Tectonics and Sedimentation: Implications for Petroleum Systems

Memoir 100

Edited by

Dengliang Gao

Published by

The American Association of Petroleum Geologists

ISBN13: 978-0-89181-381-1

AAPG Editor: Stephen E. Laubach
AAPG Geoscience Director: James B. Blankenship

COVER: Seismic facies map derived from seismic attribute analysis of a 3-D seismic survey in the Lower Congo Basin, Offshore Angola, West Africa.

This publication is available from:

The AAPG Bookstore
P.O. Box 979
Tulsa, OK U.S.A. 74101-0979
Phone: 1-918-584-2555 or 1-800-364-AAPG (U.S.A. only)
E-mail: bookstore@aapg.org
www.aapg.org

Canadian Society of Petroleum Geologists
600, 640 – 8th Avenue S.W.
Calgary, Alberta T2P 1G7
Canada
Phone: 1-403-264-5610
Fax: 1-403-264-5898
E-mail: reception@cspg.org
www.cspg.org

Geological Society Publishing House
Unit 7, Brassmill Enterprise Centre
Brassmill Lane, Bath BA13JN
United Kingdom
Phone: +44-1225-445046
Fax: +44-1225-442836
E-mail: sales@geolsoc.org.uk
www.geolsoc.org.uk

Affiliated East-West Press Private Ltd.
G-1/16 Ansari Road, Darya Gaaj
New Delhi 110-002
India
Phone: +91-11-23279113
Fax: +91-11-23260538
E-mail: affiliate@vsnl.com

The American Association of Petroleum Geologists

The American Association of Petroleum Geologists Books Refereeing Procedures

The Association makes every effort to ensure that the scientific and production quality of its books matches that of its journals. Since 1937, all book proposals have been refereed by specialist reviewers as well as by the Association's Publications Committee. If the referees identify weaknesses in the proposal, these must be addressed before the proposal is accepted.

Once the book is accepted, the Association Book Editors ensure that the volume editors follow strict guidelines on refereeing and quality control. We insist that individual papers can only be accepted after satisfactory review by two independent referees. The questions on the review forms are similar to those for the AAPG Bulletin. The referees' forms and comments must be available to the Association's Book Editors on request.

Although many of the books result from meetings, the editors are expected to commission papers that were not presented at the meeting to ensure that the book provides a balanced coverage of the subject. Being accepted for presentation at the meeting does not guarantee inclusion in the book.

More information about submitting a proposal and producing a book for The American Association of Petroleum Geologists can be found on its web site: www.aapg.org.

Acknowledgments

AAPG wishes to thank the following for their generous contributions to

Tectonics and Sedimentation: Implications for Petroleum Systems

AAPG Memoir 100

The AAPG Foundation

John J. Amoruso
Special Publications Fund

Contributions are applied toward the production cost of the publication, thus directly reducing the book's purchase price and making the volume available to a larger readership.

The editors are also grateful to the numerous reviewers acknowledged separately at the end of each paper, and the dedicated staff at the AAPG Publications Department whose useful comments and suggestions made this volume possible.

Many energy companies granted permission for the publication of their high-quality seismic data. The editor is grateful to Terri Olsen, Colin North, and Gretchen Gillis for their support for the publication of the Memoir. Many thanks go to the peer reviewers who helped improve the quality of the Memoir: Dozith Abeinomugisha, Arlene Anderson, David Barbeau, Suzanne Beglinger, Joseph Cartwright, Ian Clark, Ian Davidson, Harry Droust, Frank Ettensohn, Oscar Lopez-Gamundi, Paulo-Otávio Gomes, Daniel Harris, Dennis Harry, William Hill, Luo Hongjun, Michael Hudec, Ian Hutchinson, Hermann Lebit, John Londono, Paul Mann, Shankar Mitra, Chris Morley, Colin P. North, Maria José R. Oliveira, Terry Pavlis, Dave Pivnik, Paul Reemst, Richard Smosna, Mike Sweet, Duan Taizhong, Gabor Tari, Jaime Toro, and Qingming Yang. West Virginia University's Department of Geology and Geography and the Eberly College of Arts and Sciences were both supportive of the effort to publish this volume. This Memoir is a contribution to the West Virginia University Advanced Energy Initiative program.

About the Editor

Dr. Dengliang Gao received a PhD (1997) in geology and geophysics from Duke University, an MS (1994) in geology from West Virginia University, MS (1986) and a BS (1983) in geology from Hefei University of Technology. He is an associate professor at West Virginia University (2009–present), and was an adjunct professor at the University of Houston (2007) and a lecturer at Tongji University (1986–1991) in China. He worked at Chevron Energy Technology Company (2008–2009) as a staff geophysicist, Marathon Oil Company (1998–2008) as a geologist, and Exxon Production Research Company (1997–1998) as a post-doc fellow. He was the recipient of U.S. patents (2005 and 2001), Achievement of Company Excellence awards (2001 and 2000), and Science and Technology Advancement awards (China's State Education Commission) (1994 and 1991). Gao was recognized as an outstanding *Geophysics* associate editor (2008 and 2007) and an outstanding *Geophysics* peer reviewer (2006). He served as the *Geophysics* associate editor (2006–present), as cochair for the 31st Annual GCSSEPM Foundation Bob F. Perkins Research Conference (2011), and on the AAPG Publications Committee (2006–2009). He has extensive publications, including one AAPG Memoir 100, which received the AAPG foundation award (2012), one *Geophysics* "Bright-Spot" paper (2012), one *Geophysics Today* reprint paper (2010), two AAPG digital publication series (2007a and 2007b), and two GCAGS Grover E. Murray Best Published Papers (2006a and 2006b).

Table of Contents

Preface

Tectonics and Sedimentation: Implications for Petroleum Systems

In early 2009, in response to the American Association of Petroleum Geologists (AAPG) Publications Committee's request, I submitted a proposal for publishing an AAPG Memoir entitled "Dynamic Interplay among Tectonics, Sedimentation, and Petroleum Systems." The initially proposed chapters are primarily case studies selected from presentations at the AAPG and other professional conferences.

After the proposal was reviewed and accepted by the AAPG Publications Board, I invited several review papers per suggestions of the Board. These eventually led to this Memoir that includes both reviews and case studies, providing a snapshot of recent advances in understanding the dynamic interplay among tectonics, sedimentation, and petroleum systems in some of the most prolific sedimentary basins around the world. The recent exploration successes and great hydrocarbon potentials in these sedimentary basins, the newly acquired high-quality seismic data coupled with state-of-the-art visualization technologies, and the latest physical and/or numerical modeling capabilities provided the impetus for me to propose the publication of the Memoir.

This volume was designed to be broad in scope and, as such, covers a wide spectrum of structural styles and depositional environments in extensional, strike-slip, and contractional geologic settings. To ensure a prompt quality publication of an affordable book, the length of the volume was restricted to 18 chapters from the originally solicited 36 papers. Not all of the invited authors were able to contribute full-length manuscripts in a timely manner for rigorous peer reviews, but the final chapters well represent the latest developments in the subject area, focusing on the high-profile petroliferous sedimentary basins around the world. At the end of each chapter is a bibliography of relevant articles and books. Interested geoscientists may use it to delve into publications related to the theme of each chapter and of the whole volume.

Dengliang Gao (Editor)
July 2011
Morgantown, West Virginia

1

Gao, Dengliang, 2012, Dynamic interplay among tectonics, sedimentation, and petroleum systems: An introduction and overview, *in* D. Gao, ed., Tectonics and sedimentation: Implications for petroleum systems: AAPG Memoir 100, p. 1 – 14.

Dynamic Interplay among Tectonics, Sedimentation, and Petroleum Systems: An Introduction and Overview

Dengliang Gao

Department of Geology and Geography, West Virginia University, 98 Beechurst Ave., Morgantown, West Virginia, 26506, U.S.A. (e-mail: dengliang.gao@mail.wvu.edu)

ABSTRACT

In the past few decades, the petroleum industry has seen great exploration successes in petroliferous sedimentary basins worldwide; however, the net volume of hydrocarbons discovered each year has been declining since the late 1970s, and the number of new field discoveries per year has dropped since the early 1990s. We are finding hydrocarbons in more difficult places and in more subtle traps. Although geophysical and engineering technologies are crucial to much of the exploration success, fundamentally, the success is dependent on innovative play concepts associated with spatial and temporal relationships among deformation, deposition, and hydrocarbon accumulation.

Unraveling the dynamic interplay among tectonics, sedimentation, and petroleum systems in the subsurface is a challenge and relies on an integrated approach that combines seismic imaging, well logging, physical and/or computational modeling, as well as outcrop analogs. In recent decades, an increasing coverage of high-quality three-dimensional (3-D) seismic data, along with state-of-the-art 3-D visualization technologies, extensive well tests, sophisticated modeling capabilities, and field (outcrop) analogs, has significantly added to our understanding of subsurface complexities in structure, stratigraphy, and petroleum systems. This volume is intended to provide a snapshot of the most recent advances in petroleum exploration by presenting state-of-the-art reviews and overviews, current case studies, and the latest modeling results. The reviews and overviews offer the current status of knowledge in extensional, strike-slip, and contractional tectonic settings, as well as their influence on sedimentation and hydrocarbon accumulation. The case studies cover diverse geologic settings, with special reference to the most prolific high-profile frontier sedimentary basins, such as those in west Africa, east Africa, east Brazil, Gulf of Mexico, South China Sea, Russian Arctic, and the Mediterranean Sea. The models provide both numerical and physical simulations of basin structures as well as their spatial variation and temporal evolution in response to different tectonic processes. The objective of this volume is to contribute toward an enhanced understanding of the spatial and temporal relationships among tectonics of different structural styles, syntectonic sedimentation, and hydrocarbon accumulation. Achieving this objective is the key to

DOI:10.1306/13351546M1003528

overcoming the challenges that we face in the exploration for hydrocarbons in complex reservoirs, subtle traps, and in increasingly difficult places at a time of growing global demand for energy.

REVIEWS, CASE STUDIES, AND MODELS

In Chapter 2, Davison and Underhill (2012) provide a review of extensional tectonics and its control on sedimentation patterns and hydrocarbon generation and accumulation. In an extensional tectonic setting, extremely prolific source rocks may be produced during the rapid rifting phase, particularly if half-graben depocenters are starved and oceanographic circulation is poor. The authors cite several recent studies demonstrating that thick (unfaulted) sag basins can develop very rapidly above rifted continental crust. In these circumstances, subsidence is thought to be too rapid to have been produced by normal thermal conductive cooling and is believed to have been produced by stretching of the lower or middle crust with no observable faulting. In particular, the authors report several important observations and findings. First, hotter, highly extended shallow basins generally contain fewer hydrocarbon fields than colder rift basins because the fault spacing in these shallower basins is relatively small and also because heat maintains the rift elevation so that deep lakes (source rocks) are generally not developed. Second, a simple shear rift basin may have a rift geometry and sedimentary fill that are similar to those of an asymmetric half graben produced by pure shear, but the simple shear case will have a lesser thermal subsidence sag phase developed directly above the central rift axis. Third, the initial dip of faults can be critical because high-angle faults produce faster subsidence rates than low-angle faults for a given strain rate, thereby favoring deep rift lake development. Fourth, the largest faults will form the deepest adjacent hanging-wall half graben, and these will be the preferred sites of lake development in continental rifts. Finally, transfer faults are a preferred location for clastic sediments to enter the rift, for complex structural closures, and for natural convergent points for hydrocarbon accumulation. Consequently, they are commonly a focus for hydrocarbon prospecting.

In Chapter 3, Mann (2012) compiles and describes structural styles in four active strike-slip regions with known concentrations of giant oil and gas fields including the following strike-slip fault system: (1) the Californian San Andreas; (2) the southern Caribbean Bocono–El Pilar; (3) the Sumatra; and (4) the eastern China. In each of the four study areas, the author compiles the main geologic and tectonic parameters that include (1) regional plate motions; (2) basin types along the strike-slip faults; (3) structural styles along strike-slip faults; and (4) structural and stratigraphic traps that accommodate large accumulations of hydrocarbons.

Based on the compilation and comparison, the author states that despite the relative paucity of giant fields found associated with strike-slip faults, several active strike-slip basins have produced billions of barrels of oil and deserve special analysis to understand the regional controls on their past productivity and reserve estimates. Furthermore, the author reports the following important observations and interpretations. (1) Regional compressive stresses are commonly oriented at right angles to strike-slip faults, indicating that these faults are surfaces of low shear stress. (2) Folds are parallel to the strike-slip fault (instead of oblique or en echelon). This supports the idea of strike-slip fault normal shortening instead of the more classical view of maximum compressive stresses that are obliquely inclined to the strike-slip fault plane and controlling the generation of en echelon arrays of folds and normal faults. (3) En echelon folds and faults are present in some settings where regional compressive stresses are oblique to fault trends, but en echelon folds and faults are less prevalent than fault-normal folds and faults based on this compilation. (4) Large strike-slip faults, with their parallelism to continental margins and regional compressive stresses at high angles to continental margins, may be closely controlled by active oblique subduction of oceanic plates occurring along offshore trenches at varying distances to the strike-slip areas. (5) The hydrocarbon potential of active strike-slip areas are handicapped by seal failures related to continued activity on strike-slip faults; however, despite this problem, many transpressive basins are removed from the locus of activity along major strike-slip faults and form well-sealed environments for hydrocarbon accumulations.

In Chapter 4, Ettensohn and Lierman (2012) provide a review and discussion of tectonic controls on the origin of Paleozoic dark shale basins in the Appalachian foreland, eastern United States. Recent mapping of dark shale units in the Appalachian Basin has shown that these units are major parts of unconformity-bound sequences that form during episodes of rapid loading-related subsidence and subsequent relaxation in the foreland basin. These episodes of subsidence and relaxation are related to coeval phases of tectonism that were mediated by collisions with successive continental promontories during orogeny. Early parts of the sequences include a basal unconformity and an overlying succession of dark shale that represents accumulation of organic-rich mud at a time when clastic sources were not yet developed and when sedimentation could not keep pace with subsidence. Once active loading and deformation ceased, the lithosphere responded by relaxing in a series of steps that allowed the basin to infill progressively with flyschlike and molasselike clastic sediments. As a result, 13 such cycles were generated during the Taconian, Salinic, Acadian/Neoacadian, and Alleghanian orogenies in the Appalachian Basin. Each cycle contains a major dark shale unit and an overlying coarser clastic sequence. Every part of the cycle has potential economic importance. Dark shale units become major hydrocarbon source rocks, and many of the overlying coarser clastic units develop into reservoir rocks.

The authors also recognize the diachronous nature of tectonic loading and basin formation along the Appalachian foreland. In a simple and typical case, if the collision and transfer were synchronous along the entire continental margin, the basin and forebulge would migrate perpendicular to the strike of the orogen. However, in an oblique collisional situation, the convergence becomes diachronous along strike, and basin forebulge migration exhibits a major component parallel to strike. Consequently, lithologies in each cycle also migrated in space and time, tracking the progress of convergence along the margin.

In Chapter 5, Gomes et al. (2012) provide a description of the intrabasinal Out High of the Santos Basin that formed during an early phase of rifting and thermal subsidence (sag). The extreme extension of the continental crust in this situation led to deep crustal, and even upper mantle, exhumation along with northwest-trending transfer on a fault system that runs between and separates the extensional segments. The extensional (rift) structures were later affected by postrift strike-slip structures, which were associated with continental breakup and opening of the South Atlantic Ocean. The presalt section of the deep-water Santos Basin forms a unique and attractive exploration play that features prolific and mature source rocks, synrift structural highs as four-way closures, a regional focus for hydrocarbon migration, and an overlying evaporite seal. The upper evaporite seal is thick and continuous, which contrasts with the inboard part of the Santos and Campos basins, where salt windows provided migration avenues from presalt sources to postsalt reservoirs. However, uncertainties exist about the association between reservoir presence and deliverability, and a long and heated debate has developed about the viability of siliciclastic reservoir models versus carbonate reservoir models (with enhanced reservoir quality from karstic carbonate facies) for the presalt section.

In Chapter 6, Lopez-Gamundi and Barragan (2012) present a case study showcasing the seismic expression of extensional structures of presalt, Early Cretaceous half grabens, and synrift and postrift sag sequences in the Greater Campos Basin at the southeastern continental margin of offshore Brazil. Based on recently acquired high-quality, prestack, depth-migrated seismic data, they provide evidence for distinct reflection characters and geometries of sequences formed in response to the faulting and rotation of half grabens that preceded thermal subsidence. They interpret that thickening and/or fanning (divergence) onto the fault margins and thinning and/or onlap (convergence) onto the flexural margins of the half grabens are indicative of differential subsidence caused by an early rift phase with an episode of rotation of fault blocks bound by planar faults (domino style). Moreover, the asymmetric geometry defined by the fault and flexure margins is shown to have affected across-axis variations in thickness and depositional facies of source and reservoir rocks in the synrift and postrift sag sequences. From a fault margin that features thickening and compaction to a flexural margin that features thinning and onlapping, the seismic boundary between synrift and postrift sag sequences becomes progressively angular and recognizable. The seismic base of the sag commonly shows a pronounced impedance contrast characterized by a clear and continuous reflection event. This event also shows lateral changes in amplitude character, probably reflecting changes in lithology from margins dominated by shallow-water conditions to deep basin centers. In the deep basin centers, similar lithologies are more likely to occur above and below the base of the sag, thereby inducing a rather subdued impedance contrast.

According to the authors, structural asymmetry caused by differential subsidence created optimal conditions for the deposition of lacustrine-to-brackish source rocks in the hanging-wall area of maximum accommodation space next to the fault margins of half grabens. The fault margin itself, however, is characterized by a low potential for high-quality clastic reservoirs because of underfilled conditions and provenance. The flexural margin of a half graben is a site of low subsidence and little accommodation space suitable for the development of high-energy carbonates. In contrast, a distinctive sag-facies association commonly forms on bathymetric highs developed on underlying rift shoulders. This facies association contains microbial limestone deposited in subtidal-to-peritidal environments.

In addition, the authors discuss the structural and stratigraphic traps associated with rifting and postrift subsidence. Differential compaction at a fault margin is a key factor in creating the counterregional dip necessary to form structural four-way closures at the sag level. Combined structural-stratigraphic traps are defined by an updip pinch-out component at the base of the sag or at any horizon within the sag interval. Updip onlap and three-way closures define these combined traps. Potential for onlap and/or pinch-out traps is also present at the rift-fill level on the flexural flank of half grabens. Along the rift axis, the polarity of half grabens switch vergence across intervening accommodation zones (transfer faults) in a pattern similar to that observed in the East African rifts. This kind of architecture can contain four-way closures that provide optimal conditions for focused hydrocarbon accumulation.

In Chapter 7, Oliveira et al. (2012) present a discussion about basin slope contractional tectonics linked to upslope extensional tectonics in shale-dominated gravitational gliding systems in the Pará-Maranhão and Barreirinhas basins of Brazil. Both basins lack salt detachments, which contrast strongly with other basins on the passive continental margins of the South Atlantic Ocean. As opposed to salt-based systems, the gravitational systems reported in this chapter detach on a decollement surface of presumably overpressured shale and marl at depths of about 6 km (~3.37 mi). The known exploration successes in the contractional domains of shale-detached gravitational systems in the distal parts of deltas in Nigeria, Indonesia, Trinidad and Tobago, Egypt, India, and Brunei hold the promise of high potential at the equatorial margins of Brazil and Africa where shale-based gravitational systems are also present.

By integrating seismic interpretation, physical and/or numerical modeling, structural restoration, and field investigation, the authors demonstrate that thrusts developed in a backstepping sequence such that depocenters migrated landward through time. Their results show that the total amount of shortening exceeded the total amount of stretching in basal layers close to the detachment surface, whereas stretching exceeded shortening in the upper layers. Among the possible reasons for the excess of total shortening over total stretching in the basal layers is evidence for a greater amount of compression over shale detachments than over salt detachments. Furthermore, physical modeling that best reproduced the structural features similar to those present in the Barreirinhas Basin was the one that simulated two bordering dextral strike-slip faults, analogous to the Romanche and Saint Paul fracture zones that bounded the gravitational cell. Resulting high pore pressures in the shales and marls, which might have induced the gliding, possibly originated from hydrocarbon generation. Understanding similar spatial and temporal relationships could be instrumental in unraveling the dynamic interplay among deep-water basin slope thrusting, strike-slip tectonics, sedimentation, and fluid flow.

In Chapter 8, Tiercelin et al. (2012) report on the Mesozoic and Early–Middle Cenozoic rift basins of central and northern Kenya (East African rift) and their relationships to hydrocarbon prospects. In terms of source rock quality and reservoir properties, they review the evidence of oil potential in a suite of Cretaceous(?)–Paleogene and Paleogene–Middle Miocene basins identified in the two regions. The study provides an important set of data for potential oil exploration in these rift basins. In addition to the variation in structural styles through time, hydrocarbon potential seems to be mostly controlled by the nature of the basin-fill stratigraphy. Sedimentologic, petrologic, and geochemical studies conducted on likely reservoir and source rocks in the major basins permit the development of a provisional ranking of these basins in terms of source rock quality and quantity, reservoir potential, and hydrocarbon prospects. These can be very important in future exploration and well-bore planning in the rift basins.

In Chapter 9, Abeinomugisha et al. (2012) describe the Tertiary Albertine Graben, a part of the East African rift system, as a classic example of the processes of continental breakup. The graben has experienced several extensional and transpressional episodes, resulting in structures that are typical to both settings. They describe variations in graben geometry, sedimentation, and prospect potential along the graben segmented by intersecting accommodation zones. All the wells drilled in the graben have been on either positive flower (palm tree) structures or on rotated fault blocks. Despite the excellent hydrocarbon potential of the graben, several issues remain poorly resolved, including source rock depositional environments, migration pathways, and trapping mechanisms. A geologic section along the strike of the graben indicates two subbasins separated by an accommodation zone. The authors conclude that the Albertine Graben evolved through multiple phases of tectonic deformation, involving both extension and inversion (compression). Extensional regimes created accommodation space for the accumulation of a thick sequence of sediments necessary for the generation of hydrocarbons. However, early extensional and subsequent contractional regimes created both effective hydrocarbon migration pathways and structural traps. All the discoveries made to date are structurally controlled and are mainly in contractional anticlines developed along basin margin faults. The high potential of the petroleum system is spatially associated with transfer faults that controlled the along-axis variation in graben geometry and sedimentation.

In Chapter 10, Gao and Milliken (2012) use high-quality 3-D seismic data to discuss basin-scale lineaments in the Lower Congo Basin at the west African passive continental margin. Among the different structural grains, they focus on cross-regional intraslope lineaments and their implications for hydrocarbon accumulation in the postsalt Tertiary section. These lineaments are spatially associated with allochthonous salt bodies, turbidite channel systems, hydrocarbon field discoveries, and direct hydrocarbon indicators. Seismic evidence suggests that the cross-regional lineaments might have had a significant strike-slip component and that they are spatially associated with, but kinematically different from, regional thrusts and folds. Moreover, the authors suggest that these lineaments are transfer faults (tear faults) induced by regional gravitational sliding involving the postsalt sedimentary section. The authors further speculate that the northeast-trending lineaments could be an expression of presalt basement-involved transfer faults associated with the Early Cretaceous rifting of the continent. The obliquity of the northeast-trending (45°) continental transfer faults relative to the northeast–east-trending (80°) oceanic fracture zones of the South Atlantic can be explained by a finite 35° counterclockwise rotation of the west African continental margin. This obliquity is strikingly similar to (but with an opposite polarity) that reported at the Campos Basin along the eastern Brazilian continental margin, which indicates an opposite clockwise rotation of the east Brazilian continental margin. Such relationships and patterns from both conjugate continental margins may well be seismic evidence of the southward scissorslike opening of the South Atlantic Ocean, which in turn sheds new light on the diachronous nature and timing of the sedimentary basins along the west African passive continental margin.

Numerous major oil and gas field discoveries that apparently reflect focused commercial hydrocarbon accumulation along the lineaments underscore the economic implications of these lineaments. In the authors' discussion regarding the implications of cross-regional lineaments in the Lower Congo Basin and in other petroliferous sedimentary basins worldwide, they argue that high-angle dip and intensive shear deformation along the lineaments were favorable for effective hydrocarbon migration from the deep source to the shallow reservoirs. They additionally suggest that these cross-regional lineaments have significant strike-slip components, and that strike-slip faults could have had a different, if not more important, function in the accumulation of hydrocarbons than low-angle dip-slip faults. Such a function is particularly important in parts of basin slopes where extensive toe thrusts and folds associated with salt provide effective traps, and structural migration pathways become critical for hydrocarbon migration from deeply buried sources to shallower reservoirs. Given the uncertainty of vertical relationships between presalt basement transfer faults and the cross-regional lineaments shown in the postsalt Tertiary seismic section, the authors have stressed the importance of differentiating postsalt lineaments (detachment structures) from the presalt, deeply rooted, transfer faults (basement structures). Hence, better seismic imaging and a more complete understanding of the vertical extent of these cross-regional lineaments could be instrumental in evaluating the hydrocarbon potential of presalt and postsalt petroleum systems and the connectivity of the presalt source rock in the rift sequence to the postsalt Tertiary reservoirs of the passive-margin sequence.

In Chapter 11, Linzer and Tari (2012), for the first time, make lithologic and structural correlations between the classic Alpine folded belt of the Northern Calcareous Alps (NCA) of Austria and the Transdanubian Central Range (TCR) of Hungary, which presently is located some 200 km (~12.4 mi) away. It is widely accepted that the NCA of Austria and Germany represents a thin-skinned, fold-and-thrust belt along the northern margin of the Eastern Alps, whereas the TCR of Hungary is traditionally considered to be a simple autochthonous unit without any internal deformation. The allochthonous versus autochthonous nature of the TCR is still a subject of debate. Nonetheless, palinspastic restoration of the post-nappe deformation presented in this Chapter clearly reveals the spatial and temporal relationship between these major Alpine units despite their present-day separation and indicates clockwise rotation of the NCA and counterclockwise rotation of the TCR.

The TCR has seen several hydrocarbon exploration campaigns on its northwestern flank, but the NCA remains practically unexplored. Seismic sections illustrate the same overall Alpine characteristics in both the NCA and the TCR. The internal structures of the NCA are highlighted by a set of dextral strike-slip faults interpreted as tear faults or transfer faults caused by right-lateral oblique convergence. The strike-slip faults terminate at depth at the same detachment as that of the internal thrusts. In the clastic basins of the western NCA, deep-water sediments were separated from shallow-water sediments by these transfer faults, attesting to their syndeformational nature. In comparison with the NCA, the Eo-Alpine thrust systems of the

TCR are not as well known. However, seismic reflection data indicate that the TCR was, just as the NCA, dismembered by west–northwest-oriented dextral strike-slip faults during the Cretaceous. The TCR structural pattern of northeast-trending thrust boundaries and fold axes and the northwest orientation of the dextral strike-slip faults appear to be identical with those of the NCA in terms of geometry and timing. Hence, the similarity in structural patterns suggests a coherently deformed Eo-Alpine nappe stack along with a cross-strike, strike-slip (transfer or tear) fault. The recognition of a once continuous, regional, right-lateral strike-slip fault system in the NCA/TCR areas has important implications for pre-Tertiary kinematic reconstructions of the broader Eastern Alps and Pannonian Basin region. This reconstruction could lead to a better understanding of the regional tectonic architecture, growth history, and control on syntectonic deposition. Understanding these aspects is fundamental to future hydrocarbon exploration in this mostly unexplored region.

In Chapter 12, Verzhbitsky et al. (2012) use newly acquired and processed two-dimensional (2-D) seismic data and field geologic observations to discuss the tectonics and hydrocarbon potential of the South Chukchi Basin (Chukchi Sea, Russian Arctic). Their analysis of onshore data points to the beginning of sedimentation during the Aptian–Late Cretaceous. The geometry of the faults indicates an extensional and/or transtensional setting, although folds, thrust faults, pop-up and positive flower structures also occur, indicating local development of compressional and transpressional structural styles. The South Chukchi Basin experienced multiphase extension and/or right-lateral transtension and subsidence during the Paleocene–Quaternary. The existence of a prerift thrust system at the base of the basin suggests that the South Chukchi Basin inherited a preexisting Late Mesozoic zone of displacement and that the basin was formed during the Aptian–Late Cretaceous–Cenozoic as a result of the change in tectonic setting from compression in the Neocomian to subsequent extension. The observed right-lateral component of the youngest south-dipping normal faults and transtensional structures identified on seismic sections also indicate the function of dextral strike-slip movements in the development of the South Chukchi Basin. Structural and stratigraphic traps associated with bright spot anomalies and gas chimneys suggest that hydrocarbon potential in the South Chukchi Basin may be significantly higher than previously recognized. An investigation of the spatial and temporal relationships among strike-slip, extensional, and contractional structural styles based on seismic and outcrop observations will have important implications for future hydrocarbon exploration in this frontier Arctic sedimentary basin.

In Chapter 13, Anderson et al. (2012) examine the Oligocene–Miocene evolution of the Lower Congo Basin and how variations in structural style and history along the strike of the basin affected the distribution of reservoir sandstones on a regional scale. The authors provide examples of the interaction between active structures and syntectonic sediments, showing that evolving structures exerted a primary control on distribution and architecture of deep-water reservoirs. The Oligocene–Miocene paleogeographic evolution of the Lower Congo Basin demonstrates the interaction of active structures with deep-water depositional systems as a progradational passive margin evolved through time. Slope gradient generally became steeper through time as progradation of the shelf edge advanced and generated systematic changes in large-scale reservoir geometry, distribution, and organization. In addition, overprinting this large-scale slope gradient change were local changes in sea-floor gradient caused by active gravitational structures. This local gradient perturbation apparently resulted in rapid lateral changes in accommodation space affecting the deposition of reservoir facies. The study shows that interaction between active structures and reservoirs may be subtle and episodic and may be better detected by observing channel trends and behaviors near active faults and folds in map view instead of by relying entirely on cross sectional stratal geometries such as onlap or truncation.

In Chapter 14, Clark and Cartwright (2012) consider how the relative rates of sedimentation and uplift vary along the strike of folds in the Levant Basin of the eastern Mediterranean Sea, where deep-marine growth structures and channel-levee systems are key targets for hydrocarbon exploration. In particular, the authors provide detailed documentation on the interplay among folding, strike-slip faulting, and sedimentation in deep-water channel-levee systems. They demonstrate spatial and temporal relationships between thrust-related folding and strike-slip faulting at the fold terminations and show that multiple submarine channel-levee systems were coeval with structural development. Based on the variation in onlap observed along the strike of the fold, a zone of increased structural relief is located adjacent to the segment boundaries (lateral terminations) of strike-slip faults, which could represent the earliest onset of folding. In addition, uplift maxima are located toward the lateral terminations of folds with strike-slip faults, resulting in increased stratal thinning along those strike-slip faults where the lateral terminations act as sediment barriers. This situation contrasts with systems in which the maximum fold amplitude is located in the center and decreases toward the lateral tip regions. In addition, the authors discuss factors that affect the architecture of growth folds and their along-strike variations, emphasizing possible problems that may result from interpretations based solely on limited outcrops or 2-D seismic lines. The influence of such structures in a submarine lower fan setting is especially telling. In this setting, channel axes are mostly confined by the levees and not by incision, and increased channel aggradation may allow a channel flowing perpendicular to folds to maintain its course, particularly if the uplift rate is low relative to sedimentation rate. However, if the uplift rate is relatively high, as indicated by an increasing occurrence of onlap on the limbs of anticlines, a channel becomes blocked or diverted, indicating coeval structural uplift and sedimentation.

In Chapter 15, Tari et al. (2012) report on three deep-water exploration wells from the underexplored Moroccan salt basin in the deep water of West African Atlantic. The basin features a contractional toe thrust and fold belt and a mid-Tertiary allochthonous salt that was sourced from the latest Triassic to earliest Jurassic autochthonous salt level that was deposited during synrift margin evolution. The central part of the Moroccan Atlantic margin is adjacent to the Atlas Mountains. One would expect a much larger sedimentary influx and, therefore, more clastic reservoir facies in this segment of the African margin than, for example, farther to the south. Yet, the three deep-water exploration wells were all dry holes and basically documented the same general lack of clastic reservoirs that is observed throughout the entire Tertiary–Upper Cretaceous succession.

The authors present different scenarios for explaining the apparent lack of reservoirs in the deep-water domain of central Atlantic Morocco. First, the well locations could be inboard of the paleochannel lobe–transition zone and therefore the reservoirs might be expected farther out. Second, the wells could have been drilled outboard of reservoir facies that were trapped closer to the paleoshelf edge. Third, the well locations could have been in a paleosediment shadow zone, bypassed by potential reservoir facies. The authors also discuss and predict the existence of potential reservoir facies equivalent to other siliciclastic turbidites based on regional correlation, outcrop, and the presence of a shallow-water to onshore deltaic complex. They also suggest the presence of Jurassic reservoirs based on high-amplitude fanlike geometries identified in seismic reflection data. However, these possible reservoirs occur seaward of the toe thrust belt and appear to be relatively small and thin.

In Chapter 16, Londono et al. (2012) present a study based on modeling of lithospheric flexure and related base-level stratigraphic cycles in the Putumayo Foreland Basin, Northern Andes, showing that the basin history is different from the current evolutionary concept proposed for the basin. Their results suggest that onlapping seismic facies migrated toward the foreland predominantly during sediment-controlled flexural periods, whereas onlapping seismic facies migrated toward the hinterland when thrusts belt loads dominated. On average, tectonic-loading subsidence is thought to be responsible for about 23% of the total subsidence, whereas sediment loading is responsible for the remaining 77%. Moreover, the sandstone/shale ratio is interpreted to increase with an increasing ratio of sediment-related subsidence to tectonic-related subsidence.

According to the authors, the geometry of the effective thrust belt and the wavelength of the lithospheric deflection preclude the need to invoke dynamic topography as a downward force acting on the plate to create extra accommodation in the basin. As is manifest in their results, the geometry of the loads apparently had a strong control on the final geometry of the basin. Tectonic loads produced forebulges that were closer to the hinterland and deflections that were narrower and deeper than those created by sedimentary loads. According to the subsidence regime, the authors divide the sedimentary sequence into two facies. The first is a tectonically induced high-subsidence facies, with probably low sandstone/shale ratios, that is predicted to have been associated with onlap shifts toward the hinterland. The second is a sediment loading-induced low-subsidence facies, with probably high sandstone/shale ratios, that is predicted to have been associated with continual onlap events toward the foreland bounded by onlap shifts and/or unconformities. The authors also suggest that tectonic events produced a narrow but deep depocenter with a high subsidence rate (in relation to sediment supply) and a low sandstone/shale ratio. In the seismostratigraphic record, a tectonic-loading event would be recognized by regional (tens of kilometers) onlap shifts from the foreland toward the hinterland. Conversely, a sediment-related subsidence event (controlled mostly by the weight and dispersion of sediments) would produce a wide but relatively shallow depocenter with high sedimentation rates compared with subsidence rates and a high sandstone/shale ratio. During these sediment-related events, flexure should be expected to widen as sediments propagate toward the foreland. Discoveries of commercial heavy oil in an adjacent foreland basin could indicate that the Putumayo foreland has hydrocarbon potential, especially with traps developed in or around forebulges coeval with each loading event.

In Chapter 17, Luo and Nummedal (2012) show the results of 3-D flexural numerical modeling and a subsurface study in southwestern Wyoming to identify Late Cretaceous forebulge location and migration in response to thrust loads. They recognize three forebulges with respective amplitudes ranging from 40 to 70 m (120–210 ft) and widths ranging from 30 to 50 km (~20–30 mi). Their data indicate that the forebulges migrated eastward (basinward) in spatial and temporal association with progressively younger thrusting in the Wyoming thrust belt. The width of the foredeep (the distance from the thrust front to the associated forebulge) was between 160 and 194 km (~100–121 mi). They performed 3-D flexural numerical modeling using an elastic lithospheric model, which was achieved by solving a differential equation for the vertical deflection of a thin elastic plate over a fluid substratum with a specific applied surface load. Their modeling results are reportedly consistent with the geologic observations of forebulge migration and explain the complexity of the 3-D distribution of forebulge-related unconformities in the subsurface of the Greater Green River Basin.

In Chapter 18, Yang et al. (2012) provide an overview of different generations of the South China Sea (SCS) polyhistory basins and discuss their implications for hydrocarbon exploration. On the basis of previous studies and most recent advances in polyhistory basin classification schemes, they divide the evolution of the SCS Basin into two cycles: an early divergent-convergent cycle and a later divergent-convergent cycle. Geohistory analysis suggests that the SCS Basin has experienced multiphase deformation, creating an assemblage of numerous basins of different generations and styles.

The authors emphasize the physical condition of each basin and the overprinting (temporal) relationships of different basins. Each basin features a specific pressure, space, and temperature at a specific geologic time in the basin history, whereas overprinting of a different basin could cause changes in pressure, space and temperature with time, leading to changes (either enhancement or reduction) in the hydrocarbon potential of the basin. Therefore, the polyhistory basin analysis is helpful in evaluating hydrocarbon potential from a dynamic and historic perspective in the SCS.

STRUCTURAL STYLES IN PETROLEUM SYSTEMS

State-of-the-art reviews and/or overviews, current case studies, and recent numerical and/or physical modeling results presented in this *Memoir* demonstrate how different structural styles influence the effectiveness, thickness, and spatial distribution of source, reservoir, and seal/cap rocks, as well as migration pathways and traps in such a different manner as to enhance (or hinder) hydrocarbon generation and accumulation. Understanding the influence of different structural styles in petroleum systems can be crucial to exploration success by better predicting risk factors in more difficult basins.

Extensional structures and tectonics, particularly those at the crustal scale associated with rifting and thermal subsidence, have a major function in controlling basinwide syntectonic sedimentation in the development of petroleum systems (Brice et al., 1982; Harris et al., 2004; Guiraud et al., 2010; Versfelt, 2010; Davison and Underhill, 2012; Gomes et al., 2012; Lopez-Gamundi and Barragan, 2012; Tiercelin et al., 2012; Abeinomugisha et al., 2012). Crustal-scale extension could lead to significant thinning of the crust that is accommodated by regional grabens as well as by cross-regional strike-slip transfer faults (e.g., Wernicke, 1981; Bosworth, 1985; Barr, 1987; Talbot, 1988; Frostick and Reid, 1989; Jackson and White, 1989; Figueiredo et al., 1995; Katz, 1995; Lambiase and Bosworth, 1995; Morley, 1995; Gawthorpe and Leeder, 2000; Davison and Underhill, 2012). Major crustal extension is most effective at creating accommodation space, allowing for the influx and deposition of thick sequences of sediments that can later become prolific source, reservoir, and seal/cap rocks (Harris et al., 2004). For example, extreme stretching in highly extended terranes can lead to significant crustal thinning and denudation, causing major crustal isostatic rebound that could complicate the geothermal gradient and maturation history of intracontinental rift basins (Davison and Underhill, 2012; Abeinomugisha et al., 2012). This could be different from contractional foreland basins, where the impinging thrust front can cannibalize and disrupt any maturation that may have been occurring. This could also be different from strike-slip pull-apart basins, where focused sedimentation and limited isostatic rebound occur.

Strike-slip structures and tectonics play yet a different function in a working petroleum system and have contributed significantly to the accumulation of oil and gas (Mann et al., 2003; Mann, 2012). Strike-slip faults have three characteristics (Stone, 1969; Harding, 1973, 1990; Wilcox et al., 1973; Crowell, 1974; Fleming and Johnson, 1989; Xu, 1993) that make significant differences in controlling fluid-flow efficiency and direction. First, strike-slip faults are typically vertical or subvertical (Stone, 1969; Harding, 1973). Vertical or subvertical faults can be most straightforward migration pathways for fluid to flow efficiently from source to reservoirs. Second, strike-slip faults are typically characterized by a large lateral component of slip, producing highly strained and/or intensively fractured fault zones with enhanced fracture connectivity, porosity, and permeability. These can be important attributes controlling the efficiency and extent that fluids can flow. Third, steeply dipping strike-slip faults are oriented in such a way that the effects of gravity (overburden) on hydraulic properties along the faults are relatively insignificant compared with shallowly dipping thrusts or normal faults; whereas the effect of tectonic stress on hydraulic properties along the faults depends on the partitioning between extensional and/or contractional stress and shear stress (Wilcox et al., 1973; Mann, 2012). Such a stress state and high pressure gradient associated with vertical strike-slip faults could make the faults effective fluid pumps. However, continued deformation along active strike-slip faults may jeopardize the retention of trapped hydrocarbons because of the enhanced likelihood of fluid leakage along the faults. In addition, although strike-slip faults can also control the play fairways and create thick sequences of syntectonic deposits, the deposits are relatively focused, localized, and linearly distributed as in the pull-apart basins (Crowell, 1974; Mann et al., 2003) compared with those in extensional tectonics settings (Lambiase and Bosworth, 1995; Morley, 1995; Gawthorpe and Leeder, 2000; Harris et al., 2004) or contractional foreland basins (Ettensohn and Lierman, 2012; Londono et al., 2012; Luo and Nummedal, 2012).

Contractional structures and tectonics are yet another structural style that has had an important and different function in petroleum systems. Unlike extensional and strike-slip tectonics, crustal contraction generally creates regional distributed folds and thrusts that are commonly associated with overpressured salt and/or shale (Fox, 1959; Dahlstrom, 1970; Harris and Milici, 1977; Suppe, 1980; Gries, 1983; Mitra, 1988; Suppe et al., 1992; Weimer and Buffler, 1992; Shaw and Suppe, 1994; Rowan, 1997). These can make contractional structures effective traps and seals for oil and gas, although traps can also be found in extensional and strike-slip structures (Davison and Underhill, 2012; Mann, 2012; Verzhbitsky et al., 2012). Tectonic loading of the crust by large-scale folding and thrusting creates foreland basins in which basinwide deposition can create source and reservoir rocks (Ettensohn, 1994; Ettensohn and Lierman, 2012). The lateral distribution and vertical stacking patterns of lithofacies and their thickness are spatially and temporally associated with the flexural forebulge that migrates in response to tectonic and/or sediment loading (Ettensohn, 1994; Ettensohn and Lierman, 2012; Londono et al., 2012; Luo and Nummedal, 2012). Although contractional faults can contribute to the syntectonic deposition of source, reservoir, and seal/cap rocks on a general basis, their lateral and/or vertical distribution pattern is significantly different from those in the extensional and strike-slip settings because of their distinct deformational, depositional, and geothermal settings.

INTERACTION OF STRUCTURAL STYLES IN PETROLEUM SYSTEMS

Results presented in this *Memoir* demonstrate that it is a combination or complementation of different components of a petroleum system that controls the potential of hydrocarbons. Because of such an integrated nature, any paucities or gaps in the system, which are critical exploration risk factors to the accumulation of oil and gas, directly define and thus point to the exploration direction or target. These could change from region to region depending on the dominant structural style of the region and from basin to basin depending on the tectonic regime of the basin.

In an extensional and/or passive-margin regime, many fields that appear to be in a seemingly simple setting turn out to have very complicated spatial and temporal relations to different structural styles and syntectonic sediments (Harding and Lowell, 1979; Mann et al., 2003; Davison and Underhill, 2012; Oliveira et al., 2012). For example, in the North Sea (Rouby et al., 1996), hydrocarbons commonly occur at the intersections of north–south-trending grabens with east–west-trending transfer faults. The steeply dipping transfer faults provide effective vertical migration pathways from deeply buried source rocks, associated with rifting and thermal subsidence (sag), to shallow structural and stratigraphic traps (Rouby et al., 1996). The transfer faults also influenced the location, direction, and distribution of channels or turbidites that are highly productive reservoirs (Rouby et al., 1996; Davison and Underhill, 2012). At the west African and eastern Brazilian passive continental margins above rift and sag basins, it is well known that the downslope deep-marine areas have prolific source rocks with extensive fetch area. It is also known that this region is dominated by gravity-induced thrusts, folds, and salt structures with a high sealing capacity. In addition, many recent deep-water drilling, 3-D seismic analyses, and outcrop studies (Kolla et al., 2001; Posamentier and Kolla, 2003; M. Gardner, 2005, personal communication) reveal great potential for reservoir presence in deep-marine turbidite systems. These studies lead to the conclusion that migration pathways could be a critical gap and, as such, strike-slip faults are of primary importance and have a crucial function in filling the gap because of their effectiveness as migration pathways. For example, in the Lower Congo Basin of offshore Angola (west Africa) (Cramez and Jackson, 2000; Da Costa et al., 2001;

Hudec and Jackson, 2002; Gawenda et al., 2004; Jackson et al., 2004; Nombo-Makaya and Han, 2009), cross-regional northeast-trending faults are spatially associated with many discoveries in the Tertiary sedimentary section (Gao and Milliken, 2012). Similarly, this spatial relationship occurs in the deep water of the Campos Basin, offshore Brazil, where extensive cross-regional northwest-trending lineaments cut across regional northeast-trending folds and thrusts in the Upper Cretaceous and Tertiary sections (Gao et al., 2009). Furthermore, as indicated by recent field discoveries, the relationship also occurs in the deep water of offshore equatorial Guinea (west Africa) (Lawrence et al., 2002) (e.g., the Alba Field, Lawrence et al., 2002; Wolak and Gardner, 2008), in deep water of offshore Gabon (west Africa) (e.g., the Tchitamba Field, Kilby et al., 2004), and in the deep-water Gulf of Mexico (U.S.A.) (Weimer and Buffler, 1992; Rowan, 1997). The spatial association of hydrocarbon fields with transfer faults is primarily caused by their high migration potential that is critical to the accumulation of oil and gas in the deep-water regions of passive margins. In contrast, the upslope shallow-marine areas of passive margins are dominated by gravity-induced extensional faults (Anderson et al., 2000; Oliveira et al., 2012) along with both siliclastic and carbonate reservoirs. Because extensional faults could enhance the migration potential, effective traps become the critical component in this part of the passive margin, and the exploration focus needs to change to finding traps such as rollover anticlines on the hanging wall of listric normal faults or cross-fault juxtaposition of shale or other rocks with a high sealing capacity. This concept could be used in building an exploration strategy and risk profile from the shelf down to the basin floor across passive continental margins.

In a wrench tectonic regime, strike-slip faults provide an optimal setting for fluid migration from source rocks to reservoirs. Examples include the many giant oil and gas fields along the northwest-trending San Andreas fault in the western United States (Harding, 1973, 1974; Crowell, 1974; Mann et al., 2003; Mann, 2012) and the major fields associated with the northeast-trending Tan-Lu fault in eastern China (Xu, 1993; Mann et al., 2003; Hsiao et al., 2004; Mann, 2012). Many strike-slip faults are commonly kinematically associated with either extensional (transfer faults) or contractional (tear faults) tectonics (Linzer and Tari, 2012; Mann, 2012). Actually, many oil and gas fields are directly associated with strike-slip faults, although they are located in extensional or contractional tectonic regimes. Examples include the hydrocarbon fields in the North Sea associated with the Mesozoic northeast-trending transfer faults across the grabens (Rouby et al., 1996), the Paleozoic northwest-trending lineaments across the rift in the Sirte Basin of Libya (Guiraud and Bosworth, 1997), the Paleozoic east–west-trending and northwest-trending lineaments across the Rome trough in the Appalachian Basin (Wheeler 1980; Shumaker, 1986; Gao and Shumaker 1996; Shumaker and Wilson, 1996; Gao et al., 2000), and the Paleozoic strike-slip faults in the Anadarko Basin and the Permian Basin (Mann et al., 2003). In a strike-slip tectonic regime, the exploration focuses should be directed at traps, together with source rocks and reservoirs. Typically, deposition of source and reservoir rocks occurs in a releasing bend (Crowell, 1974) or at the intersection with rift, and creation of traps occurs at the restraining bend (Crowell, 1974) or at the intersection with thrusts or folds. In eastern China, for example, Mesozoic and Cenozoic strike-slip faulting has had a major function in controlling the scale and distribution pattern of petroliferous sedimentary basins and major oil and gas fields. These include the most recent major oil fields found in the Bohai Bay Basin (Mann et al., 2003; Hsiao et al., 2004; Mann, 2012), a major pull-apart basin spatially and temporally associated with the strike-slip Tan-Lu fault (Mann, 2012). Along the San Andreas fault in the western United States, major oil and gas fields tend to cluster in the areas of major restraining bends at distances less than 100 km (<62 mi) from the San Andreas fault zone, where crustal shortening and thrust faulting in the vicinity of the restraining bend are associated with giant oil fields (Mann et al., 2003; Mann, 2012). In the Sirte Basin of Libya (Guiraud and Bosworth, 1997), major oil and gas fields are associated with regional northwest-trending strike-slip faults, particularly at their intersection with a northeast-trending rift that is filled with a thick sequence of source rocks and is overprinted by inversion structures that form structural traps.

In a contractional tectonic regime, many oil and gas fields have been found to be associated with the intersection of thrusts and folds with transfer and/or tear faults. For example, in the central Appalachian Basin in the eastern United States, a regional northeast-trending rift basin (Shumaker, 1986; Shumaker and Wilson, 1996) was segmented by cross-regional northwest-trending basement transfer faults during the Early Cambrian (Harris, 1978; Wheeler, 1980; Shumaker, 1986; Shumaker and Wilson, 1996; Gao et al., 2000). This early rift basin was overprinted by tectonic inversion, and the latter created extensive northeast-trending detachment folds and cross-strike lineaments that have been considered to be effective traps and migration pathways (Wheeler, 1980; Gao et al., 2000). In the foreland basin setting during the late Paleozoic, cyclic sedimentation that occurred in response to tectonic loading controlled the distribution of black shale as prolific source and reservoirs (Ettensohn, 1994; Ettensohn and

Lierman, 2012). Late Paleozoic thrust faults coupled with the northwest-trending cross-regional discontinuities or tear faults controlled syntectonic sedimentation and accumulation of oil and gas (Wheeler, 1980; Gao et al., 2000). Similar interaction also occurs in the Anadarko Basin, the Powder River Basin, the Fort Worth Basin, and the Permian Basin in the United States, in which many hydrocarbon fields are associated with the intersection of strike-slip faults with folds and thrusts (Gao, unpublished data). In the Fort Worth foreland basin, in particular, strike-slip faults are spatially associated with gas chimneys (Sullivan et al., 2006), a direct hydrocarbon indicator for natural gas, suggesting that these regional intrabasin strike-slip faults in a contractional foreland basin setting might represent high-potential sweet spots as exploration targets at their intersections with folds and thrusts.

ACKNOWLEDGMENTS

Dave Pivnik and Frank Ettensohn provided constructive comments for this Chapter. This chapter is a contribution to the West Virginia University Advanced Energy Initiatives program.

REFERENCES CITED

Abeinomugisha, D., and R. Kasande, 2012, Tectonic control on hydrocarbon accumulation in the intracontinental Albertine Graben of the East African rift system, *in* D. Gao, ed., Tectonics and sedimentation: Implications for petroleum systems: AAPG Memoir 100, p. 209–228.

Anderson, A. V., D. K. Sickafoose, T. R. Fahrer, and R. R. Gottschalk, 2012, Interaction of Oligocene–Miocene deep-water depositional systems with actively evolving structures: The Lower Congo Basin, offshore Angola, *in* D. Gao, ed., Tectonics and sedimentation: Implications for petroleum systems: AAPG Memoir 100, p. 291–313.

Anderson, J. E., J. Cartwright, S. J. Drysdall, and N. Vivian, 2000, Controls on turbidite sand deposition during gravity-driven extension of a passive margin: Examples from Miocene sediments in Block 4, Angola: Marine and Petroleum Geology, v. 17, p. 1165–1203, doi:10.1016/S0264-8172(00)00059-3.

Barr, D., 1987, Structural/stratigraphic models for extensional basins of half-graben type: Journal of Structural Geology, v. 9, p. 491–500, doi:10.1016/0191-8141(87)90124-6.

Bosworth, W., 1985, Geometry of propagating continental rifts: Nature, v. 316, p. 625–627, doi:10.1038/316625a0.

Brice, S. E., M. D. Cochran, G. Pardo, and A. D. Edwards, 1982, Tectonics and sedimentation of the South Atlantic rift sequence, Cabinda, Angola, *in* J. S. Watkins and C. L. Drake, eds., Studies in continental margin geology: AAPG Memoir, ISBN: 978-0-89181-311-8, p. 5–18.

Clark, I. R., and J. A. Cartwright, 2012, A case study of three-dimensional fold and growth sequence development and the link to submarine channel-structure interactions in deep-water fold belts, *in* D. Gao, ed., Tectonics and sedimentation: Implications for petroleum systems: AAPG Memoir 100, p. 315–335.

Cramez, C., and M. P. A. Jackson, 2000, Superposed deformation straddling the continental-oceanic transition in deep-water Angola: Marine and Petroleum Geology, v. 17, p. 1095–1109.

Crowell, J. C., 1974, Origin of later Cenozoic basins in southern California, *in* W. R. Dickinson, ed., Tectonics and sedimentation: SEPM Special Publication 22, p. 190–204.

Da Costa, J. L., T. W. Schirmer, and B. R. Laws, 2001, Lower Congo Basin, deep-water exploration province, offshore west Africa: AAPG Memoir 74, p. 517–530.

Dahlstrom, C. D. A., 1970, Structural geology in the eastern margin of the Canadian Rocky Mountains: Bulletin of Canadian Petroleum Geology, v. 18, p. 332–406.

Davison, I., and J. R. Underhill, 2012, Tectonics and sedimentation in extensional rifts: Implications for petroleum systems, *in* D. Gao, ed., Tectonics and sedimentation: Implications for petroleum systems: AAPG Memoir 100, p. 15–42.

Ettensohn, F. R., 1994, Tectonic control on the formation and cyclicity of major Appalachian unconformities and associated stratigraphic sequences, *in* J. M. Dennison and F. R. Ettensohn, eds., Tectonic and eustatic controls on sedimentary cycles: SEPM Concepts in Sedimentology and Paleontology 4, p. 217–242, doi:10.2110/csp.94.04.0217.

Ettensohn, F. R., and R. T. Lierman, 2012, Large-scale tectonic controls on the origin of paleozoic dark-shale source-rock basins: Examples from the Appalachian Foreland Basin, Eastern United States, *in* D. Gao, ed., Tectonics and sedimentation: Implications for petroleum systems: AAPG Memoir 100, p. 95–124.

Figueiredo, A. M. F., J. A. E., Braga, J. C. Zabalaga, J. J. Oliveira, G. A. Aguiar, O. B. Silva, and L. F. Mato, 1995, Recôncavo Basin, Brazil: A prolific intracontinental rift basin, *in* S. M. Landon, ed., Interior rift basins: AAPG Memoir 59, p. 157–203.

Fleming, R. W., and A. M. Johnson, 1989, Structures associated with strike-slip faults that bound landslide elements: Engineering Geology, v. 27, p. 39–114, doi:10.1016/0013-7952(89)90031-8.

Fox, F. G., 1959, Structure and accumulation of hydrocarbons in southern Foothills, Alberta, Canada: AAPG Bulletin, v. 43, p. 992–1025.

Frostick, L., and I. Reid, 1989, Is structure the main control of river drainage and sedimentation in rifts?: Journal of African Earth Science, v. 8, p. 165–182, doi:10.1016/S0899-5362(89)80022-3.

Gao, D., and J. Milliken, 2012, Cross-regional intraslope lineaments on the Lower Congo Basin Slope, offshore Angola (west Africa): Implications for tectonics and petroleum systems at passive continental margins, *in* D. Gao, ed.,

Tectonics and sedimentation: Implications for petroleum systems: AAPG Memoir 100, p. 229–248.

Gao, D., and R. C. Shumaker, 1996, Subsurface geology of the Warfield structures in southwestern West Virginia: Implications for tectonic deformation and hydrocarbon exploration in the central Appalachian Basin: AAPG Bulletin, v. 80, p. 1242–1261.

Gao, D., R. C. Shumaker, and T. H. Wilson, 2000, Along-axis segmentation and growth history of the Rome trough: Central Appalachian Basin: AAPG Bulletin, v. 84, p. 75–99, doi:10.1306/C9EBCD6F-1735-11D7-8645000102C1865D.

Gao, D., L. Seidler, D. Quirk, M. Bissada, M. Farrell, and D. Hsu, 2009, Intraslope northwest-trending lineaments and geologic implications in the central Campos Basin, offshore Brazil: AAPG Search and Discovery article 90100: http://www.searchanddiscovery.com/abstracts/html/2009/intl/abstracts/gao.htm (accessed November 9, 2011).

Gawenda, P., J-M. Conne, A. Hayman, and M. Marchat, 2004, Offshore west Africa offers exceptional opportunities: West Africa—Dana Petroleum Inc.: Offshore, p. 42–44.

Gawthorpe, R. L., and M. R. Leeder, 2000, Tectonosedimentary evolution of active extensional basins: Basin Research, v. 12, p. 195–218, doi:10.1046/j.1365-2117.2000.00121.x.

Gomes, P. O., B. Kilsdonk, T. Grow, J. Minken, and R. Barragan, 2012, Tectonic evolution of the outer high of santos basin, southern São Paulo Plateau, Brazil, and implications for hydrocarbon exploration, *in* D. Gao, ed., Tectonics and sedimentation: Implications for petroleum systems: AAPG Memoir 100, p. 125–142.

Gries, R., 1983, Oil and gas prospecting beneath Pre-Cambrian of foreland thrust plates in the Rocky Mountains: AAPG Bulletin, v. 67, p. 1–28.

Guiraud, R., and W. Bosworth, 1997, Senonian basin inversion and rejuvenation of rifting in Africa and Arabia: Synthesis and implication to plate-scale tectonics: Tectonophysics, v. 282, p. 39–82, doi:10.1016/S0040-1951(97)00212-6

Guiraud, M., A. Buta-Neto, and D. Quesne, 2010, Segmentation and differential postrift uplift at the Angola margin as recorded by the transform-rifted Benguela and oblique-to-orthogonal-rifted Kwanza basins: Marine and Petroleum Geology, v. 27, p. 1040–1068.

Harding, T. P., 1973, Newport-Inglewood trend, California: An example of wrenching style of deformation: AAPG Bulletin, v. 57, p. 97–116.

Harding, T. P., 1974, Petroleum traps associated with wrench faults: AAPG Bulletin, v. 58, p. 1290–1304.

Harding, T. P., 1990, Identification of wrench faults using subsurface structural data: Criteria and pitfalls: AAPG Bulletin, v. 74, p. 1590–1609.

Harding, T. P., and J. D. Lowell, 1979, Structural styles, their plate-tectonic habitats, and hydrocarbon traps in petroleum provinces: AAPG Bulletin, v. 63, p. 1016–1058.

Harris, L. D., 1978, The Eastern Interior aulacogen and its relation to Devonian shale gas production: Second Eastern Gas Shales Symposium, U.S. Department of Energy, Morgantown Energy Technology Center, DOE/METC/SP-78/6, p. 55–72, doi:10.1306/02260403069.

Harris, L. D., and R. Milici, 1977, Characteristics of thin-skinned style of deformation in the southern Appalachians and potential hydrocarbon traps: U.S. Geological Survey Professional Paper 1018, 40 p.

Harris, N. B., K. H. Freeman, R. D. Pancost, T. S. White, and G. D. Mitchell, 2004, The character and origin of lacustrine source rocks in the Lower Cretaceous synrift section, Congo Basin, west Africa: AAPG Bulletin, v. 88, p. 1163–1184.

Hsiao, L. Y., S. A. Graham, and N. Tilander, 2004, Seismic reflection imaging of a major strike-slip fault zone in a rift system: Paleogene structure and evolution of the Tan-Lu fault system, Liaodong Bay, Bohai, offshore China: AAPG Bulletin, v. 88, p. 71–97, doi:10.1306/09090302019.

Hudec, M. R., and M. P. A. Jackson, 2002, Structural segmentation, inversion, and salt tectonics on a passive margin: Evolution of the inner Kwanza Basin, Angola: Geological Society of America Bulletin, v. 114, p. 1222–1244, doi:10.1130/0016-7606(2002)114<1222:SSIAST>2.0.CO;2.

Jackson, J. A., and N. J. White, 1989, Normal faulting in the upper continental crust-observations from regions of active extension: Journal of Structural Geology, v. 11, p. 15–36, doi:10.1016/0191-8141(89)90033-3.

Jackson, M. P. A., M. R. Hudec, and D. C. Jennette, 2004, Insights from a gravity-driven linked system in deep-water lower Congo Basin, Gabon, *in* P. J. Post, D. L. Olson, K. T. Lyons, S. L. Palmes, P. F. Harrison, and N. C. Rosen, eds., Salt-sediment interactions and hydrocarbon prospectivity: Concepts, applications, and case studies for the 21st century: 24th Annual Gulf Coast Section SEPM Foundation Bob F. Perkins Research Conference, p. 735–752.

Katz, B., 1995, A survey of rift basin source rocks, *in* J. J. Lambiase, ed., Hydrocarbon habitat in rift basins: Geological Society (London) Special Publication 80, p. 213–240, doi:10.1144/GSL.SP.1995.080.01.11.

Kilby, R. E., M. P. A. Jackson, and M. R. Hudec, 2004, Preliminary analysis of thrust kinematics in the Lower Congo Basin, deep-water southern Gabon (abs.): Geological Society of America, Abstracts with Programs, v. 36, no. 5, p. 505.

Kolla, V., Ph. Bourges, J.-M. Urruty, and P. Safa, 2001, Evolution of deep-water Tertiary sinuous channels offshore Angola (west Africa) and implications for reservoir architecture: AAPG Bulletin, v. 85, p. 1373–1405, doi:10.1306/8626CAC3-173B-11D7-8645000102C1865D.

Lambiase, J. J., and W. Bosworth, 1995, Structural controls on sedimentation in continental rifts, *in* J. J. Lambiase, ed., Hydrocarbon habitat in rift basins: Geological Society (London) Special Publication 80, p. 117–144, doi:10.1144/GSL.SP.1995.080.01.06.

Lawrence, S. R., S. Munday, and R. Bray, 2002, Regional geology and geophysics of the eastern Gulf of Guinea (Niger Delta to Rio Muni): The Leading Edge, v. 21, no. 11, p. 1112–1117, doi:10.1190/1.1523752.

Linzer, H.-G., and G. C. Tari, 2012, Structural correlation between the Northern Calcareous Alps (Austria) and the Transdanubian Central Range (Hungary), *in* D. Gao, ed., Tectonics and sedimentation: Implications for petroleum systems: AAPG Memoir 100, p. 249–266.

Londono, J., J. M. Lorenzo, and V. Ramirez, 2012, Lithospheric flexure and related base-level stratigraphic cycles in continental foreland basins: An example from the Putumayo Basin, Northern Andes, *in* D. Gao, ed., Tectonics and sedimentation: Implications for petroleum systems: AAPG Memoir 100, p. 357–375.

López-Gamundí, O., and R. Barragan, 2012, Structural framework of Lower Cretaceous half-grabens in the presalt section of the southeastern continental margin of Brazil, *in* D. Gao, ed., Tectonics and sedimentation: Implications for petroleum systems: AAPG Memoir 100, p. 143–158.

Luo, H., and D. Nummedal, 2012, Forebulge migration: A three-dimensional flexural numerical modeling and subsurface study of southwestern Wyoming, *in* D. Gao, ed., Tectonics and sedimentation: Implications for petroleum systems: AAPG Memoir 100, p. 377–395.

Mann, P., 2012, Comparison of structural styles and giant hydrocarbon occurrences within four active strike-slip regions: California, Southern Caribbean, Sumatra, and East China, *in* D. Gao, ed., Tectonics and sedimentation: Implications for petroleum systems: AAPG Memoir 100, p. 43–93.

Mann, P., L. M. Gahagan, and M. B. Gordon, 2003, Tectonic setting of the world's giant oil fields, *in* M. Halbouty and M. Horn, eds., Giant oil and gas fields of the decade, 1990–2000: AAPG Memoir 78, p. 15–105.

Mitra, S., 1988, Effects of deformation mechanisms on reservoir potential in central Appalachian overthrust belt: AAPG Bulletin, v. 72, p. 536–554.

Morley, C. K., 1995, Developments in the structural geology of rifts over the last decade and their impact on hydrocarbon exploration, *in* J. J. Lambiase, ed., Hydrocarbon habitat in rift basins: Geological Society (London) Special Publication 80, p. 1–32, doi:10.1144/GSL.SP.1995.080.01.01.

Nombo-Makaya, N. L., and C. H. Han, 2009, Pre-salt petroleum system of Vandji-Conkouati structure (Lower Congo Basin), Republic of Congo: Research Journal of Applied Sciences, v. 4, p. 101–107.

Oliveira, M. J. R., P. V. Zalán, J. L. Caldeira, A. Tanaka, P. Santarem, I. Trosdtorf Jr., and A. Moraes, 2012, Linked extensional compressional tectonics in gravitational systems in the Equatorial Margin of Brazil, *in* D. Gao, ed., Tectonics and sedimentation: Implications for petroleum systems: AAPG Memoir 100, p. 159–178.

Posamentier, H. W., and V. Kolla, 2003, Seismic geomorphology and stratigraphy of depositional elements in deep-water settings: Journal of Sedimentary Research, v. 73, p. 367–388, doi:10.1306/111302730367.

Rouby, D., H. Fossen, and P. R. Cobbold, 1996, Extension, displacement, and block rotation in the larger Gullfaks area, northern North Sea, determined from map view restoration: AAPG Bulletin, v. 80, p. 875–890.

Rowan, M. G., 1997, Three-dimensional geometry and evolution of a segmented detachment fold, Mississippi Fan fold belt, Gulf of Mexico: Journal of Structural Geology, v. 19, p. 463–480, doi:10.1016/S0191-8141(96)00098-3.

Shaw, J. H., and J. Suppe, 1994, Active faulting and growth folding in the eastern Santa Barbara Channel, California: Geological Society of America Bulletin, v. 106, p. 607–626, doi:10.1130/0016-7606(1994)106<0607:AFAGFI>2.3.CO;2.

Shumaker, R. C., 1986, Structural development of Paleozoic continental basins of eastern North America: Proceedings of the 6th International Conference on Basement Tectonics, p. 82–95.

Shumaker, R. C., and T. H. Wilson, 1996, Basement structure of the Appalachian foreland in West Virginia: Its style and affect on sedimentation, *in* B. A. van der Pluijm and P. A. Catacosinos, eds., Basement and basins of eastern North America: Geological Society of America Special Paper 308, p. 141–155, doi:10.1130/0-8137-2308-6.139.

Stone, D. S., 1969, Wrench faulting and Rocky Mountains tectonics: The Mountain Geologist, v. 6, no. 2, p. 67–79.

Sullivan, E. C., K. J. Marfurt, A. Lacazette, and M. Ammerman, 2006, Application of new seismic attributes to collapse chimneys in the Fort Worth Basin: Geophysics, v. 71, p. B111–B119, doi:10.1190/1.9781560801900.ch18.

Suppe, J., 1980, Imbricate structure of western Foothills Belts, south-central Taiwan: Petroleum Geology of Taiwan, v. 17, p. 1–16.

Suppe, J., G. T. Chou, and S. C. Hook, 1992, Rates of folding and faulting determined from growth strata, *in* K. R. McClay ed., Thrust tectonics: London, Chapman & Hall, p. 105–121.

Talbot, M. R., 1988, The origins of lacustrine oil source rocks: Evidence from lakes of tropical Africa, *in* A. J. Fleet, K. Kelts, and M. R. Talbot, eds., Lacustrine petroleum source rocks: Geological Society (London) Special Publication 40, p. 29–43, doi:10.1144/GSL.SP.1988.040.01.04.

Tari, G., H. Jabour, J. Molnar, D. Valasek, and M. Zizi, 2012, Deep-water exploration in Atlantic Morocco: Where are the reservoirs?, *in* D. Gao, ed., Tectonics and sedimentation: Implications for petroleum systems: AAPG Memoir 100, p. 337–355.

Tiercelin, J. J., P. Thuo, J. L. Potdevin, and T. Nalpas, 2012, Hydrocarbon prospectivity in Mesozoic and Early–Middle Cenozoic Rift Basins of central and northern Kenya, Eastern Africa, *in* D. Gao, ed., Tectonics and sedimentation: Implications for petroleum systems: AAPG Memoir 100, p. 179–207.

Versfelt, J. W., 2010, South Atlantic margin rift basin asymmetry and implications for pre-salt exploration: Search and Discovery article 30112: http://www.searchanddiscovery.com/documents/2010/30112versfelt/ndx_versfelt.pdf (accessed November 9, 2011).

Verzhbitsky, V. E., S. D. Sokolov, E. M. Frantzen, A. Little, M. I. Tuchkova, and L. I. Lobkovsky, 2012, The South Chukchi Sedimentary Basin (Chukchi Sea, Russian Arctic): Age, structural pattern, and hydrocarbon potential, *in* D. Gao, ed., Tectonics and sedimentation: Implications for petroleum systems: AAPG Memoir 100, p. 267–290.

Weimer, P., and R. Buffler, 1992, Structural geology and evolution of the Mississippi Fan fold belt, deep Gulf of Mexico: AAPG Bulletin, v. 76, p. 225–251.

Wernicke, B. P., 1981, Low-angle normal faults in the Basin and Range province: Nappe tectonics in an extending orogen: Nature, v. 291, p. 645–648, doi:10.1038/291645a0.

Wheeler, R. L., 1980, Cross-strike-structural discontinuities: Possible exploration tool for natural gas in Appalachian overthrust belt: AAPG Bulletin, v. 64, p. 2166–2178.

Wilcox, R. E., T. P. Harding, and D. R. Seely, 1973, Basic wrench tectonics: AAPG Bulletin, v. 57, p. 74–96, doi:10.1306/819A424A-16C5-11D7-8645000102C1865D.

Wolak, J. M., and M. H. Gardner, 2008, Synsedimentary structural growth in a deep-water reservoir, Alba Field, Equatorial Guinea: AAPG International Conference and Exhibition, Cape Town, South Africa 2008, AAPG Search and Discovery article 90082: http://www.searchanddiscovery.com/abstracts/html/2008/intl_captown/abstracts/495780.htm (accessed December 22, 2011).

Xu, J., 1993, Basic characteristics and tectonic evolution of the Tancheng-Lujiang fault zone, *in* J. Xu, ed., The Tancheng-Lujiang strike-slip fault system: New York, John Wiley and Sons, p. 17–50.

Yang, F., D. Gao, Z. Sun, Z. Zhou, Z. Wu, and Q. Li, 2012, The evolution of the South China Sea Basin in the Mesozoic–Cenozoic and its significance for oil and gas exploration: A review and overview, *in* D. Gao, ed., Tectonics and sedimentation: Implications for petroleum systems: AAPG Memoir 100, p. 397–418.

2

Davison, Ian, and John R. Underhill, 2012, Tectonics and sedimentation in extensional rifts: Implications for petroleum systems, *in* D. Gao, ed., Tectonics and sedimentation: Implications for petroleum systems: AAPG Memoir 100, p. 15–42.

Tectonics and Sedimentation in Extensional Rifts: Implications for Petroleum Systems

Ian Davison

Earthmoves Ltd., 38-42 Upper Park Rd., Camberley, Surrey, GU15 2EF, United Kingdom (e-mail: i.davison@earthmoves.co.uk)

John R. Underhill

Grant Institute of Earth Science, School of Geosciences, University of Edinburgh, The King's Buildings, West Mains Rd., EH9 3JW, Scotland (e-mail: jru@staffmail.ed.ac.uk)

ABSTRACT

Extensional rifts and their overlying sag basins host prolific hydrocarbon provinces in many parts of the world. This chapter reviews the development of rifts and the controls of tectonics on sedimentation patterns and hydrocarbon prospectivity. Rift tectonics exert the most important control on sedimentation and trap formation, and subsidence rate controls the geometry and facies of the rift fill. Rifting also controls the heat flow and burial history that determine the source rock maturation. Many rifts commence with evenly distributed small extensional faults when they are commonly characterized by closed drainage and continental sedimentation with localized lacustrine facies. As continental rifts develop, source rocks commonly accumulate in deep lakes, especially when the rifting is rapid, and organic shales are commonly located in the bottom third of the rift. Fault displacements increase and faults grow laterally to produce linked normal fault arrays. Marine deposits commonly replace the early rift continental deposition (e.g., North Sea) as the rift propagates to reach the world ocean system. Extremely prolific source rocks may be produced during rapid rifting in the marine phase, especially if half-graben depocenters are starved and oceanographic circulation is poor (e.g., Kimmeridge Clay Formation of northwestern Europe). As rifting wanes the rift fills, and fluvial sedimentation predominates, or in a passive margin, the basins enter into an extensional sag and/or postrift thermal sag phases when shallow-marine to deep-marine sediments infill and bury the former half grabens as sedimentation catches up and exceeds basin subsidence.

Recent studies in western Australia, Brazil, and west Africa indicate that thick (unfaulted) sag basins can develop very rapidly above rifted continental crust. The subsidence is too rapid to be produced by normal thermal conductive cooling and is believed to be caused by stretching of the lower/middle crust with no observable faulting (depth-dependent stretching; Driscoll and Karner, 1998). The stretching process can produce a broad passive margin (100–500 km [62–311 mi]) where the continental crust is only 5 km (3.1 mi) thick ($\beta = 7$). The sag basin

DOI:10.1306/13351547M1001556

in southern Brazil reaches about 1 to 2 km (0.6–1.2 mi) in thickness and drapes over the earlier rifted blocks to produce some of the largest oil fields discovered in the last 30 yr (e.g., Lula with 20 to 30 billion bbl of oil in place). These fields are trapped in lacustrine algal-bacterial carbonates that pinch out onto the fault block crests. Many rift basins remain unexplored in remote or inaccessible areas, and new future hydrocarbon provinces are anticipated in the African, Southeast Asian, Arctic, and Atlantic margin rift basins.

INTRODUCTION

Extensional rift systems have proven to be highly prospective petroleum provinces, especially in sedimentary basins where migration from a mature source rock occurred during the postrift phase of thermal subsidence following trap formation (e.g., Harding, 1984). It is now thought that rift basins on old passive continental margins (e.g., west Africa and Brazil) and intracontinental rift systems like the North Sea rift, host hundreds of billions of barrels of recoverable oil, and some estimates suggest that more than two thirds of the world's petroleum reserves lie in rift settings (Mann et al., 2003), making extensional systems the most significant of all tectonic regimes. In recent years, additional oil reserves have been discovered in young (Miocene–Holocene) intracontinental rifts such as the Albertine Graben in the East African rift, and more discoveries are expected to be realized in new opportunities like the previously inaccessible rifted margins adjacent to the Labrador Sea, West Greenland, the Beaufort Sea of northern Alaska and Canada, the Russian Arctic Ocean, and the Atlantic Ocean margins. Other rift exploration opportunities exist in areas that were previously off limits and lie on trend with known extensional petroleum provinces such as the offshore Sirte Basin, Libya.

Despite the evident prospectivity associated with rift systems, the study of extensional basins was a rather neglected subject in earth sciences (e.g., Gregory, 1896, 1921; McConnell, 1972) until the oil crisis in the early 1970s provoked an upsurge in hydrocarbon exploration, and seismic reflection surveys became a routine exploration tool. One of the reasons that extensional systems were neglected for so long was that most are buried and, hence, difficult to study. Furthermore, many of the ancient examples are only exposed because they have been inverted and uplifted in response to basin compression (e.g., Palymira fold and thrust belt; Wessex Basin) or now form parts of mountain belts (e.g., Tethyan continental margin of the Alpine-Himalayan Chain), and therefore it is difficult to observe the original rift geometry and facies relationships. However, important insights can be gained using seismic stratigraphy in shallowly buried rifts (e.g., the Inner Moray fifth in the North Sea; Underhill, 1991a), or by the study of well-exposed relatively undeformed rifts (e.g., Gulf of Suez, Aegean domain of Greece, and East Africa).

The amount of extension the crust has undergone is commonly expressed as the β factor (McKenzie, 1978). Continental rifts commonly develop to approximately 20% extension (β = 1.2), then strain hardening of the lithosphere occurs because of replacement of crust by cold continental mantle, and rifting ceases. However, rifts can continue to extend and develop into a passive margin with the formation of new oceanic crust. Complete opening commonly happens where thermal weakening of the lithosphere occurs because of the either focused heat source (e.g., the Parana-Etendeka plume in the South Atlantic; Turner et al., 1994) or a more diffuse thermal weakening (e.g., Central Atlantic Magmatic Province; McHone, 2000). On passive margins, the largest hydrocarbon accumulations are commonly trapped in the postrift thermal sag phase where high-quality reservoirs exist (e.g., South Atlantic margins of Brazil and west Africa). The initial sag basin in the South Atlantic developed in something like 5 m.y. and reaches up to 5 to 6 km (3.1–3.7 mi) thick on the west African margin in the vicinity of the Congo Fan (Henry et al., 2004) (Figure 1). This sag phase is located above the rifted zone and does not overlap onto the basin margins, as is expected with a thermal sag phase (Figure 1) (McKenzie, 1978). The very rapid development of this sag phase also precludes the conductive cooling model of McKenzie (1978), and the rapid subsidence is attributed to lower crustal stretching, where the upper crust remains unaffected, so that no large extensional faults or bed rotations are observed in this presalt sag phase (e.g., Karner and Gamboa, 2007).

This chapter will cover the deposition of the synrift fill and the rapid sag phase attributed to depth-dependent stretching but will not cover the later thermal subsidence phase of sedimentation. We aim to summarize a variety of basins in different locations, but will not cover the longer (postrift) thermal subsidence through time. Most of the examples we will discuss are basins that we have actually worked on ourselves. A plethora of reviews on rift tectonics and sedimentation exists, and here we recommend the following to gain an overview of the main factors to consider when addressing

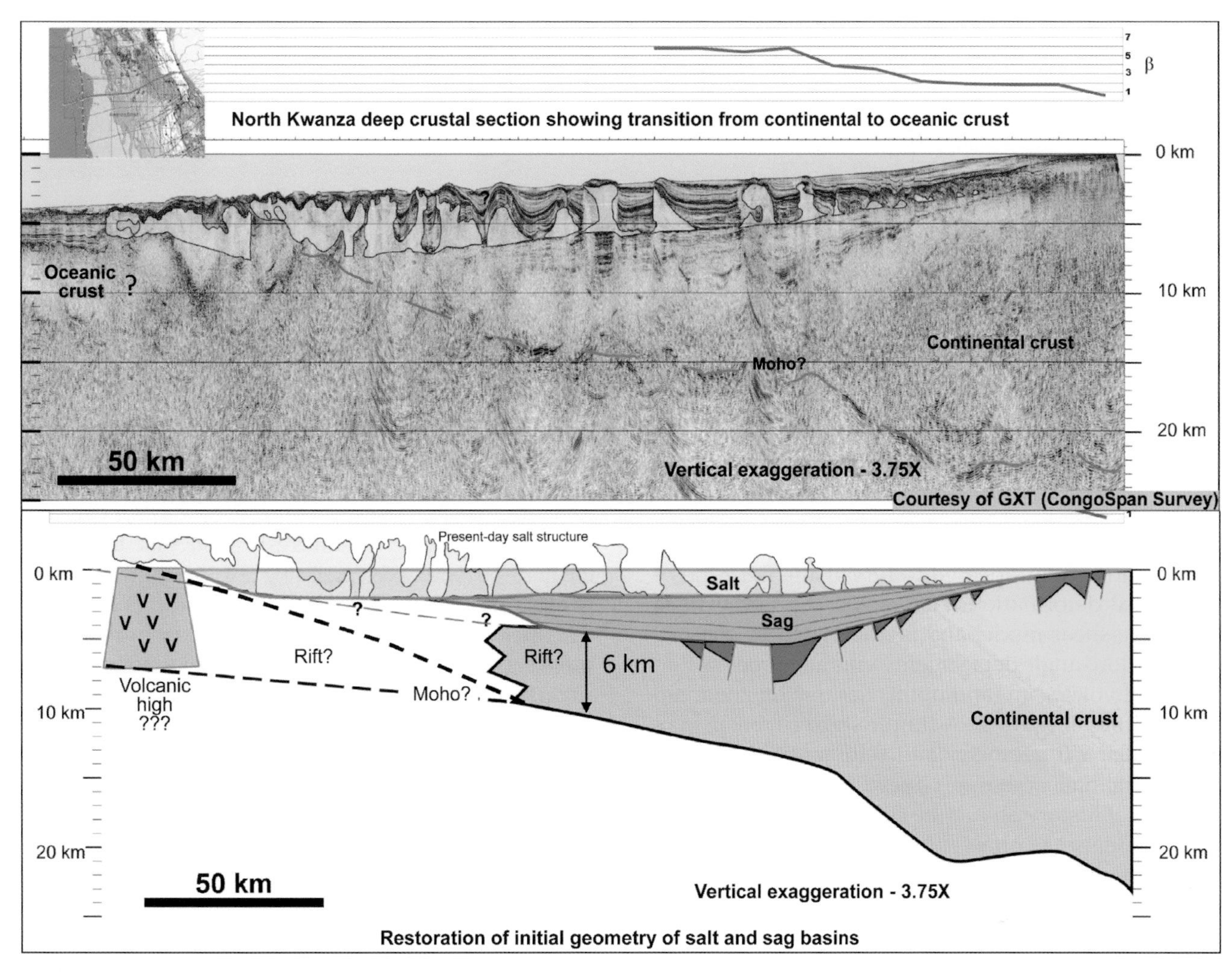

Figure 1. Seismic section and cross section flattened at the top salt level (to remove the tilt effect of thermal subsidence) through a rifted continental margin of west Africa showing a rifted series of relatively small half graben (3–4 km [1.9–2.5 mi] deep), an extensional sag basin approximately 5 km (~3.1 mi) thick, and a thermal sag phase of sedimentation (4 km [2.5 mi] deep). The β curve above the diagram indicates the calculated stretching profile across the eastern half of the section where the moho is imaged clearly. Seismic data courtesy of ION's Basin SPAN programs. 50 km (31 mi).

synsedimentary controls on petroleum play fairways: Frostick and Reid (1989), Schlische (1993), Roberts and Yielding (1994), Lambiase and Bosworth (1995), Morley (1995), Ravnas and Steel (1998), Lambiase and Morley (1999), Gawthorpe and Leeder (2000), Morley (2002b), Leeder (2007), and their impact in seismic stratigraphy (Underhill, 1991b).

RIFT TECTONICS

Rifting can be caused by lithospheric stretching and thinning without volcanic involvement, in such cases, narrow rifts are developed. In hotter heat flow regimes, broad rifted zones (e.g., Afar Triangle and Basin and Range province) develop that have complex fault patterns produced by multiple-phased extension on planar domino and/or listric faults (Morton and Black, 1975; Proffett, 1977) (Figure 2). The very large extensions produced in the Basin and Range province create steeply dipping early rift strata developed on low-angle normal faults with broad shallow basins. These hotter highly extended ($\beta > 2$) shallow basins are generally less prospective than colder rift basins because the fault spacings are relatively small, and the increased heat input maintains the rift elevation so that deep lakes are generally not developed.

Other workers, most notably Wernicke (1985, 2009), argued that rift processes were inherently asymmetric and cited cases like the Basin and Range province in the

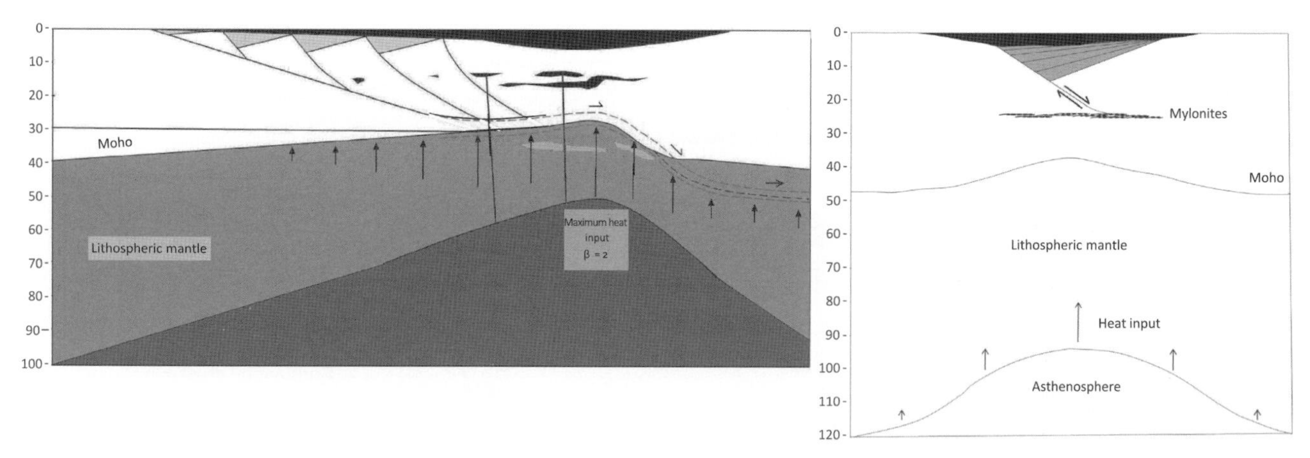

Figure 2. Different rift styles. (A) Basin and range style wide shallow rift with simple shear (Wernicke) model. (B) Deep simple rift with pure shear (McKenzie) stretching model.

western United States, where the locus of main volcanism and uplift appeared to lie outside the zone of extensional deformation. This model is referred to as the simple-shear model and is dominated by the formation of low-angle detachment faults and predicts that the axis of volcanism (and uplift) lies outside the main zone of rifting (Figure 2). A simple shear rift basin may have similar rift geometry and sedimentary fill to an asymmetric half graben produced by pure shear, but the thermal history should be different with less thermal subsidence sag phase developed directly above the central rift axis in the simple shear case (Figure 2). Continental rifting commonly occurs during a 5- to 20-m.y. period, with subsidence of up to 12 km (7.5 mi) recorded in some basins (e.g., Central Tucano Basin, northeastern Brazil; Magnavita et al., 1994; Colorado Basin, Argentina, authors' own observations). This represents a subsidence rate of 0.5 mm to 2 mma^{-1}, with the most rapid rifts reaching 5 mm^{-1} (Ravnas and Steel, 1998). Sedimentation rates can keep pace with even the most rapid subsidence rate where major river input points exist, but commonly, sedimentation rate is less than the maximum tectonic subsidence rate, so that deep lakes form, where the exposed basin floor may dip as much as 10° (Ravnas and Steel, 1998).

RIFT ARCHITECTURE

The main feature of extensional rifts is tilted half grabens where the deepest part of the half graben is adjacent to the fault (Figure 3). Rift fault blocks generally range from 5 to 25 km (3.1–15.5 mi) in width but occasionally reach up to 50 km (31 mi) (Ravnas and Steel, 1998). The deepest rifts (~10 km [~6.2 mi]) appear to develop where the crust is thickest and strongest (and the seismogenic layer thickest) such that local deep basin depocenters develop (e.g., Colorado Basin in Argentina has 10 km [6.2 mi] thickness and is located in the thickest surrounding crust (~40 km [~25 mi]) on the Argentine margin, authors' own observations). The rift is commonly a single half graben (e.g., North American Triassic rifts; Schlische, 1993; Olsen, 1997). In such cases, no structural closures exist within the basin, and any hydrocarbons generated can migrate out to the surface on the flexed rift margin. Stratigraphic traps with ponded lacustrine turbidites or fluvial and shoreline progrades off the hanging wall would be required for the petroleum system to work.

The various aspects of the fault and basin geometry are described below, and the implications for sedimentation and hydrocarbon potential are discussed at the end of this chapter.

Fault Geometry

The recognition of low-angle normal faulting in Wernicke's (1985) model, together with the apparent listricity displayed by seismic time sections, led many interpreters to produce structural interpretations that were consistent with the listric fault model (e.g., Gibbs, 1984). There appeared to be an inconsistency with neotectonic observations, however, because coseismic activity on neotectonic faults (e.g., linking surface or bathymetric rupture with earthquake epicenters) demonstrated that most normal faults in continental rifts are almost planar (straight in cross section) within the upper crust dipping at angles between 30 and 60° (Stein and Barrientos, 1985; Jackson and White, 1989) down to midcrustal levels. The increasing appreciation that the curvature and progressively shallow dip of the faults seen on seismic sections was removed when velocity

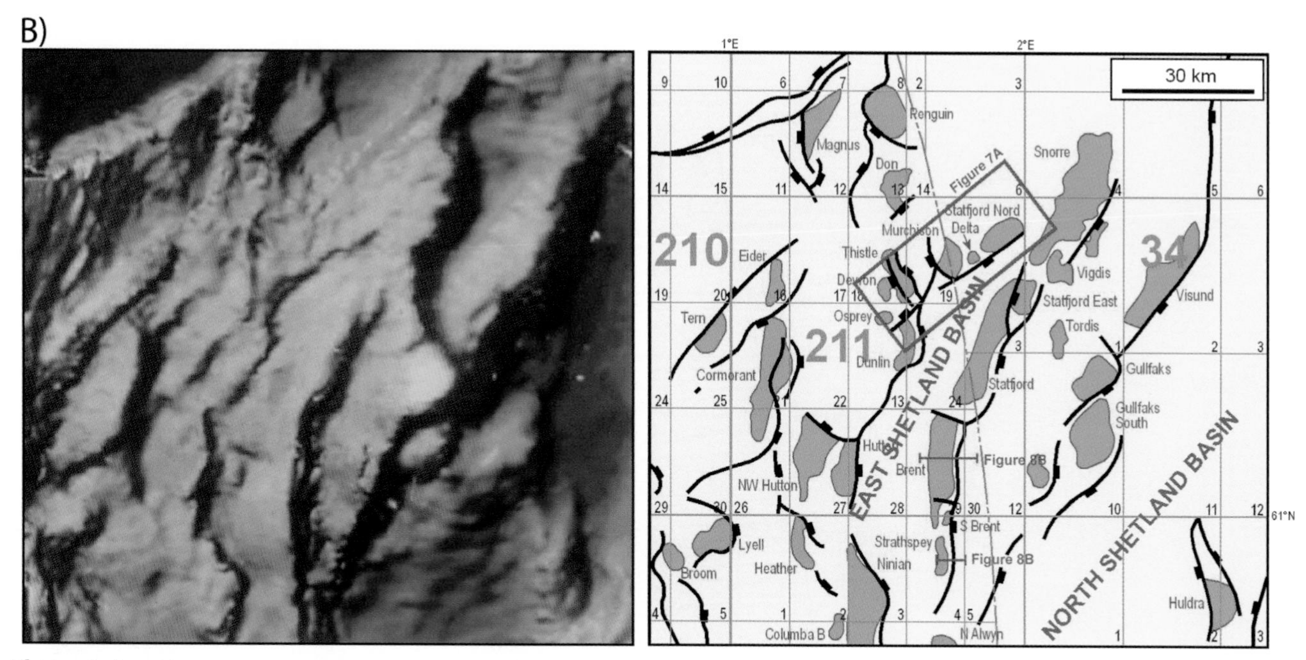

Figure 4. (cont.).

provide insights into how the normal faults nucleated, grew, and evolved to form their final patterns as they represent the end point of the rift process. Instead, insights have had to be gained from physical analog and numerical modeling (e.g., McClay, 1995), from neotectonic studies of active rifts (Cowie and Scholz, 1992a, b; Scholz et al., 1993), and from ancient examples where sufficient stratigraphic control exists to produce structural restorations through time (e.g., the Late Jurassic rift system of the northern North Sea; Dawers and Underhill, 2000; McLeod et al., 2000, 2002).

The growth of normal faults have been studied extensively (e.g., Walsh and Watterson, 1988a; Cowie and Scholz, 1992a, b; Scholz et al., 1993) and is believed to be a self-similar process that leads to fractal geometry (Scholz and Aviles, 1986). These models have predicted growth of the fault displacement to be greatest at the central fault nucleus and decrease outward following a power law decrease (Walsh and Watterson, 1988a) or simple linear displacement gradient to the fault tip (Cowie and Scholz, 1992a). However, the growth of individual faults becomes much more complex as the faults become larger, interact, and link to produce multisegmented major normal faults (Machette et al., 1991; Dawers et al., 1993; Cartwright et al., 1996; Morley, 2002a). Hence, individual depocenters are small and shallow when faulting commences, then depocenters grow and merge through time, and finally, only the largest faults remain active and localized major depocenters develop in their hanging walls. These will be the preferred sites of lake development in continental rifts (Figure 5).

The fault displacement to fault length relationship changes when the fault extends across the whole brittle crust and can no longer grow upward or downward so the fault plane is constrained to grow only laterally. Large faults are also most likely to link up with smaller faults, hence, large faults tend to have larger length/displacement ratios. Most large normal faults have horizontal lengths of 100 to 200 km (62–124 mi) and average maximum displacements approximately 5 km (~3.1 mi); however, the longest active part of a fault that breaks in a single seismic event is approximately 25 km (~15.5 mi) (Jackson and White, 1989) with coseismic slips as great as 6 m (20 ft) and earthquake recurrence intervals ranging from 500 to 7600 yr (Ravnas and Steel, 1998). One of the main controls on ultimate fault length appears to be the thickness of the seismogenic layer. East African examples demonstrate that the thicker the seismogenic layer is, the longer the faults are; and the thinner it is, the shorter the segment lengths (e.g., Jackson and Blenkinsop, 1993; Bilila-Mtakataka fault in Malawi, East Africa; Ebinger et al., 1999). Consequently, a thicker, brittle layer favors large faults and deeper half graben and source rock development. In general, the locus of extension migrates toward the center of a rift province (e.g., Goldsworthy and Jackson, 2001), leading to many faults becoming dormant as rifting proceeds (Gupta et al., 1998) whereas those that lie closest to the rift axis had the longest activity and greatest throw (Cowie et al., 2005). The process also accentuates rift flank uplift and exposure of early synrift sediments in basin margin positions.

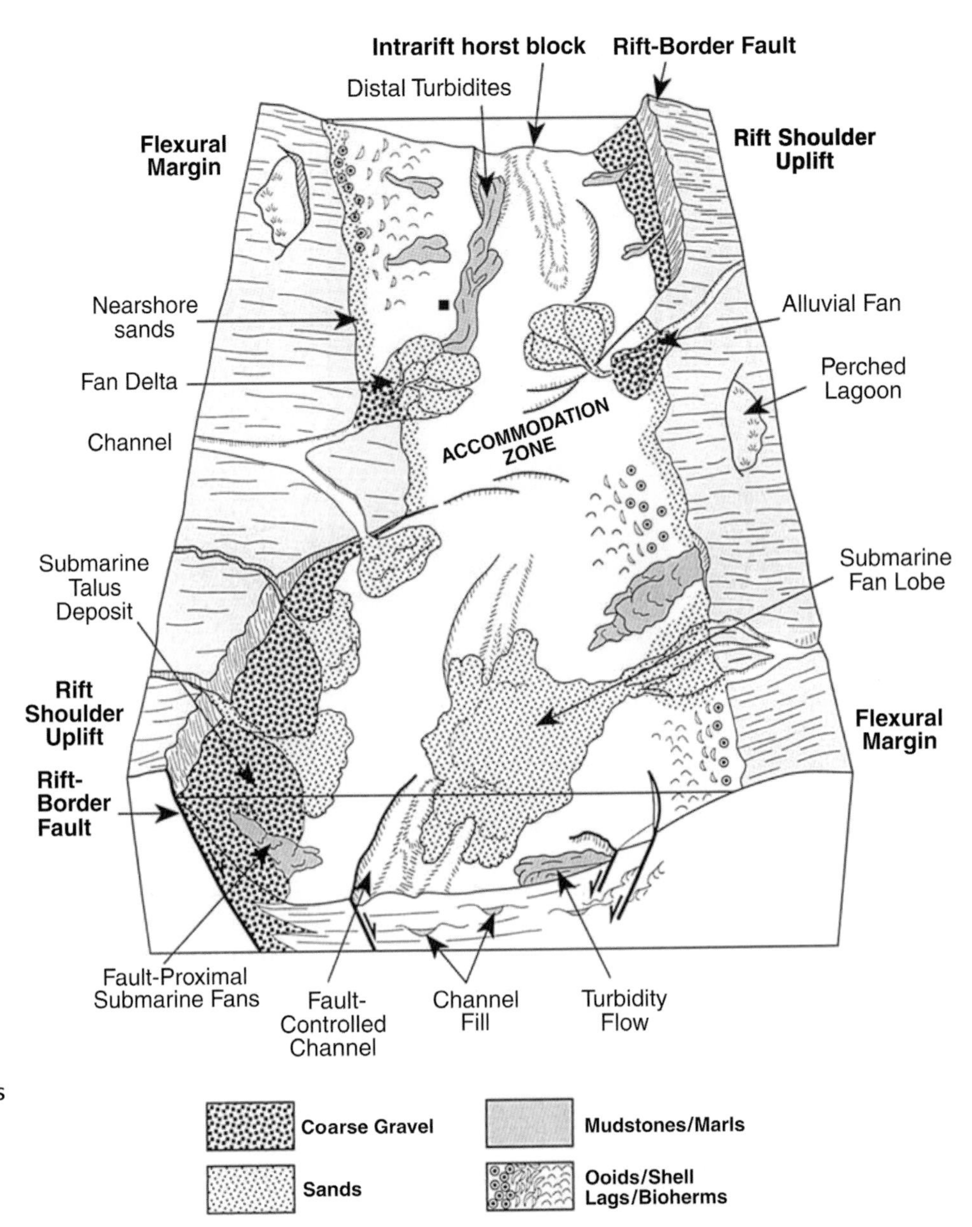

Figure 5. Block diagram of rift showing the footwall uplift zones and accommodation zone produced by the overlapping faults. Modified from Younes and McClay (2002).

Footwall Uplift

Footwall uplift and plume dynamics under a rift commonly produces surrounding rift shoulders up to 1–4 km (0.6–2.5 mi) high. These can trap large isolated lakes, 1–3 km (0.6–1.9 mi) deep, where very large fluctuations in lake level occur because of rapid evaporation and precipitation. For example, the water level of Lake Tanganyika has risen by more than 600 m (>1970 ft) in the last 25 k.y. (Scholz and Rosendahl, 1988). The zone of footwall uplift adjacent to a normal fault reaches a maximum in the center of the fault where the fault displacement reaches a maximum (Jackson and McKenzie, 1983; Yielding, 1990) and the width of the zone affected by footwall uplift can reach up to approximately 20 km (~12.4 mi) distance away from the surface fault trace (Stein and Barrientos, 1985; Stein et al., 1988). The amount of footwall uplift ranges from about 5 up to 20% of the fault throw (Roberts and Yielding, 1994), where uplift is related to the elastic properties of the lithosphere, the throw and dip on the fault, and the fault spacing (Jackson and McKenzie, 1983; Barr, 1987a, b). The amount of footwall uplift also depends on the density of the rift valley fill. Air-filled empty graben footwalls are uplifted more than water-filled or sediment-filled rifts, where the sediment loads in the hanging wall is loading the footwall of the major normal faults. Conversely, when the uplifted footwall block is eroded and the material is removed and is not deposited in the adjacent hanging-wall basin, then the basin itself will be uplifted across a zone up to 50 km (31 mi) wide away from the fault (Magnavita et al., 1994). This effect is most marked in rift shoulders of passive margins. Eventually, the footwall fault scarp will be eroded back to create a drainage catchment area so that sediment will be transported toward the rift instead of away from it. This change in sediment flow direction from the uplifted footwall blocks may take several million years.

If the footwall is uplifted above sea or lake level, subaerial erosion is important and sediment is redeposited into the surrounding hanging-wall graben. However, if the footwall blocks are submerged, subaqueous erosion is limited and the basin will be relatively more starved (e.g., Upper Jurassic of the Central Graben, North Sea).

Unconformities and surprising thickness trends are commonly produced in extensional provinces as new fault sets develop and transect older footwall uplifts (e.g., in Grand Banks, offshore Morocco, northwestern Australia) such that restoration of the later rift system is the only way to accurately assess the disposition of older plays. In the North Sea, restoration for the Late Jurassic extension has revealed prospectivity in the underlying precursor Triassic rift (Tomasso et al., 2008) with discovery of new fields (e.g., Alwyn North). This has extended the production life of the basin and allowed for a better understanding of the footwall and hanging-wall positions, the synsedimentary control on sedimentation and, hence, highlighted new exploration opportunities at that stratigraphic level.

Cross-Faults, Relay Ramps, and Accommodation Zones

Faults that form at high angles to the rift axis were termed transfer faults by Gibbs (1984) and transverse faults by Colletta et al. (1988). These faults do not form in isotropic scaled models of extension (McClay, 1995), nor are they observed in many rifts even where dip polarity changes occur along strike (e.g., Gulf of Suez and the North Sea). Instead, these areas are flexed and only lightly deformed. Where two major bounding faults of opposing polarity propagate toward each other, this produces a zone of arching between the two faults known as an accommodation zone (Bosworth, 1985; Rosendahl, 1987; Moustafa, 1997) (Figures 5, 6), and the interference of two hanging-wall flexures can produce a structural arch between the two interacting faults (Magnavita et al., 1994).

Most hard-linked transfer faults in rifts are believed to be controlled by inherited structures from the basement (e.g., Milani and Davison, 1988; McClay and Khalil, 1998) (Figures 5, 6) or reactivated precursor normal fault systems that underlie the later rift, which are activated during rifting because they are preferred zones of weakness.

Shorter cross-fault elements that branch from the master extensional fault are increasingly being recognized along normal fault arrays. They tend to be perpendicular to their controlling faults, show rapid decrease in displacement away from the master fault, and appear unrelated to preexisting faults. They have been interpreted as release faults (Destro, 1995; Destro et al., 2003) and accommodate the strain resulting from along-strike differential line stretch (Figure 7A). Although published examples have been confined to the hanging wall, where the stretch is greatest, several North Sea fields lying in the footwalls to major normal faults are believed to be either defined by such structures or compartmentalized by them (e.g., along the Brent-Statfjord and Murchison-Statfjord Nord fault arrays in the northern North Sea; McLeod et al., 2002) (Figure 7B). However, larger cross faults (>1 km [>0.6 mi] throw) could not be generated by this mechanism because the predicted differential strains along a fault are relatively small.

Relay ramps and transfer faults are preferred entry points for clastic sediments to enter the rift, and they also produce a highly structured zone with multiple structural closures along the transfer fault. They are a natural focus for oil fields to develop. In the Gulf of Suez, 4 of the 6 billion bbl of oil reserves are hosted by transfer zones (Morley, 1995), and a concentration of oil fields is apparent along the Matu-Catu Fault in the Recôncavo Basin of Brazil, where several major fields are located (Figure 6A) (Figueiredo et al., 1995).

Scarp Degradation and Failure

One major consequence of fault propagation and ever-increased displacement on the master normal fault is that a progressive increase in bathymetric or topographic relief exists in the center of fault segments as the fault arrays propagate and accumulate strain. This leads to the footwall block being exposed to subaerial weathering and erosion, or to it forming a major submarine scarp. In cases, where the footwall lithology is strong, such as with metamorphosed Mesozoic limestones in the Aegean domain, or the Mazagan Plateau in northwestern Morocco, large topographic fault scarps up to 4 km (2.5 mi) high can result (Ruellan and Auzende, 1985). However, where weaker and unconsolidated sediments make up the exposed footwall, landslides or submarine slumping commonly result. The progressive degradation of footwalls can lead to complete erosion of the scarp and subsequent stratal onlap characterizing the leading edge of many extensional fault blocks (Figure 8). In some cases, the more resistant

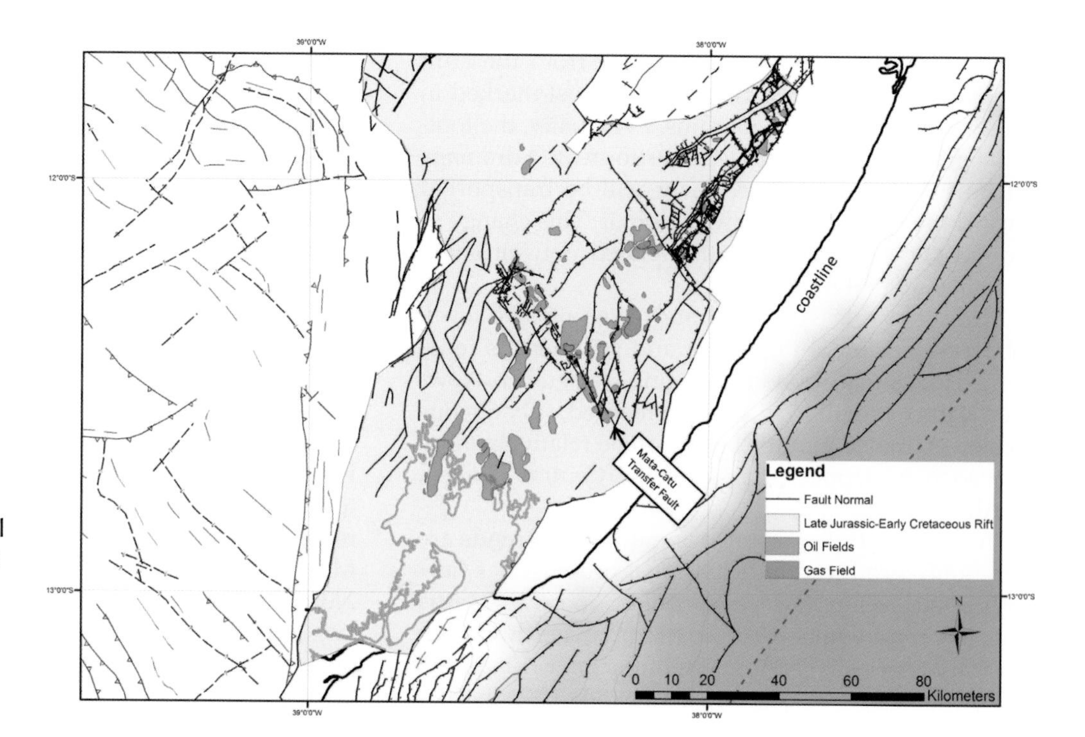

Figure 6. Maps of normal and transfer fault surface or subsurface traces. (A) Recôncavo. (B) Gulf of Suez. Modified from Younes and McClay (2002). 10 km (6.2 mi).

lithologies lead to the erosional unconformity having a distinctive stepped (ramp-flat) topographic profile; whereas in others, the footwall domain can be dominated by footwall-mounted fault-scarp degradation complexes with intense collapse faulting (Figure 8) (Underhill et al., 1997). These features are inherently difficult to image seismically and to produce oil from, being so structurally complex. However, as examples from the Brent Province in the northern North Sea demonstrate (e.g., Brent, Statfjord, Ninian, and Strathspey fields), a significant oil prize is commonly contained within them. Consequently, new methods of seismic processing are being used in an attempt to exploit these features in large North Sea oil fields affected by fault scarp degradation (e.g., Underhill et al., 1997; McLeod and Underhill, 1999).

SEDIMENTATION IN RIFTS

Tectonic subsidence is the main control on the geometry and facies of the synrift fill (Ravnas and Steel, 1998). However, in the early stages of a shallow-marine rift, glacioeustatic sea level changes can be very important. The initial subsidence can be very slow in some basins, for example, Viking Graben when the Middle Jurassic Brent Group was deposited or the Upper Jurassic Sergi and Alliance formations were deposited in the Recôncavo Basin of Brazil. Both of these occurrences represent approximately 500 m (~1640 ft) of subsidence through 10 m.y. These sediments appear as parallel-bedded units that spread evenly across the rift. When the extension rate increases, several main extensional faults dominate the system, and large asymmetric half grabens are developed. It is rare for symmetric graben to develop where the boundary faults are of equal magnitude and opposite polarity, except in zones of overlapping fault tips with opposing polarity.

Many continental rifts have a similar overall stratigraphy that commences with a basal fluvial unit dominated by coarse clastic sediment, followed by a transition to deep lacustrine sediments. When the lake fills up, the sedimentation produces a deltaic and lake shoreline and deep-water lacustrine facies and, finally, fluvial and alluvial facies (Figure 3). The thickness of the early-stage fluvial deposits will depend on the subsidence rate, and this phase may even be absent with very fast subsidence rates. This sequence development is produced by a slow initial extension with distributed faulting, followed by the accelerated subsidence and deep lake development during the main rifting phase, terminating with a waning of rift subsidence when the basin fills in.

When subsidence is more rapid than sedimentation rates, strong onlap of stratigraphic units is observed onto the flexed margin of the rift (underfilled model Figure 9A). If subsidence is equal to sedimentation rates (balanced model), stratigraphic units thicken into the hanging wall of the main graben boundary fault and the stratigraphic bounding surfaces converge to the edge of

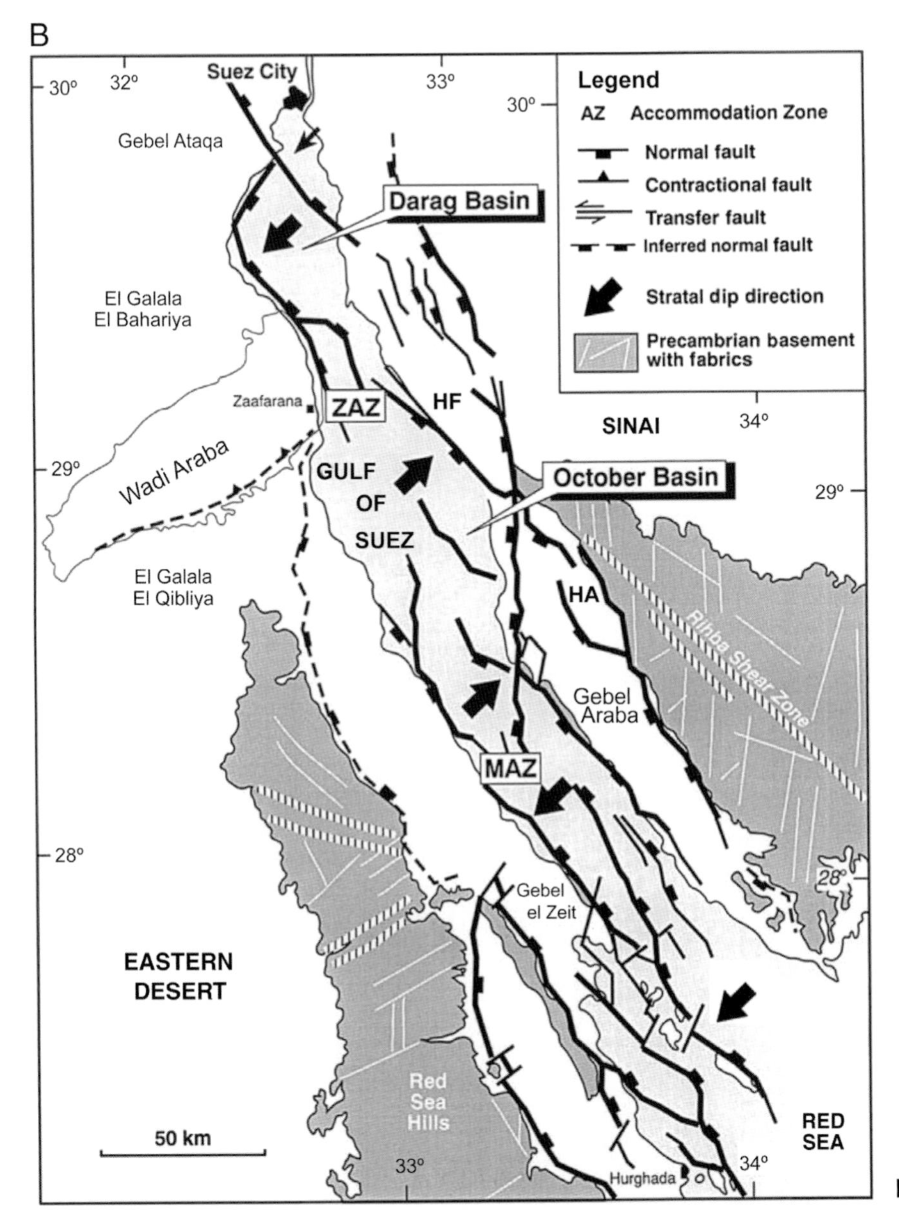

Figure 6. (cont.).

the flexed margin (Figure 9B). In rifts where sedimentation outpaces subsidence, prograding wedges of sediment will be observed that can either prograde along the rift or perpendicular to the rift axis (e.g., overfilled model). The reader is referred to an excellent discussion of facies variation and marine rift fill architecture based on North Sea examples by Ravnas and Steel (1998).

Footwall uplift produced during rifting causes sediments to initially flow away from rifts (Frostick and Reid, 1989). However, once fault scarps begin to erode these, source coarse-grained alluvial fans and deltas are distributed along the basin margins. The fans are generally limited to a zone of approximately 5 km (~3.1 mi) in width away from the border fault and reach up to 4 km (2.5 mi) in thickness (Steel, 1976; Magnavita and Da Silva, 1995). A lag of several million years commonly exists before marginal alluvial fans begin to develop as the footwall scarp has to degrade before a drainage catchment area develops that is large enough to source significant volumes of coarse conglomeratic material. Hence, early axial-derived turbidites, with good reservoir properties, can sometimes be found below the alluvial fan. These deep reservoirs are not imaged on the seismic data as the overlying alluvial fan top reflects all the energy and the top of the fan is commonly mistaken as top basement (e.g., Figure 10).

In some cases, these marginal alluvial fans do not develop because little or no erosion of the footwall exists. This can occur in areas where the rift is marine

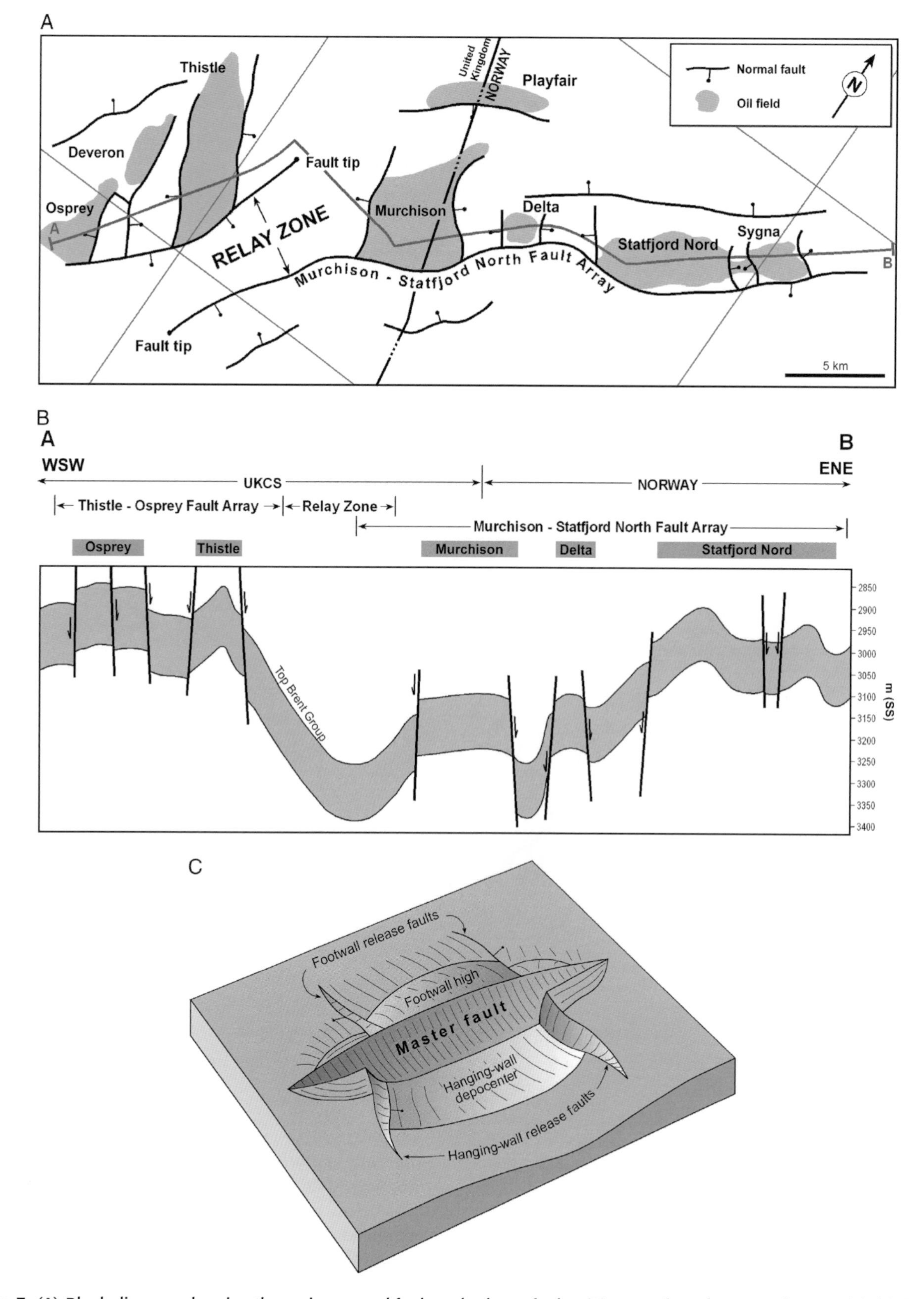

Figure 7. (A) Block diagram showing the major normal fault and release faults. (B) Map of northern North Sea oil fields located in Figure 4. (C) Longitudinal section along the major North Sea fault showing the interpreted release faults. UKCS = UK continental shelf; ss = subsea. 5 km (3 mi).

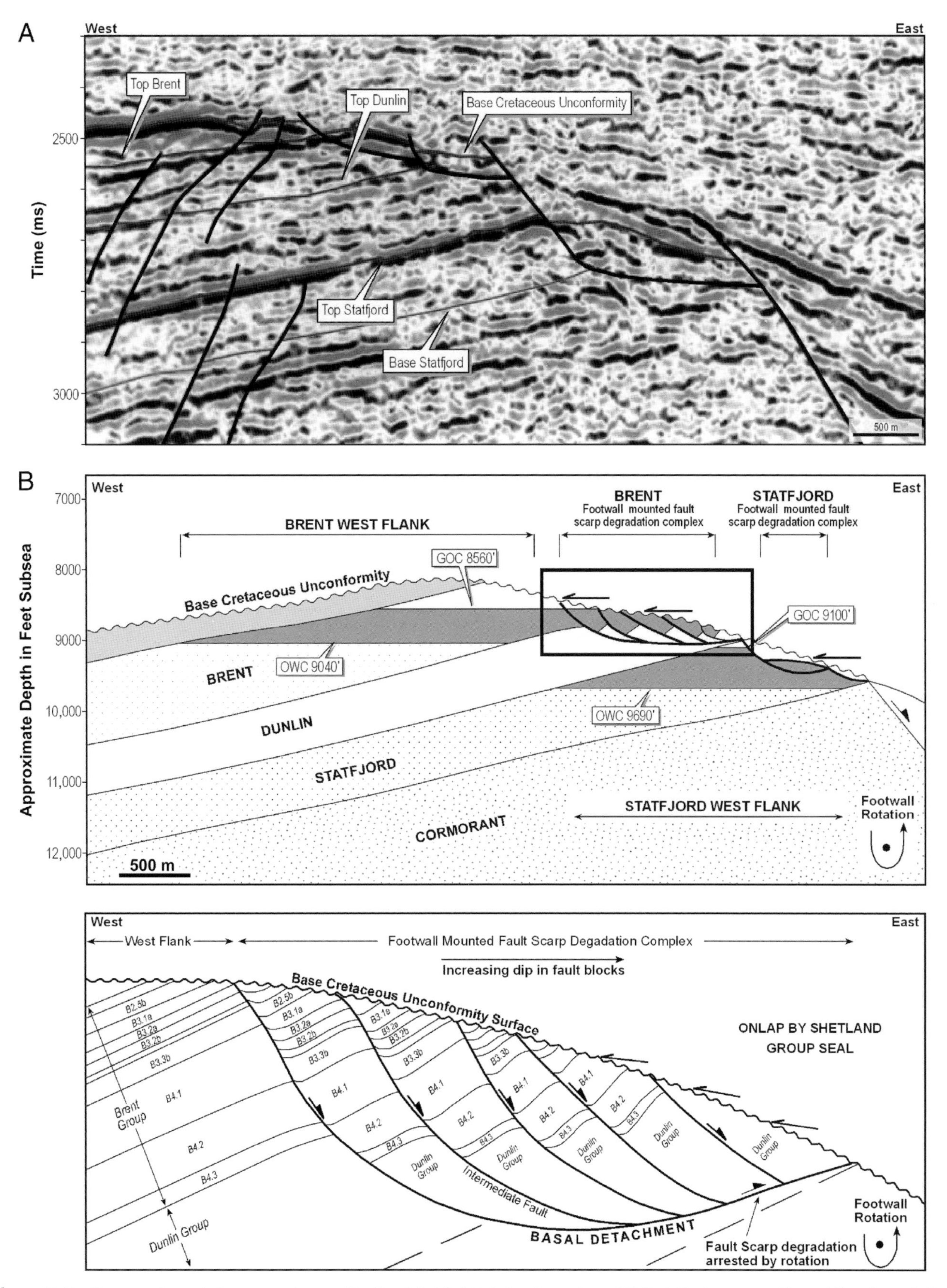

Figure 8. Fault scarp degradation complex on Statfjord field. (A) Seismic section. (B) Detailed cross section of the degradation complex. Location of the section shown in Figure 4B. Modified after Taylor et al. (2003). 500 m (1640 ft), 1000 ft (305 m).

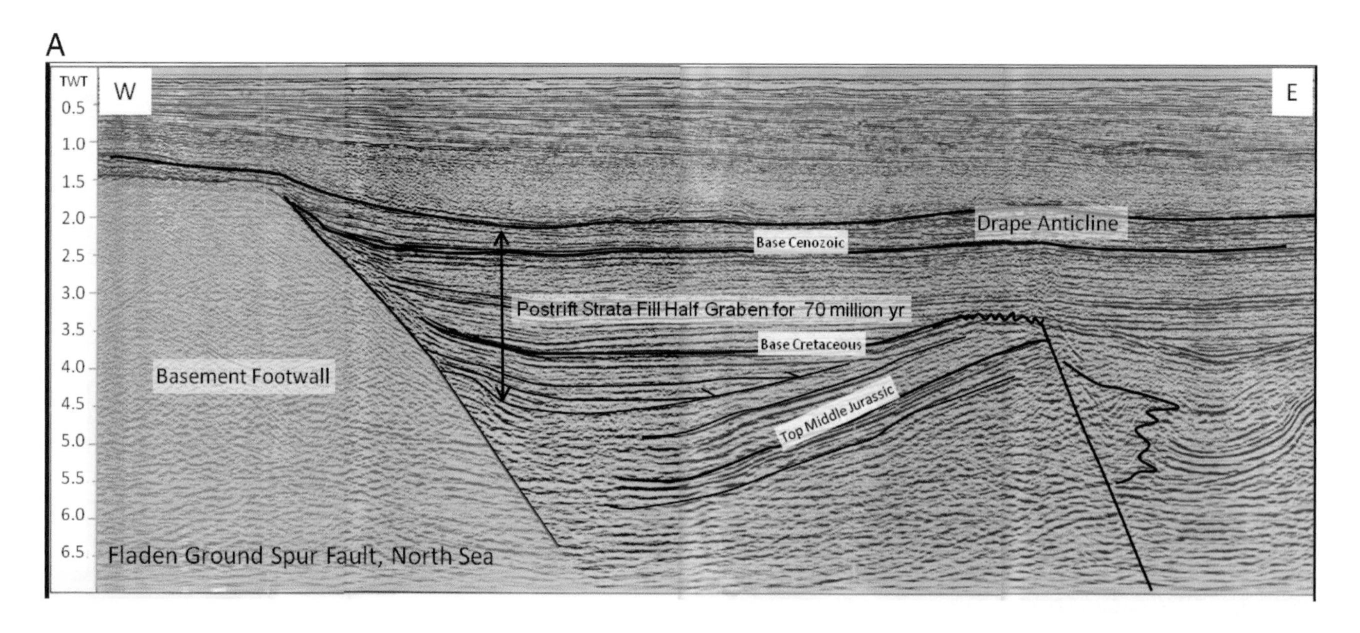

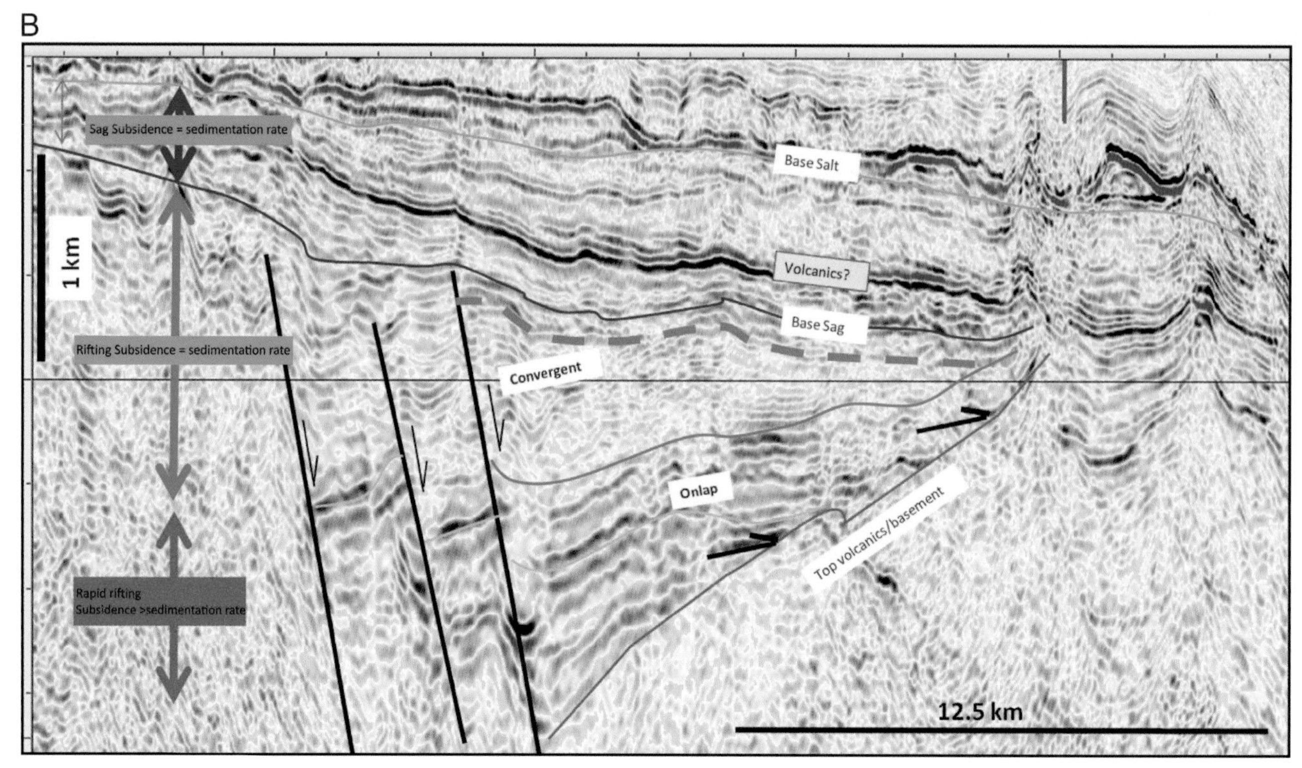

Figure 9. Stratal geometries in (A) underfilled starved half-graben basin geometry from the Fladen Ground Spur North Sea. (B) Balanced basin in mid rift phase where subsidence = sedimentation rate from the Santos Basin in Brazil. Data courtesy of ION's basin SPAN programs. 1 km (0.6 mi), 12.5 km (7.8 mi).

and the footwall is submerged, as is the case in the Central Graben of the North Sea during Late Jurassic rifting, so only deep-water shales are developed adjacent to the Coffee Soil Fault scarp (Cartwright, 1991).

In many rift systems, the main fault scarp and associated footwall uplift act as a topographic barrier for drainage, and dispersal patterns are either directed away from the adjacent half graben or are routed and focused into it through relay ramps between individual fault segments (Peacock and Sanderson, 1991). During the main rift phase in the Recôncavo and East African rift systems, the main coarse clastic sediment input to

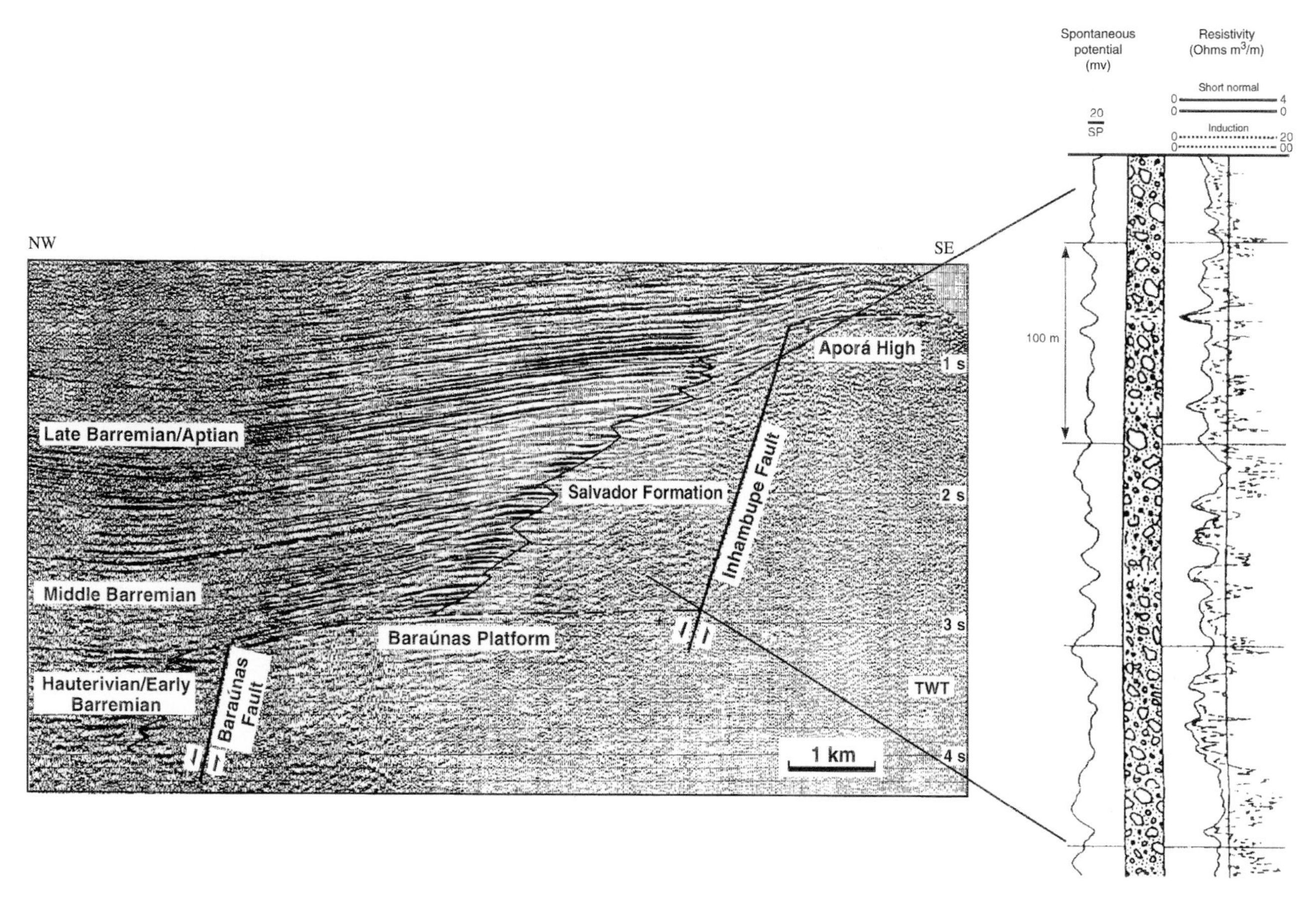

Figure 10. Section of marginal fanglomerates that developed along the rift border fault. The top of the conglomerate fan is commonly confused with the top basement seismic pick, so basin size is underestimated. High-quality turbidite sandstone reservoirs have been found below the marginal fan in the Recôncavo Basin of Brazil. TWT = two-way traveltime (Magnavita, 1992). 1 km (0.6 mi).

the basin was from the flexed margin of the half graben (Figures 11, 12) and the deep hanging-wall graben is sediment starved, with lakes forming there (Frostick and Reid, 1989) (e.g., Lake Tanganyika). As recent exploration success on the eastern shore of Lake Albert has demonstrated, the flexed margin (otherwise known as the hanging-wall dip slope) can be the site for significant sandstones and produce major stratigraphic traps for updip oil accumulations charged from more deeply buried interdigitating lacustrine source rocks (Figure 11B). A similar system characterized the Triassic rifts of the eastern parts of the United States (Schlische, 1992, 1993) and beneath the North Sea (Tomasso et al., 2008).

Carbonates in Rifts

The Toca and Lagoa Feia synrift carbonates in Brazil and Angola are developed along lake shorelines on footwall highs. These Barremian–early Aptian carbonates are composed mainly of pelecypods (called coquinas in Brazil). The shelly banks are reworked along the highs in a high-energy environment, but slumped blocks and reworked turbidites are shed off into the deeper parts of the basin. Moderate-size oil fields 50 to 200 million bbl in place have been discovered in these shelly limestones, but the presence of the reservoirs is very difficult to see on the seismic data sets.

Carbonate banks are developed along footwall highs in the Gulf of Suez where rifting occurred in a shallow-marine setting (Bosence, 2005; Cross and Bosence, 2008) (Figure 13A). Prograding carbonate wedges are observed over the fault blocks and the geometry of these prograding sequences is similar to a clastic delta, but there is no obvious source region for the clastic material in a delta developed along linear fault block crests, which are not undergoing erosion, so the presence of a prograding wedge initiating from an uneroded shallow-water fault block crest is a characteristic of in-situ carbonate growth and reworking (Figure 13B).

A

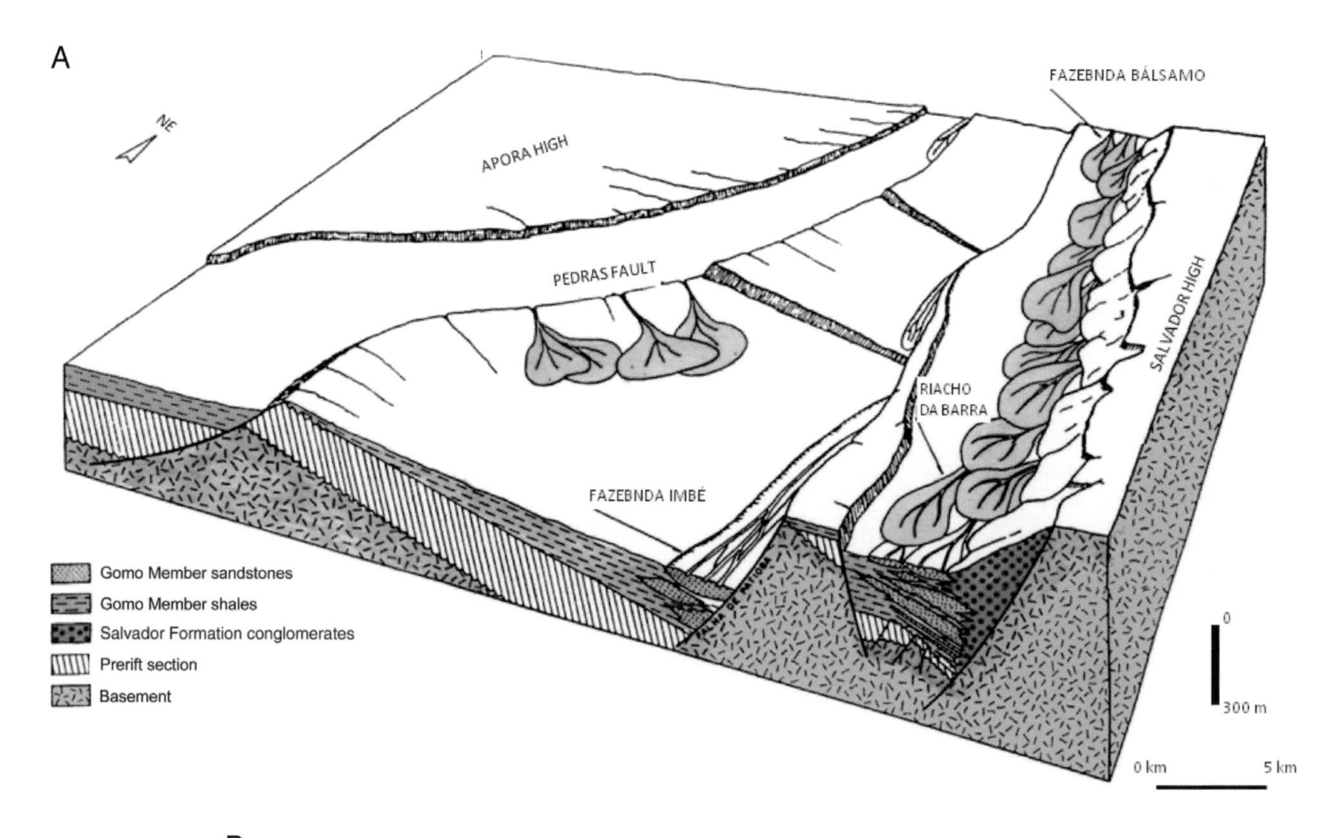

B

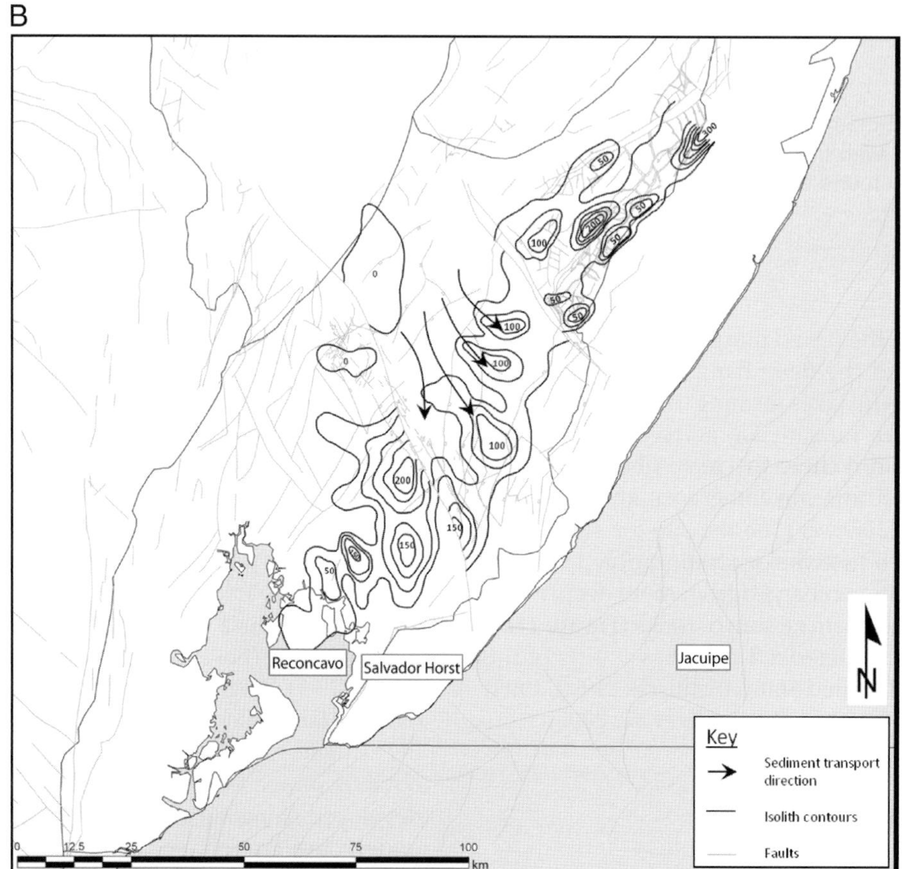

Figure 11. Block diagram (A) and map (B) showing hanging-wall reservoir sandstone deposition during the active rifting phase. Thickness of Gomo sandstones in meters. Note the possibility of turbidite sandstones below the marginal alluvial fan in (A). Modified from Bruhn (1985). 5 km (3 mi), 300 m (984 ft).

INTERACTION BETWEEN CLIMATE AND TECTONIC CONTROL ON RIFT SEDIMENTATION

Tectonics exerts the most important control on sedimentation in rifts (Ravnas and Steel, 1998; Caroll and Bohacs, 1999; Lambiase and Morley 1999), with climate being a second-order control (Olsen, 1990, 1997; Scholz et al., 1997). Milankovitch cycles produce climate variations with a cyclicity of 10 to 100 k.y. and produce high-frequency variations in the stratigraphic record generally on a scale of 1 to 10 m (see Olsen, 1997, for an excellent review of sediment cyclicity in Triassic rifts of North America). In the early stages of rifting, drainage pathways are commonly short, and closed internal drainage basins dominate. In arid regions, this commonly leads to evaporation exceeding sediment supply and runoff to produce evaporite basins (e.g., the modern lakes of East Africa).

Climate will have a control on whether lacustrine, fluvial, or eolian sedimentation dominates a rift. High rainfall in a hot climate may produce lakes with carbonates, in contrast to clastics in a cooler rift. A dry climate may also produce evaporite-filled basins such as the South Atlantic or the Gulf of Suez, where a restricted seawater entry point is present. Most evaporite deposition is very rapid (1 cm a^{-1}) and 2-km (1.2-mi)-thick evaporite sequences can develop in 200 k.y. Most large evaporite sequences are produced from seawater, but occasionally, saline lakes can locally develop with hydrothermal activity and leaching of volcanic rocks, leading to evaporite units several hundred meters thick (a possible example is the Macalungo-1 well in the Kwanza Basin, Angola).

Wind Direction

The prevailing wind direction can affect the local climate within a basin. Where the wind blows from the footwall to hanging wall of a half graben, the main runoff is away from the basin. However, if the wind blows toward the footwall, the main runoff is toward the basin, down the hanging-wall dip slope (Figure 14). Hence, wind blowing toward the footwall will favor deeper lakes and source rock development.

IMPLICATIONS FOR PETROLEUM SYSTEMS

Controls on Source Rock Deposition in Rifts

Source rocks in most hydrocarbon-producing rifts can be type I (algal) to type III (higher land plant) kerogens or, more commonly, a mixture of the two (type II) (Katz, 1995). Source rocks in tropical rifts are presently forming in the East African rift lakes today (Talbot, 1988). Demaison and Moore (1980) observed the Tanganyika Lake and suggested that organic-rich sediments will deposit in a deep anoxic lake in a warm humid climate regime with minimal seasonal contrasts, which is typical of low-latitude climate. Talbot (1988) also concluded from a study of several African rift lakes that the favorable climate conditions for source rock development are a humid climate with reduced wind conditions. In such conditions, lakes will be deep, fresh to slightly alkaline, with stable stratification (Talbot, 1988). Humid conditions will also promote dense vegetation growth, and large amounts of organic material will be washed into the basin. The dense vegetation inhibits surface erosion and clastic sediment runoff, which further enhances the source rock potential. Intense chemical weathering in a hot humid climate enhances the nutrient content of the water to promote organic productivity, especially of phytoplankton, which are common in present-day African lakes. Dry climate conditions produce shallow saline lakes, which are less favorable for major source rock development, although localized highly productive source rocks may develop, as in the Green River shales (Eugster and Hardie, 1975).

In marine rifts, oil-prone source rocks accumulate in half-graben depocenters especially those that experience poor circulation and anoxia. The main prolific source rock for the North Sea petroleum system, the Upper Jurassic Kimmeridge Clay Formation, formed in such a setting in water depths of several hundred meters to 1 km or more ($\geq$0.6 mi).

A deep narrow rift will enhance the source rock potential as the water is deeper, and this enhances the preservation potential of the organic matter. The deepest rifts form in cold stable lithosphere, where the crustal deformation localizes into a narrow rift zone. If one master fault develops along an initial weakness, the rift will be deep and narrow. If several faults of equal magnitude develop, the rift will be broader and shallower. Hence, rifts cutting through old cratons should be the deepest (e.g., the Tucano Rift cutting the Sao Francisco Craton in northeastern Brazil, which has 12 km (7.5 mi) of synrift sediment fill. Rapid rifting will also increase the depth of the rift lake where subsidence outpaces sedimentation.

Reservoirs in Rifts

In general, sandstones are the primary reservoir in rifts and carbonates are fairly rare (Bruhn et al., 1988; Morley, 2002b). The main clastic reservoir types are prerift or

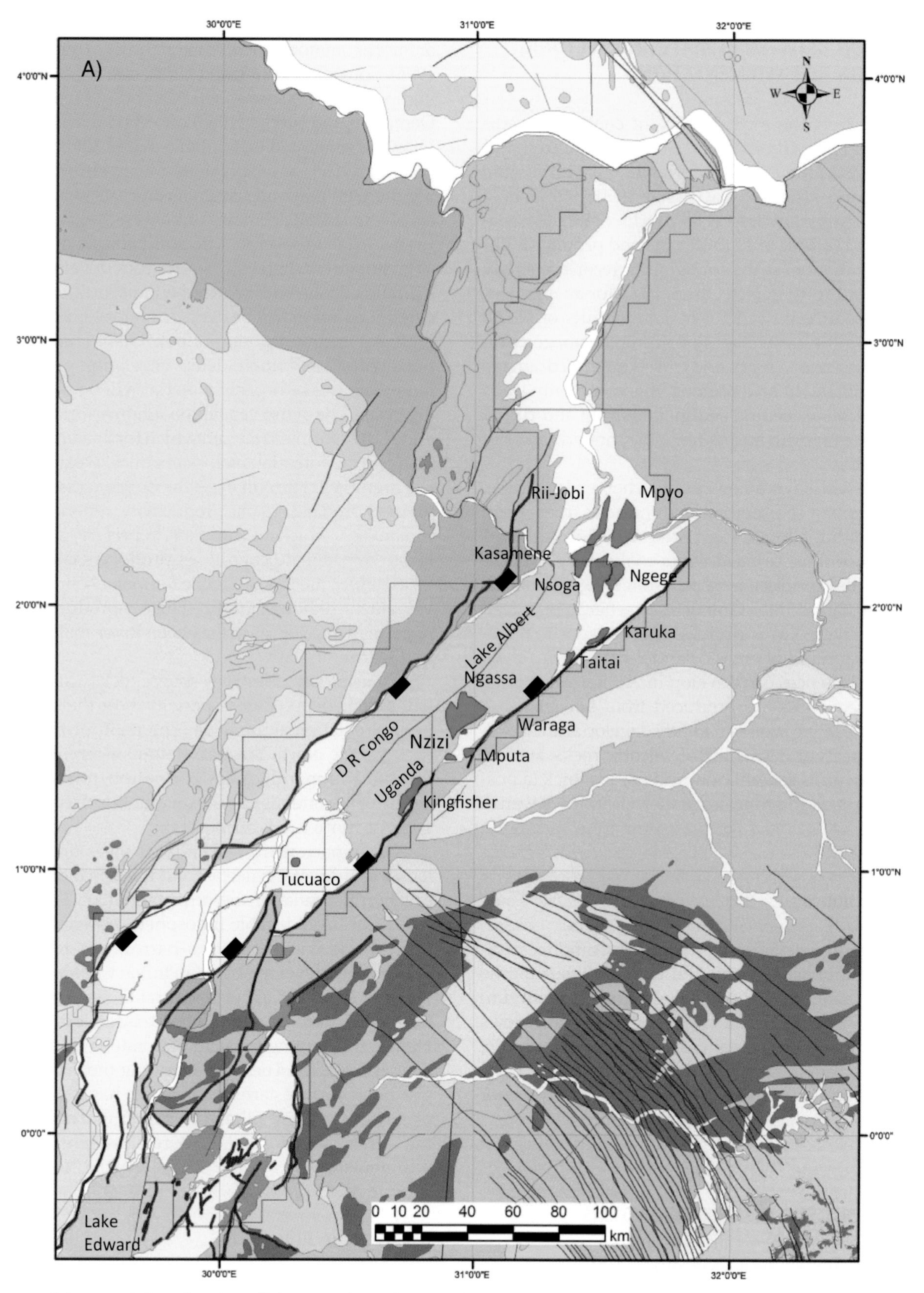

Figure 12. (A) Map of the Albertine Rift in East Africa showing the newly discovered hydrocarbon fields. (B) Block diagram showing the Albertine Rift and sediment input along the accommodation zone at the northern end of the rift. Pale yellow areas indicate Cenozoic sedimentary basin. Green areas indicate oil fields. Blue areas indicate lakes. DRC = Democratic Republic of Congo. 10 km (6.2 mi).

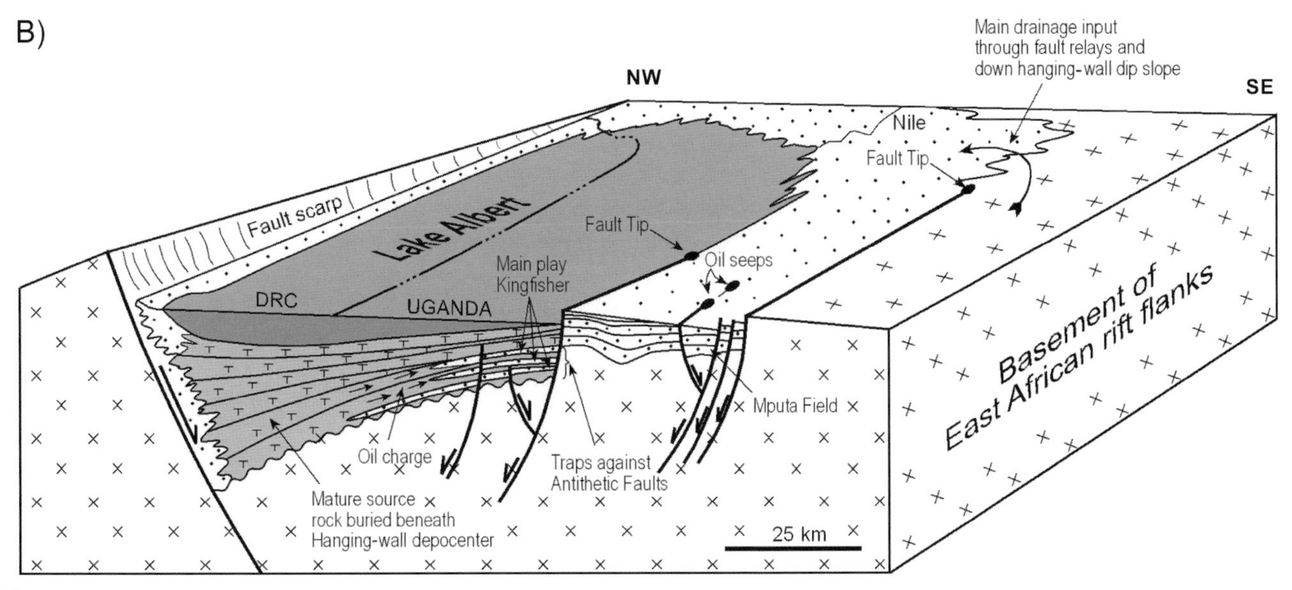

Figure 12. (cont.).

early-phase fluvial sandstones, synrift deeper water turbidite sandstones, and lacustrine deltas and late-stage fluvial or shallow-marine sandstones (Figure 3). Early-phase fluvial sandstones can be widespread and occur across the tops of rotated fault blocks except where later erosion has removed them (e.g., Lucula sandstones in west Africa). Synrift turbidite sandstones can be deposited along axially flowing canyon systems in the deep half graben or be deposited down the shallow flexural margin, but will be rarely found on fault block crests (Figure 11). Poor-quality conglomerates will occur along the hanging walls of major normal faults and may even act as seals to adjacent better quality reservoirs. Care should be taken that the border conglomerate fan is not mistakenly identified as the true basement, especially where only drill cuttings and logs are used.

Good-quality sandstones will develop where sediment provenance sources are also good quality. Examples include where high-grade granitic basement rocks or preexisting sandstones are exposed in the drainage catchment area; poor-quality reservoirs often derive from volcaniclastic or exotic terranes where mineral assemblages are such that primary depositional textures, sorting, or diagenetic effects may be detrimental to reservoir quality. Porosity, permeability and, hence, sandstone reservoir quality often deteriorate rapidly with burial as a consequence of diagenetic reactions and greater compaction. Sandstones in rifts are generally moderately sorted basement-derived arkoses, which undergo drastic reduction in porosity and permeability when feldspars break down. The porosity reduces most rapidly in the first 2 km (1.2 mi) of burial but can increase again because of secondary porosity development, where organic acids are generated as a precursor to oil expulsion from source rock shales. Hence, reservoir rocks with 20 to 25% porosity may be preserved at 4 km (2.5 mi) burial depth (Bruhn et al., 1988).

Occasionally, carbonates are important in the right climatic conditions either regionally or as fringing reef facies within what is otherwise a clastic system, such as the Gulf of Suez (Bosence, 2005) or the Santos and Campos basins in Brazil. A shallow-marine setting in a tropical climate will favor carbonate growth, and this should most likely be in the late phase of rifting when the basin is beginning to fill up, but regional subsidence and lateral rift propagation increase the likelihood of marine incursion.

Hydrocarbon Seals in Rifts

In the early stages of rift formation, closed drainage systems are commonly created. This leads to the formation of lakes that are a prerequisite for development of organic-rich shale source rocks and regional seals. The shales in a continental rift section are best developed in deep-lacustrine or -marine waters. The reason for exploration failure in many continental rift basins is commonly the lack of a good regional source rock and seal horizon, where the seal and source rocks can be the same formation.

If arid climatic conditions prevail, seawater or saline lake evaporation may lead to the formation of extensive evaporites during the extension. These act as seals for the underlying prerift and synrift reservoirs (e.g.,

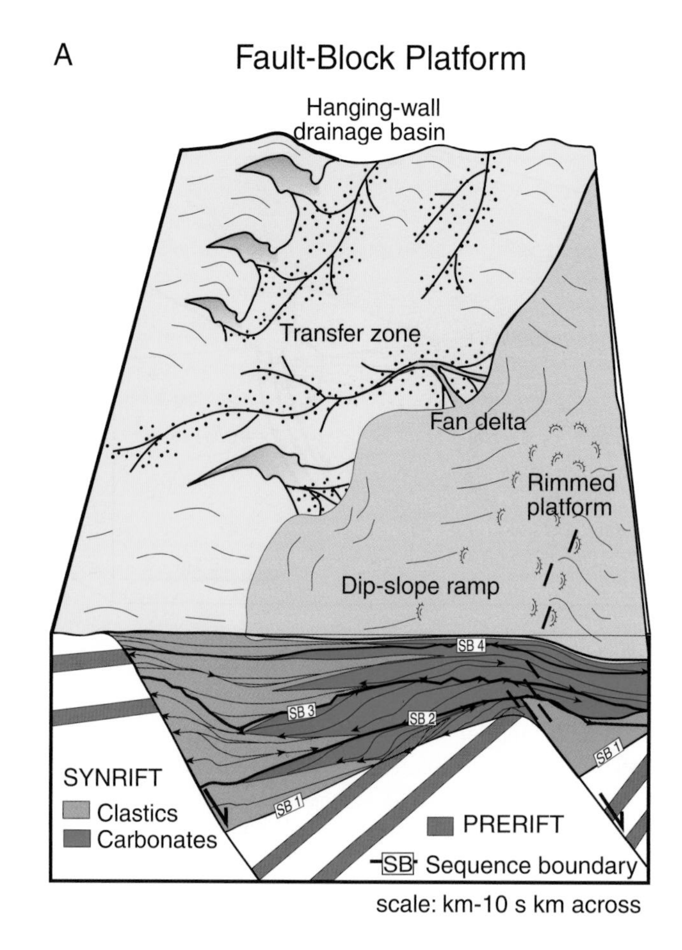

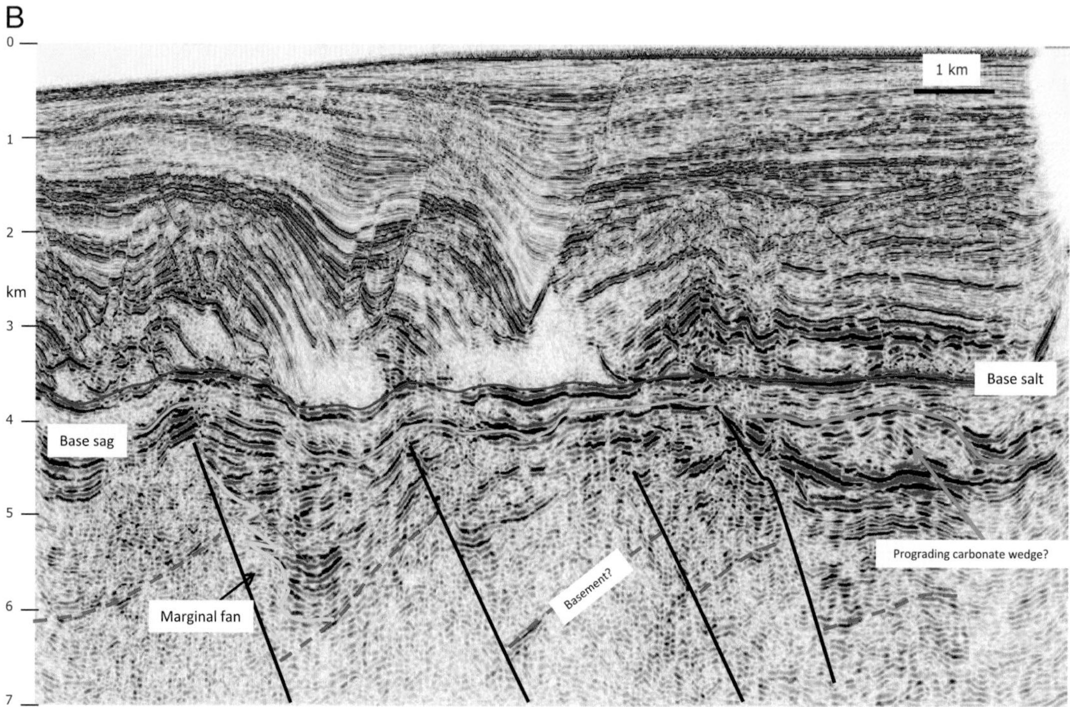

Figure 13. Examples of carbonate reservoir targets in emerging subsalt synrift plays. (A) Carbonate platform development on the footwall block crest based on Gulf of Suez examples. Modified from Bosence (2005); reproduced with the permission of Elsevier. (B) Seismic section through a west African rifted margin showing a prograding wedge (in blue) that has prograded eastward toward the continent from a narrow fault block that does not show any large erosion features on the fault block crest. This is characteristic of in-situ carbonate growth (courtesy of Maersk Oil and Sonangol). 1 km (0.6 mi).

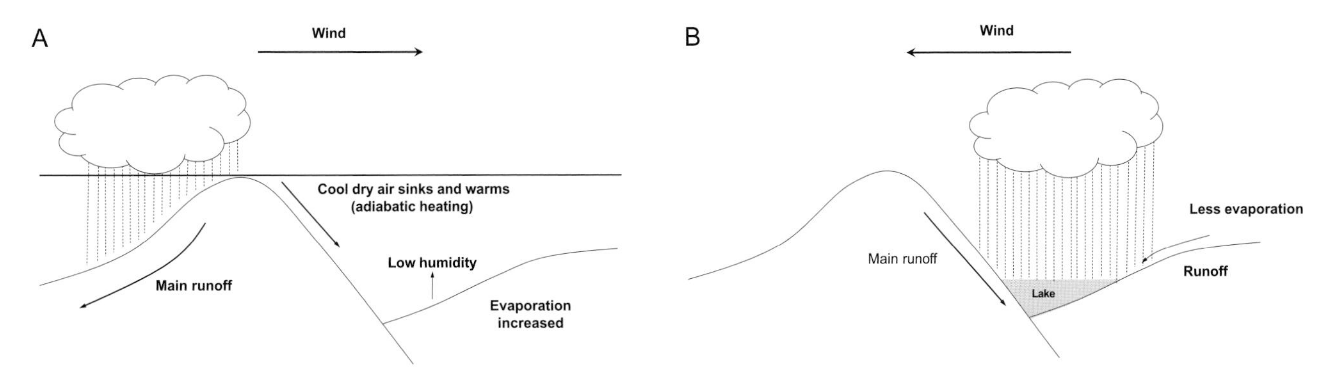

Figure 14. Cartoon sketch showing how wind direction and rift polarity control the microclimate in a half graben and the likelihood of lake formation and organic-rich source material.

existing and emerging deep-water plays of the west African and Brazilian continental margins, such as the Lula Field, and in the Gulf of Suez and northern Red Sea in Egypt).

CONCLUSIONS

Extensional rifts and their overlying sag basins contain large volumes of hydrocarbons in many part of the

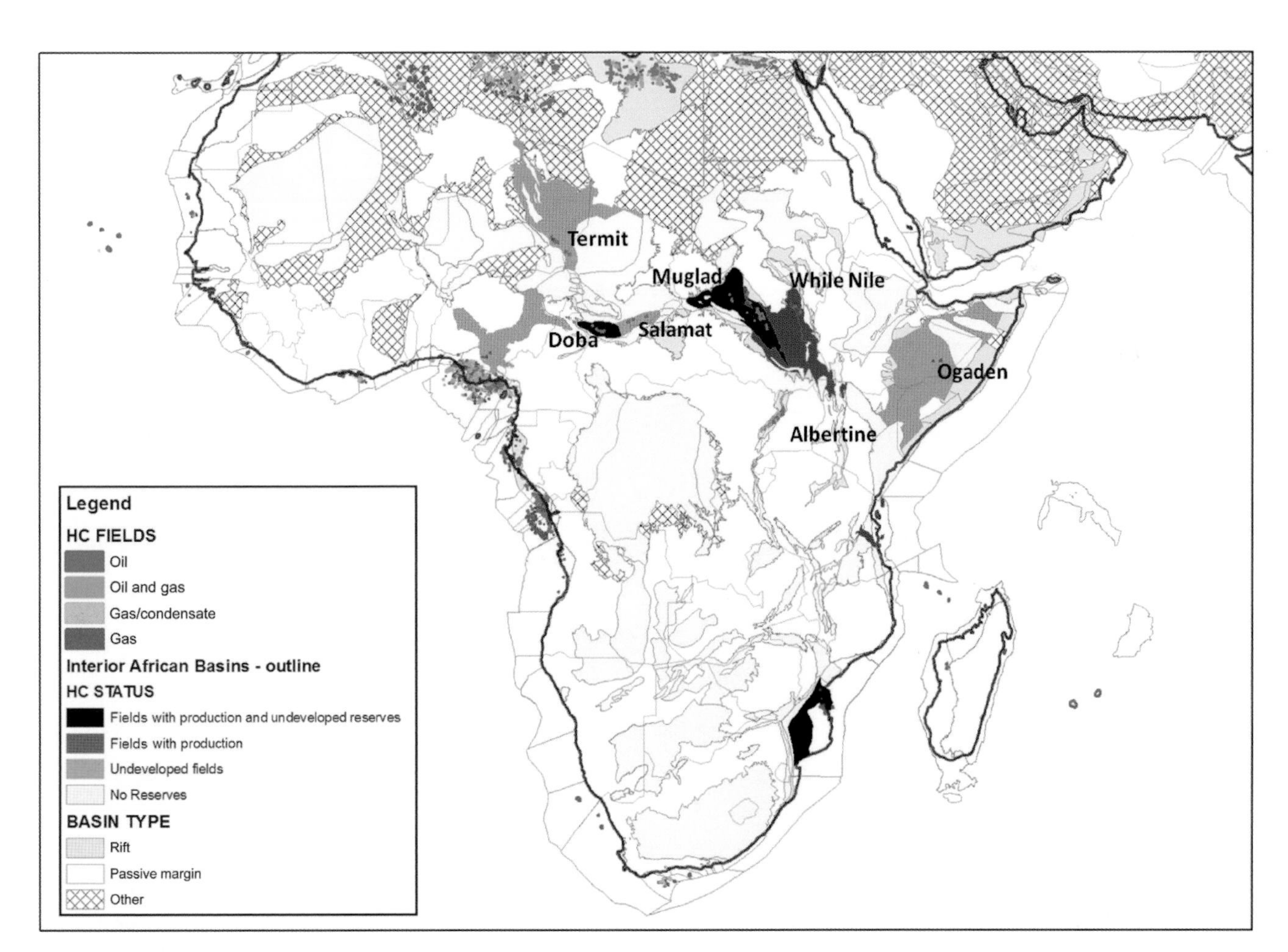

Figure 15. Map of the African rifts showing the rifts with proven potential for commercial production. Pipelines are marked as thin green lines.

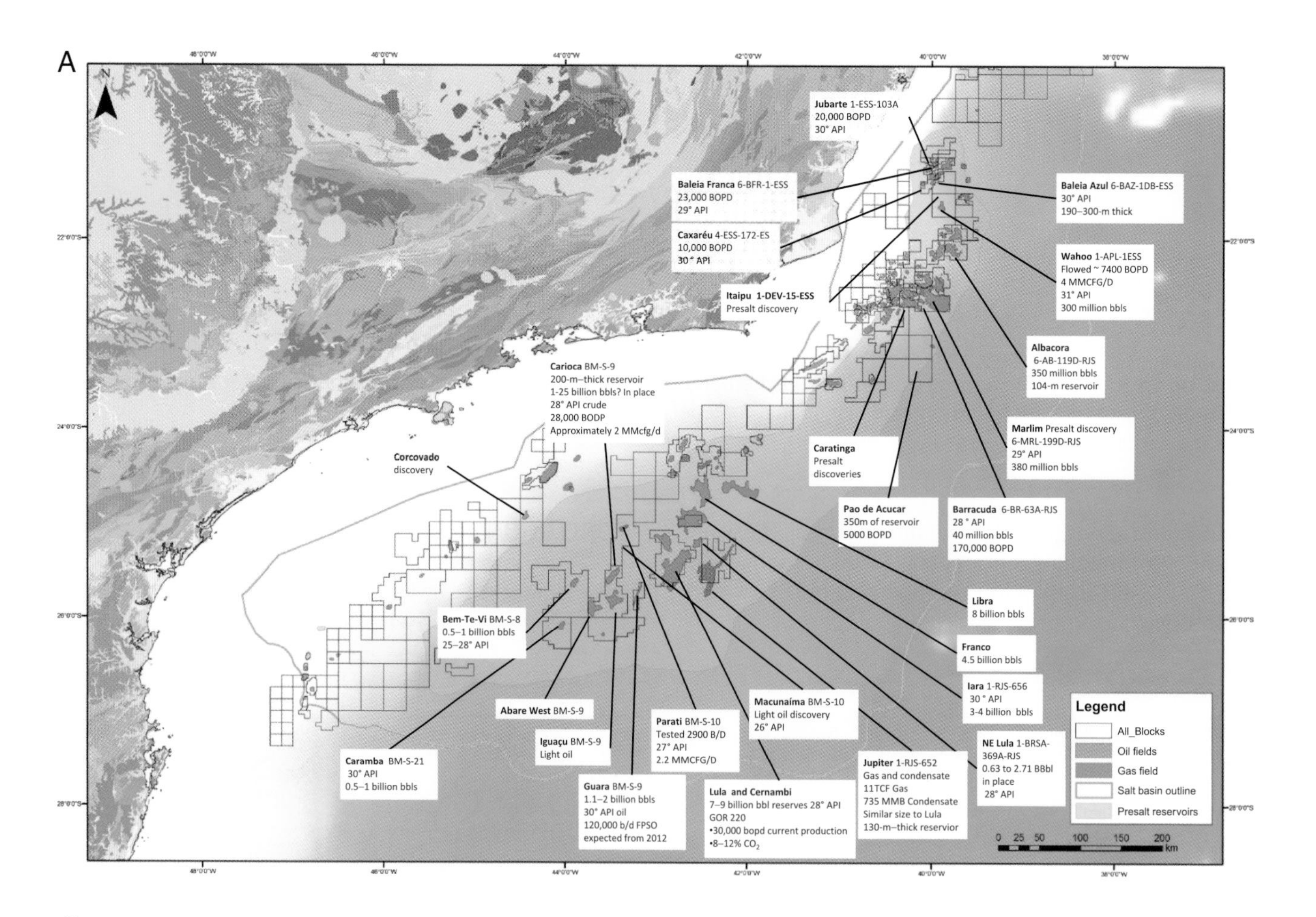

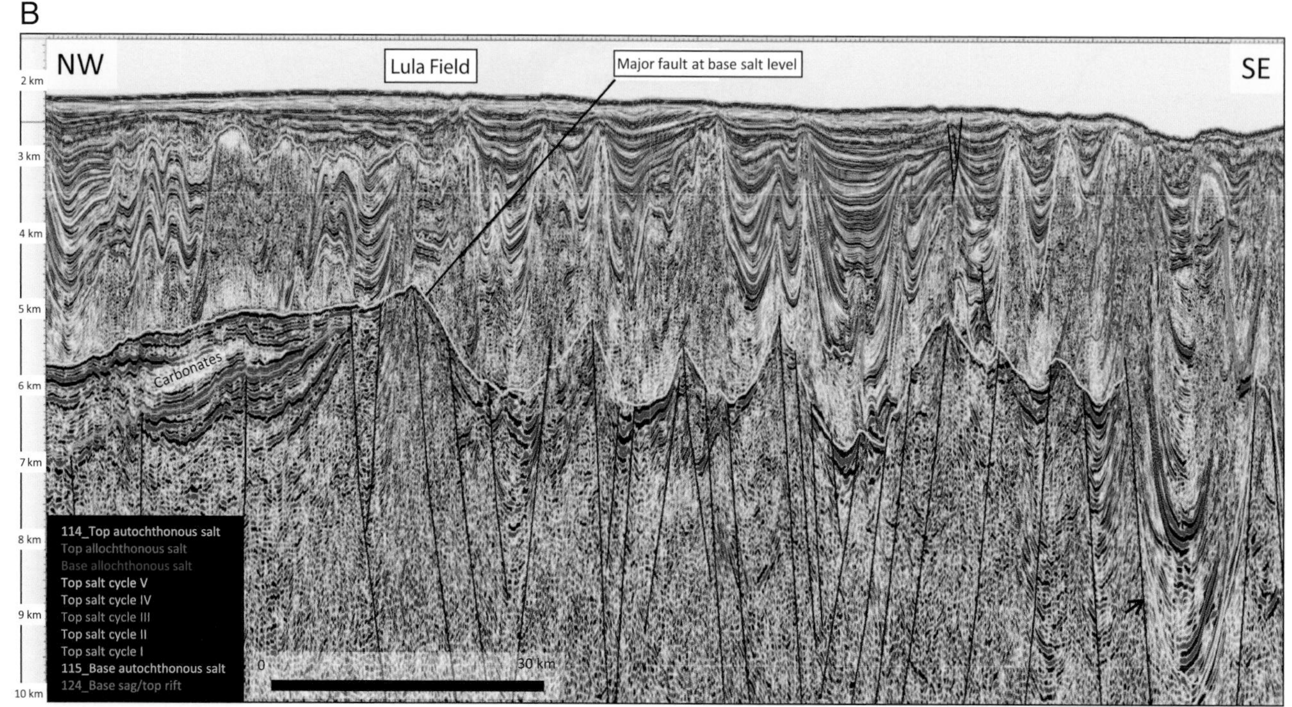

Figure 16. (A) Map of Santos and Campos basin major oil fields in presalt synextensional sag. (B) Seismic section through the Lula Field in the Santos Basin, Brazil. Seismic data courtesy of ION's Basin SPAN programs. 30 km (18.6 mi).

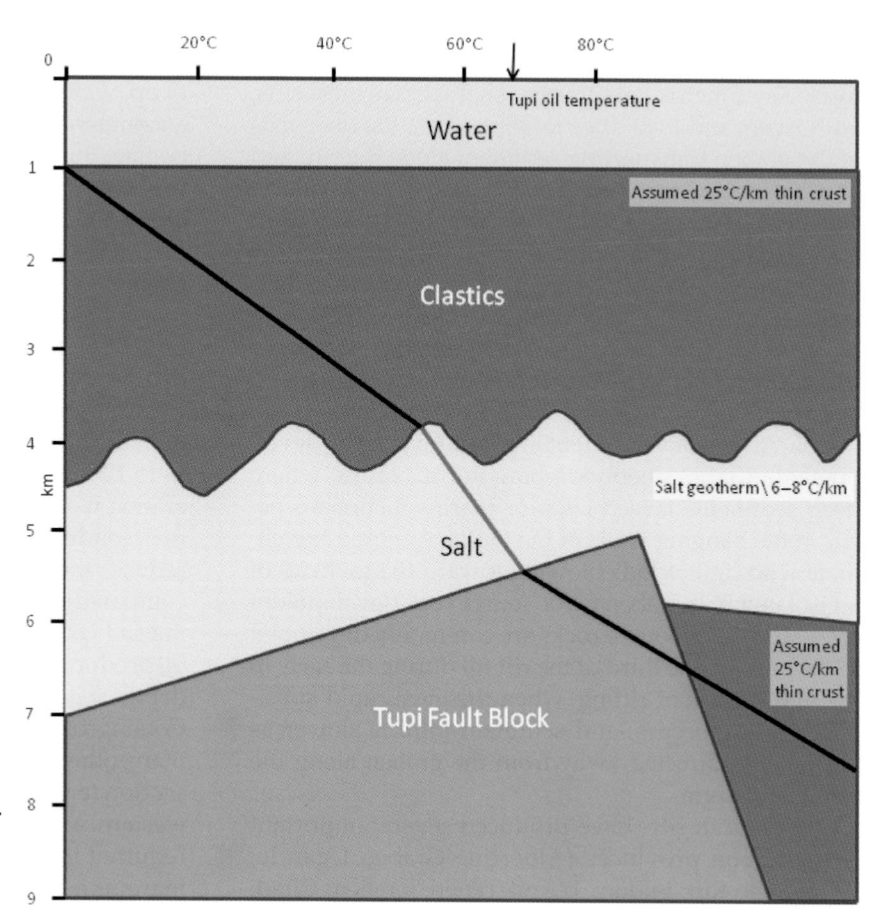

Figure 17. Cross section showing the cooling effect of the continuous salt layer. The presalt layer cools down by approximately 20°C/km of salt. Oil in the Lula Field is only 68°C. 1 km (0.6 mi).

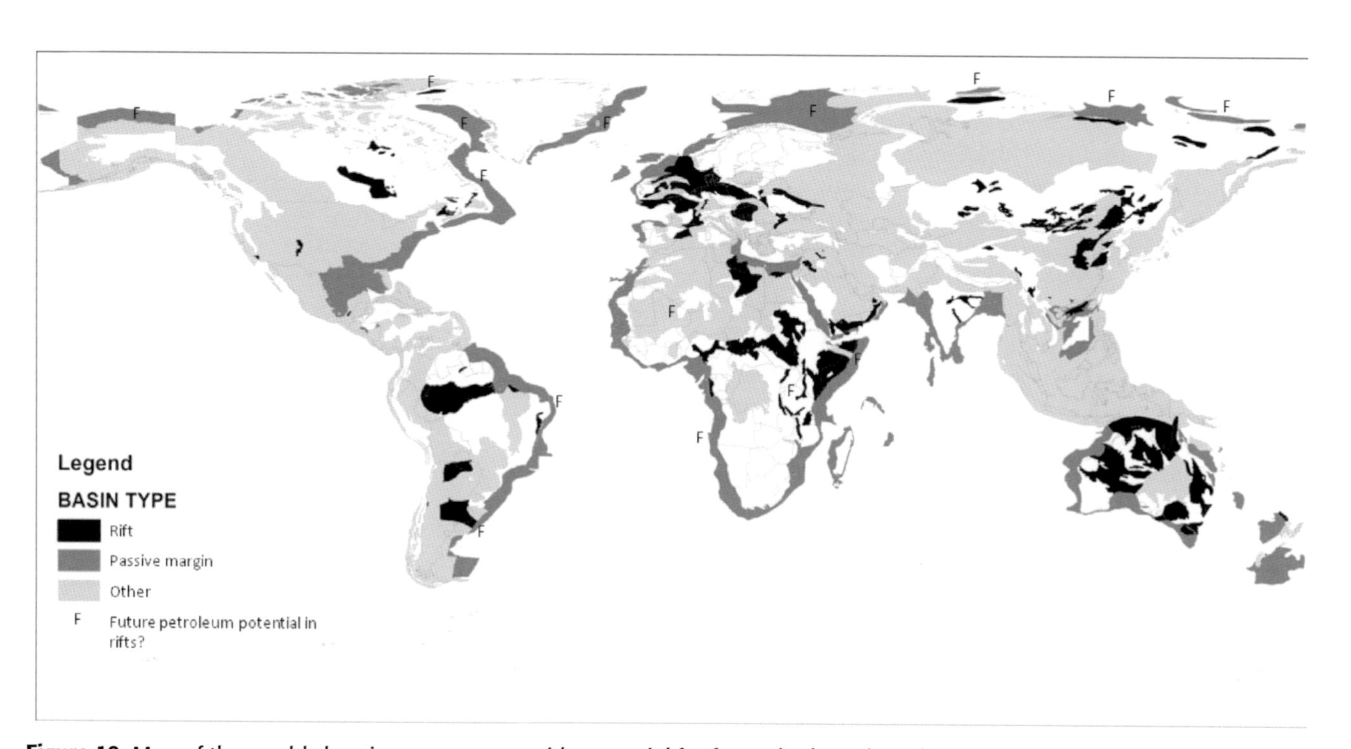

Figure 18. Map of the world showing some areas with potential for future hydrocarbon discoveries in extensional rifts, marked F (courtesy of M. K. Taylor and P. V. Baptista, 2010).

world and remain the focus for global exploration efforts. Asymmetric half grabens characterize most rifts, with rivers and turbidites flowing along the deep axis of the graben transporting sediment along the rift, and also rivers and tubidites flowing down the hanging-wall flexural margin will be important. Footwall-derived alluvial fans develop through time as the footwall is eroded back to create a drainage catchment area. Otherwise, sediment entering from the footwall side of a half graben commonly enters along accommodation or transfer zones, where a topographic low occurs along the footwall.

Source rock development in rifts is favored by development of large steeply dipping normal faults, where deep sediment-starved lakes or marine incursions occur in the hanging walls of the faults. A humid hot climate with light winds blowing toward the footwall of an isolated half graben favor source rock development (Figure 14). The source rocks are commonly developed in the bottom one third of the rift fill during the early to the main phase of rifting, when the most rapid subsidence phase occurs, and sediment infill is slower as sediment is directed away from the graben along the footwall margin.

The African rifts have produced several important hydrocarbon provinces (Albertine Graben Uganda; White Nile Rift, Sudan; Termit-Tenere Graben, Chad; Figure 15). However, exploration is still in its infancy in many rifts where no wells or seismic data exist. Gravity anomalies and surface seeps indicate that many of these undrilled African basins have hydrocarbon potential. The remote locations of many onshore rifts mean that the basins have to host large volumes of hydrocarbons in place before pipelines can be installed to tap these resources, unless they are close to an existing oil refinery. The total volume of hydrocarbons in place in the Albertine Graben now appears to be great enough to warrant building a 2000 km (1243 mi) pipeline to the East African seaboard (Figure 15).

Aptian carbonates deposited along the Brazilian margin host the largest hydrocarbon fields discovered in the last 20 yr (Figure 16A). The reservoir is a lacustrine carbonate composed of thrombolites and stromatolites, with local tufa development, dolomitization, and leaching that created porous reservoirs capable of producing 20,000 bbl of oil per day per well. These limestones are distributed over a very large area of the Santos, Campos, and Espírito Santo basins and are trapped in draped closures over large rotated fault blocks top sealed with Aptian salt (Figure 16B). The reservoirs were actually deposited after the main faulting phase but are interpreted to be part of the synextensional presalt sag caused by lower crustal stretching. The equivalent sag interval occurs along the west African margin, but the only well that appears to have drilled into this sag in deep water (>1000 m [>3280 ft]) is Falcao-1, which encountered good-quality source rock shales and carbonates but did not flow oil. Less prospective structures are present in the upper part of the sag phase, along the African margin, as the rifted fault block topography was buried by up to 6 km (3.7 mi) of synextension sag strata. These preliminary indications suggest that the presalt sag on the African margin may not be as prolific as the Brazilian margin, but we await further drilling to confirm whether this is the case.

The synrift source rocks in the Santos Basin are generating light oils, although they may be buried to 8- to 12-km (5- to 7.5-mi) depth. At this depth, gas generation would normally be predicted, but the high thermal conductivity of a 2 km (1.2 mi) thickness of salt reduces the presalt temperature by approximately 40°C compared with a shale-filled basin (Mello et al., 1995). Subsalt cooling is supported by the temperature of the oil produced in Lula, which is measured at only 68°C at a depth of 3.5 km (2.2 mi) below the seabed (Figure 17). Cooling of presalt synrift sections can be expected in many other rifts, where salt has developed late in the rift section (e.g., eastern Canada and United States, northwestern Africa, Gulf of Suez), and superdeep wells are required to test these presalt oil targets. Many passive margins have only been tested in the upper thermal sag phase of sedimentation, and the underlying deep rifts remain to be tested in deep water (Figure 18).

ACKNOWLEDGMENTS

We thank ION/GXT, PGS, and Maersk Oil and Sonangol for permission to publish seismic sections used in this chapter. Matthew Taylor and Pedro Baptista kindly provided maps for Figures 17 and 18. We thank Dengliang Gao and Scott Fraser for their reviews of this chapter. We also thank Dan Harris who kindly provided a review of our rifts chapter for this volume.

REFERENCES CITED

Armstrong, R. L., 1972, Low-angle (denudation) faults, hinterland of Sevier orogenic belt, eastern Nevada and western Utah: Geological Society of America Bulletin, v. 83, p. 1729–1754, doi:10.1130/0016-7606(1972)83[1729:LDFHOT]2.0.CO;2.

Badley, M. E., B. Freeman, A. M. Roberts, J. S. Thatcher, J. Walsh, J. Watterson, and G. Yielding, 1990, Fault interpretation during seismic interpretation and reservoir evaluation:

The integration of geology, geophysics and petroleum engineering in reservoir delineation, description and management: 1st Archie Conference, AAPG, p. 224–241.

Barr, D., 1987a, Lithospheric stretching, detached normal faulting and footwall uplift, *in* M. P. Coward, J. F. Dewey, and P. L. Hancock, eds., Continental extensional tectonics: Geological Society (London) Special Publication 28, p. 75–94, doi:10.1144/GSL.SP.1987.028.01.07.

Barr, D., 1987b, Structural/stratigraphic models for extensional basins of half-graben type: Journal of Structural Geology, v. 9, p. 491–500, doi:10.1016/0191-8141(87)90124-6.

Bosence, D. W. J., 2005, Toward a genetic classification of carbonate platforms: A proposal based on the basinal and tectonic settings of Cenozoic platforms: Sedimentary Geology, v. 175, p. 49–72, doi:10.1016/j.sedgeo.2004.12.030.

Bosworth, W., 1985, Geometry of propagating continental rifts: Nature, v. 316, p. 625–627, doi:10.1038/316625a0.

Bruhn, C. H. L., 1985, Sedimentação e evolução diagenética dos turbiditos eocretácicos do membro Gomo, Formação Candeias, no compartimento nordeste da Bacia do Recôncavo, Bahia: Master's thesis, Universidade Federal de Ouro Preto, Minas Gerais, Brazil, 203 p.

Bruhn, C. H. L., C. Cainelli, and R. M. D. Matos, 1988, Habitat do Petroleo e fronteiras exploratorias nos rifts brasileiros: Boletim de Geociencias da Petrobras, v. 2, p. 217–253.

Caroll, A. R., and K. M. Bohacs, 1999, Stratigraphic classification of ancient lakes: Balancing tectonic and climatic controls: Geology, v. 27, p. 99–102, doi:10.1130/0091-7613(1999)027<0099:SCOALB>2.3.CO;2.

Cartwright, J., 1991, The kinematic evolution of the Coffee Soil Fault, *in* A. M. Roberts, G. Yielding, and B. Freeman, eds., The geometry of normal faults: Geological Society (London) Special Publication 56, p. 29–40, doi:10.1144/GSL.SP.1991.056.01.03.

Cartwright, J. A., B. D. Trudgill, and C. M. Mansfield, 1996, Fault growth by segment linkage: An explanation for scatter in maximum displacement and trace length data for the Canyonlands Grabens of SE Utah: Journal of Structural Geology, v. 17, p. 1319–1326, doi:10.1016/0191-8141(95)00033-A.

Colletta, B., P. Le Quellec, J. Letouzey, and I. Moretti, 1988, Longitudinal evolution of the Suez rift structure (Egypt): Tectonophysics, v. 153, p. 221–233, doi:10.1016/0040-1951(88)90017-0.

Cowie, P. A., and C. H. Scholz, 1992a, Physical explanation for the displacement-length relationship of faults using a post-yield fracture mechanics model: Journal of Structural Geology, v. 14, p. 1133–1148, doi:10.1016/0191-8141(92)90065-5.

Cowie, P. A., and C. H. Scholz, 1992b, Displacement-length scaling relationship for faults: Data synthesis and discussion: Journal of Structural Geology, v. 14, p. 1149–1156, doi:10.1016/0191-8141(92)90066-6.

Cowie, P., J. R. Underhill, M. Behn, J. Lin, and C. M. Gill, 2005, Spatiotemporal evolution of strain accumulation derived from multiscale observations of Late Jurassic rifting in the northern North Sea: A critical test of models for lithospheric extension: Earth and Planetary Science Letters, v. 234, p. 401–419.

Cross, N., and D. W. J. Bosence, 2008, 3-D tectonosedimentary models for marine, rift-margin, systems, *in* J. Lakasik and T. Simo, eds., Controls on carbonate platform and reef development: SEPM Special Publication 89, p. 83–106, doi:10.2110/pec.08.89.0083.

Dawers, N. H., and J. R. Underhill, 2000, The role of fault interaction and linkage in controlling synrift stratigraphic sequences: Late Jurassic, Statfjord East area, northern North Sea: AAPG Bulletin, v. 84, p. 45–64, doi:10.1306/C9EBCD5B-1735-11D7-8645000102C1865D.

Dawers, N. H., M. H. Anders, and C. H. Scholz, 1993, Growth of normal faults: Displacement-length scaling: Geology, v. 21, p. 1107–1110, doi:10.1130/0091-7613(1993)021<1107:GONFDL>2.3.CO;2.

Demaison, G. J., and G. T. Moore, 1980, Anoxic environments and oil source bed genesis: Organic Geochemistry, v. 2, p. 9–31, doi:10.1016/0146-6380(80)90017-0.

Destro, N., 1995, Release fault: A variety of cross fault in linked extensional fault systems, in the Sergipe-Alagoas Basin, NE Brazil: Journal of Structural Geology, v. 17, p. 615–629, doi:10.1016/0191-8141(94)00088-H.

Destro, N., P. Szatmari, F. F. Alkmim, and L. P. Magnavita, 2003, Release faults, associated structures, and their control on petroleum trends in the Recôncavo rift, northeast Brazil: AAPG Bulletin, v. 87, p. 1123–1144, doi:10.1306/02200300156.

Driscoll, N., and G. Karner, 1998, Lower crustal extension across the Northern Carnarvon Basin, Australia: Evidence for an eastward-dipping detachment: Journal Geophysical Research, v. 103, p. 4975–4991, doi:10.1029/97JB03295.

Ebinger, C. J., J. A. Jackson, A. N. Foster, and N. J. Hayward, 1999, Extensional basin geometry and the elastic lithosphere: Philosophical Transactions Royal Society London A, v. 357, p. 741–765, doi:10.1098/rsta.1999.0351.

Eugster, H. P., and L. A. Hardie, 1975, Sedimentation in an ancient playa-lake complex: The Wilkins Peak Member of the Green River Formation of Wyoming: Geological Society of America Bulletin, v. 86, p. 319–334, doi:10.1130/0016-7606(1975)862.0.CO;2.

Figueiredo, A. M. F., J. A. E. Braga, J. C. Zabalaga, J. J. Oliveira, G. A. Aguiar, O. B. Silva, and L. F. Mato, 1995, Recôncavo Basin, Brazil: A prolific intracontinental rift basin, *in* S. M. Landon, ed., Interior rift basins: AAPG Memoir 59, p. 157–203.

Frostick, L., and I. Reid, 1989, Is structure the main control of river drainage and sedimentation in rifts?: Journal of African Earth Science, v. 8, p. 165–182, doi:10.1016/S0899-5362(89)80022-3.

Gawthorpe, R. L., and M. R. Leeder, 2000, Tectonosedimentary evolution of active extensional basins: Basin Research, v. 12, p. 195–218, doi:10.1046/j.1365-2117.2000.00121.x.

Gibbs, A. D., 1984, Structural evolution of extensional basin margins: Journal of the Geological Society (London), v. 141, p. 609–620, doi:10.1144/gsjgs.141.4.0609.

Goldsworthy, M., and J. Jackson, 2001, Migration of activity within normal fault systems: Examples from the

Quaternary of mainland Greece: Journal of Structural Geology, v. 23, p. 489–506.

Gregory, J. W., 1896, The Great Rift Valley: London, John Murray, 405 p.

Gregory, J. W., 1921, The rift valleys and geology of East Africa: London, Seeley Service, 479 p.

Gupta, S., P. A. Cowie, N. H. Dawers, and J. R. Underhill, 1998, A mechanism to explain rift-basin subsidence and stratigraphic patterns through fault-array evolution: Geology, v. 26, p. 595–598.

Harding, T. P., 1984, Graben hydrocarbon occurrences and structural styles: AAPG Bulletin, v. 68, p. 333–362.

Henry, S., A. Danforth, S. Ventrakaman, and C. Willacy, 2004, PSDM subsalt imaging reveals new insights into petroleum systems and plays in Angola-Congo-Gabon (abs.): Petroleum Exploration Society of Great Britain-Houston Geological Society Africa Symposium, September 7–8, 2004.

Hunt, C. B., and D. R. Mabey, 1966, Stratigraphy and structure of Death Valley, California: U.S. Geological Survey Professional Paper 494-A, 162 p.

Jackson, J. A., and T. Blenkinsop, 1993, The Bilila-Mtakataka fault in Malawi: An active, 100-km long, normal fault segment in thick seismogenic crust: Tectonics, v. 16, p. 137–150, doi:10.1029/96TC02494.

Jackson, J. A., and D. McKenzie, 1983, The geometrical evolution of normal fault systems: Journal of Structural Geology, v. 5, p. 471–482, doi:10.1016/0191-8141(83)90053-6.

Jackson, J. A., and N. J. White, 1989, Normal faulting in the upper continental crust: Observations from regions of active extension: Journal of Structural Geology, v. 11, p. 15–36, doi:10.1016/0191-8141(89)90033-3.

Karner, G. D., and L. A. P. Gamboa, 2007, Timing and origin of the South Atlantic pre-salt sag basins and their capping evaporites *in* B. C. Schreiber, S. Lugli, and M. Babel, eds., Evaporites through space and time: Geological Society (London) Special Publication 285, p. 15–35, doi:10.1144/SP285.2.

Katz, B., 1995, A survey of rift basin source rocks, *in* J. J. Lambiase, ed., Hydrocarbon habitat in rift basins: Geological Society (London) Special Publication 80, p. 213–240, doi:10.1144/GSL.SP.1995.080.01.11.

Lambiase, J. J., and W. Bosworth, 1995, Structural controls on sedimentation in continental rifts, *in* J. J. Lambiase, ed., Hydrocarbon habitat in rift basins: Geological Society (London) Special Publication 80, p. 117–144, doi:10.1144/GSL.SP.1995.080.01.0.

Lambiase, J. L., and C. K. Morley, 1999, Hydrocarbons in rift basins: The role of stratigraphy: Philosophical Transactions of the Royal Society, v. 357, p. 877–900, doi:10.1098/rsta.1999.0356.

Leeder, M. R., 2007, Cybertectonic Earth and Gais's weak hand: Sedimentary geology, sediment cycling and the earth system: Journal of the Geological Society (London), v. 164, p. 277–296.

Machette, M. N., S. F. Personius, A. R. Nelson, D. P. Schwartz, and W. R. Lund, 1991, The Wasatch fault zone, Utah: Segmentation and history of earthquakes: Journal of Structural Geology, v. 13, p. 137–149, doi:10.1016/0191-8141(91)90062-N.

Magnavita, L. P., 1992, Geometry and kinematics of the Recôncavo-Tucano-Jatobá rift, northeast Brazil: Ph.D. thesis, University of Oxford, Oxford, England, 493 p.

Magnavita, L. P., I. Davison, and N. J. Kusznir, 1994, Rifting, erosion and uplift in Recôncavo-Tucano-Jatobá Rift, NE Brazil: Tectonics, v. 13, p. 367–388, doi:10.1029/93TC02941.

Magnavita, L. P., and H. T. F. Da Silva, 1995, Rift border system: The interplay between tectonics and sedimentation in the Recôncavo Basin: AAPG Bulletin, v. 79, p. 1590–1607.

Mann, P., L. Gahagan, and M. B. Gordon, 2003, Tectonic setting of the world's oil and gas fields, *in* M. T. Halbouty, ed., Giant oil and gas fields of the decade 1990–1999: AAPG Memoir 78, p. 15–105.

McClay, K. R., 1995, 2-D and 3-D analog modeling of extensional structures: Templates for seismic interpretation: Petroleum Geoscience, v. 1, p. 163–178.

McClay, K., and S. Khalil, 1998, Extensional hard-linkages, eastern Gulf of Suez, Egypt: Geology, v. 26, p. 563–566, doi:10.1130/0091-7613(1998)0262.3.CO;2.

McConnell, R. B., 1972, Geological development of the rift system of eastern Africa: Geological Society of America Bulletin, v. 83, p. 2549–2572, doi:10.1130/0016-7606(1972)83[2549:GDOTRS]2.0.CO;2.

McHone, J. G., 2000, Nonplume magmatism and rifting during the opening of the central Atlantic Ocean: Tectonophysics, v. 316, p. 287–296, doi:10.1016/S0040-1951(99)00260-7.

McKenzie, D., 1978, Some remarks on the development of sedimentary basins: Earth and Planetary Sciences, v. 40, p. 25–32, doi:10.1016/0012-821X(78)90071-7.

McLeod, A. E., and J. R. Underhill, 1999, Footwall scarp instability and the processes and products of degradation, northern Brent Field, northern North Sea, *in* A. J. Fleet and S. A. R. Boldy, eds., Petroleum geology of northwest Europe: 5th Conference of the Geological Society of London, p. 91–106, doi:10.1144/0050091.

McLeod, A. E., N. H. Dawers, and J. R. Underhill, 2000, The propagation and linkage of normal faults: Insights from the Strathspey-Brent-Statfjord fault array, northern North Sea: Basin Research, v. 12, p. 263–284, doi:10.1111/j.1365-2117.2000.00124.x.

McLeod, A. E., J. R. Underhill, S. J. Davies, and N. H. Dawers, 2002, The influence of fault array evolution on synrift sedimentation patterns: Controls on deposition in the Strathspey-Brent-Statfjord half graben, northern North Sea: AAPG Bulletin, v. 86, p. 1061–1093, doi:10.1306/61EEDC24-173E-11D7-8645000102C1865D.

Mello, U. T., G. Karner, and R. N. Anderson, 1995, Role of salt in restraining the maturation of subsalt source rocks: Marine and Petroleum Geology, v. 12, p. 697–716.

Milani, E. J., and I. Davison, 1988, Basement control, and transfer tectonics in the Recôncavo-Tucano-Jatoba rift, northeast Brazil: Tectonophysics, v. 18, p. 41–70, doi:10.1016/0040-1951(88)90227-2.

Morley, C. K., 1995, Developments in the structural geology of rifts over the last decade and their impact on hydrocarbon exploration, *in* J. J. Lambiase, ed., Hydrocarbon habitat in rift basins: Geological Society (London) Special Publication 80, p. 1–32, doi:10.1144/GSL.SP.1995.080.01.01.

Morley, C. K., 2002a, Evolution of large normal faults: Evidence from seismic reflection data: AAPG Bulletin, v. 86, p. 961–978, doi:10.1306/61EEDBFC-173E-11D7-8645000102C1865D.

Morley, C. K., 2002b, Tectonic settings of continental extensional provinces and their impact on sedimentation and hydrocarbon prospectivity: Sedimentation in continental rifts: SEPM Special Publication 73, p. 25–55.

Morton, W. H., and R. Black, 1975, Crustal attenuation in Afar, *in* A. Pilger and A., Rosier, eds., Afar depression of Ethiopia: Inter-Union Commission on Geodynamics Scientific Report 14, p. 55–65.

Moustafa, A. R., 1997, Controls on the development and evolution of transfer zones: The influence of basement structure and sedimentary thickness in the Suez rift and Red Sea: Journal of Structural Geology, v. 19, p. 755–768, doi:10.1016/S0191-8141(97)00007-2.

Olsen, P. E., 1990, Tectonic, climatic and biotic modulation of lacustrine ecosystems: Examples from Newark Supergroup of eastern North America, *in* B. J. Katz, ed., Lacustrine basin exploration-case studies and modern analogs: AAPG Memoir 50, p. 209–224.

Olsen, P. E., 1997, Stratigraphic record of the early Mesozoic breakup of Pangea in the Laurasia-Gondwana rift system: Annual Review of Earth Planetary Science, v. 25, p. 337–401, doi:10.1146/annurev.earth.25.1.337.

Peacock, D. C. P., and D. J. Sanderson, 1991, Displacements, segment linkage and relay ramps in normal fault zones: Journal of Structural Geology, v. 13, p. 721–733, doi:10.1016/0191-8141(91)90033-F.

Proffett, J. M., 1977, Cenozoic geology of the Yerington district, Nevada, and implications for the nature of basin and range faulting: Geological Society America Bulletin, v. 88, p. 247–266, doi:10.1130/0016-7606(1977)882.0.CO;2.

Ravnas, R., and R. J. Steel, 1998, Architecture of marine rift-basin successions: AAPG Bulletin, v. 82, p. 110–146.

Roberts, A. M., and G. Yielding, 1994, Continental extensional tectonics, *in* P. Hancock ed., Continental deformation: Oxford, England, Pergamon Press, p. 223–250.

Rosendahl, B. R., 1987, Architecture of continental rifts with special reference to the eastern Africa: Annual Review of Earth and Planetary Sciences, v. 15, p. 445–503, doi:10.1146/annurev.ea.15.050187.002305.

Ruellan, E., and J.-M. Auzende, 1985, Structure and evolution of El Jadida submarine plateau (Mazagan, West Morocco): Bulletin de la Societé Géologique de France, v. 11, p. 103–114.

Schlische, R. W., 1992, Structural and stratigraphic development of the Newark extensional basin, eastern North America: Evidence for the growth of the basin and its bounding structures: Geological Society America Bulletin, v. 104, p. 1246–1263, doi:10.1130/0016-7606(1992)104<1246:SASDOT>2.3.CO;2.

Schlische, R. W., 1993, Anatomy and evolution of the Triassic–Jurassic continental rift system, eastern North America: Tectonics, v. 12, p. 1026–1042, doi:10.1029/93TC01062.

Scholz, C. A., and B. R. Rosendahl, 1988, Low lake stands in lakes Malawi and Tanganyika, East Africa, delineated with multifold seismic data: Science, v. 240, p. 1645–1648, doi:10.1126/science.240.4859.1645.

Scholz, C. H., and C. Aviles, 1986, The fractal geometry of faults and faulting, *in* S. Das, J. Boatwright, and C. Scholz, eds., Earthquake source mechanics: American Geophysical Union Monograph 37, p. 147–155.

Scholz, C. H., N. H. Dawers, J.-J. Yu, M. H. Anders, and P. Cowie, 1993, Fault growth and fault scaling laws: Preliminary results: Journal Geophysical Research, v. 98, p. 21,951–21,161, doi:10.1029/93JB01008.

Scholz, C. A., T. C. Moore, D. R. Hutchinson, A. J. Golmshtok, K. D. Klitgord, and A. G. Kurotchkin, 1997, Comparative sequence stratigraphy of low-latitude versus high-latitude lacustrine rift basins: Seismic data examples from the East African and Baikal rifts: Paleoecology, Palaeoclimatolology, Palaeoecology, v. 140, p. 401–420.

Seranne, M., and M. Seguret, 1987, The Devonian basins of western Norway: Tectonics and kinematics of an extending crust, *in* M. P. Coward, J. F. Dewey, and P. L. Hancock, eds., Continental extensional tectonics: Geological Society (London) Special Publication 28, p. 537–548, doi:10.1144/GSL.SP.1987.028.01.35.

Steel, R., 1976, Devonian basins of western Norway: Sedimentary response to tectonism and to varying tectonic context: Tectonophysics, v. 36, p. 207–224, doi:10.1016/0040-1951(76)90017-2.

Stein, J. B., G. C. P. King, and R. S. Rundle, 1988, The growth of geological structures by repeated earthquakes: 2. Field examples of continental dip-slip faults: Journal of Geophysical Research, v. 93, p. 13,307–13,318, doi:10.1029/JB093iB11p13319.

Stein, R. S., and S. E. Barrientos, 1985, Planar high angle faulting in the basin and range: Geodetic analysis of the 1983 Borah Peak, Idaho, earthquake: Journal of Geophysical Research, v. 90, p. 11,355–11,366, doi:10.1029/JB090iB13p11355.

Talbot, M. R., 1988, The origins of lacustrine oil source rocks: Evidence from lakes of tropical Africa, *in* A. J. Fleet, K. Kelts, and M. R. Talbot, eds., Lacustrine petroleum source rocks: Geological Society (London) Special Publication 40, p. 29–43, doi:10.1144/GSL.SP.1988.040.01.04.

Taylor, S. R., J. Almond, S. Arnott, D. Kemshell, and D. Taylor, 2003, The Brent field, Block 211/29, U.K. North Sea: Geological Society Memoirs 20, p. 233–250.

Tomasso, M., J. R. Underhill, R. A. Hodgkinson, and M. J. Young, 2008, Structural styles and depositional architecture in the Triassic of the Ninian and Alwyn North fields: Implications for basin development and prospectivity in the northern North Sea: Marine and Petroleum Geology, v. 25, p. 588–605.

Turner, S., M. Regelous, C. J., Hawkesworth, and M. Mantovani, 1994, Magmatism and continental break up in the South Atlantic: High-precision ^{40}Ar-^{39}Ar geochronology: Earth

Planetary Science Letters, v. 121, p. 333–348, doi:10.1016/0012-821X(94)90076-0.

Underhill, J. R., 1991a, Implications of Mesozoic–Recent basin development in the Inner Moray Firth, UK: Marine and Petroleum Geology, v. 8, p. 359–369, doi:10.1016/0264-8172(91)90089-J.

Underhill, J. R., 1991b, Controls on Late Jurassic seismic sequences, Inner Moray Firth: A critical test of Exxon's original sea level chart: Basin Research, v. 3, p. 79–98, doi:10.1111/j.1365-2117.1991.tb00141.x.

Underhill, J. R., M. J. Sawyer, P. Hodgson, M. D. Shallcross, and R. L. Gawthorpe, 1997, Implications of fault scarp degradation for Brent Group prospectivity, Ninian Field, northern North Sea: AAPG Bulletin, v. 81, p. 999–1022.

Walsh, J. J., and J. Watterson, 1988a, Analysis of the relationship between displacements and dimensions of faults: Journal of Structural Geology, v. 10, p. 239–247, doi:10.1016/0191-8141(88)90057-0.

Walsh, J. J., and J. Watterson, 1988b, Dips of normal faults in coal measures and other sedimentary sequences: Journal of the Geological Society (London), v. 145, p. 859–873, doi:10.1144/gsjgs.145.5.0859.

Wernicke, Â., 1985, Uniform-sense normal simple shear of the continental lithosphere: Canadian Journal of Earth Sciences, v. 22, p. 108–125.

Wernicke, B., 2009, The detachment era (1977-1982) and its role in revolutionizing continental tectonics, *in* U. Ring and B. Wernicke, eds., Extending a continent: Architecture, rheology and heat budget: Geological Society (London) Special Publication 321, p. 1–8, doi:10.1144/SP321.1.

Yielding, G., 1990, Footwall uplift associated with Late Jurassic normal faulting in the northern North Sea: Journal of the Geological Society (London), v. 147, p. 219–222, doi:10.1144/gsjgs.147.2.0219.

Yielding, G., M. E. Badley, and B. Freeman, 1991, Seismic reflections from normal faults in the northern North Sea, *in* A. M. Roberts, G. Yielding, and B. Freeman, eds., The geometry of normal faults: Geological Society (London) Special Publication 56, p. 79–89, doi:10.1144/GSL.SP.1991.056.01.06.

Younes, A. I., and K. R. McClay, 2002, Development of accommodation zones in the Gulf of Suez-Red Sea rift, Egypt: AAPG Bulletin, v. 86, p. 1003–1026, doi:10.1306/61EEDC10-173E-11D7-8645000102C1865D.

3

Mann, Paul, 2012, Comparison of structural styles and giant hydrocarbon occurrences within four active strike-slip regions: California, Southern Caribbean, Sumatra, and East China, *in* D. Gao, ed., Tectonics and sedimentation: Implications for petroleum systems: AAPG Memoir 100, p. 43–93.

Comparison of Structural Styles and Giant Hydrocarbon Occurrences within Four Active Strike-slip Regions: California, Southern Caribbean, Sumatra, and East China

Paul Mann

Department of Earth and Atmospheric Sciences, University of Houston, 312 Science and Research Building 1, Houston, Texas, 77204, U.S.A. (e-mail: pmann@uh.edu)

ABSTRACT

From previous compilations, active strike-slip tectonic settings include only a small fraction (~5%) of the world's basins containing giant oil and gas fields. Despite the relative global paucity of oil within strike-slip margins, several active strike-slip basins like the San Joaquin and Los Angeles basins of southern California, U.S.A. have produced billions of barrels of oil and deserve special analysis by explorationists to understand the regional geologic and stress controls on their productivity. The purpose of this chapter is to compile structural and tectonic parameters of four active strike-slip settings characterized by the greatest known concentrations of giant oil and gas fields: faults and mainly onshore basins of the San Andreas fault system in southern and central California; southern Caribbean plate boundary and related onshore and offshore basins in Venezuela and Trinidad and Tobago; onshore and offshore plate boundary strike-slip faults and related basins in the area of Sumatra and Southeast Asia; and onshore and offshore strike-slip faults and basins in the area of the Bohai Basin of eastern China. Main geologic and tectonic parameters compiled from a variety of public access databases using geographic information system technology for all four of these strike-slip and hydrocarbon settings include regional plate motions based on global positioning system (GPS)–based geodesy, basin types along the strike-slip faults are defined using compilations of depth to basement data and include localized pull-aparts and more regional transpressional basins flanking major strike-slip faults, well-described examples of structural styles along strike-slip faults based on surface and subsurface mapping mostly related to hydrocarbon exploration, and well-studied structural and stratigraphic traps that accommodate large accumulations of hydrocarbons in each of the four regional study areas. The compilation

DOI:10.1306/13351548M100861

reconfirms earlier insights into the relationship between regional stresses and structures along major strike-slip faults as well as suggesting that distant subduction zones may control the stress state in some active plate margin-parallel strike-slip faults.

INTRODUCTION

In a previous compilation of the 877 known giant oil fields at the end of the last decade, Mann et al. (2003) classified the tectonic setting of all these giants using six simplified classes for the world's basins: (1) continental passive margins fronting major ocean basins (304 giants); (2) continental rifts and overlying sag or "steer's head" basins (271 giants); (3) continental margins produced by terminal collision between two continents (173 giants); (4) collisional margins produced by continental collision related to terrane accretion, arc collision, and/or shallow subduction (71 giants); (5) strike-slip margins (50 giants); and (6) subduction margins not affected by major arc or continental collisions (8 giants). The challenges of classifying the one tectonic style that dominates the most typical stratigraphic and structural level for basins with multiphase histories are discussed by Mann et al. (2003) and are not repeated here.

The starting observation for this article is that relatively few giant oil and gas fields have been found along strike-slip margins (about 50 giants or about 5% of the total population of giants). In Mann et al. (2003), the relative paucity of giants in strike-slip settings was related to tectonic instability related to continued activity and ultimate destruction of large reservoirs needed for the preservation of giants as was previously discussed by MacGregor (1996) using many global examples. In contrast to these tectonically disrupted areas along strike-slip faults, rift and passive-margin tectonic settings account for more than two thirds of all giant oil and gas fields compiled in the Mann et al. (2003) compilation. Following the earlier ideas of Macgregor (1996), Mann et al. (2003) proposed that early rifting allowed hydrocarbons to remain undisturbed because tectonic activity had shifted toward more distant plate boundaries.

Despite the relative paucity of giants found within strike-slip margins, several active strike-slip basins have produced billions of barrels of oil and deserve special analysis by explorationists to understand their regional controls on their past productivity and reserve estimates (Figure 1). These highly productive strike-slip basins are remarkably small in area compared with much larger basins found in rift and passive-margin settings. They are also interesting in that tectonic activity continues in all four of the basins, thus, in contrast to rift setting, where tectonic activity has shifted to other areas, leaving all faults inactive. This chapter focuses on four hydrocarbon-rich basins within or adjacent to active strike-slip faults:

1) California and San Andreas strike-slip fault system: According to the U.S. Geological Survey (USGS) report on the world's hydrocarbon supply (Masters et al., 1998), cumulative production from the San Joaquin Basin adjacent to the long continental-style San Andreas fault system is 2.6 billion bbl of oil, with identified reserves of 1.16 billion bbl of oil (Figure 1). The Midway-Sunset giant oil field in the San Joaquin Basin adjacent to the San Andreas fault system is the third largest field in the United States. The Ventura Basin within the Big Bend transpressional area of the San Andreas fault system has produced 0.66 billion bbl of oil, with 0.28 billion bbl of identified reserves. The adjacent Los Angeles Basin has produced 0.76 billion bbl of oil, with 0.93 billion bbl of identified reserves. The offshore Santa Maria Basin has produced 0.67 billion bbl of oil, with 0.16 bbl of identified reserves. This is a remarkably productive region despite diffuse active strike-slip faulting and related folding that potentially disrupts major reservoirs and allows leakage of biodegradation of large oil and gas pools (i.e., "zone of shallow destruction" discussed by Macgregor, 1996).
2) Southern Caribbean and Bocono–El Pilar strike-slip fault system: Of the four strike slip-controlled hydrocarbon regions compared in this chapter, the Caribbean region is by far the richest region, with production amounts ranging from 44.6 (Masters et al., 1998) to 60 billion bbl of oil (Escalona and Mann, 2006) (Figure 1). It should be noted that, in the sixfold tectonic classification of Mann et al. (2003), only a relatively small number of the many giant fields in this region were classified in the strike-slip category because a preceding mid-Cenozoic arc-continent collision appeared more dominant in the localization of giants. Estimates by Masters et al. (1998) are more conservative than these estimates based on reports by the Venezuelan National Oil Company, Petróleos de Venezuela S.A. Recent exploration has moved from collision-related foreland areas onland to more strike slip-related areas offshore along the coast of northern Venezuela and Trinidad.

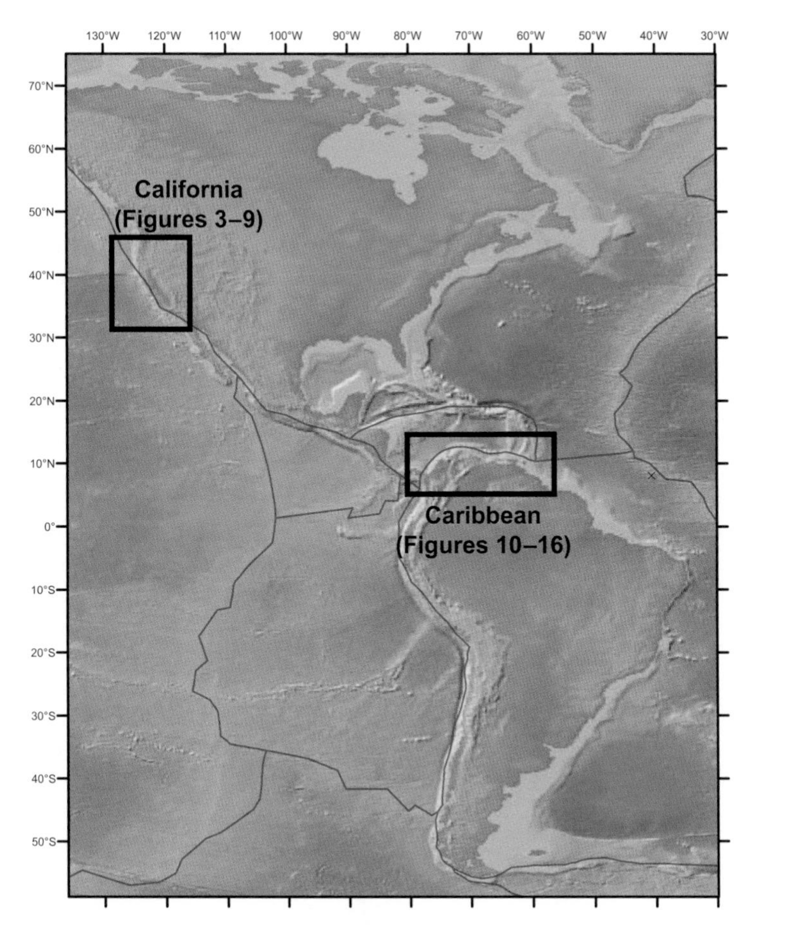

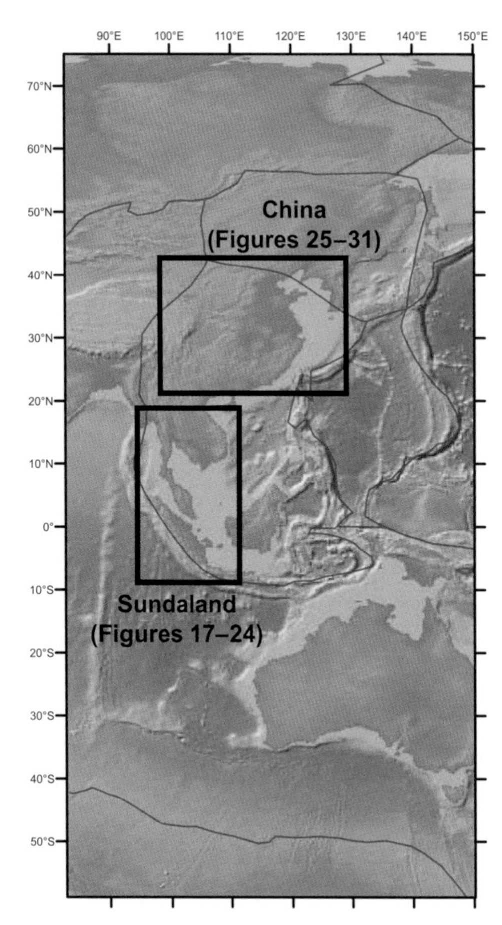

Figure 1. Global map showing active plate boundaries and four hydrocarbon-rich active strike-slip regions compared and discussed in this chapter: California, southern Caribbean, Sumatra, and eastern China. Figures indicated for each region provide detailed maps of features within these regions. Source of plate boundaries is from UTIG Plates Project (Topography and Predicted Bathymetry of the World with Present-day Plate Boundaries, 1999). Source of topography and bathymetry: Google (2009).

3) Sumatra and Sumatra strike-slip system: This region is one of the most concentrated areas of strike slip-controlled giant fields of the four regions discussed in this chapter (Figure 1). Strike-slip faulting in this area is driven by oblique subduction of the Indian oceanic plate beneath the Sumatran fore arc and mostly submerged Sunda continent and has produced a variety of both subduction-related and strike slip-related basin types. On the island of Sumatra that exhibits both strike-slip and subduction control on its basin types according to Mann et al. (2003), almost 10 billion bbl of oil have been produced, and reserve estimates are 7.82 billion bbl of oil (Masters et al., 1998). Other basins in the region include the Pattani trough, a superdeep pull-apart basin on the active Three Pagodas strike-slip fault, with 0.1 billion bbl of oil and gas equivalent produced and the West Natuna Basin, with 0.07 billion bbl of oil and gas equivalent produced (Masters et al., 1998).

4) East China and associated strike-slip faults: The onshore and offshore Bohai Basin, a pull-apart basin on active strike-slip faults in East China, has been the most productive basin in China, with 6.3 billion bbl of oil and gas equivalent produced within a relatively small basin (Masters et al., 1998) (Figure 1). This area is characterized by intraplate strike-slip motions driven by escaping continental fragments related to the collision of India and Eurasia to the southwest and/or oblique subduction of the Pacific plate beneath Eurasia.

PREVIOUS MODELS FOR THE ORIGIN AND STRUCTURAL STYLE OF MAJOR STRIKE-SLIP FAULTS AND ADJACENT SEDIMENTARY BASINS

Structural styles and orientations of secondary faults and folds formed along active strike-slip faults and

have been interpreted by previous workers in terms of two different deformational models:

1) In the classic wrench tectonics model of Wilcox et al. (1973), Harding and Wickham (1988), and Jamison (1991), en echelon folds form in a narrow zone straddling active strike-slip faults. These folds are interpreted as the product of a maximum compressive stress oriented oblique to the strike-slip fault plane. The strike-slip fault is interpreted as a high-drag tightly coupled surface along which the en echelon folds form as drag-related features (Figure 2A).
2) In the San Andreas fault model of Zoback et al. (1987) and Mount and Suppe (1987, 1992), a 50- to 100-km (~31- to 62-mi)-wide fold and thrust belt parallel to the primary strike-slip fault is interpreted as the product of a maximum horizontal stress oriented approximately 90° to the strike-slip fault plane. The strike-slip fault is interpreted as a low-drag decoupled surface along which the maximum horizontal stress is partitioned into a fault-normal component responsible for the folding and thrust faulting and a fault-parallel component responsible for strike-slip motion along the fault. Ben-Avraham (1992) and Zoback (1992) have proposed that the stresses at right angles to the fault can also be extensional and create asymmetrical rifts bounded by a normal fault on one side and a strike-slip fault on the other.

Harding and Wickham (1988) criticized Mount and Suppe's (1987) application of their low-drag San Andreas fault model to the central segment of the San Andreas strike-slip fault adjacent to the hydrocarbon-bearing San Joaquin Basin for two reasons. (1) The fault-parallel fold belt in California used as evidence for a compressive stress perpendicular to the San Andreas fault contains

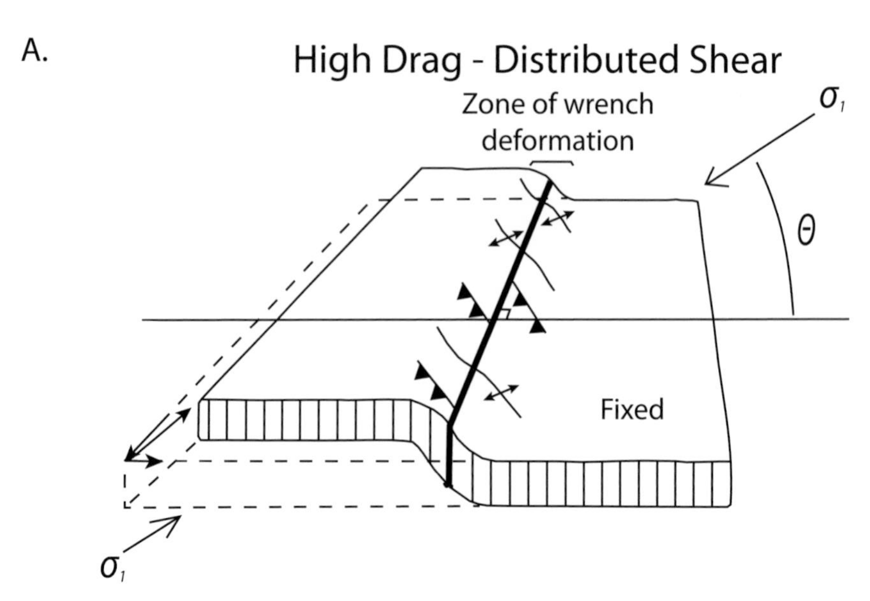

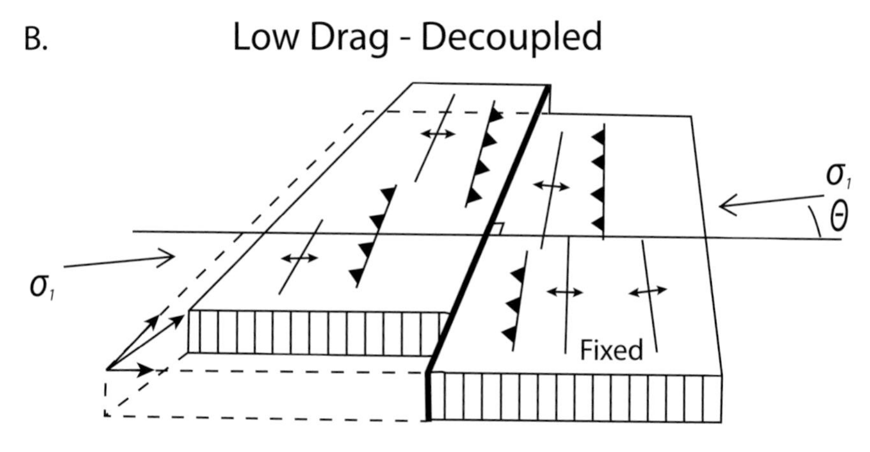

Figure 2. (A) Deformation style predicted by Wilcox et al. (1973) based on principles of wrench tectonics largely developed from results of hydrocarbon exploration in California and Sumatra. This style consists of a high-drag strike-slip zone that forms a broad zone of deformation where the angle between maximum compressive stress and fault is 30 to 45°. This broad zone of deformation includes an array of en echelon folds and faults formed by synchronous shearing. (B) Contrasting deformation style predicted by decoupled, or partitioned, strike-slip and thrust deformation along a low-drag strike-slip fault that lacks the broad zone of distributed shear seen in the model in (A). Maximum compressive stress is at a right angle to strike-slip fault and produces a wide zone of folds and thrusts that strike parallel to the main strike-slip fault trace. Figure is modified from Mount and Suppe (1987).

four possibly separate fold belts that differ in trend, age of commencement or cessation of folding, position relative to the strike-slip fault, properties of underlying basement, or thickness of the sedimentary cover, and possibly, fold style.

Harding and Wickham (1988) considers only two of the four fold belts as possible evidence to use to distinguish the function of wrench fault versus the San Andreas fault models. The other two fold belts may not be directly related to strike-slip tectonics; (2) two of the four possibly separate fold belts in California contain structures that formed as early as the late Oligocene. Because it is unlikely that the regional stress field has remained unchanged for such a long period and because older folds tend to rotate into alignment with younger strike-slip faults, the orientation of older folds are of little or no use for inferences of the direction of the present-day stress field associated with active strike-slip faults. Hardings and Wickham's (1988) observations emphasized the need for observations to be made across extensive along-strike and across-strike areas of actively deforming strike-slip faults and fault-related structures, especially in areas where the ages of deformation adjacent to major strike-slip faults are well known from both outcrop and subsurface studies.

OBJECTIVES OF THIS CHAPTER

Since the time of the controversy between Harding and Mount and Suppe more than 20 yr ago, there has been considerable progress in understanding strike-slip margins and active deformation mainly from the application of GPS-based geodesy and improved mapping of regional stress fields from the World Stress Map Project (Zoback, 1992; Heidbach et al., 2008) of the International Association of Seismology and Physics of the Earth's Interior. These GPS and stress data now span the width of most of the active strike-slip zones discussed in this chapter and provide many new insights into the deformation that were not available to earlier workers. Results of GPS-determined plate motions in active plate boundary zones are relatively uniform in direction (as opposed to less organized motions more suggestive of simultaneous rotations of many smaller blocks). Stresses compiled from the World Stress Project (2008) and reproduced on many of the maps used in this chapter show that the maximum compressive stress direction remains parallel over large areas of the Earth's crust. The widespread uniformity of stresses indicates that tectonics is a likely cause with possible mechanisms such as slab pull and ridge push (Zoback, 1992). Regional stresses tend to be orthogonal to both collisional and strike-slip boundaries but slightly more oblique along strike-slip boundaries. The degree of obliquity of regional stresses along strike-slip boundaries remains at the center of the controversy between Harding and Mount and Suppe.

Using these and other observations compiled from the published literature and from internet resources, the objectives of this chapter include:

1) To compile maps of the four regional hydrocarbon-bearing strike-slip basins that characterize their active plate motions based on GPS-based geodesy, their depth to acoustic basement and overall basin geometry known from wells and interpretation of seismic reflection data, their structural geology from geologic maps, regional stress fields from the World Stress Map Project, and the distribution of hydrocarbons in these basins from a variety of published and Internet resources. A similar style is used for the basin maps from all four regions to allow for comparisons to be made between the geographically separate strike-slip settings.
2) Use these data to revisit the controversy of the late 1990s between Harding and Mount and Suppe. Can the newer compilation presented here distinguish between the deformational mechanisms of the "classic wrench tectonics" model versus the "San Andreas fault model"? Moreover, what is the function of uniform or slightly varying regional stress fields now known from the World Stress Project in controlling the geometry of the major strike-slip faults along with associated folds and secondary faults?
3) To place the regional structures and production areas in the four strike-slip zones into their active regional stress fields known from the World Stress Project. The present-day maximum horizontal stress orientation is a primary control on fluid flow in the subsurface both in fractured and unfractured rocks (Tingay et al., 2005). The direction of normal or strike-slip faults commonly strike subparallel or about 30° to the present-day regional maximum horizontal stress orientation. This information is important for exploration because these faults can breach reservoirs and because fluids released in reservoir flooding operations will move preferentially in the direction of maximum horizontal stress.
4) Because each of the four strike-slip zones described in this chapter are found in different tectonic settings, one goal is to see whether structures can be related to specific strike-slip settings or whether all strike-slip faults and structures exhibit similar geometries regardless of their tectonic settings. Once common characteristics emerge, it might be

possible to show typical structural settings for the large hydrocarbon deposits in each area.

METHODS AND DATA

The main tool used to prepare regional maps for the four strike-slip areas was a geographic information system that allows all maps and cross sections compiled from a variety of scanned or digital sources to be georeferenced. Data sources used in these map compilations were public access appearing in the published literature or taken from online resources. All data sources including data Web sites are cited in the list of references at the back of this chapter. For most regions, uniform coverage of key basin-defining elements lacked—including depth to basement or GPS vectors—so the maps shown is the best effort at showing all of the data that could be located from public sources.

INTERCONTINENTAL STRIKE-SLIP FAULT: SAN ANDREAS FAULT SYSTEM IN CALIFORNIA

Geographic Location of Study Area

The curvilinear San Andreas fault system of California forms a prominent topographic lineament 1300 km (~808 mi) in the length from the Mexican border with the United States in the south to the northern border of California and Oregon (Figure 3A). The San Andreas is undoubtedly the best studied active strike-slip fault system in the world because of the seismic hazard it poses to the 37 million residents of the state of California and an additional 3.8 million residents living near the U.S. border in northern Mexico. Moreover, the fault is spatially associated with some of the largest oil fields in the world (Figure 3B). The regional description of fault deformation will focus on the area from the city of Fresno in the San Joaquin Basin to San Diego on the Pacific coast near the border of California with Mexico.

Classification of the San Andreas Strike-slip Fault System

Woodcock (1986) made the first comprehensive global classification of active strike-slip faults based on their tectonic origin and setting. Mann (2007) slightly modified Woodcock and Daly's (1986) classification into five major classes of faults that are used in this chapter to classify faults in each of the four plate boundaries discussed. In this classification scheme, the San Andreas fault system is a continental boundary strike-slip fault characterized by a long fault trace separating continental plates with most plate motion parallel to the fault trace as shown from GPS studies by Feigl et al. (1993) (Figure 3A).

Crustal Motions Determined from Global Positioning System-based Geodesy

Global positioning system studies in California (UNAVCO, 2009) show that plate motions of the Pacific plate relative to stable North America are subparallel to the main trace of the San Andreas strike-slip plate boundary fault that is shown as the heavy red line in Figure 3B. The maximum area of plate convergence along the San Andreas fault corresponds to the topographically elevated Transverse Ranges (Big Bend area north and northwest of the city of Los Angeles) (Figure 3A). Argus and Gordon (2001) conclude from their regional study of GPS results that the San Andreas fault accommodates most of the plate motion by plate boundary-parallel strike-slip faulting at rates of 39 ± 2 mm/yr (Figure 3A). Global positioning system shows that fault-normal motion is small and occurs at much slower rates of 3.3 ± 2 mm/yr than the fault-parallel motions.

Plate reconstructions of the San Andreas fault system show that a brief period of greater fault normal-motion that can explain the present-day topographic elevations along the length of the San Andreas fault zone began approximately 8 to 6 Ma (Argus and Gordon, 2001) (Figure 3A). The parallelism of GPS vectors over wide areas indicates that large-scale rotation of fault blocks is not presently occurring (Feigl et al., 1993). Geologic and paleomagnetic studies show that these complex block rotations may have occurred in the area of the Transverse Ranges during the Miocene (Hornafius et al., 1986).

Depth to Basement Map

Depth to basement information for the areas adjacent to the San Andreas fault system was found from various sources including Harding (1973) and Exxon Tectonic Production Research Company (1985) (Figure 3B). The basement is deformed asymmetrically with steeper dipping and deeper basement in the San Joaquin Basin, whereas basement contours in the offshore area of California show a more gently dipping, westward-dipping surface (with the exception of isolated deeps near an offshore thrust front that roughly parallels both the coast and the Pacific-North America direction of relative plate motion (Figure 3B). Giant oil and gas fields shown in Figure 3B cluster in the area of the Transverse Ranges' restraining bend at distances less than 100 km (<62 mi) from the San Andreas fault zone. Basement deepening that is likely associated with crustal shortening and thrust faulting in the vicinity of the Transverse Ranges' restraining bend acts to localize nine giant oil fields (Figure 3B).

Seismic profiling across the thrust front in central California by McIntosh et al. (1991) reveals the presence of oceanic crust or thickened oceanic crust subducting beneath the margin, an accretionary prism, and complex structures in the basins on the overriding plate. Basement deepening in the fore-arc area may be related to the formation of fore-arc basins that obviously lack an active arc to the very slow amount of plate-boundary normal/subduction beneath the California margin (~3.3 mm/yr) (Argus and Gordon, 2001). These authors interpret the 5- to 6-km (3.1- to 3.7-mi)-thick subducted oceanic crust as either part of the Pacific plate or, more likely, a subducted fragment derived from the Farallon plate. Seismic data show that the accretionary prism reaches a thickness of 10 to 15 km (6.2–9.3 mi) and therefore formed as a result of a significant amount of subduction beneath the margin. Basins and depocenters near the trench and prism have been inverted by compression in the past 3 to 5 m.y. Compressional features and inverted normal faults are attributed to a change in Pacific-North America plate motion from strike-slip to more compressional about 3 to 5 Ma (Argus and Gordon, 1991).

In the San Francisco Bay area of northern California, deep seismic reflection and refraction profiling by the BASIX project showed subducting oceanic crust dipping beneath the northern California margin and extending to a depth of 13 km (~8 mi) beneath the onland San Andreas fault (Brocher et al., 1994) (Figure 3B). These authors also interpreted the dipping reflector beneath the margin as the top of a subducted slab because the reflector is prominent and can be tracked to a prominent and active frontal thrust at the seaward edge of the continental margin (Figure 3B). In places, the subducted layer appears uncommonly thick (~11 km [~6.8 mi]) to be normal oceanic crust, and in these areas, the crust may have thickened by thrust imbrication. The prominent dipping reflector observed on the BASIX lines is taken as the contact between overlying silicic (Franciscan) rocks and the underthrusting mafic oceanic slab. Deep seismic profiling and refraction by Henstock et al. (1997) in northernmost California revealed a similar landward-dipping subducted area of oceanic crust that they inferred was subducted beneath the California margin about 2 Ma. The subducted plate shows some vertical offset on the downdip projection of the San Andreas fault zone.

Regional Stress Map

Regional stresses from the Heidbach et al. (2008) are compiled in Figure 3C and are color coded to the stress indicator used in the lower part of Figure 3C. As discussed by Mount and Suppe (1987), Zoback (1992), and Townend and Zoback (2004), the state of stress is remarkably uniform across southern and central California. In the San Francisco Bay area of northern California, the trajectory of the maximum compressive stress measured as close as 1 km (~0.62 mi) from several methods, including earthquake focal mechanisms and wellbore breakouts. Adjacent to the main trace of the San Andreas fault, dense measurements show that the maximum compressive stress is oriented at right angles to the fault trace (average value of about 85°). This observation supports the earlier hypothesis of Mount and Suppe (1987) for a low-drag and decoupled San Andreas fault model, as shown schematically in Figure 2B.

Farther to the south in southern California, the orientation of the maximum compressive stress compiled from various methods is at an angle of approximately 65° and rotated in a clockwise direction relative to the main trace of the San Andreas fault zone (Townend and Zoback, 2004) (Figure 3C) and more compatible with the high-drag and distributed shear model of Harding and Wickham (1988). Larger discrepancies between the trajectory of the maximum compressive stress and the trace of the fault occur at restraining bends in the fault trace in the Big Bend area of the fault and the San Bernardino bend area to the south (Figure 3C).

Strike-Slip Related Structures and Relation to Hydrocarbons

The Los Angeles Basin

The Los Angeles Basin of southern California is a complexly faulted elongate basin adjacent to the Transverse Ranges' gentle restraining bend on the San Andreas fault zone (Figure 3A). Global positioning system measurements from Feigl et al. (1993) show that North America-Pacific plate motions closely parallel the overall northwestward strike of the San Andreas fault zone and do not deviate in the area of the Transverse Ranges' restraining bend that defines the northern edge of the Los Angeles Basin (Figure 3A). Geodetic studies by Feigl et al. (1993) and Griffith and Cooke (2005) have

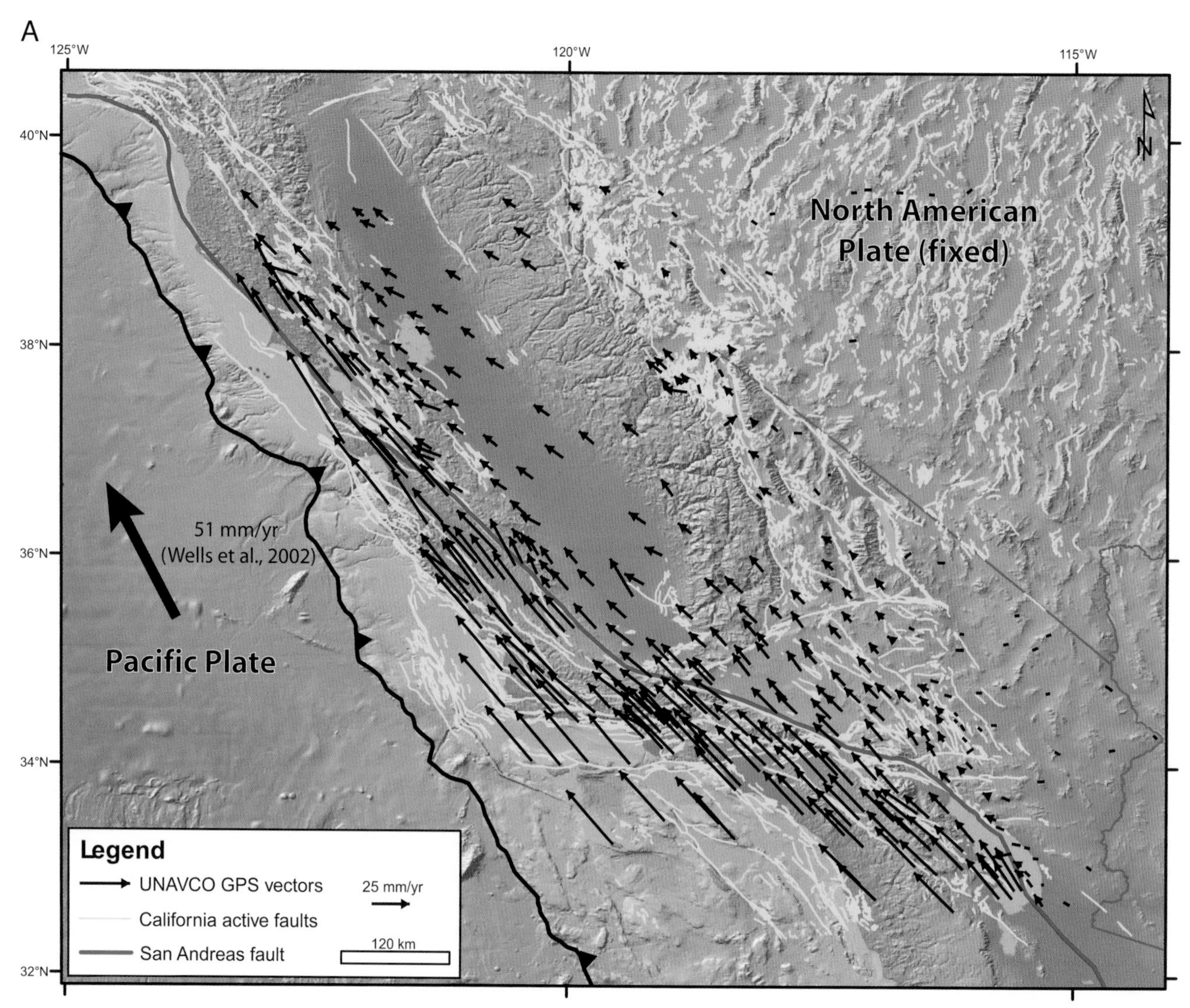

Figure 3. (A) Map showing global positioning system (GPS) vectors and the motion of the Pacific plate relative to a fixed point in the interior of the North American plate (UNAVCO, 2009). The heavy red line is the main trace of the San Andreas right-lateral strike-slip fault zone the yellow lines are other active faults taken from fault compilation by the State of California Department of Conservation (Bryant, 2005). Direction and rate of Pacific plate relative to fixed North America plate is from Wells et al. (2002). (B) Map showing the depth of the igneous/metamorphic/acoustic basement in feet beneath the onland San Joaquin and Los Angeles basins and offshore California borderlands of southern California. Sources of depth to basement information are taken from Exxon Tectonic Production Research Company (1985); depth to basement maps are modified from Harding (1973, 1974, 1985). Giant oil fields are shown in green stars, and giant gas fields are shown in red stars (data from Mann et al., 2003). (C) Stress map of California modified from the Heidbach et al. (2008) of the International Association of Seismology and Physics of the Earth's Interior (IASPEI). Various types of stress indicators are color coded on the map and summarized in the map key. Oil and gas giants are shown by green and red stars, respectively, and oil fields from the Zucca (2001) are shown in brown. The heavy red line is the main trace of the San Andreas right-lateral strike-slip fault zone, and the yellow lines are active faults taken from the State of California Department of Conservation. White boxes indicate the locations of more detailed map areas shown in Figures 4 and 7. Stress indicators on the map, including strikeslip (SS), normal faults (NF), thrust faults (TF), and undefined (U), are color coded and summarized in the map keys.

determined that contraction ranges from north-south to northeast-southwest in the Los Angeles Basin and is roughly parallel to the direction of maximum compressive stress (Figure 3C). Geologic studies have shown that strain in the Los Angeles Basin is accommodated by active systems of east–west-oriented reverse faults

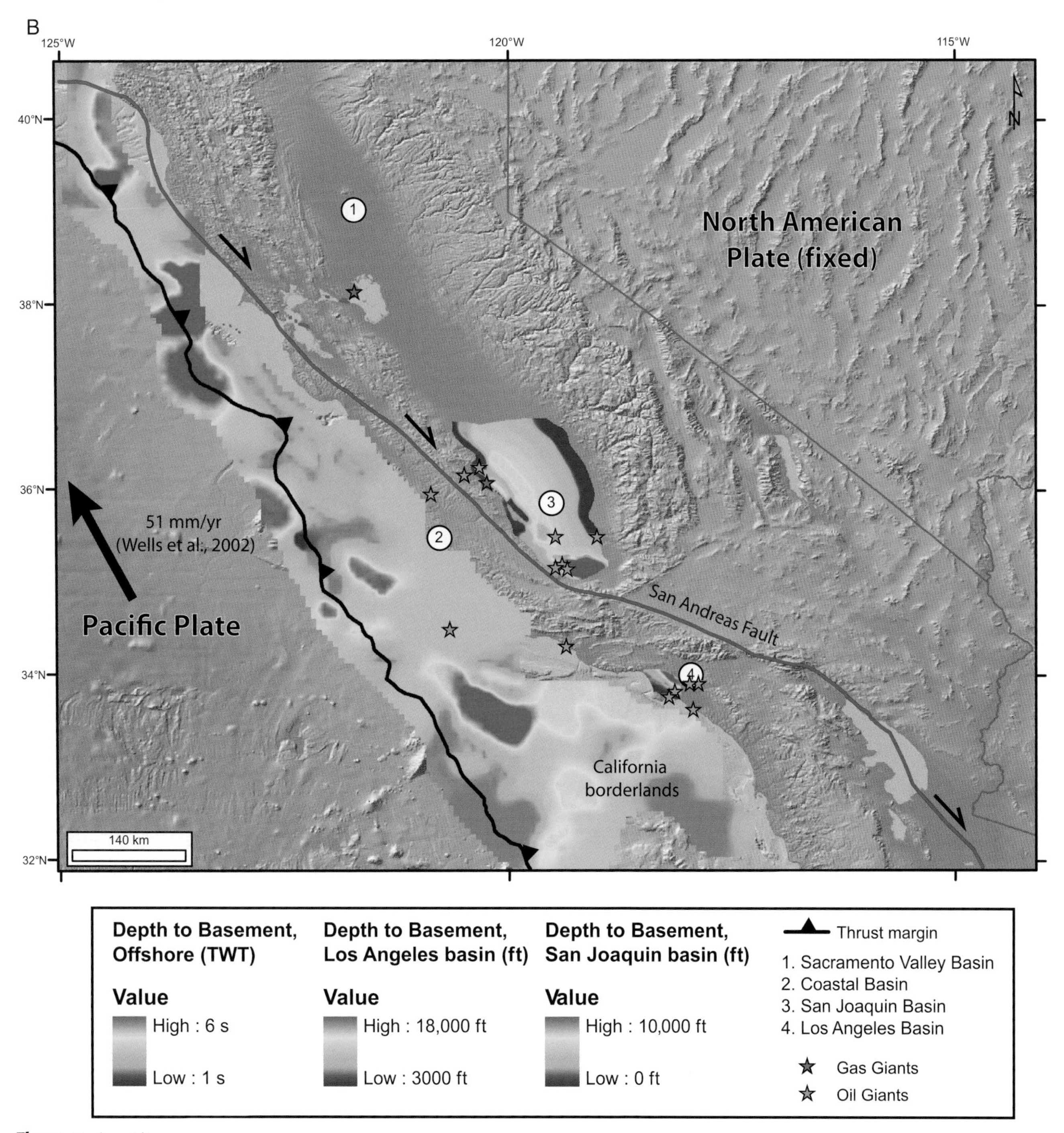

Figure 3. (cont.).

and southwest–northeast and northwest–southeast striking strike-slip faults (Figures 3C, 4).

Geologic studies of the Los Angeles Basin that make use of the extensive subsurface database from the oil industry have shown a two-stage history consisting of (1) an early stage of northeast–southwest rifting accompanied by volcanic activity from the middle Miocene (14 Ma) and the late Miocene (8 Ma) (Bjorklund et al., 2002); these rifts formed deep-marine basins suited for the deposition of source rocks and turbiditic reservoir rocks (Wright, 1991; Peters et al., 1994) and (2) a later tectonic phase between the late Miocene to Quaternary of north-south to northeast-oriented shortening that inverted the rifts to form blind thrust faults and overlying anticlinal uplifts that are now expressed as topographic highlands (Bjorklund and Burke, 2002)

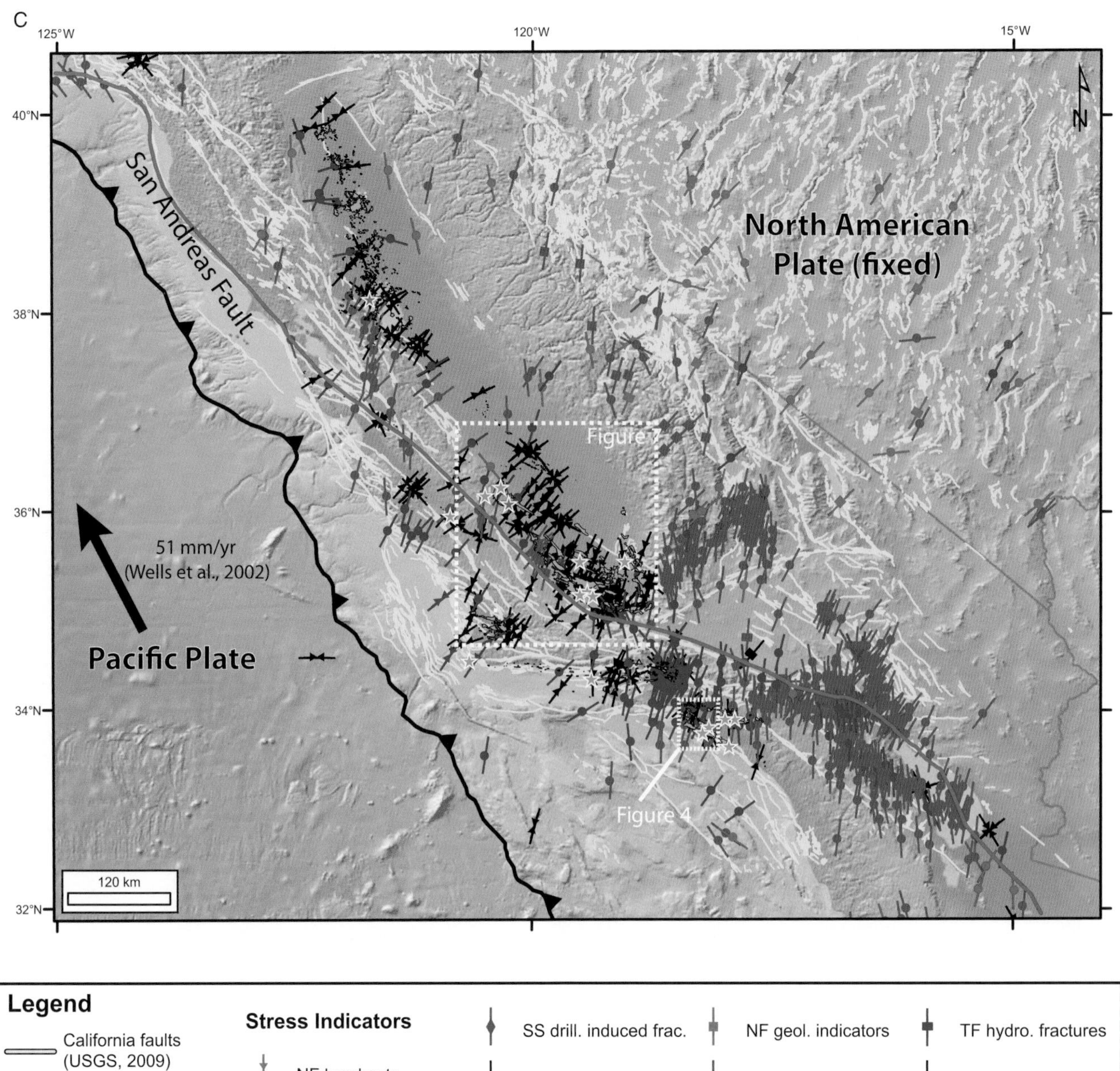

Figure 3. (cont.).

(Figure 4). During the early rift phase, the close association of volcanic and intrusive rocks with the basin suggests that the basin may have been part of a fore-arc system related to eastward subduction of the Pacific or Farallon plate that was similar to the San Joaquin fore arc basin to the north (Moxon and Graham, 1987).

The global compilation by Mann et al. (2003) of giant oil and gas fields shows that the Los Angeles Basin

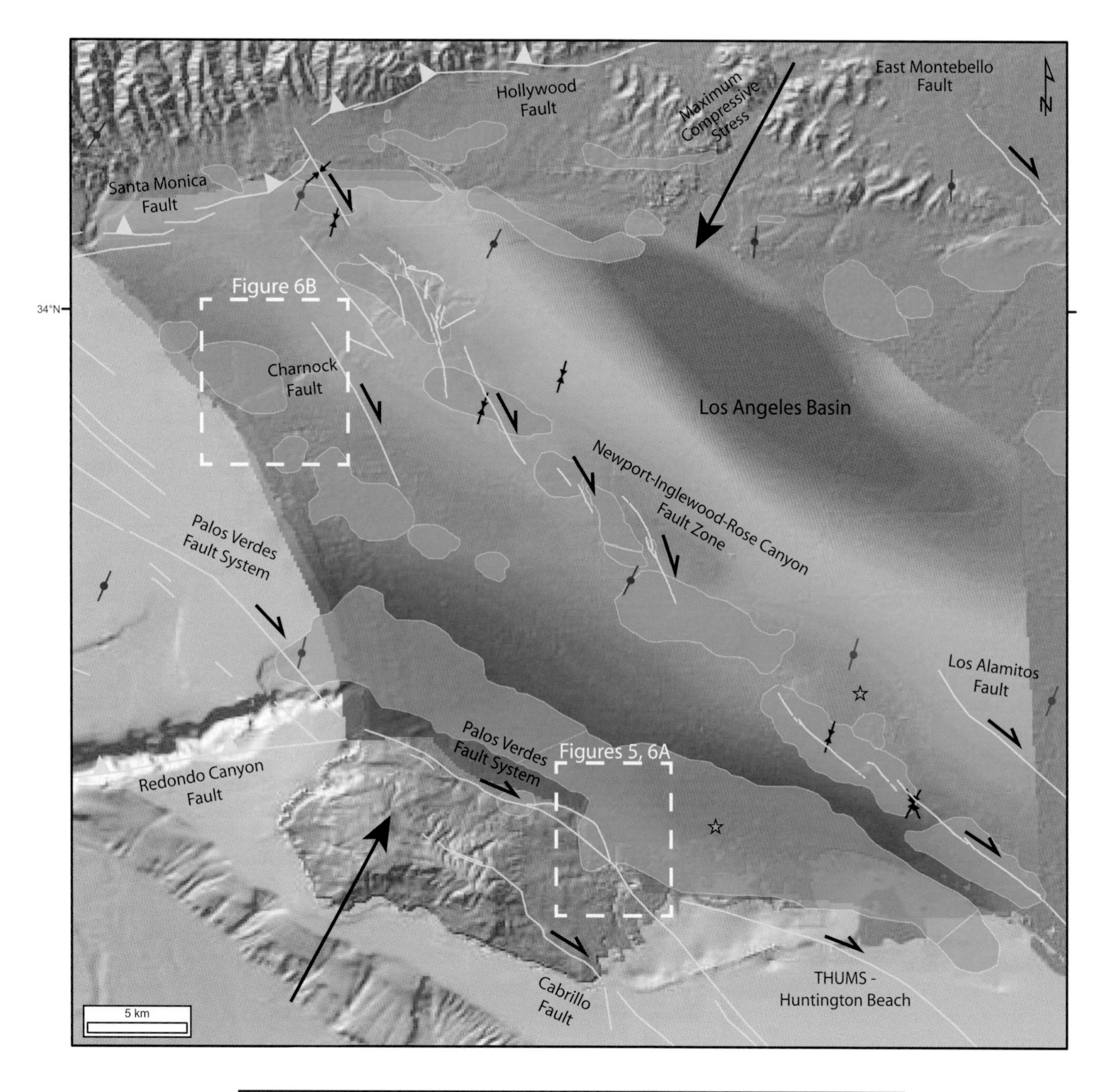

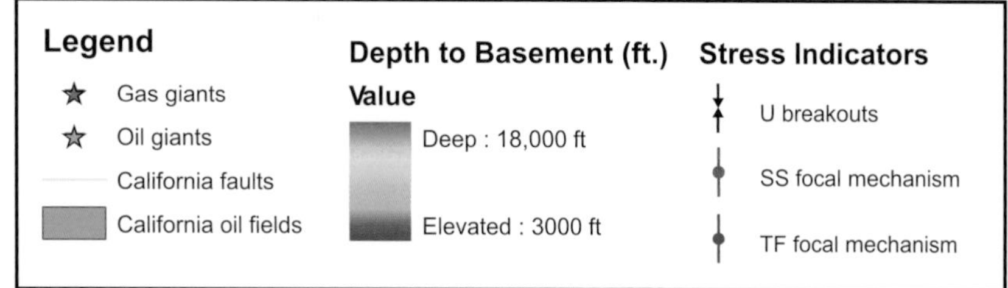

Figure 4. Detailed map showing faults, depth to basement in feet, and hydrocarbon occurrences in supergiant Los Angeles Basin of southern California. Sources of depth to basement information include Harding (1973) and Exxon Tectonic Production Research Company (1985). The giant oil fields are shown in green stars, and the giant gas fields are shown in red stars (data from Mann et al., 2003). Oil fields from compilation by the Zucca (2001) are shown in green. Various types of stress indicators are color coded on the map and summarized in the map key. The white boxes indicate the locations of more detailed map areas shown in Figures 5 and 6. U = undefined; SS = strikeslip; TF = thrust fault.

hosts five giant oil fields, with the Wilmington field adjacent to the Palos Verdes fault zone (Figure 4) being the third largest oil field in the United States. Discovered in 1932, the Wilmington field originally contained about 3 billion bbl of reserves, with only about 300 million bbl still in place. Smaller fields in the Los Angeles Basin are formed in anticlinal traps that are either parallel or oblique to presently active right-lateral strike-slip faults (Figure 4). Source rocks in the Los Angeles Basin include Eocene and Miocene units, including the well-known Miocene Monterey diatomaceous shale with oil-prone kerogen derived from marine plankton (Wright, 1991; Peters et al., 1994). Oil traps and the five giant fields of the basin are mainly found along strike-slip faults and include anticlinal closures modified by updip fault closures (Harding, 1973).

The depth to basement map shows that the deepest area of the Los Angeles Basin reaches depths of 550 m (~18,000 ft) in the northeastern part of the basin and rises to a shallow northwest-striking ridge structurally controlled by the Newport-Inglewood-Rose Canyon right-lateral strike-slip fault zone (Harding, 1973) (Figure 4). The deeper northeastern part of the basin is controlled by overthrusting of the adjacent transpressional ranges to the east that are controlled by the San Andreas fault zone (Figure 3B).

Described in detail using subsurface data by Harding (1973), the Newport-Inglewood fault zone contains an alignment of six major oil fields, each with more than 50 million bbl of cumulative production and a total cumulative production from 1932 through 1972 of 2.22 billion bbl of oil. Harding (1973) presents evidence for right-lateral strike-slip displacements along the elongate trend of the Newport-Inglewood fault zone that include (1) laterally offset fold axes and fold flanks, (2) horizontal slickensides on individual faults within the zone, (3) juxtaposed dissimilar stratigraphies across the fault zone, and (4) en echelon fold and fault pattern (Figure 4). The north–south orientation of maximum compressive stresses in the Los Angeles Basin is consistent with right-lateral shear along faults like the Newport-Inglewood fault zone (Figure 4).

The Inglewood field illustrates the subsurface structure of one of the larger oil-bearing anticlines along the Newport-Inglewood fault zone (Harding, 1973). The upright Inglewood anticline is symmetrical and plunges away from the obliquely bisecting right-lateral fault zone. The fault zone offsets the fold axis in a direction consistent with right-lateral shear along the fault. The structural trap is a combination of the early formed fold culmination with subsequent updip strike-slip faulting that combines to make a large and effective trap. The direction of maximum compressive stress from Zoback (1992) and Mount and Suppe (1992) is consistent with active right-lateral shear on the bisecting strike-slip fault zone.

The large doubly plunging anticline hosting the Long Beach field forms a structural high parallel to the main right-lateral strike-slip fault that runs along its fold axis (Figure 5). A right step in the trace of this right-lateral fault along the fold axis creates a small restraining bend or "push-up block" (Mann, 2007) near the plunging southeastern end of the fold (Figure 5A). Localized shortening in the restraining bend area has elevated the fold axis as a central block and has formed three compartmentalized structural traps with updip closures along the steeply dipping strike-slip faults (Figure 5B).

In comparison, the Potrero field is located on a wider restraining bend and fold linked by oblique reverse faults in the stepover area (Figure 6). The wider bend produces a wider fold structure with dipping limbs that are shallower than the Long Beach structure (Figure 6). The direction of maximum compressive stress shown by the large areas in Figure 6 is oblique to both structures and consistent with right-lateral shear along the central strike-slip faults.

The San Joaquin Basin

The San Joaquin Basin of California is the southern termination of a large asymmetrical fore-arc basin, the Great Valley, formed in a Mesozoic–Cenozoic Andean-type arc-trench system between an accretionary prism in the coastal area of California and the volcanic arc of the now deeply exhumed Sierra Nevada Mountains (Moxon and Graham, 1987) (Figure 7). GPS work by Feigl et al. (1993) shows that modern North America–Pacific plate motions are remarkably parallel to the San Andreas fault zone that forms a structurally controlled linear ridge along the southeastern edge of the San Joaquin Basin (Figure 7).

Geologic and fission track studies by Moxon and Graham (1987) show that the Great Valley fore-arc basin (including the San Joaquin) formed in two stages: (1) between 100 and 80 Ma, rapid subduction of a steeply east-dipping Pacific or Farallon slab was manifested by voluminous magmatism in the Sierra Nevada volcanic arc east of the San Joaquin Basin; and (2) between 40 and 45 Ma, shallow east-dipping subduction and very rapid Pacific or Farallon convergence led to a lull in the Sierra Nevada magmatism and convergent deformation of the Laramide orogeny across the western United States.

The compilation by Mann et al. (2003) shows the presence of nine oil and gas fields in the Great Valley, with five giants clustered in the San Joaquin Basin

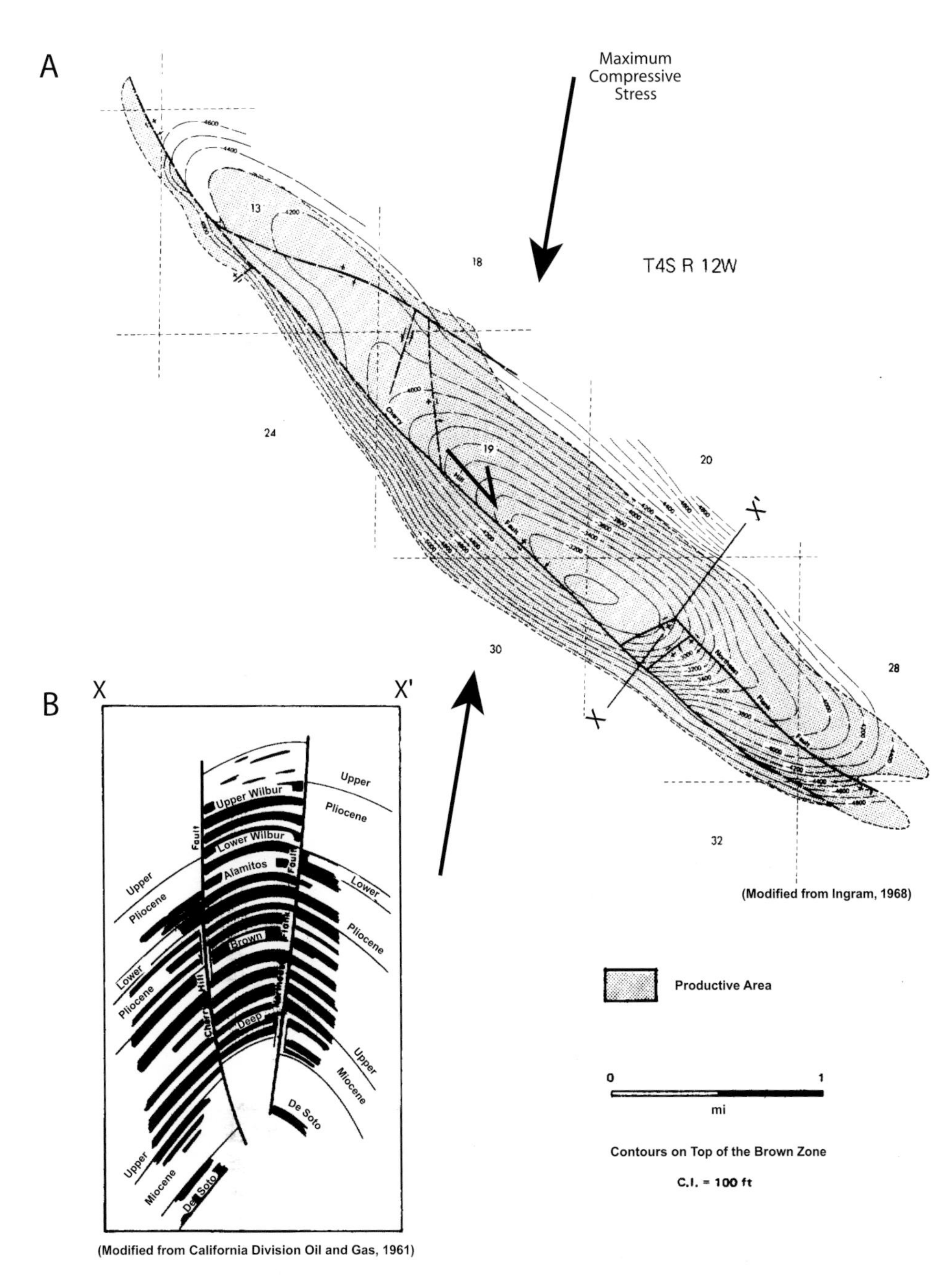

Figure 5. Inglewood oil-bearing strike-slip-related fold in Los Angeles Basin showing the upright symmetrical anticline plunging away from bisecting right-lateral strike-slip fault that offsets the fold axis by 1500 ft (457 m). A more regional map of this fold is shown in Figure 4. The closure is defined by the earlier formed fold culminations and the later formed up-plunge strike-slip fault. The productive area is in gray. The figure is modified from Harding (1973).

(Figure 7). The San Joaquin Basin includes the Midway-Sunset oil field, the largest oil field in California and the second largest field in the United States. This extensive oil field aligned in a large anticline adjacent to the San Andreas fault zone near the southeastern edge of the San Joaquin Basin (Figure 7) has a cumulative production of 3 billion bbl of oil at the end of 2006 and has an estimated remaining reserve of 532 million bbl of oil. This field and others surrounding it formed by overthrusting of the California coastal ranges over the San Joaquin Basin (Figure 8A). Northeast-directed thrusting accounts for the asymmetrical depth to basement in the San Joaquin Basin, with the deeper part of the basin depressed along this thrust system (Figure 7).

In the San Joaquin Basin, most of the folds (and their associated oil fields) are parallel, not oblique to the San Andreas fault zone (Zoback et al., 1987; Mount and Suppe, 1992) (Figure 8A). The San Andreas fault at this latitude is transpressional in nature with bidirectional thrusting: northeastward thrusting and associated blind thrusts in the San Joaquin Basin and southwestward overthrusting of the San Andreas fault-controlled ridge over the adjacent and less deep Cuyama Basin (Figure 8B). The bidirectional component of thrusting

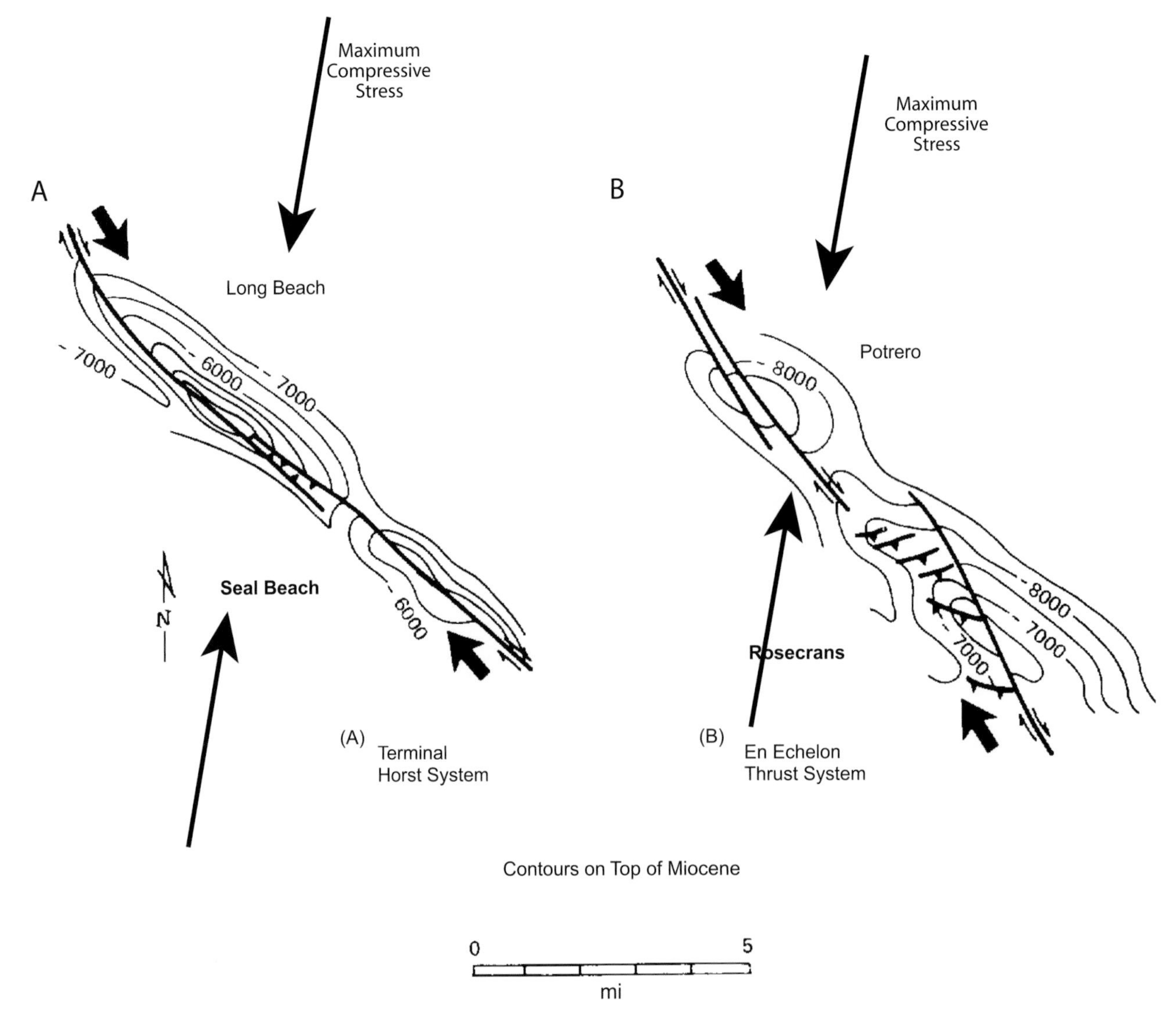

Figure 6. Comparison of schematic views of Long Beach left-stepping restraining bend folds with wider left-stepping faults and connecting thrust faults at the Rosecrans field. Contours on top of the Miocene. 5 mi (8 km).

is driven by the relatively small component of fault-normal compression across the San Andreas fault as measured by GPS (Figure 3A).

The age of the faulting and folding on both sides of the fault is Pliocene–Pleistocene (Zoback et al., 1987). All of the large structural traps hosting giants of the San Joaquin Basin are folds produced by overthrusting related to shortening along the San Andreas fault zone (Figure 7). From the compilation map in Figure 7, it is clear that the folds shown are parallel to the San Andreas fault zone. Harding and Wickham (1988) criticized Mount and Suppe's (1987) application of their low-drag San Andreas fault model to the central segment of the San Andreas strike-slip fault adjacent to the hydrocarbon-bearing San Joaquin Basin for two reasons: (1) the fault-parallel fold belt in California used as evidence for a compressive stress perpendicular to the San Andreas fault contains four possibly separate fold belts that differ in trend, age of commencement or cessation of folding, position relative to the strike-slip fault, properties of underlying basement, or thickness of the sedimentary cover, and possibly, fold style. Harding and Wickham (1988) regarded two of the fold belts to be not directly related to strike-slip motions along the San Andreas fault zone and instead formed before the formation of the San Andreas in the late Oligocene. Moreover, he argued that an unchanging stress field would tend to rotate all older (possibly unrelated) folds into alignment with younger strike-slip faults, so the present orientations of folds would be of

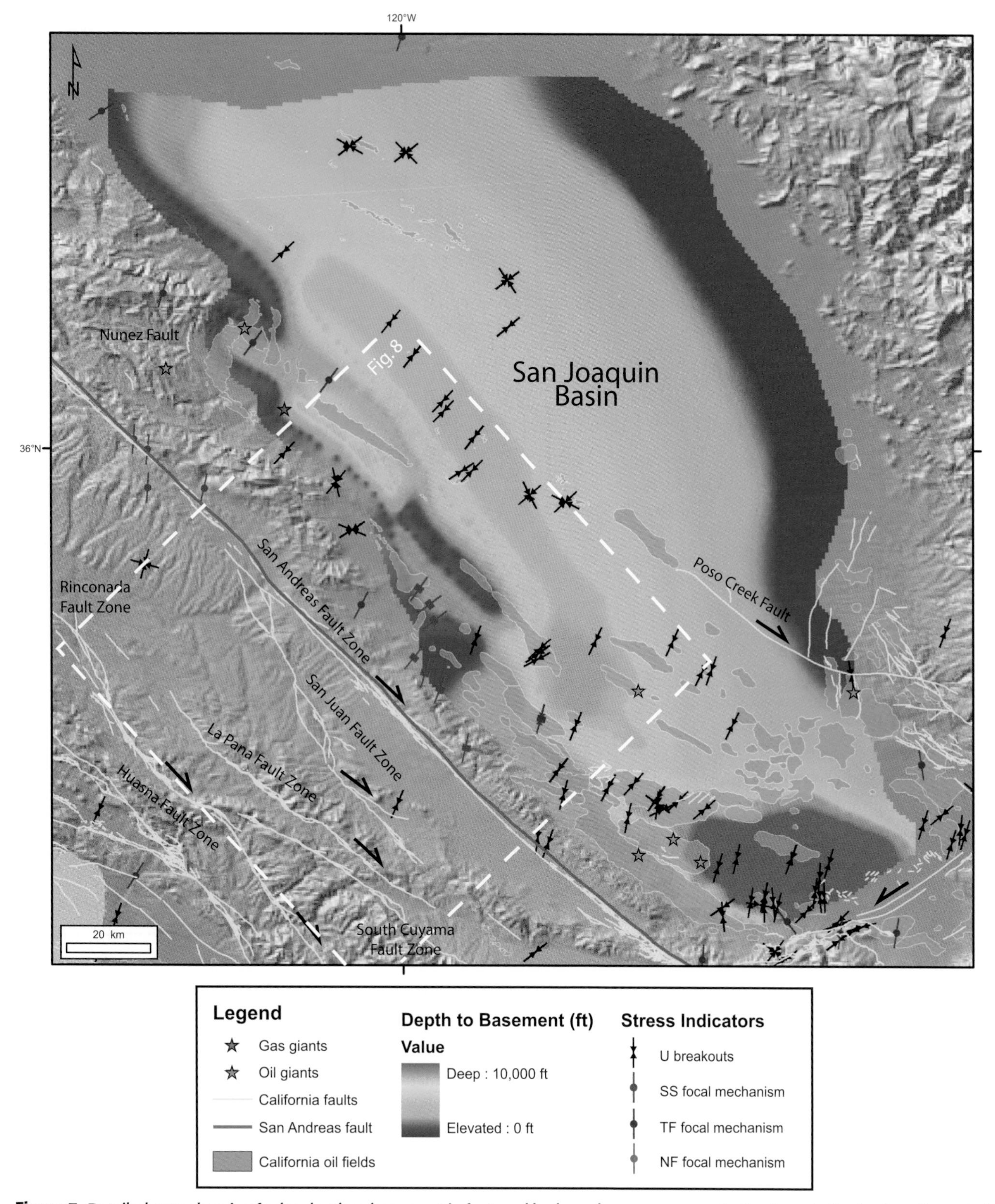

Figure 7. Detailed map showing faults, depth to basement in feet, and hydrocarbon occurrences in San Joaquin Basin of south-central California. Sources of depth to basement information from Harding (1974). The giant oil fields are shown in green stars, and the giant gas fields are shown in red stars (data from Mann et al., 2003). Oil fields from the Zucca (2001) are shown in green. Stress indicators on the map, including strike-slip (SS), normal faults (NF), thrust faults (TF), and undefined (U), are color coded and summarized in the map key. The white box indicates the location of a more detailed map shown in Figure 8. 20 km (12.4 mi).

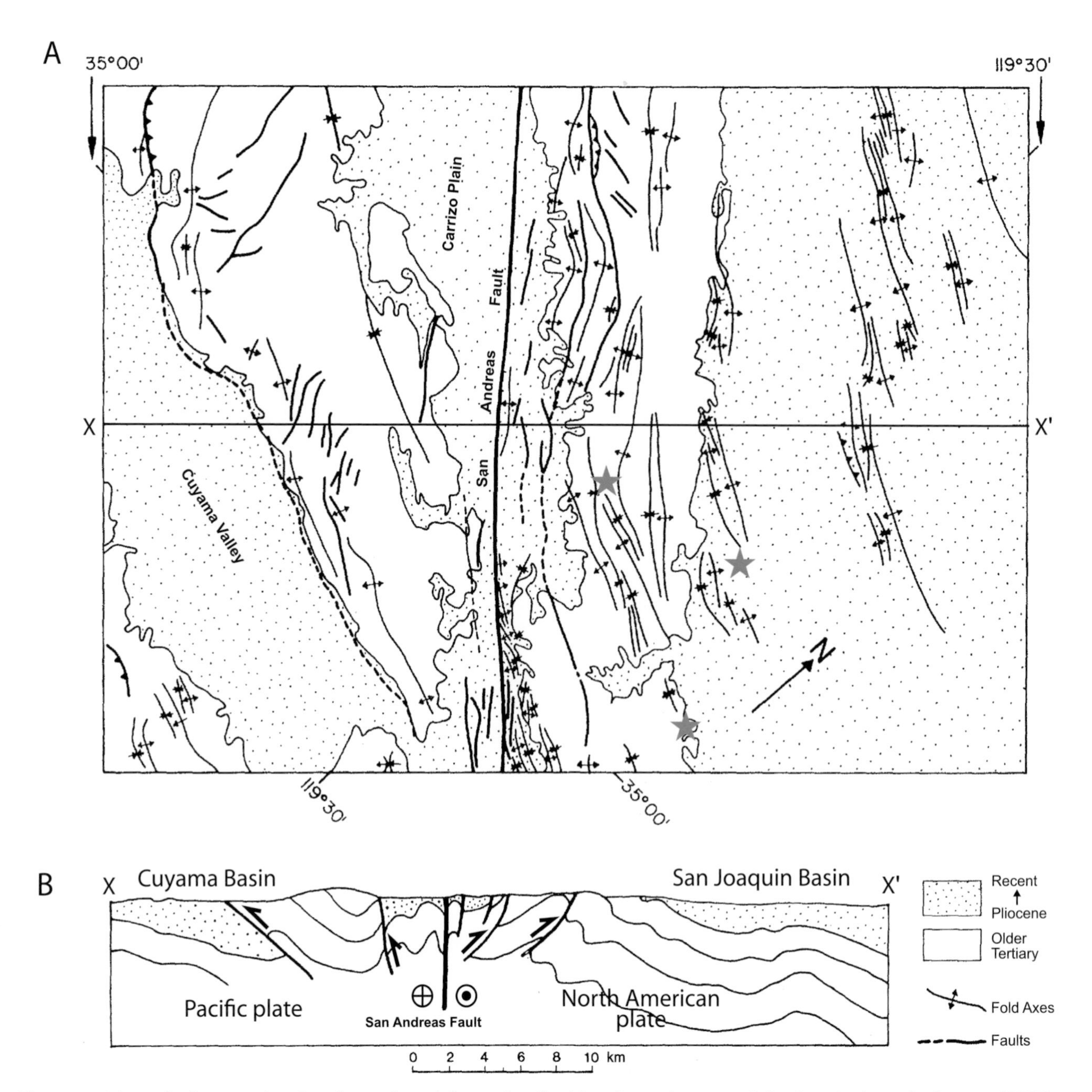

Figure 8. (A) Geologic map showing the surface deformation flanking the main trace of the San Andreas fault zone (SAFZ) in the Carrizo plain area adjacent to the flanking Cuyama and San Joaquin basins (more regional map shown in Figure 7). Most folds deforming rocks of the Pliocene to the Holocene are parallel to the main trace of the SAFZ, indicating a large degree of fault-normal convergence consistent with the regional stress pattern shown in Figure 7. The giant oil fields are shown in green stars, and the giant gas fields are shown in red stars (data from Mann et al., 2003, updated in 2009). Additional oil fields from the Zucca (2001) are shown in green. (B) Cross section along line XX′ in (A) showing the symmetrical positive flower structure along the main trace of SAFZ and flanking Cuyama and San Joaquin basins. 10 km (6.2 mi).

little or no use for inferences of the direction of the present-day stress field associated with active strike-slip faults.

Mount and Suppe (1987) and Zoback et al. (1987) have both concluded that motion along the San Andreas fault is partitioned into (1) a low-shear stress strike-slip fault component (~39 mm/yr) parallel to the San Andreas fault; and (2) a high-stress thrust component (~3.3 mm/yr) normal to the San Andreas fault. The strike-slip component is accommodated along a

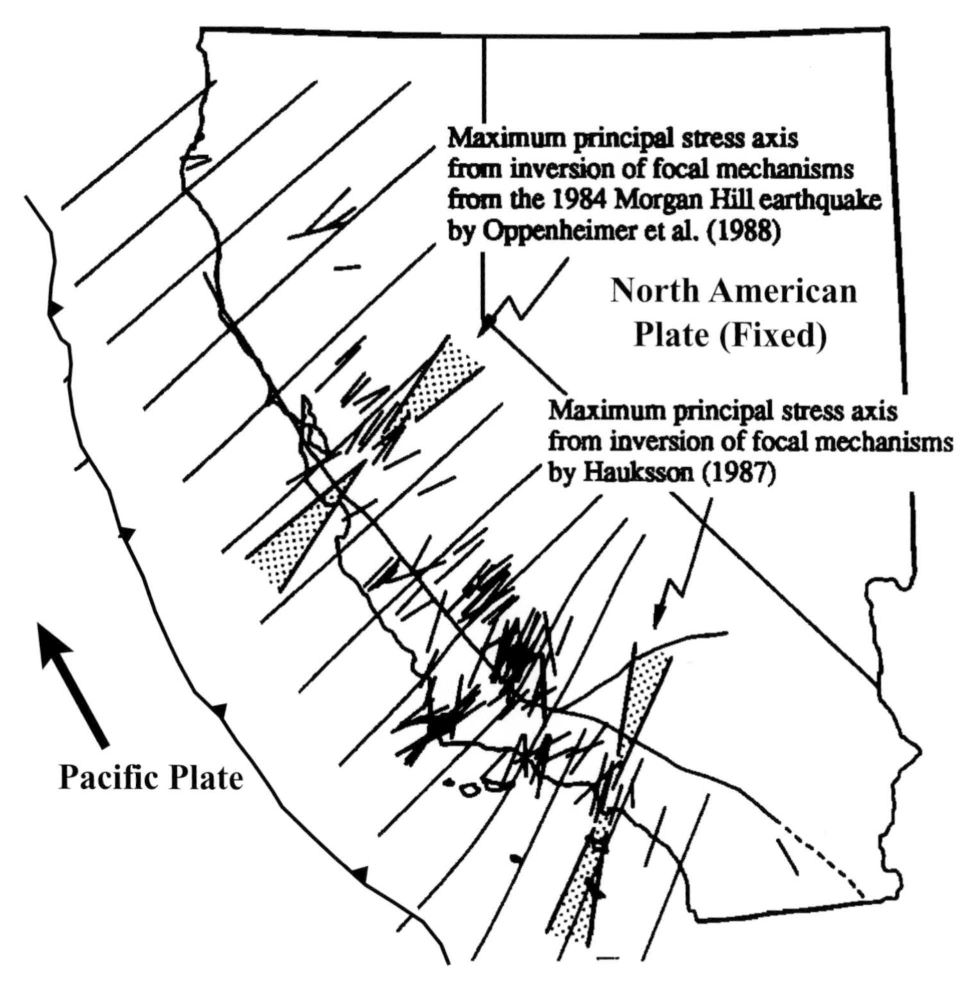

Figure 9. Lines show trajectories of regional maximum compressive stress in California based on well breakout data compiled by Mount and Suppe (1992). Stress trajectories are near San Andreas fault-normal and support the observed deformation styles of folds in the Pliocene to recent rocks that parallel the main trace of the San Andreas fault zone (SAFZ) shown in Figure 8B. Directions of maximum compressive stresses derived from inversion of earthquake focal mechanism data support stress trajectories from well breakout data (Webb and Kanamori, 1985; Hauksson, 1987; Oppenheimer et al., 1988). Offshore thrust accommodates overthrusting of coastal California over the area of deeper water in the Pacific Ocean and is based in part on McIntosh et al. (1991). Note that the shape of the offshore thrust zone closely parallels the shape of the onland main trace of the SAFZ.

narrow (3–10 km [1.9–6.2 mi] wide) zone of nearly pure strike-slip faulting. This narrow zone of shearing is expressed as the center part of the topographically elevated and symmetrical flower zone of deformation seen on the cross section in Figure 8B. The compressive component is accommodated in a wider (50–100 km [31-62] mi) zone of thrust faults, which propagate in directions almost perpendicular to the San Andreas fault (Figure 2B). These authors concluded that the broad zone of folding is not a consequence of distributed shear along the high-strength San Andreas fault as envisioned by Harding (1976) (Figure 2A).

Tectonic Origin of the Regional Stress Field of California and Its Effects on Petroleum in the Los Angeles and San Joaquin Basins

The map in Figure 9 shows a compilation of all maximum horizontal stress trajectories in California computed from breakout data by Mount and Suppe (1992). The map shows an overall homogeneous stress field more than 45,000 km^2 ($\sim$27,962 mi^2) with stress axes at high angles to the folds, thrusts, and transpressive strike-slip faults forming giant oil and gas traps in both the Los Angeles (Figure 4) and the San Joaquin basins (Figure 7). The remarkable uniformity of the stress field over such a large area suggests that the origin of the stress field is related to a plate tectonic driving force such as slab pull, ridge push, collisional resistance, trench suction, and basal drag (Zoback, 1992). The stress pattern is clearly unrelated to the present directions of North America–Pacific plate motions as shown by GPS work compiled in Figure 3A.

In Figure 9, the position of the offshore California subduction zone based on studies by McIntosh et al. (1991), Brocher et al. (1994), and Henstock et al. (1997) is superimposed on the onland stress map of Zoback (1992). The slightly arcuate shape of this offshore subduction zone parallels the same shape seen in the arcuate California coastline and the shape of the San Andreas fault itself. One possibility is that subduction of an eastwardly subducting thickened slab is still occurring at slow rates perhaps on the order of fault-normal

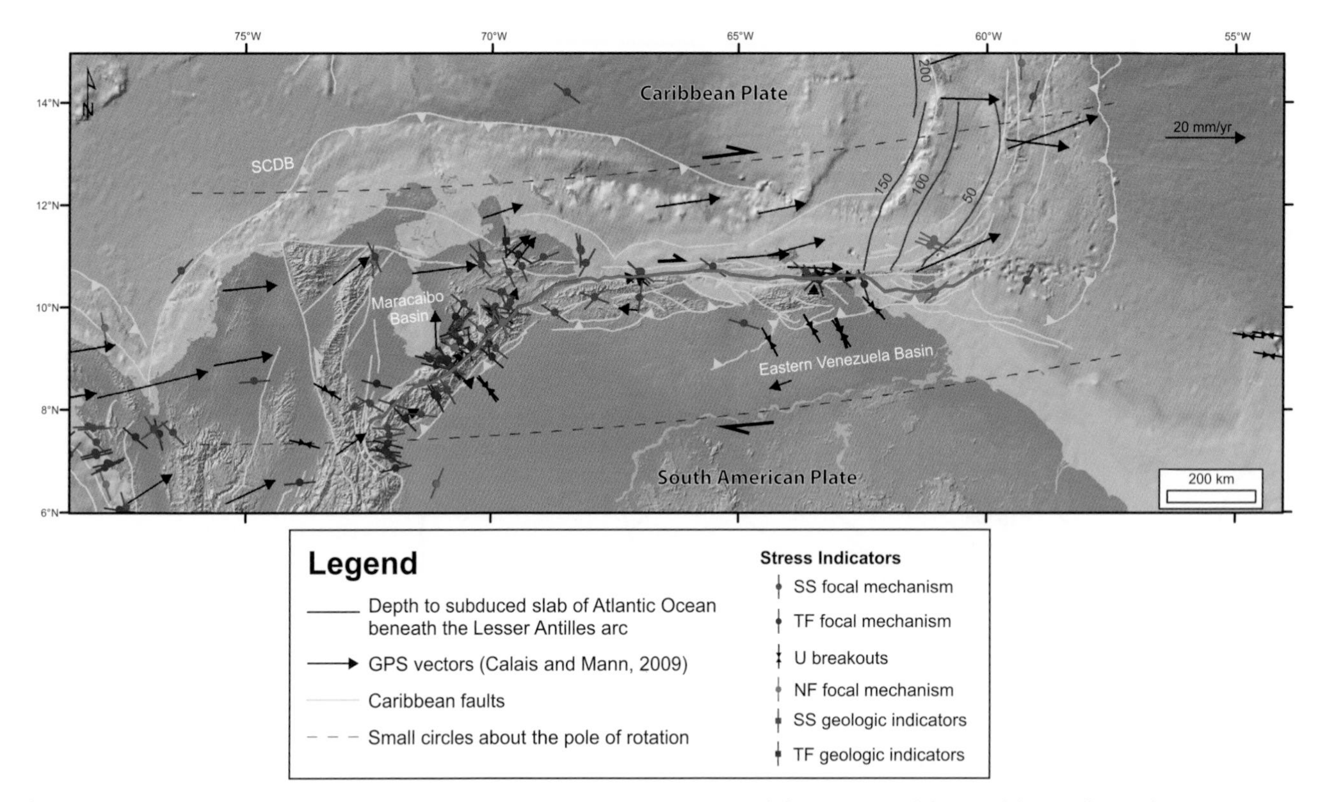

Figure 10. (A) A map showing global positioning system (GPS) vectors and the motion of the Caribbean plate relative to a fixed point in the interior of the South American plate (Calais and Mann, 2009). The heavy red line is the main trace of the Bocono–El Moron–El Pilar–Central Range right-lateral strike-slip fault zone of Venezuela and Trinidad and Tobago. Direction and rate of Caribbean plate relative to the fixed South America plate is from Calais and Mann (2009). The stress map of the southern Caribbean is from the Heidbach et al. (2008) of the International Association of Seismology and Physics of the Earth's Interior (IASPEI). Various types of stress indicators are color coded on the map and summarized in the map key. (B) Map showing the depth of the igneous/metamorphic/acoustic basement in feet beneath onshore and offshore basins of Venezuela and Trinidad and Tobago (numbers show major basin names in the legend below the map). Sources of depth to basement information include depth to basement compilation in Mann et al. (2010). The giant oil fields are shown in green stars, and the giant gas fields are shown in red stars (data from Mann et al., 2003). (C) Oil and gas giants are shown by green and red stars, respectively, and oil fields from the U.S. Geological Survey (USGS, 1999) are shown as orange dots. The heavy red line is the main trace of the Bocono–El Moron–El Pilar–Central Range right-lateral strike-slip fault zone, and the yellow lines are active faults from Mann et al. (2006). The white boxes indicate the locations of more detailed map areas shown in Figures 11 and 13. SCDB = southern Caribbean deformed belt.

shortening in California (3.3 mm/yr) (Argus and Gordon, 2001) along this offshore subduction feature and actively controls the regional stress field by the process of collisional resistance. This is a very low angle and shallowly dipping subduction system, as determined in the study by Brocher et al. (1994) in northern California. For this reason, this shallow subduction process may involve a very wide zone of coupling and collisional resistance that would likely affect the stress state of the uppermost crust in the overriding plate. Similar zones of slightly active and shallowly dipping subducted slabs are discussed later to explain similar patterns of regional stress and hydrocarbon distribution in the strike-slip zones of the southern Caribbean, Sumatra, and eastern China.

STRIKE-SLIP FAULT FORMED AT THE SITE OF ARC-CONTINENT COLLISION: SOUTHERN CARIBBEAN PLATE BOUNDARY

Geographic Location of Study Area

The active southern Caribbean-South America plate boundary is defined by the Bocono–El Pilar right-lateral strike-slip fault system that extends as a single continuous fault zone approximately 1500 km (~932 mi) from Colombia in the west to Trinidad and Tobago in the east (Figure 10A). This single fault system accommodates most of the Caribbean–South America interplate motion (Calais and Mann, 2009) and poses a major seismic hazard to the populations of these three countries it traverses

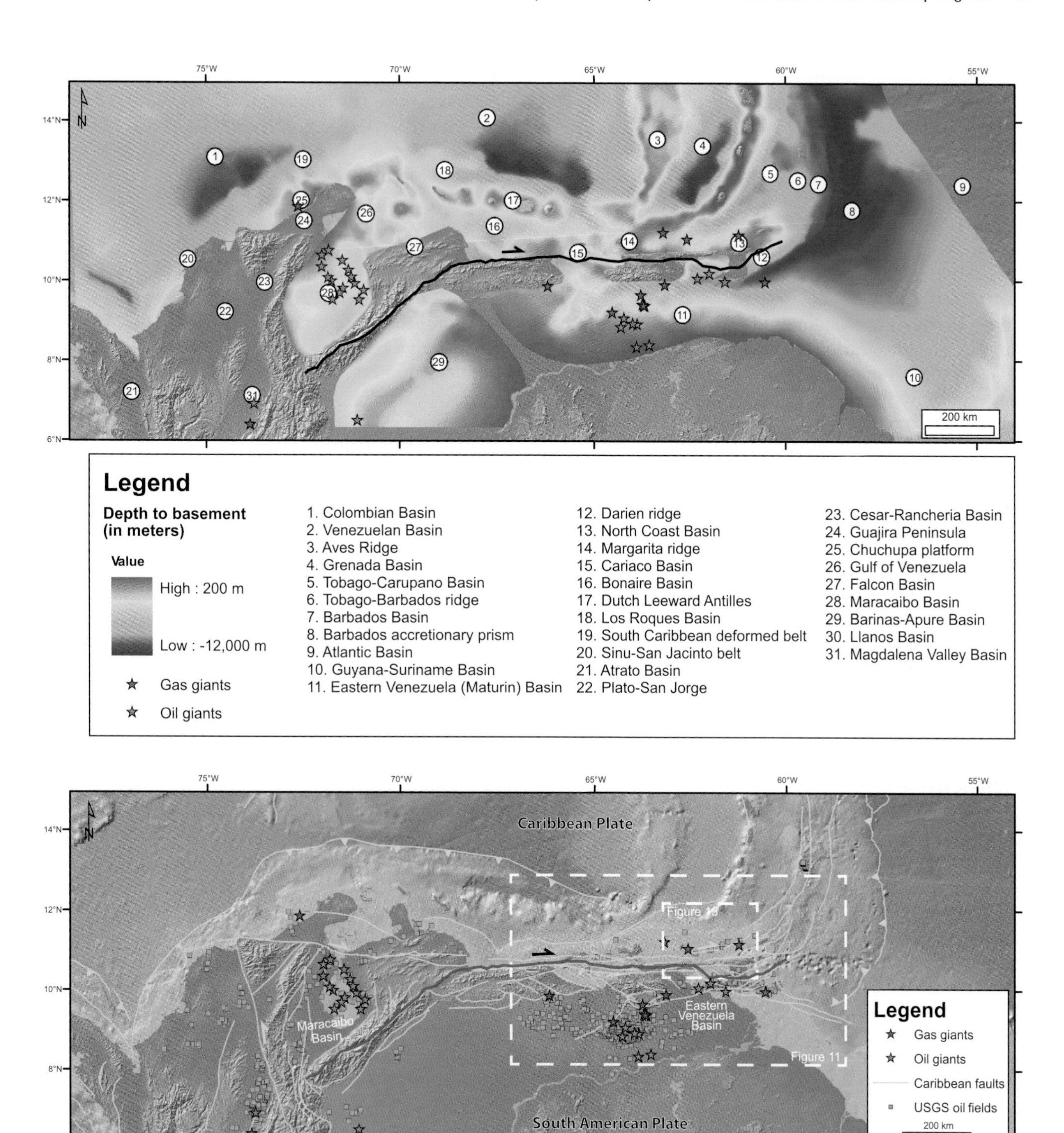

Figure 10. (cont.).

(Figure 10A). This description of the fault system will focus on a few of its better studied basinal areas of hydrocarbon accumulations and will attempt some comparisons between its oil and gas structural traps with those seen in the previous examples from California and the later examples from Sumatra and eastern China.

Classification of the Bocono–El Pilar Strike-slip Fault System

In the classification scheme of Mann (2007), the Bocono–El Pilar fault system is a continental boundary strike-slip fault characterized by a long fault trace separating a continental plate (South America) from an obliquely collided plate composed of arc and oceanic terranes (Caribbean) (Escalona and Mann, 2011) with most interplate motion concentrated along and parallel to the single fault trace (Figure 10A).

Crustal Motions Determined from Global Positioning System-based Geodesy

Global Positioning System studies in northern South America and the Caribbean (UNAVCO, 2009; Calais and Mann, 2009) show that the motion of the Caribbean plate relative to the stable South America is subparallel to the main trace of the Bocono–El Pilar strike-slip plate boundary fault that is shown as a heavy red line in Figure 10A. Weber et al. (2001) concluded from a regional study of GPS results that the Bocono–El Pilar system accommodates most of the plate motion by plate boundary-parallel strike-slip faulting at rates of 20 mm/yr (Figure 10A). Global positioning system shows that fault-normal motion is small and occurs at much slower rates than the fault-parallel motion. The nonparallelism and varying magnitudes of GPS vectors over wide areas indicate that some large-scale rotations of fault blocks may be occurring (Escalona and Mann, 2011) (Figure 10A).

Depth to Basement Map

Depth to basement information for the areas adjacent to the Bocono–El Pilar fault system was found from various sources, including many unpublished studies of the CBTH (Caribbean Basins, Tectonics and Hydrocarbons) Project at the University of Texas (Figure 10B). As observed in California, the basement is deformed asymmetrically with steeper dipping and deeper basement flanking the thrust-controlled and elevated ridge parallel to the Bocono–El Pilar strike-slip fault system (Figure 10B). Giant oil and gas fields shown in Figure 10B cluster in the deeper basinal areas flanking the fault system, especially in the superdeep foreland basin area of eastern Venezuela and Trinidad and Tobago (Figure 10B). This pattern of a deeply depressed basement with giant fields is similar to the setting of the cluster of giant fields in the Big Bend area of the San Andreas fault system (Figure 3B).

Seismic profiling and tomographic imaging across the offshore thrust front shown in Figure 11B (South Caribbean deformed belt) reveals the presence of subducted oceanic crust of oceanic plateau crust extending southeastward and southward beneath the margin of northern South America (van der Hilst and Mann, 1994; Clark et al., 2007; Magnani et al., 2009). The length of the subducted slab imaged on refraction and reflection data beneath the South Caribbean deformed belt exceeds 300 km (186.5 mi) beneath Colombia (van der Hilst and Mann, 1994) and 100 km (62 mi) in northwestern South America (Magnani et al., 2009) but disappears entirely in the east near Trinidad (Clark et al., 2007) (Figure 10A). This pattern of subduction can be explained by the direction of plate motion known from GPS vectors (Figure 10A). As one moves from west to east along the plate boundary, plate motion becomes increasingly oblique to the point that insufficient plate boundary normal stress exists to drive subduction.

Regional Stress Map

Regional stress directions by Zoback (1992), Colmenares and Zoback (2003), and the Heidbach et al. (2008) are compiled in Figure 10A, with the type of stress indicator used color coded in the lower part of Figure 3D. As discussed by Colmenares and Zoback (2003), these directions of the maximum compressive stresses change their orientation to remain perpendicular to the trace of the offshore South Caribbean deformed belt and its subducted slab (Figure 10A). In the Ecuadorian Andes, where the maximum compression is approximately northwest–southeast, the stress magnitudes are greatest and were interpreted as manifesting the process of collisional resistance of the fast-subducting and shallow-dipping slab in that area. Stresses become less compressive in the area of eastern Venezuela and Trinidad in northeastern South America and show a northeast–southwest direction of extension. This change in magnitude and direction may reflect the absence of the shallowly subducting Caribbean slab in that area. As in California, the regional stress directions remain almost perpendicular to the Pacific and Caribbean coastlines and the Bocono–El Pilar strike-slip fault system over much of its length.

Strike Slip-related Structures in Eastern Venezuela and Trinidad and Relation to Tectonics and Hydrocarbons

Hydrocarbon Distribution

The best area for seeing the relationship between the strike-slip fault system and oil and gas fields is in the area of eastern Venezuela and Trinidad and Tobago (Figure 10C). Volumetrically, the Maracaibo Basin of western Venezuela is by far the largest contributor to the reserves of the Gulf of Mexico–Caribbean region by contributing 37% of the ultimate hydrocarbon reserves

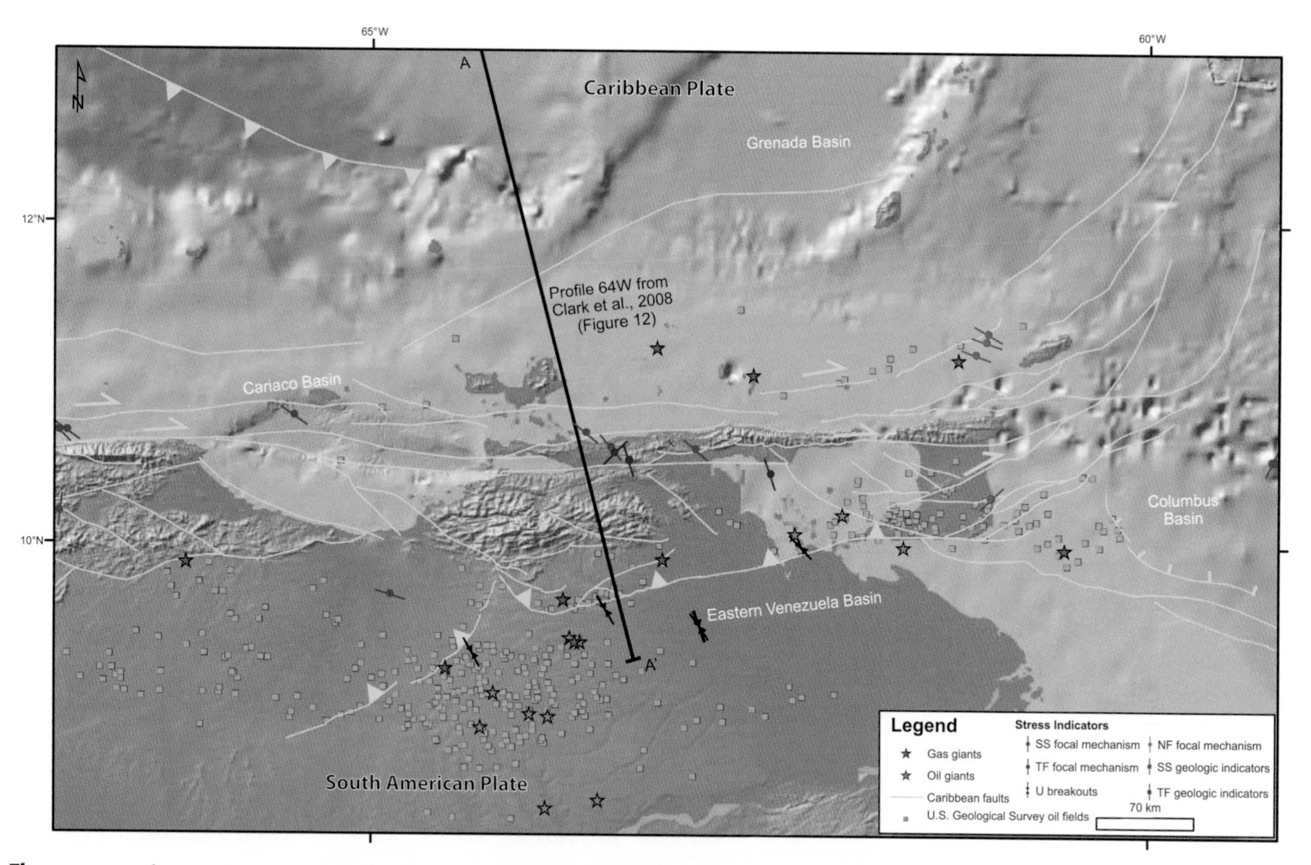

Figure 11. Bathymetric-topographic map of eastern Venezuela showing stress indicators and major faults and giant oil and gas fields as green and red stars, respectively. Orange dots represent oil and gas fields from the U.S. Geological Survey source (USGS, 1999) cited in Figure 10. Line of cross section for section AA′ shown in Figure 12 is indicated. 70 km (43.5 mi).

of the region (USGS, 1999; Escalona and Mann, 2006). The second largest contributor in the region is the eastern Venezuelan basin, with 19% of the ultimate reserves, whereas Trinidad and Tobago occupy a distant third place, with 1% of the ultimate reserves. Conventional oil and gas deposits in the eastern Venezuelan foreland basin are greatly increased by the addition of nonconventional resources of the Venezuelan Heavy Oil belt along its southern fringe in the vicinity of the Orinoco River (Figure 10C), adding 316 billion bbl of reserves, making Venezuela control the second largest hydrocarbon reserve in the world, with about 524 million bbl of oil equivalent. These estimates do not include the recent La Perla giant gas discovery off the northwestern coast of Venezuela, which is considered to be 9 to 10 tcf and, therefore, the largest single-field gas discovery in the past 10 yr.

Eastern Venezuelan Foreland Basin

The largest depocenter in northeastern South America is the Eastern Venezuela foreland basin that extends across eastern Venezuela into the Columbus Basin of offshore Trinidad, with more than 9 s of two-way traveltime in the thickest part of the offshore Columbus Basin (Garciacaro et al., 2011) (Figure 10B). Most of these clastic sediments are sourced by the Orinoco River system. A cluster of oil and gas giants parallels this deep elongate basin from Venezuela into Trinidad and its offshore area (Columbus Basin) (Figure 10B). Clark et al. (2008) have studied the crustal structure underlying this hydrocarbon-rich area using seismic refraction and reflection methods on a long transect that includes the deeper area of the eastern Venezuelan basin (Figure 11). Their refraction profile reveals the three tectonically juxtaposed crustal components underlying these basins: (1) an oceanic plateau area (~20 km [~12.5 mi]) underlying the Venezuelan basin; (2) an intermediate thickness (~30 km [~18.5 mi]) area of island arc crust that was accreted to the northern margin of South America in the Oligocene and the Miocene; and (3) a thick (~50 km [~31 mi]) area of continental crust underlying the South American continent (Figure 12). The suture zone between the arc crust and the continental crust coincides exactly with the

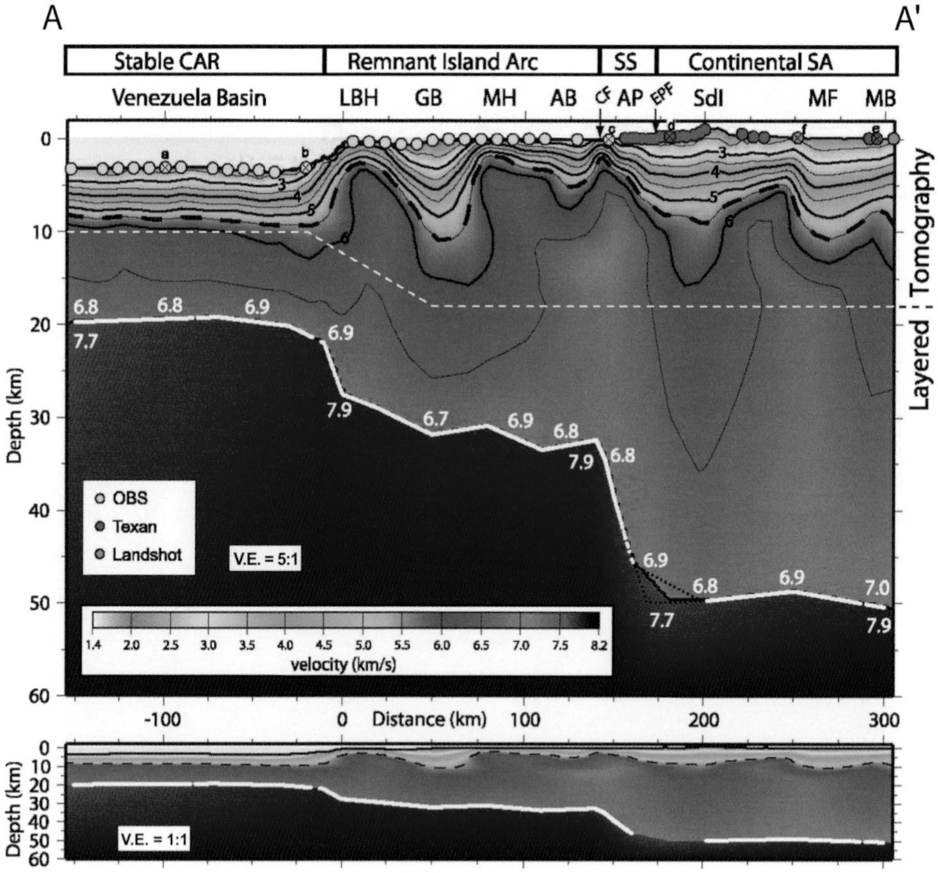

Figure 12. Crustal refraction section along line AA′ in offshore and onshore eastern Venezuela in the southeastern Caribbean shown in figure 12 of Clark et al. (2008), with the lowermost crust and uppermost mantle velocities (km/s) labeled above and below the Moho. LBH = La Blanquilla High; GB = Grenada Basin; MH = Margarita High; AB = Araya Basin; CF = Coche fault; A-P = Araya-Paria peninsula; EPF = El Pilar fault; SdI = Serrania del Interior; MF = Monagas foothills; MB = Maturin Basin; SS = strike-slip system; CAR = Caribbean plate; OBS = Offshore Business Solutions well; VE = vertical exaggeration; SA = South American plate.

surficial expression of the active El Pilar strike-slip fault system (Clark et al., 2008). The El Pilar fault zone, therefore, marks the suture zone between the accreted arc and oceanic terranes of the Caribbean plate and the continental rocks of northern South America (Escalona and Mann, 2011). An important point for hydrocarbons is that each of the three basement types has a unique overlying stratigraphy that controls the source and reservoir potential for the overlying basins. The South American continental basement has the most prolific overlying hydrocarbon basins.

Cariaco Pull-apart Basin

The Cariaco Basin of offshore Venezuela forms the largest pull-apart at a 30 km (~18.5 mi) wide right stepover in the Bocono–El Pilar fault system (Figure 13). Structural and isopach time maps based on an integration of all the seismic reflection data available from the basin in Figure 13 reveal the shape and structural control of the basin (Escalona and Mann, 2006). The Cariaco Basin geometry is characterized by two main depocenters (West and East Cariaco basins) elongated in an east–west direction, with a main clastic infill of Neogene age. These maps indicate that the basin formed in three main phases: (1) from the middle Miocene to Pliocene, the West Cariaco Basin formed as a rhomboidal pull-apart at a 30 km (~18.5 mi) wide stepover between the northern branch of the San Sebastian fault and the El Pilar fault zone; (2) during the early Pliocene, the southern branch of the San Sebastian fault and the El Pilar fault zone transected the West Cariaco Basin and caused extension to cease; and (3) during the early Pliocene to Holocene, a "lazy-Z" shaped pull-apart formed as the gently curving connection between the San Sebastian and El Pilar fault zones. Oil fields have been found around the margins of the basin but not within the basin itself.

Eastern Offshore Area of Trinidad and Tobago

The eastern offshore area of Trinidad and Tobago is dominated by the presence of the Central Range fault zone that cuts obliquely across the island of Trinidad and elevates the Central Range (Soto et al., 2011) (Figure 14A).

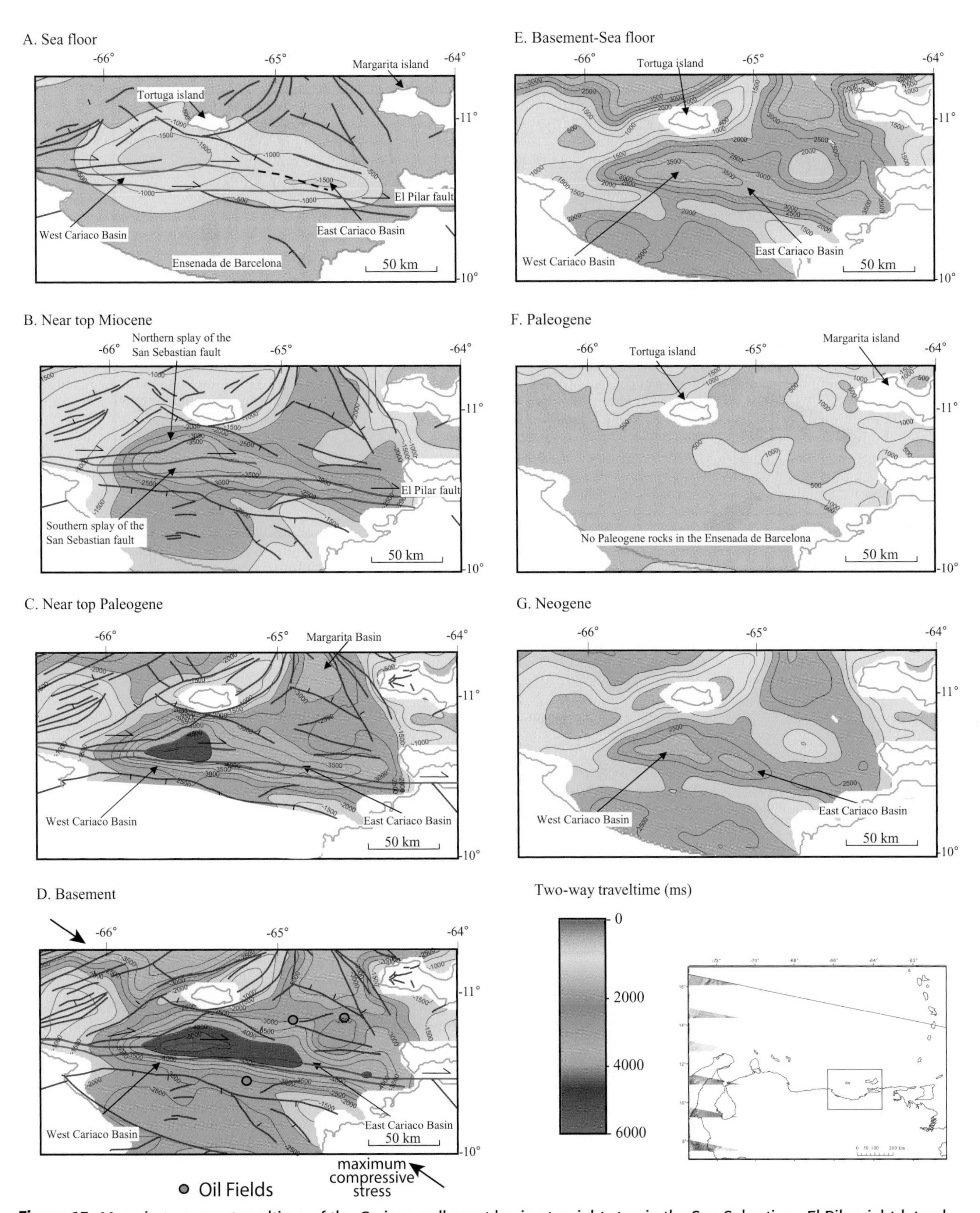

Figure 13. Maps in two-way traveltime of the Cariaco pull-apart basin at a right step in the San Sebastian–El Pilar right-lateral strike-slip fault zone from Escalona et al. (2011). (A) Sea-floor horizon. (B) Structure near the top Miocene horizon. (C) Structure near the top Paleogene. (D) Structure of basement horizon. (E) Basement to sea-floor isochron map. (F) Paleogene isochron. (G) Neogene isochron. Major oil fields around the flanks of the pull-apart basin are shown in Figure 16D. CRFZ = central range fault zone. 50 km (31 mi).

A

maximum compressive stress
62°W
61°W
60°W
11°N
10°N
N
Oil Giants
Gas Giants
U.S.Geological Survey oil fields
Capital
3-D seismic
2-D seismic
Wells
SRTM topography (m)
0 - 10
10 - 20
20 - 30
30 - 40
40 - 50
50 - 60
60 - 70
70 - 80
80 - 90
90 - 100
100 - 200
200 - 300
300 - 400
400 - 500
500 - 600
600 - 800
800 - 1000
10
100
1000
Bathymetry (m)
20 km
Emerald shoal
Prospector patch
N1
CRFZ
Emerald 2
Nothern Range
Venezuela
Port of Spain
Nothern basin
Darien ridge
Darien rock
Red Snapper
Gulf of Paria
Central Range
Southern basin
Radix point
Trinidad
Block 2ab 3-D survey
Galeota ridge
Southern Regna
Columbus basin
Wells in block 2AB 3-D survey
Atlantic Ocean
Columbus channel
Venezuela

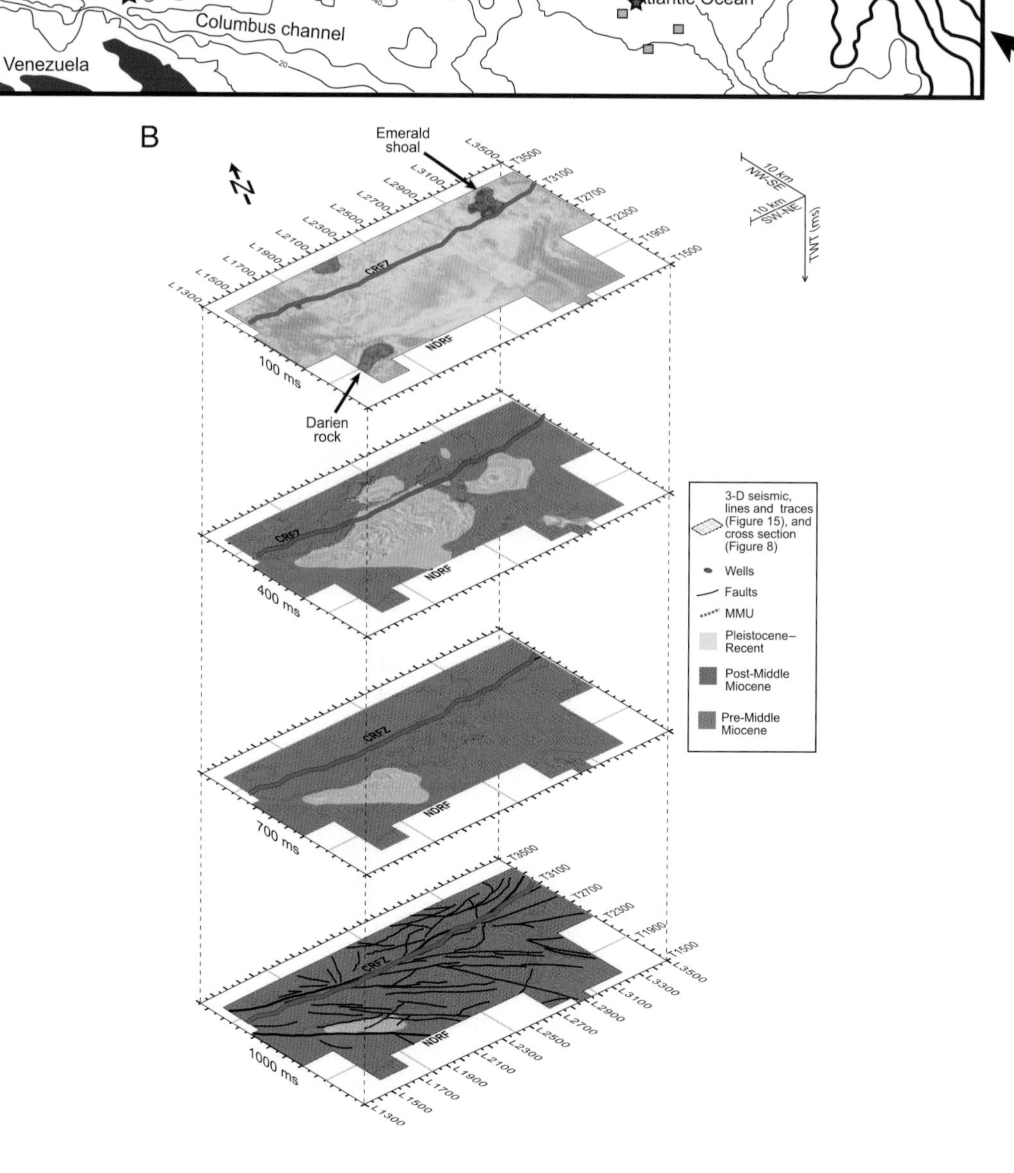

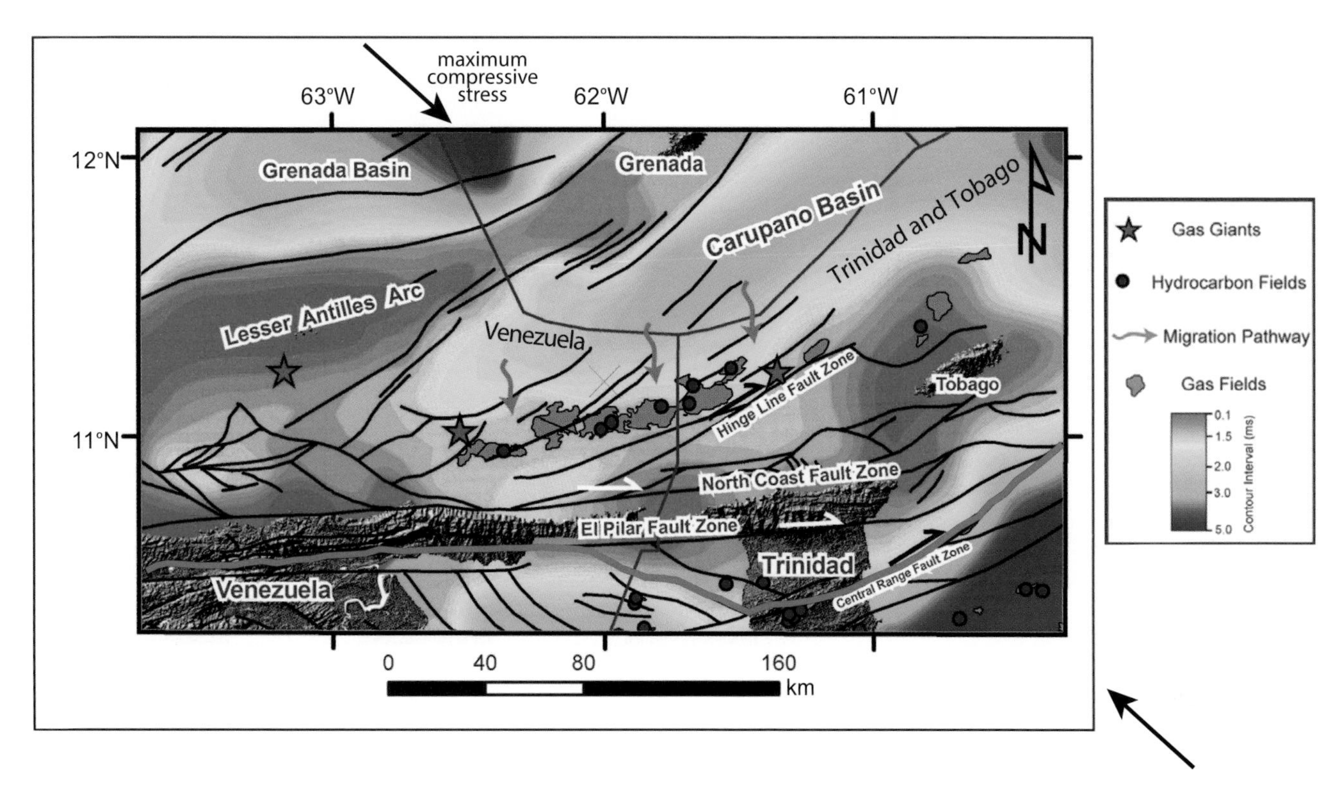

Figure 15. Map showing the depth of the igneous/metamorphic/acoustic basement in two-way traveltime (in milliseconds) beneath offshore Trinidad and Tobago (from Punnette, 2010). Sources of depth to basement information include Robertson and Burke (1989). The giant oil fields are shown in green stars, and the giant gas fields are shown in red stars (data from Mann et al., 2003). The heavy red line is the main trace of the El Pilar–Central Range right-lateral strike-slip fault zone. 40 km (24.8 mi).

Oil fields have been discovered with the complex strands of the 20 km (~12.5 mi) wide Central Range fault zone on the shallow eastern shelf of Trinidad. The fault zone affects a pre-middle Miocene section of highly folded and faulted rocks related to a middle Miocene event from a less folded post-middle Miocene section.

North Coast of Trinidad and Tobago

The north coast marine area of northern Trinidad is characterized by major gas fields aligned along the Hinge Line fault zone (Figure 15). The trend of structures in this fault indicates the occurrence of active right-lateral shear (Punnette, 2010) (Figure 15). The gas may be of a shallow thermogenic origin or may be derived from a deeper kitchen to the north in the Carupano Basin (Figure 15). The strike-slip faults form elevated ridges bounding the deeper parts of the adjacent basin as seen for the Los Angeles Basin (Figure 4).

Tectonic Origin of the Regional Stress Field of Northern South America and Its Effects on Petroleum in Eastern Venezuela and Trinidad

Colmenares and Zoback (2003) determined that the regional stress field of northern South America reflects the east–west convergence of the Nazca plate and Panama arc in northwestern South America and the more northwest to southeast subduction of the Caribbean plate beneath northeastern South America (Figure 16). As proposed for California, subduction of thickened slabs on the Nazca plate, the Panama arc, and the

Figure 14. (A) Topographic-bathymetric map of Trinidad and Tobago region showing the location of three-dimensional (3-D) seismic data used from study area in eastern offshore exploration block shown in gray from Soto et al. (2011). (B) Time slices through 3-D seismic survey of exploration block showing strike-slip fault controls on basin formation along Central Range fault zone (CRFZ) (heavy red line) and North Darien Ridge fault zone (NDRFZ). The giant oil fields are shown in green stars, and the giant gas fields are shown in red stars (data from Mann et al., 2003). MMU = middle Miocene unconformity.

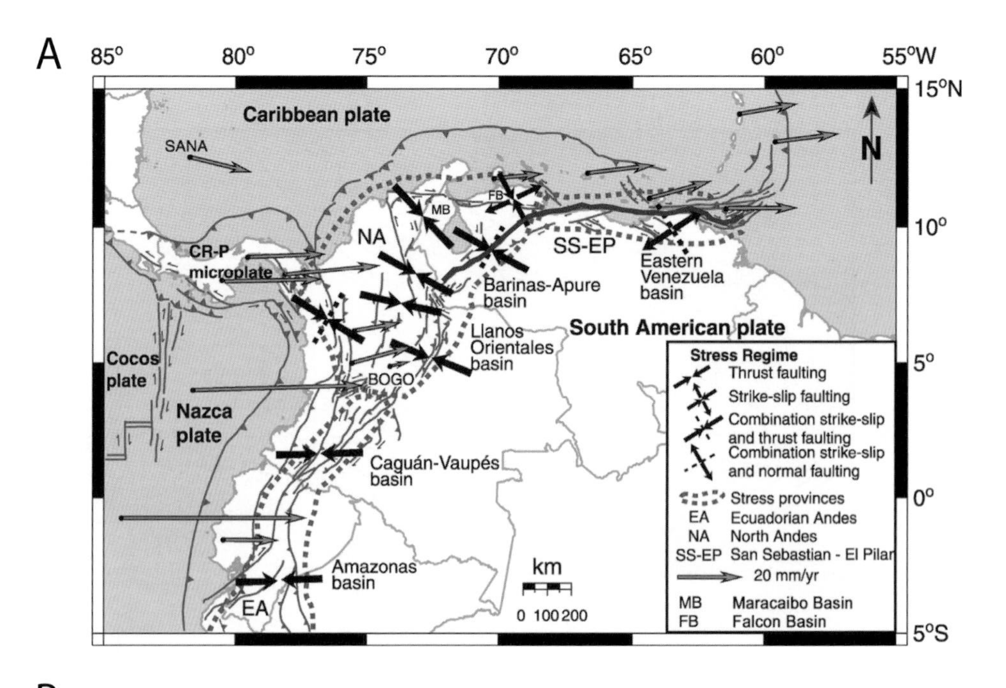

Figure 16. (A) Generalized tectonic map of northern South America showing major stress provinces attributed by Colmenares and Zoback (2003) to the southward component of subduction beneath northern South America. (B) Map showing the depth of regional seismicity with contour lines on top of inclined seismic zone of underthrusting Caribbean plate from Colmenares and Zoback (2003). The heavy black arrows show generalized stress directions linked to shallow subduction. SCDB = southern Caribbean deformed belt.

Caribbean plate are still occurring along offshore subduction features and may actively control the regional stress field by the process of collisional resistance. This is a very low angle and shallowly dipping subduction system, as determined by the tomographic study of van der Hilst and Mann (1994) in northern California. For that reason, the subduction process would involve a very wide zone of coupling and collisional resistance that would likely affect the regional stress state of the uppermost crust in the overriding plate and likely have major effects on the hydrocarbon-bearing structures over a wide region.

STRIKE-SLIP FAULT FORMED IN ZONE OF OBLIQUE SUBDUCTION: SUMATRAN AND SOUTHEAST ASIAN SUBDUCTION AND STRIKE-SLIP FAULT SYSTEMS

Geographic Location of Study Area

The curvilinear right-lateral Sumatran fault system (Sieh and Natawidjaja, 2000) extends 1900 km (~1181 mi) from its southern terminus as a Z-shaped pull-apart basin in the Sunda Strait between the islands of Sumatra and Java (Lelgemann et al., 2000) to an extreme rhomboidal pull-apart basin floored by oceanic crust in the Andaman

Sea to the north (Curray, 2005) (Figure 17A). The rate of slip on the fault passes from zero at its terminus in the Sunda Sea to a rate of 60 mm/yr in the Andaman Sea. The fault poses a large seismic hazard to the 50 million inhabitants living on the island of Sumatra, although the most recent destructive earthquakes and tsunamis that affected the region in 2004, 2005, 2007, and 2010 originated on the Sumatran subduction zone located about 300 km (~186.5 mi) to the west of the Sumatran fault system (Figure 17A). Other right-lateral strike-slip faults in this region in areas north of Sumatra in Southeast Asia include the northern extension of the Sumatra fault into Myanmar (Sagaing fault) and faults to the northeast in Thailand and Malaysia (Figure 17A). All of these faults accommodate oblique subduction of the Australian oceanic plate beneath continental and arc crust of the Sunda plate (Figure 17A).

Classification of the Sumatran Strike-slip Fault System

According to Mann's (2007) slight modification of Woodcock and Daly's (1986) classification, five major classes of faults are used in this chapter to classify faults in each of the four plate boundaries discussed. In this classification scheme, the Sumatran fault system is a trench-linked strike-slip fault accommodating the strike-slip component of an oblique subduction zone (Figure 17A). The basic principle is that a vertical strike-slip fault can concentrate shear in a much more efficient way than distributing shear across a much larger sub-horizontal subduction zone interface. Fitch (1972) noted that trench-linked strike-slip faults probably nucleate on weak zones in the crust formed by a heated and thinned crust of the volcanic arc itself. Jarrard (1986) summarized all aspects of trench-linked strike-slip faults and introduced the term "fore-arc sliver plate" to refer to the microplate defined by the trench axis and the trench-parallel, strike-slip fault (Figure 17B). Because most subduction zones have an oblique component of subduction, fore-arc sliver plates are present along many subduction zones (Mann, 2007).

A 300 km (~186 mi) wide fore-arc sliver plate occupies the space between the Sumatran trench and deformation front and the Sumatran fault (Figure 17A). The overall shape of the Sumatran fault system is sinusoidal with the northern half of the fault concave to the southwest and the southern half of the fault concave to the northeast (Figure 17A). Near its southern end, the Sumatran fault system curves toward the offshore trench and makes the fore-arc sliver plate progressively narrower in map view (Figure 17A). More orthogonal subduction in this southern area does not require the formation of a trench-linked strike-slip fault along this part of the trench and the subducted slab has a lesser dip and a greater penetration into the mantle than in the obliquely subducting area (Figure 17A) (Mann, 2007).

Sieh and Natawidjaja (2000) noted that the sinusoidal trace of the Sumatran fault is mimicked by the sinusoidal trace of the Sumatran subduction front, indicating a close genetic link between the active subduction process and the geometry of the trench-linked fault. The question is whether deformation of the subducted slab influences the fault geometry in the overriding plate, or whether the fault geometry of the overriding plate has influenced the subducting plate. As described for the California (Figure 9) and Caribbean (Figure 16) strike-slip margins above, I propose that the deformation of subducted slab is more significant based on the observed regional stress pattern.

Crustal Motions Determined from Global Positioning System–based Geodesy

Global positioning system studies in Sumatra from Barber and Crow (2005b) show motion of the fore-arc sliver plate relative to a fixed point in the stable interior of the Sunda plate (Figure 17A). The colored contours on Figure 17A indicate the dip of the subducting Indian plate. Note that the tectonic process of strain partitioning of the type originally proposed by Fitch (1972) for this same region shows only partial strain partitioning with the direction of the vectors on the fore-arc sliver plate having a significant landward component of motion. This pattern of vectors would indicate a degree of coupling between the subducting Indian plate and the fore-arc sliver plate in this region of highly oblique subduction. This larger degree of coupling may reflect the presence of incoming bathymetric ridges on the subducting plate that would act to sweep the length of the subduction zone. Higher coupling would also mean large earthquakes such as the series of large destructive events that affected the region in 2004, 2005, 2007, and 2010.

Depth to Basement Map

The depth to basement compilation map shows depth to igneous/metamorphic/acoustic basement in feet beneath numbered onshore and offshore basins of Southeast Asia (Figure 17C). Several types of basins are apparent on the compilation. The fore-arc sliver plate along the Sumatran fault system is characterized by the presence of a relatively narrow but continuous fore-arc basin defined by the structural high of the accretionary prism to the west and the volcanic arc and Sumatran fault system to the east. An isolated elongated deep basin in the Andaman Sea is formed at the pull-apart

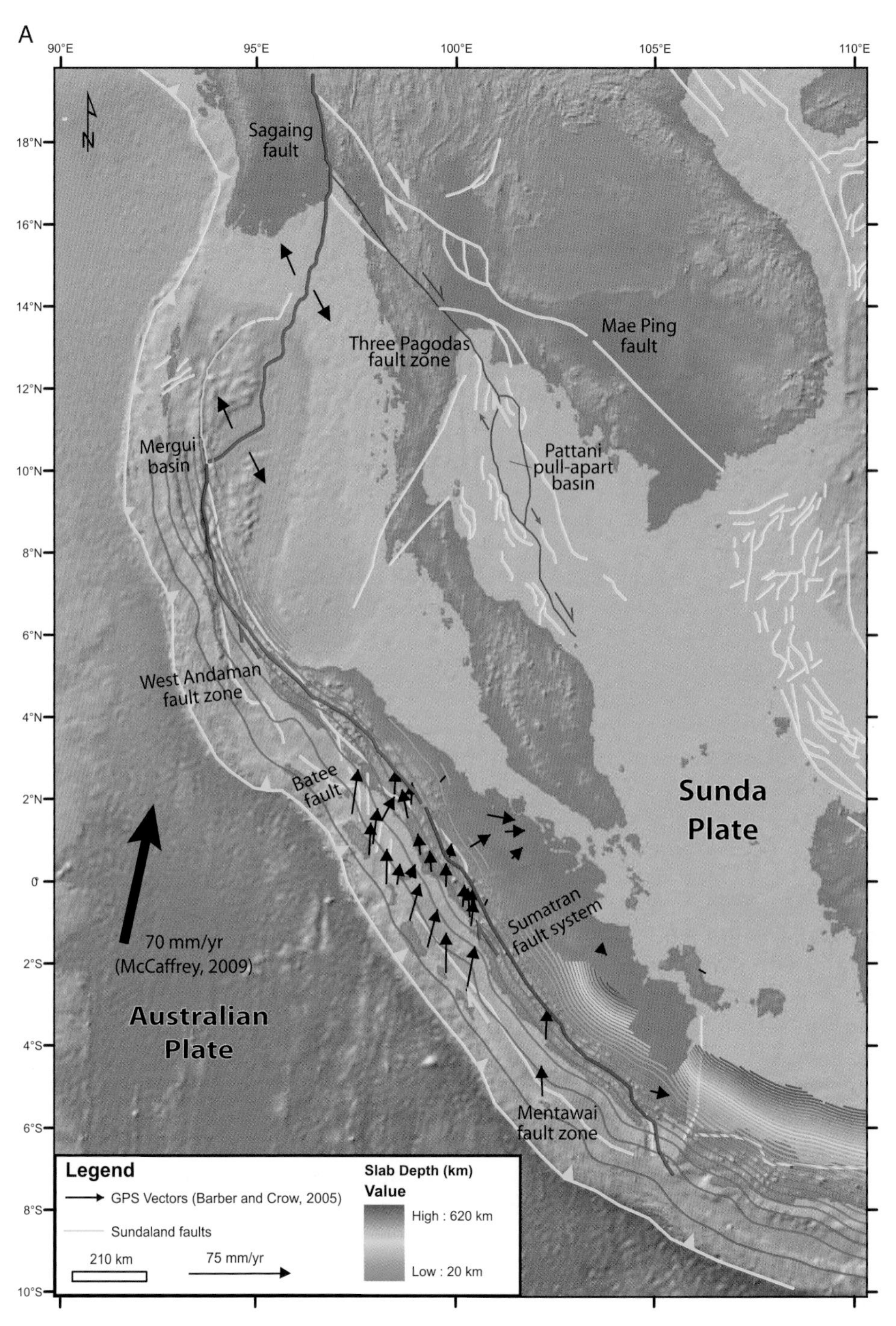

Figure 17. (A) A map showing global positioning system (GPS) vectors from Barber and Crow (2005b) and the motion of the fore-arc area and volcanic areas of Sumatra relative to a fixed point in the stable interior of the Sunda plate (McCaffrey, 2009). The heavy red line is the main trace of the Sumatra right-lateral strike-slip fault zone. The direction and rate of the Australian plate relative to the fixed Sunda plate is from McCaffrey (2009). The depth to the top of the subduction zone is from the U.S. Geological Survey slab models for subduction zones (Hayes et al., 2012). (B) Block diagram from McCaffrey (2009) showing the conceptual model of mechanism for northward transport of the Sumatran fore arc as a "sliver plate" between the right-lateral Sumatra strike-slip fault zone and the trench. Velocity diagram shows angles and magnitudes of velocities of the sliver plate. (C) A map showing the depth of igneous/metamorphic/acoustic basement in feet beneath numbered onshore and offshore basins of Southeast Asia. Source of depth to basement information includes Exxon Tectonic Production Research Company (1985). The giant oil fields are shown in green stars, and the giant gas fields are shown in red stars (data from Mann et al., 2003). (D) Stress map of Southeast Asia from the Heidbach et al. (2008) of the International Association of Seismology and Physics of the Earth's Interior (IASPEI). Types of stress indicators are shown in the key. The white boxes indicate the location of detailed maps shown in Figures 18–20, 22, and 23. 200 km (124 mi).

segment in that area as is the deep elongate Malay Basin formed along the Three Pagodas strike-slip fault zone off the coast of Malaysia. The much greater depth to basement in the Andaman and Three Pagodas pull-apart basins indicates their origin by localized stretching in a pull-apart setting that contrasts to the limited crustal extension in the case of the fore-arc basin.

Giant oil fields are shown in green stars and giant gas fields are shown in red stars (Mann et al., 2003). Giant gas fields flank the Andaman Sea, Pattani Basin, and Malay Basin (Figure 17C). Giants are also present on continental crust on the northeastern side of Sumatra but are absent along the entire length of the narrow Sumatran fore-arc basin. The lack of giants in the Sumatran fore-arc basin is consistent with the generally poor hydrocarbon setting of fore-arc basins that generally have poor reservoir rocks related to the proximity of volcanic arc sources that clog porosity and permeability.

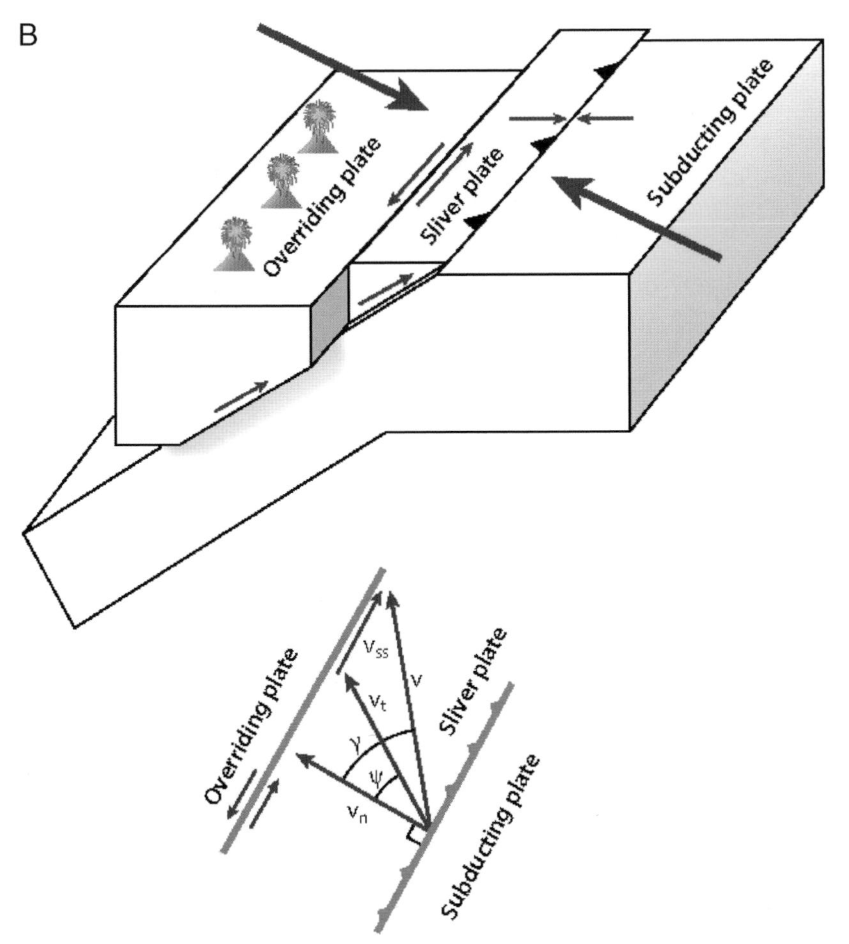

Figure 17. (cont.).

Regional Stress Map

Regional stresses from the Heidbach et al. (2008) are compiled in Figure 17D and are colored coded to the stress indicator used in the lower part of Figure 17D. As previously discussed by Mount and Suppe (1992) and Tingay et al. (2010a), stresses are roughly perpendicular to the subduction boundary in Sumatra (northeast–southwest) and Java (north-northeast to south-southwest) as was observed for the California (Figure 3C, 9) and southern Caribbean (Figure 10A) margins. The perpendicular compressive stresses across the subduction boundary and the Sumatran fault system indicate that the two plates are strongly coupled (Mount and Suppe, 1992) and that perhaps a component of gravitational collapse of the high topography on these islands exists that contributes to the subduction-related forces (Tingay et al., 2010a). Stresses in Thailand and near the Pattani and Malay basins tend to trend north–south and have been interpreted by Tingay et al. (2010a) as an expression of the southward extrusion of Southeast Asia as a result of the India-Asia collision with perhaps a contribution of rollback of the Sumatra-Andaman subducted slab (Figure 17D).

Strike Slip-related Structures in the Sumatra Region and Relation to Tectonics and Hydrocarbons

In this complex region with many local variations in its stress field, hydrocarbon-bearing structures vary from folds and faults generated within strike-slip fault zones, to fore arc-related structures, to deep pull-apart basins along major strike-slip faults, to reactivated rift structures.

Andaman Sea, Offshore Indonesia

The detailed map of the Sumatran fault zone terminating on the Andaman pull-apart basin is from Harding (1985) (Figure 18). The main strike-slip fault is characterized by changes in the dip of the fault over distances of 25 km (~15.5 mi) and the formation of anticlines in areas where the main fault changes its strike. Normal faults to the northeast of the main fault manifest extension in the southern part of the Andaman pull-apart basin that locally has extended to form several active short spreading ridges (Figure 17D). In this area, maximum compression directions are highly oblique to the main trend of the fault and likely reflect

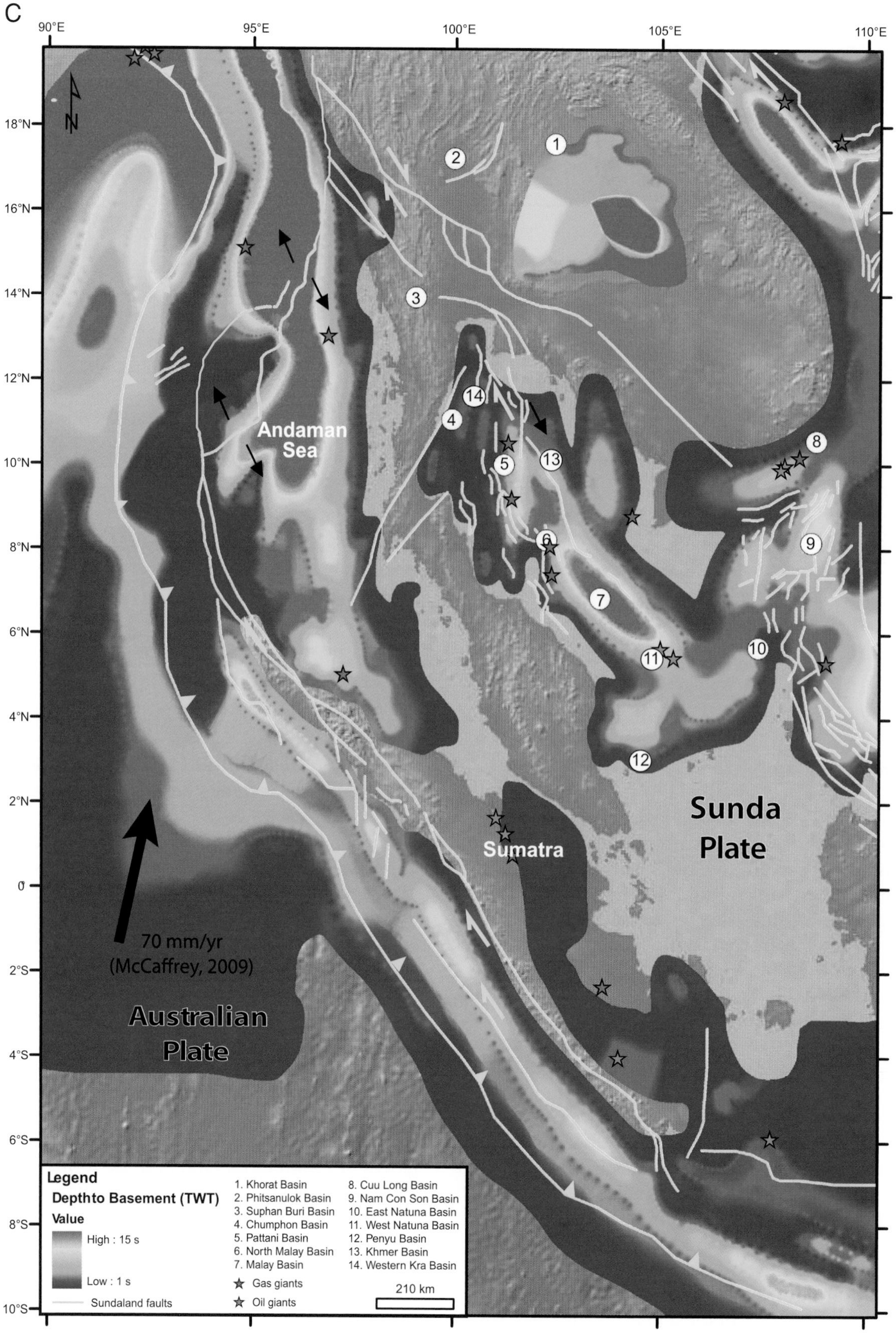

C
90°E
95°E
100°E
105°E
110°E
18°N
16°N
14°N
12°N
10°N
8°N
6°N
4°N
2°N
0
2°S
4°S
6°S
8°S
10°S
N
Andaman Sea
Sumatra
Sunda Plate
Australian Plate
70 mm/yr
(McCaffrey, 2009)
Legend
Depthto Basement (TWT)
Value
High : 15 s
Low : 1 s
Sundaland faults
1. Khorat Basin
2. Phitsanulok Basin
3. Suphan Buri Basin
4. Chumphon Basin
5. Pattani Basin
6. North Malay Basin
7. Malay Basin
Gas giants
Oil giants
8. Cuu Long Basin
9. Nam Con Son Basin
10. East Natuna Basin
11. West Natuna Basin
12. Penyu Basin
13. Khmer Basin
14. Western Kra Basin
210 km

Figure 17. (cont.).

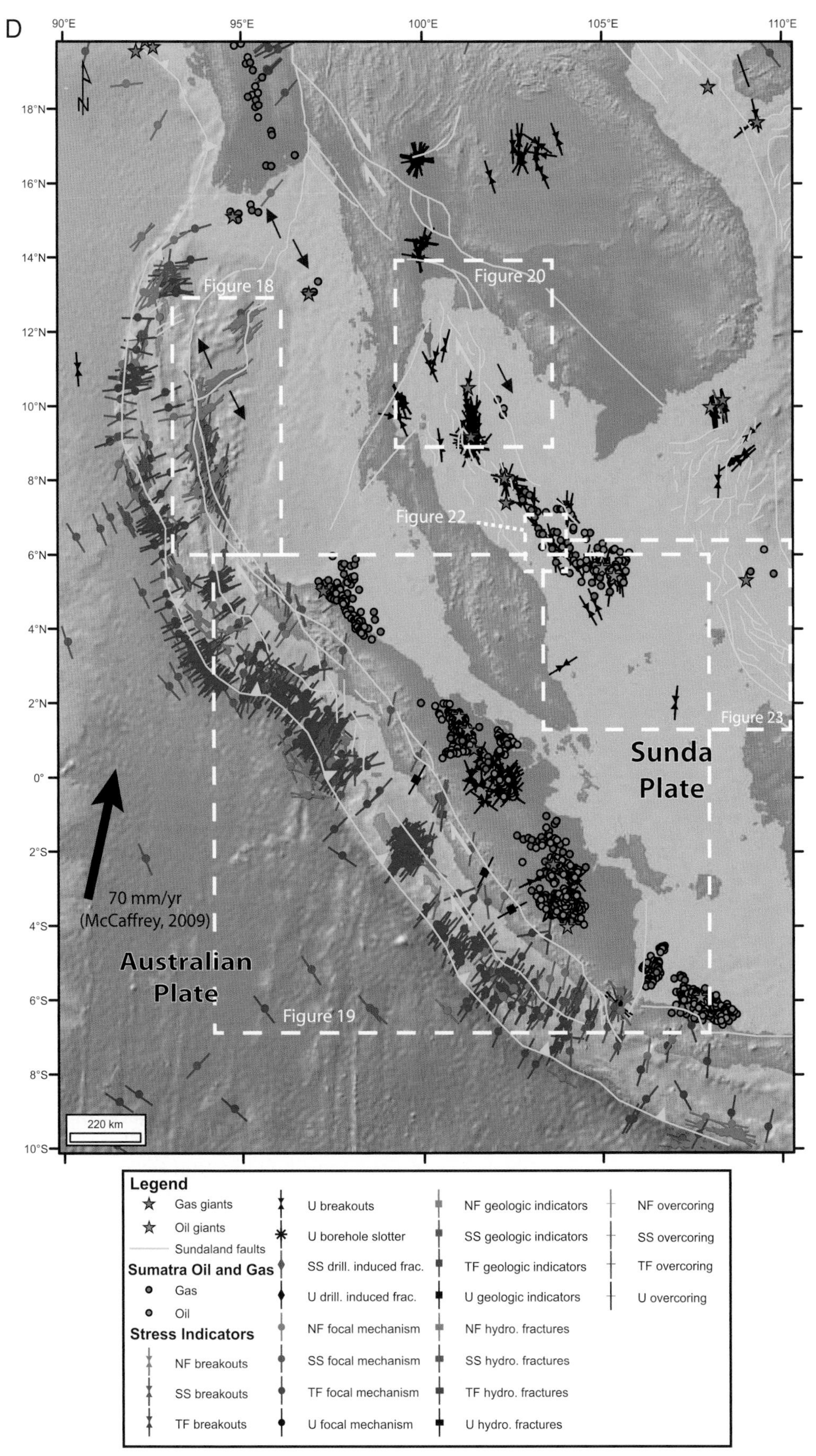

Figure 17. (cont.).

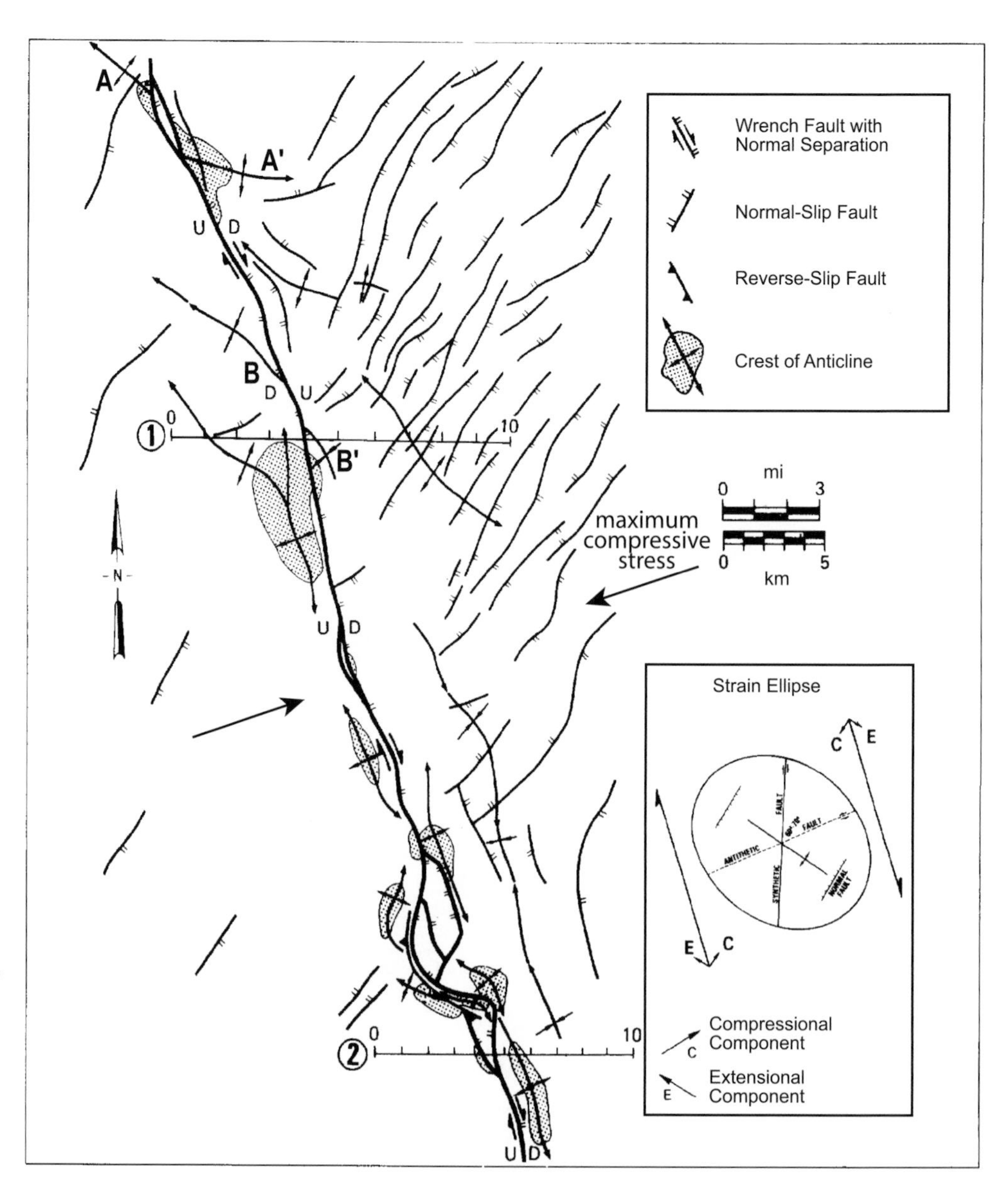

Figure 18. A map showing offshore extension of the right-lateral Sumatra fault adjacent to the southern edge of the Andaman Sea pull-apart basin (from Harding, 1985). The location of map area is shown by a white box in Figure 17D. U = upward; D = downward.

stress perturbations related to the presence of the pull-apart basin. Similar stress perturbations are observed in the vicinity of the San Andreas fault in the Big Bend region of southern California (Figure 3C). The activity of the Sumatran fault is likely to compromise reservoirs in this area and may explain why no production from these fault-related folds exists.

Sumatran Fore Arc, Arc, and Back Arc

Strong coupling of arc-perpendicular stresses is observed and is consistent with a broad zone of folding north of the Sumatran fault zone (Figure 19). Many of these folds contain oil and gas, with an ultimate recoverable reserve estimate of 3.1 billion bbl of oil (Bishop, 2001). Most of these fields have sandstone reservoirs with some limestone and calcareous sandstone. The anticlines resulted from compression that began as early as the Miocene but was most pronounced between 2 and 3 Ma (Bishop, 2001). Stratigraphic pinch-outs and carbonate buildups locally combine with folds and anticlines to enhance the effectiveness of the primary trap type. Recoverable reserves of 178 million bbl of oil are found in bioherms and carbonate buildups and 668 million bbl in fault traps. Much of the oil in Sumatra is paraffinic and low in sulfur content. Both lacustrine and terrigenous facies on the margins of the lacustrine environments have been interpreted as the sources for the oils. The presence of preexisting rift basins may explain why fields are concentrated in several major clusters in Sumatra (Figure 19). These deeper rifts, active from the late Paleocene to the early Miocene (Harding, 1985), may provide kitchens for oils to mature and move updip into shallower reservoirs found in large anticlines.

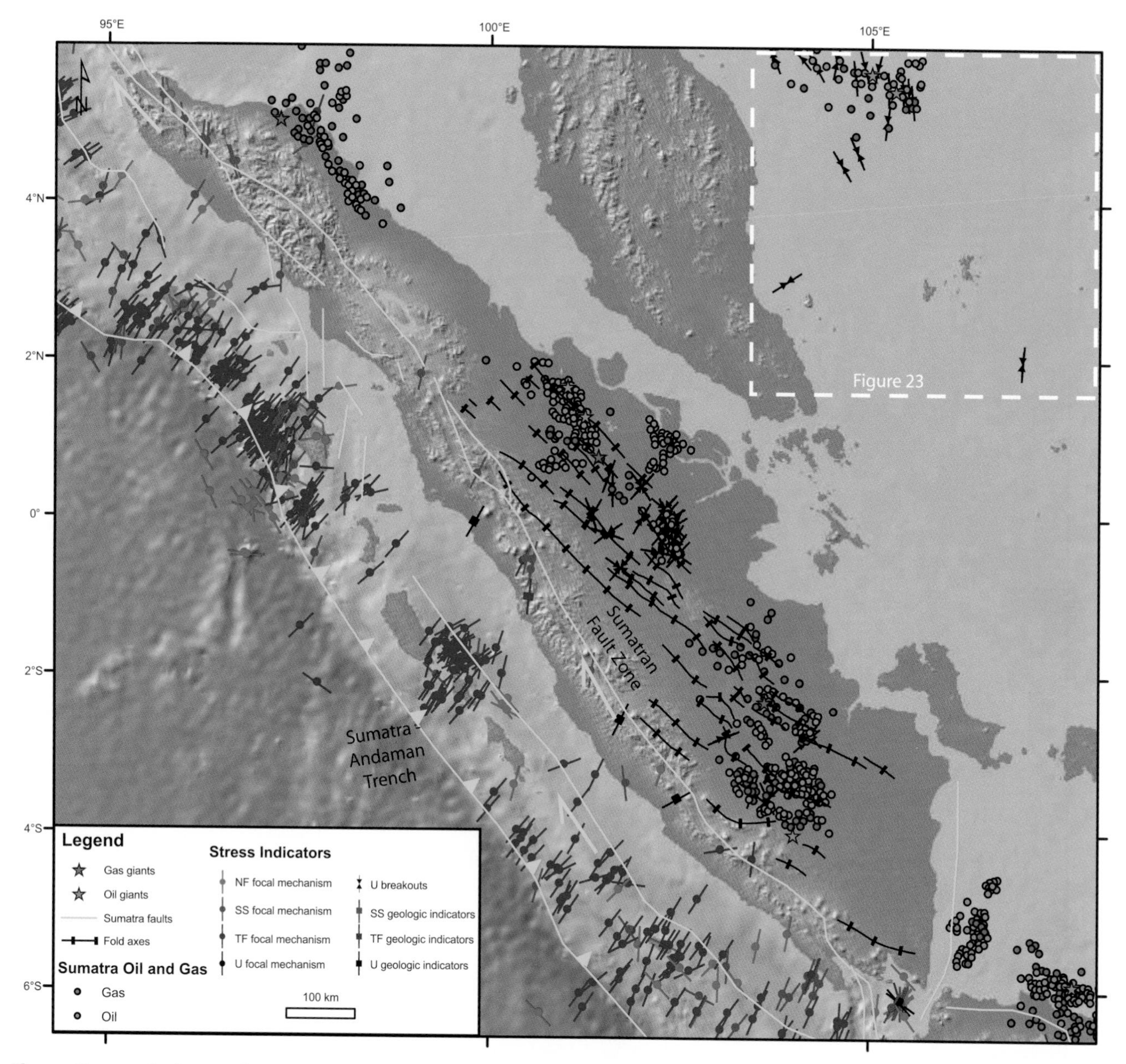

Figure 19. Detailed map showing stress indicators, faults, and hydrocarbon occurrences in the Sumatra region. The giant oil fields are shown in green stars, and the giant gas fields are shown in red stars (data from Mann et al., 2003). Oil fields from the U.S. Geological Survey (USGS) are shown as green dots; gas fields are shown as red (USGS, 2003, Description and results: various reports/geologic provinces within assessment [accessed 2010]). Various types of stress indicators are color coded on the map and summarized in the map key. The white box indicates the location of a more detailed map shown in Figure 23. NF = normal fault; SS = strikeslip; TF = thrust fault; U = undefined. 100 km (62 mi).

Pattani and Malay Basins, Gulf of Thailand

The Pattani and Malay troughs form large pull-apart basins along the right-lateral Three Pagodas fault zone (Figure 17C). The basins are both localized areas of extension filled by up to 12 km (~7.5 mi) of sediment mostly from early Miocene to Holocene in age (Morley and Westaway, 2006). Giant gas fields are present within and on the flanks of the Pattani and northern Malay basins, whereas several oil giants are found in the West Natuna area southeast of the Malay Basin (Figure 17C). The alignment of maximum compressive stress may represent reorientation of stresses within the pull-apart environment instead of southward transport of material related to the India collision (Figure 20). Morley and Westaway (2006) describe both basins as synrift basins of Eocene–Oligocene age (~12 km [~7.5 mi]) and early Miocene to Holocene postfill fill (~4 km

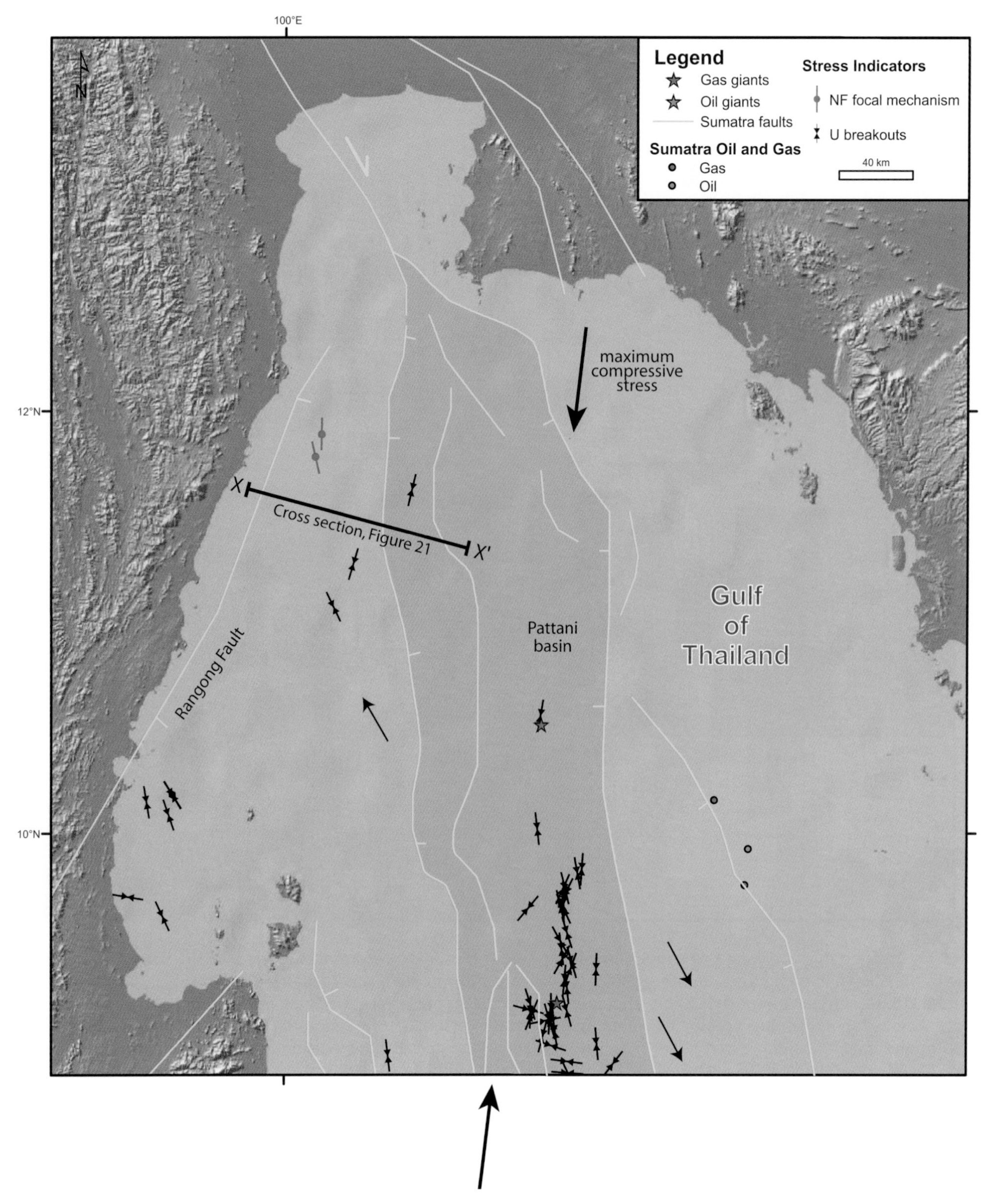

Figure 20. A detailed map showing faults and hydrocarbon occurrences in the area of the northern Pattani pull-apart basin in the northern Gulf of Thailand (regional map shown in Figure 18E). The giant oil fields are shown in green stars, and the giant gas fields are shown in red stars (data from Mann et al., 2003). Oil fields from the U.S. Geological Survey are shown as green dots, with gas fields shown as red dots (USGS, 2003, Description and results: various reports/geologic provinces within assessment [accessed 2010]). Various types of stress indicators are color coded on the map and summarized in the map key of Figure 17D. 40 km (24.8 mi).

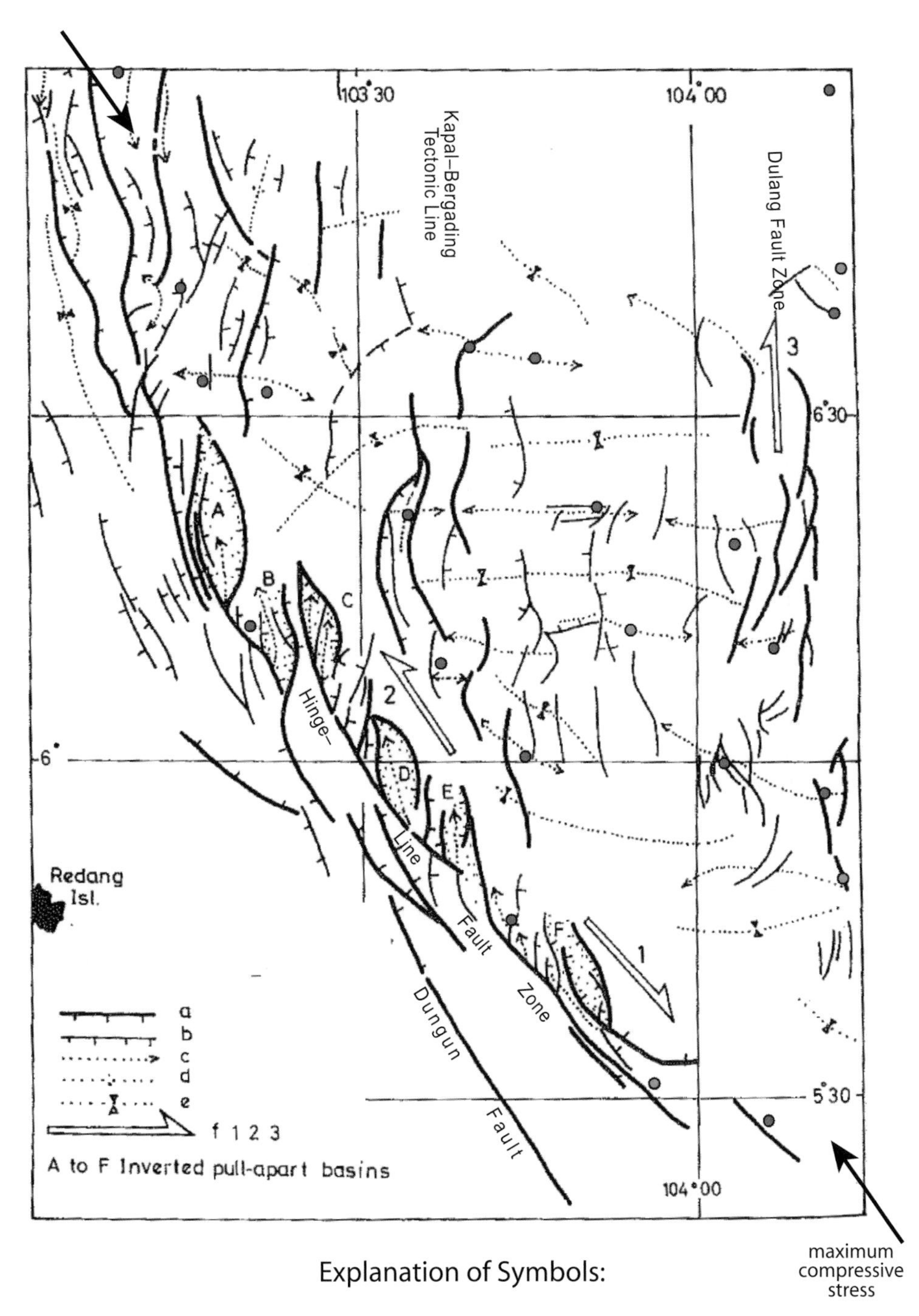

Figure 21. The structural map of the Hinge Line strike-slip fault zone from Bishop (2001). Oil fields are shown by green dots and gas fields are shown by red dots (USGS, 1999).

[~2.5 mi]). The cross section of the Pattani Basin located in Figure 20 appears to be more of a deep sag basin than a rift. Normal faults have small throws.

Older pre-middle Miocene rift faults have been reactivated by right-lateral shearing along the southern extension of the Three Pagodas fault zone (Figure 21).

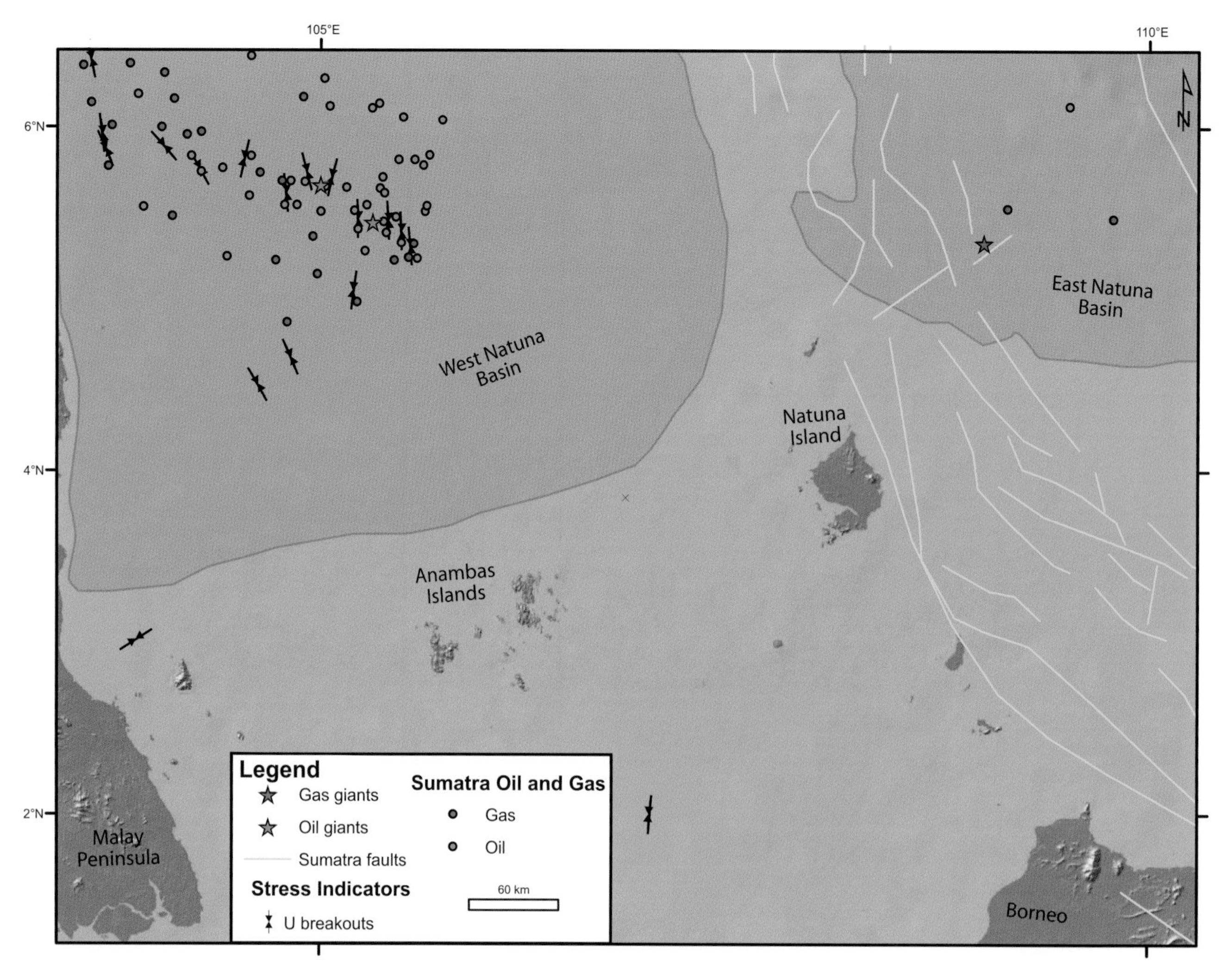

Figure 22. A detailed map showing faults and hydrocarbon occurrences in the oil-rich area of Natuna Basin (regional map shown in Figure 17D). The giant oil fields are shown in green stars, and the giant gas fields are shown in red stars (data from Mann et al., 2003). Oil fields from the U.S. Geological Survey are shown as green dots, with gas fields shown as red dots (USGS, 2003, Description and results: various reports/geologic provinces within assessment [accessed, 2010]). Various types of stress indicators are color coded on the map and summarized in the map key of Figure 17D. 60 km (37 mi).

This process is illustrated by the Hinge Line fault zone (Bishop, 2001) along the edge of the Malay Basin. Shearing causes the underlying rift faults to link up to form more continuous and straighter faults at higher levels in the basin (Tingay et al., 2010b) (Figure 21).

West and East Natuna Basins, Indonesia

This is an area of active exploration with many wells and one gas giant (Figure 22). Ngah et al. (1996) have proposed that this region represents a triple-armed Late Cretaceous rift system that was later slightly inverted in the middle to late Miocene. The concentration of fields presumably overlies the deeper rifts. Most workers agree that the most probable cause of the mild inversion is subduction-related stresses transmitted more than 1000 km (>621.5 mi) from the trench (Figure 23).

Tectonic Origin of the Regional Stress Field of the Sumatra Region and Its Effects on Petroleum

Subduction-related stresses have a major function in forming folds and hydrocarbon traps on the island of Sumatra (Figure 24). Note that subduction-related compression in Sumatra is approximately normal to the trench, implying that thrust deformation is directed

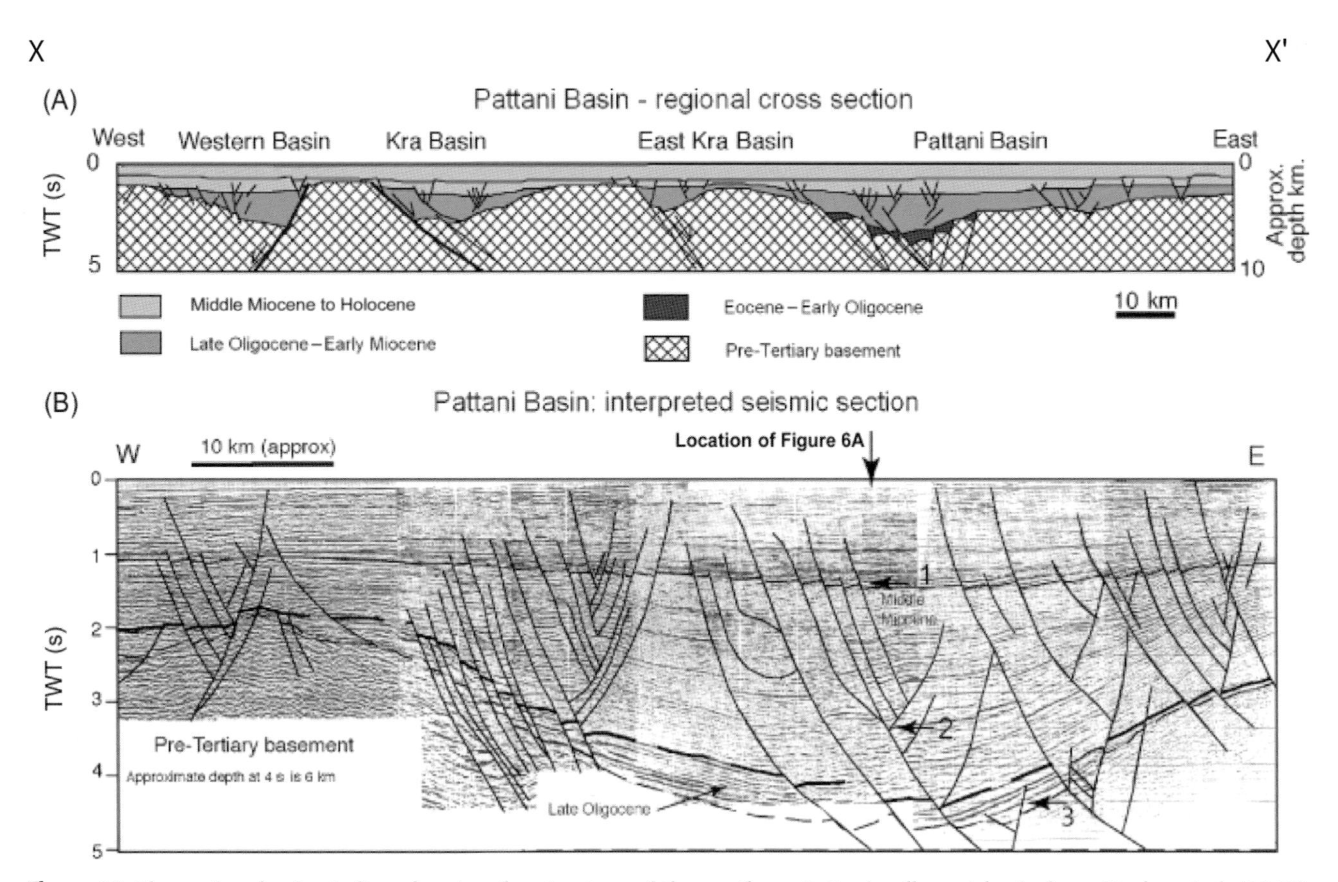

Figure 23. The regional seismic line showing the structure of the northern Pattani pull-apart basin from Morley et al. (2006) (approximate location of line shown on Figure 20). TWT = two-way traveltime. 10 km (6.2 mi).

nearly normal to the trench and not parallel to the oblique plate vector (Mount and Suppe, 1992). This observation supports the model by Fitch (1972) that predicted that oblique subduction of the Indian plate is partitioned into strike-slip faulting along the Sumatra fault zone and subduction. Pliocene–Pleistocene folds in the Sumatra area at right angles to these stress directions are likely the result of this partitioning (Figure 19). In summary, the low-drag decoupled model by Mount and Suppe (1992) (Figure 2B) explains the regional folding pattern better than the high-drag coupled model of Harding (1973) (Figure 2B).

STRIKE-SLIP FAULT FORMED IN INTRAPLATE DEFORMATIONAL ZONE IN EASTERN CHINA

Geographic Location of Study Area

The Bohai Basin is a shallow-marine embayment and hydrocarbon-bearing basin located in northeastern China (Figure 25) about 200 km (~124 mi) east of Beijing (Figure 26). The onshore extension of the Bohai Basin to the southwest is known as the Hebei Plain. Water depths in the Bohai Basin range from less than 5 m (<16 ft) in the coastal area of the western part to more than 30 m (>99 ft) in the deeper eastern part of the basin. The delta of the Yellow River empties into the southwestern part of the basin.

After 40 yr of exploration, the Bohai Basin is currently China's third largest producing area for crude oil. The net proved reserves in the basin is 1000 million bbl of oil. Average daily production from the basin is about 187,000 bbl of oil. The Bohai Basin is bounded to the east by the active right-lateral Tan-Lu fault zone and to the west and by the active oblique-slip Taihang fault zone forming the eastern front of the Taihang Mountains (Zhang et al., 2005) (Figure 26A).

Classification of the Strike-slip System

The Tan-Lu and Taihang strike-slip fault zones bounding the Bohai Basin are cratonic strike-slip faults because they are not part of a well-defined continental

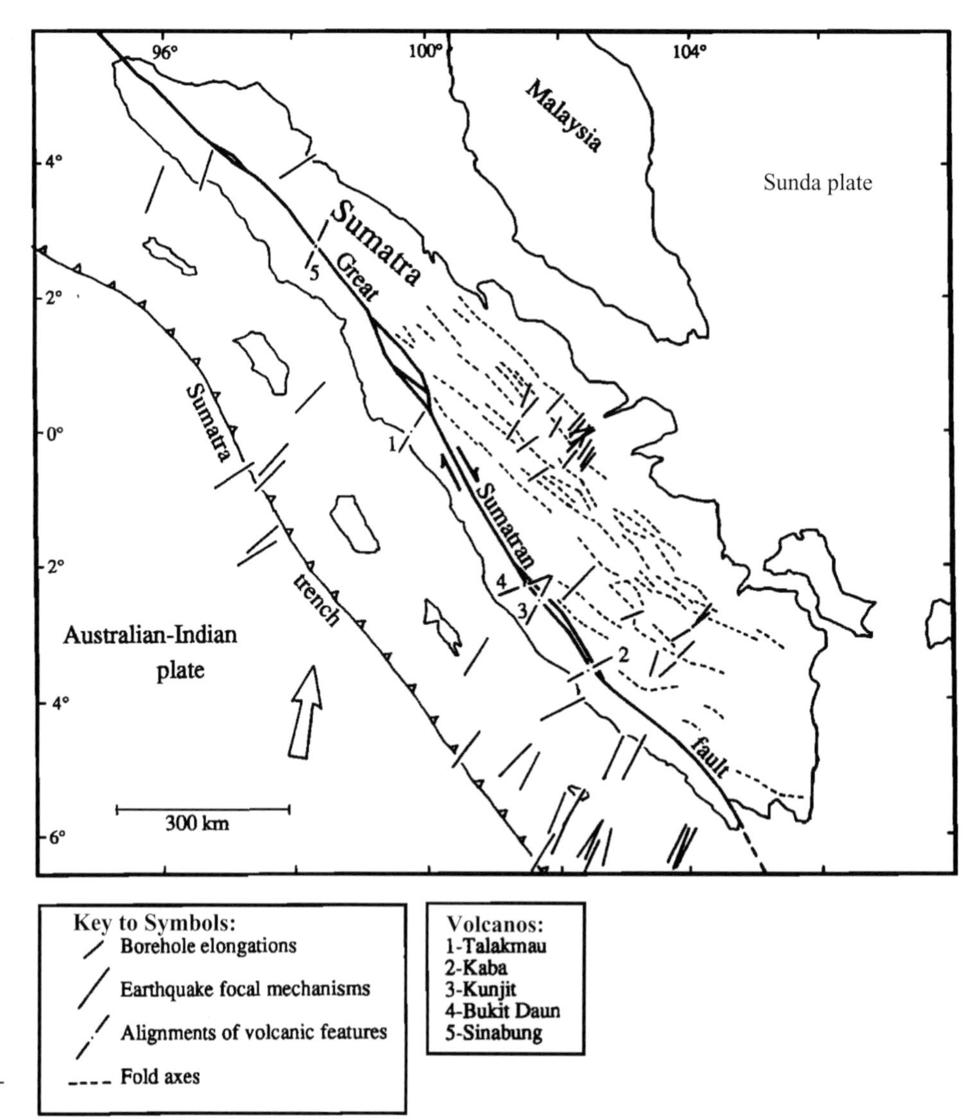

Figure 24. Maximum horizontal stress directions (short bars on map) in Sumatra compiled by Mount and Suppe (1987) using borehole breakouts, alignments of volcanic flank eruptions, and earthquake focal mechanisms. The arrow shows the relative motion of subducting Australian-Indian plate beneath Sumatra.

plate boundary; they are well within a cratonic interior of east Asia, and they do not appear to be either indent-linked or trench-related. The north–northeast-striking Tan-Lu fault is one of the most prominent strike-slip faults of eastern Asia and extends 3500 km (~2175 mi) from southern China to eastern Siberia (Figure 26A) (Castellanos, 2007). During the middle Mesozoic, the sense of shear on the Tan-Lu fault was left lateral and resulted in about 700 to 800 km (~435–497 mi) of lateral displacement (Xu, 1993). The Tan-Lu fault was quiescent from the middle Cretaceous to the beginning of the Cenozoic. During the early Cenozoic, the Tan-Lu fault exhibited renewed right-lateral strike-slip displacement with about 40 km (~25 mi) of offset (Zhu et al., 2005).

Frequent earthquakes and deformation on Quaternary strata along the fault demonstrate continued right-lateral displacement of the Tan-Lu fault (Figure 26A). Geodetic studies using GPS have been unable to detect measurable right-lateral slip along the Tan-Lu fault, although it is possible that fault motion is episodic or occurring at a slow rate within the error of the GPS measurements.

Crustal Motions Determined from Global Positioning System-based Geodesy

Global Positioning System data show that the North and East China blocks move eastward as extrusion blocks away from the India-Asia collision zone (Figure 25). To the east of the basin Pacific subduction zones dip westward beneath the Bohai Basin. A fundamental question is the relative importance of extrusion versus subduction

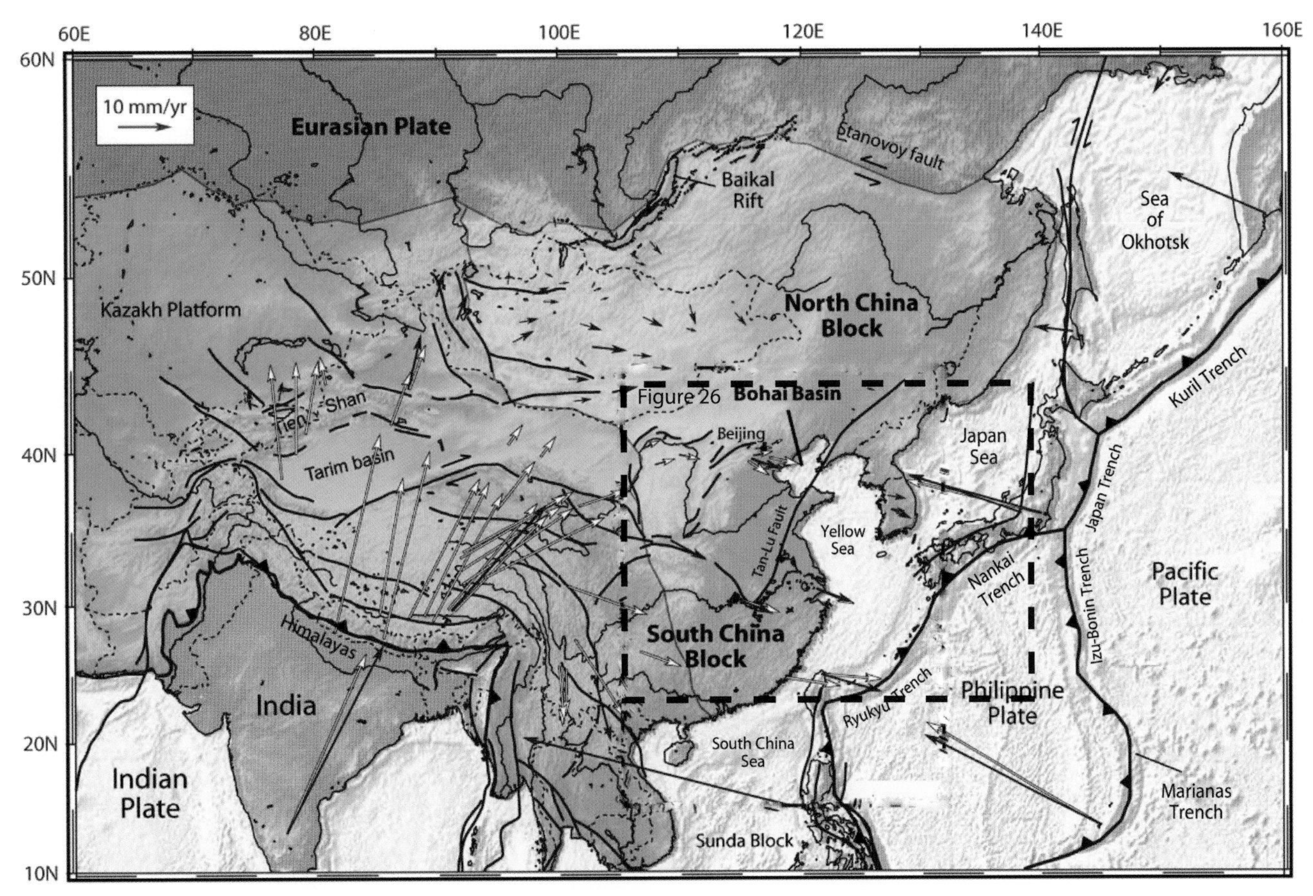

Figure 25. Location and active tectonic setting of strike-slip faults in eastern China from global positioning system (GPS) velocities with respect to a fixed Eurasia plate shown in red (modified from Calais et al., 2003, and Castellanos, 2007). In the green area, from the Himalayas to the Tien Shan mountain range, the GPS vectors trend nearly north–south. Northward-directed GPS vectors show northward displacement of India and the Himalayas into Asia. In the blue area, the vectors are roughly east-southeast–west-northwest. These vectors show eastward and southeastward displacement of extruding continental areas of Southeast Asia in response to the India collision to the southwest. Subduction of the Pacific plate may also have a major function in the eastward motion of south and north China. The black triangle shows the strike-slip region shown in more detail in Figure 26A.

tectonics on the late Cenozoic origin and development of the Bohai Basin.

Depth to Basement Map

Beginning in the Late Cretaceous–early Oligocene, East Asia underwent a period of widespread Cenozoic extension (Allen et al., 1997, 1998; Yang and Xu, 2004) (Figure 26). It is possible that the Bohai Basin was one of many basins generated during this time as a consequence of this regional extension. One tectonic mechanism for extension is rollback of a subducted Pacific slab underlying the basin. Alternatively, the lazy-Z map pattern of the Bohai Basin depocenter indicates the importance of right-stepping pull-apart control on the early rifting and later sag basin formation.

Eocene–Oligocene marks the period of highest subsidence in the Bohai Basin and the period of lacustrine source rock deposition in the restricted rift basins (Hu et al., 2001). Up to 10 km (~6.2 mi) of continental clastic sediments were deposited in the northeast-striking faulted depressions (Hu et al., 1989; Allen et al., 1997; Hsiao et al., 2004) (Figure 26). Almost all of these rifts or depressions within the Bohai Basin, so-called in the Chinese geologic literature, were formed during this period. Each depression in the Bohai Basin had its own sedimentary system that was isolated from the surrounding depressions. Major faulting culminated in the Bohai Basin by the end of the Oligocene. Regional denudation of the basin at the end of the Oligocene was

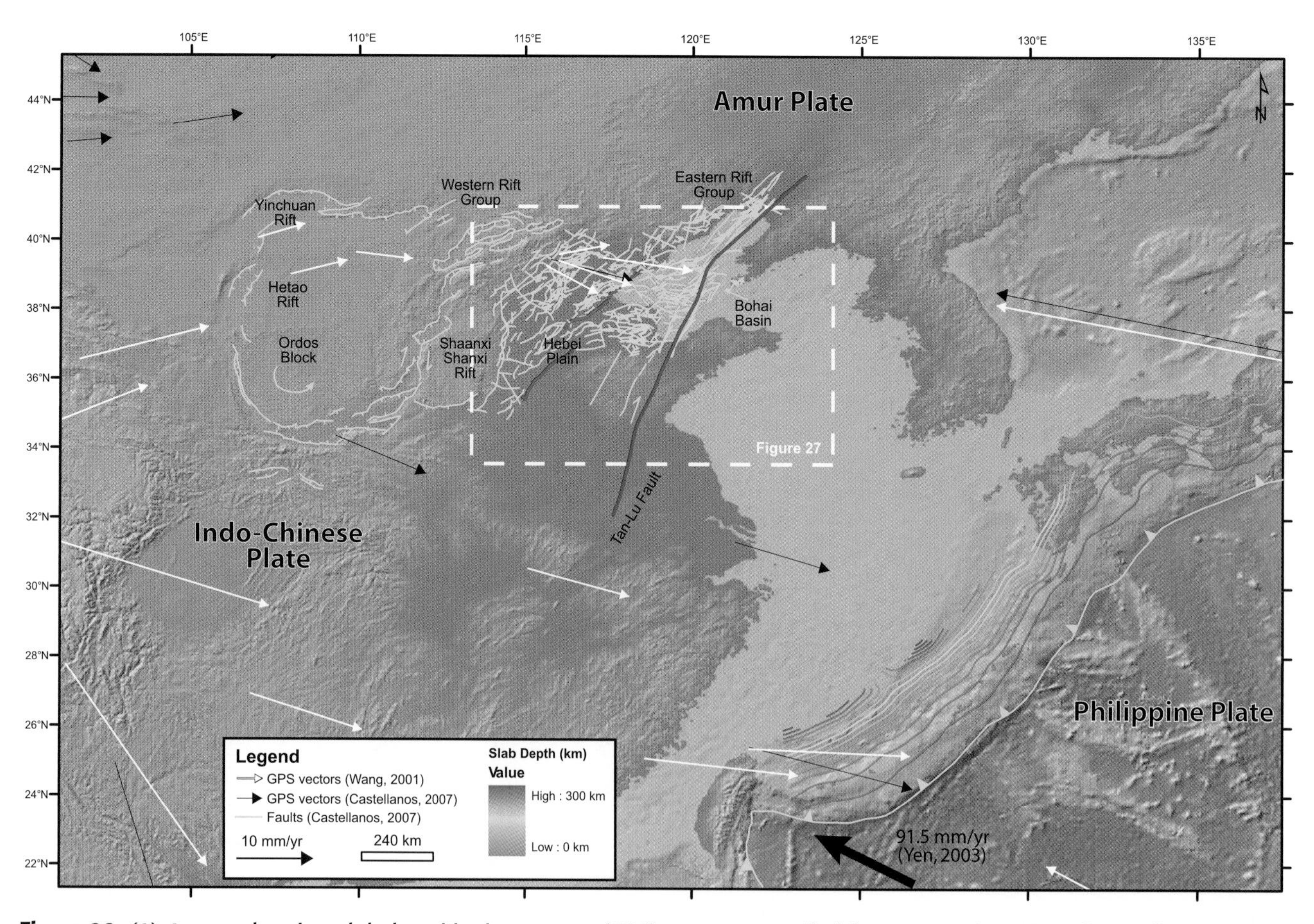

Figure 26. (A) A map showing global positioning system (GPS) vectors compiled from Wang (2011) and Castellanos (2007) showing the motion of eastern China relative to the Pacific plate. The heavy red lines are the main traces of the Hebei and Tanlu right-lateral strike-slip fault zones. The depth to the top of the subduction zone is from U.S. Geological Survey slab models for subduction zones (Hayes et al., 2012). (B) A map showing the depth of the igneous/metamorphic/acoustic basement in feet beneath numbered onshore and offshore basins of eastern China and offshore areas taken from Pfeiffer, 2002, and Castellanos, 2007. Sources of depth to basement information include Exxon Tectonic Production Research Company (1985). The giant oil fields are shown in green stars, and the giant gas fields are shown in red stars (data from Mann et al., 2003). (C) The stress map of Southeast Asia from the Heidbach et al. (2008) of the International Association of Seismology and Physics of the Earth's Interior (IASPEI), with stress indicators shown in the key below.

followed by more uniform and less fault-controlled Neogene sedimentation. Neogene sediments form a thick widespread blanket or sag basin of continental clastic deposits that are up to 3 km (~1.9 mi) thick in the central part of the basin.

Regional Stress Map

The pattern of regional stresses in China shows a rough east–west orientation related to either the eastward extrusion of blocks away from the site of India-Eurasia collision in central Asia or to farfield subduction effects from subduction of the Pacific plate to the east (Figure 26C). As in the case of California, the Caribbean, and Sumatra, regional stresses are approximately orthogonal to the major strike-slip fault of eastern China, the Tan-Lu fault zone (Figure 26C).

Strike-slip-related Structures in the Bohai Basin Region and Their Relation to Hydrocarbons

As common to many rifted areas, hydrocarbons are widespread across the Bohai Basin (Figure 27). The fault pattern in the subsurface of the basin is complex with east–west and southeast-trending faults (Figure 27). In the pull-apart basin model of Allen et al. (1997),

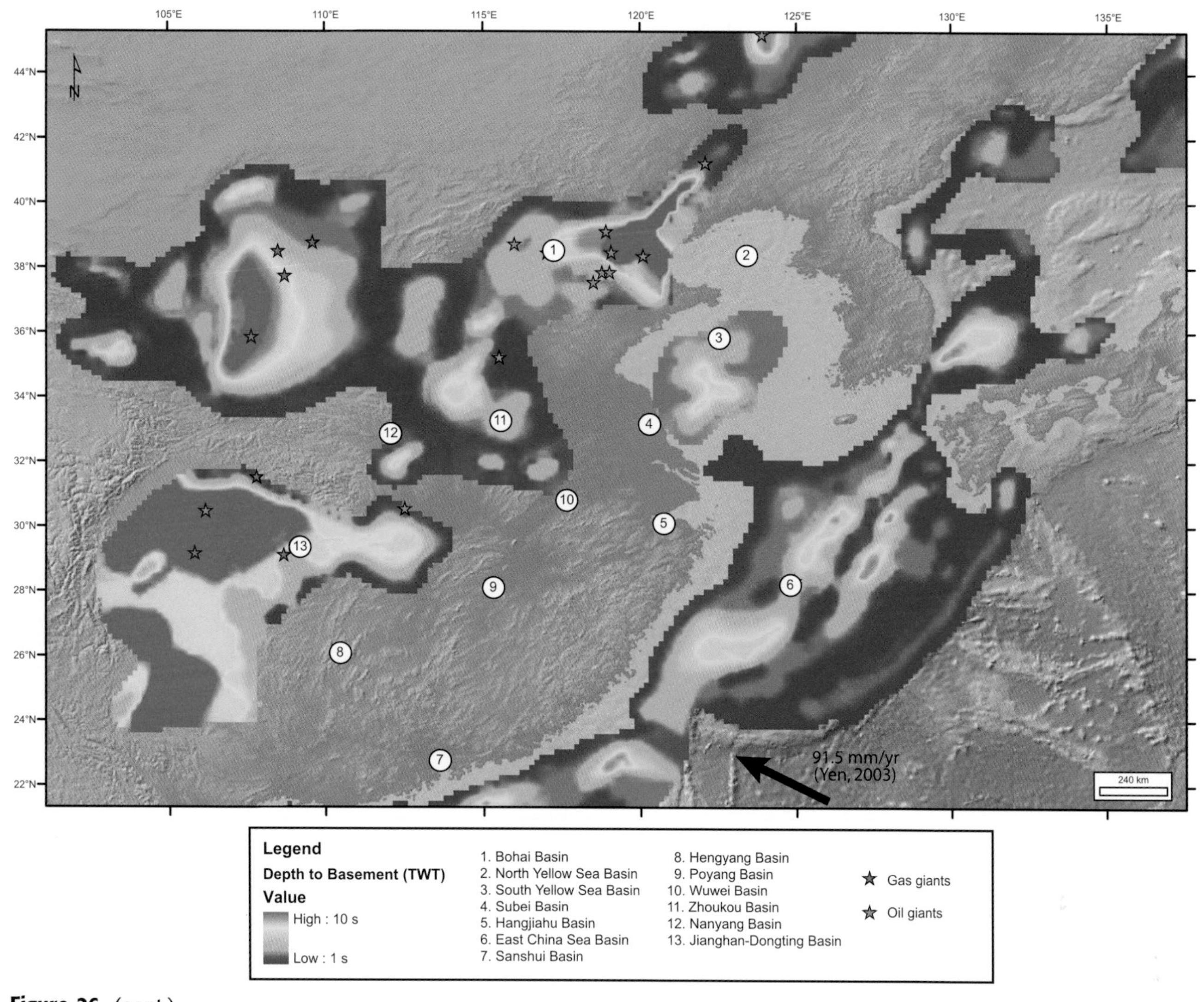

Figure 26. (cont.).

these trends of faults would connect the Taihang and Tan-Lu fault zones to form the deepest part of the basin (Figure 26B) and would likely combine strike-slip faults with normal faults.

The Tan-Lu fault zone in the central part of the basin consists of two parallel strike-slip faults separated by a distance of about 3 km (~1.9 mi) (Figure 28). Many normal faults intersect but do not crosscut the Tan-Lu fault and show that the Tan-Lu fault is either a younger feature or that all normal faults terminate on the Tan-Lu fault (Figure 28). Despite the recency of deformation, one giant oil field is present within 1 km (0.6 mi) of the two main fault strands. In cross section, the Tan-Lu fault in the central Bohai Basin appears in most areas as a large faulted anticline (Figure 29). In cross section and on three-dimensional (3-D) seismic data in the northern Bohai Basin (Liaodong Bay), the fault reactivated older normal faults related to rifting as positive flower structures (Hsiao et al., 2004) (Figures 30, 31).

Structures outside the Tan-Lu fault include inversion structures (Zhu et al., 2009), shearing-related folds (Chen et al., 2010), and low-angle normal faults (Qi and Yang, 2010).

Tectonic Origin of the Regional Stress Field of Eastern China and Its Effects on Petroleum

It is not clear whether subduction or extrusion is a dominant factor in forming folds and hydrocarbon traps within the Bohai Basin because both processes would create the observed pattern of east–west-oriented stresses (Figure 26C). Because the area is a submarine pull-apart

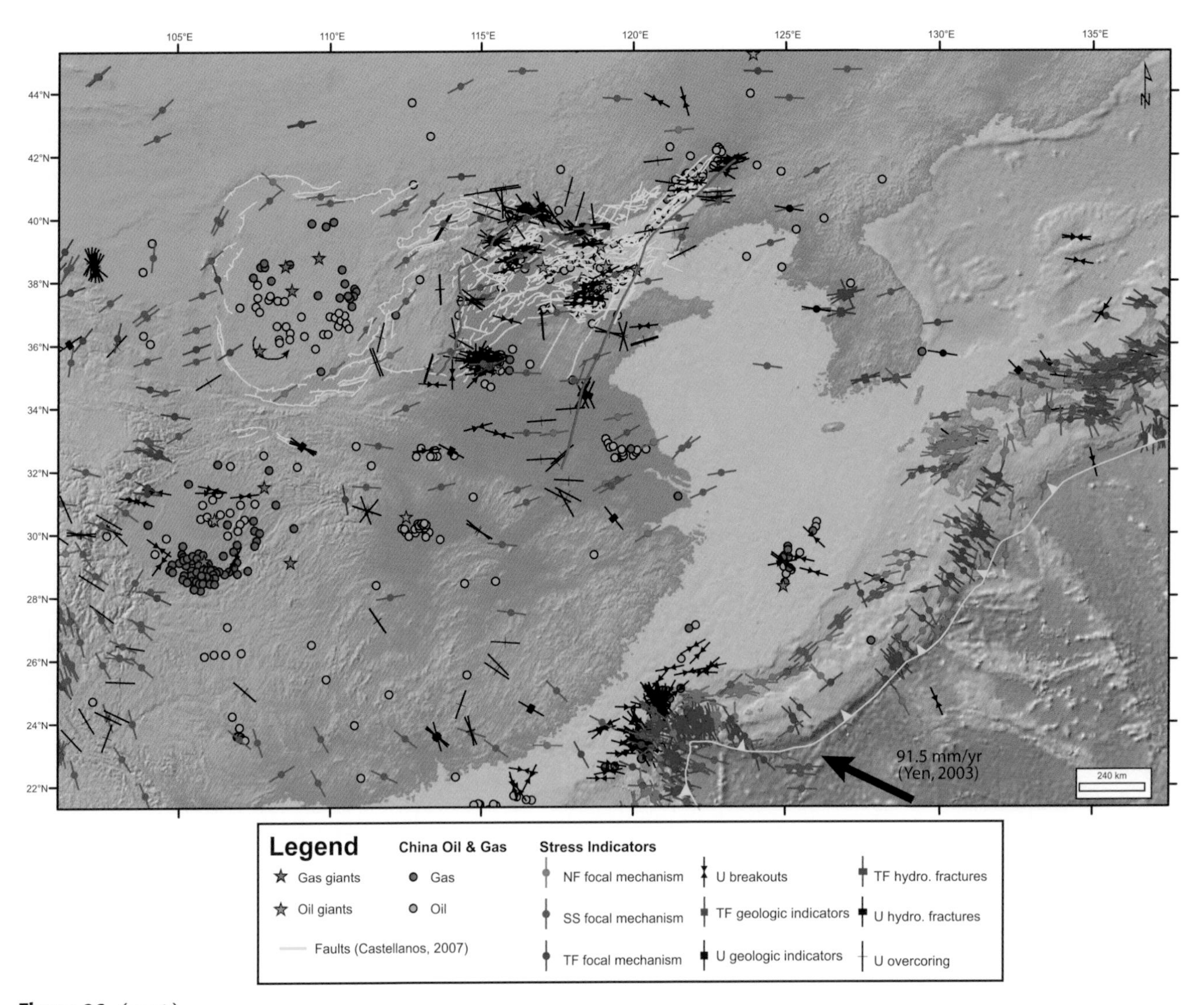

Figure 26. (cont.).

basin, a few folds are present probably as a result of lessening of compressive stress in the area of the localized extensional pull-apart basin. Structural traps for petroleum mainly include inversion structures and various types of fault traps.

DISCUSSION

This chapter has compiled a variety of publicly available data on several characteristics of strike slip-controlled areas with large hydrocarbon reserves: (1) regional plate motions based on a GPS-based geodesy; (2) basin types along the strike-slip faults are defined using compilations of depth to basement data and include localized pull-aparts and more regional transpressional basins flanking major strike-slip faults; (3) well-described examples of structural styles along strike-slip faults based on surface and subsurface mapping mostly related to hydrocarbon exploration; and (4) well-studied structural and stratigraphic traps that accommodate large accumulations of hydrocarbons in each of the four strike slip-dominated tectonic settings.

The compilation reconfirms earlier insights into the relationship between regional stresses and structures along major strike-slip faults. In all four areas, the maximum compressive stress is almost at right angles to the major strike-slip faults. This relationship suggests that the low-drag uncoupled model of Mount and Suppe (1992) (Figure 2B) can explain the regional folding pattern better than the high-drag coupled model of Harding (1973) (Figure 2A). In all four areas, distant subduction zones may control the relatively

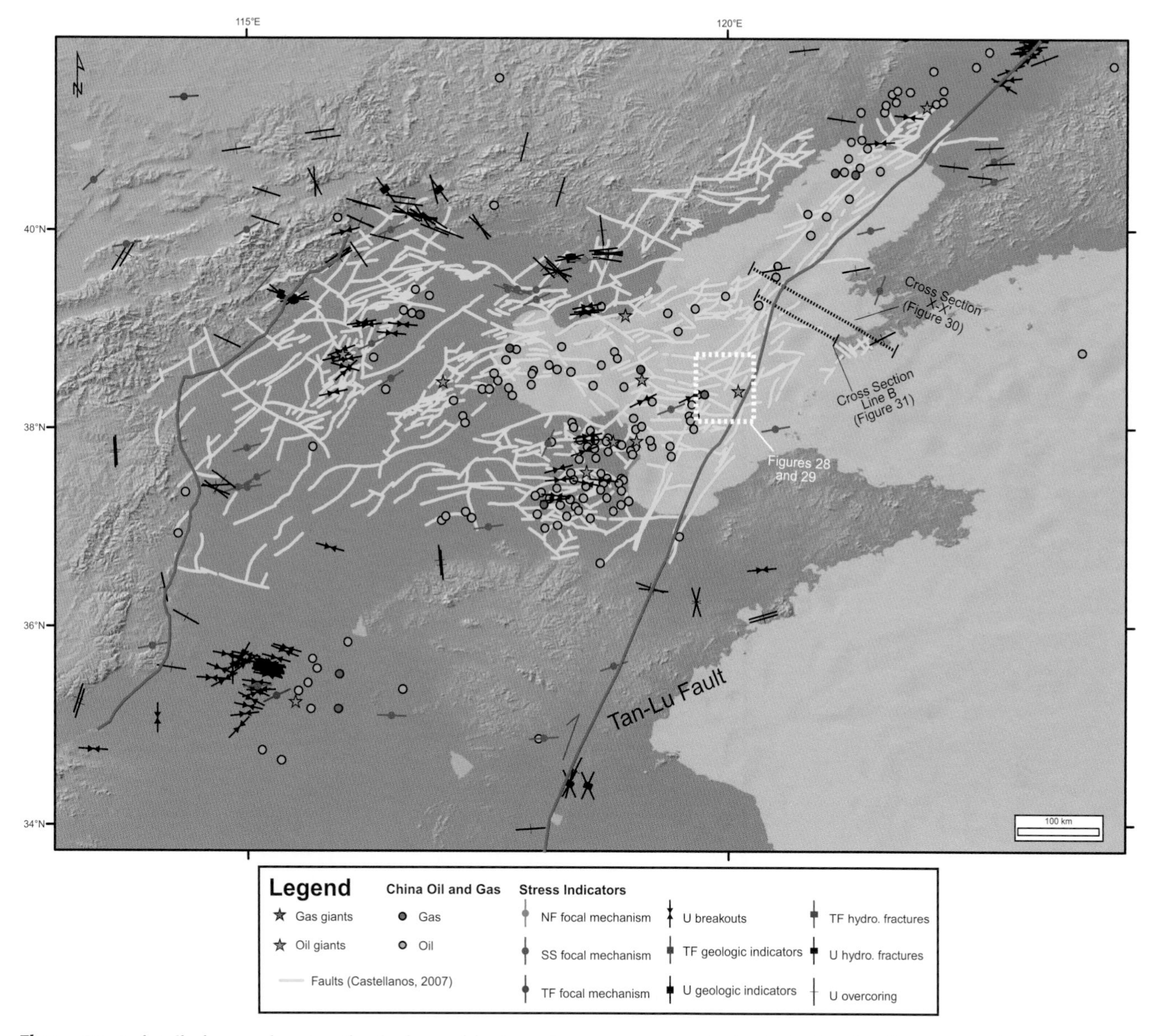

Figure 27. A detailed map showing the faults and hydrocarbon occurrences in the Bohai Basin of eastern China, as modified from Castellanos (2007). The giant oil fields are shown in green stars, and the giant gas fields are shown in red stars (data from Mann et al., 2003). Oil fields shown in green are from Steinshouer et al. (2002). Various types of stress indicators are color coded on the map and summarized in the map key. The white box indicates the locations of more detailed maps shown in Figures 28 and 29 and cross sections in Figures 30 and 31. NF = normal fault; SS = strike slip; TF = thrust fault; U = undefined. 100 km (62 mi).

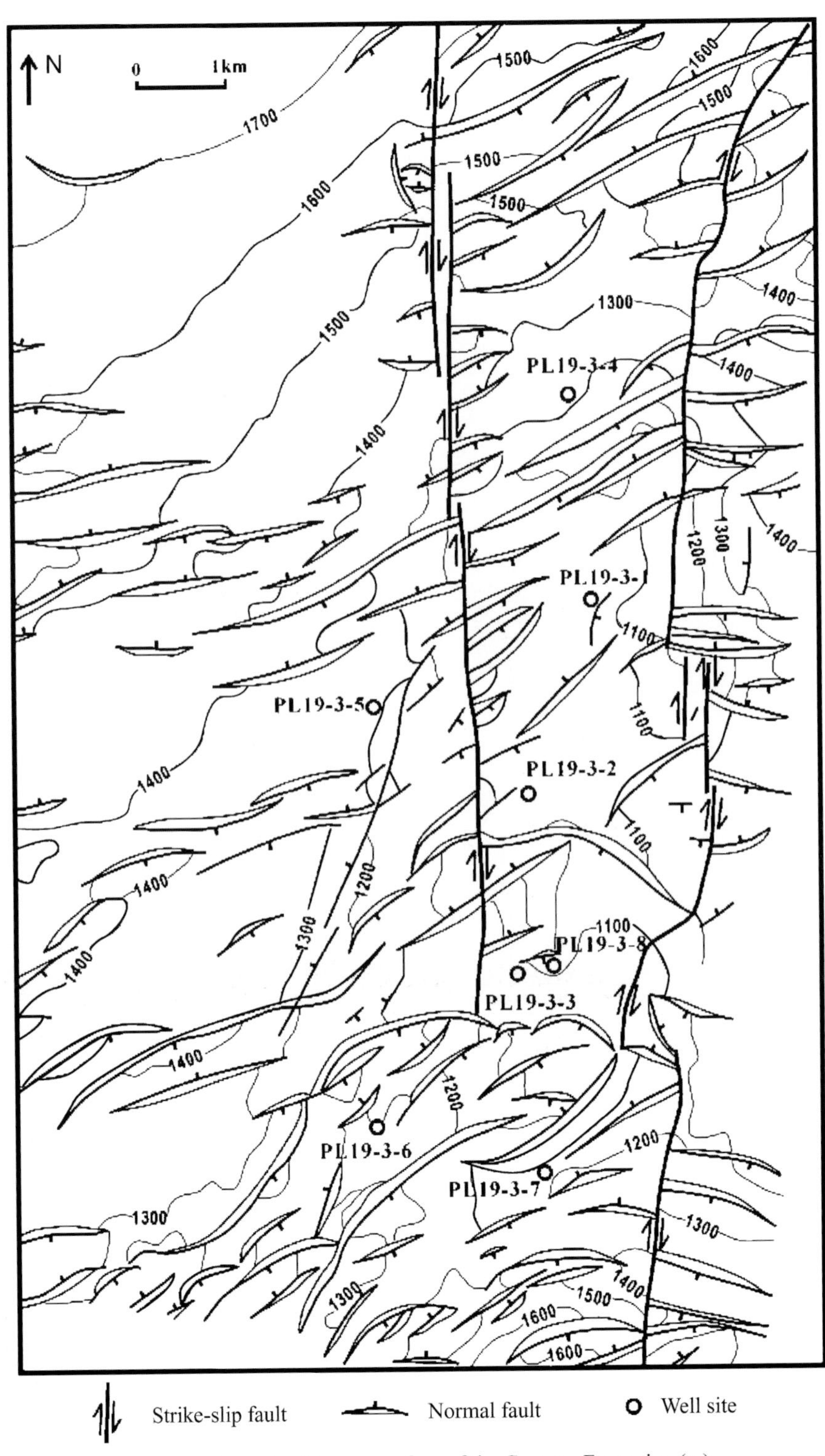

Figure 28. A detailed map from Yang and Xu (2004) of normal faults in the stepover areas of major strands of the right-lateral Tanlu fault zone. The area is located on the map in Figure 27.

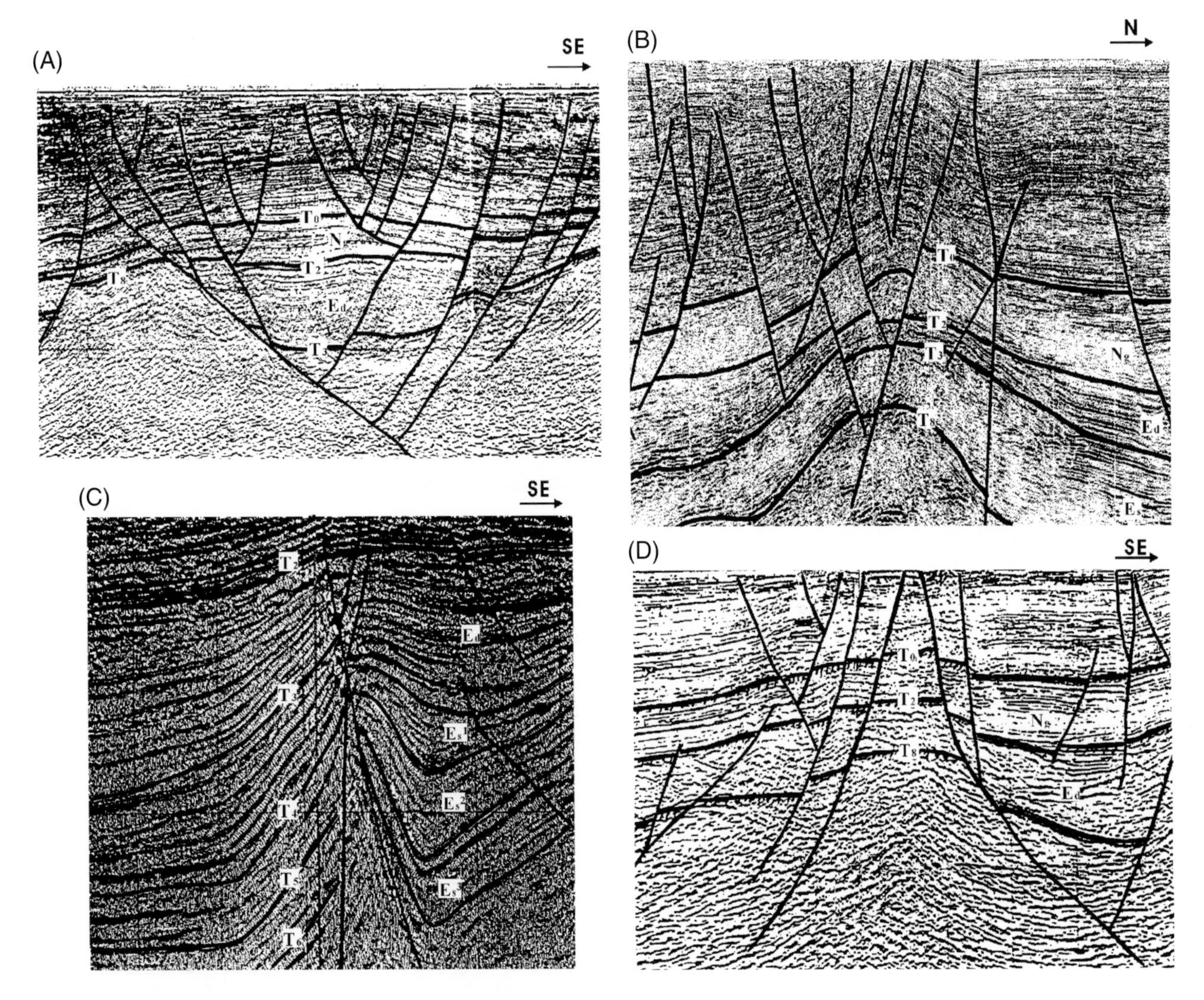

(A) Transtensional faulted anticline of the PL14-3 structure

(B) Transcompressional faulted anticline of the PL25-6 structure

(C) Diapiric anticline of the JZ31-2 structure

(D) Drape anticlines complicated by shallow faults of the PL19-3 structure

Figure 29. Examples of structures in the Bohai Basin from Yang and Xu (2004). Sections are located on the map. (A) Transtensionally faulted anticline of the PL14-3 structure. (B) Transpressionally faulted anticline of the PL25-6 structure. (C) Diapiric anticline of the JZ31-2 structure. (D) Drape anticlines complicated by shallow faults of the PL19-3 structure.

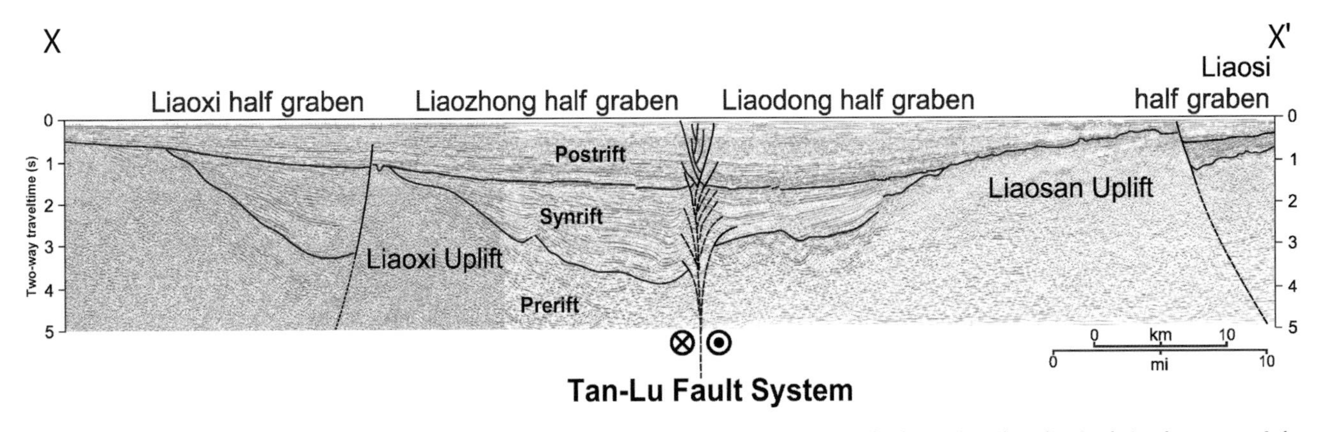

Figure 30. A regional seismic cross section from Hsiao et al. (2004) of the Tanlu right-lateral strike-slip fault in the area of the northern Bohai Basin (line is located in Figure 27).

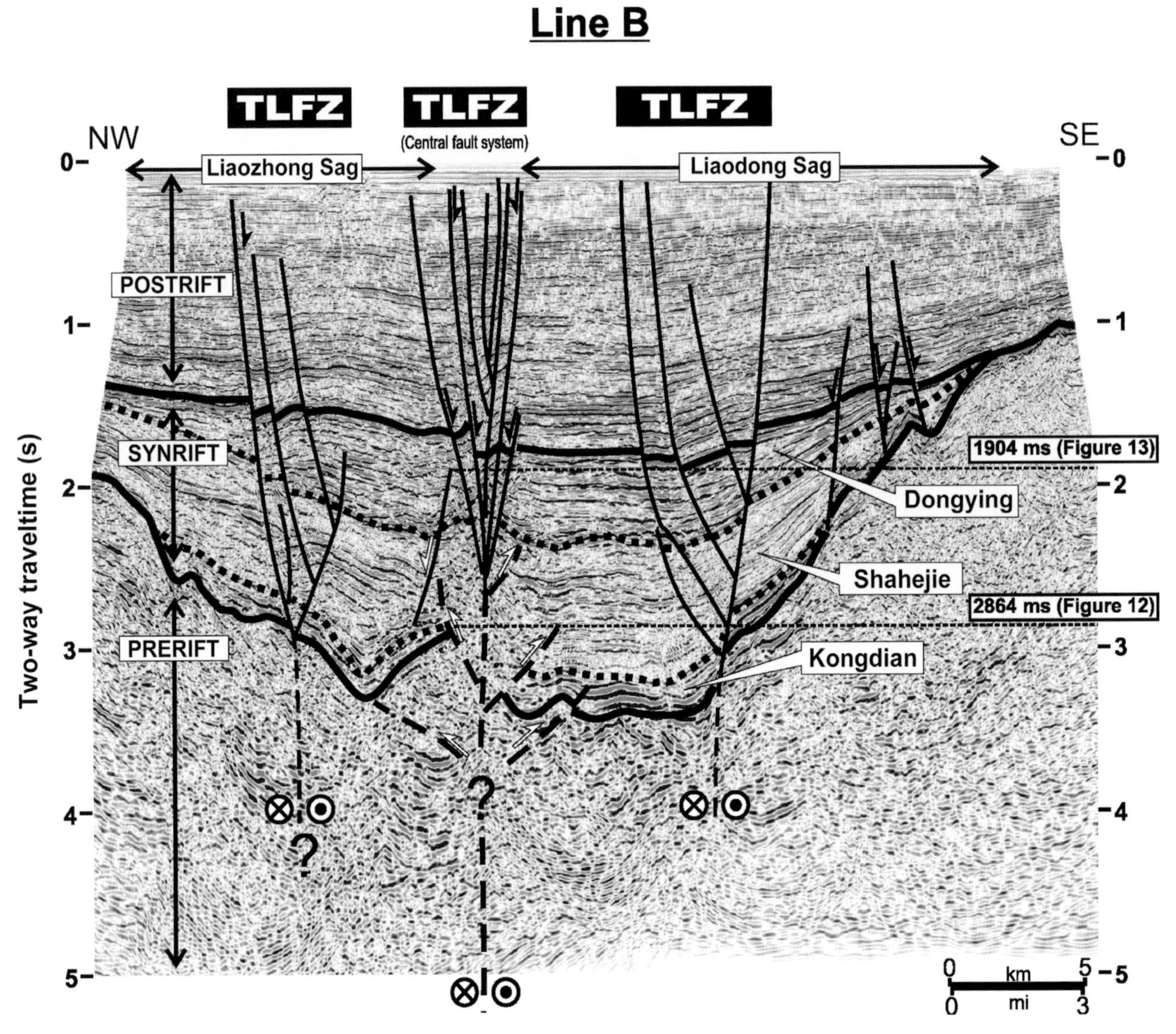

Figure 31. A regional seismic cross section from Hsiao et al. (2004) of the Tanlu right-lateral strike-slip fault in the area of the northern Bohai Basin (line is located in Figure 27). TLFZ = Tan Lu fault zone.

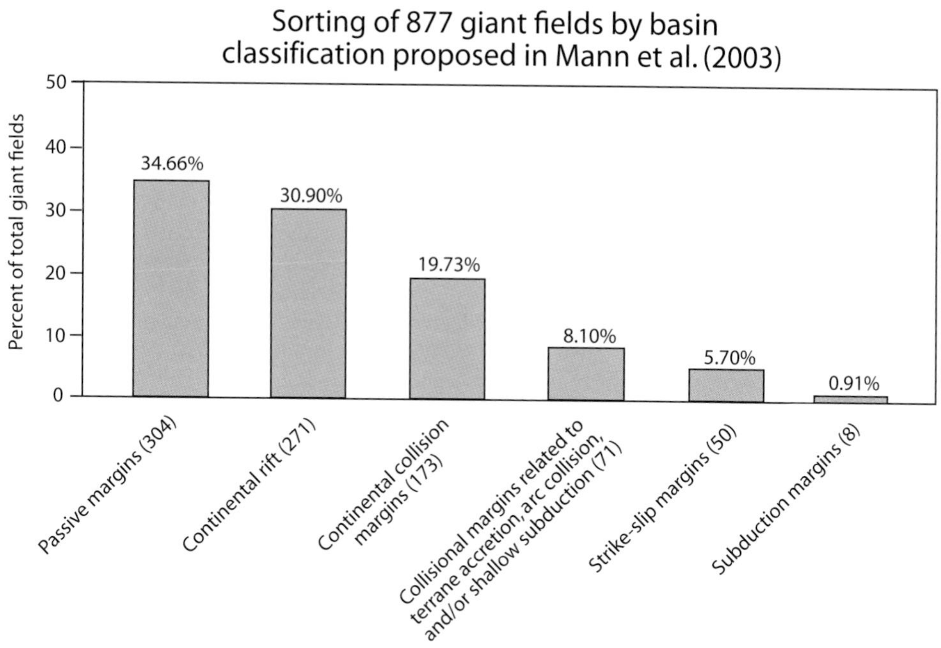

Figure 32. A histogram showing the sorting of 877 giant oil and gas fields according to their tectonic setting by Mann et al. (2003). See text for discussion.

uniform stress state by a process of subduction resistance. This process may be important because slow subduction of ancient areas of thicker-than-average oceanic crust may remain aseismic and go unnoticed as a major influence along many margins.

Giant clusters in the four strike-slip margins commonly occur in deep basins directly adjacent to the major active strike-slip fault in the region. These deep basins may be inherited and reactivated from their previous function as either arc-related basins within subduction systems (California, Caribbean) or as rift basins (Sumatra, Bohai).

CONCLUSIONS

Based on the previous study by Mann et al. (2003), the numbers of giants were subdivided according to their tectonic setting (Figure 32). Strike-slip faults showed the second lowest number of giants after subduction settings. One of the main conclusions of this chapter is that subduction may have a larger function in areas of large strike-slip faults than previously realized. Subduction is likely to be waning because of the increased motion on the overlying strike-slip fault system. Waning subduction means less volcanism that leads to an improvement in reservoir sands. During the waning phases of subduction, shallow subduction acts to uplift and deform the overlying fore arc and volcanic arc, with most fold axes oriented parallel to the plate boundary. Deformation is, therefore, a combination of both plate boundary normal stresses related to subduction and more localized stresses along major strike-slip faults.

ACKNOWLEDGMENTS

I appreciate the patience of the editor of this volume, Dengliang Gao. I thank Jeff Storms for helping with all graphic design, map compilations in geographic information system, and text editing. Financial support to Mann and Storms was provided by the sponsoring companies of the CBTH (Caribbean Basins, Tectonics and Hydrocarbons) industry consortium. UTIG (University of Texas Institute for Geophysics) contribution no. 2391. Dengliang Gao and an anonymous reviewer provided peer reviews for this chapter.

REFERENCES CITED

Allen, M. B., D. I. M. Macdonald, S. J. Vincent, C. Brouet-Menzies, and Z. Xun, 1997, Early Cenozoic two-phase extension and late Cenozoic thermal subsidence and inversion of the Bohai Basin, northern China: Marine and Petroleum Geology, v. 14, p. 951–972, doi:10.1016/S0264-8172(97)00027-5.

Allen, M. B., D. I. M. Macdonald, Z. Xun, S. J. Vincent, and C. Brouet-Menzies, 1998, Transtensional deformation in the evolution of the Bohai Basin, northern China:

Geological Society (London) Special Publications, v. 135, p. 215–229.
Argus, D. F., and R. G. Gordon, 2001, Present tectonic motion across the Coast Ranges and San Andreas fault system in central California: Geological Society of America Bulletin, v. 113, p. 1580–1592.
Barber, A. J., and M. J. Crow, 2005a, Pre-Tertiary stratigraphy, *in* A. J. Barber, M. J. Crow, and J. S. Milsom, eds., Sumatra: Geology, resources and tectonic evolution: Geological Society of London Memoir 31, p. 24–53.
Barber, A. J., and M. J. Crow, 2005b, Structure and structural history, *in* A. J. Barber, M. J. Crow, and J. S. Milsom, eds., Sumatra: Geology, resources and tectonic evolution: Geological Society of London Memoir 31, p. 175–233.
Barber, A. J., M. J. Crow, and M. E. M. De Smet, 2005, Tectonic Evolution, *in* A. J. Barber, M. J. Crow, and J. S. Milsom, eds., Sumatra: Geology, resources and tectonic evolution: Geological Society of London Memoir 31, p. 234–259.
Ben-Avraham, Z. M., 1992, Transform-normal extension and asymmetric basins: An alternative to pull-apart models: Geology, v. 20, p. 423–426, doi:10.1130/0091-7613(1992)020<0423:TNEAAB>2.3.CO;2.
Bishop, M., 2001, South Sumatra Basin Province, Indonesia: The Lahat/Talang Akar-Cenozoic Total Petroleum System: U.S. Geological Survey Open-File Report, p. 1–22.
Bjorklund, T., and K. Burke, 2002, Four-dimensional analysis of the inversion of a half graben to form the Whittier fold-fault system of the Los Angeles Basin: Journal of Structural Geology, v. 24, p. 1369–1387.
Bjorklund, T., K. Burke, H.-w. Zhou, and R. S. Yeats, 2002, Miocene rifting in the Los Angeles Basin: Evidence from the Puente Hills half graben, volcanic rocks, and P-wave tomography: Geology, v. 30, p. 451–454, doi:10.1130/0091-7613(2002)030<0451:MRITLA>2.0.CO;2.
Brocher, T., J. McCarthy, P. Hart, S. Holbrook, K. Furlong, T. McEvilly, J. Hole, and S. Klemperer, 1994, Seismic evidence for a lower-crustal detachment beneath San Francisco Bay, California: Science, v. 265, p. 1436–1439.
Bryant, W. A., compiler, 2005, Digital database of Quaternary and younger faults from the fault activity map of California, version 2.0: California Geological Survey Web Page: http://www.consrv.ca.gov/CGS/information/publications/QuaternaryFaults_ver2.htm, (accessed July 8, 2010).
Calais, E., and Mann, P., 2009, A combined GPS velocity field for the Caribbean plate and its margins: AGU Fall Meeting, San Francisco, California, December 14–18, 2009.
Calais, E., M. Vergnolle, V. San'kov, A. Lukhnev, A. Miroshnitchenko, S. Amarjargal, and J. Deverchere, 2003, GPS measurements of crustal deformation in the Baikal-Mongolia area (1994–2002): Implications for current kinematics of Asia: Journal of Geophysical Research, v. 108, 13 p., doi:10.1029/2002JB002373.
Castellanos, H. A., 2007, Sequence stratigraphy and tectonics of the Guantao and Minghuazhen formations, Zhao Dong Field, Bohai Bay, eastern China: Ph.D. thesis, University of Texas, Austin, Texas, 242 p.
Chen, J. S., B. C. Huang, and L. S. Sun, 2010, New constraints to the onset of the India-Asia collision: Paleomagnetic reconnaissance on the Linzizong Group in the Lhasa Block, China: Tectonophysics, v. 489, p. 189–209.
Cheng-yong, C., 1991, Geological characteristics and distribution patterns of hydrocarbon deposits in the Bohai Bay Basin, east China: Marine and Petroleum Geology, v. 8, p. 98–106, doi:10.1016/0264-8172(91)90048-6.
Clark, S. A., A. Levander, F. Niu, C. A. Zelt, M. Sobiesiak, and M. B. Magnani, 2007, Seismic evidence for a vertical tear in the South American lithosphere offshore Venezuela: American Geophysical Union Fall Meeting, San Francisco, California, December 10–14, 2007.
Clark, S. A., C. A. Zelt, M. B. Magnani, and A. Levander, 2008, Characterizing the Caribbean-South American plate boundary at 64°W using wide-angle seismic data: Journal of Geophysical Research, v. 113, no. B07401, 16 p., doi:10.1029/2007JB005329.
Colmenares, L., and M. D. Zoback, 2003, Stress field and seismotectonics of northern South America: Geology, v. 31, p. 721–724, doi:10.1130/G19409.1.
Curray, J. R., 2005, Tectonics and history of the Andaman Sea region: Journal of Asian Earth Sciences, v. 25, p. 187–232.
Escalona, A., and P. Mann, 2006, Tectonic controls of the right-lateral Burro Negro tear fault on Paleogene structure and stratigraphy, northeastern Maracaibo Basin: AAPG Bulletin, v. 90, p. 479–504.
Escalona, A., and P. Mann, 2011, Tectonics, basin subsidence mechanisms, and paleogeography of the Caribbean-South American plate boundary zone: Marine and Petroleum Geology, v. 28, p. 8–39, doi:10.1016/j.marpetgeo.2010.01.016.
Exxon Production Research Company, 1985, The tectonic map of the world (scale varies with panel): Exxon Production Research Company, 17 sheets.
Feigl, K. L., et al., 1993, Space geodectic measurement of crustal deformation in central and southern California 1984–1992: Journal of Geophysical Research, v. 98, p. 21,677–21,712.
Fitch, T. J., 1972, Plate convergence, transcurrent faults, and internal deformation adjacent to Southeast Asia and the Western Pacific: Journal of Geophysical Research, v. 77, p. 4432–4460.
Fuchs, K., and B. Muller, 2001, World Stress Map of the Earth: A key to tectonic processes and technological applications: Naturwissenschaften, v. 88, p. 357–371.
Garciacaro, E., P. Mann, and A. Escalona, 2011, Regional structure and tectonic history of the obliquely colliding Columbus foreland basin, offshore Trinidad and Venezuela: Marine and Petroleum Geology, v. 28, p. 126–148.
Google Inc., 2009, Google Earth (version 5.1.3533.1731): http://earth.google.com (accessed July 2010).
Griffith, W. A., and M. L. Cooke, 2005, How sensitive are fault-slip rates in the Los Angeles Basin to tectonic boundary conditions?: Bulletin of the Seismological Society of America, v. 95, p. 1263–1275, doi:10.1785/0120040079.
Harding, T., 1990, Identification of wrench faults using

subsurface structural data: Criteria and pitfalls: AAPG Bulletin, v. 74, p. 1590–1609.

Harding, T. P., 1973, Newport-Inglewood trend, California: An example of wrenching style of deformation: AAPG Bulletin, v. 58, p. 97–116.

Harding, T. P., 1974, Petroleum traps associated with wrench faults: AAPG Bulletin, v. 58, p. 1290–1304.

Harding, T. P., 1976, Tectonic significance and hydrocarbon trapping consequences of sequential synchronous with San Andreas faulting, San Joaquin Valley, California: AAPG Bulletin, v. 60, p. 356–378.

Harding, T. P., 1983, Divergent wrench fault and negative flower structure, Andaman Sea, *in* T. P. Harding, ed., Seismic expression of structural styles: A picture and work atlas: AAPG Studies in Geology 15, p. 4.2-1–4.2-8.

Harding, T. P., 1985, Seismic characteristics and identification of negative flower structures, positive flower structures, and positive structural inversion: AAPG Bulletin, v. 69, p. 582–600.

Harding, T. P., and J. S. Wickham, 1988, Comments on "State of stress near the San Andreas fault: Implications for wrench tectonics": Geology, v. 16, p. 1151–1153.

Hauksson, E., 1987, Seismotectonics of the Newport-Inglewood fault zone in the Los Angeles Basin, southern California: Bulletin of the Seismological Society of America, v. 77, no. 2, p. 539–561.

Hauksson, E., 1990, Earthquakes, faulting, and stress in the Los Angeles Basin: Journal of Geophysical Research, v. 95, no. B10, p. 15,365–15,394, doi:10.1029/JB095iB10p15365.

Hayes, G. P., D. J. Wald, and R. L. Johnson, 2012, Slab 1.0: A three-dimensional model of global subduction zone geometries: Journal of Geophysical Research, v. 117, no. B01302, doi:10.1029/2011JB008524.

Hefu, L., 1986, Geodynamic scenario and structural styles of Mesozoic and Cenozoic basins in China: AAPG Bulletin, v. 70, p. 377–395.

Heidbach, O., J. Reinecker, M. Tingay, B. Müller, B. Sperner, K. Fuchs, and F. Wenzel, 2007, Plate boundary forces are not enough: Second- and third-order stress patterns highlighted in the World Stress Map database: Tectonics, v. 26, p. TC6014, doi:10.1029/2007TC002133.

Heidbach, O., M. Tingay, A. Barth, J. Reinecker, D. Kurfess, and B. Müller, 2008, The 2008 release of the World Stress Map: Helmholtz Centre Potsdam, Potsdam, Germany: http://www.world-stressmap.org (accessed July 2010).

Heidbach, O., M. Tingay, A. Barth, J. Reinecker, D. Kurfeß, and B. Müller, 2010, Global crustal stress pattern based on the World Stress Map database release 2008: Tectonophysics, v. 482, p. 3–15, doi:10.1016/j.tecto.2009.07.023.

Henstock, P., A. Levander, and J. Hole, 1997, Deformation in the lower crust of the San Andreas fault system in northern California: Science, v. 278, p. 650–653, doi:10.1126/science.278.5338.650.

Hornafius, J., B. Luyendyk, R. Terres, and M. Kamerling, 1986, Timing and extent of Neogene tectonic rotation in the western Transverse Ranges, California: Geological Society of America Bulletin, v. 97, p. 1476–1487, doi:10.1130/0016-7606(1986)972.0.CO;2.

Hsiao, L., S. Graham, and N. Tilante, 2004, Seismic reflection imaging of a major strike-slip fault zone in a rift system: Paleogene structure and evolution of the Tan-Lu fault system, Liaodong Bay, Bohai, offshore China: AAPG Bulletin, v. 88, p. 71–97, doi:10.1306/09090302019.

Hu, J. Y., S. B. Xu, X. G. Tong, and H. Y. Wu, 1989, The Bohai Bay Basin: Chinese sedimentary basins: Amsterdam, Elsevier, p. 89–105.

Hu, S., P. B. O'Sullivan, A. Raza, and B. P. Kohn, 2001, Thermal history and tectonic subsidence of the Bohai Basin, northern China: A Cenozoic rifted and local pull-apart basin: Physics of the Earth and Planetary Interiors, v. 126, p. 221–235.

Jamison, W. R., 1991, Kinematics of compressional fold development in convergent wrench terranes: Tectonophysics, v. 190, p. 209–232.

Jarrard, R. D., 1986, Terrane motion by strike-slip faulting of forearc slivers: Geology, v. 14, p. 780–783.

Lelgemann, H., M. Gutscher, J. Bialas, E. Flueh, and W. Weinrebe, 2000, Transtensional basins in the western Sunda Strait: Geophysical Research Letters, v. 27, p. 3545–3548, doi:10.1029/2000GL011635.

Macgregor, D., 1996, Factors controlling the destruction or preservation of giant oil fields: Petroleum Geoscience, v. 2, p. 197–217, doi:10.1144/petgeo.2.3.197.

Magnani, M. B., C. A. Zelt, A. Levander, and M. Schmitz, 2009, Crustal structure of the South America-Caribbean plate boundary at 67°W from controlled-source seismic data: Journal of Geophysical Research, v. 114, p. 1–23.

Mann, P., 2007, Global catalog, classification and tectonic origins of restraining and releasing bends on active and ancient strike-slip fault systems: Geological Society (London) Special Publication, v. 290, p. 13–142.

Mann, P., L. Gahagan, and M. Gordon, 2003, Tectonic setting of the world's giant oil and gas fields, *in* M. Halbouty, ed., Giant oil fields of the decade 1990–1999: AAPG Memoir 78, p. 15–105.

Mann, P., A. Escalona, and CBTH Research Group, 2006, Caribbean basins, tectonics and hydrocarbons (CBTH) phase I: Atlas volume 2006: Austin, Texas, University of Texas, Jackson School of Geosciences, Institute for Geophysics, p. 91.

Mann, P., A. Escalona, and CBTH Research Group, 2010, Caribbean basins, tectonics and hydrocarbons (CBTH) phase II: Atlas volume 2010: Austin, Texas, University of Texas, Jackson School of Geosciences, Institute for Geophysics, p. 85.

Masters, C., D. Root, and R. Turner, 1998, World conventional crude oil and natural gas: Identified reserves, undiscovered resources, and futures: U.S. Geological Survey Open-File Report 98-468: http://pubs.usgs.gov/of/1998/of98-468/text.htm (accessed July 2010).

McCaffrey, R., 2009, The tectonic framework of the Sumatran Subduction Zone: Annual Review of Earth and Planetary Science, v. 37, p. 24, doi:10.1146/annurev.earth.031208.100212.

McCarthy, A. J., and C. F. Elders, 1997, Cenozoic deformation in Sumatra: Oblique subduction and the development of

the Sumatran Fault System: Geological Society (London) Special Publication 126, p. 355–363.
McIntosh, K., D. Reed, and E. Silver, 1991, Deep structure and structural inversion along the central California continental margin from EDGE seismic profile RU-3: Journal of Geophysical Research, v. 96, p. 6459–6473, doi:10.1029/89JB01172.
Morley, C. K., 2009, Evolution from an oblique subduction back-arc mobile belt to a highly oblique collisional margin: The Cenozoic tectonic development of Thailand and eastern Myanmar: Geological Society (London) Special Publication 318, p. 373–403.
Morley, C. K., and R. Westaway, 2006, Subsidence in the super-deep Pattani and Malay basins of Southeast Asia: A coupled model incorporating lower-crustal flow in response to postrift sediment loading: Basin Research, v. 18, p. 51–84, doi:10.1111/j.1365-2117.2006.00285.x.
Mount, V., and J. Suppe, 1987, State of stress near the San Andreas fault: Implications for wrench tectonics: Geology, v. 15, p. 1143–1146, doi:10.1130/0091-7613(1987)152.0.CO;2.
Mount, V. S., and J. Suppe, 1992, Present-day stress orientations adjacent to active strike-slip faults: California and Sumatra: Journal of Geophysical Research, v. 97, p. 11,995–12,013, doi:10.1029/92JB00130.
Moxon, I. W., and S. A. Graham, 1987, History and controls of subsidence in the Late Cretaceous–Tertiary Great Valley fore-arc basin, California: Geology, v. 15, p. 626–629, doi:10.1130/0091-7613(1987)152.0.CO;2.
Ngah, K., M. Mazlan, and H. D. Tjia, 1996, Role of pre-Tertiary fractures in formation and development of the Malay and Penyu basins, *in* R. Hall and D. Blundell, eds., Tectonic evolution of Southeast Asia: Geological Society Special Publication 106, p. 281–289.
Oppenheimer, D., P. Reasenberg, and R. Simpson, 1988, Fault-plane solutions for the 1984 Morgan Hill, California, earthquake sequence: Evidence for the state of stress on the Calaveras fault: Journal of Geophysical Research, v. 93, no. B8, p. 9007–9026, doi:10.1029/88JB00071.
Pfeiffer, D. A., 2002, Sizing up the competition: Is China the end game?: Wilderness Publications: http://www.fromthewilderness.com/free/ww3/092502_endgame.html (accessed July 2010).
Punnette, S., 2010, Structural framework and its influence on the Quaternary-age sequence architecture of the northern shelf of Trinidad and Tobago: M.S. thesis, University of Texas at Austin, Austin, Texas, 187 p.
Qi, J. F., and Q. Yang, 2010, Cenozoic structural deformation and dynamic processes of the Bohai Bay Basin province, China: Marine and Petroleum Geology, v. 27, no. 4, p. 757–771.
Robertson, P., and K. Burke, 1989, Evolution of southern Caribbean Plate boundary, vicinity of Trinidad and Tobago: AAPG Bulletin, v. 73, p. 490–509.
Schlüter, H., C. Gaedicke, H. Roeser, B. Schreckenberger, H. Meyer, C. Reichert, Y. Djajadihardja, and A. Prexl, 2002, Tectonic features of the southern Sumatra-western Java fore arc of Indonesia: Tectonics, v. 21, p. 1–15, doi:10.1029/2001TC901048.
Sieh, K., and D. Natawidjaja, 2000, Neotectonics of the Sumatran fault, Indonesia: Journal of Geophysical Research, v. 105, p. 28,295–28,326.
Soto, D., P. Mann, and A. Escalona, 2011, Miocene-to-recent structure and basinal architecture along the Central Range strike-slip fault zone, eastern offshore Trinidad: Marine and Petroleum Geology, v. 28, p. 212–234, doi:10.1016/j.marpetgeo.2010.07.011.
Steinshouer, D. W., J. Qiang, P. McCabe, and R. Ryder, 1997, Maps showing geology, oil and gas fields, and geologic provinces of the Asia-Pacific region: U.S. Geological Survey Open-File Report 97-470F, 1 CD-ROM.
Susilohadi, S., C. Gaedicke, and A. Ehrhardt, 2005, Neogene structures and sedimentation history along the Sunda forearc basins off southwest Sumatra and southwest Java: Marine Geology, v. 219, p. 133–154, doi:10.1016/j.margeo.2005.05.001.
Tingay, M. R. P., B. Müller, J. Reinecker, O. Heidbach, F. Wenzel, and P. Fleckenstein, 2005, Understanding tectonic stress in the oil patch: The World Stress Map Project: The Leading Edge, p. 1276–1282, doi:10.1190/1.2149653.
Tingay, M. R. P., B. Müller, J. Reinecker, and O. Heidbach, 2006, State and origin of present-day stress fields in sedimentary basins: New results from the World Stress Map Project: The 41st U.S. Symposium on Rock Mechanics (USRMS), p. 1–10.
Tingay, M. R. P., C. K. Morley, R. R. Hillis, and J. Meyer, 2010a, Present-day stress orientation in Thailand's basins: Journal of Structural Geology, v. 32, p. 235–248, doi:10.1016/j.jsg.2009.11.008.
Tingay, M. R. P., C. K. Morley, R. King, R. R. Hillis, D. Coblentz, and R. Hall, 2010b, Present-day stress field of Southeast Asia: Tectonophysics, v. 482, p. 92–104, doi:10.1016/j.tecto.2009.06.019.
Topography and Predicted Bathymetry of the World with Present-day Plate Boundaries, 1999, PLATES Project digital data compilation: University of Texas Institute for Geophysics.
Townend, J., and M. D. Zoback, 2004, Regional tectonic stress near the San Andreas fault in central and southern California: Geophysical Research Letters, v. 31, p. L15S11.
UNAVCO (University Navstar Consortium), 2009, PBO GPS velocities in southern California: http://geon.unavco.org/unavco/IDV_for_GEON_gps.html (accessed August 2010).
USGS (U.S. Geological Survey), 1999, Maps showing geology, oil and gas fields, and geologic provinces of South America: U.S. Geological Survey Open-File Report 97-470-D http://certmapper.cr.usgs.gov/rooms/utilities/full_metadata.jsp?docId=%7B6BE604B3-7118-4653-AFFF-B6B2804508FB%7D&loggedIn=false (accessed January 2011).
USGS (U.S. Geological Survey), 2003, U.S. Geological Survey world petroleum assessment 2000: New estimates of undiscovered oil and natural gas, natural gas liquids, including reserve growth, outside the United States: U.S. Department of the Interior, U.S. Geological Survey, Reston, Virginia, p. 1–2.
van der Hilst, R. D., and P. Mann, 1994, Tectonic implications of tomographic images of subducted lithosphere

beneath northwestern South America: Geology, v. 22, p. 451–454.

Wang, Q., et al., 2001, Present-day crustal deformation in China constrained by Global Positioning System measurements: Science, v. 294, p. 574–577, doi:10.1126/science.1063647.

Webb, T. H., and H. Kanamori, 1985, Earthquake focal mechanisms in the eastern Transverse Ranges and San Emigdio Mountains, southern California, and evidence for a regional decollement: Bulletin of the Seismological Society of America, v. 75, p. 737–757.

Weber, J., T. Dixon, C. DeMets, W. Ambeh, P. Jansma, G. Mattioli, R. Bilham, J. Saleh, and O. Perez, 2001a, A GPS estimate of the relative motion between the Caribbean and South American plates, and geologic implications for Trinidad and Venezuela: Geology, v. 29, p. 75–78.

Wells, R. E., R. J. Blakely, and C. S. Weaver, 2002, Cascadia microplate models and within-slab earthquakes, *in* S. Kirby, K. Wang, and S. Dunlop, eds., The Cascadia subduction zone and related subduction systems: Seismic structure, intraslab earthquakes and processes, and earthquake hazards: U.S. Geological Survey Open-File Report 02-328 and Geological Survey Canada Open-File Report 4350, p. 17–23.

Wilcox, R. E., T. P. Harding, and D. R. Seely, 1973, Basic wrench tectonics: AAPG Bulletin, v. 57, p. 74–96, doi:10.1306/819A424A-16C5-11D7-8645000102C1865D.

Woodcock, N. H., and M. C. Daly, 1986, The role of strike-slip fault systems at plate boundaries (with discussion): Philosophical Transactions of the Royal Society of London, Series A, Mathematical and Physical Sciences, v. 317, p. 13–29.

Wright, T. L., 1991, Structural geology and tectonic evolution of the Los Angeles Basin, *in* K. T. Biddle, ed., Active margin basins: AAPG Memoir 52, p. 35–134.

Xu, J., 1993, Historical review and present setting, *in* J. Xu, ed., The Tancheng-Luijiang Wrench Fault System: New York, Wiley, p. 3–15.

Xu, Z.-h., 2001, A present-day tectonic stress map for eastern Asia region: Acta Seismologica Sinica, v. 14, p. 524–533, doi:10.1007/BF02718059.

Yang, Y., and T. Xu, 2004, Hydrocarbon habitat of the offshore Bohai Basin, China: Marine and Petroleum Geology, v. 21, p. 691–708, doi:10.1016/j.marpetgeo.2004.03.008.

Zhang, S., Y. Wang, M. Li, X. Pang, D. Shi, and Dongxia, 2005, Fault-fracture mesh petroleum plays in the Zhanhua Depression, Bohai Bay Basin: Part 1. Source rock characterization and quantitative assessment: Organic Geochemistry, v. 36, p. 183–202, doi:10.1016/j.orggeochem.2004.08.003.

Zhu, G., Y. Wang, G. Liu, M. Niu, C. Xie, and C. Li, 2005, $^{40}Ar/^{39}Ar$ dating of strike-slip motion on the Tan-Lu fault zone, east China: Journal of Structural Geology, v. 27, p. 1379–1398, doi:10.1016/j.jsg.2005.04.007.

Zoback, M. D., et al., 1987, New evidence on the state of stress of the San Andreas Fault System: Science, v. 238, p. 1105–1111, doi:10.1126/science.238.4830.1105.

Zoback, M. L., 1992, First- and second-order patterns of stress in the lithosphere: The World Stress Map Project: Journal of Geophysical Research, v. 97, p. 11,703–11,728, doi:10.1029/92JB00132.

Zucca, A. J., compiler, 2001, Oil, gas, and geothermal fields in California: California Department of Conservation, scale 1,500,000, 1 sheet: ftp://ftp.consrv.ca.gov/pub/oil/maps/Map_S-1.pdf (accessed July 14, 2010).

4

Ettensohn, Frank R., and R. Thomas Lierman, 2012, Large-scale tectonic controls on the origin of paleozoic dark-shale source-rock basins: Examples from the Appalachian Foreland Basin, Eastern United States, *in* D. Gao, ed., Tectonics and sedimentation: Implications for petroleum systems: AAPG Memoir 100, p. 95–124.

Large-scale Tectonic Controls on the Origin of Paleozoic Dark-shale Source-rock Basins: Examples from the Appalachian Foreland Basin, Eastern United States

Frank R. Ettensohn

Department of Earth and Environmental Sciences, University of Kentucky, 101 Sloan Bldg., Lexington, Kentucky, 40506, U.S.A. (e-mail: f.ettensohn@uky.edu)

R. Thomas Lierman

Department of Geography and Geology, Eastern Kentucky University, 521 Lancaster Ave., Richmond, Kentucky, 40475, U.S.A. (e-mail: tom.lierman@eku.edu)

ABSTRACT

Recent plays like the Middle Devonian Marcellus Shale and possible prospects like the Upper Ordovician Utica Shale point out the significance of dark-shale source rocks in the Appalachian Basin. Mapping the distribution of such shales in space and time throughout the basin shows that periods of dark-shale deposition coincided with orogenies and the related formation of foreland basins. The fact that foreland basins form and become repositories for organic-rich dark-shale source rocks is mostly the result of deformational loading in the adjacent orogen. Tectonism mostly exerts its control through the flexural effects of deformational loading and subsequent relaxation in the orogen. These flexural processes generate sedimentary responses in the foreland basin that are reflected in a seven-part unconformity-bound cycle, of which dark shales are a major component. Because orogenies comprise a series of smaller deformational events, or tectophases, and each tectophase generates a similar cycle, many foreland basins typically exhibit a cyclic array of dark-shale and intervening clastic units, called tectophase cycles. Thirteen such third-order tectophase cycles, formed during four orogenies, are present in the Appalachian Basin. Using examples of foreland-basin dark-shale units formed during the Ordovician–Silurian Taconian and Devonian–Mississippian Acadian/Neoacadian orogenies, the timing of cycles and migration of successive dark-shale units within them relative to the progress of orogeny are presented as evidence of causal relationships between tectonism and dark-shale sedimentation. However, tectonic influence may extend well beyond the confines of the foreland basin in the form of far-field tensional and compressional forces. This may impel the yoking of foreland and intracratonic basins as well as the reactivation of foreland basement

DOI:10.1306/13351549M1003529

structures—the former allowing dark-shale depositional conditions to move from one basin to the other, and the latter, inaugurating new basins for dark-shale accumulation.

INTRODUCTION

Dark shales, as hydrocarbon source and reservoir rocks, comprise some of the most prominent and economically significant stratigraphic intervals in epicontinental settings. In such settings, most dark shales occur in marine compressional or tensional basins that are related to tectonic events. The most extensive and variable of these basins are the compressional retroarc and peripheral foreland basins that mostly form in response to deformational loading in an adjacent orogen (Quinlan and Beaumont, 1984). In fact, the mapping of major Paleozoic and Mesozoic dark-shale units across North America has shown that periods of dark-shale deposition are commonly coeval with orogenies and with the concomitantly developing foreland basins that become repositories for these organic-rich source rocks (Ettensohn, 1997). Units like the Devonian Marcellus and Mississippian Barnett dark shales are clear examples.

Like the Marcellus Shale, other Appalachian, Devonian–Mississippian, foreland-basin, dark shales are important because of their hydrocarbon potential and widespread occurrence. These shales underlie approximately 725,000 km^2 (279,924 mi^2) of the eastern United States, and if the dark shales and related rocks are considered to be total petroleum systems, they are estimated to contain between 577 and 1131 tcf of in-place gas (Charpentier et al., 1993), of which about 34.1 tcf are probably recoverable (Milici et al., 2003; Milici and Swezey, 2006), accounting for approximately 44% of undiscovered Appalachian gas. In addition, these same shales have been considered to be world-class oil shales (Roen and Kepferle, 1993) and sources of uranium (Conant and Swanson, 1961). Moreover, the Devonian–Mississippian dark shales are estimated to contain up to 2.8 trillion bbl of in-place oil (Russell, 1990), which, if properly exploited, might produce shale-oil resources estimated to be between 423 and 800 billion bbl (Nowacki, 1981; Dyni, 2005). Although not as well known at this time, Ordovician dark shales in the Appalachian Basin are also potential hydrocarbon sources and reservoir rocks (e.g., Ryder et al., 1998). With this kind of potential, it is particularly important to understand the origin of such shales and the controls on their distribution in time and space, as these aspects have clear implications for exploration in other foreland basins. Inasmuch as the Appalachian Basin is effectively called the type foreland basin, its stratigraphy is well understood, and it has a production history extending back to the 1800s; it provides an ideal setting in which to understand the origin and development of dark-shale source and reservoir rocks.

The Appalachian Basin (Figure 1) is composite foreland basin that formed in stages during a period of nearly 340 Ma from latest Neoproterozoic (~570 Ma) to the Late Triassic (~230 Ma). Although influenced strongly by relict, Precambrian, continental-margin architecture left from Iapetan rifting (Figure 1), the basin was mostly produced during four nearly continuous orogenies (Taconian, Salinic, Acadian/Neoacadian, and Alleghanian) along the southeastern Laurentian/Laurussian (Appalachian) margin, and it contains 13 distinct, unconformity-bound, sedimentary cycles that begin with marine dark shales (Ettensohn, 2005). Because most of the Appalachian cycles are lithologically similar, we will examine only Ordovician–Silurian and Devonian–Mississippian shale units because these shales are the most extensive and best developed, and they are the only ones currently being actively prospected or produced.

In this chapter, the term Laurentia is used to describe early Paleozoic nuclear parts of North America. After the collision of Laurentia and Baltica (western Europe) in the late Early Silurian, the combined continent is called Laurussia.

PROCEDURES

Tectonics mostly exerts its control on dark-shale origins through the lithospheric flexure that accompanies deformational loading in the orogen (Quinlan and Beaumont, 1984). Such flexural processes in turn generate large, basin-scale, wavelike, deformational responses in the lithosphere (Figure 2) that migrate in time with changing loads. It is, of course, the broad troughlike part of the waveform near the orogen (Figure 2) that provides the accommodation space for deposition of dark shales and related sediments. Given the large size of the trough-like basins, it is only through mapping the distribution of distinctive basin fills that we can discern the generation and migration of the basins in time and space. Tephra geochronology and biostratigraphy provide the temporal framework, whereas surface and subsurface mapping provide the spatial distribution of the shales.

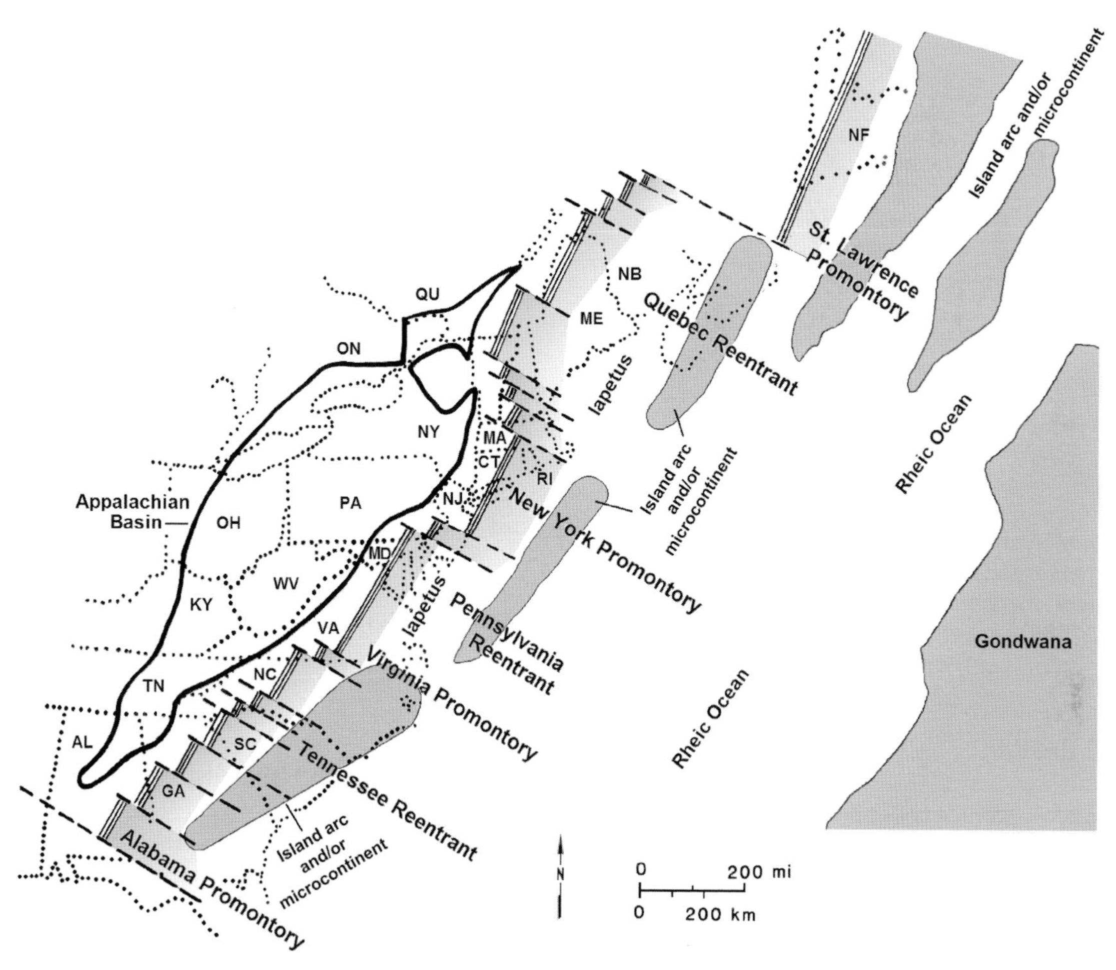

Figure 1. A generalized map of eastern United States and Maritime Canada showing preserved parts of the Appalachian foreland basin relative to promontories and reentrants on the late Precambrian – Early Cambrian rifted margin of Laurentia. Hypothetical island arcs and/or microcontinents are shown off the coast because accretion of such bodies to Laurentia during closure of the Iapetus Ocean constituted the Caledonian orogenic supercycle (Taconian and Salinic orogenies). Closure of the Rheic Ocean during convergence of Gondwana with Laurussia constituted the Variscan-Hercynian orogenic supercycle (Acadian/Neoacadian and Alleghanian orogenies).

Because dark shales are distinctive lithologies and commonly exhibit characteristic gamma-ray signatures, their distribution is easily mapped in the subsurface. Moreover, many dark-shale units are separated by nondark-shale clastic or carbonate units, making unit identification even more obvious; however, once these intervening units have thinned, pinched out, or changed colors, individual organic- and clastic-rich units commonly merge into even larger, composite, dark-shale units on adjacent parts of the craton. Units like the Devonian and/or Mississippian Ohio, Chattanooga, Woodford, Huoy, and New Albany shales are distal composites of several thinned Appalachian dark-shale units in which intervening clastic and carbonate units have pinched out or changed colors (Ettensohn et al., 1988). In these composite, cratonic, dark-shale units, discerning included units requires much care in examining gamma-ray logs or—in surficial exposures—using artificial gamma-ray logs (Ettensohn et al., 1979) that can be correlated into the subsurface. Hence, the dark-shale distribution patterns used in this chapter were derived from thousands of subsurface gamma-ray correlations and stratigraphic charts (Patchen et al., 1985a, b); these correlations were subsequently informed by comparison with existing Devonian (Pepper and deWitt, 1950, 1951; Pepper et al., 1956; deWitt and Colton, 1959, 1978; Wallace et al., 1977, 1978; Kepferle et al., 1978; Roen et al., 1978; Wilson et al., 1981; Woodrow et al., 1988; Ryder et al., 2009) and Ordovician/Silurian (Ryder, 1991, 2002a, b, 2008; Ryder et al., 1992, 1996, 2008) stratigraphic cross sections through the basin. The geologic time scale of Gradstein et al. (2004) was used as a temporal framework for the shales.

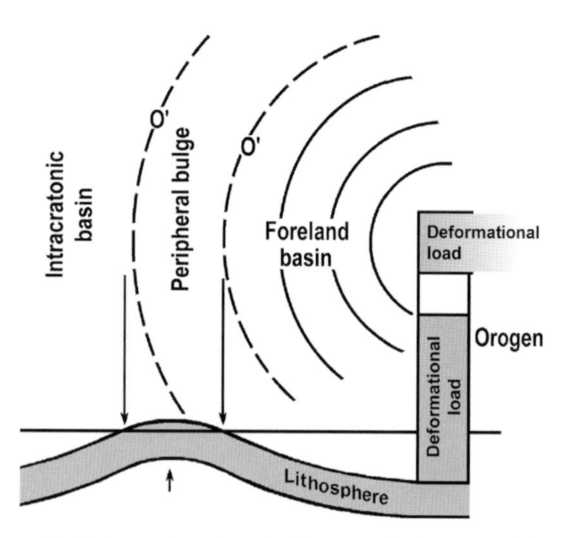

Figure 2. Cartoon showing the large-scale, long-wavelength deformational response of the lithosphere to loading in an adjacent craton-margin orogen. Modified from Quinlan and Beaumont (1984).

FORELAND BASINS

Discussion of orogenies commonly brings to mind the external parts of an orogen elevated by lateral thrusting, nappes, faulting, and folding, as well as those internal parts of an orogen affected at depth by metamorphism, plastic folding, and plutonism. However, other orogenic effects may extend more than a thousand kilometers beyond the actual orogenic highlands. The influence of these effects may extend far onto the craton or foreland, and through interaction with preexisting structures may greatly affect the nature of sedimentation there. In his seminal 1920 work on the nature of mountain building, Stille (1920) called these two aspects of tectonism, "alpinotype" and "germanotype," respectively, and development of foreland basins and related distal deformation is clearly a milder, more epeirogenic aspect of germanotype tectonism. Stille's terms are rarely used today, but in place of germanotype tectonism, the term "far-field tectonics" (Klein, 1994) has been substituted, and far-field processes have been reported at least 1300 km (808 mi) from the originating orogen (e.g., Ziegler, 1978).

In the dominantly compressive regimes that characterized Appalachian orogens, foreland-basin development is mostly thought to reflect the isostatic effects of flexural deformational loading. This loading results from the emplacement of subsurface (obducted crustal blocks or flakes) or surface (folds, thrust sheets, and nappes) loads on adjacent parts of the crust. To compensate for the load, the lithosphere deforms into a downwarping flexural foreland basin just cratonward of the alpinotype deformation and a subtle peripheral bulge, about as wide as the foreland basin, on the cratonward margin of the basin (Figure 2); both the bulge and basin migrate in space and time relative to the nature and position of the load (Figure 3) (Quinlan and Beaumont, 1984). Although several models exist for explaining the nature of the flexure relative to lithospheric rheology, we use the models of Quinlan and Beaumont (1984), Beaumont et al. (1987, 1988), and Jamieson and Beaumont (1988), based on the loading and unloading of a temperature-dependent viscoelastic lithosphere, because the predictions of these models accurately match the sedimentary sequences observed in the Appalachian Basin and elsewhere.

Emplacement of a subsurface load, which may result in very little or no surface deformation and relief (Figure 3A), marks the initiation of subsidence and foreland-basin formation; thrusting and nappe formation are typically later secondary sources of loading (Figure 3B). Deformation provides most of the load, whereas sediment loading is ancillary.

TECTOPHASES AND FORELAND-BASIN CYCLICITY

The term tectophase was first used by Johnson (1971) to describe a pulse of orogenic activity that occurred during a distinct time span—time spans that he recognized were almost wholly coeval with certain Devonian transgressive and regressive cycles on parts of North America. In fact, Johnson (1971) was able to correlate these Devonian sedimentary cycles in the Appalachian Basin with three phases of Acadian orogeny recognized by Boucot et al. (1964) in Maine based on times of intrusion, metamorphism, and clastic-wedge formation in the orogen. Although Johnson's cycles were based on successions of transgressive and regressive lithologies, the association between cyclic dark shales and orogeny via lithospheric flexure was not made until 1985 in two articles on the Catskill delta complex (Ettensohn, 1985a, b). In these articles, Ettensohn showed that Appalachian Basin, Devonian–Mississippian, dark shales occur as parts of third-order unconformity-bound sequences related to rapid subsidence that accompanied phases of major structural deformation concentrated at changing locations along the Acadian/Neoacadian orogen. These locations, moreover, coincided with promontories (Figure 1), or structurally controlled protuberances on the continental margin left from earlier Iapetan rifting (Thomas, 2006), which successively localized convergence for a distinct time interval and generated an episode of orogeny or tectophase. Ettensohn (1985a, 1987) also showed that the diachronous nature of orogeny was related to distinct

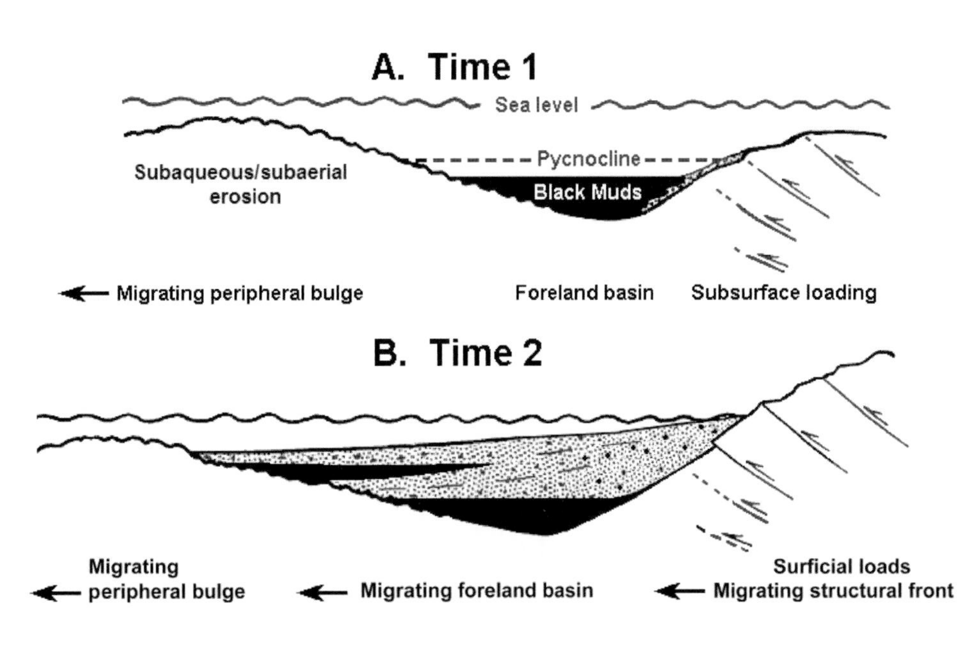

Figure 3. A schematic diagram showing the relationships between foreland-basin generation, sediment infill, and deformational loading. (A) Basin-bulge generation and subsurface loading with little clastic influx; most dark shales develop during this short part of the cycle. (B) Situation after two loading-relaxation cycles during which deformation, basin, and bulge have migrated cratonward. The pycnocline reflects thermohaline density stratification in deeper basins with declining O_2 content. Dark color = organic-rich muds; large stipple = coarser clastic sediments; wavy lines = unconformities. Modified from Ettensohn (1997).

periods of convergence with successive promontories along a scissors-like or obliquely convergent subduction zone and that dark-shale depositional basins in the foreland tracked the progress of convergence. This work not only provided the structural and tectonic bases for Johnson's tectophases but also provided a direct mechanism for explaining the connection between dark-shale deposition and episodic tectonism. Although all Appalachian tectophases were apparently mediated by continental promontories, it is now understood that most orogenies progress via short-lived pulses of convergence and deformation, or tectophases, which persist for several million to a few tens of millions (10^6–10^7) of years (Jamieson and Beaumont, 1988).

From the Early Ordovician (~472 Ma) through the Permian (~251 Ma), the Appalachian margin was the site of four nearly continuous orogenies, organized into two larger, second-order (10^7–10^8 yr), orogenic supercycles: the Middle Ordovician–Early Devonian Caledonian cycle (Tippecanoe sequence, 472–411 Ma) and the Early Devonian–Permian Variscan-Hercynian cycle (Kaskaskia and Absaroka sequences, 411–251 Ma) (Figure 4). The Caledonian cycle includes the Taconian and Salinic orogenies and reflects the closure of the Iapetus Ocean, an oceanic basin between the rifted margin of Laurentia and a series of outboard rifted Grenvillian blocks, peri-Gondwanan terranes, and island arcs (Figure 1); in contrast, the Variscan–Hercynian cycle includes the Acadian/Neoacadian and Alleghanian orogenies (Figure 4) and represents the closure of the Rheic Ocean, an oceanic basin between the newly accreted margin of Laurussia and Gondwana (Figure 1). These orogenies not only generated nearby sediment sources in adjacent orogenic highlands but also generated enough foreland-basin accommodation space because of the isostatic effects of flexural deformational loading to accumulate more than 13,700 m (44,948 ft) of sediment in the Appalachian Basin. This sediment accumulated in 13 third-order unconformity-bound cycles (Figure 4), each of which shows a similar progression of lithologies, includes dark shales, and corresponds to an Appalachian tectophase (Ettensohn, 2005).

Tectophase cycles in the Taconian orogeny reflect the convergence of island arcs and terranes closing obliquely from south to north; the Acadian/Neoacadian orogeny, in contrast, reflects the convergence of terranes closing obliquely from north to south. During each orogeny, oblique convergence at successive promontories generated phases of deformational loading that produced coeval periods of foreland-basin subsidence, substantial and rapid enough to enable dark-shale deposition. Hence, each convergence event, or tectophase, generated its own dark-shale basin, and these basins migrated in time, tracking the course of convergence. The dark shales, moreover, are part of a distinct sequence of lithologies that reflects the generation and relaxation of flexural stresses, causing adjacent parts of the foreland to rise or subside in long-wavelength undulations apparent in the foreland basin and bulge (Figures 2, 3). These undulations and the related phases of erosion and basin infilling are clearly responses to loading-related processes. These processes are outlined below, emphasizing the occurrence and significance of dark shales.

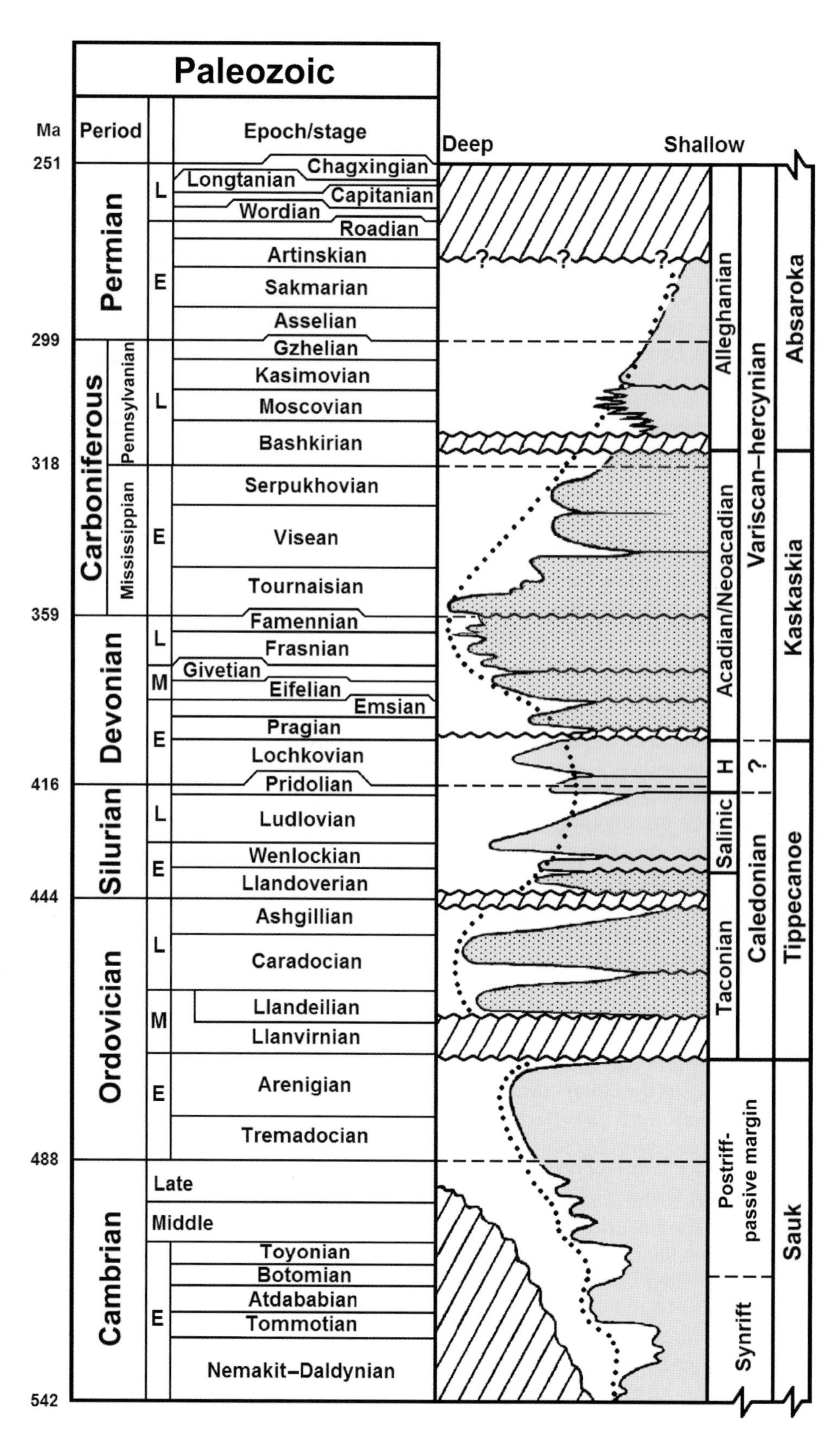

Figure 4. Paleozoic time scale showing the occurrence of 13 unconformity-bound tectophase cycles as a relative sea level curve from the Appalachian Basin. Third-order tectophase cycles occur as parts of two orogenic supercycles (Caledonian and Variscan–Hercynian) and are shown relative to Appalachian orogenies and Sloss sequences. Curves for the Taconian and Acadian/Neoacadian tectophases examined in this chapter are infilled with coarse stipple. Undulating lines = unconformities; diagonal hatching = missing time on unconformities; H = Helderberg cycles. Modified from Ettensohn (2005).

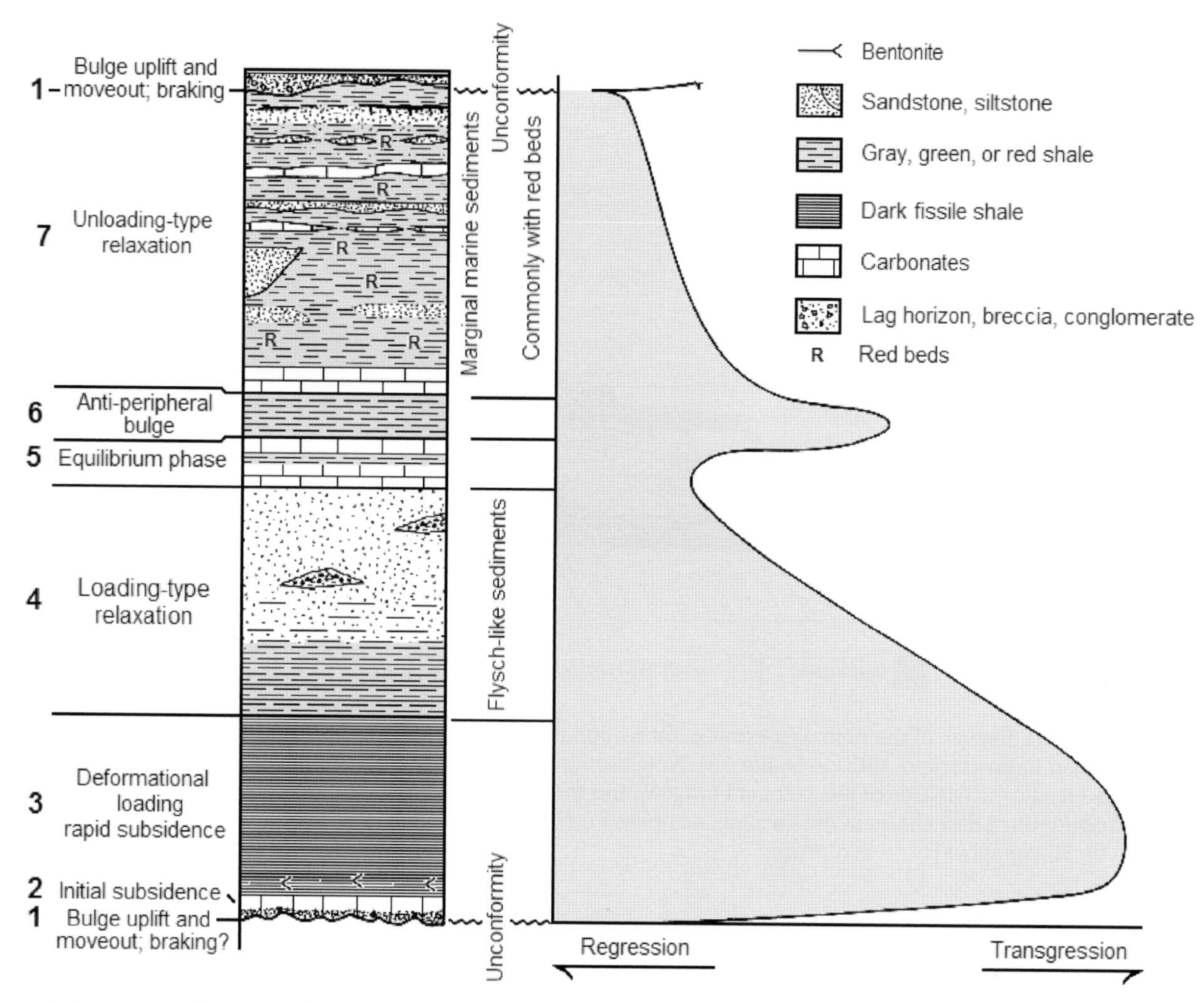

Figure 5. Parts of an ideal tectophase cycle for subduction-type orogenies and the related sequence of flexural events, lithologies, and relative sea level. Modified from Ettensohn (1994).

FLEXURAL, FORELAND-BASIN, PROCESS-RESPONSE MODELS

A fully developed tectophase cycle consists of seven parts and represents a complete transgressive-regressive cycle bound by unconformities (Figure 5) (Ettensohn, 2005, 2008). Although the cycles are primarily limited to foreland basin, the overall transgressive-regressive regime inherent in each cycle should be apparent in platform and ramp settings on adjacent parts of the foreland, mainly in the form of changing carbonate depositional environments (Figures 6, 7). Each cycle begins with the transfer of a load from the margin of the subducted plate or from the subducting plate onto more cratonward parts of the subducted or subducting plate, generating a distal bulge and a rapidly subsiding retroarc or peripheral basin, respectively (Figures 2, 3). If the collision and transfer are synchronous along the entire continental margin, the basin and bulge will only migrate perpendicular to the strike of the orogen. However, if convergence is oblique and diachronous along strike, as was typical along the Laurentian/Laurussian (Appalachian) margin, basin-and-bulge migration will not only exhibit a component of migration perpendicular to strike but also a major component of shift parallel to strike (Ettensohn, 1985a, 1987). Accordingly, lithologies in each cycle will also migrate in space and time (Figures 6, 7), tracking the progress of convergence along the margin. Initial parts of each tectophase are characterized by major loading, which results in a rapidly subsiding foreland basin, so that lower parts of each cycle are transgressive and the basin is mostly underfilled (Figures 3A, 5). Once active loading has abated, the lithosphere responds by relaxing stress through a series of stages that result in a basin that is filling to overfilled with postorogenic clastic wedges; these latter parts of the cycle are mostly

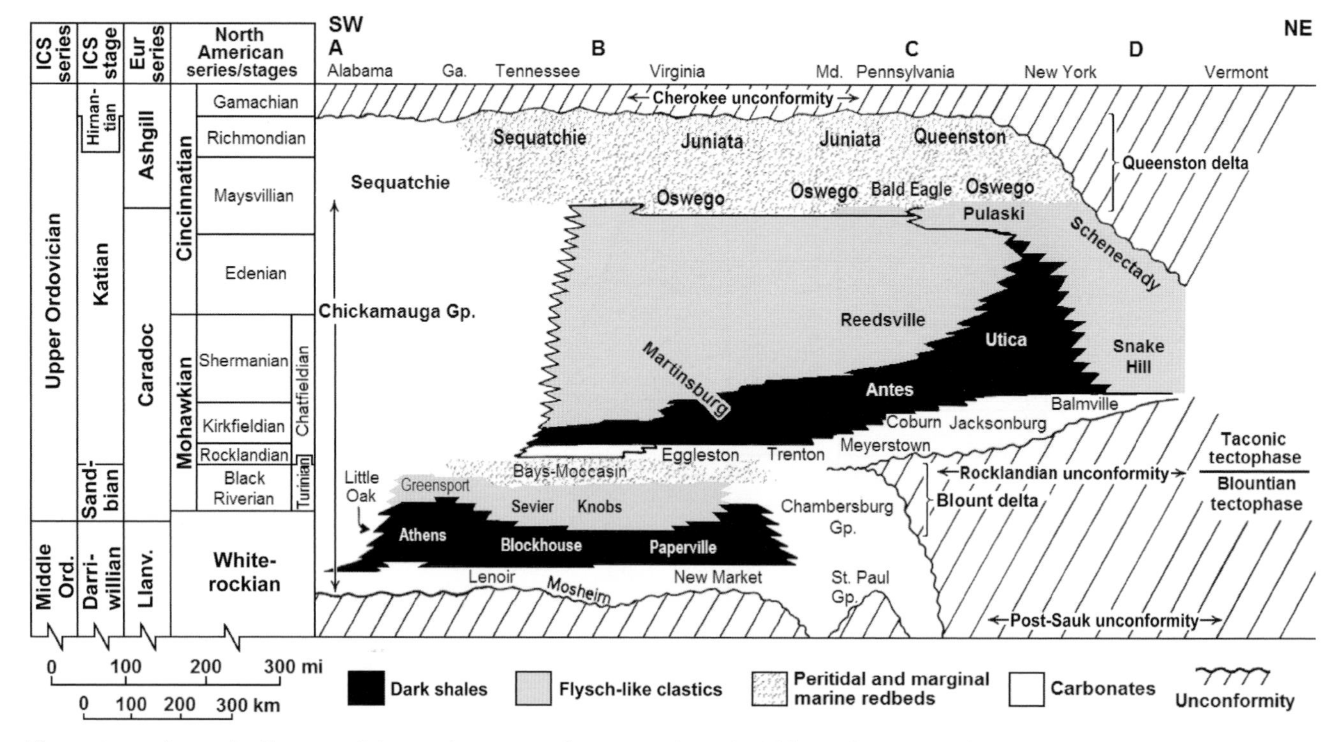

Figure 6. A schematic diagram of the northeast–southwest section of Middle and Upper Ordovician rocks, parallel to the strike of the Appalachian Basin (section ABCD in Figure 12). Note the repetition of two unconformity-bound tectophase cycles for Blountian and Taconic tectophases of the Taconian orogeny and the northeastward migration of successive tectophase cycles and of Martinsburg-Utica dark shales in the Taconic tectophase cycle. Modified from Ettensohn (1991). ICS = International Commission on Stratigraphy; Eur. = European.

regressive with a small transgressive subcycle (Figures 3B, 5). However, it is only in its early, deep, underfilled stages that the foreland basin provides an ideal setting for dark-shale accumulation (Figure 3A); in its later stages, the basin fills with postorogenic clastic wedges (Figure 3B) that help protect the underlying organic-rich rocks and provide many settings for future reservoir development.

Part 1: Unconformity Formation

Once subduction begins and a deformational load accumulates on the continental margin, the lithosphere responds by generating a compensating foreland basin and peripheral bulge (Figure 2). As convergence continues and the load moves cratonward, the basin and uplifted bulge also migrate cratonward, and erosion, either subaerial or submarine, on the uplifted bulge generates the bounding lower unconformity (Figure 3A).

Part 2: Transgressive Shallow-water Deposition

After bulge moveout, continued deformational loading induces rapid subsidence of the foreland basin. Subsidence is typically so rapid that shallow transgressing seas are soon replaced by deeper water deposits. Hence, part 2 of the tectophase cycle, transgressive shallow-water deposition (Figure 5), is typically very short-lived and represented by a thin carbonate or clastic unit or a condensed lag deposit that records rapid flooding and reworking of the underlying unconformity surface. The nature of the deposit that does finally develop depends on the climate, presence or absence of preuplift residual debris, and the pace of subsidence and transgression. In subtropical settings with little clastic influx, carbonates typically develop, and in carbonate settings with slow to moderate rates of subsidence, an entire carbonate transgressive sequence from intertidal to deeper subtidal may develop. For example, the Early–Middle Ordovician transitions into the Taconian orogeny (lower Chickamauga Group of Tennessee and the Mosheim, Lenoir, New Market, and related formations of Virginia; Figures 6, 7), as well as the Onondaga and Tully limestones (Figure 8) in the Acadian orogeny, represent more gradual, complete, transgressive transitions. In contrast, transitions from the peritidal Moccasin and Bays formations into the deeper water Martinsburg Shale (Figures 6, 7) are very abrupt, and the presence of very thin, sandy, lag horizons, locally only millimeters or centimeters thick, at the bases of Upper Devonian

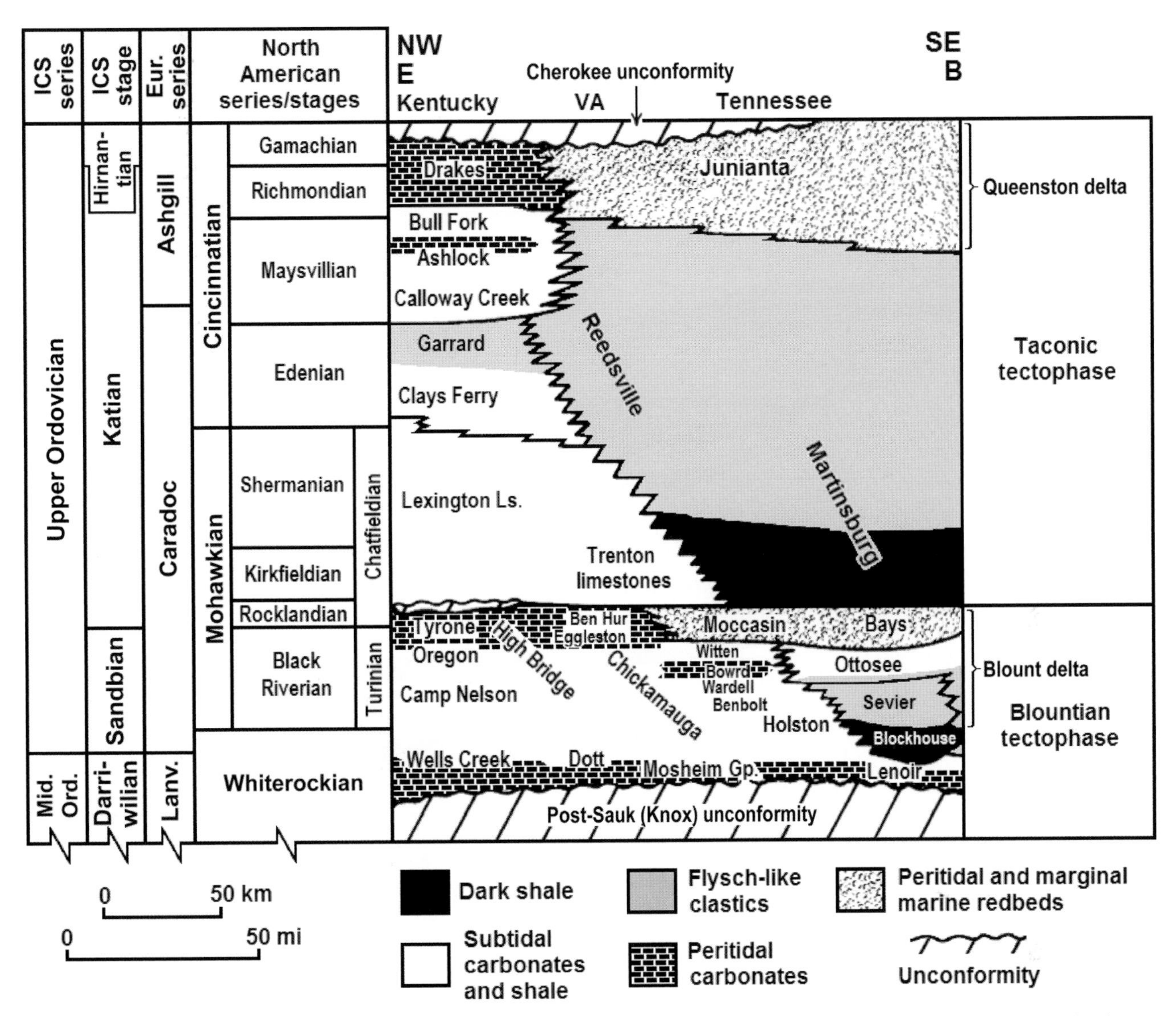

Figure 7. A schematic diagram of the northwest–southeast section of Middle and Upper Ordovician rocks, perpendicular to the strike of the Appalachian Basin (section EB in Figure 12). Note the repetition of tectophase cycles and cratonward migration of dark-shale basins in time and space. Modified from Ettensohn (1991). ICS = International Commission on Stratigraphy; Eur. = European.

and Lower Mississippian dark shales of the third and fourth tectophases (Figure 8) suggests very rapid subsidence and transgression.

Part 3: Rapid Subsidence and Dark-mud Sedimentation

As deformational loading continues, the third part of the cycle, dark-mud sedimentation, ensues. Initial loading is typically related to crustal shortening and load transfer along a steep basement ramp during subduction. During early development of an orogen, it is likely that such ramps can accommodate up to 20 km (12.4 mi) of vertical deformation without creating major subaerial topography or source areas (Jamieson and Beaumont, 1988), so that much of the initial deformation occurs in the subsurface or in subaqueous environments that generate little or no subaerial relief (Figure 3A). As a result, no major source of externally derived sediment is available to fill the subsiding foreland basin during early parts of the orogeny, so that once shallow-water deposits are drowned, the foreland basin becomes underfilled and sediment starved. In the absence of nearby clastic sources, organic matter from the water column, as well as suspended clays and silts from wind-blown and distal terrestrial sources, becomes the predominant input. Because of ongoing subduction, wind-blown ash, preserved as bentonites, may also be common in this part of the section (Figure 5). Furthermore, Ettensohn

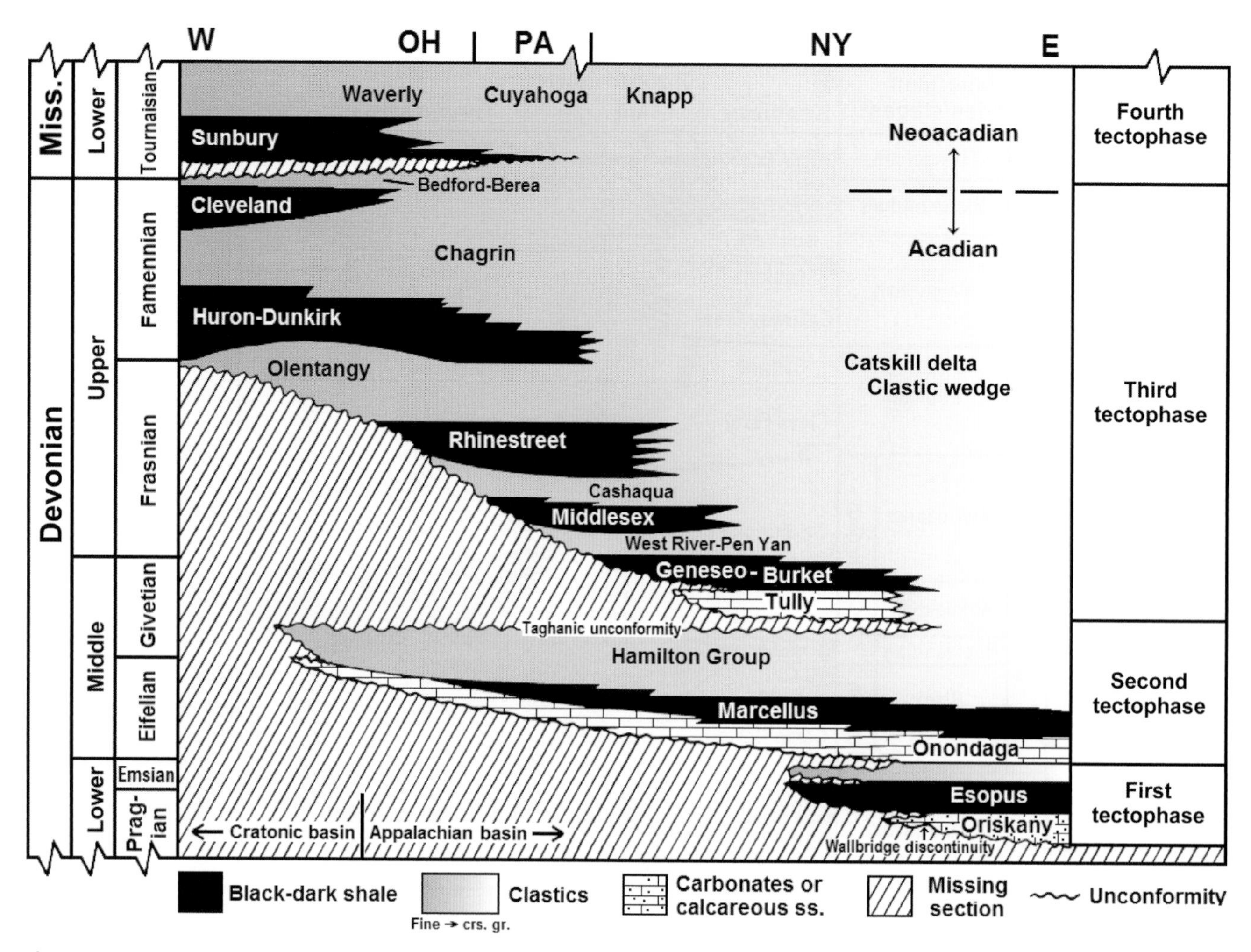

Figure 8. A highly schematic east–west section from east-central New York to north-central Ohio in the northern Appalachian Basin showing the relative timing and distribution of dark-shale basins formed during the four tectophases of the Acadian/Neoacadian orogeny. Modified from Ettensohn (1987). crs. gr. = coarse grained (color gradation in clastic symbol box).

(1997) has suggested that the paleoclimatic and paleogeographic settings of many foreland basins may create situations that promoted upwelling and sequestering of terrestrial nutrients, which greatly enhance organic productivity. All the while, the basin is experiencing rapid subsidence with which the scant sedimentation cannot keep pace. Once the basin attains depths more than 150 m (>492 ft), a stratified water column, or pycnocline, typically develops (Rhoads and Morse, 1971), and the predominant organic matter is buried and preserved as dark muds on the resulting deep oxygen-poor (dysoxic and anoxic) basin floor. Hence, early parts of a typical subduction-related foreland basin are nearly always disposed toward dark-shale deposition.

Dark-shale sedimentation is most prominent in the rapidly subsiding proximal and central parts of a foreland basin, and although subsidence-related transgression also occurs in distal parts of the basin and on adjacent parts of the foreland, the effects of subsidence are reduced. Hence, in more distal parts of the basin and adjoining foreland, the dark-shale sequences typically grade into coeval transgressive carbonates and lighter colored shales and coarser clastic deposits (Figures 6, 7).

During times of maximum subsidence and dark-shale deposition in the foreland basin, flexural interactions with adjacent intracratonic basins may contribute to coeval subsidence and dark-shale deposition in these basins as well. Rheological modeling shows that if basins and intervening arches are considered to be long-wavelength flexures of the lithosphere, continued cratonward loading of the lithosphere from the continental margin can lead to basin yoking through destructive interference and the lowering of intervening bulges (Figure 9) (Quinlan and Beaumont, 1984). This effectively leads to the migration of dark-shale depositional conditions from the foreland basin into adjacent intracratonic basins, and both modeling and mapping dark-shale distribution (Beaumont et al., 1988; Coakley

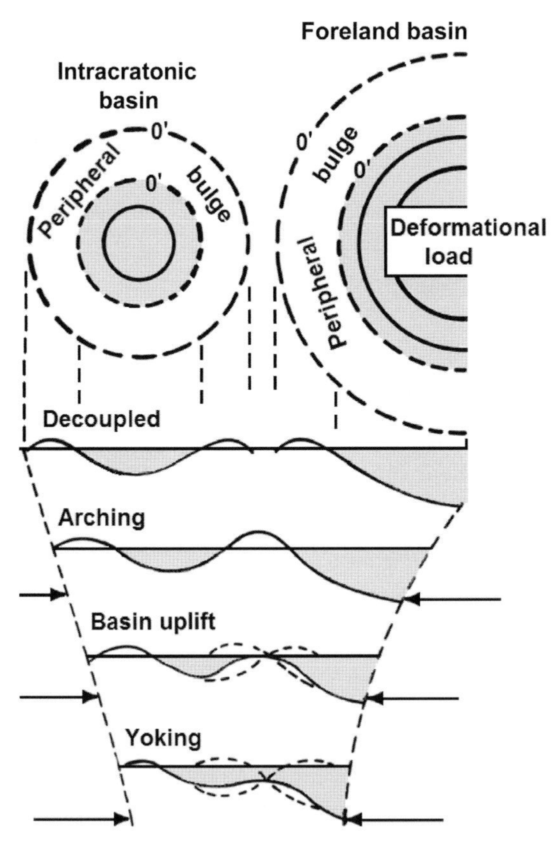

Figure 9. A schematic diagram showing hypothetical constructive-destructive interactions between long-wavelength deflections of the lithosphere that lead to the yoking of foreland and intracratonic basins during deformational loading in a craton-margin orogen. Modified from Quinlan and Beaumont (1984).

and Gurnis, 1995) show that such yoking did occur during some early Appalachian tectophases.

Part 4: Loading-type Relaxation and Flysch-like Sedimentation

As long as active orogeny and cratonward movement of deformational loads continue, subsidence should be sufficient to ensure deposition of dark muds in a cratonward-expanding foreland basin. However, once active thrusting declines and tectonic quiescence ensues, the deformational load becomes static. Lithospheric response to the now static load is a relaxation of stress such that the proximal foreland basin subsides while the peripheral bulge is uplifted and migrates toward the subsiding load (Figure 10A). Moreover, by this time, convergence has generated a substantial, subaerial load in the form of high, alpinotype, fold-thrust belt deformation, and surface drainage nets have had time to develop. Consequently, coarser grained clastic debris are eroded from adjacent tectonic highlands and transported into the deepening foreland basin in the form of deeper water deltaic deposits, turbidites, contourites, debris flows, and storm deposits (Figure 3B). Because basin sedimentation at this stage is rapid and includes darker, deeper water, commonly immature, siliciclastic components, sediments in part 4 of the cycle are typically flysch-like (Figure 5). As the basin fills with flysch-like sediments, the bulge is uplifted and shifts basinward (Figure 10A), possibly generating a local or regional unconformity that truncates the flysch-like infill and/or regressive carbonates that form atop the bulge, depending on the nature of sea level. Ettensohn (1994) has called this process "loading-type relaxation," and it generates a large part of the generally shallowing-upward, regressive, foreland-basin infill (Figures 5–7). At this stage, the basin will typically have two paleoslopes, but most of the basin infill comes from the adjacent, rapidly eroding, unloading tectonic highlands.

Part 5: Equilibrium and Regressive Shallow-water Blanket of Carbonates and/or Shales

With erosion of adjacent tectonic highlands, the rate of clastic influx begins to exceed subsidence rates until the basin is filled to capacity or overflows. The infilled foreland basin, combined with greatly lowered source areas and a waning supply of clastic sediment, sets the stage for the deposition of an extensive blanket of shallow-water carbonates or mixed carbonates and shales in part 5 of the cycle (Figure 5). These carbonates and shales mark consummation of the regression and shallowing that began with the cessation of active loading during loading-type relaxation. At this stage, the carbonate and/or shale blanket may briefly overstep large parts of the foreland basin and adjacent areas because the filled basin and lowered source areas briefly approach elevational equilibrium, facilitating the widespread expansion of shallow seas.

Part 6: Peripheral Sag and a Thin, Infilling, Transgressive Sequence

Approximate elevational equilibrium, however, is short-lived because the nearly leveled area of former alpinotype deformation and adjacent parts of the foreland basin begin to rebound upward in response to the lost load (Figure 10B), the beginning of unloading-type relaxation. As a result, previously eroded upland areas and nearby parts of the foreland basin respond through compensating uplift and an anti-peripheral bulge, or peripheral sag, which moves toward the rebounding area (Figure 10B) (Beaumont et al., 1988). The peripheral sag is reflected in a short-lived transgressive sequence of

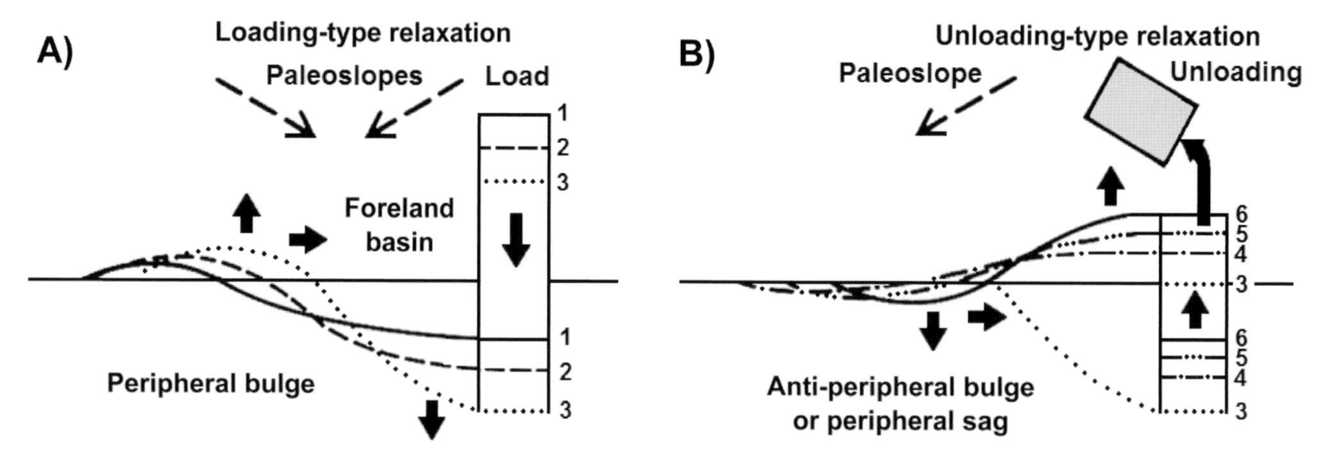

Figure 10. Schematic diagrams showing successive flexural responses to lithospheric stress relaxation. Modified from Beaumont et al. (1988). (A) Loading-type relaxation, wherein the now static load causes subsidence-type relaxation in the basin in part 4 of a typical cycle (Figure 5). (B) Unloading-type relaxation, wherein erosional unloading causes rebound and erosion of old uplands and proximal parts of the foreland basin in parts 5 to 7 of a typical cycle (Figure 5).

open-marine shales or mixed shales and carbonates, which is part 6 in a complete cycle (Figure 5).

Part 7: Unloading-type Relaxation and Forming a Marginal-marine to Terrestrial Clastic Wedge

Part 7 in the cycle reflects the uplift and erosion of the rebounding load (Figure 10B), which consists of formerly beveled tectonic highlands and cannibalized parts of the foreland basin; it is manifested in the basin by a cratonward-prograding wedge of marginal-marine to terrestrial siliciclastic sediments, which are commonly labeled post-orogenic, molasse-like, or deltaic (Figure 5). Because parts of the new source area have already been through earlier phases of erosion and deposition, sediments eroded from the rebounded area will mostly be finer grained, consisting of fine-grained sand, silt, silty muds, argillaceous muds, and muddy carbonates, commonly interspersed with accumulations of organic matter. Because flexural processes at this stage start from conditions of approximate elevational equilibrium at or near sea level, a single cratonward-dipping paleoslope is formed, and it supports a similarly dipping clastic-wedge complex that may be several thousand meters thick, extend hundreds of kilometers across the basin, and parallel basin strike for more than 1000 km (>621 mi); these are the so-called "tectonic delta complexes" of Friedman and Johnson (1966). Low gradients, proximity to the sea, poorly developed systems of energy dispersal, local topographic and structural features, and autogenic sedimentary fluctuations typically generate complex vertically and laterally changing depositional environments in this setting. Hence, this part of the cycle is characterized by a facies mosaic, commonly composed of redbeds, paleosols, coals, and thin carbonates in close association with various siliciclastic components (Figure 5). Redbeds are particularly common (Figure 5), and because of their common association with the terrestrial parts of deltas, many Appalachian Basin redbed units in this part of the cycle, like the Queenston, Bloomsburg, Catskill, Bedford-Berea, Pennington-Mauch Chunk, and Conemaugh, have been called deltas.

Model Implications

Although the typical cycle has seven parts (Figure 5), not all cycles exhibit the complete sequence of parts, especially at the cycle top, because of erosion at the base of the subsequent cycle, because initiation of the following cycle began before the earlier one was complete, or because the area of observation during deposition was too distant from the center of active deformation and subsidence. It is also common for subcycles of dark-shale and flysch-like clastic sediments to repeat several times in a single tectophase (Figure 8, third tectophase), perhaps reflecting the episodic movement of major thrust systems during the tectophase. In fact, the simplest cycles and subcycles in the Appalachian Basin consist of an unconformity or subtle discontinuity, dark shales with a very thin, basal, lag horizon, and an overlying flysch-like sequence (Figure 8).

The duration of tectophase cycles in the Appalachian Basin ranges from about 4 to 41 m.y. (Figure 4), although the single Mississippian cycle, at 41 m.y. in length, is unusually long and probably represents an atypical tectonic regime that will be discussed later. Based on the extrapolation of absolute ages from conodont zones (e.g., Webby, 1995; Gradstein et al., 2004; Engelder et al.,

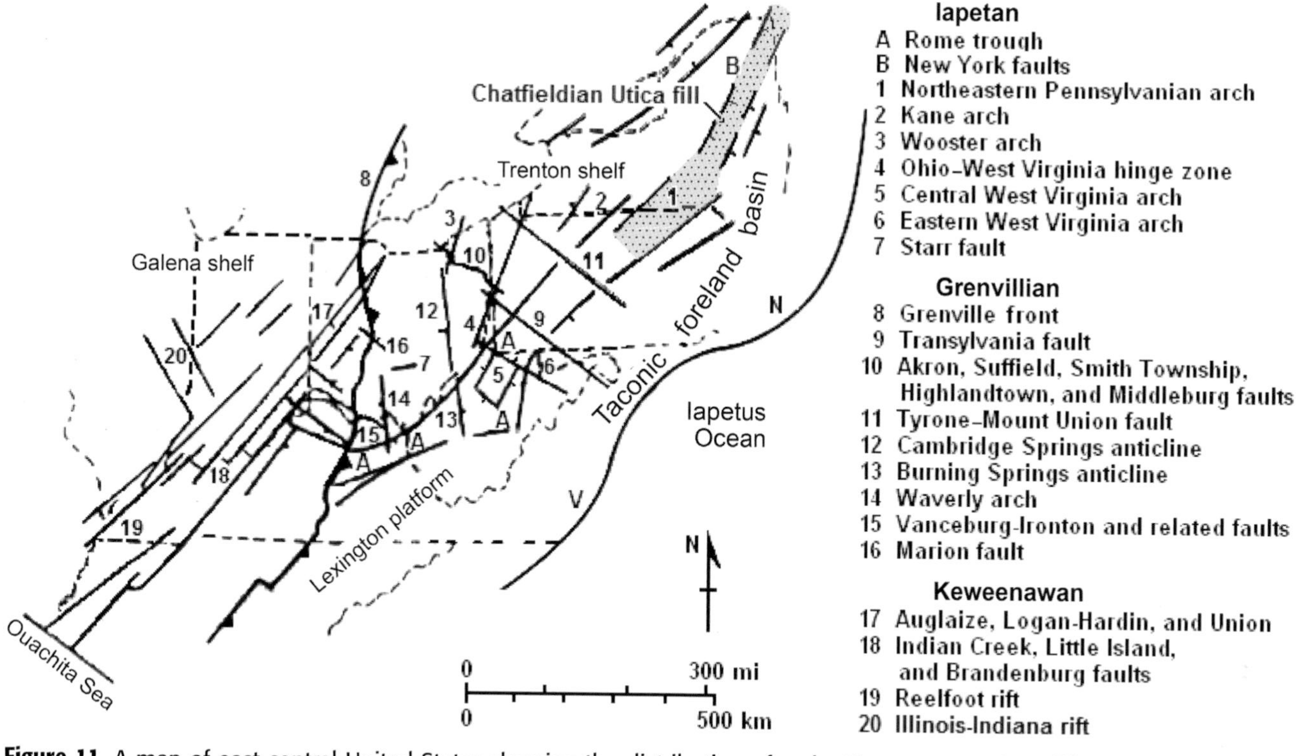

Figure 11. A map of east-central United States showing the distribution of major Keweenawan, Grenvillian, and Iapetan basement structures in the area of the Appalachian foreland basin and adjacent parts of the foreland relative to the New York (N) and Virginia (V) promontories. The stippled areas in New York and Pennsylvania reflect structures that were reactivated to form the initial Utica foreland basin and its dark shales during the Taconic tectophase (see Figures 13, 14A, 15B). Modified from Ettensohn et al. (2002).

2009), the dark-shale parts of a typical cycle range in duration from about a few hundred thousand years to 1.5 m.y., although the 5- to 6-m.y. duration of the Sunbury dark shale is considerably longer than, but proportional to, the duration of its tectophase. These estimates suggest that the dark-shale part of a cycle is relatively rapid, and that most of the time in a tectophase cycle is expended in the various flexural and sedimentational aspects of relaxation (Figure 5).

FAR-FIELD EFFECTS

Aside from the broad, long-wavelength, lithospheric undulations that give rise to foreland and intracratonic basins (Figures 2, 10), the upper parts of the lithosphere are characterized by widespread brittle deformation in the form of faults that reflect previous episodes of lithospheric tension and compression. Before any Paleozoic Appalachian deformation, the eastern parts of Laurentia had already been deformed by Keweenawan extension (~1.1 Ga), Grenvillian compression (~1.0 Ga), and Iapetan extension (~740 Ma), and each event left zones of structural weakness with particular orientations (Figure 11). More distal parts of the craton show the effects of Keweenawan deformation with mostly southwest-to-northeast-trending basement dislocations, whereas areas in and adjacent to the foreland basin show the effects of Grenvillian and Iapetan deformation with predominantly northwest-to-southeast and northeast-to-southwest orientations, respectively (Figure 11) (e.g., Denison et al., 1984; Shumaker, 1986; Furer, 1996). These and other studies show that the basement is inhomogeneous and rife with dislocations of various ages and that these dislocations remain zones of weakness, which can be reactivated when subjected to new stresses. Large-scale compression and tension and related deformational loading at the Appalachian margin during the Paleozoic closure of the Iapetus and Rheic oceans were just such stresses, and not only were they involved in the formation and yoking of foreland and intracratonic basins but also in various far-field structural displacements, or germanotype deformation, more than 1300 km (808 mi) from the originating orogen (Mitrovica et al., 1989; Coakley and Gurnis, 1995). Similar displacements may occur both in more proximal areas, where they can enhance foreland-basin development and partitioning and in more distal areas where they can form large fault-related basin complexes in which dark-shale source rocks can accumulate.

TACONIAN OROGENY: ORDOVICIAN–SILURIAN FORELAND-BASIN AND FAR-FIELD DEVELOPMENTS

The Taconian orogeny marks initiation of the closure phase of the Iapetan or Appalachian Wilson cycle and the subsequent tectonic mobilization of the Laurentian Appalachian margin. Although orogeny did not begin until the Early–Middle Ordovician transition at about 472 Ma, island arcs and terranes that would later collide with the margin were already present by the Late Cambrian (Figure 1). On the foreland, the beginning of the orogeny is marked by the widespread post-Sauk or Knox unconformity (Figures 6, 7). This unconformity also marks the transition from a passive to active margin, inception of the Appalachian foreland basin, and the tectonic/sedimentary differentiation of the foreland from a passive-margin, shallow-water, carbonate bank (Knox, Beekmantown, and equivalent units) to a mosaic of deep-to-shallow clastic-to-carbonate units that reflect foreland flexure and the far-field reactivation of many basement structures. The Taconian foreland basins and their substantial dark-shale infillings (Figures 6, 7, 12, 13) were the direct results.

Taconian Foreland Basins

Blountian Tectophase and the Sevier Foreland Basin

In the southern and central Appalachians, the Taconian orogeny was diachronous and heterogeneous in kinematic style and reflects a progressive northeastward shift in convergence through time. Accordingly, Taconian foreland basins show a similar northeastward migration in space and time (Figures 6, 12). The first Taconian tectophase represents convergence of an island arc, now represented by the Dahlonega terrane, and/or crustal fragments and microcontinents (Faill, 1997; McClellan et al., 2005) at the Alabama and Virginia promontories (Figure 12). This part of the Taconian orogeny has been called the Blountian tectophase, and it formed a deep narrow basin from Alabama to northern Virginia, the first Appalachian foreland basin, now called the Sevier Basin (Figure 12). The basin hosts a complete tectophase cycle (Figures 6, 7) that begins with the post-Sauk or Knox unconformity and is followed by a transgressive carbonate sequence represented by the lower Chickamauga and St. Paul groups and basal carbonate units like the Lenoir, Mosheim, New Market, and Lincolnshire formations. The carbonates abruptly give way to deep-water, dark, graptolitic shales represented by the Athens, Blockhouse, Paperville and equivalent dark shales in the Rockmart, Columbiana, Liberty Hall and Rich Valley formations. These dark shales typically range in thickness from 100 to 1300 m (328–4265 ft) and grade upward into flysch-like clastic sediments of the Sevier, Knobs, Greensport, and equivalent units. The flysch-like units in turn grade upward into the peritidal and marginal-marine, carbonate- and clastic-rich redbeds of Bays and Moccasin formations (Figures 6, 7); together, clastic units overlying the dark shales have been termed the Blount delta. In the section perpendicular to the basin (Figure 7), the regressive-to-transgressive sequence inherent in the upper Benbolt-to-Witten formations represents a migrating peripheral bulge and peripheral sag before unloading-type relaxation (Figures 5, 10B). Poor development of the Blountian tectophase cycle behind the New York promontory, represented by thin Black River and lower Trenton (lower Rocklandian) carbonates (Figure 13), reflects the relatively great distance between New York and southern locus of Blountian tectonism.

Taconic Tectophase and the Martinsburg Foreland Basin

Although the second, or Taconic tectophase, of the Taconian orogeny was concentrated at the New York promontory, convergence began at the Virginia promontory and is reflected in the spatial overlap of the Middle Ordovician (Blountian) Sevier and the Late Ordovician (Taconic) Martinsburg foreland basins (Figure 12). Docking at the Virginia promontory was followed shortly by collision of the Shelburne Falls and Bronson Hill arcs at the New York promontory (Karabinos et al., 1998), where the best manifestations of the Taconic tectophase occur. The initiation of these collisions is represented by the Rocklandian unconformity (Figures 6, 7, 13), which is best developed in the foreland basin behind the New York promontory. The initial transgressive carbonates (Martinsburg, Eggleston, Meyerstown, Jacksonburg, and Balmville; Figure 6) are relatively thin and, with overlying Martinsburg, Antes, and Utica dark shales, become progressively younger toward the northeast (Figure 6), tracking the northeastward advance of orogeny. Bentonites in lower parts of this succession from Alabama to New York also mark the inception of the tectophase. Compared with the Sevier Basin, however, the Martinsburg Basin shows a broader, more cratonward development and progressive northeastward migration (Figure 12); it also exhibits a complete seven-part tectophase cycle (Figures 6, 7, 13).

In the Martinsburg Basin, dark shales are the most prominent constituents of the tectophase cycle with thicknesses ranging from 300 to 700 m (984–2297 ft); however, in the most proximal parts of the basin behind the New York promontory in eastern Pennsylvania,

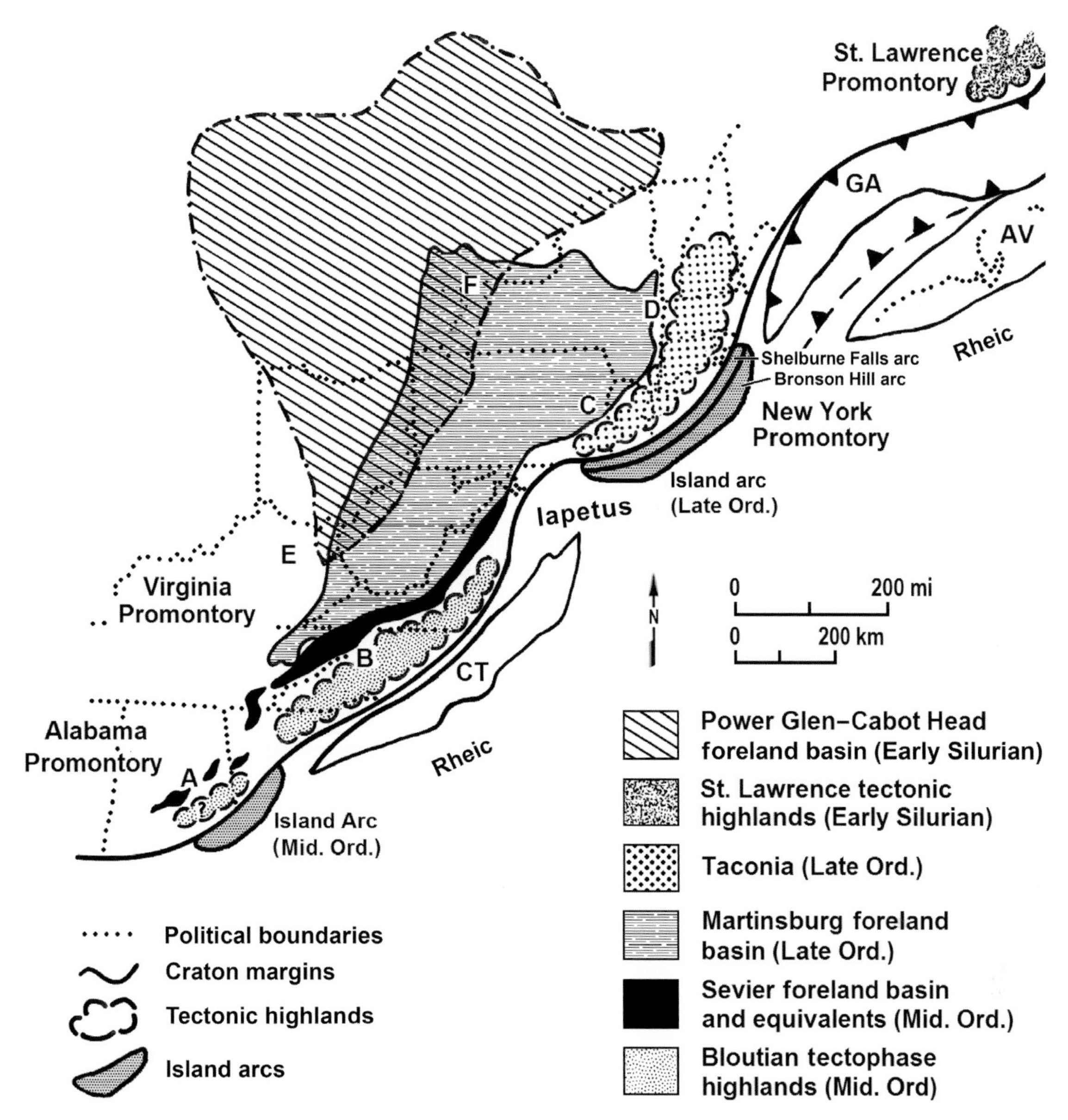

Figure 12. A schematic map showing the relative positions of Taconian tectonic highlands, migrating dark-shale basins, continental promontories, as well as colliding arcs and peri-Gondwanan microcontinents during the Middle Ordovician to Early Silurian Taconian orogeny. The extent of successive foreland basins (Sevier, Martinsburg, and Power Glen-Cabot Head) reflects the distribution of dark shales in the lower parts of each tectophase cycle. CT = Carolina terrane; GA = Ganderia; AV = Avalonia. Modified from Ettensohn (2008).

thicknesses ranging from 2400 to 3800 m (7874–12467 ft) are more typical (Patchen et al., 1985a, b). In more proximal eastern and distal southern parts of the basin, interbedded clastic and carbonate units are common, but the most organic-rich shales occur in the northern part (Utica Shale) of the basin in central and western New York and adjacent parts of Pennsylvania; these areas were more distal from major clastic sources and yet subject to substantial subsidence behind the area of greatest Taconic deformation and loading, the New York promontory. Moreover, paleocurrent analyses (McBride, 1962; McIver, 1970) indicate that the northern parts of the basin, where the Utica Shale was deposited, were partitioned into a separate sub-basin by reactivated Grenvillian basement structures (Figure 11, no. 9 and 11). Reactivation of other basement structures in the central and western parts of New York was probably responsible for the multiple unconformities in the Taconic tectophase cycle there (Figure 13), but the unusual distribution of the Utica and its overall migration and younging to the west (Figure 13) require additional explanation.

In fact, the initial unconformity development below the Utica probably represents uplift on local structures

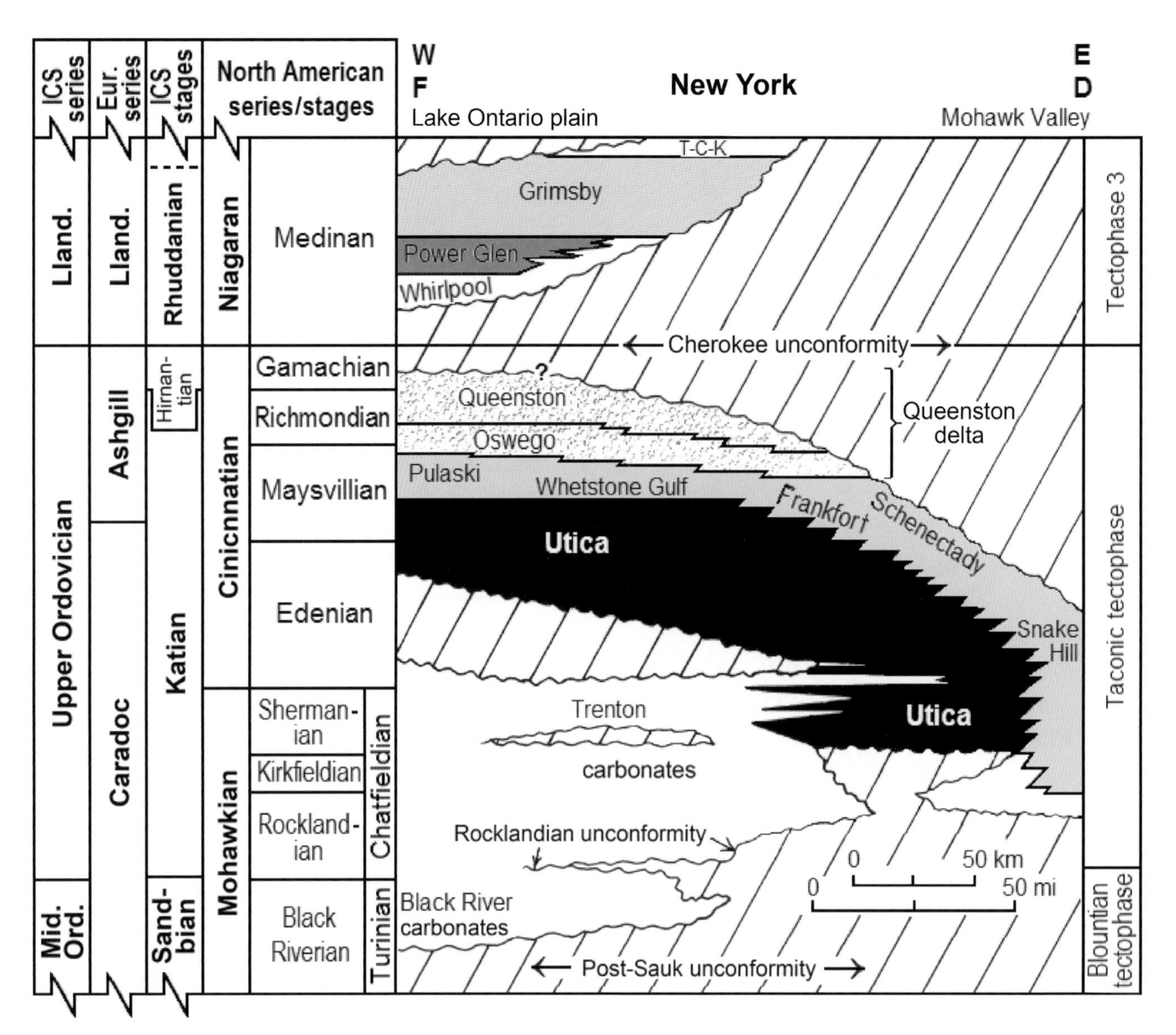

Figure 13. A schematic east–west section, nearly perpendicular to the strike of the Appalachian foreland basin (section FD in Figure 12), showing the nature and disposition of three Taconian tectophase cycles. Note the younging of Utica dark shales and the underlying unconformities. Modified from Ettensohn (2008). ICS = International Commission on Stratigraphy; Eur. = European; T-C-K = Thorold, Cambria, and Kodek formations.

accompanying initiation of convergence and loading to the south. The few-million-year duration of the unconformity from inception until flooding by deeper Utica seas in the early Shermanian (Figure 13) likely reflects uplift on other local structures during the northward shift to full convergence at the New York promontory. For another few million years until the latest Shermanian (latest Chatfieldian) and the early Edenian, rapid subsidence behind the promontory produced a stable Utica dark-shale basin (Figure 13), but by the early Edenian, the bulge-related unconformity and the Utica shale basin itself migrated westward, expanding the size of the foreland basin (Figures 13, 14). Ettensohn (2008) has suggested that this change in Utica distribution and the westward expansion of the shale basin were related to changes in subduction vergence direction at the time. According to Karabinos et al. (1998), early parts of the Taconic tectophase reflect obduction and buildup of an accretionary wedge at the New York promontory because of eastward subduction below the Shelburne Falls arc (Figure 14A); at about 455 Ma, when the westward expansion of the Utica Shale occurred, obduction and slab breakoff are interpreted to have halted eastward subduction and created a new westward-dipping subduction zone below the new Bronson Hill arc (Figure 14B). This change in vergence

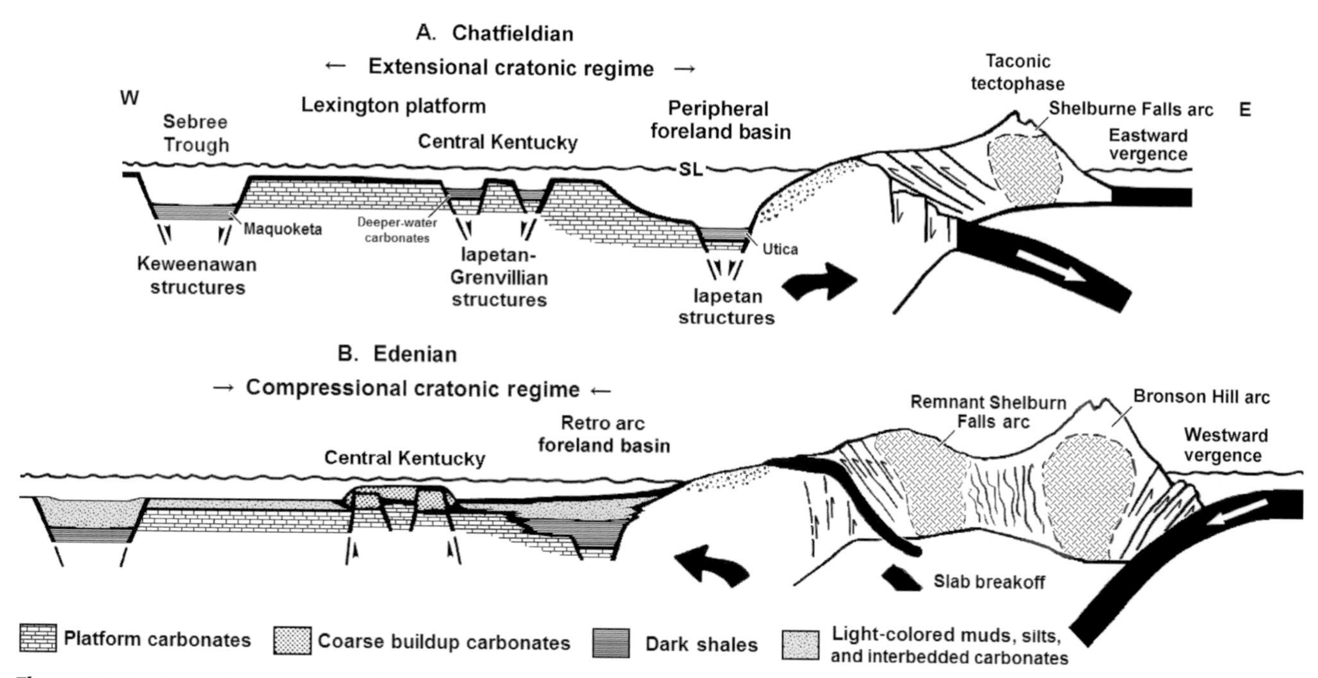

Figure 14. A schematic, east–west, interpretive section, running from the early Taconic subduction zone in northern New York to distal cratonic areas in western Kentucky, showing the likely influence of subduction vergence direction on the development of Taconic foreland and cratonic (Sebree trough) basins; no scale intended. (A) East-verging subduction zone and establishment of an extensional foreland and cratonic regime; (B) West-verging subduction zone and establishment of a compressional foreland and cratonic regime.

direction apparently had significant impacts on cratonic sedimentation because during eastward subduction, the lithosphere experienced an overall tensional regime that allowed subsidence to be accommodated by downdrop on local structures (Figure 14A); hence, for a few million years in the Shermanian–earliest Edenian, the early Utica foreland basin (northern parts of the Martinsburg Basin) and dark-shale deposition within it were localized in a narrow trough (Figures 13, 14A) defined by basement structures (Figure 11, no. 1 and B). Here, dark shales accumulated to thicknesses of more than 400 m (>1312 ft).

With the change to westward subduction, the lithospheric stress regime altered to compression, so that some structural highs were uplifted, forming local unconformities, whereas deformational loads moved westward onto the craton, generating a migrating bulge and foreland basin that expanded Utica deposition (10–100 m [33–328 ft]) westward (Figures 13, 14, 15B), eventually yoking the Appalachian and Michigan basins (Coakley and Gurnis, 1995). So the expansion of the Martinsburg (Utica) Basin in the north (Figure 12) was not only related to the westward change in subduction direction but also to the more prominent westwardly advancing deformational load and its far-field interactions with basement structures.

By the late Chatfieldian, loading-type relaxation, accompanied by flyschlike sedimentation in coarser clastic-rich parts of the upper Martinsburg and Reedsville formations, had reached the southern parts of the basin (Figures 6, 7). However, similar relaxation and flysch-like sedimentation did not occur in the northern parts of the basin until the Edenian and Maysvillian (Figures 6, 13). Following equilibrium-phase transitions (Ashlock/Bull Fork formations; Figure 7), the basin was subsequently filled with a widespread, marginal-marine to terrestrial, molasse-like sequence represented by the non-red Oswego and Bald Eagle formations and the red Queenston, Juniata, and Sequatchie formations, commonly called the Queenston delta (Figures 6, 7, 13). These redbeds represent unloading-type relaxation and are present throughout most of the basin from Tennessee to south-central Canada. Where not present, they are represented by regressive peritidal to shallow-marine carbonates and shales in the Sequatchie, Drakes, Whitewater, and Saluda formations.

Third Tectophase and the Power Glen-cabot Head Foreland Basin

The Late Ordovician redbed sequence in most of the Appalachian Basin is truncated by the widespread Ordovician-Silurian Cherokee Unconformity (Figures 6, 7, 13), an unconformity that has been widely ascribed to global, glacio-eustatic, sea level drawdown (e.g., Dennison, 1976). Although the global nature of the

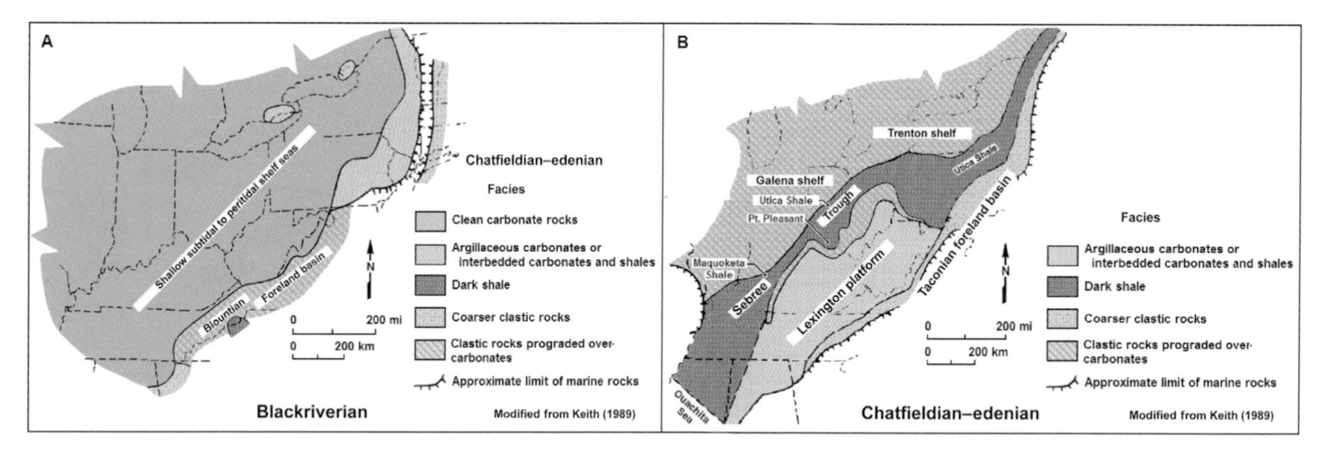

Figure 15. Paleoenvironmental interpretations of east-central Laurentia during the Blackriverian (A) and the Chatfieldian–Edenian (B), showing the stratigraphic and structural differentiation of the area caused by far-field forces during the Taconic tectophase of the Taconian orogeny. (A) Extensive Black River carbonate platform during the late Blountian tectophase. (B) Stratigraphic and structural differentiation of the Black River platform into Lexington, Trenton, and Galena platforms/shelves, as well as the Sebree Trough. Modified from Ettensohn et al. (2002).

unconformity supports this interpretation (McKerrow, 1979), the unconformity becomes increasingly angular in the northeastern parts of the basin, suggesting also a major tectonic component (e.g., Dorsch et al., 1994). Moreover, the presence of an overlying flexural sequence (Figure 13) and its infill of a basin strongly asymmetrical toward the St. Lawrence promontory (Figure 12), together with evidence for the Late Ordovician–Early Silurian convergence and deformation at the St. Lawrence promontory caused by the accretion of Ganderia (Figure 12) (van Staal, 1994), points to a third and final Taconian tectophase (Ettensohn and Brett, 2002). This phase of tectonism generated the largest and most distal of the Taconian basins; not only did the basin migrate farther north than previous Taconian basins but also moved farther cratonward (Figure 12), reflecting the cratonward movement of deformation. The initial transgressive part of the cycle is represented by marginal-marine to shoreface sands in the Tuscarora, Clinch, and Whirpool sandstones, and overlying the sands, are dark basinal shales of the Power Glen and Cabot Head formations, whose distribution marks the limits of the basin in Figure 12. The shales are typically not more than 25 m (>82 ft) thick and the overlying, relaxational, clastic units are typically poorly preserved, reflecting deep erosion that accompanied inception of the following Salinic orogeny. The entire tectophase cycle is rarely more than 200 m (656 ft) thick.

Taconian Far-field Effects and the Sebree Trough

Although Taconian orogeny and related foreland-basin development continued on the eastern parts of the Laurentian margin, the cratonward transmission of far-field stresses, focused on preexisting zones of basement weakness (Figure 11), had a major impact on more distal, foreland, depositional environments and stratigraphy. This was particularly the case during the Taconic tectophase, when a linear basin of deeper water dark shales, called the Sebree Trough (Schwalb, 1980), abruptly developed between shallow-water carbonate shelves, 400 to 500 km (249–311 mi) west of the foreland basin (Figures 14A, 15B).

Before the inception of the Taconic tectophase, the east-central U.S. foreland was characterized by an extensive peritidal carbonate platform, called the Black River Platform (Figure 15A) (Keith, 1989). The carbonates (Black River and High Bridge groups) reflect an overall regressive regime that coincided with regional regression during final unloading-type relaxation in the Blountian tectophase (Figure 7). However, in the late Rocklandian (early Chatfieldian) at the beginning of the Taconic tectophase, parts of the platform suddenly collapsed and were differentiated into the rectilinear Lexington carbonate platform to the southeast and the Galena and Trenton carbonate shelves to the northwest, separated by the linear Sebree Trough (Keith, 1989), in which little sedimentation initially occurred (Figures 14A, 15B). The facts that the trough mostly coincides with Keweenawan and Grenvillian structures (Figure 11) and that renewed downdrop along these structures coincides with the beginning of an overall tensional regime related to eastward subduction at the inception of the Taconic tectophase (Figure 14A) strongly suggest that the structural and stratigraphic differentiation of the Black River platform (Figure 15) was a response to Taconic far-field tensional stresses.

In the late Shermanian to the early Edenian, this situation again abruptly changed when dark organic-rich muds began to fill the trough. The muds are the same age as the late Shermanian and the early Edenian parts of the Utica Shale in New York and have been called Utica Shale as far south as southeastern Indiana; beyond southwestern Indiana, the dark shales are considered to be parts of the Maquoketa Formation (Figures 14A, 15B). At the same time, evidence supports deepening and titling of the entire area toward the southeast (Mitrovica et al., 1989; Coakley and Gurnis, 1995). All of these events were concurrent with the westwardly change in subduction direction and the already noted westwardly expansion of the northern Martinsburg (Utica) foreland basin (Figures 13, 14) and its included dark-shale (Utica) depositional setting (Figure 14). In fact, it is likely that subsidence on Iapetan and Grenvillian structures in northeastern Ohio and northwestern Pennsylvania effectively yoked the expanded foreland basin with the Sebree Trough in north-central Ohio, allowing direct connection between the foreland basin and the trough (Figure 15B). That all of these changes were synchronous with the change to a westward subduction direction is probably no accident. Westward subduction and the presence of a new westwardly directed load apparently created a net cratonic compressional regime that allowed for regional flexural subsidence, reactivation of structures, and a more proximal source of sediments to fill the foreland basin and structural lows beyond during loading- and unloading-type relaxation (Figure 14B). Furthermore, Ettensohn et al. (2002) have suggested that the Late Ordovician paleoclimatic and paleogeographic setting promoted quasi-estuarine circulation across the area, which enhanced anoxia in the trough and upwelling onto adjacent platforms so that the dark-shale source rocks in the trough developed in close proximity to potential carbonate reservoirs (Trenton, Lexington) on adjacent platforms. The above scenario illustrates how proximal tectonic events might control the development of dark-shale source rocks—even in areas beyond the foreland basin—through far-field responses.

ACADIAN/NEOACADIAN OROGENY: DEVONIAN–MISSISSIPPIAN FORELAND-BASIN DARK SHALES AND BASIN YOKING

The Acadian/Neoacadian orogeny began in the Early Devonian (~411 Ma) and marks the beginning of the Variscan–Hercynian orogenic supercycle (Figure 4). The orogeny mostly reflects impending closure of the Rheic Ocean (Figures 1, 16), involving the reorganization of island arcs and/or microcontinents (terranes), which had docked earlier on or near the eastern margin of Laurussia (Figure 16), and a change in the oblique translation direction of terrane convergence under the influence of impact by Meguma and Gondwana (Figure 16). The initial convergence of the Carolina terrane with the Laurussian margin appears to have had an oblique sinistral aspect such that translational movement of the terrane was to the northeast relative to Laurussia (Figure 18A) (e.g., Dennis, 2007). After the Early Devonian, however, translational movement switched to a dominantly dextral aspect (Merschat and Hatcher, 2007) such that terranes moved to the southwest relative to Laurussia (Figure 16), and it was this southwestward, oblique, translational movement of the Carolina terrane toward the New York and Virginia promontories that was mostly responsible for the Acadian/Neoacadian orogeny (Figure 16B–D) (Ettensohn 1985a, 1987; Merschat and Hatcher, 2007). Oblique convergence occurred in four distinct tectophases that resulted in four distinct sets of dark-shale basins in the foreland (Figures 8, 17). These dark shales are parts of unconformity-bound tectophase cycles, which migrated parallel to strike and tracked the progress of orogeny. The beginning of the orogeny is manifest in the foreland basin by the prominent Walbridge discontinuity (Figure 8). The three following tectophase cycles are incomplete, consisting of a shallow-water transgressive unit, a basal dark-shale unit, and an overlying flysch-like sequence (Figure 8), which is subdivided into several subcycles (e.g., Rickard, 1975). These three early cycles range in duration from approximately 10 to 30 m.y., and their incomplete nature indicates that convergence and deformation were effectively nonstop, such that convergence at one promontory was still ongoing during the onset of convergence at the next one (Figure 16). The fourth tectophase, at about 41 m.y. in duration, is the longest tectophase and the only one to go to completion.

First Tectophase Dark-shale Basins

By the Early Devonian, the northeastward translation and dispersal of the Carolina terrane (Dennis, 2007) had apparently halted perhaps because of its encroachment on the southern end of Avalonia, which was obliquely docking in a southwestward direction with the St. Laurence promontory area (Figure 16A). As a result, Carolina was left sitting astride both the New York and Virginia promontories. By the Early Devonian, however, the full docking of Avalonia with the St. Lawrence promontory initiated the first tectophase of the Acadian orogeny. This tectophase is what many workers consider to be the traditional Acadian

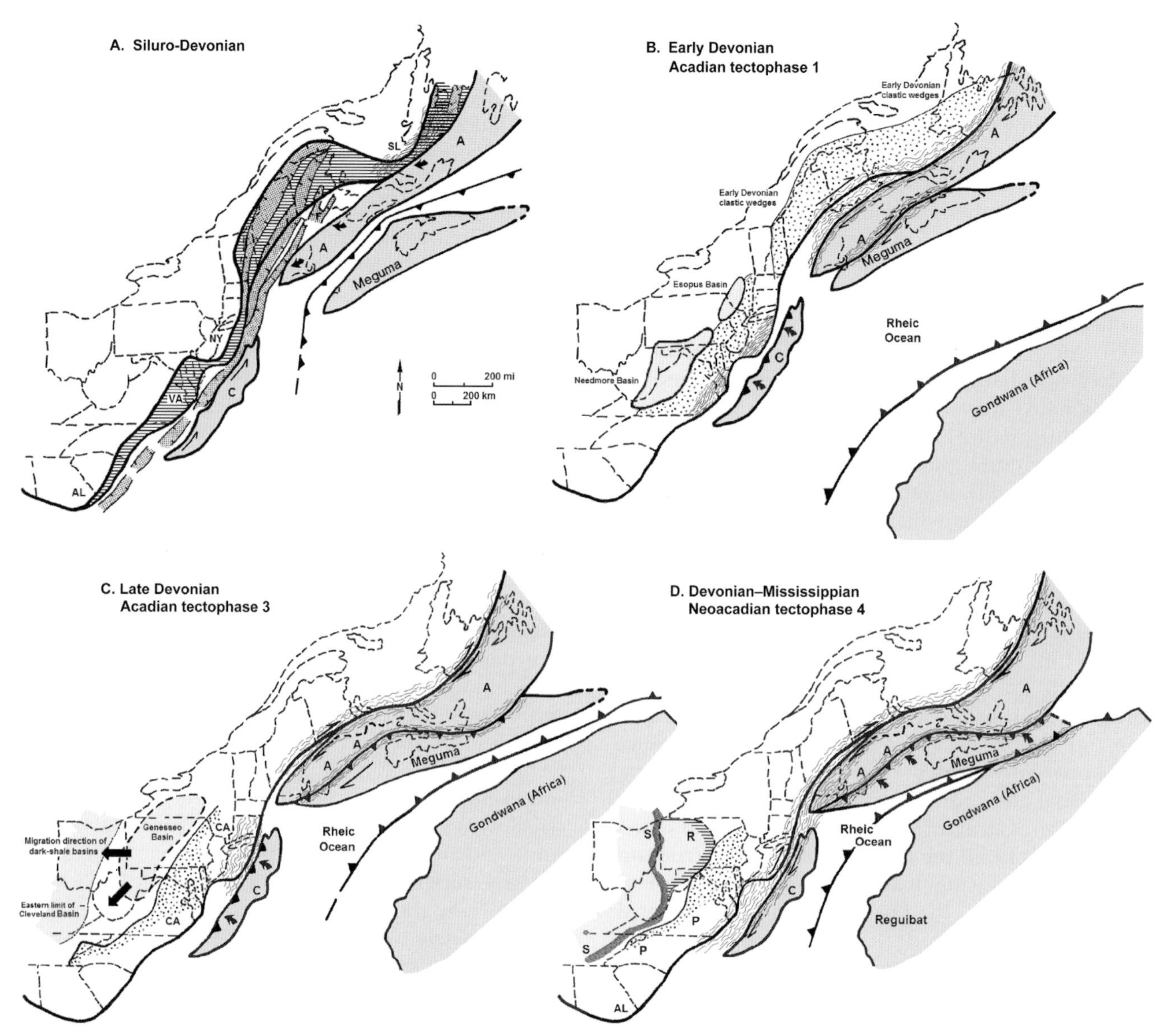

Figure 16. Schematic representations of foreland and proforeland tectonic settings before and during the Acadian/Neoacadian orogeny to show the influence of tectonic setting on the formation of dark-shale basins. (A) The Silurian–Devonian pre-Acadian setting: collapse of Taconian and Salinic orogens showing newly accreted areas (dark, lined pattern) and formation of successor basins (light stipple). AL = Alabama promontory; VA = Virginia promontory; NY = New York promontory; SL = St. Lawrence promontory; C = Carolina terrane; A = Avalonia. (B) The Early Devonian Acadian orogeny, first tectophase: Carolina converges at New York and Virginia promontories, and dark-shale basins develop behind the promontories. (C) The Middle–Late Devonian Acadian orogeny, third tectophase: Carolina docks in a north–south zipper-like fashion with New York and Virginia promontories, generating Catskill delta wedge (CA) and six southwardly migrating dark-shale foreland basins (Figures 8, 17C, 18). (D) The latest Devonian–Early Mississippian Neoacadian orogeny, fourth tectophase: Avalonia and Meguma collide with Carolina and New York promontory; the Price-Pocono-Borden-Grainger clastic wedge (P).

orogeny. It was concentrated at the St. Lawrence promontory and generated a series of dark shales and an extensive Early–Middle Devonian clastic wedge in the New England–Maritime province area (Figure 16B), but most of the dark shales and clastic wedge were structurally dispersed, severely deformed, or destroyed by subsequent tectonism.

Farther south in the central Appalachian Basin, the presence of dark-shale foreland basins behind the New York and Virginia promontories (Figures 16B, 17A) indicates that a similar convergence regime—with deformation concentrated at the promontories—had begun. Here, the Carolina terrane collided obliquely with outboard parts of the respective promontories.

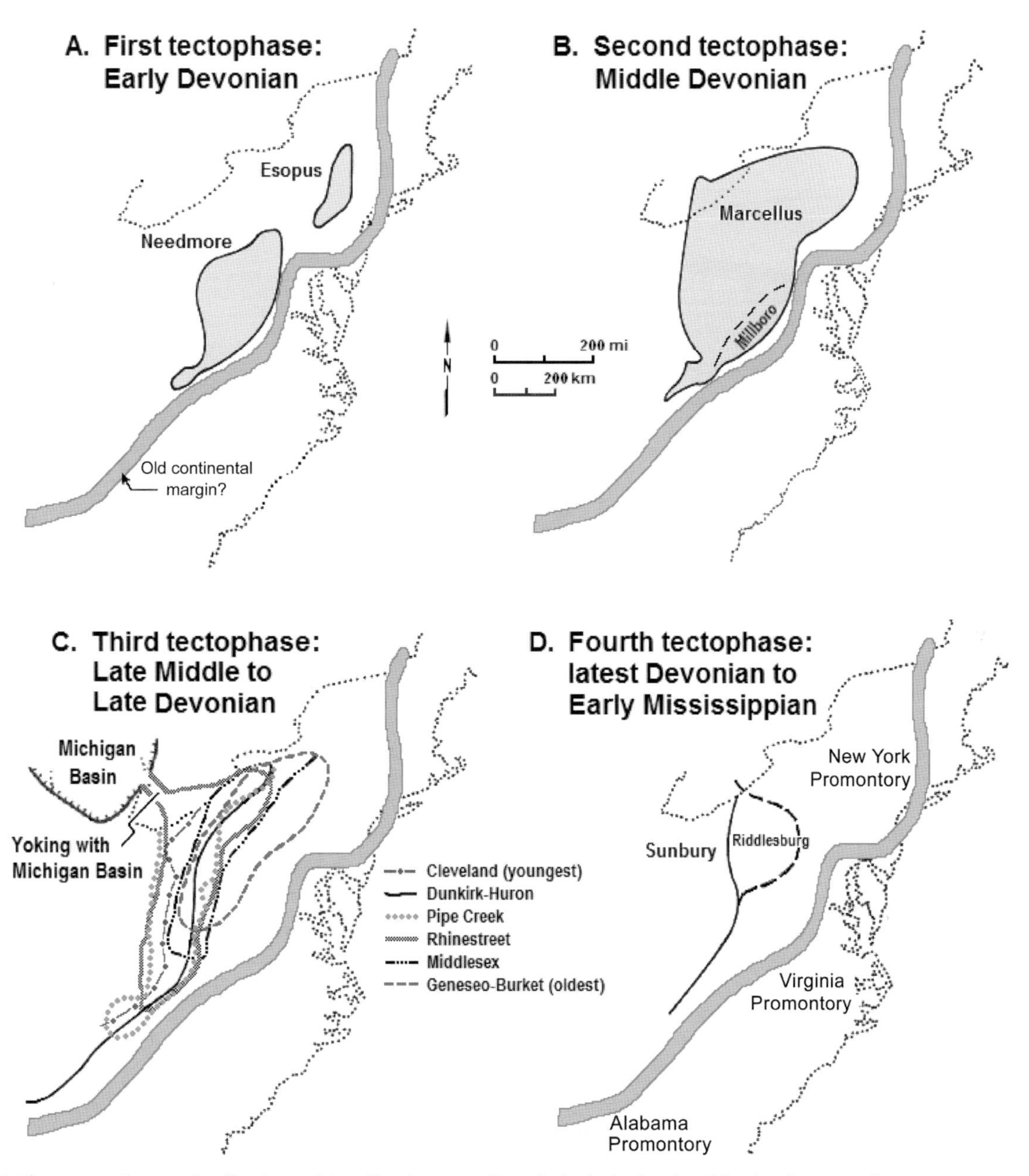

Figure 17. The approximate distribution of Acadian/Neoacadian dark-shale foreland basins by tectophase. By Huron-Dunkirk time, most of the foreland basin had been infilled with coarser clastic sediments, forcing deeper water, dark-shale, basinal environments onto the craton (Figure 18) so that only the proximal eastern margins of these basins are shown. Modified from Ettensohn (1994).

Deformation at the New York and Virginia promontories was sufficient to generate an unconformity behind the involved promontories (Figure 8), which was followed by deposition of the transgressive Oriskany/Ridgeley sandstones and the subsequent development of two small dark-shale basins, the Esopus and Needmore, behind the respective promontories (Figures 8, 16B, 17A). Although laminated black shales, representing anoxic environments, did develop in parts of the Needmore Basin (Newton, 1979), the gray sandy shales of the Esopus indicate that the basin was only deep enough to support dysoxic environments. The dark shales are overlain by gray calcareous shales, sandstones, and argillaceous carbonates of the Carlisle Center and Schoharie formations and their equivalents (Rickard, 1975), but these units are not the expected flysch-like sediments, suggesting that basin subsidence was not great and that loading-type relaxation was incomplete. Overall, dark shales and the tectophase cycles in these two basins are not extensive or well developed,

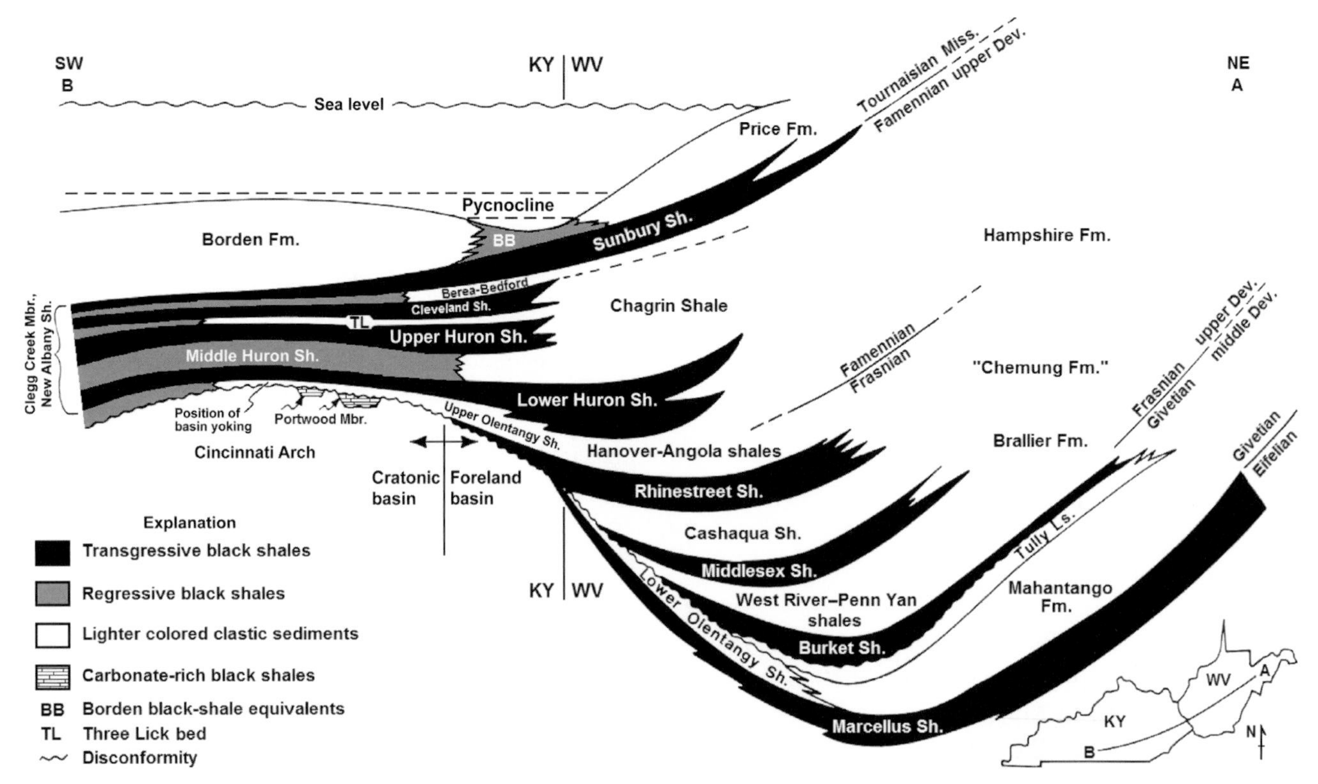

Figure 18. A schematic northeast–southwest section (AB) through the central Appalachian Basin in West Virginia and Kentucky, showing positions of the Middle Devonian to Early Mississippian dark-shale basins during the second-to-fourth Acadian/Neoacadian tectophases. By early Famennian time, the Appalachian Basin had filled with coarser clastic sediments, pushing deeper, sediment-depleted conditions for dark-shale deposition farther cratonward into extensive cratonic basins, marking the time and location of yoking with the Illinois Basin. Modified from Ettensohn et al. (1988).

meaning that deformational loading and the resulting subsidence were relatively insignificant at this time.

Second Tectophase Marcellus Dark-shale Basin

The second tectophase in the central Appalachians was a wholly Middle Devonian event (Figure 8) that represents a phase of more intense head-on collision between Carolina and the New York and Virginia promontories; it is apparent as an incomplete tectophase cycle. The initial pre-Onondaga unconformity is widespread throughout the basin (Figure 8), and it is overlain by a partial tectophase cycle that includes the dark Marcellus Shale. The cycle begins with the transgressive Onondaga Limestone, which grades up into and intertongues with dark Marcellus shales (Figure 8), indicating slow but steady subsidence as dark shales progressively overlapped carbonates in a westward direction (Ettensohn, 1985a). The Marcellus dark-shale basin is one of the largest Appalachian foreland basins; compared with the previous Esopus and Needmore basins, it fully overlapped all of the area between them and extended westwardly well beyond them (Figure 17B). This distribution connotes that the second tectophase deformation was not only more intense and extensive across the New York–Virginia area but had also migrated substantially westward across the promontories.

The Marcellus Shale contains evidence of considerable facies variation, reflecting differences in the degree of basin oxygenation at different times and places. Oxic, dysoxic, and anoxic conditions were present, but based on the overall color and fissility of the shales, anoxic conditions seem to have been more prevalent in northern parts of the basin behind the New York promontory, whereas dysoxic conditions seem to have prevailed in more sandy Millboro equivalents in southern parts of the basin (Figure 17B). These differences may reflect partitioning of the larger basin along basement structures (Figure 11) and the fact that collision and deformation at the New York promontory were more intense.

The Marcellus and overlying parts of the Hamilton Group are clearly cyclic. The Marcellus itself contains two dark-shale cycles (Union Springs and Oatka Creek shales), separated by the Cherry Valley Limestone (e.g., Rickard, 1975). The overlying parts of the Hamilton Group, which represent incomplete loading-type relaxation, contain three subcycles represented by the Skaneateles, Ludlowville, and Moscow formations in New York (e.g., Rickard, 1975), as well as equivalent subcycles to the south (Hasson and Dennison, 1977). Each cycle generally begins with dark-gray shale, succeeded by blue-gray or green-gray silty shales, and followed by a limestone; and the cycles represent transition from marginally anoxic to dysoxic and oxic environments, accompanying a sudden pulse of subsidence and basin infilling. Although the Marcellus and overlying parts of the Hamilton Group clearly represent a single tectophase cycle, the presence of subcycles throughout demonstrate the continued presence of episodic deformational loading probably through movement on distinct thrust complexes.

Migrating Third Tectophase Dark-shale Basins

The third tectophase begins at the pre-Tully or Taghanic unconformity (Figure 8) and includes the development of five to six consecutive dark-shale basins (Figures 8, 17C). The tectophase begins in the late Middle Devonian, but most of the dark shales are Late Devonian in age (Figures 8, 18). Mapping the distribution of the dark-shale basins shows that each successive basin not only migrates cratonward in time (Figures 8, 18) but also southwestward, parallel to the strike of the larger Appalachian Basin and Acadian orogen (Figure 17C). This migration pattern can only mean that the dark-shale basins were tracking the progress of orogeny, and that the orogeny had a dextral transpressive component involving the oblique subduction below Carolina, such that any gap between Carolina and the Laurussian margin (between the New York and Virginia promontories) zipped shut in a scissor-like fashion (Figure 16C) (Ettensohn, 1987; Merschat and Hatcher, 2007). This was effectively a terminal collisional-type tectophase that ended in the latest Devonian with the full collision of Carolina against the restraining Virginia promontory. Unlike many dark shales in the first and second tectophase basins, shales in third tectophase basins are mostly black, pyritic, and lack major transgressive deposits at their bases. These characteristics suggest episodic rapid subsidence, which quickly generated sediment starvation and anoxia throughout the respective basins—conditions that are consistent with a cadence of abrupt cratonward-moving thrust complexes that successively loaded the crust.

Each of the six dark-shale units is overlain by a flysch-like clastic unit that was abruptly truncated by the next dark-shale basinal unit (Figures 8, 18). Each of these flysch-like clastic units, moreover, grades successively into shallow-marine, marginal-marine, and terrestrial clastic units in a shoreward direction, building the Catskill clastic wedge to thicknesses of more than 3000 m (>9843 ft) behind the New York promontory and to more than 2400 m (7874 ft) behind the Virginia promontory. By the early Famennian, however, the combination of cratonward-moving flexural subsidence and increasing clastic influx overfilled the foreland basin and pushed individual dark-shale basins onto the craton, forming what Ettensohn et al. (1988) called a cratonic dark-shale basin (Figures 8, 18). Although northern parts of the Appalachian Basin had yoked earlier with the Michigan Basin during deposition of Rhinestreet dark shales (Figure 17C), westwardly migrating subsidence and basin overflow forced the central parts of the Appalachian Basin to yoke with the Illinois Basin during Huron-Dunkirk deposition. Subsequently, all third-tectophase dark-shale deposition was pushed westward to areas beyond the foreland basin during Huron and Cleveland deposition (Figures 17C, 18).

The lower Huron-Dunkirk, upper Huron, and Cleveland dark shales all represent episodes of rapid tectonically related subsidence, and they are separated by clastic-rich intervals like all the underlying third tectophase dark shales. However, in the cratonic basin, subsidence was apparently great enough and the clastic influx so reduced that even during intervening periods of flysch-like deposition, the sea bottom never aggraded high enough to move much above the pycnocline into more oxic waters. As a result, in contrast to the alternating dark shales and intervening lighter colored clastics in the foreland basin (Figure 18, right side), most shale units on the craton are mostly dark colored, whether clastic or organic rich. This situation has given rise to dark-shale units beyond the foreland basin like the Ohio, New Albany, Woodford, and Chattanooga shales, which are distal composites of several larger units from the foreland basin (Figure 18, left side). Even in these mostly dark-shale units, it is still possible to distinguish dark shales of different origins. Shales that formed during episodes of tectonic subsidence are more radioactive and organic rich, forming pulp-like paper shales (Figure 19A) that have been called transgressive black shales. Those that formed during intervening episodes of rapid clastic influx are less radioactive and contain more clastic constituents that are apparent as ribbed regressive black shales (Figure 19B) (Ettensohn et al., 1988).

Fourth Tectophase Dark-shale Basin and the Acadian/Neoacadian Orogeny

The fourth tectophase began in the Early Mississippian (Tournaisian), although it took the entirety of the

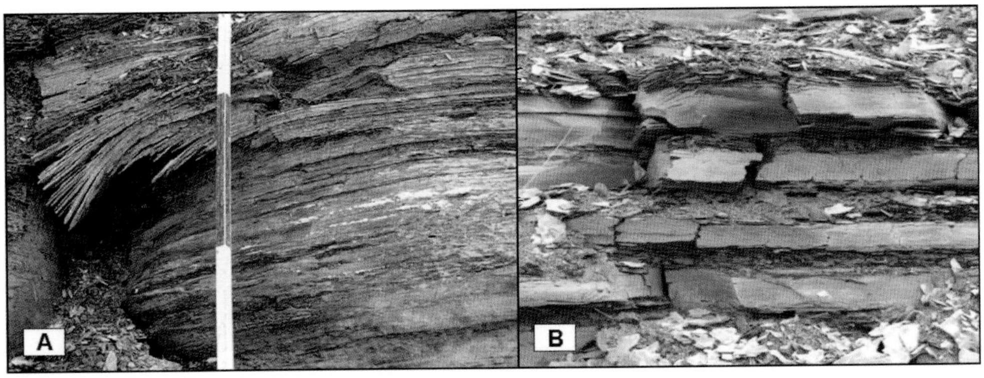

Figure 19. (A) Transgressive, pulpy, paper-like dark shales. (B) Regressive ribbed dark shales. Ohio Shale from northeastern Kentucky; meter stick for scale.

Mississippian Period to fully develop. The tectophase is represented by a single dark-shale unit, the Sunbury Shale and its equivalents (Figures 8, 17D, 18). The Sunbury occurs in the western parts of the Appalachian Basin as a single, transgressive, dark shale. Behind each promontory, the shale unconformably (Figure 8) overlies the lighter sands and shales of the Bedford-Berea sequence (Figures 8, 18), and it is overlain by the lighter sands, silts, and shales of the Borden-Price-Pocono-Grainger clastic wedge and its equivalents (Figures 8, 18), which form the most extensive of the Acadian/Neoacadian clastic wedges (Ettensohn, 2004). The Sunbury is the darkest, most organic-rich, and most radioactive of all Appalachian Basin dark shales; it also has the largest distribution of any Acadian/Neoacadian dark shale because its equivalents are included as uppermost parts of the composite Chattanooga, New Albany, Woodford, Houy, and Antrim shales throughout the central and south-central United States (Mankin, 1987; Ettensohn et al., 1988; Matthews, 1993). Estimates of water depth during the Acadian/Neoacadian dark-shale deposition indicate that Sunbury dark shales were deposited in the deepest of the Acadian/Neoacadian basins (Ettensohn, 1984).

The earlier Devonian tectophases of the Acadian orogeny were effectively the consequence of northwestward Early–Middle Devonian docking of the Carolina terrane at the New York and Virginia promontories (Figure 16B) and the final, north-to-south, zipperlike accretion of the terrane to the Laurussian margin (Figure 16C); and among other lines of evidence, this tectonic mechanism is clearly supported by the migration of Acadian dark-shale basins in the foreland (Figure 17). However, the unusual distribution pattern of the dark Sunbury Shale and its equivalents, along with new compilations of evidence from metamorphic, magmatic, and deformational sources, strongly suggests that what has been called the fourth tectophase of the Acadian orogeny (Ettensohn, 1985a), or more recently, the Neoacadian orogeny (e.g., Robinson et al., 1998; Merschat and Hatcher, 2007), represents a mostly different tectonic scenario.

In contrast to the earlier Late Devonian dark-shale basins that had migrated progressively westward in time (Figures 8, 17, 18), by the earliest Mississippian, basin migration shifted abruptly to the east and is represented by the Sunbury Shale with a conspicuous asymmetrical embayment toward the east in the coeval Riddlesburg dark shales just behind the New York promontory (Figures 16D, 17D). Considering the flexural mechanisms necessary to switch the direction of basin migration, this change must reflect a major transition in the tectonic regime, and such a transition was noted by Robinson et al. (1998) as an episode of intense latest Devonian–Early Mississippian magmatism, metamorphism, and deformation in central Massachusetts—an episode that they first called the Neo-Acadian orogeny. At the same time, in the central Appalachians, evidence supports major dextral shear involving the Carolina terrane near the Virginia promontory (e.g., Hibbard et al., 2002; Trupe et al., 2003) and peak metamorphism (Dennis, 2007; Merschat and Hatcher, 2007), all of which has similarly been termed Neoacadian.

In particular, the prominent eastward expansion and amplification of subsidence required to develop the Sunbury-Riddlesburg basin behind the New York promontory (Figures 16D, 17D) can only mean that the New York promontory was the site of major deformational loading in latest Devonian–Early Mississippian time. As Carolina had already accreted to Laurussia, the most likely source of the intense deformation was the collision of the Avalon and Meguma terranes with the New York promontory and Carolina under the impact of Gondwana (Figure 16D). Clearly, the prominent eastward embayment in the Sunbury-Riddlesburg shale distribution, the source areas of the overlying Borden-Grainger-Price-Pocono clastic wedge, as well as deformational and magmatic evidence, all suggest that the area of the New York promontory was the site of major collision (Figure 16D). Nonetheless, the continuing southward distribution of the Sunbury basin and overlying Borden-Grainger-Price-Pocono clastic wedge (Figures 16D, 17D) shows that both continued their southward expansion to areas behind and beyond the Virginia promontory. This expansion can only indicate that deformation must have also migrated southward, and this is most easily explained through the development of major dextral shear between the Carolina terrane and the margin of Laurussia near the Virginia promontory. The dextral shear probably reflects the ejection of the Avalon and Meguma terranes in a pincer movement between the margin of Laurussia and Gondwana (Figure 16D). As the terranes were ejected, they first collided with the New York promontory and the northern end of Carolina, initiating the major pulse of tectonism now called the Neoacadian orogeny, and with it, the Sunbury-Riddlesburg dark-shale basin. However, as terrane ejection continued, dextral shear was transferred to the recently accreted Carolina terrane, causing it to move southwestwardly in a transpressional regime that generated new deformational loading and clastic sources on and near the Virginia promontory. As the deformation moved southwestward, the Sunbury Basin and overlying Borden-Grainger-Price-Pocono wedge expanded in a similar direction at least as far south as Tennessee, where the uppermost parts of the Chattanooga Shale are Sunbury equivalents of Mississippian age (Glover, 1959; Roen et al., 1964; Ettensohn et al., 1988).

The intensity of the fourth tectophase, or Neoacadian, tectonic event was apparently much greater than any of the other Acadian tectophases, resulting in a deeper, more extensive dark-shale basin and in the largest and most extensive of the Acadian/Neoacadian clastic wedges. The Sunbury and its equivalents in the upper Chattanooga, New Albany, Woodford, and Huoy shales form the largest of the cratonic dark-shale basins (Figure 18), extending across much of the central and south-central United States. In addition to the Neoacadian subsidence already noted, part of the great extent of this unit may also be related to the initiation of Ouachita subsidence, which had begun by the Early Mississippian along the southern margin of Laurussia (Arbenz, 1989), and to Early Mississippian sea level rise (e.g., Dennison, 1989).

Other measures of the intensity and influence of this tectophase are the facts that it alone among Acadian/Neoacadian tectophases exhibits a complete seven-part tectophase cycle, and at about 41 m.y. in duration, or the entirety of the Mississippian Period, it is the longest of any Appalachian tectophase (Figure 4). Although major collisional and loading parts of the tectophase cycle, represented by the Sunbury and its equivalents, were completed in the earliest 5 to 6 m.y. of the Mississippian (early–middle Tournaisian; Kinderhookian), it took the remaining 35 m.y. of the Mississippian for relaxational parts of the cycle to finish. The long duration of the cycle may reflect the oblique nature of transpression and the fact that lateral translation involved the entire 650 km (404 mi) length of Carolina. Overlying relaxational parts of the cycle contain a loading-type relaxational clastic wedge (Borden, Grainger, Price, Pocono, and equivalent formations), a thick equilibrium interval of carbonates (Greenbrier, Newman, Slade, Tuscumbia, Bangor, Monteagle, and Maxville limestones), and an overlying unloading-type relaxational wedge (Pennington and Mauch Chunk groups).

Of particular interest in understanding the intensity of this tectophase is the unusual extent of the Borden-Grainger-Price-Pocono clastic wedge, a product of loading-type relaxation. This complex emanated from behind both the New York and Virginia promontories as an extensive subaerial (Price and Pocono) and subaqueous (Grainger and Borden) delta complex and reflects the production of so much clastic influx that the wedge migrated beyond the Appalachian Basin and into the Illinois Basin (Ettensohn, 2004). In areas just behind the Virginia promontory, subsidence was locally great enough to continue Sunbury-type dark-shale deposition into the Middle Mississippian (Tournaisian–Visean; Osagean) (Figure 18, BB).

CONCLUSIONS

Marine dark shales, which are important hydrocarbon source and reservoir rocks, are natural constituents in most foreland basins. Their presence in foreland basins is partly a problem of accumulation and preservation and partly a problem of water chemistry and organic-matter generation. Aspects like eustasy (Werne et al., 2002), types of organic matter (Algeo and Scheckler, 1998; Rimmer et al., 2004), water circulation (Perkins et al., 2008), and glaciation (Armstrong et al., 2005; Ettensohn et al., 2009) are clearly important in understanding the geochemical framework of such shales. Although the timing and paleogeography of foreland basins may also contribute to certain geochemical conditions (Ettensohn, 1997), it is the large-scale tectonic controls, which generate foreland- and intracratonic-basin repositories, that are responsible for the accumulation and preservation of organic matter in quantities great enough to form major units of dark-shale source rocks. The Appalachian Basin and nearby areas clearly provide an ideal framework for understanding these processes during the Paleozoic.

Mapping the distribution of dark shales in time and space throughout the basin has shown that temporal and spatial patterns of dark-shale deposition strongly coincide with the timing and formation of foreland basins during orogenies on the eastern margin of Laurentia/Laurussia. It was already known that foreland basins form as flexural responses to deformational loading in the adjacent orogen, but recent work in the Appalachian Basin has shown that dark shales are major parts of unconformity-bound sequences that form during episodes of rapid loading-related subsidence and subsequent relaxation in the foreland basin. These episodes of subsidence and relaxation are clearly related to coeval phases of tectonism or tectophases that were mediated by collisions with successive continental promontories during orogeny. Using Ordovician and Devonian–Mississippian dark-shale sequences from the Appalachian Basin, this chapter illustrates how intense deformation and deformational loading initiate a foreland-basin cycle with periods of rapid subsidence and subsequent relaxation that both generate the basin and fill it with sediments. Early parts of the cycle include a basal unconformity and an overlying succession of dark shales, which represent accumulation of organic-rich muds at a time when clastic sources were not yet developed and in a basin subsiding so fast that sedimentation could not keep pace with subsidence; once active loading and deformation ceased, the lithosphere responded by relaxing in a series of steps that allow the basin to infill progressively with flyschlike and molasse-like clastic sediments. Each successive collision with promontories on the Laurentian/Laurussian margin generated a tectophase cycle that was superimposed on earlier cycles. As a result, 13 such cycles were generated during the Taconian, Salinic, Acadian/Neoacadian, and Alleghanian orogenies in the Appalachian Basin, and each cycle contains a major dark-shale unit and an overlying coarser clastic sequence. Every part of the cycle has potential economic importance inasmuch as dark shales become major hydrocarbon source rocks, whereas many of the overlying coarser clastic units develop into reservoir rocks. The influence of these same flexural forces, however, may extend well beyond the foreland basin itself through the reactivation of distal basement structures, which can generate intracratonic basinal settings that also favor dark-shale accumulation. The same processes and controls have almost certainly operated in other foreland-basin settings and may provide models for understanding foreland-basin accumulations of dark-shale source rocks elsewhere (see Su et al., 2009).

ACKNOWLEDGMENTS

We thank Richard Smosna and Dengliang Gao for their helpful reviews of the manuscript.

REFERENCES CITED

Algeo, T. J., and S. E. Scheckler, 1998, Terrestrial-marine teleconnections in the Devonian: Links between the evo lution of land plants, weathering processes, and marine anoxic events: Philosophical Transactions of the Royal So ciety of London Series B, v. 353, p. 113–130, doi:10.1098/rstb.1998.0195.

Arbenz, J. K., 1989, The Ouachita system, *in* A. W. Balley and A. R. Palmer, eds., The geology of North America: An overview: Geological Society of America, The Geology of North America, v. A, p. 371–396.

Armstrong, H. A., B. R. Turner, I. M. Makhlouf, G. P. Weedon, M. Williams, A. Al Smadi, and A. Abu Salah, 2005, Origin, sequence stratigraphy and depositional environment of an Upper Ordovician (Hirnantian) deglacial black shale, Jordan: Palaeogeography, Palaeoclimatology, Palaeoecology, v. 220, p. 273–289, doi:10.1016/j.palaeo.2005.01.007.

Beaumont, C., G. M. Quinlan, and J. Hamilton, 1987, The Alleghanian orogeny and its relationship to the evolution of the Eastern Interior, *in* C. Beaumont and A. J. Tankard, eds., Sedimentary basins and basin-forming mechanisms: Canadian Society of Petroleum Geologists Memoir 12, p. 425–445.

Beaumont, C., G. M. Quinlan, and J. Hamilton, 1988, Orogeny and stratigraphy: Numerical models of the Paleozoic in the Eastern Interior of North America: Tectonics, v. 7, p. 389–416, doi:10.1029/TC007i003p00389.

Boucot, A. J., M. T. Field, R. Fletcher, W. H. Forbes, R. Naylor, and L. Pavlides, 1964, Reconnaissance bedrock geology of the Presque Isle Quadrangle, Maine: Augusta, Maine Geological Survey Quadrangle Mapping Series 2, 123 p.

Charpentier, R. R., W. de Witt Jr., G. E. Claypool, L. D. Harris, R. F. Mast, J. D. Megeath, J. B. Roen, and J. W. Schmoker, 1993, Estimates of unconventional natural gas resources of the Devonian shales of the Appalachian basin, *in* J. B. Roen and R. C. Kepferle, eds., Petroleum geology of the Devonian and Mississippian black shales of eastern North America: U.S. Geological Survey Bulletin 1909, p. N1–N20.

Coakley, B., and M. Gurnis, 1995, Far-field tilting of Laurentia during the Ordovician and constraints on the evolution of a slab under an ancient continent: Journal of Geophysical Research, v. 100, p. 6313–6327, doi:10.1029/94JB02916.

Conant, L. C., and V. E. Swanson, 1961, Chattanooga Shale and related rocks of central Tennessee and nearby areas: U.S. Geological Survey Professional Paper 357, 91 p.

Denison, R. E., E. G. Lidiak, M. E. Bickford, and E. V. Kisvarsanyi, 1984, Geology and geochronology of Precambrian rocks in the Central Interior Region of the United States: U.S. Geological Survey Professional Paper 1241-C, 20 p.

Dennis, A. J., 2007, Cat Square Basin, Catskill clastic wedge: Silurian–Devonian orogenic events in the central Appalachians and the crystalline southern Appalachians, *in* J. W. Sears, T. A. Harms, and C. Evenchick, eds., Whence the mountains? Enquiries into the evolution of orogenic belts: A volume honoring Raymond A. Price: Geological Society of America Special Paper 433, p. 313–329.

Dennison, J. M., 1976, Appalachian Queenston Delta related to eustatic sea level drop accompanying Late Ordovician glaciation centered in Africa, *in* M. G. Bassett, ed., The Ordovician System: Palaeontological Association Symposium, Birmingham, September, 1974: Cardiff, University of Wales Press and National Museum of Wales, p. 107–120.

Dennison, J. M., 1989, Paleozoic sea level changes in the Appalachian Basin: 28th International Geological Congress Field Trip Guidebook T354: Washington, D. C., American Geophysical Union, 56 p.

DeWitt Jr., W., and G. W. Colton, 1959, Revised correlations of lower Upper Devonian rocks in western and central New York: AAPG Bulletin, v. 43, p. 2810–2828.

DeWitt Jr., W., and G. W. Colton, 1978, Physical stratigraphy of the Genesee Formation (Devonian) in western and central New York: U.S. Geological Survey Professional Paper 1032-A, 22 p.

Dorsch, J., R. K. Bambach, and S. G. Driese, 1994, Basin-rebound origin for the "Tuscarora unconformity" in southwestern Virginia and its bearing on the nature of the Taconic orogeny: American Journal of Science, v. 294, p. 237–255, doi:10.2475/ajs.294.2.237.

Dyni, J. R., 2005, Geology and resources of some world oil-shale deposits: U.S. Geological Survey Scientific Investigations Report 2005-5294, 42 p.

Engelder, T., G. C. Lash, and R. S. Uzcátegui, 2009, Joint sets that enhance production from Middle and Upper Devonian gas shales of the Appalachian Basin: AAPG Bulletin, v. 93, p. 857–889, doi:10.1306/03230908032.

Ettensohn, F. R., 1984, The nature and effects of water depth during deposition of the Devonian–Mississippian black shale: 1984 Eastern Oil Shale Symposium: Lexington, University of Kentucky Institute for Mining and Minerals Research and Kentucky Energy Cabinet, IMMR84/124, p. 333–346.

Ettensohn, F. R., 1985a, The Catskill delta complex and the Acadian orogeny: A model, *in* D. L. Woodrow and W. D. Sevon, The Catskill delta: Geological Society of America Special Paper 201, p. 39–49.

Ettensohn, F. R., 1985b, Controls on development of Catskill delta complex basin-facies, *in* D. L. Woodrow and W. D. Sevon, The Catskill delta: Geological Society of America Special Paper 201, p. 65–75.

Ettensohn, F. R., 1987, Rates of relative plate motion during the Acadian orogeny based on the spatial distribution of black shales: Journal of Geology, v. 95, p. 572–582.

Ettensohn, F. R., 1991, Flexural interpretation of relationships between Ordovician tectonism and stratigraphic sequences, central and southern Appalachians, U.S.A., *in* C. R. Barnes and S. H. Williams, eds., Advances in Ordovician geology: Geological Survey of Canada Paper 90-9, p. 213–224.

Ettensohn, F. R., 1994, Tectonic control on the formation and cyclicity of major Appalachian unconformities and associated stratigraphic sequences, *in* J. M. Dennison and F. R. Ettensohn, eds., Tectonic and eustatic controls on sedimentary cycles: SEPM Concepts in Sedimentology and Paleontology 4, p. 217–242.

Ettensohn, F. R., 1997, Assembly and dispersal of Pangea: Large-scale tectonic effects on coeval deposition of North American, marine, epicontinental black shales: Journal of Geodynamics, v. 23, p. 287–309, doi:10.1016/S0264-3707(96)00045-2.

Ettensohn, F. R., 2004, Modeling the nature and development of major Paleozoic clastic wedges in the Appalachian Basin, U.S.A.: Journal of Geodynamics, v. 37, p. 657–681, doi:10.1016/j.jog.2004.02.009.

Ettensohn, F. R., 2005, Chapter 5, The sedimentary record of foreland-basin, tectophase cycles: Examples from the Appalachian Basin, *in* J. M. Mabesoone and V. H. Neuman, eds., Cyclic development of sedimentary basins: Developments in Sedimentology 57: Amsterdam, Elsevier, p. 139–172.

Ettensohn, F. R., 2008, Chapter 4, The Appalachian foreland basin in the eastern United States, *in* A. Miall, ed., The Sedimentary basins of the United States and Canada: Sedimentary basins of the world: Amsterdam, Netherlands, Elsevier, p. 105–179.

Ettensohn, F. R., and C. E. Brett, 2002, Stratigraphic evidence from the Appalachian Basin for continuation of the Taconian orogeny into Early Silurian time, *in* C. E. Mitchell and R. Jacobi, eds., Taconic convergence: Orogen, foreland basin, and craton: Physics and Chemistry of the Earth, v. 27, p. 279–288, doi:10.1016/S1474-7065(01)00010-9.

Ettensohn, F. R., L. P. Fulton, and R. C. Kepferle, 1979, Use of scintillometer and gamma-ray logs for correlation and stratigraphy in homogeneous black shales: Geological Society of America Bulletin, v. 90, pt. I, p. 421–432, doi:10.1130/0016-7606(1979)902.0.CO;2.

Ettensohn, F. R., M. L. Miller, S. B. Dillman, T. D. Elam, K. L. Geller, D. R. Swager, G. Markowitz, R. D. Woock, and L. S. Barron, 1988, Characterization and implications of the Devonian–Mississippian black-shale sequence, eastern and central Kentucky, U.S.A.: Pycnoclines, transgression, regression, and tectonism, *in* N. J. McMillan, A. F. Embry, and G. J. Glass, eds., Devonian of the world: Second International Symposium on the Devonian System: Canadian Society of Petroleum Geologists Memoir 14, v. 2, p. 323–345.

Ettensohn, F. R., J. C. Hohman, M. A. Kulp, and N. Rast, 2002, Evidence and implications of possible far-field responses to Taconian orogeny: Middle–Late Ordovician Lexington Platform and Sebree Trough, east-central United States: Southeastern Geology, v. 41, p. 1–36.

Ettensohn, F. R., R. T. Lierman, C. E. Mason, A. J. Dennis, and E. D. Anderson, 2009, Kentucky dropstone "ices" the case for Late Devonian alpine glaciation in the central Appalachians: Implications for Appalachian tectonics and black-shale sedimentation, *in* F. R. Ettensohn, R. T. Lierman, C. E. Mason, and G. Clayton, eds., Changing physical and biotic conditions on eastern Laurussia, Field Trip 2: Cincinnati, North American Paleontological Convention, p. 44–51.

Faill, R. T., 1997, A geologic history of the north-central Appalachians: Part 1. Orogenesis from the Mesoproterozoic through the Taconic orogeny: American Journal of Science, v. 297, p. 551–619, doi:10.2475/ajs.297.6.551.

Friedman, G. M., and K. T. Johnson, 1966, The Devonian Catskill deltaic complex of New York, type example of a "tectonic delta complex," *in* M. L. Shirley, ed., Deltas and their geologic framework: Houston, Houston Geological Society, p. 172–188.

Furer, L. C., 1996, Basement tectonics in the southeastern part of the Illinois Basin and its effect on Paleozoic sedimentation, *in* B. A. van der Pluijm and P. A. Catacosinos, eds., Basement and basins of eastern North America: Geological Society of America Special Paper 308, p. 109–126.

Glover, L., 1959, Stratigraphy and uranium content of the Chattanooga Shale in northeastern Alabama, northwestern Georgia, and eastern Tennessee: U.S. Geological Survey Bulletin 1087-E, p. 133–168.

Gradstein, F. M., J. G. Ogg, and A. G. Smith, 2004, A geologic time scale: Cambridge, Cambridge University Press, 589 p.

Hasson, K. O., and J. M. Dennison, 1977, Devonian Harrell and Millboro shales in parts of Pennsylvania, Maryland, West Virginia, and Virginia, *in* G. L. Schott, W. K. Overby Jr., A. E. Hunt, and C. A. Komar, eds., First Eastern Gas Shales Symposium: U.S. Department of Energy, Morgantown Energy Research Center, MERC/SP-77/5, p. 634–636.

Hibbard, J. P., E. F. Stoddard, D. T. Secor, and A. J. Dennis, 2002, The Carolina Zone: Overview of Neoproterozoic to early Paleozoic peri-Gondwanan terranes along the eastern flank of the southern Appalachians: Earth-Science Reviews, v. 57, p. 299–339, doi:10.1016/S0012-8252(01)00079-4.

Jamieson, R. A., and C. Beaumont, 1988, Orogeny and metamorphism: A model for deformation and pressure-temperature-time paths with applications to the central and southern Appalachians: Tectonics, v. 7, p. 417–445, doi:10.1029/TC007i003p00417.

Johnson, J. G., 1971, Timing and coordination of orogenic, epeirogenic, and eustatic events: Geological Society of America Bulletin, v. 82, p. 3263–3298, doi:10.1130/0016-7606(1971)82[3263:TACOOE]2.0.CO;2.

Karabinos, P., S. D. Samson, J. C. Hepburn, and H. M. Stoll, 1998, The Taconian orogeny in New England: Collision between Laurentia and the Shelburne Falls arc: Geology, v. 26, p. 215–218, doi:10.1130/0091-7613(1998)026<0215:toitne >2.3.co;2.

Keith, B. D., 1989, Regional facies of the Upper Ordovician Series of eastern North America, *in* B. D. Keith, ed., The Trenton Group (Upper Ordovician Series) of eastern North America: Deposition, diagenesis, and petroleum: AAPG Studies in Geology 29, p. 1–16.

Kepferle, R. C., E. N. Wilson, and F. R. Ettensohn, 1978, Preliminary stratigraphic cross section showing radioactive zones in the Devonian black shales in the southern part of the Appalachian Basin: U.S. Geological Survey Oil and Gas Investigations Chart OC-85, 1 sheet.

Klein, G. D., 1994, Depth determination and quantitative distinction of the influence of tectonic subsidence and climate on changing sea level during deposition of mid-continent Pennsylvanian cyclothems, *in* J. M. Dennison and F. R. Ettensohn, eds., Tectonic and eustatic controls on sedimentary cycles: SEPM Concepts in Sedimentology and Paleontology, v. 4, p. 35–50.

Mankin, C. J., 1987, Texas-Oklahoma tectonic region: AAPG COSUNA Chart TOT, 1 sheet.

Matthews, R. D., 1993, Review and revision of the Devonian–Mississippian stratigraphy in the Michigan Basin, *in* J. B. Roen and R. C. Kepferle, eds., Petroleum geology of the Devonian and Mississippian black shale of eastern North America: U.S. Geological Survey Bulletin, v. 1909, p. D1–D85.

McBride, E. F., 1962, Flysch and associated beds of the Martinsburg Formation (Ordovician), central Appalachians: Journal of Sedimentary Petrology, v. 32, p. 39–91, doi:10.1306/74D70C45-2B21-11D7-8648000102C1865D.

McClellan, E. A., M. G. Steltenpohl, C. Thomas, and C. F. Miller, 2005, Isotopic age constraints and metamorphic history of the Talladega belt: New Evidence for timing of arc magmatism and terrane emplacement along the southern Laurentian margin, *in* M. G. Steltenpohl, ed., Southernmost Appalachian terranes, Alabama and Georgia: Field Trip Guidebook for the Geological Society of America Southeastern Section 2005 Annual Meeting: Tuscaloosa, Alabama Geological Society, p. 19–50.

McIver, N. L., 1970, Appalachian turbidites, *in* G. W. Fisher, F. S. Pettijohn, J. C. Reed, and K. N. Weaver, eds., Studies of Appalachian geology: New York, Interscience Publishers, p. 69–81.

McKerrow, W. S., 1979, Ordovician and Silurian changes in sea level: Journal of the Geological Society (London), v. 16, p. 17–145.

Merschat, A. J., and R. D. Hatcher Jr., 2007, The Cat Square terrane: Possible Siluro–Devonian remnant ocean basin in the Inner Piedmont, southern Appalachians, U.S.A., *in* R. D. Hatcher Jr., M. P. Carlson, J. H. McBride, and J. R. Martínez Catalán, eds., 4-D framework of the continental crust: Geological Society of America Memoir 200, p. 553–565.

Milici, R. C., and C. S. Swezey, 2006, Assessment of Appalachian basin oil and gas resources: Devonian Shale—Middle and Upper Paleozoic Total Petroleum System: U.S. Geological Survey Open-File Report Series 2006-1237, 70 p.

Milici, R. C., R. T. Ryder, C. S. Swezey, R. R. Charpentier, T. A. Cook, R. A. Crovelli, T. R. Klett, R. M. Pollastro, and C. J. Schenk, 2003, Assessment of undiscovered oil and gas resources of the Appalachian basin province, 2002: U.S. Geological Survey Fact Sheet FS-009-02, 2 p.

Mitrovica, J. X., C. Beaumont, and G. T. Jarvis, 1989, Tilting of continental interiors by the dynamical effects of subduction: Tectonics, v. 8, p. 1079–1094, doi:10.1029/TC008i005p01079.

Newton, C. R., 1979, Aerobic, dysaerobic and anaerobic facies within the Needmore Shale (Lower to Middle Devonian), *in* J. M. Dennison, K. O. Hasson, D. M. Hoskins, R. M. Jolly, and W. D. Sevon, eds., Devonian shales in south-central Pennsylvanian and Maryland: Guidebook, 44th Annual Field Conference of Pennsylvania Geologists, p. 56–60.

Nowacki, P., 1981, Oil shale technical data handbook: Park Ridge, Noyes Data Corporation, 309 p.

Patchen, D. G., K. L. Avary, and R. B. Erwin, 1985a, Southern Appalachian region: AAPG COSUNA Chart SAP, 1 sheet.

Patchen, D. G., K. L. Avary, and R. B. Erwin, 1985b, Northern Appalachian region: AAPG COSUNA Chart NAP, 1 sheet.

Pepper, J. F., and W. deWitt Jr., 1950, Stratigraphy of the Upper Devonian Wiscoy Sandstone and equivalent

Hanover Shale in western and central New York: U.S. Geological Chart OC-37, 1 sheet.

Pepper, J. F., and W. deWitt Jr., 1951, Stratigraphy of the Late Devonian Perrysville Formation in western and west-central New York: U.S. Geological Chart OC-45, 1 sheet.

Pepper, J. F., W. deWitt Jr., and G. W. Colton, 1956, Stratigraphy of the West Falls Formation of Late Devonian age in western and west-central New York: U.S. Geological Chart OC-55, 1 sheet.

Perkins, R. B., D. Z. Piper, and C. E. Mason, 2008, Trace-element budgets in the Ohio/Sunbury shales of Kentucky: Constraints on ocean circulation and primary productivity in the Devonian–Mississippian Appalachian Basin: Palaeogeography, Palaeoclimatology, Palaeoecology, v. 265, p. 14–29, doi:10.1016/j.palaeo.2008.04.012.

Quinlan, G. M., and C. Beaumont, 1984, Appalachian thrusting, lithospheric flexure, and the Paleozoic stratigraphy of the Eastern Interior of North America: Canadian Journal of Earth Science, v. 21, p. 973–996, doi:10.1139/e84-103.

Rhoads, D. C., and J. W. Morse, 1971, Evolutionary and ecological significance of oxygen-deficient marine basins: Lethaia, v. 4, p. 413–438, doi:10.1111/j.1502-3931.1971.tb01864.x.

Rickard, L. V., 1975, Correlation of Silurian and Devonian rocks in New York state: New York State Museum and Science Service, Geological Survey Map and Chart Series 24, 4 plates.

Rimmer, S. M., J. A. Thompson, S. A. Goodnight, and T. L. Robl, 2004, Multiple controls on the preservation of organic matter in the Devonian–Mississippian marine black shales: Geochemical and petrographic evidence: Palaeogeography, Palaeoclimatology, Palaeoecology, v. 215, p. 125–154, doi:10.1016/j.palaeo.2004.09.001.

Robinson, P., R. D. Tucker, D. Bradley, H. N. Berry IV, and P. H. Osberg, 1998, Paleozoic orogens in New England: Geologiska Föreningens I Stockholm Förhandlingar (GFF), v. 120, p. 119–148.

Roen, J. B., and R. C. Kepferle, eds., 1993, Petroleum geology of the Devonian and Mississippian black shale of eastern North America: U.S. Geological Survey Bulletin, v. 1909.

Roen, J. B., R. L. Miller, and J. W. Huddle, 1964, The Chattanooga Shale (Devonian and Mississippian) in the vicinity of Big Stone Gap, Virginia: U.S. Geological Survey Professional Paper 501-B, p. B43–B48.

Roen, J. B., L. G. Wallace, and W. deWitt Jr., 1978, Preliminary stratigraphic cross section showing radioactive zones in the Devonian black shales in the central Appalachian Basin: U.S. Geological Survey Oil and Gas Investigations Chart OC-87, 1 sheet.

Russell, P. L., 1990, Oil shales of the world: Their origin, occurrence, and exploration: Oxford, Pergamon Press, 753 p.

Ryder, R. T., 1991, Stratigraphic framework of Cambrian and Ordovician rocks in the central Appalachian Basin from Richland County, Ohio, to Juniata County, Pennsylvania: U.S. Geological Survey Miscellaneous Investigations Series Map I-2264, 1 sheet.

Ryder, R. T., 2002a, Stratigraphic framework of Cambrian and Ordovician rocks in the Central Appalachian Basin from Lake County, Ohio, through Juniata County, Pennsylvania (revised and digitized by R. D. Crangle Jr.): U.S. Geological Survey Miscellaneous Investigations Series Map I-2200, 1 section.

Ryder, R. T., 2002b, Stratigraphic framework of Cambrian and Ordovician rocks in the Central Appalachian Basin from Richland County, Ohio, to Rockingham County, Virginia (revised and digitized by R. D. Crangle Jr.): U.S. Geological Survey Miscellaneous Investigations Series Map I-2264, 1 section.

Ryder, R. T., 2008, Stratigraphic framework of Cambrian and Ordovician rocks in the Appalachian Basin from Sequatchie County, Tennessee, through eastern Kentucky, to Mingo County, West Virginia: U.S. Geological Survey Scientific Investigations Map 2994, 1 sheet.

Ryder, R. T., A. G. Harris, and J. E. Repetski, 1992, Stratigraphic framework of Cambrian and Ordovician rocks in the central Appalachian Basin from Medina County, Ohio, through southwestern and south-central Pennsylvania to Hampshire County, West Virginia: U.S. Geological Survey Bulletin, v. 1839-K, p. K1–K32.

Ryder, R. T., J. E. Repetski, and A. G. Harris, 1996, Stratigraphic framework of Cambrian and Ordovician rocks in the central Appalachian Basin from Fayette County, Ohio, to Botetourt County, Virginia: U.S. Geological Survey Miscellaneous Investigations Series Map I-2495, 1 sheet.

Ryder, R. T., R. C. Burruss, and J. R. Hatch, 1998, Black shale source rocks and oil generation in the Cambrian and Ordovician of the central Appalachian Basin, U.S.A.: AAPG Bulletin, v. 82, p. 412–441.

Ryder, R. T., C. S. Swezey, R. D. Crangle Jr., and M. H. Trippi, 2008, Geologic cross section EE′ through the Appalachian Basin from the Findlay Arch, Wood County, Ohio, to the Valley and Ridge Province, Pendleton County, West Virginia: U.S. Geological Survey Scientific Investigations Map 2985, 2 sheets.

Ryder, R. T., R. D. Crangle Jr., M. H. Trippi, C. S. Swezey, E. E. Lentz, E. L. Rowan, and R. S. Hope, 2009, Geologic cross section DD′ through the Appalachian Basin from the Findlay Arch, Sandusky County, Ohio, to the Valley and Ridge province, Hardy County, West Virginia: U.S. Geological Survey Scientific Investigations Map 3067, 2 sheets.

Schwalb, H., 1980, Hydrocarbon entrapment along a Middle Ordovician disconformity: Kentucky Geological Survey, Series XI, Special Publication 2, p. 35–41.

Shumaker, R. C., 1986, Structural development of Paleozoic continental basins of eastern North America: 6th International Conference on Basement Tectonics, p. 82–95.

Stille, H., 1920, Über Alter und Art der Phasen varischer Gebirgsbildung: Nachrichten der königlichen Gesellschaft der Wissenschaften zu Göttingen, mathematische-physikalische Klasse, v. 1920, p. 218–224.

Su, W., W. D. Huff, F. R. Ettensohn, X. Liu, and Z. Li, 2009, K-bentonite, black-shale, and flysch successions at the Ordovician–Silurian transition, south China: Possible sedimentary responses to the accretion of Cathaysia to the Yangtze Block and its implications for the evolution of Gondwana: Gondwana Research, v. 15, p. 111–130, doi:10.1016/j.gr.2008.06.004.

Thomas, W. A., 2006, Tectonic inheritance at a continental margin: Geological Society of America Today, v. 16, p. 4–11.

Trupe, C. H., K. G. Stewart, M. G. Adams, C. L. Waters, B. V. Miller, and L. K. Hewitt, 2003, The Burnsville fault: Evidence for the timing and kinematics of southern Appalachian Acadian dextral transform tectonics: Geological Society of America Bulletin, v. 115, p. 1365–1376, doi:10.1130/?B25256.1.

van Staal, C. R., 1994, Brunswick subduction complex in the Canadian Appalachians: Record of Late Ordovician to Late Silurian collision between Laurentia and the Gander margin of Avalon: Tectonics, v. 13, p. 946–962, doi:10.1029/93TC03604.

Wallace, L. G., J. B. Roen, and W. deWitt Jr., 1977, Preliminary stratigraphic cross section showing radioactive zones in the Devonian black shales in the western part of the Appalachian Basin: U.S. Geological Survey Oil and Gas Investigations Chart OC-80, 1 sheet.

Wallace, L. G., J. B. Roen, and W. deWitt Jr., 1978, Preliminary stratigraphic cross section showing radioactive zones in the Devonian black shales in southeastern Ohio and west-central West Virginia: U.S. Geological Survey Oil and Gas Investigations Chart OC-83, 1 sheet.

Webby, B. D., 1995, Toward and Ordovician time scale, *in* J. D. Cooper, M. L. Droser, and S. C. Finney, Ordovician odyssey: Short papers for the Seventh International Symposium on the Ordovician System: Fullerton, Pacific Section, Society for Sedimentary Geology (SEPM), p. 5–9.

Werne, J. P., B. B. Sageman, T. W. Lyons, and D. J. Hollander, 2002, An integrated assessment of a ''type euxinic'' deposit: Evidence for multiple controls on black-shale deposition in the Middle Devonian Oatka Creek Formation: American Journal of Science, v. 302, p. 110–143, doi:10.2475/ajs.302.2.110.

Wilson, E. N., J. S. Zafar, and F. R. Ettensohn, 1981, Southern tie section: A stratigraphic section through the Devonian–Mississippian black-shale sequence in Ohio, Kentucky, West Virginia, and Virginia: U.S. Department of Energy METC/EGSP Series 501, 1 sheet.

Woodrow, D. L., J. M. Dennison, F. R. Ettensohn, W. D. Sevon, and W. T. Kirchgasser, 1988, The Middle and Upper Devonian of the Appalachian Basin, United States, *in* N. J. McMillan, A. F. Embry, and D. J. Glass, eds., Devonian of the world: Second International Symposium on the Devonian System: Canadian Society of Petroleum Geologists Memoir 14, v. 1, p. 277–301.

Ziegler, P. A., 1978, Northeastern Europe: Tectonics and basin development: Geologie en Mijnbouw, v. 57, p. 589–626.

5

Gomes, Paulo Otávio, Bill Kilsdonk, Tim Grow, Jon Minken, and Roberto Barragan, 2012, Tectonic evolution of the Outer High of Santos Basin, southern São Paulo Plateau, Brazil, and implications for hydrocarbon exploration, *in* D. Gao, ed., Tectonics and sedimentation: Implications for petroleum systems: AAPG Memoir 100, p. 125–142.

Tectonic Evolution of the Outer High of Santos Basin, Southern São Paulo Plateau, Brazil, and Implications for Hydrocarbon Exploration

Paulo Otávio Gomes

Hess Brasil Petróleo Limitada, Praia de Botafogo 501, Torre Corcovado, 2° andar, Rio de Janeiro, Brazil (e-mail: paulo-otavio@hotmail.com)

Bill Kilsdonk and Tim Grow

Hess Corporation, 1501 McKinney St., Houston, Texas, 77010, U.S.A. (e-mails: bkilsdonk@hess.com; tgrow@hess.com)

Jon Minken

Hess Exploration Australia, 77 St. Georges Terr., Perth, 6600, Western Australia (e-mail: jminken@hess.com)

Roberto Barragan

Hess Exploration Malaysia, 207 Jalan Tun Razak, 50400, Kuala Lumpur, Malaysia (e-mail: rbarragan@hess.com)

ABSTRACT

Multiple geologic elements combine in the presalt section of deep-water Santos Basin, forming a quite unique exploration play: prolific and mature source rocks are present, synrift structures include huge intrabasinal highs, and the overlying evaporite seal extends throughout most of the area. Significant uncertainties related to reservoir presence and deliverability, which still persist as the key risks on this emerging presalt play, have been progressively reduced throughout a continued drilling campaign, which started in 2006.

The most prominent and extensive intrabasinal high in the region is the Outer High of Santos Basin, a regional basement structure that forms a 12,000 km^2 (4633 mi^2) four-way closure at the Aptian level. The geologic history of the Outer High involves multiepisodic uplift and erosion of a series of rift fault-block shoulders during the Barremian. At that time, regional uplift resulted from a failed sea-floor spreading process that emplaced proto-oceanic crust in the southern Santos Basin. Concurrently, magmatic underplating is postulated as the mechanism responsible for locally thickening the crust and isostatically holding the Outer High as a present-day

DOI:10.1306/13351550M1003530

positive feature above its surroundings. Because of the extreme extension of the continental crust in Santos Basin, zones of deep crustal, or even upper mantle exhumation, are also expected near the transition from continental to oceanic crust.

Before continental breakup, the Outer High was roughly located 200 km (124 mi) away from both the African and Brazilian hinge lines. This distal setting, coupled with a positive relief, limited siliciclastic input from the margins. The presence of a long-lived paleohigh, in such a clastic-starved environment, favored the development of a broad carbonate platform, during the Lower Aptian. Tectonically controlled water-level fluctuations affected the evolving platform, serving an important function on reservoir facies development.

The Outer High has been the core region of a deep-water presalt exploration outbreak, after a pioneering drilling campaign that started around the middle of this decade. The hydrocarbon potential of this vast frontier area is yet to be fully unraveled.

INTRODUCTION

The deep-water region of Santos Basin, located in the heart of the São Paulo Plateau (Figure 1), has been the focus of an extensive exploration campaign in recent years. As a classical salt basin, this area is characterized by the presence of a thick evaporite layer deposited during the breakup of the Gondwana supercontinent (Pereira and Macedo, 1990; Karner and Gamboa, 2007) (Figure 2). Because of the similarities in terms of petroleum systems and tectonostratigraphic evolution, Santos Basin has been assembled together with other salt basins to the north, the Campos and Espírito Santo basins, forming a single geologic province, commonly known as the Greater Campos Basin (Mello et al., 2002).

The exploration activity in the Greater Campos Basin has been tied to the development of technology in drilling and well engineering, as well as to remarkable advances in seismic imaging and geoscience interpretation. Exploration efforts moved into progressively deeper waters since the 70s to both the 80s and 90s. Application of leading-edge technology, coupled with the introduction of new or alternative play concepts, has brought another significant shift in the present decade, focusing the search on deeper and/or unconventional targets in the Greater Campos area (Carminatti et al., 2009).

These technological advances, combined with an economic environment resulting from record highs in both realized and forecast oil prices, opened a window of opportunity within the exploration business and created a favorable scenario for the investigation of challenging high-risk, high-reward exploration plays. In this environment, it finally became possible to reach the ultra-deep-water presalt play of Santos Basin (Note 1).

To test this frontier play, a pioneering drilling campaign was conducted over a broad region known as the Outer High of Santos Basin (Gomes et al., 2002) or the Santos External High (Carminatti et al., 2008). This campaign tested, for the first time, the presalt section of the basin in ultra-deep water (Note 2). The results have been multiple discoveries and recoverable hydrocarbon reserves in multibillion barrel volumes.

The huge potential of the Outer High was recognized early by those companies who acquired exploration acreage in the region between 2000 and 2001. By that time, regional geologic evaluations had identified several characteristics of the Outer High region that were uniquely combined to form a very attractive presalt exploration play: (1) the likely presence of prolific and mature source rocks; (2) synrift structures including multiple, large intrabasinal highs, in the order of hundreds of square kilometers, which may have both trapped hydrocarbons and focused migration; and (3) the overlying evaporite seal present throughout most of the area.

In contrast to the recognized lower risk play elements enumerated above, reservoir presence and deliverability persisted as poorly quantified risk elements in the area. In fact, different approaches to regional geologic evaluation and seismic interpretation fueled a long and heated debate between proponents of siliciclastic reservoir models and proponents of carbonate reservoir models for the presalt section.

Although the initial presalt drilling campaign established a deep-water carbonate trend and reduced some of the uncertainties on the reservoir front (Carminatti et al., 2008), key issues remain to be addressed for the commercial development of this play. Notwithstanding the remaining uncertainties, the exploration success to date has fostered great expectations for untapped oil and gas resources in this vast and underexplored frontier exploration acreage (Mello et al., 2006). Note 1: A presalt play is here defined as an exploration play where the reservoir targets are older than the overlying evaporite package, which has never detached away from its original depositional interval. This is significantly different from the classical Gulf of Mexico subsalt play, where the targets are younger than the overlying allochthonous salt. Note 2: Although the presalt sequence in Campos Basin was reasonably investigated in the neighboring

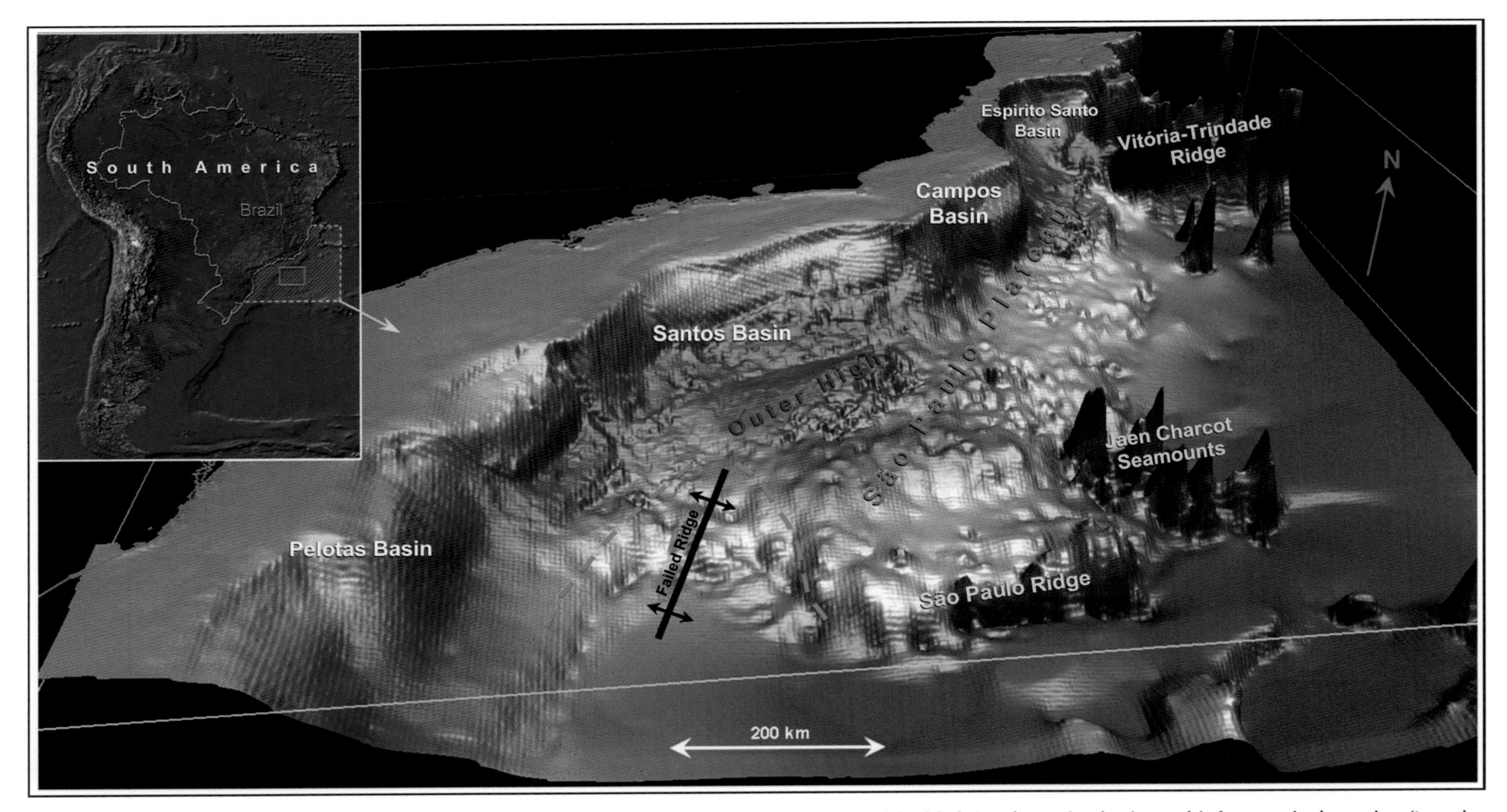

Figure 1. Three-dimensional bathymetric diagram of the southeastern continental margin off Brazil, highlighting the main physiographic features in the region (Location is indicated by the dashed yellow polygon on the map. Vertical exaggeration, 50×). The São Paulo Plateau, extending from Santos to Espírito Santo Basin, is mostly underlain by stretched continental crust. The present-day morphology of the plateau is strongly affected by halokinesis. The Outer High of Santos Basin, as a long-lived structural feature, can be also identified in the diagram. The Failed Ridge represents the region associated to the emplacement of a protooceanic crust, which was strongly affected by postbreakup subsidence, resulting in a present-day starved basin. The smaller red rectangle on the map indicates the focus area for this study. 200 km (124 mi).

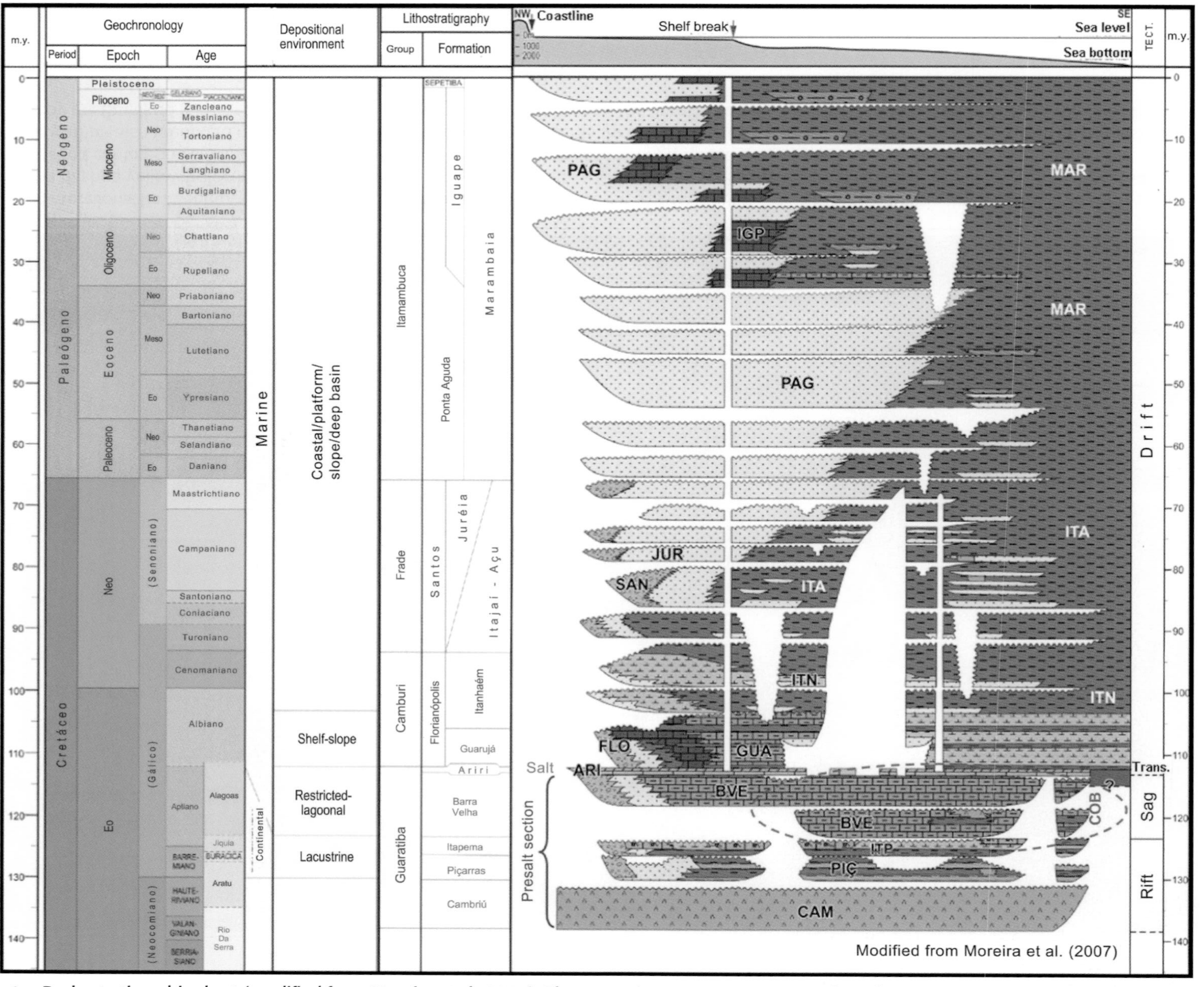

Figure 2. Santos Basin stratigraphic chart (modified from Moreira et al., 2007). The evaporite sequence corresponds to the Aptian Ariri Formation, where a thickness of roughly 2 km (6562 ft) of salt was deposited in approximately 0.6 m.y. (Dias, 2005). The focus of this study is the presalt section from Late Barremian to Late Aptian, indicated by the red ellipse. COB = continent-ocean boundary; Tect. = Tectonics; Trans. = Transitional.

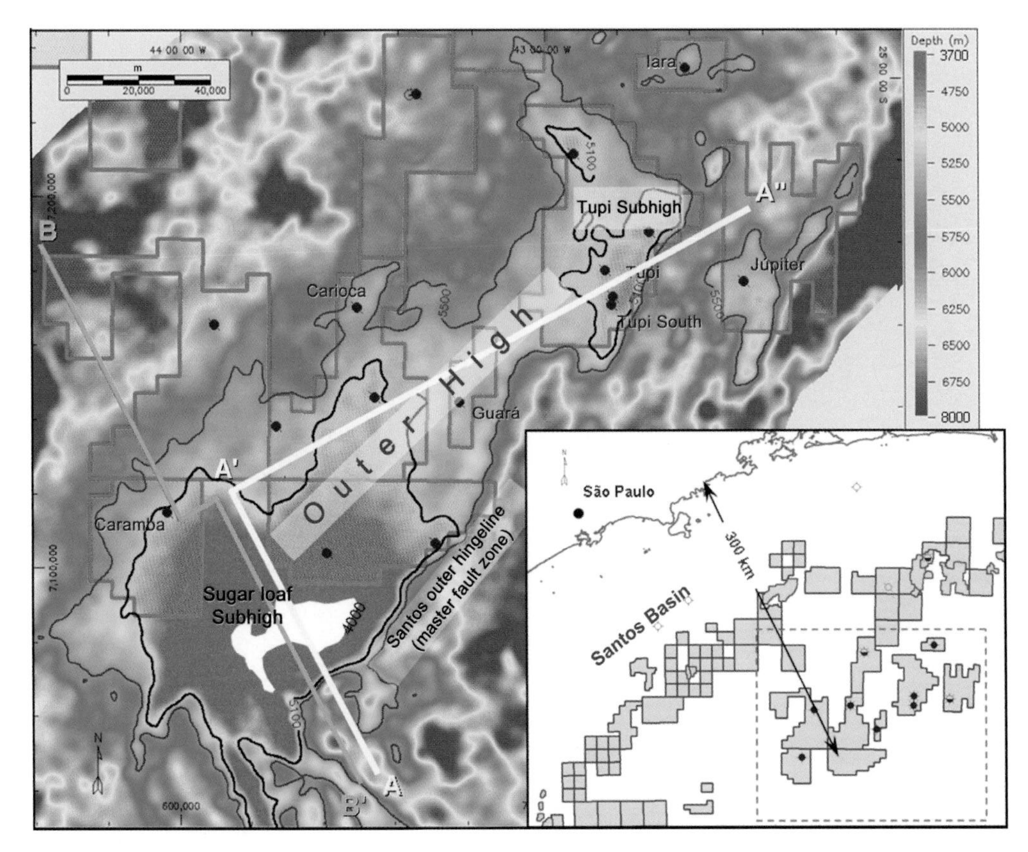

Figure 3. Base salt structure map of the Outer High of Santos Basin (Aptian, ca. 113 m.y.), showing some of the key presalt wells in the region. Selected contours highlight both the Outer High structure (blue contour, 5500 m [18,000 ft] subsea) and the two main individual culminations within the megaclosure—the Tupi and Sugar Loaf subhighs (black contours, 5100 m [16,700 ft] subsea). The depth to the crest of Sugar Loaf is approximately 3700 m (~12,100 ft). The mapped area represents the dashed red rectangle on the lower right figure (red rectangle on Figure 1). Seismic transects A–A′–A″ and BB′ are displayed on Figures 4 and 5, respectively. 300 km (186 mi).

Campos Basin, to the north, with approximately 150 wells drilled up to 1990 (Figueiredo and Martins, 1990), only a couple of wells were drilled in water depths greater than 4900 ft (>1500 m) in the region. In the Santos area, specifically, only four wells had previously reached the presalt section of the basin, before the recent Outer High campaign, and all of them were drilled in a shallow-water, proximal setting.

THE OUTER HIGH OF SANTOS BASIN

The base of the evaporite package at the Greater Campos province is marked by a regional unconformity, with an estimated age between 112 and 113 m.y. (França et al., 2007; Moreira et al., 2007). Its present-day structural expression was largely inherited from its Aptian paleotopography, which was shaped by crustal extension and related phases of mechanical and thermal subsidence. The seismic horizon associated with this geologic boundary can be mapped across the entire Greater Campos, showing a series of very large structural closures, which are associated to the presence of basement paleohighs. These structures developed during the extension and rifting of the continental crust and were later affected by postbreakup epeirogenic events and/or periodic strike-slip reactivation of the basement fabric (Fetter, 2009).

The most prominent and extensive intrabasinal high within the Greater Campos region is the Outer High of Santos Basin (Gomes et al., 2002; Gomes et al., 2008). Located in the distal deep-water part of Santos Basin, the Outer High forms a 12,000 km^2 (4633 mi^2) four-way closure at the base salt level, which is mostly covered by a thick layer of evaporite seals (Figures 3–5). This regional basement structure likely represents a segmented series of rift fault-block shoulders that were uplifted and eroded during the Late Barremian.

In fact, the Outer High contains several individual structures within a larger megaclosure, including two main presalt culminations (Modica and Brush, 2004). The lower one of these two culminations, known as the Tupi structure, after the homonym hydrocarbon discovery, forms a closure with an area of approximately 1100 km^2 (~425 mi^2) at the base salt level. The higher (crestal) and larger culmination is widely referred to as the Sugar Loaf structure. Sugar Loaf has a closure covering an area of approximately 6000 km^2 (~2317 mi^2) at the base salt level (Figure 3).

Both the Tupi and Sugar Loaf are long-lived structures that focused the migration of oil and gas generated in the thick presalt section over large fetch areas. The focusing character of these structures, probably enhanced by the presence of regional carrier beds, led the Outer High to be considered as the most favorable

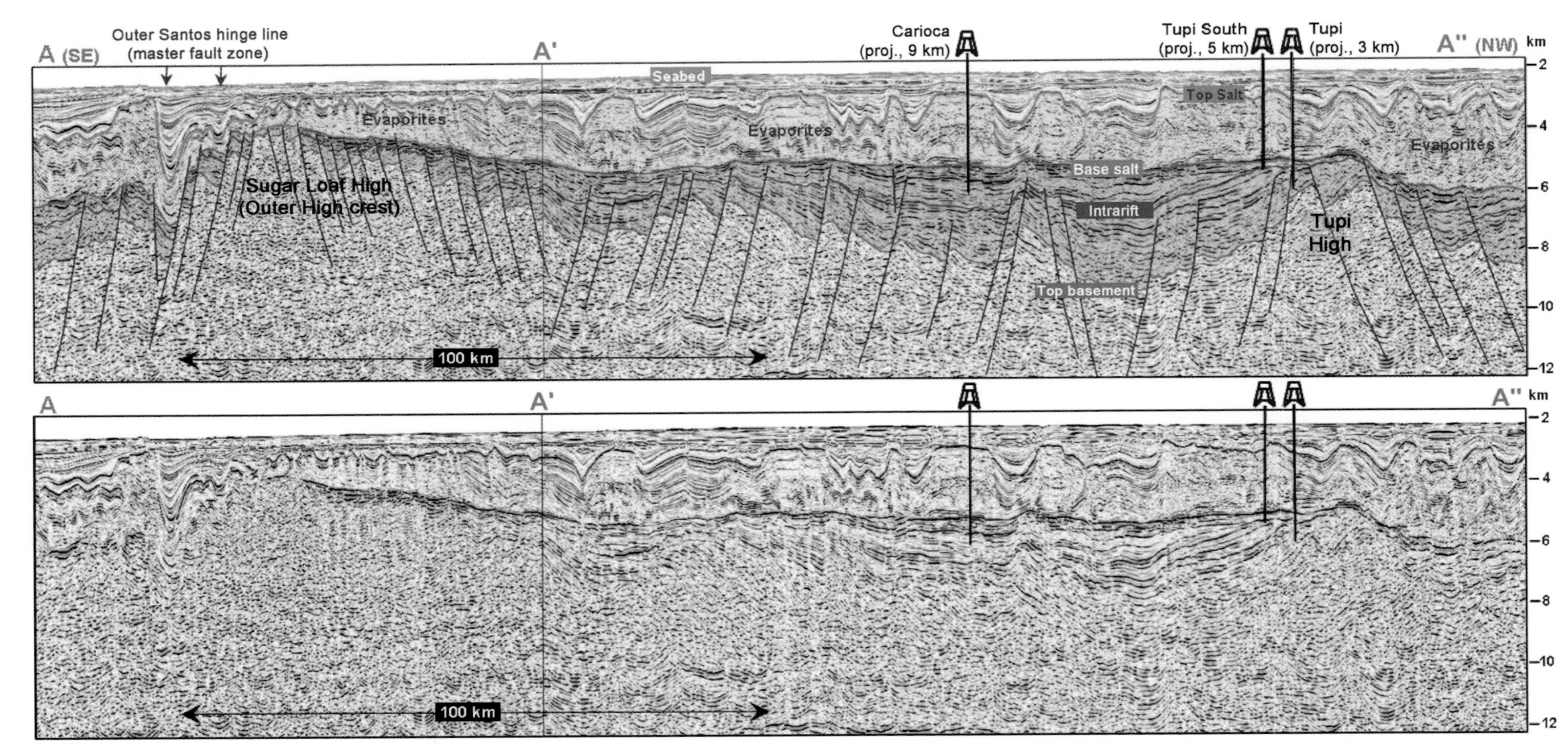

Figure 4. Regional seismic section A-A′-A″ across the Outer High of Santos Basin, with interpreted (top) and noninterpreted versions, showing the Tupi and Sugar Loaf subhighs, and the projected position of three key presalt wells (see Figure 3 for location). Note the thinning of both the synrift and sag sections against the basement culminations. The purple section, composed of upper synrift and sag, is expected to be dominated by carbonates. If not bald, the crest of the Sugar Loaf structure may be formed by early rift volcanic infill. Note also the outer Santos hinge line, a master fault region at the southeastern flank of the Outer High, showing a major throw at base salt level. The estimated ages for the horizons are as follows: top salt, approximately 112 m.y.; base salt, approximately 113 m.y.; intrarift, approximately 123 m.y.; top basement, approximately 132 m.y. (ages based on Moreira et al., 2007). Two-dimensional wave equation pre-stack depth migration data were used with the kind permission of TGS-Nopec Geophysical Company. proj. = projected. 100 km (62 mi).

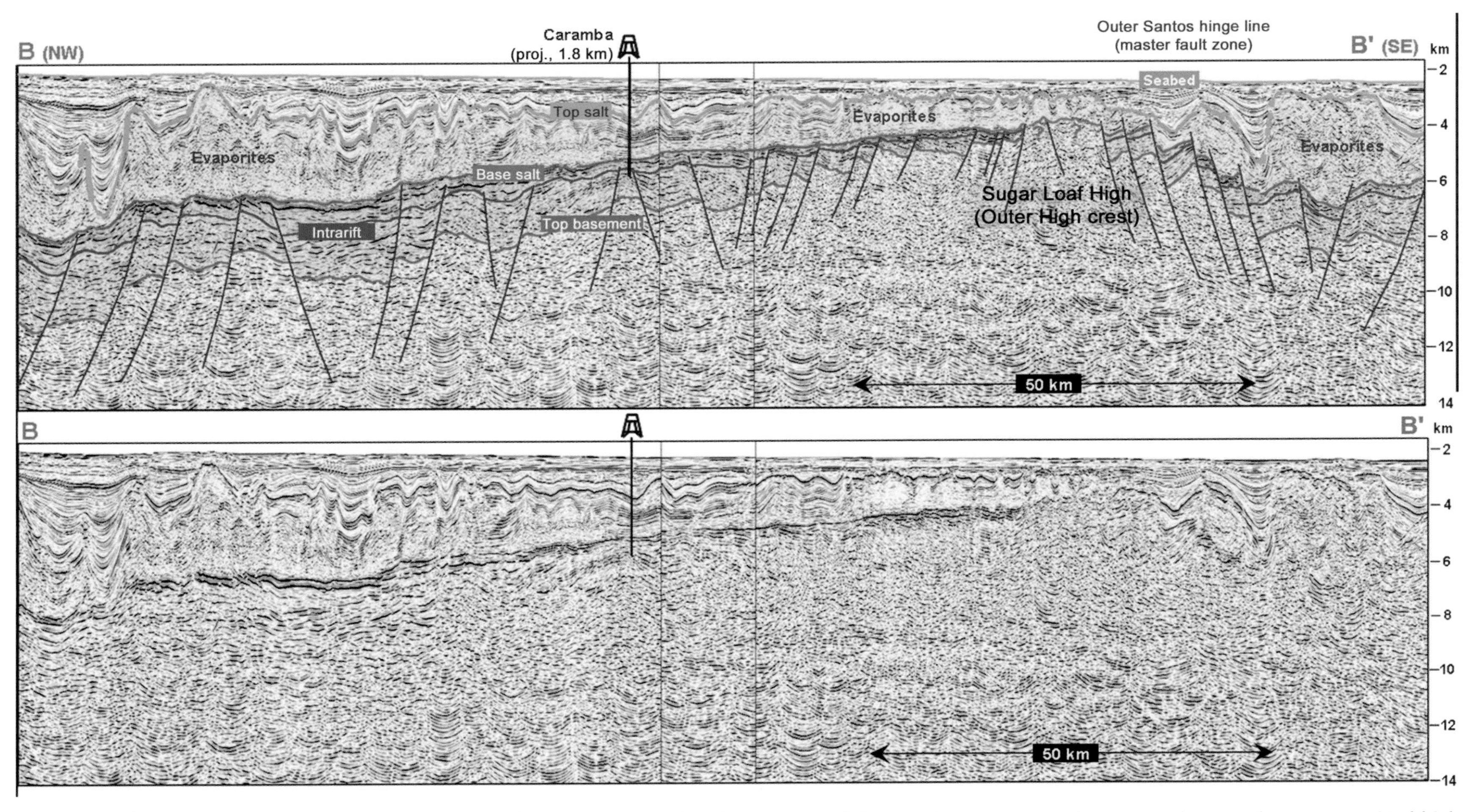

Figure 5. Regional seismic section BB′ across the Outer High of Santos Basin, with interpreted (top) and noninterpreted versions, showing the Sugar Loaf Subhigh, and the projected position of a key presalt well (see Figure 3 for location). Again, note the thinning of both the synrift and sag sections against the basement culmination and the expansion of the presalt section to the southeast of the outer Santos hinge line. The estimated ages for the horizons are as follows: top salt, approximately 112 m.y.; base salt, approximately 113 m.y.; intrarift, approximately 123 m.y.; top basement, approximately 132 m.y. Two-dimensional wave equation pre-stack depth migration data were used with the kind permission of TGS-Nopec Geophysical Company. proj. = projected. 50 km (31 mi).

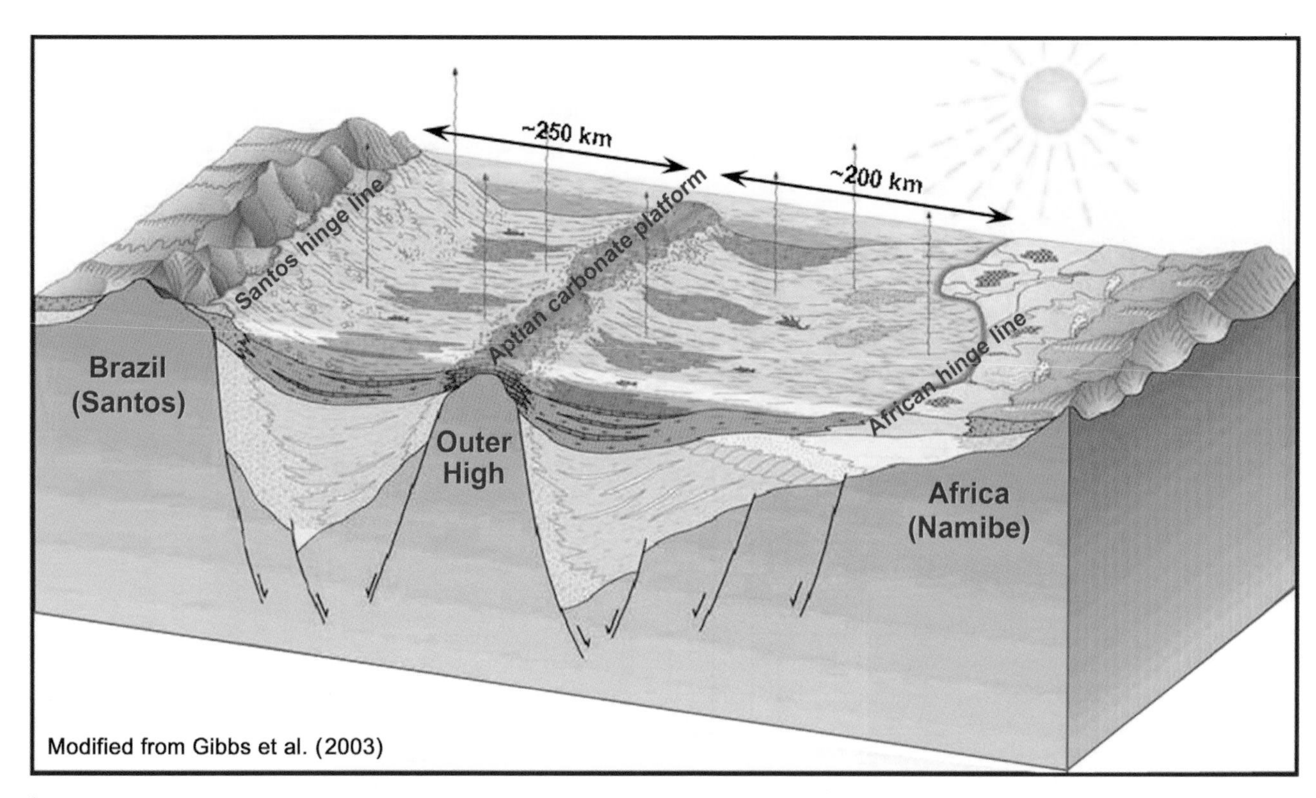

Figure 6. Simplified schematic diagram showing the depositional environment in the Santos-Namibe conjugate margin before salt deposition and continental breakup, with the relative position of the Outer High to the basin hinge lines (modified from Gibbs et al., 2003). 200 km (124 mi).

region to test the presalt play concept in deep-water Santos Basin. Another attractive element of the petroleum system in the Outer High region is seal efficiency, which is linked to the presence of a thick and continuous evaporite layer throughout most of the area. This situation contrasts with the inboard parts of the Santos and Campos basins, where salt windows are generally more common.

Notwithstanding critical reservoir uncertainties, which remain as strong risk factors, the combination of the remaining elements created a positive scenario for the presence of presalt hydrocarbon accumulations in the Outer High region. This assumption is particularly valid for the various individual four-way closures located within the megastructure and also applicable to some satellite structures along its northern and southwestern flanks. Those were the primary structures tested in the pioneering drilling campaign launched in 2006.

PALEOGEOGRAPHIC SETTING AND RESERVOIR IMPLICATIONS

Based on plate tectonic reconstructions of the Gondwana supercontinent (Davison, 1999; Scotese, 2001; Moulin et al., 2010), and taking the ocean-continent boundaries interpreted in both sides of the South Atlantic (Karner, 2000; Davison, 2007; Karner and Gamboa, 2007), it is possible to estimate that the Outer High of Santos Basin was located at least 200 km (124 mi) away from both the African and the Brazilian hinge lines, before continental breakup (Figure 6). This distal setting, coupled with the paleotopographic relief of the Outer High, resulted in a clastic-starved environment, which persisted from the Late Barremian throughout evaporite deposition. Whereas sediment input from perennial rivers was unlikely in this scenario, local sediment provenance may have been generated from erosion of the crestal part of the high. However, the provenance area would have been restricted to the exposed part of the structure, yielding a relatively small supply of clastic material. In addition, the volcanic nature of the crest of the Outer High, as indicated by gravity and magnetic modeling, would source sedimentologically immature strata, deteriorating the reservoir quality of siliciclastic facies.

In contrast, the bathymetric expression of a series of distal, intermittently exposed basement highs, in the clastic-starved environment described above, was ideal for the development of a broad carbonate platform in the Lower Aptian (Figure 6). Tectonically controlled water-level fluctuations, linked to the extension, faulting, and thermal sag, would dictate the development of this carbonate platform, leading to periods of intense growth during subsidence, and karstification as a result of uplift and subaerial exposure. The latter of these processes may have played an important function in the improvement of reservoir quality facies.

Public data from key wells, drilled in various locations over the Outer High province, confirm the presence of a carbonate-dominated upper presalt section in the ultra-deep-water region of Santos Basin (Figure 7). The carbonate environment probably spanned from the final stages of the Barremian to the Late Aptian. The most likely depositional model during this period involves a transition from a brackish-to-saline lacustrine environment, dominated by bioclastic carbonate facies (coquinas), to a broad saline epicontinental lake, likely affected by sporadic marine incursions and dominated by microbial limestones (Carminatti et al., 2008). These microbialites formed extensive deposits on the broad Aptian carbonate platform, just before evaporite deposition. A similar transition is observed at the distal part of Campos Basin, where the lacustrine Coquinas sequence (Carvalho et al., 2000) is overlain by marginal-marine carbonate deposits mainly associated with microbial limestones (Dias, 2005).

A key challenge to overcome in the Santos presalt carbonate play is related to the building of consistent models for reservoir facies distribution. Reservoir prediction requires the understanding of a complex diagenetic history of the carbonate system, which may involve different dolomitization models. Wright and Racey (2009) have reported multiphase diagenetic effects on the presalt carbonates of Santos Basin, which would include late-stage corrosion as a factor for primary porosity enhancement.

KEY STRUCTURAL ASPECTS OF THE OUTER HIGH

The Outer High contains two main culminations, each the high edge of a large uplifted footwall fault block (Figures 3–5). A northwest–southeast transfer system runs between and separates the structures. The northern high, Tupi, is segmented by a series of synthetic synrift faults. In contrast, the southern high, Sugar Loaf, is segmented mostly by antithetic faults (Figure 5). The eastern flank of each structure is bounded by a major southwest–northeast down-to-the-east normal fault system, here called the outer hinge line of the basin (Figures 3–5). The throw on this fault system is roughly 1 to 2.5 km (0.6–1.6 mi) at the base salt level. In addition, the upper Cretaceous section expands significantly on the downthrown side of this fault system, suggesting repeated reactivation and footwall uplift (cf. Weissel and Karner, 1989).

The main tectonic pulse leading to the uplift of the Outer High probably occurred at the end of the early synrift phase (Late Barremian?). However, several later, additional events seem to have reactivated structures throughout evaporite deposition (Late Aptian). The reactivation history has affected both the evolution of the outer hinge line and of related footwall uplift.

The Outer High remained a positive feature throughout its long tectonic history and persisted as a basin-wide culmination from the early stages of rifting to the continental breakup and drift. It impacted depositional thicknesses and patterns of both the transitional evaporites and the overlying post-breakup sediment packages from Late Cretaceous to Neogene. Regional seismic data, flattened at the base salt level, clearly indicate this long-lived structural influence. In Figure 8, the maximum thinning of the presalt section coincides with the maximum thinning of both the evaporite sequence and the overlying drift sediment package. In contrast, some of the Campos outer basin highs to the northeast, which form the Campos External High (Dias et al., 1990), experienced significant subsidence after continental breakup. There, the locus of maximum sedimentary thinning migrates landward with time, from rift to drift (Figure 9).

DISCUSSION: TECTONIC EVOLUTION OF THE OUTER HIGH

The regional processes that generated the differential uplift of fault-block structures, leading to the formation of the Outer High, are directly linked to the continental breakup and evolution of the South Atlantic. The proposed model involves a failed sea-floor-spreading center in the Santos Basin (Meisling et al., 2001), which was responsible for the emplacement of a proto-oceanic crust at the southern São Paulo Plateau (Mohriak, 2001; Gomes et al., 2002) (Figures 1, 10–12).

It has been widely recognized that the São Paulo Plateau, although mostly underlain by stretched continental crust (Kowsmann et al., 1982; Macedo, 1990; Souza et al., 1993; Karner, 2000), contains volcanic-floored subbasins (Modica and Brush, 2004), like the region that surrounds the so-called Avedis and Abimael ridges (Demercian, 1996; Mohriak, 2001). We believe that this volcanic wedge represents the onset of a period of sea-floor spreading (Kumar and Gamboa, 1979; Meisling et al., 2001; Mohriak et al., 2008), which was aborted early in its development (Scotchman et al., 2006), probably during the Early Aptian. The abandonment of this tectonomagmatic process, possibly related to a ridge-jump event, ultimately led to the formation of the São Paulo Plateau

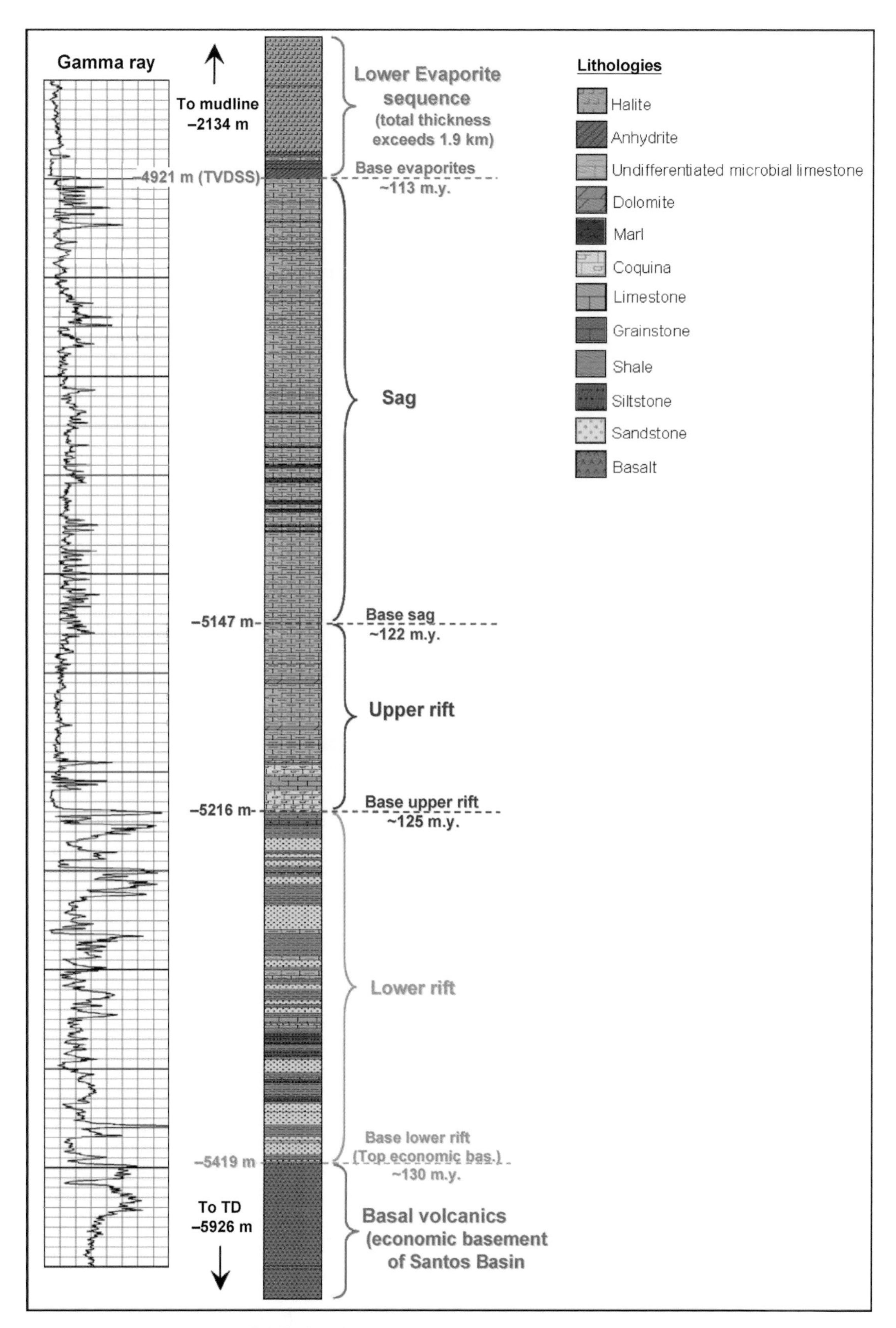

Figure 7. Presalt section from a key well drilled in the Outer High of Santos Basin (the ages of the markers are based on Moreira et al., 2007). TD = total depth; TVDSS = true vertical depth subsea; bas = basement.

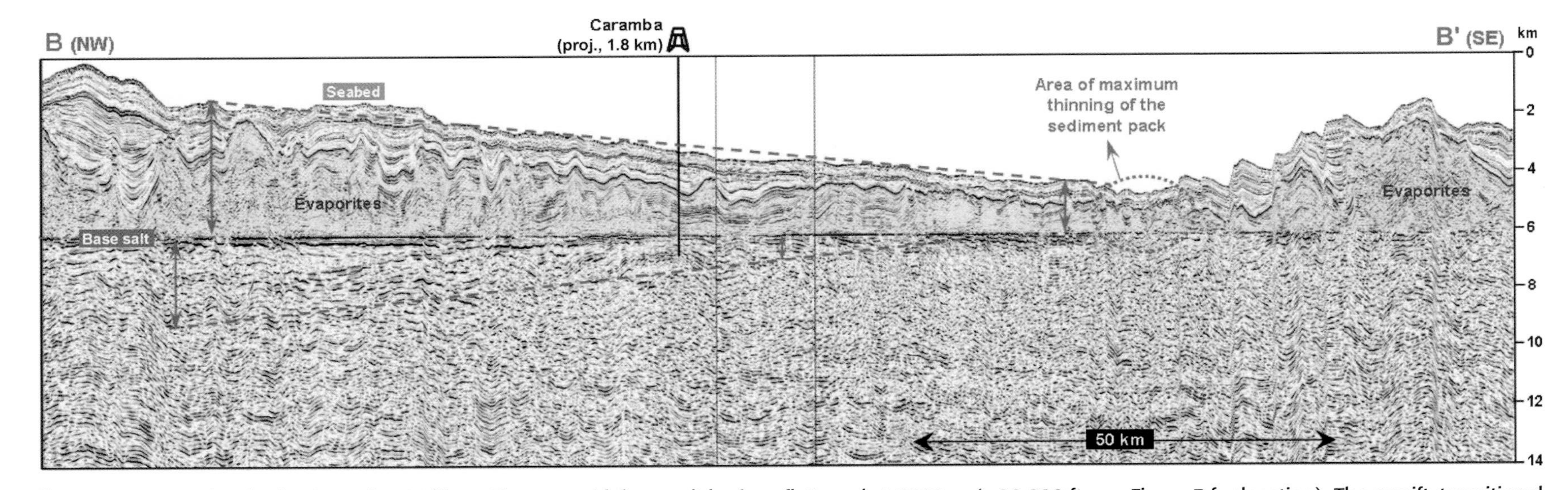

Figure 8. Same regional seismic section BB′ from Figure 5, with base salt horizon flattened at 6000 m (~20,000 ft; see Figure 3 for location). The synrift, transitional, and postrift sections thin onto the structure, indicating that the Outer High is early and long-lived. The Outer High affected the entire evolution of Santos Basin, acting as a buttress during halokinetic processes. The sediment record from Albian to present is represented by a condensed section that caps the Outer High. (Two-dimensional wave equation pre-stack depth migration data were used with the kind permission of TGS-Nopec Geophysical Company.) proj. = projected. 50 km (31 mi).

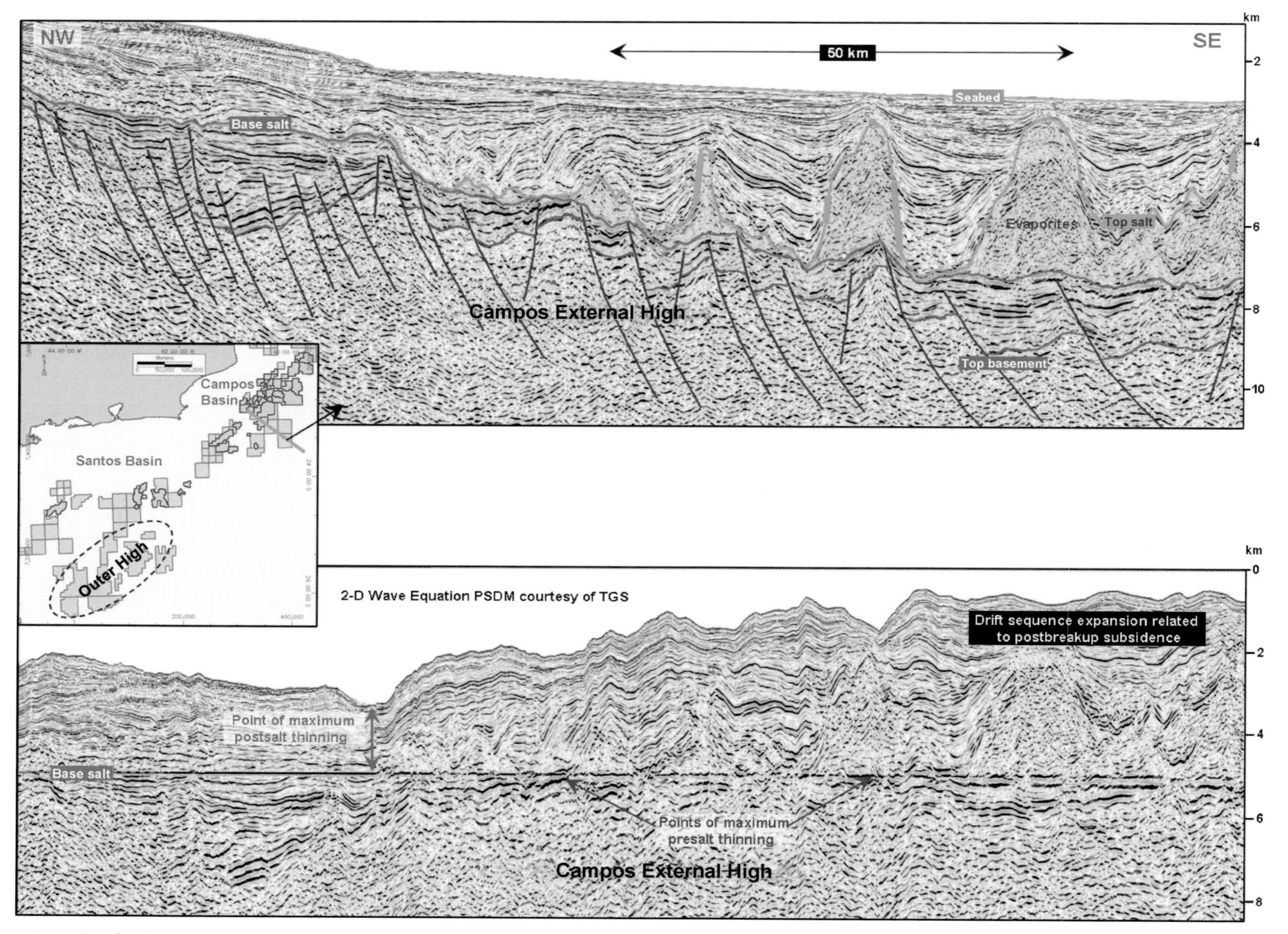

Figure 9. Regional seismic section over the southern Campos Basin, with simplified interpretation (see map for location). The bottom version is flattened at the base salt horizon. From synrift to drift, a landward shift in the zone of maximum thinning of the sediment packages exists, which is related to the subsidence history of the basin. Part of the so-called Campos External High has subsided significantly after continental breakup and no longer represents a prominent culmination at base salt level (two-dimensional wave equation pre-stack depth migration data were used with the kind permission of TGS-Nopec Geophysical Company). PSDM = prestack depth migration. 50 km (31 mi).

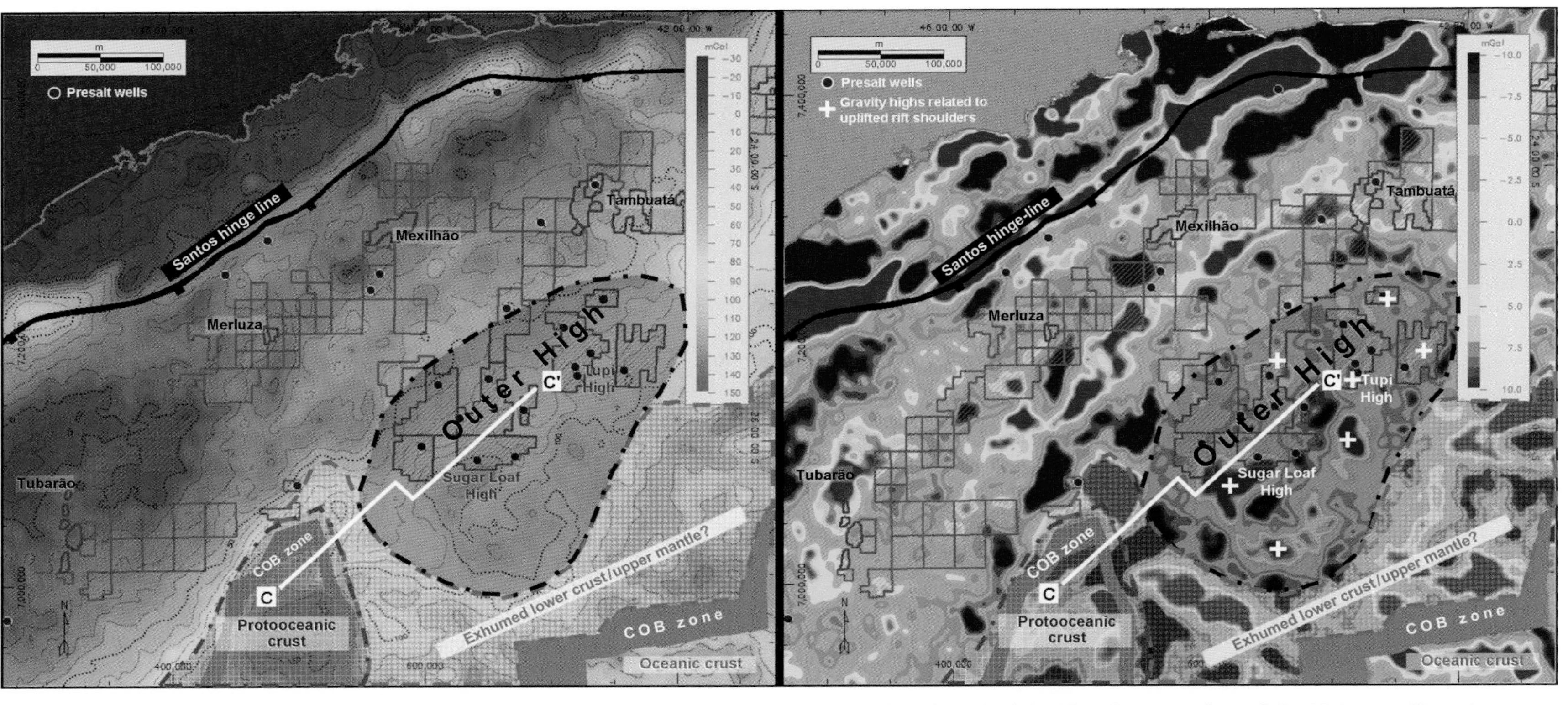

Figure 10. Gravity maps of the southern São Paulo Plateau, with the Bouguer anomalies displayed on the left side (ci. 10 mGal), and the high-pass filtered Bouguer anomalies (75 mi [120 km]) on the right side. The maps show the geographic relationship between the proto-oceanic crust (purple outline) and the Outer High. Whereas the former is emphasized on the Bouguer map, the Outer High structures are better delineated on the filtered map. The red outline between the Outer High and the highly subsided proto-oceanic crust indicates a possible zone of exhumed lower crust and/or upper mantle. Regional seismic section CC′ is displayed on Figure 12. Gravity data are used with the kind permission of GETECH. COB = continent-ocean boundary.

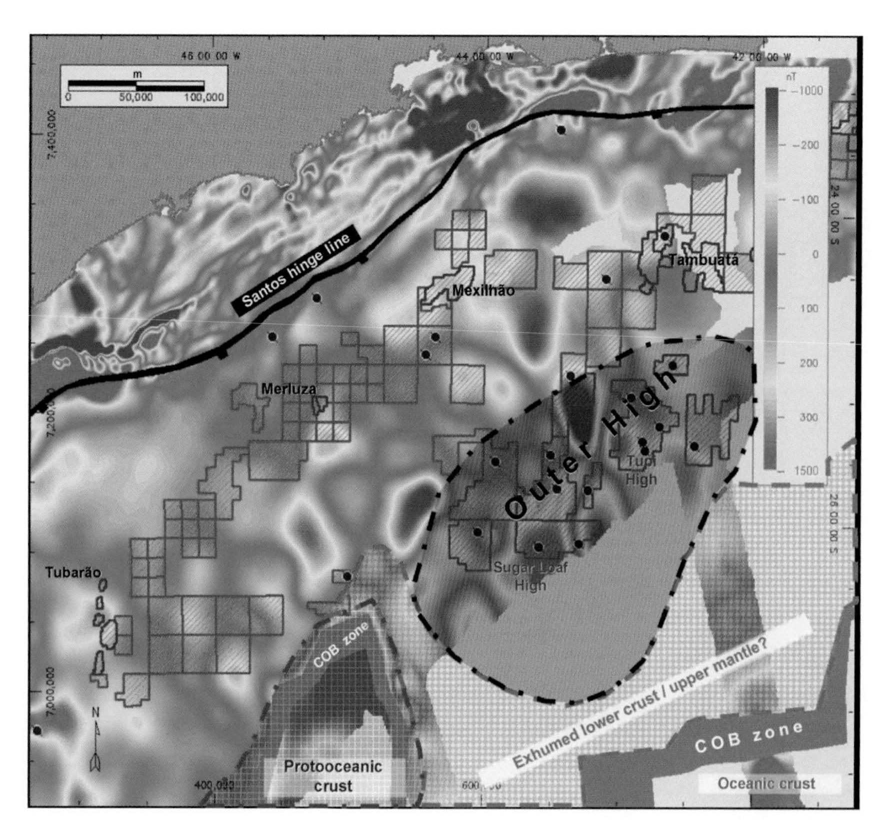

Figure 11. Map of magnetic intensity (data rotated to pole) over the southern São Paulo Plateau. The proto-oceanic crust is delineated by strong negative anomalies in the lower part of the map. Colored outlines are as described on Figure 10. Magnetic data are used with the kind permission of GETECH. COB = continent-ocean boundary.

(Figure 1) as a segment of stretched continental crust that remained attached to the South American plate (Carminatti et al., 2008).

We postulate that the thermal event that emplaced proto-oceanic crust in the southern Santos Basin was also a key event for the formation of the Outer High. Sea-floor spreading led to the inception of early volcanics to the south and caused a general uplift of the highly stretched continental crust to the north (central Santos) as part of the evolving continental breakup. During, or shortly after spreading, magmatic underplating added material to the base of the crust below the Outer High. The additional crust, fractionated from the mantle by partial melting, formed a deep root with lower density than its underlying parent. From that point forward, the low-density crustal root of the Outer High held it isostatically above its nonunderplated environs. Consequently, the Outer High region remained a positive structure, in striking contrast to the subsiding proto-oceanic wedge just south (Figure 12).

Although no deep seismic refraction data are currently available to support the underplating concept, the anomalous shallow elevation of the basement based on integrated seismic, gravity, and magnetic modeling strongly suggests the presence of a supporting crustal root derived from magmatic underplating. The short-wavelength gravity response of the elevated high density basement can be observed on the high-pass-filtered Bouguer anomaly map (Figure 10) and also in the Bouguer anomaly profile displayed on Figure 12.

It is possible that the distal São Paulo Plateau contains areas of exhumed lower crust or even upper mantle resulting from extreme extension of the continental crust (Figure 10), as suggested by Mohriak et al. (2008), Gomes et al. (2008), Gardiner et al. (2009) and Zalán et al. (2009). This interpretation is analogous to the West Iberia continental margin, where results from the Ocean Drilling Program (ODP) indicate the presence of upper mantle rocks directly below oceanic sediments (Whitmarsh and Wallace, 2001). If applicable to Santos Basin, the isostatic interplay of local crustal thickening by underplating with regional crustal thinning by hyperextension might well have amplified the relief of the outer high. Recently acquired two-dimensional seismic data in the São Paulo Plateau, aiming to image deep structures, may shed some light on the tectonic evolution and extension history of this region.

CONCLUSIONS

The Outer High is a broad structural province in the distal deep-water region of Santos Basin, represented by a segmented series of rift fault-block shoulders that

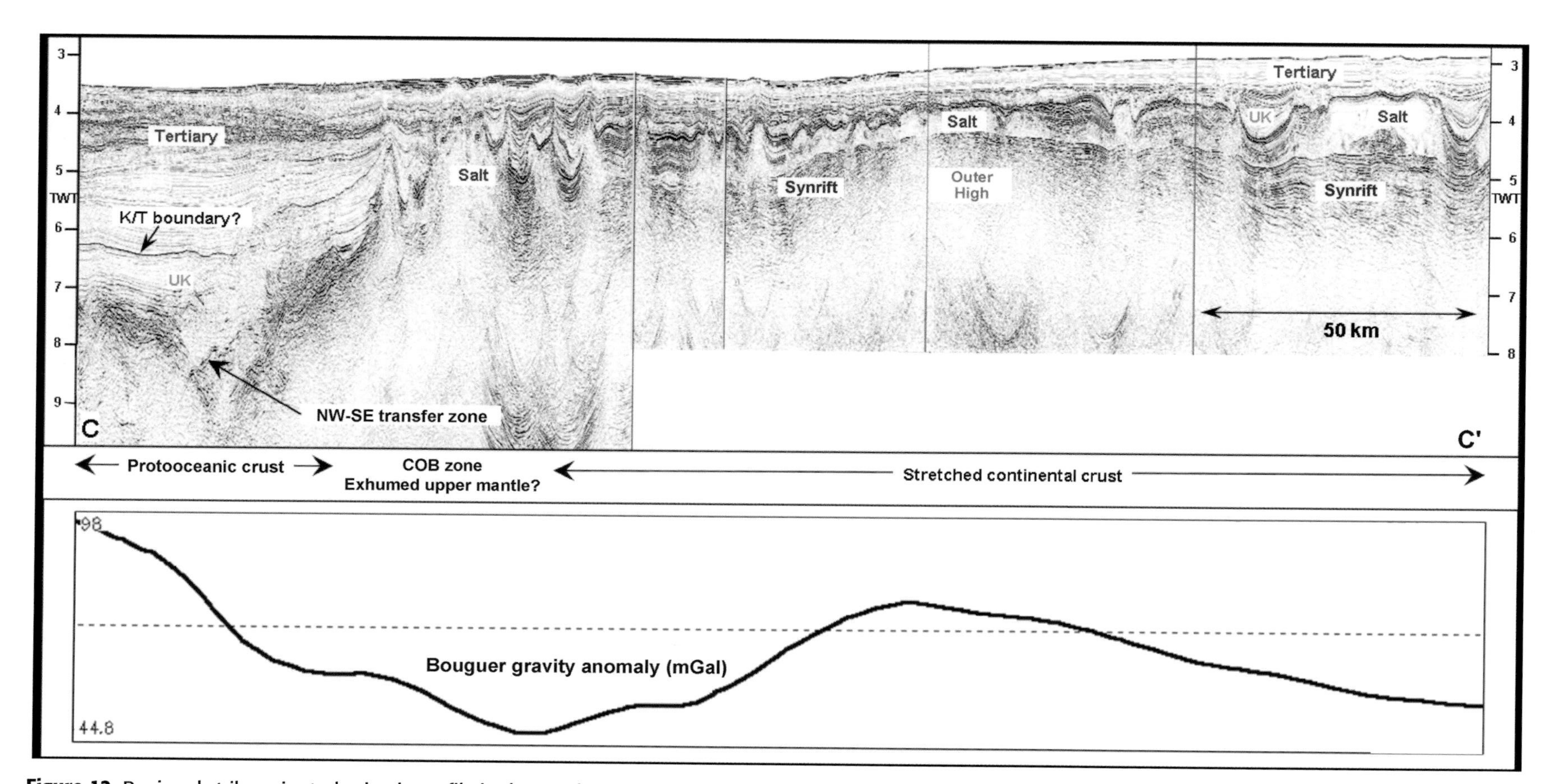

Figure 12. Regional strike-oriented seismic profile in the southern São Paulo Plateau (see Figure 10 for location), showing the basement structure of the Outer High of Santos Basin. Note that the presalt section onlaps and pinches out toward both flanks of the high, which acted as a regional focus for oil migration. A structural low can be observed at the southwestern edge of the seismic image. This region is related to the steep gradient observed in the Bouguer gravity anomaly profile and is interpreted as the edge of the proto-oceanic crust region indicated in Figures 1, 10, and 11. The contrast between the Outer High and the adjacent proto-oceanic crust is at least partly caused by isostatic adjustment: magmatic underplating locally thickened the crust beneath the Outer High, resulting in its present-day thickness and elevation (modified from Gomes et al., 2002). COB = continent-ocean boundary; UK = Upper Cretaceous; K/T boundary = Cretaceous–Tertiary boundary. 50 km (31 mi).

were uplifted and eroded during the Late Barremian. The largest culmination within this province forms a 12,000 km^2 (4633 mi^2) megaclosure at base salt level, which includes the Tupi and Sugar Loaf highs, among other satellite structures.

The regional uplift event that generated the Outer High structures is linked to a failed sea-floor-spreading process that emplaced proto-oceanic crust in the southern Santos Basin. Magmatic underplating is postulated as the mechanism responsible for locally thickening the crust and isostatically holding the Outer High as a positive feature throughout the post-breakup geologic history of Santos Basin.

As a long-lived feature, mostly covered by an efficient evaporite seal, the Outer High acted as a regional focus for hydrocarbon migration, i.e., a gathering area for the oil and gas generated in the thick presalt section of Santos Basin. Individual base salt structures of various sizes, located along this migration pathway, were progressively filled with hydrocarbons and represent the dominant trapping mechanism for several discoveries made to date in the Outer High region.

Based on plate tectonic reconstructions, we estimate that the Outer High of Santos Basin was located at least 200 km (124 mi) away from both the African and the Brazilian hinge lines, before continental breakup. This distal setting, coupled with the paleotopographic relief of the Outer High, resulted in a clastic-starved environment, which spanned from the final stages of the Barremian up to the Late Aptian. The depositional model during this period involves a transition from a brackish-to-saline lacustrine environment, dominated by bioclastic carbonate facies (coquinas), to a broad saline epicontinental lake, likely affected by sporadic marine incursions and dominated by microbial limestones. These microbialites, which formed extensive deposits on a broad Aptian carbonate platform, just before evaporite deposition, represent the main reservoirs in the deep-water Santos presalt exploration play.

ACKNOWLEDGMENTS

This study was originally presented at the 2008 AAPG international conference. It was expanded to be included in this memoir, with well information, two-dimensional seismic data, and literature available as of the first quarter of 2010. The manuscript does not incorporate, however, the most recent information in the area from the active exploration effort that was conducted in Santos Basin since April 2010.

The authors thank Hess Corporation for the encouragement to prepare and submit this paper. We also thank our colleagues Scott Pluim, Peter Mullin, Thomas Melgaard, Tom West, Grant Crandall, Ryan Mann, Robert Handford, and John Hohman for their technical input and valuable discussions. The senior author thanks his former colleagues, Dr. Webster U. Mohriak, João A. Bach de Oliveira, Stefano Santoni, Jonathan Parry, and Wisley Martins, for their collaboration on previous studies in the same area. Finally, we thank the anonymous AAPG reviewers for their valuable feedback and suggestions, which greatly improved the manuscript. Seismic data on Figures 4, 5, 8, and 9 is used by kind permission of TGS-Nopec Geophysical Company. Potential field data on Figures 10 and 11 were used after authorization from GETECH company.

REFERENCES CITED

Carminatti, M., J. L. Dias, and B. Wolff, 2009, From turbidites to carbonates: Breaking paradigm in deep waters: Offshore Technology Conference, Houston, Texas, May 4–7, 2009.

Carminatti, M., B. Wolff, and L. A. P. Gamboa, 2008, New exploratory frontiers in Brazil: 19th World Petroleum Congress, Madrid, Spain, June 29–July 3, 2008.

Carvalho, M. D., U. M. Praça, A. C. Silva-Telles, R. J. Jahnert, and J. L. Dias, 2000, Bioclastic carbonate lacustrine facies models in the Campos Basin (Lower Cretaceous), Brazil, *in* E. H. Gierlowski-Kordesch and K. R. Kelts, eds., Lake basins through space and time: AAPG Studies in Geology, v. 46, p. 245–256.

Davison, I., 1999, Tectonics and hydrocarbon distribution along the Brazilian South Atlantic margin, *in* N. R. Cameron, R. H. Bate, and V. S. Clure, eds., The oil and gas habitats of the South Atlantic. Geological Society (London) Special Publication, v. 153, p. 133–151, doi:10.1144/GSL.SP.1999.153.01.09.

Davison, I., 2007, Geology and tectonics of the South Atlantic Brazilian salt basins, *in* A. C. Ries, R. W. H. Butler, and R. H. Graham, eds., Deformation of the continental crust: The legacy of Mike Coward: Geological Society (London) Special Publication, v. 272, p. 345–359, doi:0.1144/GSL.SP.2007.272.01.18.

Demercian, L. S., 1996, A halocinese na evolução do Sul da Bacia de Santos do Aptiano ao Cretáceo Superior: Master's thesis, Universidade Federal do Rio Grande do Sul, Porto Alegre, Brazil, 201 p.

Dias, J. L., 2005, A Tectônica, estratigrafia e sedimentação no Andar Aptiano da margem leste brasileira: Boletim de Geociências da Petrobrás, v. 13, no. 1, p. 7–25.

Dias, J. L., J. C. Scarton, L. C. Guardado, F. R. Esteves, and M. Carminatti, 1990, Aspectos da evolução tectono-sedimentar e a ocorrência de hidrocarbonetos na Bacia de Campos, *in* G. P. Raja Gabaglia and E. J. Milani, eds., Origem e evolução de bacias sedimentares: Rio de Janeiro, Brazil, Petrobrás, p. 333–360.

Fetter, M., 2009, The role of basement tectonic reactivation on the structural evolution of Campos Basin, offshore Brazil:

Evidence from 3-D seismic analysis and section restoration: Marine and Petroleum Geology, v. 26, no. 6, p. 873–886, doi:10.1016/j.marpetgeo.2008.06.005.

Figueiredo, A. M. F., and C. C. Martins, 1990, 20 Anos de Exploração da Bacia de Campos e o Sucesso nas Águas Profundas: Boletim de Geociências da Petrobrás, v. 4, no. 1, p. 105–123.

França, R. L., A. C. Del Rey, C. V. Tagliari, J. R. Brandão, and P. R. Fontanelli, 2007, Bacia do Espírito Santo (Carta Estratigráfica): Boletim de Geociências da Petrobrás, v. 15, no. 2, p. 501–509.

Gardiner, W., A. Tudoran, G. Karner, C. Johnson, and I. Norton, 2009, Synrift sedimentation and timing of continental breakup of Santos and Campos basins, Brazil: AAPG International Conference and Exhibition, Rio de Janeiro, Brazil, November 15–18, 2009.

Gibbs, P. B., E. R. Brush, and J. C. Fiduk, 2003, The evolution of the synrift and transition phases of the central/southern Brazilian and west African conjugate margins: The implications for source rock distribution in time and space, and their recognition on seismic data: VIII Congresso Internacional Sociedade Brasileira de Geofísica, Rio de Janeiro, Brazil, September 14–18, 2003, p. 1–6.

Gomes, P. O., J. Parry, and W. Martins, 2002, The Outer High of the Santos Basin, southern São Paulo Plateau, Brazil: Tectonic setting, relation to volcanic events and some comments on hydrocarbon potential: AAPG Search and Discovery Article 90022: http://www.searchanddiscovery.com/abstracts/pdf/2002/hedberg_norway/extended/ndx_gomes.pdf (accessed November 9, 2011), p. 1–5.

Gomes, P. O., B. Kilsdonk, J. Minken, T. Grow, and R. Barragan, 2008, The Outer High of the Santos Basin, southern São Paulo Plateau, Brazil: Presalt exploration outbreak, paleogeographic setting, and evolution of the synrift structures: AAPG International Conference and Exhibition, Cape Town, South Africa, October 26–29, 2008, p. 1–6.

Karner, G. D., 2000, Rifts of the Campos and Santos basins, southeastern Brazil: Distribution and timing, *in* M. R. Mello and B. J. Katz, eds., Petroleum systems of South Atlantic margins: AAPG Memoir 73, p. 301–315.

Karner, G. D., and L. A. P. Gamboa, 2007, Timing and origin of the South Atlantic presalt sag basins and their capping evaporites, *in* B. C. Schreiber, S. Lugli, and M. Babel, eds., Evaporites through space and time. Geological Society (London) Special Publication, v. 285, p. 15–35, doi:10.1144/SP285.2.

Kowsmann, R. O., M. P. A. Costa, M. P. Boa Hora, and P. P. Guimarães, 1982, Geologia estrutural do Platô de São Paulo: Sociedade Brasileira de Geologia, v. 4, p. 1558–1569.

Kumar, N., and L. A. P. Gamboa, 1979, Evolution of the São Paulo Plateau (southeastern Brazilian margin) and implications for the early history of the South Atlantic: Geological Society of America Bulletin, v. 90, p. 281–293.

Macedo, J. M., 1990, Evolução tectônica da Bacia de Santos e áreas continentais adjacentes, *in* G. P. Raja Gabaglia and E. J. Milani, eds., Origem e evolução de bacias sedimentares: Rio de Janeiro, Brazil, Petrobrás, p. 361–376.

Meisling, K. E., P. R. Cobbold, and V. S. Mount, 2001, Segmentation of an obliquely rifted margin, Campos and Santos basins, southeastern Brazil: AAPG Bulletin, v. 85, no. 11, p. 1903–1924, doi:10.1306/8626D0A9-173B-11D7-8645000102C1865D.

Mello, M. R., J. M. Macedo, R. Requejo, and C. Schiefelbein, 2002, The great Campos: A frontier for new giant hydrocarbon accumulations in the Brazilian sedimentary basins: AAPG Bulletin, v. 85, no. 13.

Mello, M. R., P. W. Brooks, J. M. Moldowan, and B. Wygrala, 2006, The giant forgotten subsalt hydrocarbon province in the Greater Campos Basin, Brazil: A journey from a great mistake to a learning experience: AAPG Bulletin, v. 90.

Modica, C. J., and E. R. Brush, 2004, Postrift sequence stratigraphy, paleogeography, and fill history of the deepwater Santos Basin, offshore southeast Brazil: AAPG Bulletin, v. 88, p. 923–946, doi:10.1306/01220403043.

Mohriak, W. U., 2001, Salt tectonics, volcanic centers, fracture zones and their relationship with the origin and evolution of the South Atlantic Ocean: VII Congresso Internacional Sociedade Brasileira de Geofísica, Salvador, Bahia, Brazil, October 28–31, 2001.

Mohriak, W. U., M. Nemcok, and G. Enciso, 2008, South Atlantic divergent margin evolution: Rift-border uplift and salt tectonics in the basins of SE Brazil, *in* R. J. Pankhurst, R. A. J. Trouw, B. B. Brito Neves, and M. J. De Wit, eds., West Gondwana: Pre-Cenozoic correlations across the South Atlantic region: Geological Society (London) Special Publication, v. 294, p. 365–398.

Moreira, J. L. B., C. V. Madeira, J. A. Gil, and M. A. P. Machado, 2007, Bacia de Santos (Carta Estratigráfica): Boletim de Geociências da Petrobrás, v. 15, no. 2, p. 531–549.

Moulin, M., D. Aslanian, and P. Unternehr, 2010, A new starting point for the South and Equatorial Atlantic Ocean: Earth-Science Reviews, v. 98, no. 1–2, p. 1–37, doi:10.1016/j.earscirev.2009.08.001.

Pereira, M. J., and J. M. Macedo, 1990, A Bacia de Santos: Perspectivas de uma nova província petrolífera na plataforma continental sudeste brasileira: Boletim de Geociências da Petrobrás, v. 4, p. 3–11.

Scotchman, I. C., G. Marais-Gilchrist, F. G. Souza, F. F. Chaves, L. A. Atterton, A. Roberts, and N. J. Kusznir, 2006, A failed sea-floor spreading center, Santos Basin, Brazil: Rio Oil and Gas Conference, Instituto Brasileiro do Petróleo, Rio de Janeiro, Brazil, September 11–14, 2006, Paper IBP1070_06.

Scotese, C. R., 2001, Atlas of Earth history: Paleogeography: Arlington, Texas, PALEOMAP Project, 52 p.

Souza, K. G., R. L. Fontana, J. Mascle, J. M. Macedo, W. U. Mohriak, and K. Hinz, 1993, The southern Brazilian margin: An example of a South Atlantic volcanic margin, *in* III Congresso Internacional da Sociedade Brasileira de Geofísica, Rio de Janeiro, Brazil, November 7–11, 1993, v. 2, p. 1336–1341.

Weissel, J. K., and G. D. Karner, 1989, Flexural uplift of rift flanks due to mechanical unloading of the lithosphere during extension: Journal of Geophysical Research, v. 94, p. 13919–13950, doi:10.1029/JB094iB10p13919.

Whitmarsh, R. B., and P. J. Wallace, 2001, The rift-to-drift development of the west Iberia nonvolcanic continental margin: A summary and review of the contribution of Ocean Drilling Program Leg 173: Ocean Drilling Program Results, v. 173, p. 1–36.

Wright, P. V., and A. Racey, 2009, Presalt microbial carbonate reservoirs of the Santos Basin, offshore Brazil: AAPG Annual Convention and Exhibition, Denver, Colorado, June 7–10, 2009.

Zalán, P. V., M. C. G. Severino, J. A. B. Oliveira, L. P. Magnavita, W. U. Mohriak, R. C. Gontijo, A. R. Viana, and P. Szatmari, 2009, Stretching and thinning of the upper lithosphere and continental-oceanic crustal transition in southeastern Brazil: AAPG International Conference and Exhibition, Rio de Janeiro, Brazil, November 15–18, 2009.

6

López-Gamundí, Oscar, and Roberto Barragan, 2012, Structural framework of Lower Cretaceous half grabens in the presalt section of the southeastern continental margin of Brazil, *in* D. Gao, ed., Tectonics and sedimentation: Implications for petroleum systems: AAPG Memoir 100, p. 143 – 158.

Structural Framework of Lower Cretaceous Half Grabens in the Presalt Section of the Southeastern Continental Margin of Brazil

Oscar López-Gamundí

C&C Energy, Carrera 4 72-35, Bogota, Columbia (e-mail: olopezgamundi@ccenergy.com.co)

Roberto Barragan

Hess Exploration Malaysia, 207 Jalan Tun Razak, 5400, Kuala Lumpur, Malaysia (e-mail: rbarragan@hess.com)

ABSTRACT

Recently acquired and processed prestack depth-migrated seismic data have helped to identify the key elements of the asymmetric Lower Cretaceous half grabens in the presalt, synrift-to-postrift transitional (sag) section of the Greater Campos Basin (Santos, Campos, and Espirito Santo basins), offshore Brazil. Evidence of such a structural configuration is provided by seismic reflection geometries, such as fanning (strongly divergent internal configuration) on fault borders, thinning (convergent internal configuration), and onlap onto flexural margins. Moreover, compaction synclines over basement footwall cutoff points have been identified. In poorly imaged areas, the termination of the divergent seismic configuration can be used to place the master fault of the half graben. Differential compaction at half-graben border fault margins caused by the contrasting nature of rift-fills and adjacent basement highs is postulated to have been a critical factor for the creation of counterregional dips necessary to form structural four-way closures at the sag level. Although the sag sequence extends beyond the underlying rift fill, commonly onlapping or draping over the basement, the key risk in these types of traps is the possibility that the overlying salt layer may rest directly on the basement. Fault-plane reflections indicate the predominance of planar fault-plane geometries. This is consistent with the absence of rollover anticlines or hanging-wall antiforms, which are a direct function of nonplanar listric faults. The final configuration of the traps may also be modified by important basin-scale factors (i.e., uplift resulting from magmatic underplating in the Santos Basin).

INTRODUCTION

Recent success in exploration efforts in the Cretaceous (Barremian to Aptian) presalt rift and sag sequences of offshore Brazil (Santos, Campos, and Espirito Santo basins, collectively known as Greater Campos; Figure 1) has revived interest in the difficulty in properly imaging a section characterized by asymmetric half grabens and

[1] *Previous address*: Hess Corporation, 1501 McKinney, Houston, Texas, 77010, U.S.A.

DOI:10.1306/13351551M1003531

covered by a conspicuous salt layer of variable thickness. Although the sedimentary velocity variation is considered relatively smooth, depth migration is required for properly imaging the presalt section (Huang et al., 2009). Additional complicating factors are isolated igneous intrusions (dikes and sills) and seismically fast carbonates above and below the salt layer (Gerrard et al., 2009). The importance of having an adequate imaging of the presalt sequence is key to understanding not only the recent discoveries in this section, but also the probable extent of the prolific presalt source rocks that charged the pre- and postsalt reservoirs in Santos, Campos, and Espirito Santo basins (Mello et al., 2006). Newly acquired seismic reflection data, particularly data processed using prestack depth migration (PSDM), allow us to unveil some details of the structural framework of these individual half grabens that, in turn, may shed light on the basin-fill evolution of the synrift (half graben) and postrift (sag) stages in the offshore basins of Brazil. The tectonic setting of the rifting during the opening of the South Atlantic has been amply documented elsewhere (Rabinowitz and LaBrecque, 1979; Zanotto and Szatmari, 1987; Conceição et al., 1988; Szatmari, 2000; Rosendahl et al., 2005). This contribution deals with the influence that the tilt block nature of the presalt section had on (1) the sedimentary evolution of both the syn- and postrift presalt section and on (2) the associated trap geometries. Seismic examples provided in this contribution are from the Santos, Campos, and Espirito Santo basins (Figure 1). It is expected that the characteristics highlighted below, derived from selected dip lines with very good quality particularly for the presalt section, will help in areas with mild to poor imaging caused mainly by extreme

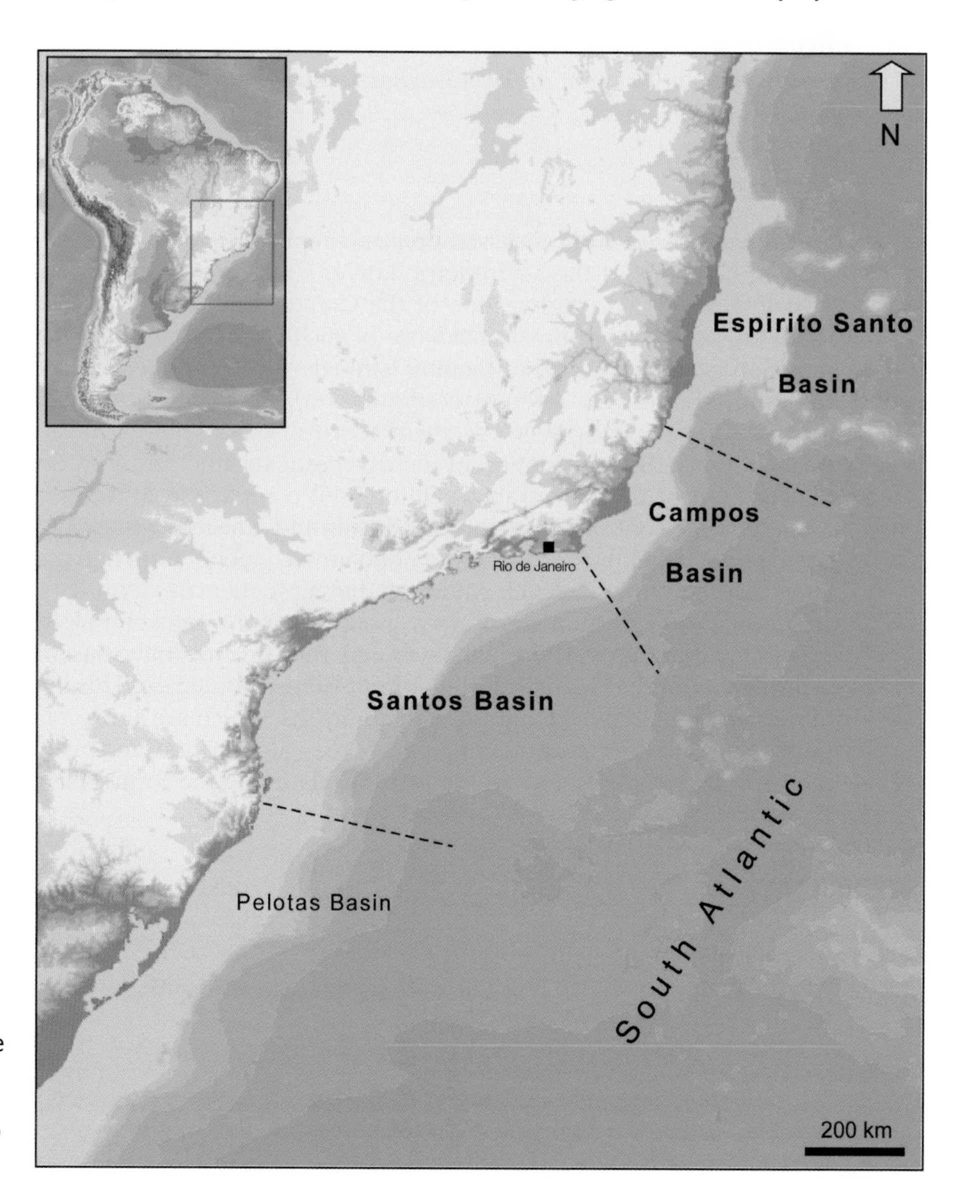

Figure 1. Offshore basins of central Brazil. Salt does not extend south into the Pelotas Basin. The presence of salt over the synrift and sag sections is widespread in the Santos, Campos, and Espirito Santo basins. 200 km (124 mi).

salt thickness variations (between 0 and 4000 m [0–13,123 ft] for most of the deep-water part of the basins) and/or a thick salt layer. In this regard, the seismic examples presented here are oriented in an approximately dip direction, orthogonal to the inferred direction of extension that occurred during rifting. This orientation is optimal to evaluate the characteristics of the half grabens and their sedimentary fills.

PRESALT RIFTING IN THE CONTEXT OF THE SOUTH ATLANTIC OPENING

The initial structural style of the South Atlantic opening is dominated by rifting. A key element in this tectonic stage is the presence of asymmetric half grabens (Szatmari, 2000; Gibbs et al., 2003; Bueno, 2004; Dias, 2005). This seems to be the dominant individual geometry on both the South American and African margins. This rifting was diachronous, beginning in the south during the Jurassic and propagating northward to the equatorial segment during the Early Cretaceous (Conceição et al., 1988; Bueno, 2004) as the South American continent rotated clockwise relative to Africa about a pole located in northeast Brazil (Rabinowitz and LaBrecque, 1979; Szatmari et al., 1985). The initial phase of opening is characterized by intracontinental rifting associated with Late Jurassic–Early Cretaceous tholeiitic basalts (Cobbold et al., 2001; Meisling et al., 2001; Mohriak et al., 2002). In the central offshore of Brazil (Santos, Campos, and Espirito Santo basins; Figure 1) and its African counterpart (Kwanza and Lower Congo basins), this first stage of opening was followed by a period initially dominated by nonmarine (primarily lacustrine) facies during the Barremian that evolved to a shallow-water engulfment, culminating with evaporite sedimentation in the Aptian. It has been suggested that rifting in the Santos and Kwanza basins is strongly asymmetric with oblique extension (transtension), resulting in rhombochasmic subbasins and listric fault geometries (Lentini and Fraser, 2008; Wilson et al., 2008). The Campos and Lower Congo basins are characterized by orthogonal stretching and simple half-graben geometries with planar master faults (Lentini and Fraser, 2008). Dias (2005) identified an eastward displacement of the rift and the exposure of the proximal areas during the Early Aptian, with significant expansion of the size of individual half grabens basinward (eastward) followed by a sag phase characterized by thermal subsidence and predominantly marine sedimentation over the pre-Late Aptian times (Figure 2). Inboard, the initial fill phase in these asymmetric half grabens was dominated by alluvial fan

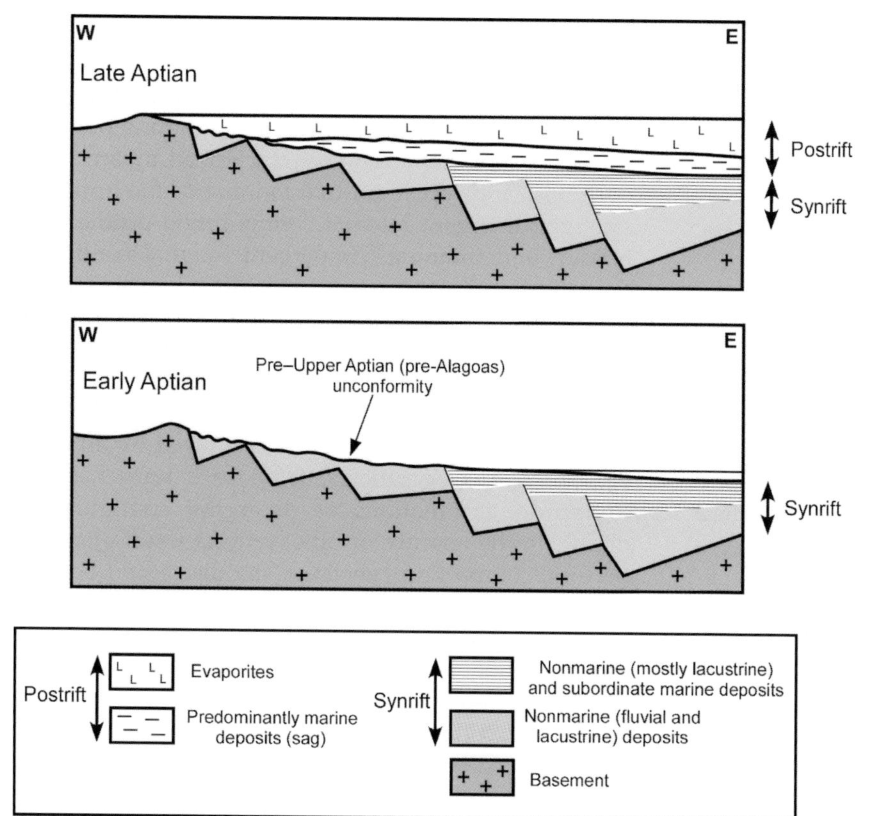

Figure 2. Schematic cross sections showing evolution of the offshore basins during the transition from the synrift stage to the sag stage. Note the increasing marine influence toward the top of the section and eastward in the synrift and postrift (sag) sections. The pre-Upper Aptian unconformity (pre-Alagoas unconformity) separates the synrift section dominated by tectonically driven subsidence and the overlying sag section characterized by regional thermal subsidence. The pre-Upper Aptian unconformity is equivalent to the base of sag marker as defined in this contribution. Marine sedimentation over the pre-Upper Aptian unconformity occurred under predominantly clastic-starved conditions toward the eastern part of the basins during the sag phase in the Late Aptian. Note the regional basinward thickening of the sag section caused by differential thermal subsidence (modified from Dias, 2005).

deposits associated with rift-generated local topography; locally, coquinas were deposited in rift shoulders isolated from high clastic sediment input (i.e., presalt coquinas in the Pampo and Badejo fields of Campos Basin; Bertani and Carozzi, 1985a,b; Horschutz and Scuta, 1992; Rangel and Carminatti, 2000). Outboard, the lower clastic sequences of Early Barremian tend to be made up of sandy to silty fine-grained lacustrine facies (Guardado et al., 1990), whereas the upper units (mid- to Late Barremian) consist of lacustrine carbonates and prolific organic-rich black shales (Dias et al., 1988; Rangel and Carminatti, 2000), which constitute the source rocks for all of the presalt, and most postsalt, hydrocarbon accumulations in offshore Brazil (Mello and Maxwell, 1990; Guardado et al., 2000). The synrift section exhibits increasing marine influence eastward and is separated by a regional unconformity (pre-Upper Aptian unconformity, also known as the pre-Alagoas unconformity) from the overlying sag deposits (Figure 2). This unconformity was produced during a phase of uplift and regional truncation and has been considered to be responsible for the development of restricted environments conducive to the formation of evaporites (Chang et al., 1992; Azevedo, 2004; Mohriak et al., 2008).

The sag phase (transitional megasequence; Guardado et al., 1990) was dominated by carbonate sedimentation in shallow-water environments (Dias, 1998, 2005; Carvalho et al., 2000; Gibbs et al., 2003), which probably evolved from a brackish-to-saline lacustrine environment to a broad saline lake, possibly punctuated by marine ingressions, and dominated by microbial limestones (Dias, 1998, 2005; Wright and Racey, 2009).

ARCHITECTURE AND FILL EVOLUTION OF THE ASYMMETRIC HALF GRABENS: EVIDENCE FROM SEISMIC DATA

Time and depth reflection seismic data have been used in this contribution to analyze the presalt half grabens of the Greater Campos basins. Although time seismic provides a good imaging of the presalt section in some areas and can highlight the general structural and stratigraphic framework (Figure 3), recently acquired PSDM seismic has been preferentially used in this contribution to constrain the geometry and internal configuration of the half-grabens because of its exceptional quality. The seismic was shot with a basin strike-and-dip orientation using an airgun source, a streamer length of 8100 m (26,574 ft), and a record length of 12 s. The spacing ranges approximately between 8 and 20 km (~5–10 mi) for the strike lines and 3 and 4 km (~2–2.5 mi) for the dip lines.

The external configuration of the half grabens is evidenced by several characteristics present on the border fault margin, namely fault-plane reflections. Fault-plane reflections (Figure 3) indicate planar geometries for the master fault of the half grabens, particularly in the Campos and Espirito Santo basins. Most boundary faults of the half-grabens are clearly defined as single fault systems (sensu Rosendahl et al., 1986); in some cases, the displacement is partitioned in distributary border fault systems (Figures 4, 5), a feature also common in the East African rift basins (Rosendahl et al., 1986). This feature may indicate migration of the boundary fault that typically occurs late in the evolution of a half graben (Morley, 2002). This migration expands the basin on both the fault and flexural margins (late-stage half graben of Morley, 2002). At the fault margin, the basin expands until the boundary fault is reestablished and footwall uplift and erosion occur. On the flexural margin, basin expansion is achieved by progressive onlap (Figures 3, 6). Onlap is herein defined as the stratal termination against a reflection of greater dip (cf. Mitchum, 1977). Onlap surfaces are most easily recognizable in the highly convergent zone, toward the flexural or ramp margin of the tilt blocks. This progressive onlap has been labeled as flexural margin transgressions by Morley and Wescott (1999) in the half-grabens of the East African rifts system. It has been related to a displacement increase on the boundary fault system (flexural margin rollback; Morley and Wescott, 1999).

Two distinct seismic sequences can be discerned in the presalt section based on the nature of the bounding surfaces and their seismic internal configurations. The synrift seismic sequence is characterized by a concordant base and top, a gradual fanning of the horizons, strongly divergent internal configuration on the fault border, and thinning (convergent internal configuration) on the flexural (ramp) margin (Figures 3–6). A hanging-wall onlap of successive horizons (progressive onlap) is common on the flexural margin of the half grabens (Figures 3, 6). This divergent internal configuration of the half grabens has been interpreted as the product of differential subsidence caused by an early episode of block rotation (rift phase 1; Karner, 2000). Reflection terminations of diverging configuration define, in the absence of other criteria (fault-plane reflectors, compaction synclines; see discussion below), the approximate location of the planar master fault of the half graben (cf. Roberts and Yielding, 1991). Commonly, chaotic to hummocky seismic facies can be identified close to the footwall scarp in areas with good seismic resolution. The continuity and concordance of the synrift divergent stratal reflections suggest that sediment supply and fault-controlled subsidence were

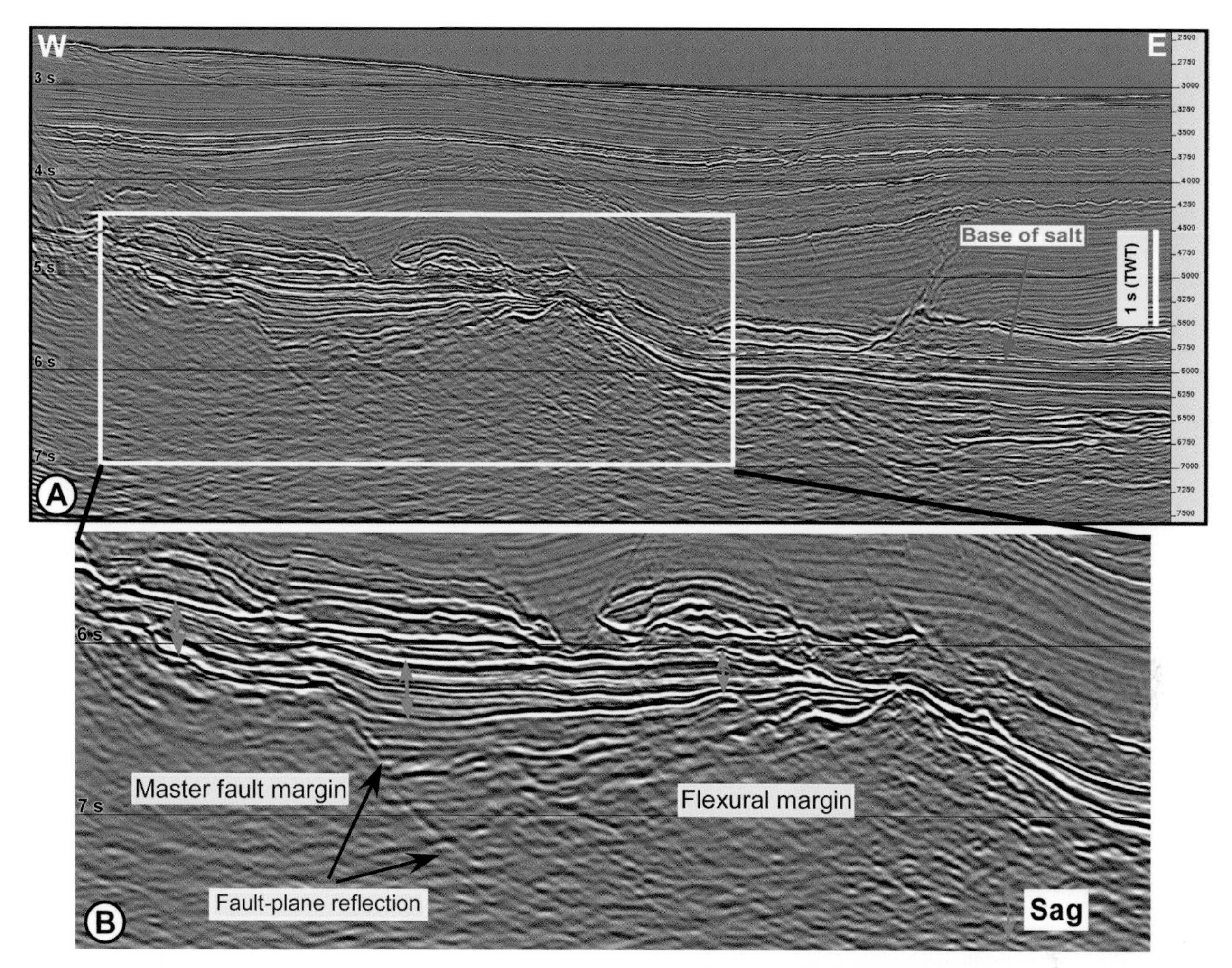

Figure 3. Regional time dip-oriented section in the central Campos Basin. General characteristics of the presalt section. (A) Two main basin-fill stages can be discerned in the presalt section: an early rift phase dominated by tectonic subsidence (laterally confined) and asymmetric half-graben geometries (ramp flexural margin and opposite fault margin) with fanning stratal patterns toward the border fault, followed by the sag phase dominated by thermal subsidence (more laterally extensive), which expanded beyond the rift fill over highs. The regional line is 38 km (24 mi) long. (B) Detail of presalt section. Note the fault-plane reflector on the master fault of the half graben and the thinning of sag section on the flexural margin and the rift shoulder beyond the master fault. Seismic is courtesy of Petroleum Geo-Services (PGS). TWT = two-way traveltime.

closely matched for most of the period of the active development of the half graben (Cartwright, 1991).

Compaction synclines (Figures 6, 7) occur in the hanging wall over the basement footwall cutoff point close to the border fault margin (Figure 8). When compaction occurs in a thick sequence with high initial fluid content at the time of deposition on the hanging wall of a normal fault, the differential compaction against the more rigid footwall gives rise to a syncline (White et al., 1986; Gibson et al., 1989). This folding causes dip reversals in the hanging wall close to the master fault (Figure 8). Compaction synclines have been identified from seismic data in many half-graben settings like the Viking Graben (see Bradley et al., 1988; Frost, 1989; Roberts and Yielding, 1991) and the Central Graben in the Danish sector of the North Sea (Cartwright, 1991), the Sirte Basin in offshore Lybia (Skuce, 1996), and the Tertiary (Paleogene) rifts of Indonesia (Atkinson et al., 2006).

Because compaction synclines are subsurface deformation features, with compaction decreasing to zero toward the top of the sedimentary pile, they cannot influence the distribution of sediments on the sea floor and hence cannot generate stratal terminations (i.e., onlap) but differential thicknesses of individual beds (Butler et al., 1998). Seismic examples and results from forward modeling (Barr, 1991) show geometric similarities and confirmed two key features of compaction synclines: the steepness of the dips is greater on deeper horizons and the synclinal axial plane is approximately

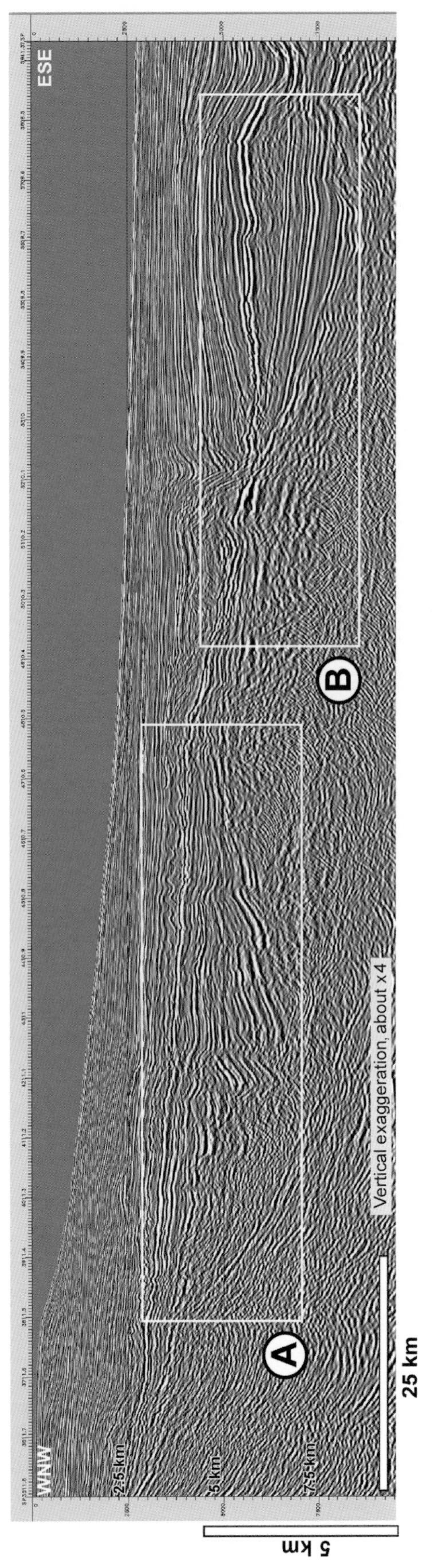

Figure 4. Regional dip-oriented, depth-converted section from the northern Campos Basin with conjugate half-grabens. Note that the opposite polarity of the half grabens created a high flank by the flexural margins of the half grabens and distributary fault systems. See Figure 5 for details in A and B inserts. The prestack depth migration seismic is courtesy of TGS-NOPEC Geophysical Company.

vertical (Skuce, 1996). Significant sag in the Miocene postrift section above Paleogene synrift depocenters has been identified in the rifts of Indonesia and related to differential compaction by Atkinson et al. (2006).

The top of the synrift section is identified by a marker (base of sag; Figures 3, 6B, 7) that separates the areally confined synrift seismic sequence from a more regionally extensive sag seismic sequence characterized by progressive onlapping on the basement. This sag infill takes the form of a more passive, parallel onlap-fill geometry (cf. Cartwright, 1991). The contact between the synrift fill and the base of sag seismic sequence can be classified as a strong angular unconformity, a minor angular unconformity, or a disconformity (Figures 3, 5A, 9). These variations reflect different structural positions within the basin at the transition from the synrift to the postrift (sag) stages (cf. Kyrkjebø et al., 2004). Strong angular unconformities are identified when the contact is located on the flexural margins of the half graben where onlap is the most common seismic configuration (Figures 3, 9). Similar angular relationships have been identified in the half grabens of the Central Sumatra Basin by Shaw et al. (1997). These authors noted that the angularity between the synrift and overlying sag strata gradually decreases toward the hanging-wall depocenters. Consequently, strata above (sag section) and below (synrift section) the unconformities are concordant in the deeper parts of the half grabens.

Seismically, the base of the sag commonly shows a pronounced impedance contrast characterized by a clear and continuous reflection. However, it exhibits in places lateral changes in seismic character, bright on both flanks (shoulders and ramps of the underlying rifts) and dim toward the depocenters, probably reflecting the change in lithology from margins dominated by shallow-water conditions to deeper basin centers where similar lithologies are more likely to occur below and above the base of the sag, inducing a subdued impedance contrast instead. The upper sag is dominated by parallel, reflection-free internal seismic facies (Figures 3, 6, 7, 9). Wells that reached the presalt section in the basin show a range of facies associations ranging from clastic (alluvial-fluvial in the rift section) to calcareous (ramp margin of the rift and sag; Dias, 2005). The latter facies association is the key objective of the emerging presalt prospectivity in offshore Brazilian basins.

IMPLICATIONS FOR FACIES DEVELOPMENT

The structural asymmetry of half-grabens exerts a fundamental control on sedimentary facies (Gibbs, 1984; Leeder and Gawthorpe, 1987; Schlische and Olsen, 1990; Lambiasse, 1990; Leeder, 1995; Morley and Wescott,

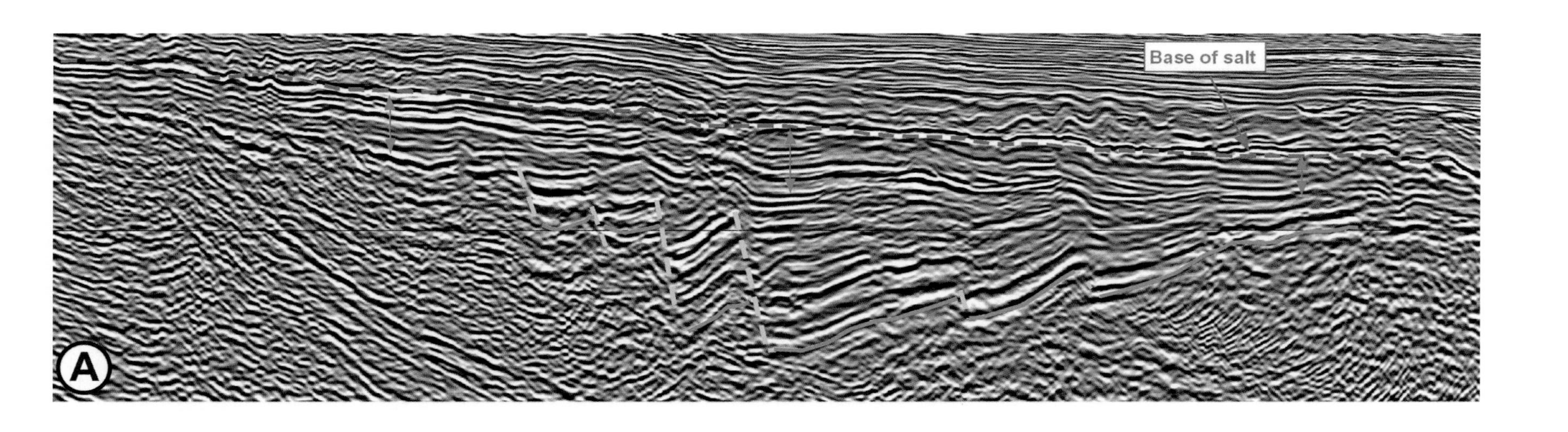

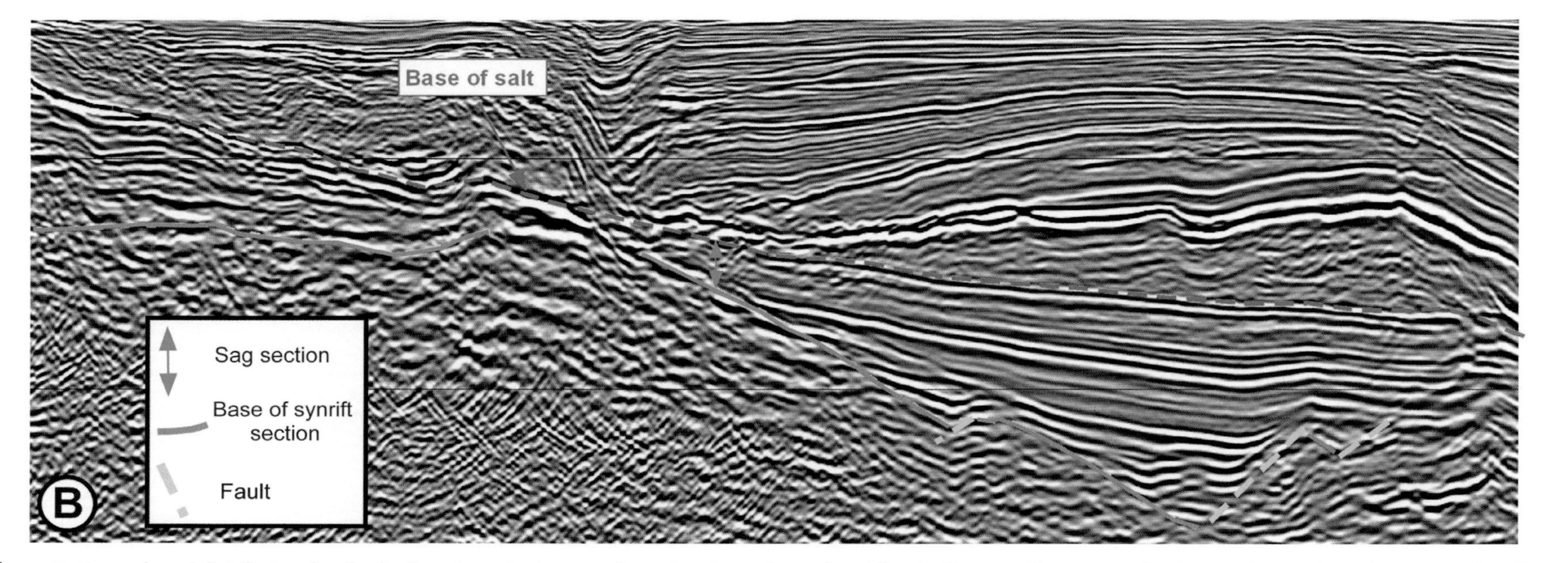

Figure 5. Examples of distributary border fault systems in the presalt section from the regional line in Figure 4. The sag section in panel A tends to drape over the rift shoulder whereas the sag section in panel B appears to pinch out on the flexural margin of the half graben. The prestack depth migration seismic is courtesy of TGS-NOPEC Geophysical Company.

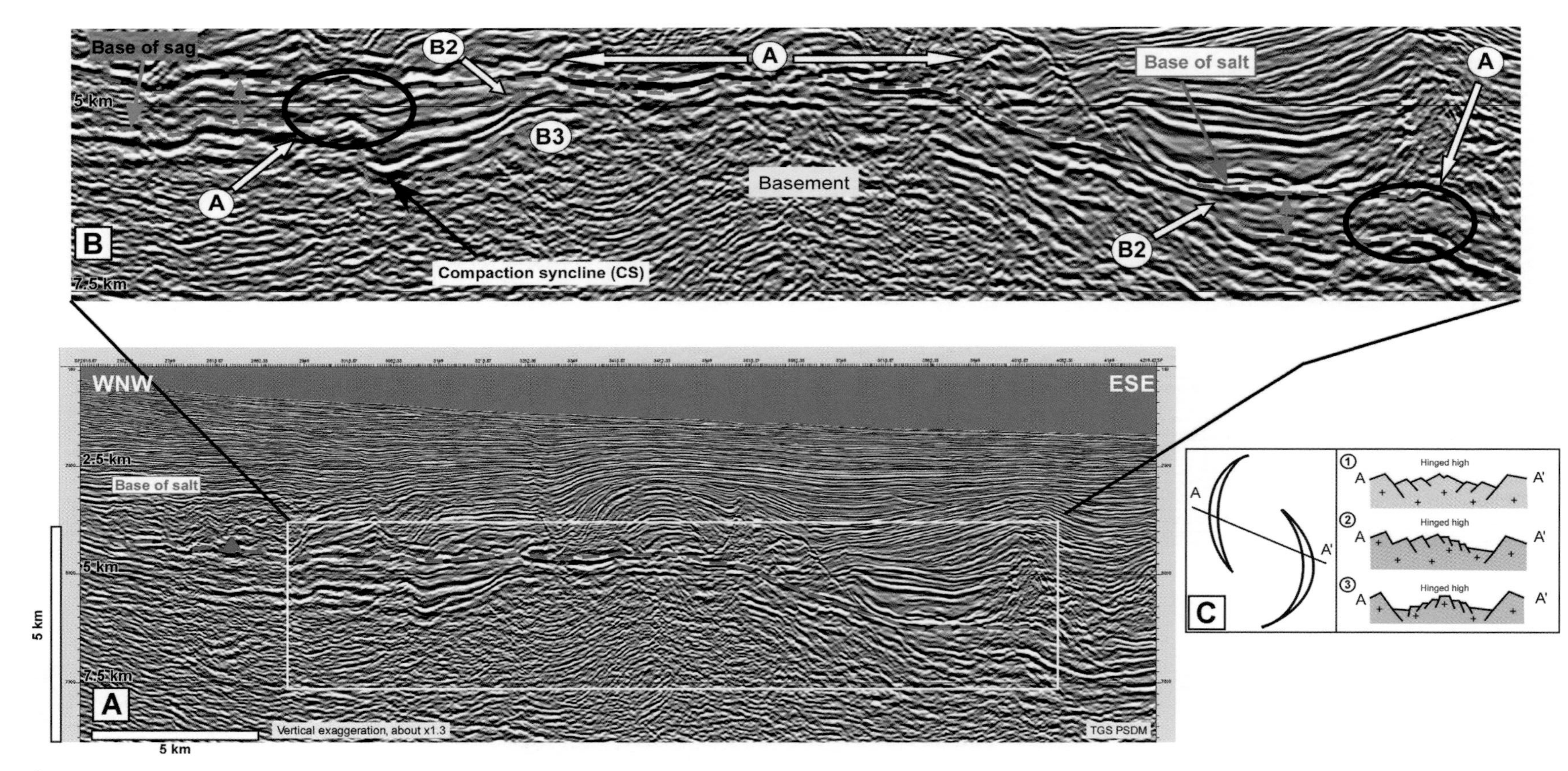

Figure 6. Trap types in the synrift and postrift (sag) sections illustrated in a depth-converted, dip-oriented section from the northern Campos Basin. (A) Regional configuration of presalt high with opposing half grabens on both flanks. (B) Detail of regional line. Differential compaction on the flanks of large basement high created counterregional dip necessary to form structural (four-way) closures at the sag level (trap type A). Moreover, the smaller subsidiary A traps on the flanks of the large four-way closure are controlled by differential compaction evidenced by compaction synclines at the rift level and more tenuously expressed at the sag level. Updip onlap and three-way dips define combined structural-stratigraphic traps (trap type B). Examples of these geometries can be identified on preexisting half-graben shoulders (half-graben ramp margins; B2) and on the ramp flank of the half graben (B3). See text for discussion. The prestack depth migration seismic is courtesy of TGS-NOPEC Geophysical Company. (C) Linking mode with overlapping half grabens. Note the similarity between cross section A and the seismic section in panel B (based on Rosendahl et al., 1986).

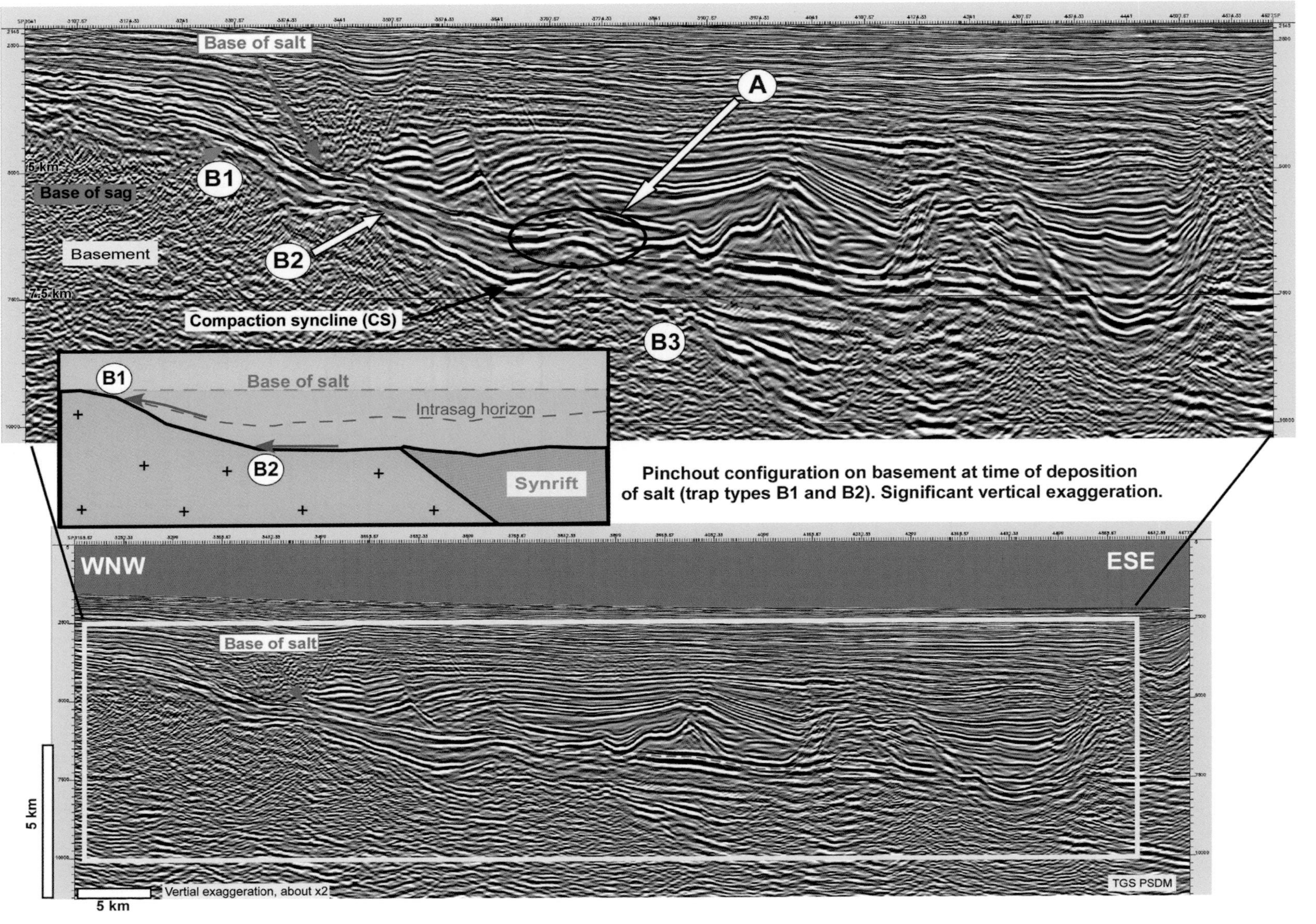

Figure 7. Trap types in the synrift and postrift (sag) sections illustrated in the dip section of central Espirito Santo Basin. Hanging wall compaction syncline created counterregional dip necessary to form structural (four-way) closures at the sag level (trap type A). Combined traps (B trap family) are illustrated: trap type B2 defined by an updip pinchout component at the base of the sag or at any horizon within the sag interval in the case of trap type B1. Examples of these geometries are common inboard (B1) and are also on preexisting half-graben ramp margins basinward (B2). A potential for onlap/pinchout traps is also present at the rift-fill level on the ramp flank of the half graben (trap type B3). See text for discussion. The prestack depth migration seismic is courtesy of The Geological Society (TGS). PSDM = prestack depth migration seismic.

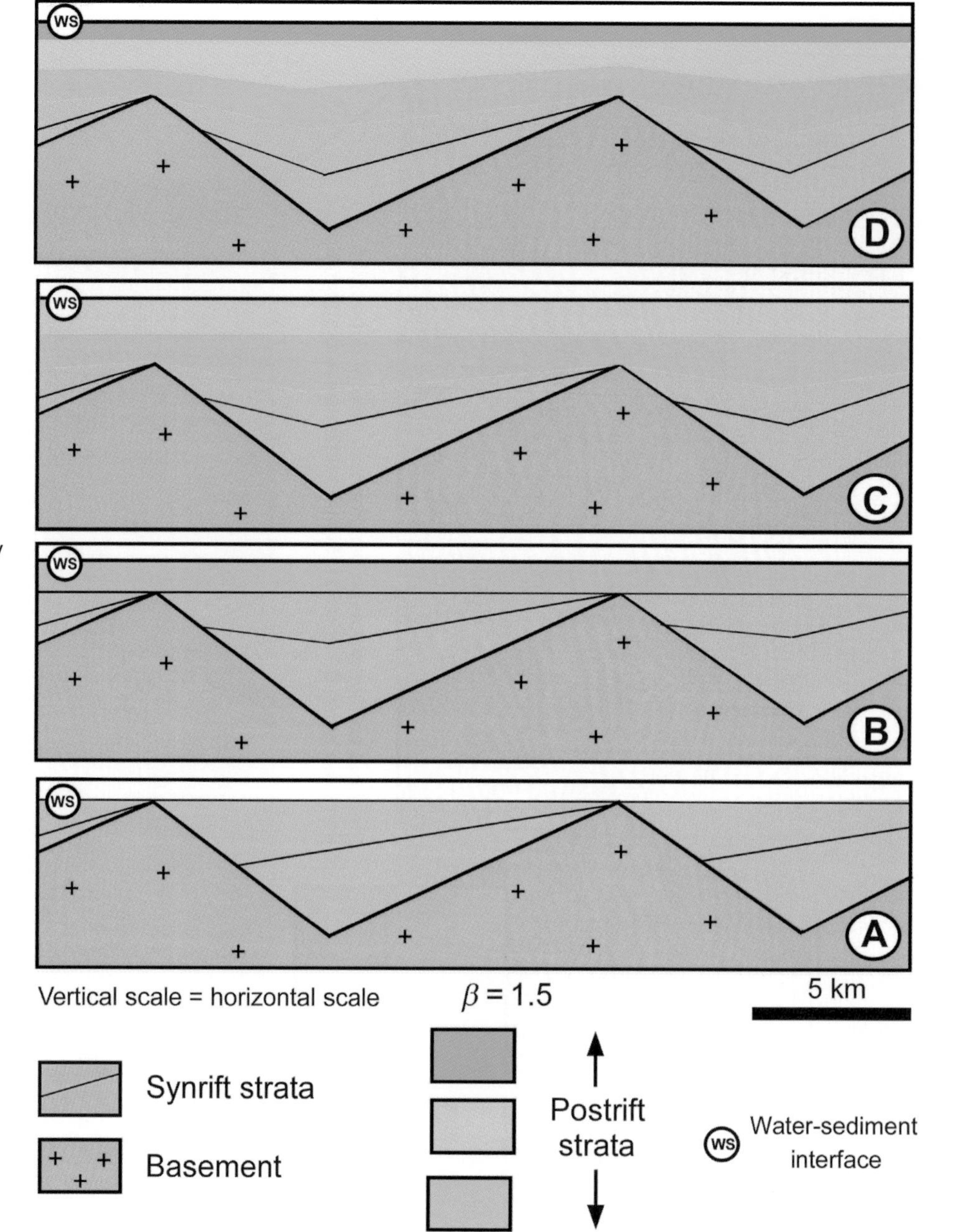

Figure 8. Evolution of sedimentary fill from initial synrift phase, characterized by asymmetric half-graben configuration, to postrift (sag) phase and development of hanging-wall compaction synclines (based on the forward modeling of North Sea examples by Barr, 1991). Basement is defined as any noncompactable or precompacted sedimentary, metamorphic, or igneous rocks such as the prerift. Note that the steepness of the dips is greater on deeper horizons, the location of the synclinal fold axis immediately above the hanging-wall cutoff of the basement and approximately vertical to the synclinal axial plane. Moreover, note the increasing steepness through evolution from the initial stage (A) to the final stage (D). A stretching factor of $B = 1.5$ was used.

1999). The fault margin is characterized by a low potential for good-quality clastic reservoirs because of underfilled conditions and provenance (volcanic basement). The flexural margin of the half graben is a site of low subsidence, is likely to undergo little deposition, and can even be affected by uplift and subsequent erosion. This area is characterized by low accommodation space, suitable for the development of high-energy carbonates (oolitic calcarenites and/or coquinas).

The half grabens of the East African rift system provide excellent modern analogs for the depositional systems and facies distribution that could have been present in the presalt half-grabens of the Greater Campos area. The open waters of Lake Tanganyika are, in practical terms, fresh but are more saline than those of Lake Malawi and have an uncommonly high Mg:Ca ratio. Lake Tanganyka is characterized by the presence of several carbonate facies, all exhibiting high Mg:Ca ratios, composed of *Chara* beds in sheltered shallows, ooid shoals, stromatolites, shelly bioclastic debris sheets, and algal bioherms in the deeper waters (Cohen and Thouin, 1985; Casanova, 1986).

Although the regional thermal subsidence that characterizes the sag phase tends to finally drown and

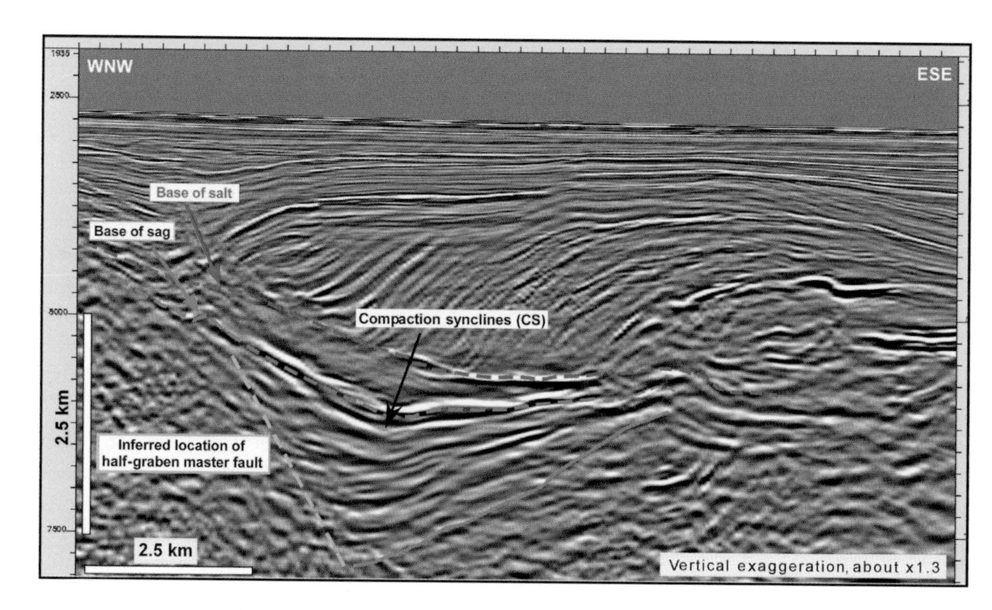

Figure 9. Dip-oriented, depth-migrated section of presalt section in the northernmost Santos Basin. The location of the half-graben master fault is defined by the termination of the diverging seismic horizons within the half-graben fill. The intersection of the inferred master fault and the projection of compaction syncline axes define the geometry of the half graben. Note the transparent seismic character of the sag section and its thinning toward the flexural margin of the underlying half graben. The prestack depth-migrated seismic is courtesy of TGS-NOPEC Geophysical Company.

smooth the relief of the synrift stage, rift shoulders remained as important paleogeographic elements that influenced the distribution of depositional environments in the early phase of the sag. Differential compaction is evidenced by the presence of hanging-wall synclines (Figures 6B, 7, 8). There is abundant seismic evidence that the sag phase expanded beyond the limits of the half-grabens (Figures 3, 6). Facies distribution in the lower part of the sag section was still controlled by highs inherited from the underlying asymmetric half graben (rift shoulders and ramps). Moreover, at this stage, the rift shoulders could have acted as efficient bathymetric highs that define shallow high-energy areas prone to the deposition of coquinas and oolitic shoals. These rift shoulders still had a positive bathymetric expression during the late sag phase as a series of distal, intermittently exposed basement highs in the clastic-starved environments prone to the development of a broad carbonate platform in the Early Aptian.

A distinctive sag facies association formed on bathymetric highs developed on underlying rift shoulders. This association was deposited in subtidal to peritidal environments where microbially mediated precipitation of carbonates produced microbial limestones (cf. Pratt et al., 1992). Subtidal stromatolites composed of peloidal discontinuous lamination are inferred to have formed by the trapping and binding of loose carbonate sediments in microbial mats. These subtidal stromatolites produced by in-situ precipitation are considered part of a carbonate-to-evaporite transition (Pope et al., 2000). Microbiolites (stromatolites and microbial laminites) have been described in the sag sequence of the Campos Basin (Dias, 1998). Calcareous organic shales and marls deposited in broad saline lakes are also present in the sag section of the Santos and Campos basins (Gibbs et al., 2003). In some extreme cases, the rift shoulders may have persisted through time as highs exposed to nondeposition and/or erosion during the sag phase, eventually leading to bald highs where salt was deposited directly over the basement (Figure 3).

IMPLICATIONS FOR TRAP GEOMETRIES

The rift section in Santos, Campos, and Espirito Santo basins shows a clear asymmetric configuration with a flexural border or ramp generally in the east and a steep fault border commonly in the west. This structural asymmetry caused by differential, tectonically induced subsidence created optimal conditions for the development of lacustrine-to-brackish source rocks in the hanging-wall area of maximum accommodation space next to the fault-margin of the half grabens (Gawthorpe et al., 1994; Howell and Flint, 1996). Internally, the asymmetric half grabens are characterized seismically by fanning of the horizons (strongly divergent internal configuration) on the fault border and thinning (convergent internal configuration) and onlap on the flexural (ramp) margin. In contrast to this gradual thickening toward the bounding fault observed in the half grabens analyzed herein, when the extension in half grabens is accommodated by normal faults that commonly flatten with depth, a collapse of the hanging wall and a formation of inclined rollover panels occur (Hamblin, 1965). This geometry leads to rollover traps that are absent in the half grabens of offshore Brazil. Seismically, the curved rollover above a nonplanar listric normal fault is evidenced by an abrupt thickening of strata toward the fault above the rollover panels (Xiao and Suppe,

1992). The nonplanar (flattening with depth) geometry of border faults and the consequent development of rollover structures are common in other rift settings (Nunns, 1991; Shaw et al., 1997) but are conspicuously absent in the half-grabens of offshore Brazil. The absence of rollover anticlines or hanging-wall antiforms suggests that the overall structure of the half grabens of Greater Campos is of tilted fault blocks bounded by planar faults in some form of a domino model, either of rigid dominoes (Barr, 1987; Jackson et al., 1988; Yielding, 1990) or soft dominoes (Gibson et al., 1989; Kusznir and Egan, 1989). Geometries similar to those interpreted here as compaction synclines have been assigned to fault-bend folds by Neto Cavalcanti de Araújo et al. (2009). These authors relate these synclines in the presalt synrift section of the Campos and Santos basins to the half-graben development above a normal fault that flattens to horizontal through bends although acknowledge the function of differential compaction in the formation of such synclines as well.

The effects of the differential postrift subsidence (and thus the creation of accommodation space) have been modeled for asymmetric half grabens by Barr (1991). Configurations similar to compaction synclines resulted from forward modeling based on North Sea data (Barr, 1991) assuming a basement (prerift) domino-style fault block generated by rotated planar faults. The results show the development of hanging-wall compaction synclines with marked thickness variations and substantial compaction-induced dips into the basin center (Figure 8) after significant thermal subsidence. He further showed that hanging-wall compaction synclines affected the subsidence of overlying sag sediments. This was evidenced by marked thickness variations and substantial compaction-induced dips into the basin center. Consequently, the thickest section of synrift sediments is overlain by the thickest section of postrift deposits. This relationship is also present in the half grabens of offshore Brazil (Figures 6, 7). As warned by Barr (1991), compaction-induced differential subsidence would probably give rise to faulting within the sag fill (the drape-slip faulting of Bertram and Milton, 1989) or, in ductile regimes, simply to a compaction syncline. In either case, uncritical evaluation of isopach maps and footwall–hanging-wall sediment thickness ratios may erroneously lead to the inference of renewed rifting without taking into consideration the effects of differential compaction.

Internal seismic configurations within the half-graben fill can be used to infer the presalt architecture in poorly imaged areas and consequently help reconstruct the basin geometry and predict the location of possible traps. Where fault-plane reflections are poorly imaged, the structural interpretation of the master fault in the half-graben can be based mainly on identifying divergent stratal terminations against the fault. The termination of the fanning seismic package indicates the approximate position of the fault plane (cf. Prosser, 1993; Figure 9). Compaction synclines can also be used to determine the footwall cutoff with the downward projection of their synclinal axes (cf. Prosser, 1993; Figure 9).

The overlying sag phase expands beyond the limits of the half grabens. The distribution of facies in the lower section is still controlled by highs inherited from the underlying asymmetric half graben (rift shoulders and ramps). Onlap-pinchout traps at the sag level are also indirectly controlled by preexisiting half-graben geometries and have a tendency to develop on the flexural margins or on the inboard shoulders (Figure 7). This type of stratigraphic onlap onto eroded basin margins is common on previous uplifted footwalls.

The influence of differential compaction in the creation of base-salt structural highs has been recently highlighted by López-Gamundí and Barragan (2008) for the Santos, Campos, and Espirito Santo basins and by Teasdale and Jensen (2008) for the Santos Basin. Differential compaction at the half-graben border fault margin is a key factor to create counterregional dip necessary to form structural (four-way) closures at the sag level (A trap type in Figure 7). Combined structural-stratigraphic traps are defined by an updip pinchout component at the base of the sag or at any horizon within the sag interval (Figures 6B, 7). Updip onlap and three-way dips define these combined traps (B trap type in Figure 7). Examples of these trap geometries at the sag level are common inboard (B1 type) as part of the regional updip onlap of the sag interval and also for intermediate levels within the sag section on preexisting half-graben shoulders (half-graben ramp margins) basinward (B2 type). A potential for onlap-pinchout traps is also present at the rift-fill level on the ramp flank of the half graben (B3 type).

Differential compaction seems to act at different scales as exemplified by the presence of a large four-way closure (trap type A) with a thin sag section draping the high (Figure 6B) and additional smaller four-way closures on both flanks of the large structure. The polarity of the half grabens on both flanks suggests an opposite alternate cross section linked by an intervening accommodation zone (Figure 6C) in a similar pattern as the one described by Rosendahl et al. (1986) for the East African rifts. Rosendahl et al. (1986) describe three families of half-graben linking modes. Family 1 comprises overlapping opposing half grabens (Figure 6C) linked by an intervening accommodation zone. This architecture is also identified in offshore Brazil (Figures 4; 6A, B) and is characteristic of four-way closures with optimal conditions of focusing for hydrocarbon charging.

CONCLUSIONS

- New two-dimensional PSDM seismic seems to be a viable tool to at least partially solve the problem of identifying the key elements of the basic motif of asymmetric half grabens: a border fault margin and a flexural or ramp margin. Evidence of this asymmetric half-graben nature is provided by seismic signatures like fanning (strongly divergent internal configuration) on fault borders, thinning (convergent internal configuration), and onlap on flexural margins. After block tilting ceased (synrift stage), the half grabens were onlapped (locally downlapped) by the postrift (sag) sequence, and this angular discordance provides a good criterion for locating on seismic data the base of the sag sequence.
- Compaction synclines over basement footwall cutoff points have been identified. Differential compaction at half-graben border fault margins caused by the contrasting nature of rift fills and adjacent basement highs is postulated to have been a critical factor for the creation of counterregional dips necessary to form structural four-way closures at the sag level, where most of the recent discoveries have been made. Consequently, the sag sediments expand over the shoulders of the fault margins and low-angle ramps of the flexural margins of the preexisting asymmetric rifts, creating potential for traps. Alternatively, the sag sequence, the key objective of the emerging presalt prospectivity, commonly extends beyond the underlying rift-fill draping over the basement. The key risk in this type of traps is the possibility that the overlying salt layer may rest directly on basement highs restricting the areal extent of the traps.
- Rift shoulders were topographic highs during the synrift stage and remained as lower relief, bathymetric highs during the sag phase. Hence, these were sites where shallow-water high-energy facies may have developed (Figure 11). Seismically, the base of the sag is concordant and exhibits lateral changes in its seismic character, bright on both flanks (shoulders and ramps of the underlying rifts) and dim toward the depocenters, probably reflecting the change in lithology from margins dominated by shallow-water conditions to deeper basin centers. The final configuration of the traps may also be modified by some important basin-scale factors such as uplift resulting from magmatic underplating like in the Santos Basin (Gomes et al., 2008).

ACKNOWLEDGMENTS

We thank the management of Hess for authorizing the publication of this contribution. The discussions with Hess colleagues Thomas Melgaard and Willy Barreda provided insight into the details of the presalt geology of offshore Brazil. The publication of depth seismic reflection data was kindly authorized by TGS-NOPEC Geophysical Company. Time seismic reflection data are courtesy of Petroleum Geo-Services (PGS). Ian Clark and an anonymous reviewer provided peer reviews that helped improve the quality and presentation of the paper.

REFERENCES CITED

Atkinson, C., M. Reynolds, and O. Hutapea, 2006, Stratigraphic traps in the Tertiary rift basins of Indonesia: Case studies and future potential, *in* M. R. Allen, G. P. Goffey, R. K. Morgan, and I. M. Walker, eds., The deliberate search for the stratigraphic trap: Geological Society (London) Special Publication, v. 254, p. 105–126.

Azevedo, R. L. M., 2004, South Atlantic paleoceanography and evolution during the Albian: Boletim de Geociencias de Petrobras, v. 12, p. 231–249.

Barr, D., 1987, Structural/stratigraphic models for extensional basins of half-graben type: Journal of Structural Geology, v. 9, p. 491–500, doi:10.1016/0191-8141(87)90124-6.

Barr, D., 1991, Subsidence and sedimentation in semistarved half graben: A model based on North Sea data, *in* A. M. Roberts, G. Yielding, and B. Freeman, eds., The geometry of normal faults: Geological Society (London) Special Publication, v. 56, p. 17–28.

Bertani, R. Z., and A. V. Carozzi, 1985a, Lagoa Feia Formation (Lower Cretaceous), Campos Basin, offshore Brazil: Rift valley stage carbonate reservoirs. Part I: Journal of Petroleum Geology, v. 8, p. 37–58, doi:10.1111/j.1747-5457.1985.tb00190.x.

Bertani, R. Z., and A. V. Carozzi, 1985b, Lagoa Feia Formation (Lower Cretaceous), Campos Basin, offshore Brazil: Rift-valley-stage carbonate reservoirs. Part II: Journal of Petroleum Geology, v. 8, p. 199–220, doi:10.1111/j.1747-5457.1985.tb01011.x.

Bertram, G. T., and N. J. Milton, 1989, Reconstructing basin evolution from sedimentary thickness: The importance of paleobathymetric control, with reference to the North Sea: Basin Research, v. 1, p. 247–257.

Bradley, M. E., J. D. Price, C. Rambech Dahl, and T. Agdestein, 1988, The structural evolution of the northern Viking Graben and its bearing upon extensional modes of basin formation: Journal of the Geological Society (London), v. 145, p. 455–472, doi:10.1144/gsjgs.145.3.0455.

Bueno, G. V., 2004, Diacronismo de eventos no rifte Sul-Atlântico: Boletim de Geociências de Petrobras, v. 12, p. 203–229.

Butler, R. W. H., M. Grasso, W. Gardiner and D. Sedgeley,

1998, Depositional patterns and their tectonic controls within the Plio-Quaternary carbonate sands and muds of onshore and offshore SE Sicily, Italy: Marine and Petroleum Geology, v. 14, p. 879–892.

Cartwright, J., 1991, The kinematic evolution of the Coffee Soil Fault, *in* A. M. Roberts, G. Yielding, and B. Freeman, eds., The geometry of normal faults: Geological Society (London) Special Publication, v. 56, p. 29–40, doi:10.1144/GSL.SP.1991.056.01.03.

Carvalho, M. D., U. M. Praça, A. C. Silva-Telles, R. J. Jahnert, and J. L. Dias, 2000, Bioclastic carbonate lacustrine facies models in the Campos Basin (Lower Cretaceous), Brazil, *in* E. H. Gierlowski-Kordesch and K. R. Kelts, eds., Lake basins through space and time: AAPG Studies in Geology 46, p. 245–256.

Casanova, J., 1986, East African rift stromatolites, *in* L. E. Frostick, R. W. Renaut, I. Reid, and J. Tiercelin, eds., Sedimentation in the African rifts: Geological Society (London) Special Publication, v. 25, p. 201–210.

Chang, H. K., R. O. Kowsmann, A. M. F. Figueiredo, and A. A. Bender, 1992, Tectonic and stratigraphy of the East Brazil rift system: An overview: Tectonophysics, v. 213, no. 1–2, p. 97–138, doi:10.1016/0040-1951(92)90253-3.

Cobbold, P. R., K. E. Meisling, and V. S. Mount, 2001, Reactivation of an obliquely rifted margin, Campos and Santos basins, southeastern Brazil: AAPG Bulletin, v. 11, p. 1925–1944, doi:10.1306/8626D0B3-173B-11D7-8645000102C1865D.

Cohen, A. S., and C. Thouin, 1985, Nearshore carbonate deposits in Lake Tanganyka: Geology, v. 15, p. 414–418, doi:10.1130/0091-7613(1987)15<414:NCDILT>2.0.CO;2.

Conceição, J. C. J., P. V. Zalán, and S. Wolff, 1988, Mecanismo, evolução e cronologia do rifte Sul Atlântico: Boletim de Geociências da Petrobras, v. 2, p. 255–265.

Dias, J. L., 1998, Análise sedimentológica e estratigráfica do Andar Aptiano em parte da Margem Leste do Brasil e no Platô das Malvinas: Considerações sobre as primeiras incursões e ingressões marinhas do Oceano Atlântico Sul Meridional: Ph.D. thesis, Universidade Federal do Rio Grande do Sul, Brazil, 208 p.

Dias, J. L., 2005, Tectônica, estratigrafia e sedimentação no Andar Aptiano da margem leste brasileira: Boletim de Geociências de Petrobras, v. 13, p. 7–25.

Dias, J. L., J. Q. Oliveira, and J. C. Vieira, 1988, Sedimentological and stratigraphic analysis of the Lago Feia Formation, rift phase of the Campos Basin, offshore Brazil: Revista Brasileira de Geociências, v. 18, p. 252–260.

Frost, R. E., 1989, Discussion on the structural evolution of the northern Viking Graben and its bearing on extensional modes of basin formation: Reply by M. E. Badley, J. D. Price, C. Rambech Dahl, and T. Agdestein: Journal of the Geological Society (London), v. 146, p. 1035–1040.

Gawthorpe, R. L., A. J. Fraser, and R. E. L. Collier, 1994, Sequence stratigraphy in active extensional basins: Implications for the interpretation of ancient basin fill: Marine and Petroleum Geology, v. 11, p. 64–658, doi:10.1016/0264-8172(94)90021-3.

Gerrard, C., J. Cramer, K. Sherwood and N. Weber, 2009, Rapid, large-scale depth imaging in the Santos Basin: 79th Society of Exploration Geophysicists International Exposition and Annual Meeting, Houston, Texas, October 25–30, 2009, p. 603–607.

Gibbs, A. D., 1984, Structural evolution of extensional basin margins: Journal of the Geological Society (London), v. 141, p. 609–620, doi:10.1144/gsjgs.141.4.0609.

Gibbs, P. B., E. R. Brush, and J. C. Fiduk, 2003, The evolution of the synrift and transition phases of the central/southern Brazilian and west African conjugate margins: The implications for source rock distribution in time and space and their recognition on seismic data: 8th International Congress of the Brazilian Geophysical Society, Rio de Janeiro, Brazil, September 14–18, 2003, 6 p.

Gibson, J. R., J. J. Walsh, and J. Watterson, 1989, Modeling of bed contours and cross sections adjacent to planar normal faults: Journal of Structural Geology, v. 11, p. 317–328, doi:10.1016/0191-8141(89)90071-0.

Gomes, P. O., B. Kilsdonk, J. Minken, T. Grow, and R. Barragan, 2008, The Outer High of the Santos Basin, southern São Paulo Plateau, Brazil: Presalt exploration outbreak, paleogeographic setting, and evolution of the synrift structures: AAPG International Conference and Exhibition, Cape Town, South Africa, October 26–29, 2008, p. 1–6.

Guardado, L. R., L. A. P. Gamboa, and C. F. Lucchesi, 1990, Petroleum geology of the Campos Basin, Brazil: A model for a producing Atlantic-type basin, *in* J. D. Edwards and P. A. Santogrossi, eds., Divergent/passive margin basins: AAPG Memoir 48, p. 3–80.

Guardado, L. R., A. R. Spadini, J. S. L. Brandão, and M. R. Mello, 2000, Petroleum system of the Campos Basin, *in* M. R. Mello and B. J. Katz, eds., Petroleum systems of South Atlantic margins: AAPG Memoir 73, p. 317–324.

Hamblin, W. K., 1965, Origin of reverse drag on the downthrown side of normal faults: Geological Society of America Bulletin, v. 76, p. 1145–1164.

Horschutz, P. M. C., and M. S. Scuta, 1992, Facies-perfis e mapeamento de qualidade do reservatório de coquinas da Formação Lagoa Feia do Campo de Pampo: Boletim Geociências Petrobras, v. 6, p. 45–58.

Howell, J. A., and S. S. Flint, 1996, A model for high-resolution sequence stratigraphy within extensional basins, *in* J. A. Howell and J. E. Aitken, eds., High-resolution sequence stratigraphy: Innovations and applications: Geological Society (London) Special Publication, v. 104, p. 129–137, doi:10.1144/GSL.SP.1996.104.01.09.

Huang, Y., D. Lin, B. Bai, and C. Ricardez, 2009, Presalt depth imaging of Santos Basin, Brazil: 79th Society of Exploration Geophysicists International Exposition and Annual Meeting, Houston, Texas, October 25–30, 2009, p. 2869–2873.

Jackson, J. A., N. J. White, Z. Garfunkel, and H. Anderson, 1988, Relations between normal-fault geometry, tilting, and vertical motions in extensional terrains: An example from the southern Gulf of Suez: Journal of Structural Geology, v. 10, p. 155–170, doi:10.1016/0191-8141(88)90113-7.

Karner, G. D., 2000, Rifts of the Campos and Santos basins, southeastern Brazil: Distribution and timing, *in* M. R.

Mello and B. J. Katz, eds., Petroleum systems of South Atlantic margins: AAPG Memoir 73, p. 301–315.

Kusznir, N. J., and S. Egan, 1989, Simple-shear and pure-shear models of extensional sedimentary basin formation: Application to the Jeanne d'Arc Basin, Grand Banks of Newfoundland, *in* A. J. Tankard and H. R. Balkwill, eds., Extensional tectonics and stratigraphy of the North Atlantic margins: AAPG Memoir 46, p. 305–322.

Kyrkjebø, R., R. H. Gabrielsen, and J. I. Faleide, 2004, Unconformities related to the Jurassic–Cretaceous syn-rift-postrift transition of the northern North Sea: Journal of the Geological Society (London), v. 161, p. 1–17, doi:10.1144/0016-764903-051.

Lambiase, J. J., 1990, A model for tectonic control of lacustrine stratigraphic sequences in continental rift basins, *in* B. J. Katz, ed., Lacustrine basin exploration: Case studies and modern analogs: AAPG Memoir 50, p. 265–276.

Leeder, M. R. 1995, Continental rifts and protooceanic rift troughs, *in* C. J. Busby and R. V. Ingersoll, eds., Tectonics of sedimentary basins: Oxford, Blackwell Science, p. 119–148.

Leeder, R. M., and R. L. Gawthorpe, 1987, Sedimentary models for extensional tilt-block/half-graben basins, *in* M. P. Coward, J. F Dewey, and P. L. Hancock, eds., Continental extensional tectonics: Geological Society (London) Special Publication, v. 28, p. 139–152, doi:10.1144/GSL.SP.1987.028.01.11.

Lentini, M., and S. Fraser, 2008, Observations from the South Atlantic: How conjugate is the conjugate margin?, *in* S. Fraser et al., eds., Rifts renaissance: Stretching the crust and extending exploration frontiers: Geological Society of London and SEPM International Conference, Houston, Texas, August 18–21, 2008, p. 68–69.

López-Gamundí, O. R., and R. Barragan, 2008, Rift architecture and its control from presalt plays in the South Atlantic: Lessons from offshore Brazil: AAPG International Conference and Exhibition, Cape Town, South Africa, October 26–29, 2008.

Meisling, K. E., P. R. Cobbold, and V. S. Mount, 2001, Segmentation of an obliquely rifted margin, Campos and Santos basins, southeastern Brazil: AAPG Bulletin, v. 85, p. 1903–1924, doi:10.1306/8626D0A9-173B-11D7-8645000102C1865D.

Mello, M. R., P. W. Brooks, J. M. Moldowan, and B. Wygrala, 2006, The giant forgotten subsalt hydrocarbon province in the Greater Campos Basin, Brazil: A journey from a great mistake to a learning experience: AAPG Convention Program Abstracts (Digital), Houston, Texas, April 9–12, 2006.

Mello, M. R., and J. R. Maxwell, 1990, Organic geochemical and biological marker characterization of source rocks and oils derived from lacustrine environments in the Brazilian continental margin, *in* B. J. Katz, ed., Lacustrine basin exploration: Case studies and modern analogs: AAPG Memoir 50, p. 77–99.

Mitchum Jr., R. M., 1977, Seismic stratigraphy and global changes of sea level. Part 1: Glossary of terms used in seismic stratigraphy, *in* C. E. Payton, ed., Seismic stratigraphy: Applications to hydrocarbon exploration: AAPG Memoir 26, p. 205–212.

Mohriak, W. U., B. R. Rosendahl, J. P. Turner, and S. C. Valente, 2002, Crustal architecture of South Atlantic volcanic margins, *in* M. A. Menzies, S. L. Klemperer, C. J. Ebinger, and J. Baker, eds., Volcanic rifted margins: Geological Society of America Special Paper 362, p. 159–202.

Mohriak, W. U., M. Nemcok, and G. Enciso, 2008, South Atlantic divergent margin evolution: Rift-border uplift and salt tectonics in the basins of SE Brazil, *in* R. J. Pankhurst, R. A. J. Trouw, B. B. Brito Neves, and M. J. De Wit, eds., West Gondwana: Pre-Cenozoic correlations across the South Atlantic region: Geological Society (London) Special Publication, v. 294, p. 365–398.

Morley, C. K., 2002, Evolution of large normal faults: Evidence from seismic reflection data: AAPG Bulletin, v. 86, p. 961–978, doi:10.1306/61EEDBFC-173E-11D7-8645000102C1865D .

Morley, C. K., and W. A. Wescott, 1999, Sedimentary environments and geometry of sedimentary bodies determined from subsurface studies in East Africa, *in* C. K. Morley ed., Geoscience of rift systems: Evolution of East Africa: AAPG Studies in Geology, v. 44, p. 211–231.

Neto Cavalcanti de Araújo, M., P. C. Santarem da Silva, G. Correa de Matos, and R. Dias Lima, 2009, Conceitos, feições diagnósticas e exemplos sísmicos de dobras associadas a falhas distensionais na seção rifte das bacias de Campos e Santos: Boletim de Geociências de Petrobras, v. 17, no. 1, p. 17–30

Nunns, A. G., 1991, Structural restoration of seismic and geologic sections in extensional regimes: AAPG Bulletin, v. 75, p. 278–297.

Pope, M. C., J. Grotzinger, and C. Schreiber, 2000, Evaporitic subtidal stromatolites produced by in-situ precipitation: Textures, facies associations, and temporal significance: Journal of Sedimentary Research, v. 70, p. 1139–1151, doi:10.1306/062099701139.

Pratt, B. R., N. P. James, and C. A. Cowan, 1992, Peritidal carbonates, *in* R. G. Walker and N. P. James, eds., Facies models: Response to sea level change: St. John's, Newfoundland, Geological Association of Canada, p. 303–322.

Prosser, S., 1993, Rift-related linked depositional systems and their seismic expression, *in* G. D. Williams and A. Dobb, eds., Tectonics and seismic sequence stratigraphy: Geological Society (London) Special Publication 71, p. 35–66.

Rabinowitz, P. D., and V. LaBrecque, 1979, The Mesozoic South Atlantic Ocean and evolution of its continental margin: Journal of Geophysical Research, v. 84, p. 5973–6002, doi:10.1029/JB084iB11p05973.

Rangel, H. D., and M. Carminatti, 2000, Rift lake stratigraphy of the Lagoa Feia Formation, Campos Basin, Brazil, *in* E. H. Gierlowski-Kordesch and K. R. Kelts, eds., Lake basins through space and time: AAPG Studies in Geology 46, p. 225–244.

Roberts, A. M., and G. Yielding, 1991, Deformation around basin-margin faults in the North Sea/mid-Norway rift, *in* A. M. Roberts, G. Yielding, and B. Freeman, eds., 1991, The geometry of normal faults: Geological Society (London) Special Publication 56, p. 61–78.

Rosendahl, R. B., D. J. Reynolds, P. M. Lorber, C. F. Burgess, J. McGill, D. Scott, J. J. Lambiase, and S. J. Derksen, 1986, Structural expression of rifting: Lessons from Lake Tanganyika, Africa, *in* L. E. Frostick, R. W. Renaut, I. Reid, and J. J. Tiercelin, eds., Sedimentation in the African rifts: Geological Society (London) Special Publication 25, p. 29–44.

Rosendahl, B. R., W. U. Mohriak, M. Nemèoc, M. E. Odegard, J. P. Turner, and W. G. Dickson, 2005, West African and Brazilian conjugate margins: Crustal types, architecture, and plate configurations, *in* 4th Houston Geological Society–Petroleum Exploration Society of Great Britain International Conference on African Exploration and Production, Houston, Texas, September 7–8, 2005, p. 3.

Schlische, R. W., and P. E. Olsen, 1990, Quantitative filling model for continental extensional basins with applications to the early Mesozoic rifts of eastern North America: Journal of Geology, v. 98, p. 135–155.

Shaw, J. H., S. C. Hook, and E. P. Sitohang, 1997, Extensional fault-bend folding and synrift deposition: An example from the Central Sumatra Basin, Indonesia: AAPG Bulletin, v. 81, p. 367–379.

Szatmari, P., 2000, Habitat of petroleum along the South Atlantic margins, *in* M. R. Mello and B. J. Katz, eds., Petroleum systems of South Atlantic margins: AAPG Memoir 73, p. 69–75.

Szatmari, P., E. Milani, M. Lana, J. Conceição, and A. Lobo, 1985, How South Atlantic rifting affects Brazilian oil reserves distribution: Oil and Gas Journal, v. 83, p. 107–113.

Skuce, A. G., 1996, Forward modeling of compaction above normal faults: An example from the Sirte Basin, Libya, *in* R. G. Buchanan and D. A. Nieuwland, eds., Modern developments in structural interpretation: Validation and modeling: Geological Society (London) Special Publication 99, p. 135–146.

Teasdale, J. P., and L. Jensen, 2008, A new structural model for the presalt Santos Basin, Brazil, based on bottom-up basin analysis, *in* S. Fraser et al., eds., Rifts renaissance: Stretching the crust and extending exploration frontiers: Geological Society of London and SEPM International Conference, Houston, Texas, August 18–21, 2008, p. 88.

Yielding, G., 1990, Footwall uplift associated with Late Jurassic normal faulting in the northern North Sea: Journal of the Geological Society (London), v. 147, p. 219–222, doi:10.1144/gsjgs.147.2.0219.

White, N., J. Jackson, and D. P. McKenzie, 1986, The relationship between the geometry of normal faults and that of the sedimentary layers in their hanging walls: Journal of Structural Geology, v. 8, p. 897–909, doi:10.1016/0191-8141(86)90035-0.

Wilson, R. W., M. Thompson, K. Boyd, K. J. W. McCaffrey, and R. E. Holdsworth, 2008, Structure and evolution of the Santos Basin, SE Brazil: Oblique extension, strain partitioning and implications for South Atlantic rift evolution *in* S. Fraser et al., eds., Rifts renaissance: Stretching the crust and extending exploration frontiers: Geological Society of London and SEPM International Conference, Houston, Texas, August 18–21, 2008, p. 90.

Wright, P. V., and A. Racey, 2009, Presalt microbial carbonate reservoirs of the Santos Basin, offshore Brazil: AAPG Annual Convention, Denver, Colorado, June 7–10, 2009.

Xiao, H., and J. Suppe, 1992, Origin of rollover: AAPG Bulletin, v. 76, p. 509–525.

Zanotto, O., and P. Szatmari, 1987, Mecanismo de rifteamento da porção ocidental da Margem Equatorial: Revista Brasileira de Geociências, v. 17, p. 189–195.

7

Oliveira, Maria José R., Pedro V. Zalán, João L. Caldeira, Arnaldo Tanaka, Paulo Santarem, Ivo Trosdtorf Jr., and Anderson Moraes, 2012, Linked extensional-compressional tectonics in gravitational systems in the Equatorial Margin of Brazil, *in* D. Gao, ed., Tectonics and sedimentation: Implications for petroleum systems: AAPG Memoir 100, p. 159–178.

Linked Extensional-compressional Tectonics in Gravitational Systems in the Equatorial Margin of Brazil

Maria José R. Oliveira, Paulo Santarem, and Anderson Moraes

Petrobras Research Center, Avenida Horácio Macedo, 950, Cidade Universitária, Rio de Janeiro, Brazil (e-mails: mjoliveira@petrobras.com.br; paulosantarem@petrobras.com.br; andersonmoraes@petrobras.com.br)

Pedro V. Zalán, João L. Caldeira, Arnaldo Tanaka, and Ivo Trosdtorf Jr.

Petrobras/E&P (Exploration and Production)-Exp, Avenida República do Chile 330/16th floor, Rio de Janeiro, Brazil (e-mails: pvzalan@uol.com.br; joao.caldeira@yahoo.com.br; atanaka@petrobras.com.br; trosdtorf@petrobras.com.br)

ABSTRACT

The Pará-Maranhão and Barreirinhas basins in the equatorial margin of Brazil contain gravitational gliding systems composed of three structural domains: a proximal extensional, a distal contractional, and a transitional (or translational) domain between the two others. The main faults of these domains detach on a decollement surface of shales and marls, presumably overpressured. Several methods were applied to investigate these thin-skinned tectonics systems, including interpretation of seismic sections, physical modeling, numerical modeling, restoration of cross sections, and integration with field data. These methods indicated that thrusts developed in a classical backstepping sequence with younger thrusts developing in the hanging wall and with landward migration of depocenters through geologic time. Out-of-sequence thrusts were observed locally. The results of cross section restorations suggested that the total amount of shortening exceeded the total amount of stretching in the basal layers, close to the detachment surface, whereas stretching exceeded shortening in the upper layers. Our conclusions point out that gravitational gliding was caused by the combined effect of sedimentary loading, slope gradient, and probably, pore fluid overpressure, with gliding events being triggered by episodic reactivations of the intervening Romanche and Saint Paul fracture zones.

INTRODUCTION

Linked extensional-compressional systems, developed in response to gravitational tectonics, are a common feature along the equatorial margins of Africa and Brazil, the latter constituting a new frontier for petroleum exploration. In the Pará-Maranhão (here abbreviated as PAMA) and Barreirinhas basins (Figure 1), discrete

DOI:10.1306/13351552M1003532

linked extensional-compressional systems formed individual gravitational gliding cells (Figure 2). Internally, each cell contains a proximal extensional domain, an intermediate translational domain, and out in the deep and ultradeep waters, a distal contractional domain (Figure 3). The extensional domain is characterized mainly by synthetic (seaward dipping) listric normal faults, as well as minor antithetic (landward dipping) planar normal faults. The contractional domain, however, shows thrusts and folds (detachment folds, fault-propagation folds, and fault-bending folds) as described by Zalán (1998, 2005). The transitional (or translational) domain may contain, in its upper levels, both normal and thrust faults, some of them reaching the sea floor. This zone commonly displays transitional or superposed styles of deformation. The main faults sole on a detachment surface composed of shales and marls, presumably overpressured, defining a shale-dominated tectonic system. The detachment surface for both basins is positioned at depths of about 6 km (~3.37 mi), in similar depths to those described from the Niger Delta (Morley and Guerin, 1996; Corredor et al., 2005). The present chapter documents the results presented by Oliveira et al. (2008).

Gravitational Systems

Gravitational mass movements may be generated by a sloping sea floor with rapidly deposited unstable sediments. The movements occur when fine sediments or fractured rocks are submitted to the effects of hurricanes, big storm waves, or high pore pressures. Submarine gliding occurs where the downslope component of stress exceeds the rock strength, causing movements along many planar and concave rupture surfaces (Hampton et al., 1996). Another cause of submarine mass movements is the flow of methane from beds rich in gas hydrates (Xu et al., 2001; Paull et al., 2003).

The driving mechanisms of gravitational systems may be classified as gravity gliding and gravity spreading (Rowan et al., 2004). Gravity gliding is defined as the rigid translation of a body down a slope, with the displacement vectors parallel to the detachment plane. Gravity spreading is the vertical collapse and lateral spreading of a rock body with a sloping upper surface under its own weight. Sometimes it is difficult to differentiate between gravity gliding and gravity spreading, and along many margins, the two mechanisms are mixed.

Shale Tectonics Versus Salt Tectonics

Gravitational systems may detach above either salt or shale. The structural styles are mostly dependent on the nature of the detachment surface. Salt-cored fold belts are dominated by symmetric polyharmonic detachment folds. Shale-cored fold belts, in contrast, are dominated by asymmetric basinward-vergent thrust imbricates, multiple detachment levels, and fault-bend and fault-propagation folds (e.g., Wu and Bally, 2000). The difference is caused by the greater resistance of shale and its plastic behavior. The deformation happens only if the loading by the sedimentary overburden is sufficiently big; being triggered by a high pore pressure (Rowan et al., 2004). Whereas the contact between salt and sedimentary loading is commonly along a well-defined interface, in the case of shales, this interface may be diffuse and progressive, difficult to be exactly defined in outcrops and seismic sections, and may migrate through the section (Vendeville, 2003). The literature on mobile shale deformation is sparse as compared with salt tectonics publications (Morley, 2003).

Salt mobility is a fundamental material property whereas shales are mobile only if overpressured fluids are present. Dewatering of shales will stop their mobility, but renewed burial and the rise of internal overpressure (e.g., diagenetic release of water or hydrocarbon generation) may renew mobility. The mobility of shales may change with time, depending on burial rate, the amount of overburden, and internal overpressuring conditions (Morley and Guerin, 1996). The volume of shale also varies, depending on the increase or decrease of pore space (Vendeville, 2003). In contrast to salt, the rheological behavior of overpressured shales is independent of deformation rates (Hubbert and Rubey, 1959; Weijermars et al., 1993).

Wu and Bally (2000) conducted comparative studies in tectonic styles generated by gliding over salt and shale in the Gulf of Mexico and offshore Nigeria, respectively. They concluded that allochthonous structures on a large scale are unique to salt basins, and salt appears to be more mobile in the Gulf of Mexico than the overpressured shales of the Niger Delta. The greater density contrast between the salt and surrounding sediments favors the rising of salt to higher structural levels than shale because of larger buoyancy forces.

Major causes of fluid overpressure in shales are (1) rapid sedimentation (Rowan et al., 2004; Nemcok et al., 2005); (2) hydrocarbon generation (Smith and Thomas, 1971; Spencer, 1987; Morley and Guerin, 1996; Cobbold et al., 2004); (3) oil cracking with gas generation (Barker, 1990; Morley and Guerin, 1996); (4) dehydration during the smectite-illite transition (Bruce, 1984; Morley and Guerin, 1996); (5) tectonic compaction during shortening (Koyi, 1995; Rowan et al., 2004; Nemcok et al., 2005); (6) strain softening caused by rotation of clay minerals during shear (Rowan et al., 2004); (7) collapse caused by removal of lithostatic support of sedimentary

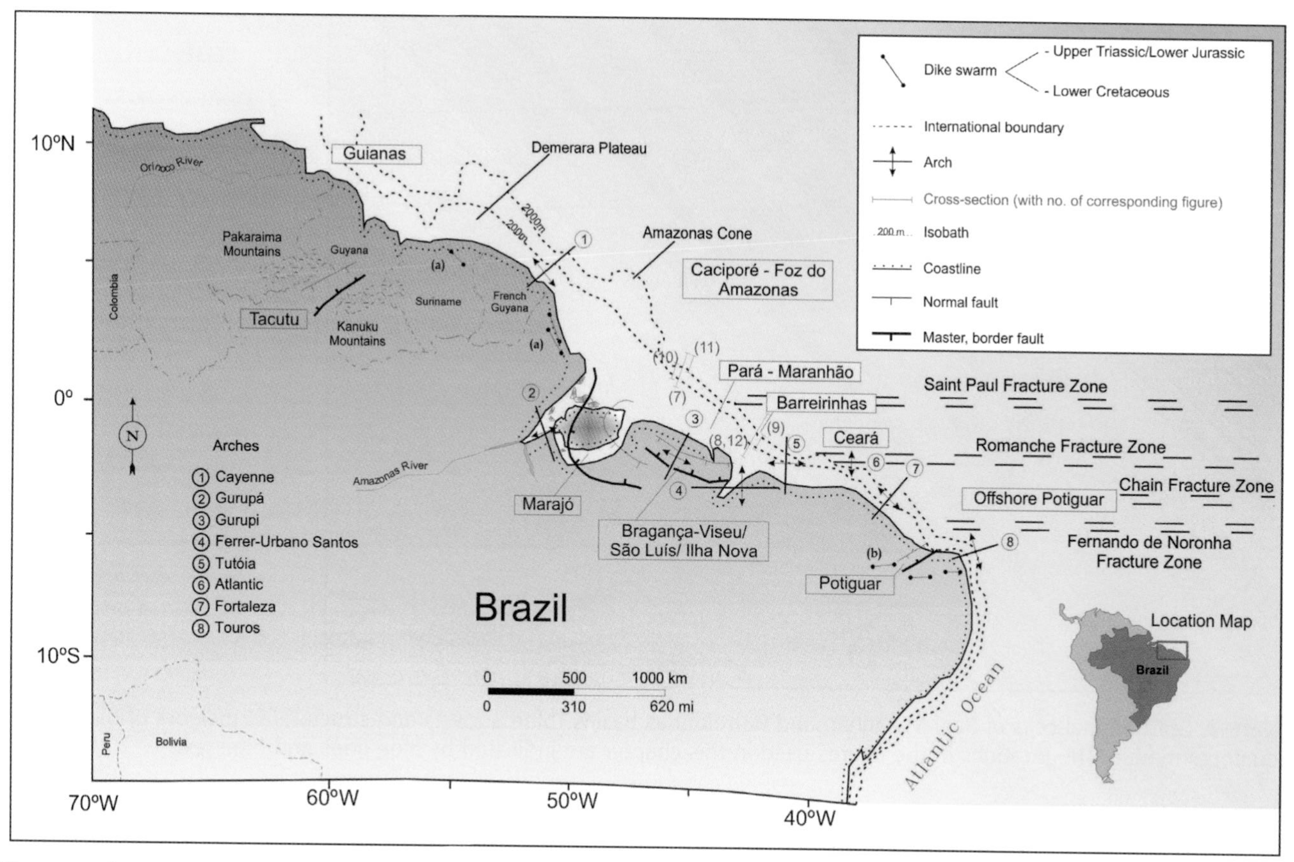

Figure 1. The Equatorial Atlantic segment of South America (from Milani and Thomaz Filho, 2000). The locations of the figures used in this chapter are indicated by blue lines and blue numbers.

loading, increasing the function of fluids (Rowan et al., 2004).

Zalán (2009) presented a scheme differentiating two subtypes of shale-detached gravitational systems (Figure 3). In the first subtype, the gliding occurs along a very distinct layer of the shale package, which commonly is very close to its top; and, thus, the detachment material (shale) does not interfere or participate in the deformation in the gliding allochthonous volume of rock. This is the case of the gravitational systems presented in this work. In the second subtype, the detachment level is involved in the deformation via shale diapirs or pillows or via shale slivers along thrust faults. In this case, the deformation resembles that of salt-cored gravitational systems. This type commonly occurs in extremely overpressured shale beds associated with prodeltaic environments, such as in the Niger (Corredor et al., 2005) and the Amazon (Cobbold et al., 2004) cones.

Hydrocarbon Potential

Deltas are known to constitute significant hydrocarbon provinces (Morley, 2003). Deformation caused by loading of overpressured, undercompacted (mobile) shales associated with deltas produces spectacular rapidly evolving structures, thick rapidly forming depocenters, and massive fluid flows in basins. The Mississippi Delta (Gulf of Mexico) and Niger Delta are the most classical examples of these petroleum-rich provinces in a passive-margin setting.

Both extensional and compressional domains present in shale gravitational systems may contain hydrocarbon accumulations. In extensional domains, large growth fault depocenters are remarkable for the amount of sediment they can accumulate in a short period. The distribution of growth faults is linked to relative rates of deposition versus subsidence. Major ways in which growth faults influence sedimentation are subsidence rapid enough to stall the lateral migration of facies, creation of bathymetric highs and lows influencing sediment pathways from the shelf to deep water, and creation of sea floor depressions that trap sediments under gravity (Morley, 2003). In the Niger Delta, for example, the biggest growth faults, both synthetic and antithetic (seaward and landward dipping) were developed in the proximal domain, each with its associated depocenter.

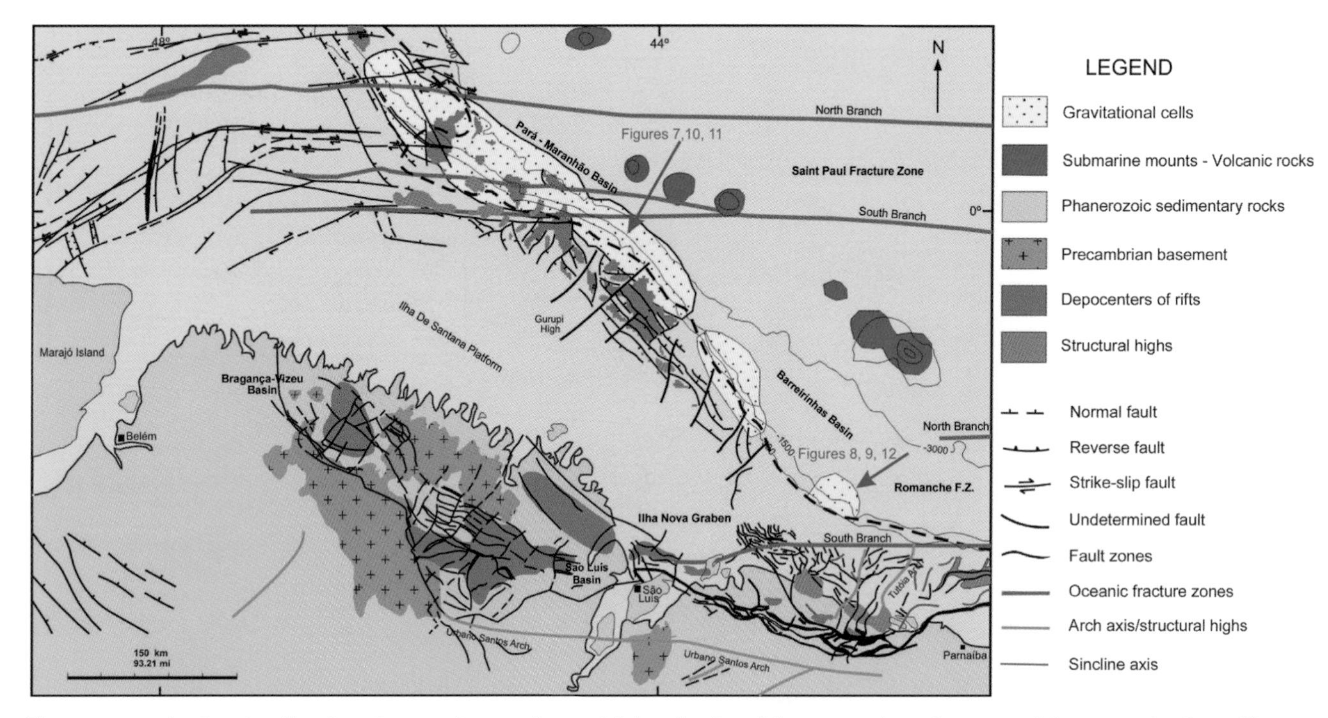

Figure 2. Gravitational cells of Pará-Maranhão and Barreirinhas basins (blue arrows) and structural framework of Brazil's equatorial margin. The locations of the figures used in this chapter are indicated by blue lines and blue text.

In the contractional domain, fold and thrust belts provide exploration plays that are highly prospective but of high risk (Nemcok et al., 2005). The authors pointed out the following major features of these tectonic settings: alteration of hydrocarbon maturation during thrust formation caused by rapid burial and uplift of source rocks; petroleum systems that are commonly highly compartmentalized; and thrust surfaces that act as conduits but also as seals, controlling petroleum migration. Hydrocarbon accumulation and preservation are disturbed by uplift and erosion of traps, by out-of-sequence thrust faults that may alter already existing traps, and by deep circulation of meteoric waters caused by topographic elevation (in the case of orogenic belts).

According to Peel (2003), the fold belts along the passive margins are intimately related to the patterns of deposition. A fold belt along a passive margin might progress continuously, episodically, or even stay stationary. A major part of the increase in thickness is caused by deposition at the top of the wedge and not, necessarily, to structural thickening. Peel (2003) notes that where a long history of fold development exists, the oldest folds of shorter wavelengths become inactive and they become involved in younger folds of longer wavelengths, generally three times those of the initial folds. He points out the importance of precursor structures that may control the style of younger folds. Reservoir distribution in the lowermost levels of the structures may be controlled by the geometry of the oldest folds and not of the more obvious younger folds. The first folds may have a critical function in shaping the migration pathways of hydrocarbons.

The exploration of petroleum in the compressional domains of shale-detached gravitational systems, always situated in deep and ultradeep waters, has met high rates of success in the distal parts of deltas in Nigeria, Indonesia, Trinidad and Tobago, Egypt, India and Brunei and is presently being pursued in the equatorial margins of Brazil and Africa.

With respect to hydrocarbon exploration in Brazilian basins with gravitational tectonics, the works by Schaller and Dauzacker (1986) and Zalán (1998) deserve particular attention.

GEOLOGIC SETTING

The equatorial Atlantic Ocean was initially formed by the fragmentation and breakup of northwestern Gondwana during the Aptian–Albian interval. According to Azevedo (1991), a transtensional shear corridor with a dextral sense of displacement was developed along the present-day northern equatorial continental margin of Brazil, forming the PAMA and Barreirinhas

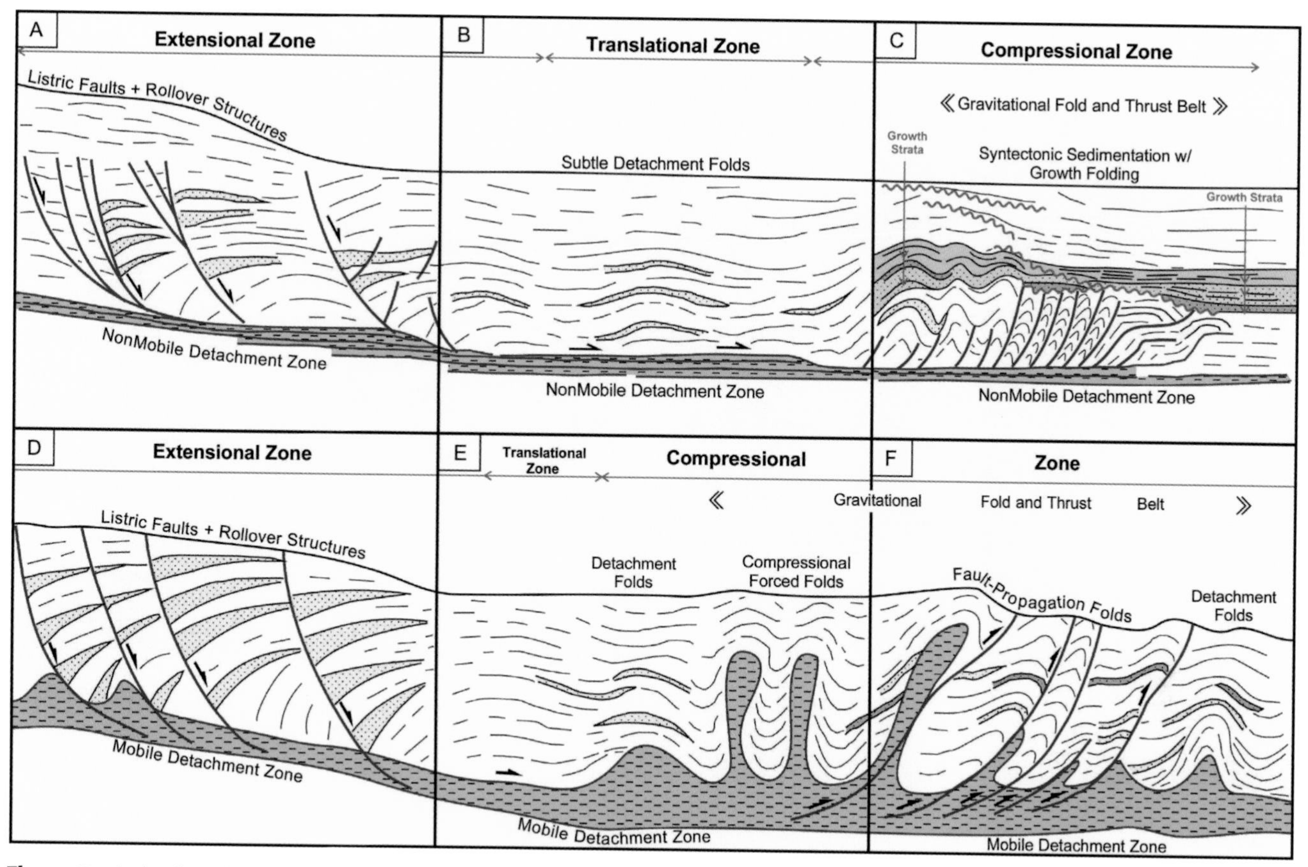

Figure 3. Shale-detached gravitational systems present two subtypes. (A, B, C) developed on nonmobile detachment level; shale does not take part in the deformation of the allochthon in any of the three realms. (D, E, F) associated with mobile detachment level; shale is folded and faulted together with the allochthonous gliding package (Zalán, 2009).

marginal basins. Although these basins could be classified as passive-margin basins (Soares et al., 2007; Trosdtorf, Jr. et al., 2007) containing typical sedimentary sequences deposited during thermal subsidence in the drift phase, Matos (2004) prefers not to classify them as conventional rifts because of the strong influence of transform faults in their evolution. The author suggests that the tectonic evolution of the margin occurred in three stages: (1) pretransform stage (before Barremian–Aptian), mainly transtensive, when the lithospheric stretching generated an en echelon array of depocenters oriented northwest–southeast (Figure 4); (2) syntransform stage (Albian–Cenomanian), with four kinematic/dynamic events. In this stage, the Barreirinhas Basin developed as a predominantly transtensive system, whereas the PAMA Basin developed through wrench-dominated transpression; and (3) posttransform stage (Cenomanian–Holocene), when drift characterized the passive margin, developing mainly in an extensional environment. The geometric relationship between the Brazilian and west African equatorial margins, reconstructed for the Aptian, is shown in Figure 4. The most prominent sea-floor features of these basins are the east–west-trending Romanche and Saint Paul fracture zones (Figures 1, 2). The PAMA and Barreirinhas basins do not contain Aptian salt, in sharp contrast with basins situated in both sides of the South Atlantic Ocean.

The offshore PAMA Basin (Figure 1), has an area of approximately 48,000 km^2 (18,533 mi^2) and was explored by PETROBRAS during the 1970s and 1980s, when subcommercial occurrences of light oil in Cenozoic carbonates (well 1-PAS-11) and Upper Cretaceous turbidites (well 1-MAS-5A) were discovered. The last well in this basin was drilled in 1993. The limits of this basin, located between the Barreirinhas Basin (to the southeast) and the Amazonas Fan (to the northwest), are not very clear because no major tectonic features bordering them exist (Soares et al., 2007). The Saint Paul fracture zone projects to the interior of the continental crust near the northern limit of the basin in two very well defined branches. The stratigraphy of the PAMA is complex (Figure 5). Paleozoic deposits may occur at its base as prerift sequences preserved inside initial

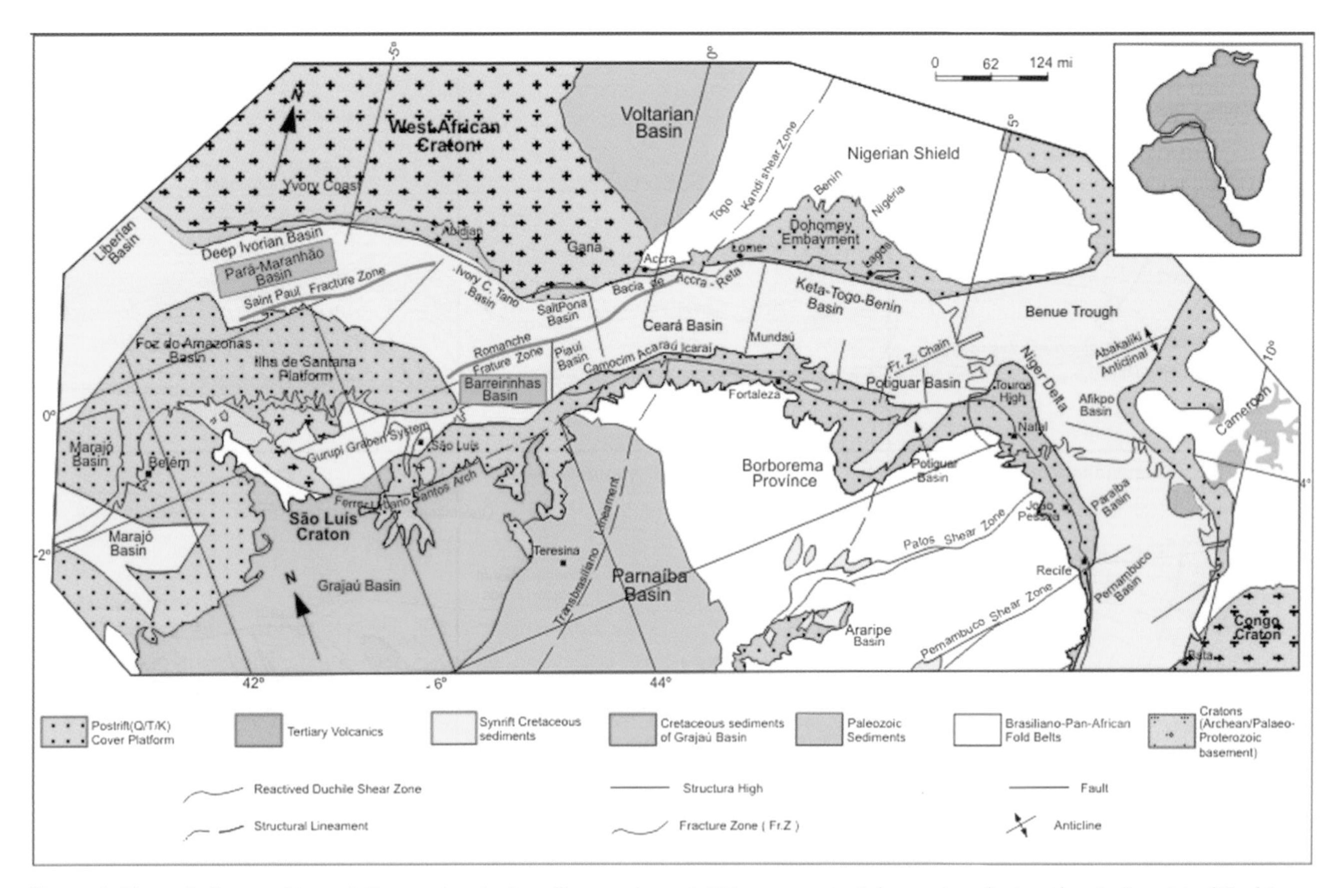

Figure 4. The relative position of the conjugate Brazilian and west African equatorial margins during the Aptian. Modified from Matos (2004). 124 mi (199.5 km).

grabens. The Mesozoic basin comprises three sequences (Brandão and Feijó, 1994): rift sequence (late Aptian–late Albian) formed by fluvial/deltaic/lacustrine sediments of the Canárias Group; transitional sequence (late Albian–mid-Cenomanian) composed of progradational sediments of the Caju Group; and drift sequence (mid-Cenomanian–Holocene) represented by the Humberto de Campos Group made up mainly of shales and carbonate rocks deposited during transgressive-regressive cycles. In a more modern interpretation, two rift events can be recognized: the first one during the Aptian, and the second one during the Albian (Soares et al., 2007). The authors interpreted a sag basin, represented by the horizontal Aptian sediments of the Codó Formation separating these two events. Basaltic magmatism occurred during the Late Cretaceous, the Eocene, and the Miocene. A particularity of the PAMA is a thick package of Cenozoic carbonate rocks that predominates in the shallow waters.

The gravitational linked extensional-compressional systems studied in this work were deposited during the drift phase. The major challenges for the exploration of petroleum in the plays associated with gravitational systems in the PAMA are (1) the decollement surface may represent an obstacle to fluid migration from structurally deeper source rock levels; and (2) the shallow depths at which thrusts form may endanger the quality of the oil that may have accumulated (Zalán, 1998).

The Barreirinhas Basin (Figure 1) comprises an area of approximately 46,000 km^2 (17,761 mi^2), of which 8500 km^2 (3282 mi^2) are situated onshore and covered by Neogene and Quaternary sediments. It is limited to the east by the Tutóia Arch and to the south by shallow basement. To the west, the Ilha de Santana Platform forms a barrier between the wide Barreirinhas Basin to the east/southeast spread from onshore to ultradeep waters and the narrow PAMA Basin that runs northwest–southeast parallel to this mighty shallow basement shelf. The Romanche fracture zone runs into the basin, dividing it in the middle (Trosdtorf, Jr. et al., 2007). The basin was explored in the 1960s, 1970s, and 1980s, but the results were not encouraging, with minor production having been established for a short period in onshore fields. At present, PETROBRAS is exploring this basin once again. The Barreirinhas Basin contains a basal prerift sequence composed of Paleozoic rocks, similar to the PAMA. The rift sequence spans in age from late Aptian to late Albian (Figure 6). In a manner similar to the PAMA Basin, two phases of rifting are interpreted, separated by a sag basin (represented by Aptian sediments of the Codó Formation) (Trosdtorf, Jr. et al., 2007). The drift sequence is composed of sediments from late Albian/Cenomanian to Pliocene/Pleistocene, the last showing retrogradational fluvial/deltaic rocks at their base (Feijó, 1994). A very thick drift section (>10 km [>5.61 mi]) was deposited on oceanic crust in deep and ultradeep waters (Trosdtorf, Jr. et al., 2007).

GRAVITATIONAL GLIDING CELLS IN PARÁ-MARANHÃO AND BARREIRINHAS BASINS

The pioneer work about the gravitational systems of PAMA and Barreirinhas basins was published by Caldeira et al. (1991). These systems are composed of three structural domains: extensional, transitional, and compressional. These domains contain faults and associated folds, which sole on detachment surfaces composed of shale and/or marl, presumably overpressured. In some places, it is possible to identify more than one detachment surface, but the main decollement is positioned within the Late Cretaceous (Barreirinhas) and Paleocene (PAMA) sediments of the Travosas Formation. Extension in the proximal regions of the basins originates spectacular compressional structures downdip in the distal areas. The geometry of the structures varies laterally, from strictly downdip tectonic transport in the middle of the linked extensional-compressional system, to laterally spreading in the margins. Locally and rarely, tectonic transport generated shale diapirism. In map view, the limits of the domains are arcuate, and their widths are variable: the extensional and the compressional domain could be alternately wider than each other.

The extensional domain is characterized mainly by synthetic listric normal faults, as well as minor antithetic planar normal faults. In the PAMA Basin, few large normal synthetic listric faults are associated with spectacular rollovers at the shelf edge (Figure 7). In the Barreirinhas Basin, numerous smaller normal listric faults, with less developed associated rollover structures, predominate (Figure 8). The compressional domain comprises folds of various styles: detachment folds, fault-bend folds, and fault-propagation folds (Figure 9). The geometric analysis of faults, folds, and growth strata observed in the seismic profiles allow characterizing the propagation of the normal listric faults and thrust faults as a backstep sequence, with normal and contractional faults becoming younger landward. Some seismic sections show out-of-sequence thrusts (Figure 7). In the PAMA Basin, igneous rocks have locally buttressed the gliding of the sediments, triggering the contractional deformation. The transitional domain contains, in its upper levels, normal faults and thrusts, some of them reaching up to the sea floor; these being the youngest faults of the gravitational system. Growth strata and structural highs could be recognized in both extensional and compressional domains (Figures 7–11) and may have acted as traps for hydrocarbon accumulation. Multiple methods, described below, were used to investigate the evolution of the gravitational systems of these basins.

TECTONICS AND SEDIMENTATION IN THE GLIDING CELLS

It is a well-known fact that most of the deformation that occurs in these gravitational systems is submarine because it commonly happens in the deep waters of the slope realms of the continental margins. As such, sedimentation occurs at the same time as deformation, giving rise to very complex interrelationships between growth strata and structural features (Zalán, 2005). In the extensional domain, growth strata are typically related to listric normal faults, thickening on the flanks and thinning onto the tops of the rollover structures. In the PAMA and Barreirinhas basins, growth strata consistently display a younging-landward distribution; thus, reflecting a back stepping pattern for the propagation of the normal listric faults (Figures 7, 10). The resulting feature is a large composite rollover structure displaying several minor structural closures, subsidiary to the main anticline, each displaying thickening growth strata toward listric faults situated landward (Figure 10).

In the compressional domain, growth strata are related to folds and thrust faults. Growth strata are thinner on the crest of folds and always thicken downward along the back limb toward the next younger thrust fault (load flexure mechanism similar to foreland basins) (Figure 9). The forelimb of the frontal fold is covered by onlapping post-tectonic strata not displaying internal diverging reflections. Figure 11 displays a superb view of the internal structure of the compressional domain of a gravitational system in the PAMA Basin. It is clear that the growth strata (I, II, and III) become younger landward, the youngest being strongly bent and cut by thrust faults toward the southwest. This pattern indicates a backstepping pattern for the propagation of the thrust faults in a coherent way to the propagation of the normal listric faults in the extensional domain (Figure 10). The same relationship between tectonics and sedimentation has been deduced

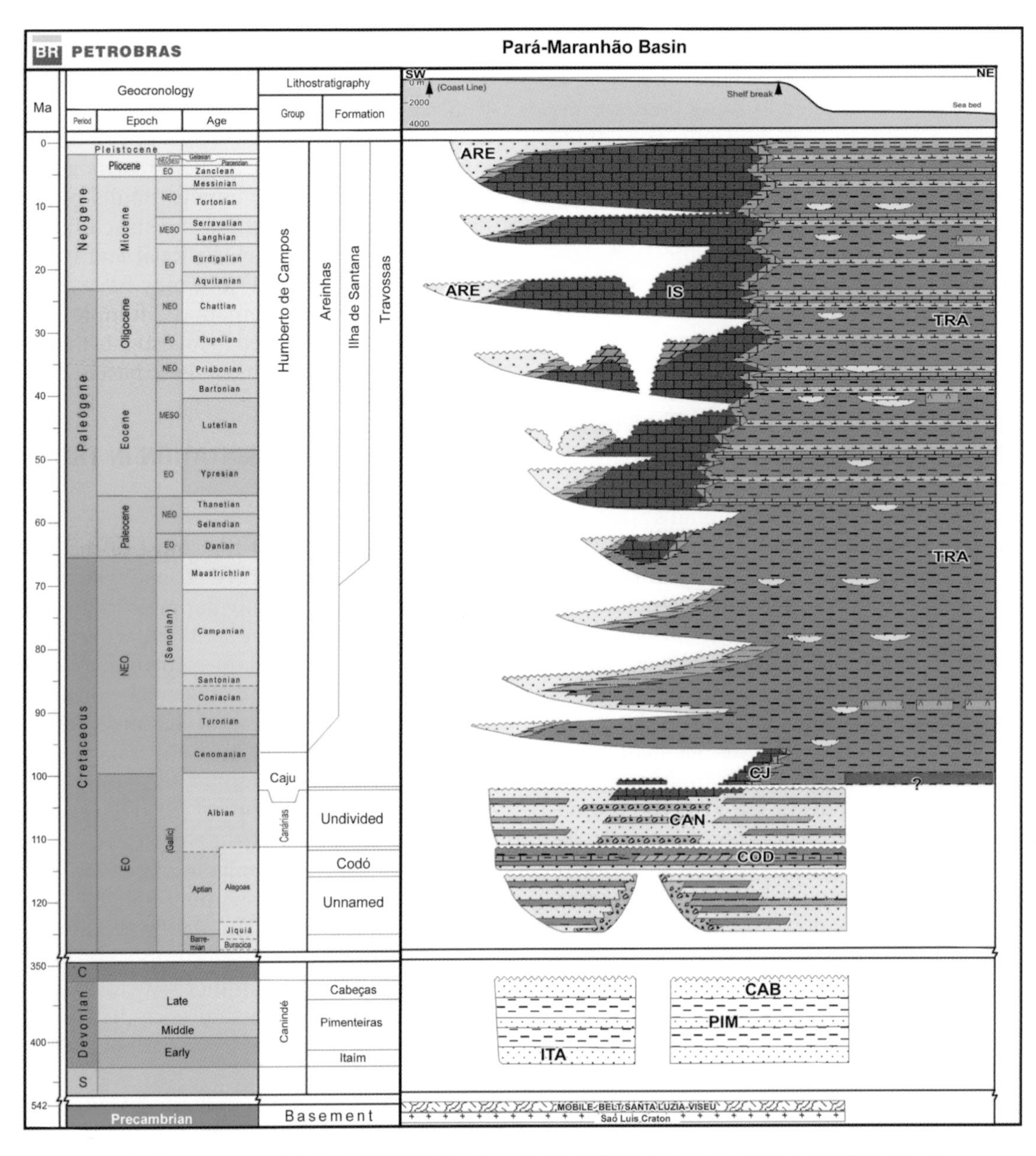

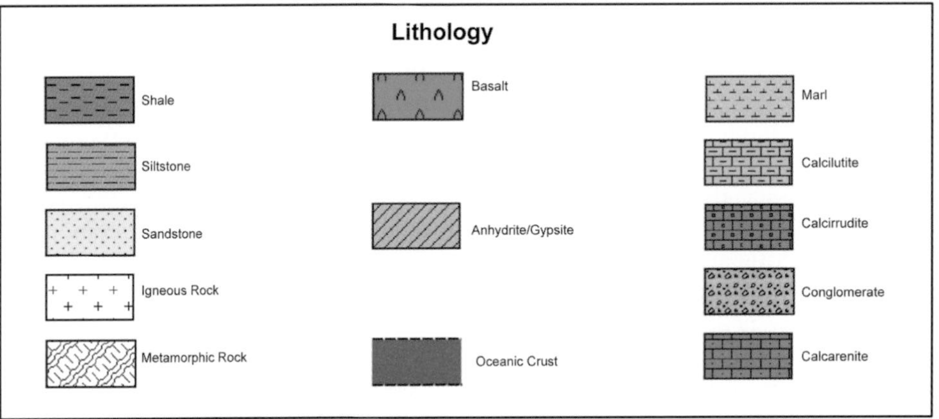

Figure 5. Stratigraphic chart of the Pará-Maranhão Basin. Modified from Soares et al. (2007). ARE = Areinhas; IS = Ilha de Santana; TRA = Travosas; CJ = Caju; CAN = Canárias; COD = Codó; CAB = Cabeças; PIM = Pimenteiras; ITA = Itaim.

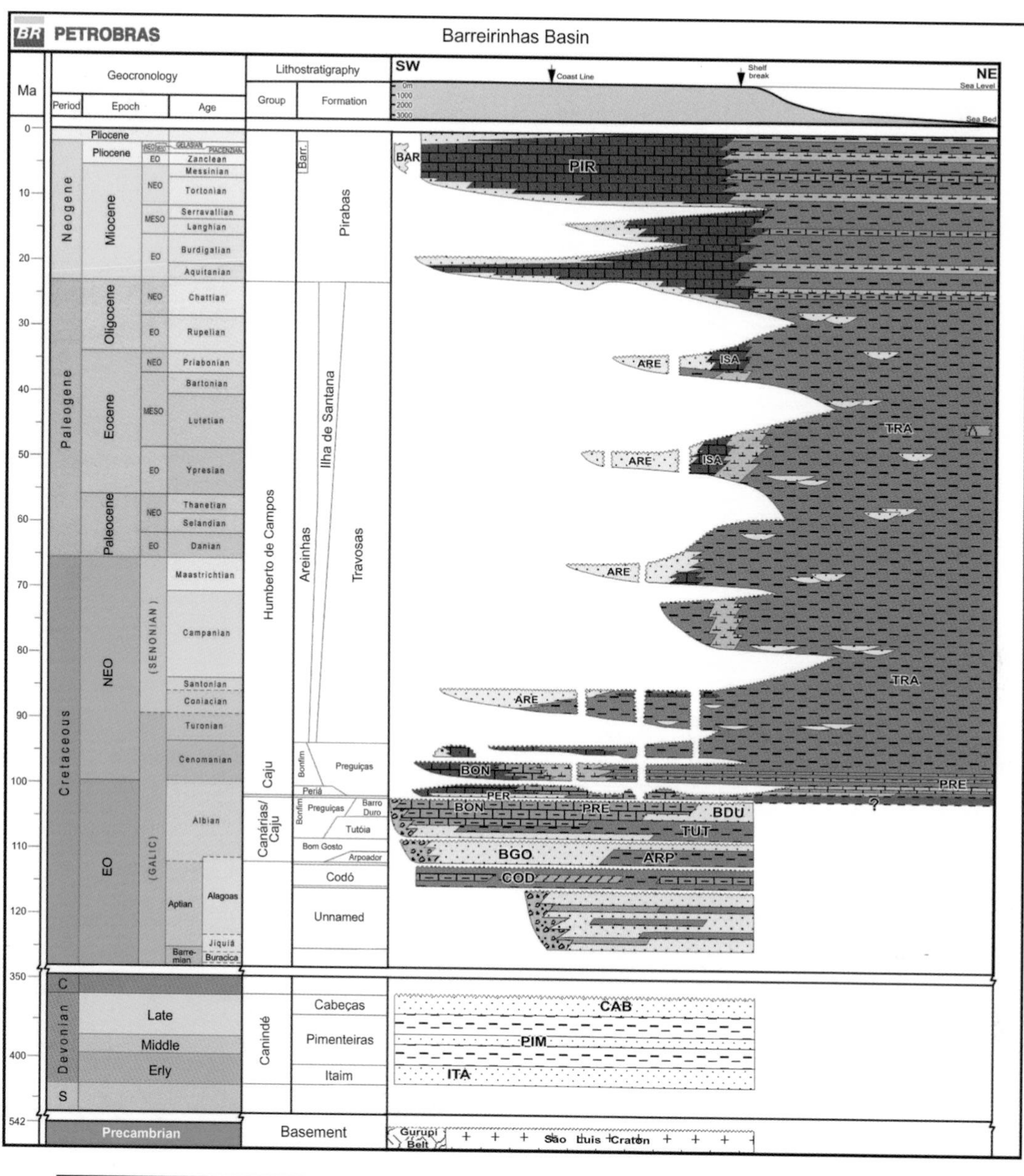

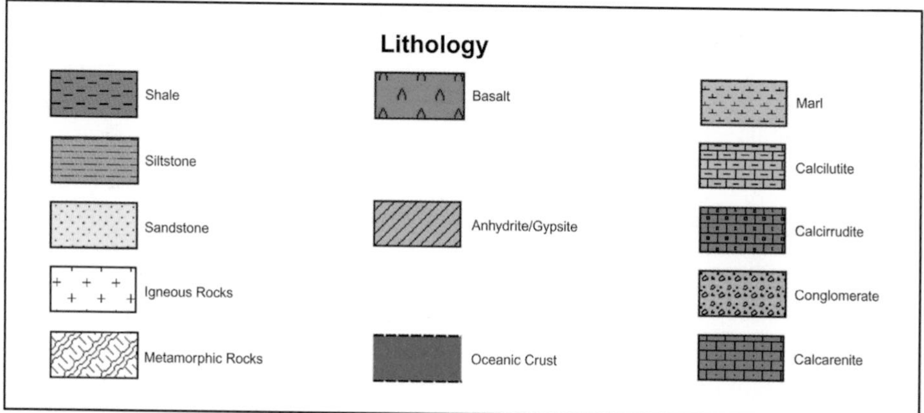

Figure 6. Stratigraphic chart of the Barreirinhas Basin. Modified from Trosdtorf, Jr. et al. (2007). BAR = Barrieras; PIR = Pirabas; ARE = Areinhas; IS = Ilha de Santana; TRA = Travosas; BON = Bonfim; PER = Periá; PRE = Preguiças; BDU = Barro Duro; TUT = Tutóia; BGO = Bom Gosto; ARP = Arpoador; COD = Codó; CAB = Cabeças; PIM = Pimenteiras; ITA = Itaim.

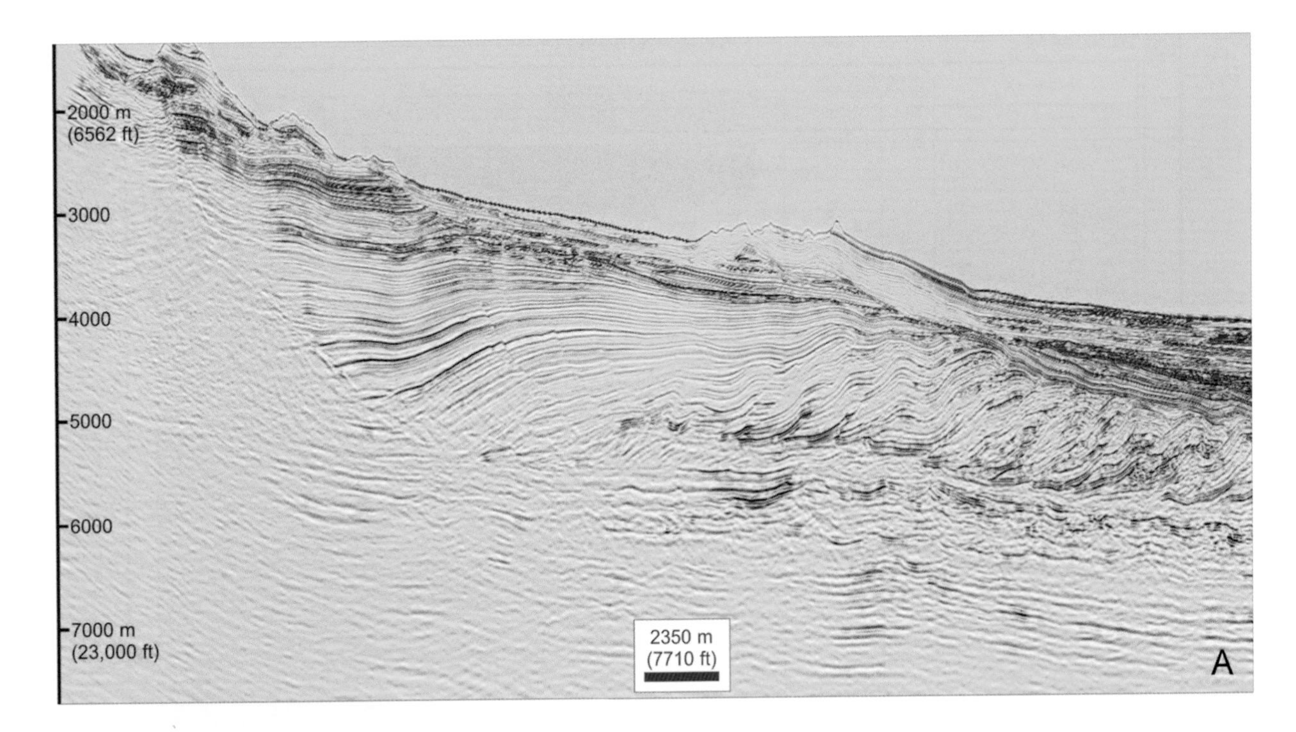

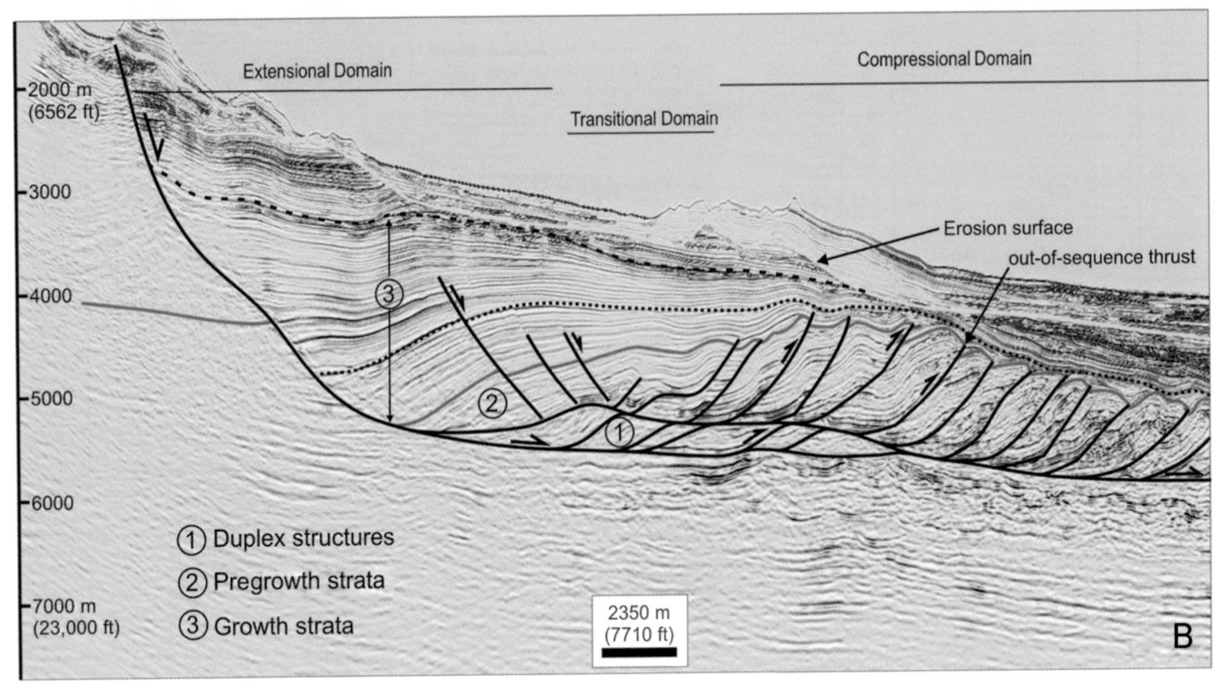

Figure 7. (A) Depth-migrated seismic section of the Pará-Maranhão Basin. Noninterpreted. (B) Interpreted seismic section. Courtesy of CGG VERITAS.

for other surrounding gliding cells. It is possible to suggest that in the equatorial margin of Brazil, the dominant pattern for the development and evolution of gravitational systems is backstepping, that is, normal listric faults and associated thrust faults become younger landward.

RESTORATION OF STRUCTURAL CROSS SECTIONS

Twelve depth-migrated interpreted seismic sections were selected to be restored by three different software packages: two commercial products and one PETROBRAS proprietary software. For the Barreirinhas Basin,

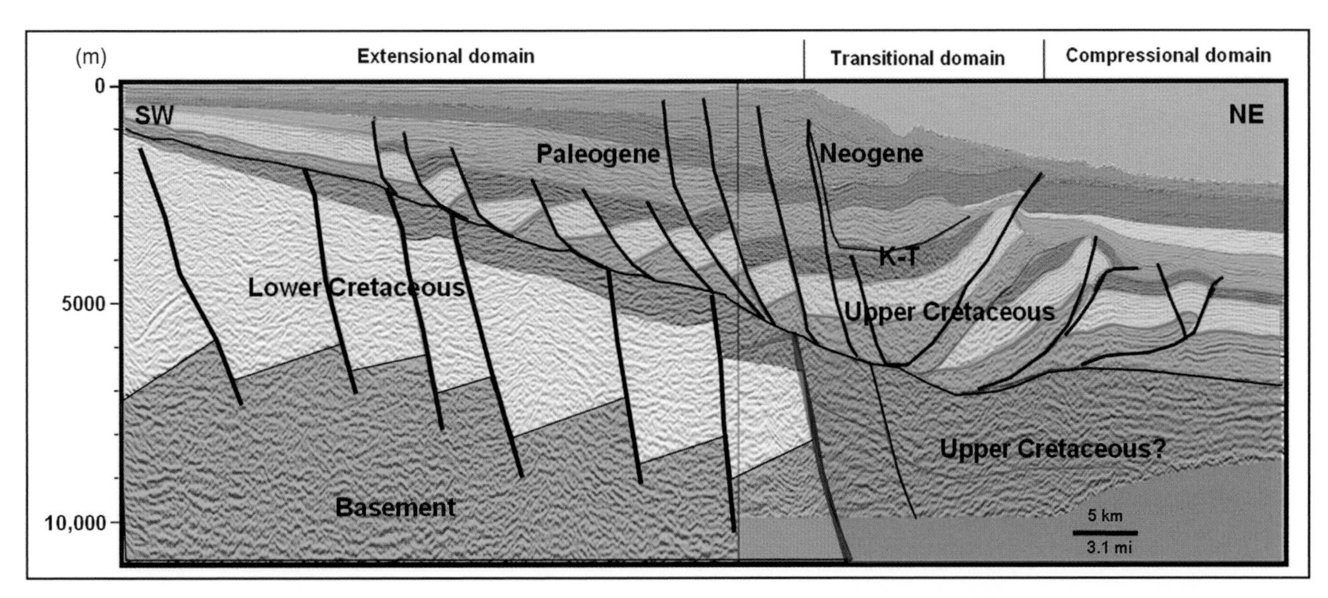

Figure 8. Interpreted seismic section of the Barreirinhas Basin (5000 m [16,404.2 ft]). Courtesy of CGG VERITAS. K-T = Cretaceous-Tertiary boundary.

we used a functionality of the PETROBRAS in-house software that allows the visualization of many parallel restored sections. The use of different software products to restore the same geologic section allowed us to compare the results, which were very similar. The restoration was done considering only the layers above the detachment surface, so that the gliding was decoupled from the layers below.

In the restorations, we applied inclined shear, movement along fault plane, rotation and translation of blocks, and unfolding. The method also considered decompaction and isostatic parameters. Decompaction is based on Athy's (1930) law: $\Phi = \Phi_0 e^{-cz}$, where Φ = porosity, Φ_0 = initial porosity, e = neperian logarithm base (= 2.71828. . .), and c = decay constant. The porosity data were extracted from available wells and from the PETROBRAS database. The isostatic parameters used were T_e = 20 km (crustal elastic thickness) and density of the mantle = 3.33 g/cm^3. The elongation of each layer was calculated according to Ramsay (1967): $e = (l_f - l_0)/l_0$, where e = elongation, l_0 = initial length of the layer, and l_f = final length of the layer. When the elongation is positive, we have stretching. If negative, we have shortening. The elongation percentage is obtained by multiplying the result by 100, as shown in Figure 12.

Assuming plane deformation orientation of the sections approximately perpendicular to the main direction of mass transport and total connection between normal folds and thrusts during the gliding, we expected that the offsets of the normal faults would be balanced by the offsets of the thrusts folds; and that present-day layers would have the same lengths as the restored layers. However, this did not occur, that is, stretching and shortening did not balance. Shortening was commonly greater than stretching in the lower layers situated closer to the detachment surface, whereas stretching exceeded shortening in the upper layers. This mismatch does not mean, necessarily, that the seismic interpretation is incorrect. Some of the possible reasons for the excess of total shortening over total stretching of the basal layers are the presence of normal faults oblique to the main displacement direction, the lack of perfect coupling between extensional and compressional domains, internal deformation on subseismic scales, and loss of mass transported obliquely to the sections. These, as well as the presence of eroded zones in the translational domain, could hinder the propagation of normal faults and amplify the compression. Furthermore, the restored sections did not include the most proximal normal faults in the shelf, thus underestimating stretching. Evidence exists for a greater amount of compression in extensional-compressional linked systems developed over shale surfaces than the systems developed over salt surfaces (Wu and Bally, 2000). Peel (2007) studied the linked extensional-compressional gravitational systems of offshore Namibia and concluded that these systems also do not balance. In this case, the amount of extension was bigger than compression. The author mentioned that "it is clear that the assumption of bed length conservation does not universally apply."

The structural restoration validated the thin-skinned tectonic style model expected for the development of

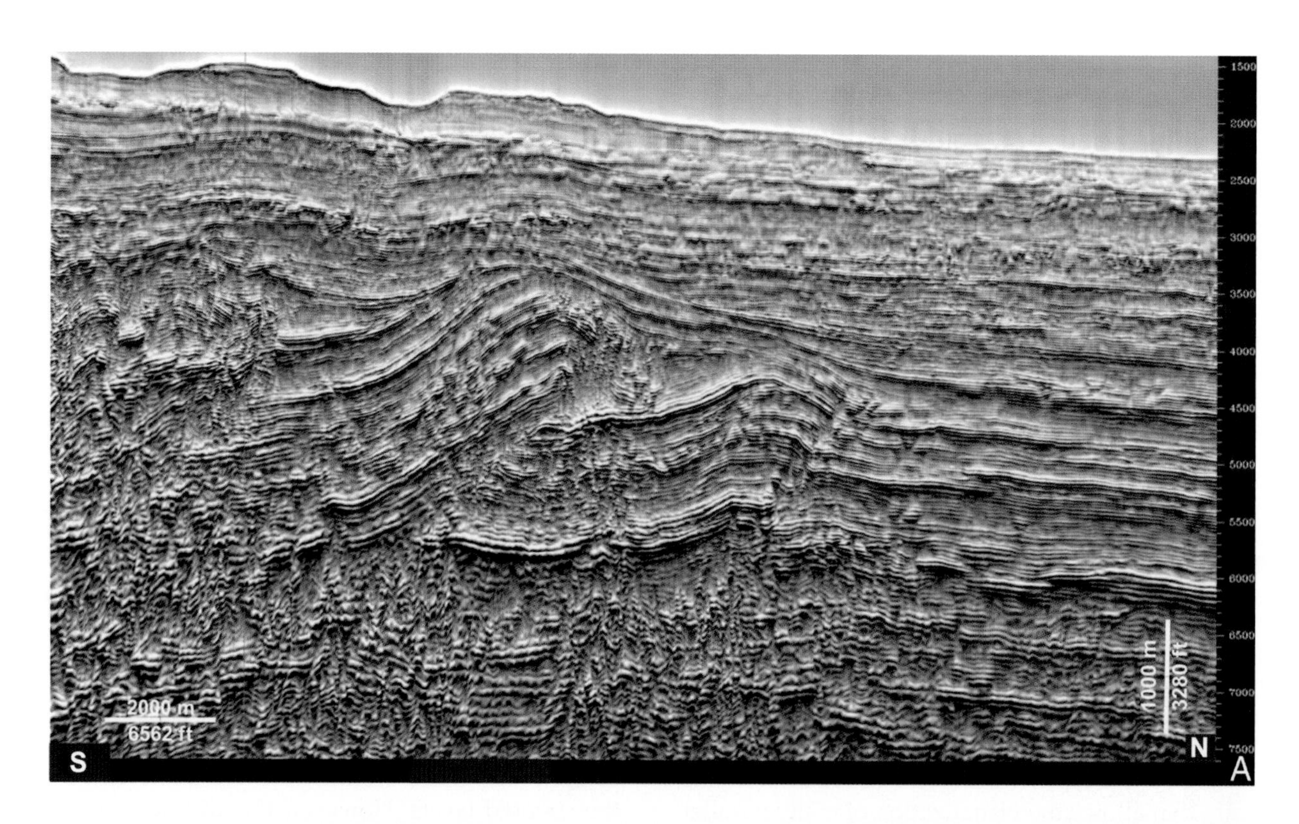

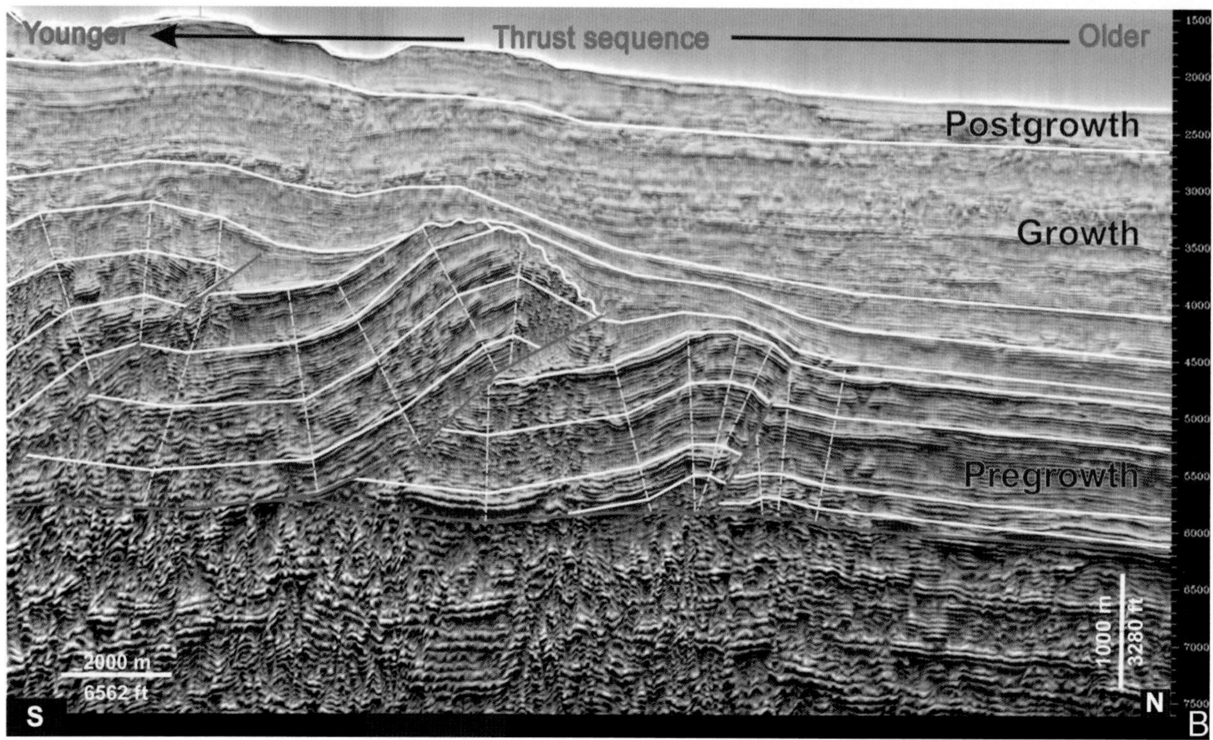

Figure 9. (A) Depth-migrated seismic section of the Barreirinhas Basin (TECVA section, PETROBRAS in-house seismic-enhancing patented display). Noninterpreted. (B) Interpreted seismic section. Courtesy of CGG VERITAS.

these gravitational cells. Structural restoration of cross sections confirmed the generation of a classical backstepping thrust system downslope (younger thrusts formed in the hanging wall of the faults) and a consequent landward migration of depocenters during geologic time. Out-of-sequence thrusts occurred occasionally

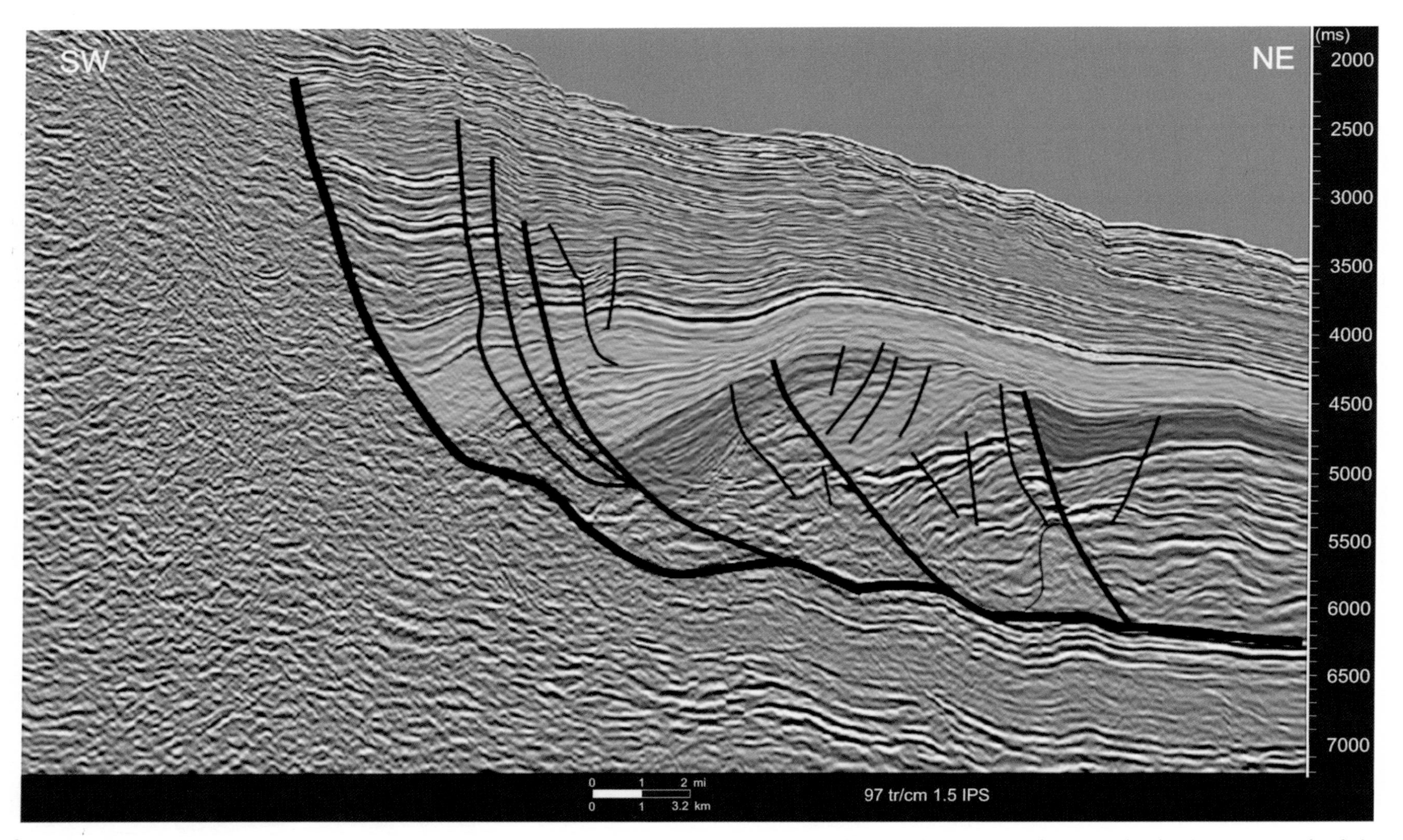

Figure 10. Time-migrated seismic section from the Pará-Maranhão Basin displaying the extensional domain of a gravitational system. The dominant structural style is a major rollover structure composed of multiple subsidiary anticlines and growth strata (in different colors) associated with several listric faults that merge into a single detachment zone. The distribution of the growth strata indicates a younging-landward pattern, reflecting a backstepping pattern for the propagation of the listric faults. A possible shale diapir is interpreted in green (3000 m [9842.52 ft]).

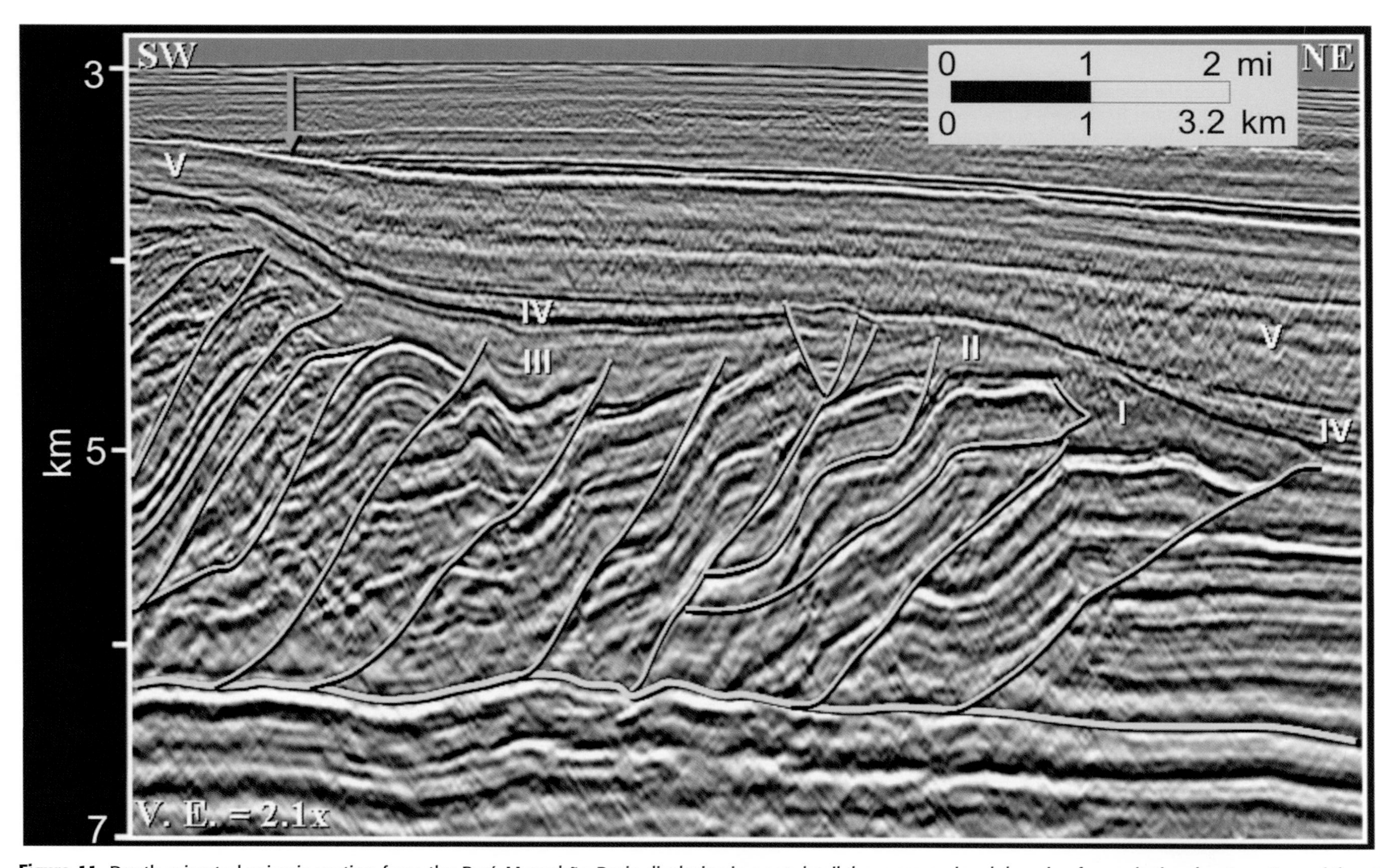

Figure 11. Depth-migrated seismic section from the Pará-Maranhão Basin displaying in great detail the compressional domain of a gravitational system. Several thrust faults (in yellow) splay upward from a single detachment level. Typical syncompression growth strata (colored packages I–III) are associated with the thrusts and display a younging-landward (to the southwest) pattern, reflecting a backstepping pattern for the propagation of the thrust faults. Filling the structural depressions, post-tectonic sequences (IV–V) simply onlap the existing structural highs. Subtle divergence of reflectors within these sequences (in the right side of the section for both IV and V and below the blue arrow for V) may indicate minor growth, probably associated with out-of-sequence reactivation of the compressional features. (3 km [1.86 mi]). V.E. = vertical exaggeration.

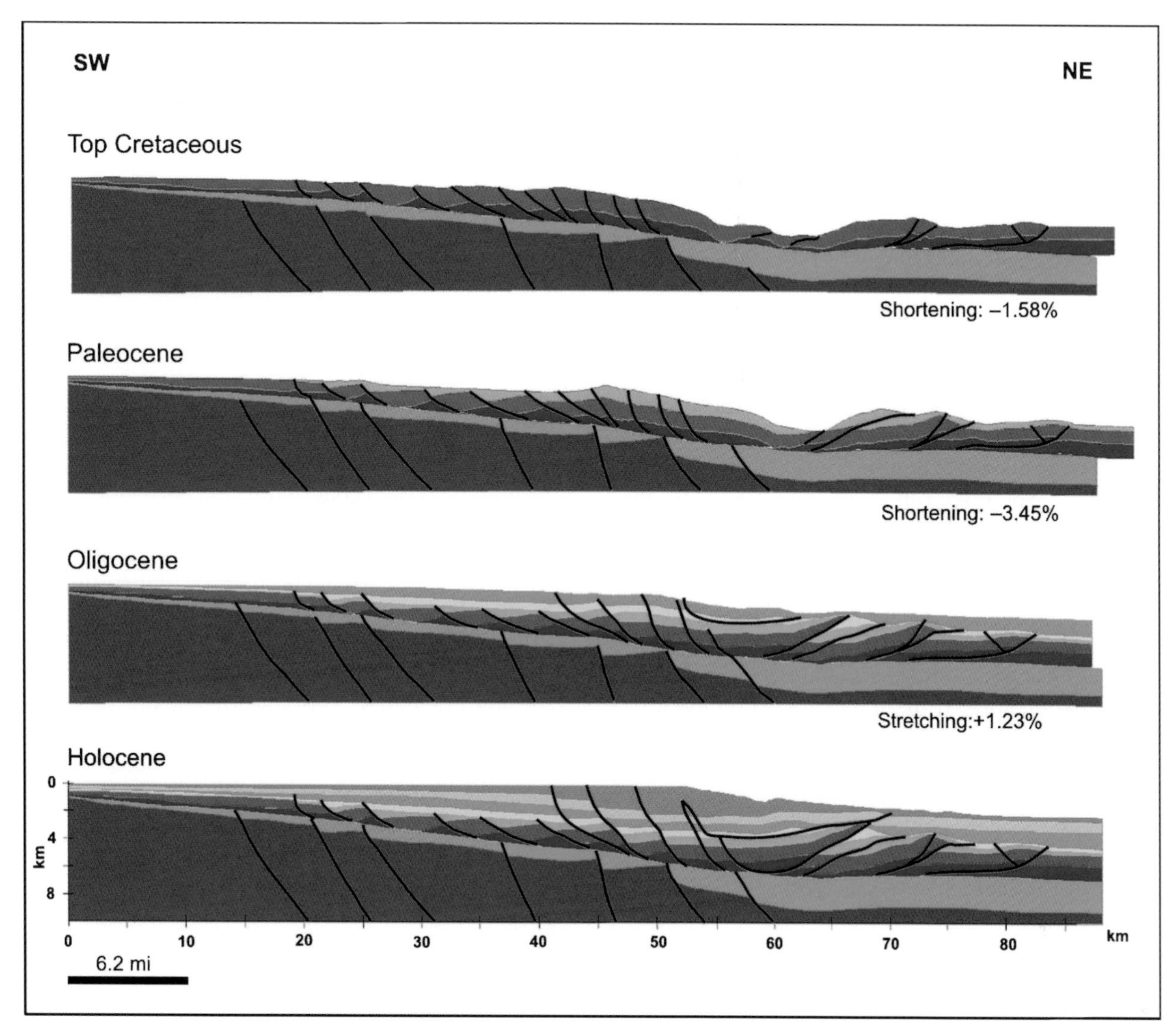

Figure 12. Cross section restoration of the Barreirinhas Basin. Note the amount of shortening of the lower layers (top Cretaceous–Paleocene) in contrast to the amount of stretching of the upper layers (Oligocene and younger) in relation to the final length of the section (Holocene). No vertical exaggeration.

(Figure 12). Another conclusion obtained from the structural restoration is that the faults present in the transitional zone are the younger ones. This fact was confirmed by numerical modeling.

The reconstructed paleostructural scenarios used in petroleum system modeling resulted in a better definition of the synchronism between petroleum generation, migration, and entrapment, thereby permitting better planning of exploration wells and reducing of risks in the exploration process in these basins.

PHYSICAL MODELING

Seven experiments were run at the Università degli Studi di Parma, Italy, attempting to reproduce the structures observed in the Barreirinhas Basin. The analog materials used were transparent silicone putty to simulate the overpressured shale (detachment surface) and colored sand and glass microbeads to simulate the overlying sediments. To reproduce the movements along the slope, the table positioned at the base of the model was inclined at 6°. Two laser scanners were located, respectively, over and under the table to allow the visualization of the morphology of the top and base of the model while the experiment was running.

The model that best reproduced the structural features present in the Barreirinhas Basin was the one that simulated two bordering dextral strike-slip faults, analogous to the two branches of the Romanche fracture zone, bounding the gravitational cell and simulating multiple tectonic pulses during geologic time

(Figure 13). Such activity was recognized in some reflection seismic sections and was described in the literature by Lehner and Bakker (1983). The simulation tried to model pregliding, synkinematic and postkinematic layers. Grains of magnetite were mixed with the prekinematic layers, and before the deformation, the model was magnetized by two permanent magnets that created a quasi-linear magnetic field, using a technique described by Costa and Speranza (2003). After finishing the experiment, the model was cut in vertical sections, and oriented samples of the model were sent to the Instituto Nazionale di Geofisica in Rome, where natural remnant magnetization of these samples was measured. This method allowed visualizing the path of deformation over time. The results are shown in Figure 13C, where we can observe the rotations suffered by the particles during gliding. More compression occurs in the northern part of the model and more extension in the southern part, as we have observed in one gravitational cell in the Barreirinhas Basin.

The physical modeling also confirmed the thin-skinned tectonic style and the backstep sequence of thrusts. Furthermore, it demonstrated the development of normal faults oblique to the main displacement direction. Such faults were recognized in seismic sections. Physical models also allowed constraining the deformation over a unique and thin detachment surface composed of the silicone putty, analogous to the shale and marl horizons interpreted in the seismic sections. A section obtained from physical modeling was chosen for structural balancing using both PETROBRAS in-house software and one commercial two-dimensional software. The restoration showed excess of shortening in the basal layers, just as observed in the structural reconstruction of the interpreted seismic sections previously described.

NUMERICAL MODELING

Numerical models were performed based on the geometry of physical models using an in-house PETROBRAS finite-element system software, which considers gravity load, elastoplastic behavior, and the physical parameters of the materials used in physical models. Two models were run and compared with two physical models, considering the following parameters to the silicone putty and sand, respectively: $E = 5 \times 10^3$ e 5×10^5 Pa (Young's modulus), $\nu = 0{,}48$ e 0,25 (Poisson ratio), $\rho = 965$ e 1560 kg m^{-3} (density), $\lambda = 0$ (pf = 0 Pa) (porous pressure/lithostatic tension), $\sigma_0 = 0.1$ e 10 Pa (coesion), $\psi = 20°–8°$ (dilatance angle), and $\Phi = 1°$ for silicone putty and 30°–26° and 29°–22° sand (internal friction angle).

The results from the numerical modeling were compatible with that obtained by physical modeling on gravitational gliding. They constrained the regions most suitable for the generation of faults and showed that most of them nucleated at the contact between the detachment surface and the overburden before propagating upward. In addition, the models demonstrated that the younger faults developed in the upper levels of the translational domain as can be observed in the structural restoration. The images related to one of the numerical models are shown in Figure 14.

ONSHORE FIELD DATA

Onshore field data collected in the correlative São Luis Basin (Figure 1) indicated the existence of a transtensional structural deformation, with the predominance of extensional or strike-slip features depending on the site. These features include high-angle normal faults, negative flower structures, and subvertical fractures. The faults present predominantly north–northwest-trending orientation, dipping toward northeast. Faults with northeast-trending strike orientation were also observed, showing similarly dextral kinematics, indicating that these two sets of faults were not generated in the same tectonic event as a conjugate pair.

Most of the tectonic structures observed were of subseismic scale. Based on the analysis of the field data, it was possible to deduce a model where S_{Hmax} and S_{Hmin} are produced by a major east–west-trending dextral shear couple rotating clockwise as the basin formed, just as it would be expected for the kinematics of the Romanche fracture zone.

CONCLUSIONS

The various methods applied to understand the gravitational linked extensional-compressional systems in the PAMA and Barreirinhas Basins indicated thin-skinned tectonics as the mechanism of deformation, with the sedimentary layers gliding over a discrete detachment surface composed of shales and marls. The thrusts developed according to a backstepping sequence, with consequent landward migration of depocenters during geologic time. As reproduced by physical models, we propose that the gravitational gliding occurred in response to sedimentary loading and increasing slope gradient during thermal subsidence, triggered and accelerated by episodic activities of the bounding Romanche and Saint Paul fracture zones (Lehner and Bakker, 1983). Another factor that may have favored the gliding was the high pore pressure in shales and marls that

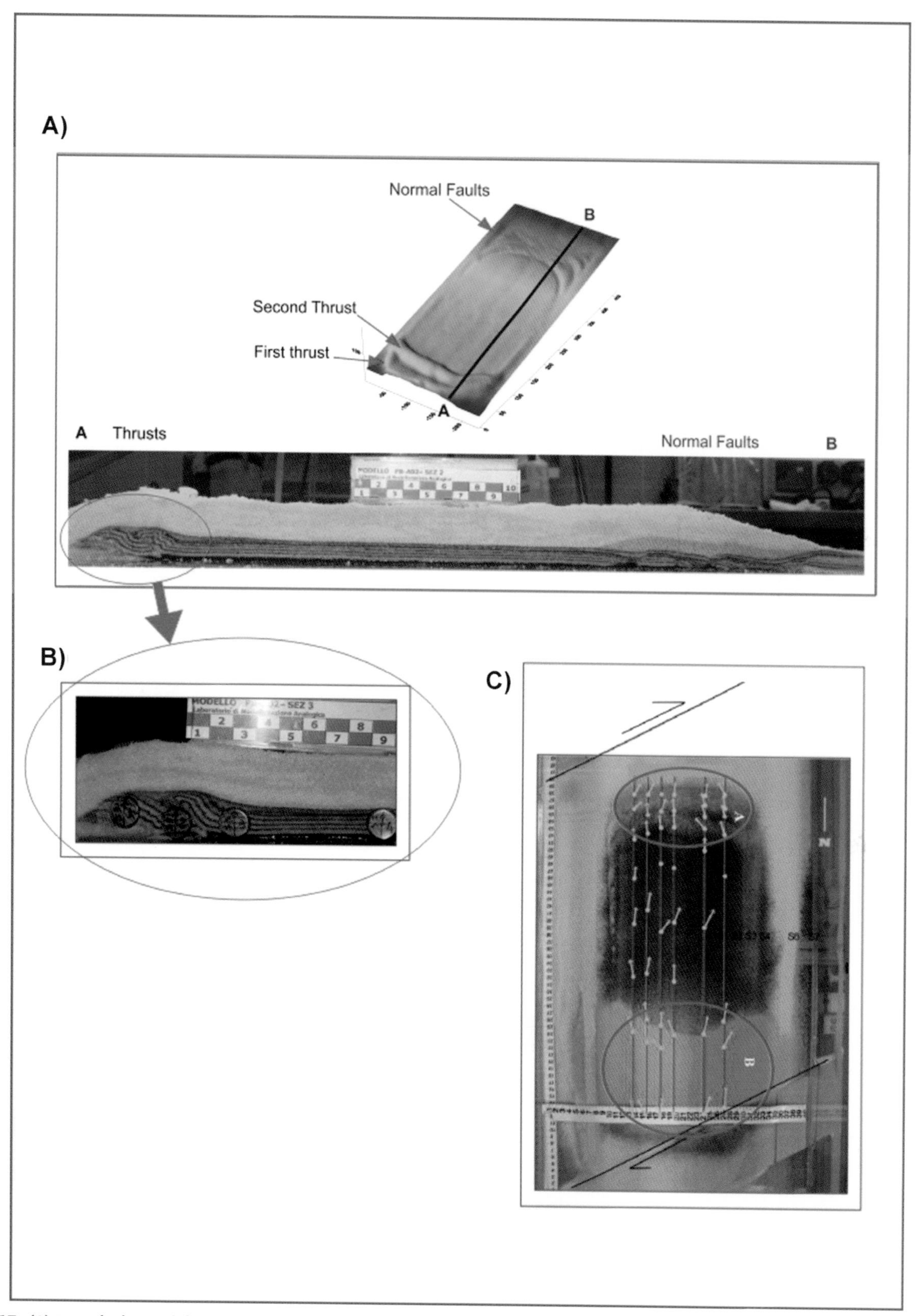

Figure 13. (A) Morphology of the top of the analog model obtained by laser scanner and cross section of the model showing the sequence of generation of thrusts and normal faults. (B) Oriented samples of one cross section of the model were submitted to paleomagnetic analysis. (C) Results of the magnetic analysis, showing the rotation of the particles during the deformation.

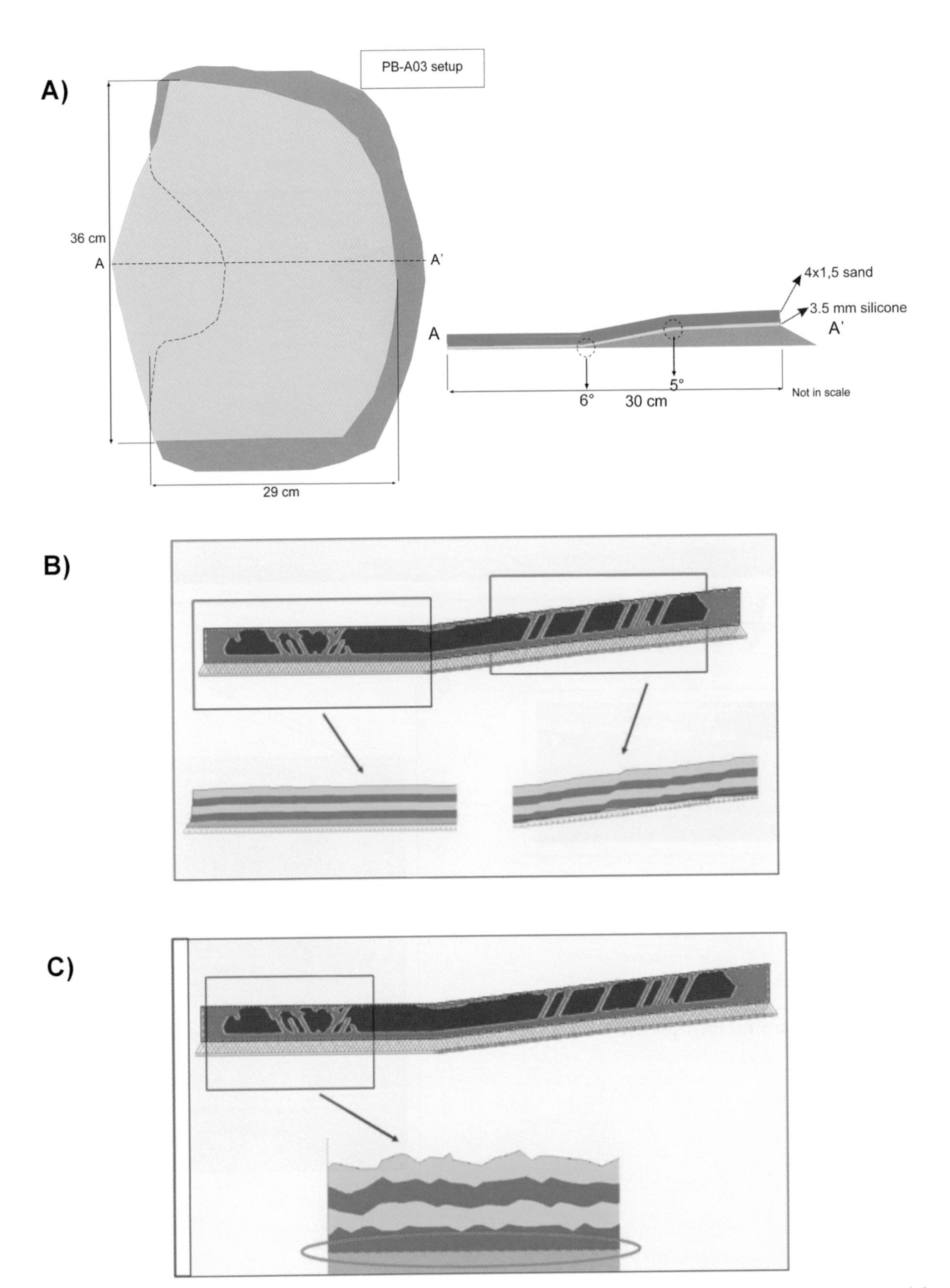

Figure 14. Numerical modeling with finite elements compared with analog modeling. (A) Outlines of the analog model used for comparison with the numerical modeling. (B) Final results of numerical modeling. Right = normal faults; left = reverse faults. (C) Effect of displacement over ductile layer.

possibly originated from generation of hydrocarbons simultaneously with the gravitational movements (Cobbold et al., 2004).

Assuming that generation of hydrocarbons could have constituted one of the major factors in causing the failure and gliding of the allochthonous units, the matter of petroleum potential immediately comes to mind. It is fair to assume that the generated fluids would concentrate along the detachment zone and be expelled upward through the faults that acted as relief valves. Localized pods of maximum subsidence and maturation would dictate what domain would receive, collect, and retain larger quantities of petroleum, either the extensional or the compressional domains, or eventually both. In the studied area, the extensional domain in the PAMA Basin and the compressional domain in the Barreirinhas Basin seem to be more favored in terms of large and well-defined structural closures. Other aspects related to petroleum systems such as the presence of effective reservoirs within the closures and impervious seals covering them were not treated in this work. The unraveling of the formation of structural traps and their timing in relation to hydrocarbon generation and migration constitutes the main contribution of this study to the petroleum exploration of the equatorial margin of Brazil.

ACKNOWLEDGMENTS

We thank PETROBRAS for permission to publish this chapter and especially Paulus Van der Ven, Edison J. Milani, Sérgio Michelucci, Otaviano Pessoa, and José G. Rizzo for the support during this work. We also thank CGG VERITAS for permission to publish the seismic lines. We thank Elisabetta Costa (in memorian, Università degli Studi di Parma) for running the physical models described here and Fabio Speranza (Instituto Nazionale di Geofisica) for paleomagnetic analysis. We also thank Gabor Tari, Ian Davison, Dengliang Gao, and Peter Szatmari for their peer reviews; Mário Neto C. de Araújo for discussions; and Elcio R. de Lima and Thiago Falcão for the drawings.

REFERENCES CITED

Athy, L. F., 1930, Density, porosity and compaction of sedimentary rocks: AAPG Bulletin, v. 14, p. 1–24.

Azevedo, R. P., 1991, Tectonic evolution of Brazilian equatorial continental margin basins: Ph.D. thesis, Imperial College, London, United Kingdom, 455 p.

Barker, C., 1990, Calculated volume and pressure changes during the thermal cracking of oil and gas in reservoirs: AAPG Bulletin, v. 74, p. 1254–1261.

Brandão, J. A. S. L., and F. J. Feijó, 1994, Bacia do Pará-Maranhão: Boletim de Geociências da Petrobras, v. 8, no. 1, p. 103–109.

Bruce, C. H., 1984, Smectite dehydration: Its relation to structural development and hydrocarbon accumulation in Northern Gulf of Mexico Basin: AAPG Bulletin, v. 68, no. 6, p. 673–683.

Caldeira, J. L., L. F. C. Coutinho, and M. F. B. Moraes, 1991, Aspectos estruturais e sismoestratigráficos da seção neocretácea e terciária da Bacia de Barreirinhas–águas profundas: Proceedings of the II Congresso da Sociedade Brasileira de Geofísica, Salvador, Bahia, v. 2, p. 667–672.

Cobbold, P. R., R. Mourgues, and K. Boyd, 2004, Mechanism of thin skinned detachment in the Amazon Fan: Assessing the importance of fluid overpressure and hydrocarbon generation: Marine and Petroleum Geology, v. 21, p. 1013–1025, doi:10.1016/j.marpetgeo.2004.05.003.

Corredor, F., J. H. Shaw, and F. Bilotti, 2005, Structural styles in the deep-water fold and thrust belts of the Niger Delta: AAPG Bulletin, v. 89, no. 6, p. 753–780, doi:10.1306/02170504074.

Costa, E., and F. Speranza, 2003, Paleomagnetic analysis of curved thrust belts reproduced by physical models: Journal of Geodynamics, v. 36, p. 633–654, doi:10.1016/j.jog.2003.08.003.

Feijó, F. J., 1994, Bacia de Barreirinhas: Boletim de Geociências da Petrobras, v. 8, no. 1, p. 101–102.

Hampton, M., H. J. Lee, and J. Locat, 1996, Submarine landslides: Reviews of Geophysics, v. 34, p. 33–59.

Hubbert, M. K., and W. Rubey, 1959, Role of fluid pressure in mechanics of over-thrust faulting, Parts I and II: Geological Society of America Bulletin, v. 70, p. 115–205.

Koyi, H., 1995, Mode of internal deformation in model accretionary wedges: Journal of Structural Geology, v. 17, no. 2, p. 293–300, doi:10.1016/0191-8141(94)00050-A.

Lehner, P., and G. Bakker, 1983, Strike-slip tectonics in an equatorial fracture zone (Romanche Fracture), *in* A. W. Bally, ed., Seismic expression of structural styles: A picture and work atlas: AAPG Volume Studies in Geology 15, v. 3, p. 25–29.

Matos, R. M. D., 2004, Petroleum systems related to equatorial transform margin Brazilian and west African conjugate basin, *in* P. J. Post, N. C. Rosen, D. L. Olson, S. L. Palmes, K. T. Lyons, and G. B. Newton, eds.: 25th Annual Gulf Coast Section, SEPM Foundation Bob F. Perkins Research Conference, p. 807–831.

Milani, E. J., and A. Thomaz-Filho, 2000, Sedimentary basins of South America, *in* U. G. Cordani, E. J. Milani, A. Thomaz-Filho, and D. A. Campos, eds., Tectonic evolution of South America: 31st International Geological Congress, p. 389–449.

Morley, C. K., 2003, Mobile shale related deformation in large deltas developed on passive and active margins, *in* P. Van Rensbergen, R. R. Hillis, A. J. Maltman, and C. K. Morley, eds., Subsurface sediment mobilization:

Geological Society (London) Special Publication 216, p. 335–357.

Morley, C. K., and G. Guerin, 1996, Comparison of gravity-driven deformation styles and behavior associated with mobile shales and salt: Tectonics, v. 15, p. 1154–1170, doi:10.1029/96TC01416.

Nemcok, M., S. Schamel, and R. Gayer, 2005, Thrust belts: Structural architecture, thermal regimes, and petroleum systems: Cambridge University Press, Cambridge, 541 p.

Oliveira, M. J. R., J. L. Caldeira, A. Tanaka, P. Santarem, I. Trosdtorf Jr., and P. V. Zalán, 2008, Linked extensional-compressional tectonics in gravitational systems of Brazil's equatorial margin (abs.): AAPG International Conference and Exhibition, Cape Town, South Africa, October 26–29, 6 p., CD-ROM.

Paull, C. K., P. G. Brewer, W. Usler III, E. T. Peltzer, G. Rehder, and D. Glague, 2003, An experiment demonstrating that marine slumping is a mechanism to transfer methane from sea-floor gas-hydrate deposits into the upper ocean and atmosphere: Geo-Marine Letters, v. 22, p. 198–203, doi:10.1007/s00367-002-0113-y.

Peel, F., 2003, Styles, mechanisms and hydrocarbon implications of syndepositional folds in deep-water fold belts: Examples from Angola and Gulf of Mexico (abs.): Houston Geological Society International Joint Dinner Meeting: www.hgs.org (accessed November 24, 2010).

Peel, F., 2007, What deep-water passive margin thrust belt tell us above thrust geometry (abs.): Geological Society of London, Continental Tectonics and Mountain Building Conference, The Peach and Horne Meeting, Ullapool: www.see.leeds.ac.uk/peachhorne (accessed November 24, 2010).

Ramsay, J. G., 1967, Folds and fracturing of rocks, 1st ed.: New York, McGraw Hill, 568 p.

Rowan, M. G., F. J. Peel, and B. C. Vendeville, 2004, Gravity-driven fold belts on passive margins, *in* K. R. MacClay, ed., Thrust tectonics and hydrocarbon systems: AAPG Memoir 82, p. 157–182.

Schaller, H., and M. V. Dauzacker, 1986, Tectônica Gravitacional e sua aplicação na exploração de hidrocarbonetos: Boletim de Geociências da Petrobras, v. 29, no. 3, p. 193–206.

Smith, N. E., and H. G. Thomas, 1971, The origin of abnormal fluid pressures in abnormal subsurface pressures: A study group report: Houston Geological Society, p. 4–19.

Spencer, C. H., 1987, Hydrocarbon generation as a mechanism for overpressuring in Rock Mountain region: AAPG Bulletin, v. 71, no. 4, p. 368–388.

Soares, E. F., P. V. Zalán, J. J. P. de Figueiredo, and I. Trosdtorf Jr., 2007, Bacia do Pará-Maranhão: Boletim de Geociências da Petrobras, v. 15, no. 2, p. 321–330.

Trosdtorf Jr., I., P. V. Zalán, J. J. P. de Figueiredo, and E. F Soares, 2007, Bacia de Barreirinhas: Boletim de Geociências da Petrobras, v. 15, no. 2, p. 331–340.

Vendeville, B., 2003, Similarities and differences between salt and shale tectonics (abs.): AAPG Annual Convention, South Lake City, 1 p., CD-ROM.

Weijermars, R., M. P. A. Jackson, and B. C. Vendeville, 1993, Rheological and tectonic modeling of salt provinces: Tectonophysics, v. 217, p. 143–174, doi:10.1016/0040-1951(93)90208-2.

Wu, S., and A. W. Bally, 2000, Slope tectonics: Comparisons and contrasts of structural styles of salt and shale tectonics on the northern Gulf of Mexico with shale tectonics of offshore Nigeria in Gulf of Guinea, *in* W. Mohriak and M. Talwani, eds., Atlantic rifts and continental margins: American Geophysical Union Geophysical Monograph 115, p. 151–172.

Xu, W., R. P. Lowell, and E. T. Peltzer, 2001, Effect of seafloor temperature and pressure variations on methane flux from a gas-hydrate layer: Composition between current and late Paleocene climate conditions: Journal of Geophysical Research, v. 106, no. 26, p. 413–426.

Zalán, P. V., 1998, Gravity-driven compressional structural closures in Brazilian deep-waters: A new frontier play (abs.): AAPG Annual Convention, South Lake City, 5 p., CD-ROM.

Zalán, P. V., 2005, End members of gravitational fold and thrust belts (GFTBs) in deep waters of Brazil, *in* J. H. Shaw, C. Connors, and J. Suppe, eds., Seismic interpretation of contractional fault-related folds: AAPG Seismic Atlas, Studies in Geology 53, p. 147–156.

Zalán, P. V., 2009, Structural styles in petroleum exploration: Short-Course 6, AAPG International Conference and Exhibition, Rio de Janeiro, Brazil, November 14, 2009, 335 p.

8

Tiercelin, Jean-Jacques, Peter Thuo, Jean-Luc Potdevin, and Thierry Nalpas, 2012, Hydrocarbon Prospectivity in Mesozoic and Early–Middle Cenozoic Rift Basins of Central and Northern Kenya, Eastern Africa, *in* D. Gao, ed., Tectonics and sedimentation: Implications for petroleum systems: AAPG Memoir 100, p. 179–207.

Hydrocarbon Prospectivity in Mesozoic and Early–Middle Cenozoic Rift Basins of Central and Northern Kenya, Eastern Africa

Jean-Jacques Tiercelin and Thierry Nalpas

UMR 6118 CNRS (Centre National de la Recherche Scientifique) Géosciences Rennes, Université de Rennes 1, Campus de Beaulieu, Bât. 15, 35042 Rennes Cedex, France
(e-mails: jean-jacques.tiercelin@univ-rennes1.fr, thierry.nalpas@univ-rennes1.fr)

Peter Thuo

National Oil Corporation of Kenya, Aon Minet Building, Mamlaka Rd., Off Nyerere Rd., PO Box 58567-00200 Nairobi, Kenya (e-mail: pthuo@erhc.com)

Jean-Luc Potdevin

Université de Lille 1, Sciences et Technologies, UFR des Sciences de la Terre, UMR 8157 CNRS (Centre National de la Recherche Scientifique) Géosystèmes, 59655 Villeneuve d'Ascq Cedex, France
(e-mail: jean-luc.potdevin@univ-lille1.fr)

ABSTRACT

The northern (NKR) and central (CKR) segments of the Kenya Rift are among the most important areas of the East African rift system for hydrocarbon prospecting because they offer the oldest and longest lived sedimentary basins and they are a crossover area between Cenozoic and Cretaceous rifts. During the 1970s and 1980s, the interest of oil companies focused in the Turkana depression and the northeastern region of Kenya. Seismic reflection surveys and several exploration wells enabled the identification of several deeply buried basins: (1) In the NKR, three strings of north–south-oriented half grabens, the oldest known basins being of Cretaceous?–Paleogene to middle Miocene age; (2) In the CKR, two north–south half grabens, the Baringo-Bogoria Basin (Paleogene–Present), and the Kerio Basin (Paleogene–upper Miocene). All basins are filled by up to 8 km (5 mi) thick sediments of alluvial, fluviodeltaic, or lacustrine origin and volcanics of late Eocene to Neogene age.

New studies have focused on reservoir and/or source rock quality in several of these basins. In terms of hydrocarbon potential, arkosic sandstones in CKR or NKR demonstrate a fair to good reservoir quality, with porosity up to 25%. Strong changes in terms of diagenetic alteration relate to deformation events or change in sediment source as a result of tectonic activity and hydrothermal fluid circulation associated with volcanism. High-quality source

DOI:10.1306/13351553M1001742

rocks were deposited in freshwater lake environments under a tropical climate. Such environments have been identified during the Paleogene in the NKR and lower Neogene in the CKR. The combination of reservoir and source rock characteristics results in a provisional classification of each studied basin, in terms of very high to medium potential for hydrocarbons.

INTRODUCTION

The northern (NKR) and central (CKR) segments of the Kenya Rift are important areas for hydrocarbon prospecting in East Africa. Not only because they contain the oldest and longest lived group of sedimentary basins related to the Cenozoic East African rift system (EARS), but they also crosscut preexisting rifts of Cretaceous and earliest Cenozoic age (Figure 1). In this chapter, we review the evidence of petroleum potential in a suite of Cretaceous?–Paleogene and Paleogene–Middle Miocene basins identified in these two regions in terms of source rock quality and reservoir properties. This study provides an important basic set of data for potential oil explorers in northern and central Kenya. During the 1970s and early 1980s, high oil prices led to major exploration activity in the northern region of Kenya. The interest of the oil companies focused mainly on the Anza Rift, in the northeastern part of Kenya (Figure 1), where several kilometers of fluvial sandstones, carbonates, and shales of lacustrine and deep-water marine origin were investigated by seismic reflection surveys and several exploration wells. The Anza Rift represents a large multiphase rift basin that has been active from the Lower Cretaceous up to the Paleogene (Winn et al., 1993; Morley et al., 1999b) and has been interpreted to be the eastern termination of the Central African rift system (CARS) (Schull, 1988) (Figure 1).

In the mid-1980s, following the discovery of large hydrocarbon accumulations in the Cretaceous–Paleogene rifts of southern Sudan (Peterson, 1986; Schull, 1988; Mohamed et al., 1999, 2002; Obaje et al., 2004), a phase of aggressive exploration was developed by Amoco and Shell companies in the western part of the Turkana depression of northern Kenya (Figure 2). In 1981–1983, the oil industry-supported Project PROBE (Proto-Rifts and Ocean Basin Evolution; Duke University, United States of America) undertook an extensive reflection seismic program on Lake Malawi and Lake Tanganyika, the two largest lake basins of the western branch of the EARS (Rosendahl et al., 1986, 1988, 1992; Rosendahl, 1987). In 1984, Project PROBE conducted an academic seismic reflection campaign covering the offshore areas of Lake Turkana (Dunkelman et al., 1988, 1989). This was followed by a series of intense onshore seismic reflection surveys acquired to the immediate west and southwest of Lake Turkana by the Amoco Kenya Petroleum Company (AKPC) in 1985–1986. Two exploration wells, Loperot-1 and Eliye Springs-1, drilled by Shell E&P Kenya B.V. in 1990–1991 (Figures 1, 3), enabled the geometry and lithostratigraphy of a suite of five buried half-graben basins to be ascertained. These basins, called the Lokichar, North Kerio, North Lokichar, Lothidok, and Turkana basins, contain up to 7 km (4.4 mi) of sedimentary fill and are among the oldest and deepest known basins in the EARS, being of Paleogene to middle Miocene age (Boschetto et al., 1992; Morley et al., 1992, 1999a; Vincens et al., 2006; Ducrocq et al., 2010). Gravity data and five reflection seismic lines were acquired by AKPC in 1985 in the northwest of Lake Turkana and indicated the presence of two other deeply buried north–south-oriented sedimentary basins: the Lotikipi Basin to the west and the smaller Gatome Basin to the east, separated by a structural high called the Lokwanamoru Range (Figure 1). Maximum depth to basement was interpreted to be 4 km (2.5 mi) for the Lotikipi Basin and up to 6 km (3.7 mi) in the Gatome Basin (Wescott et al., 1999; Desprès, 2008) (Figure 1).

Farther south, between lat. 2°N and the equator, the CKR is a particularly well-known segment of the EARS in terms of rift development and tectonic and morphologic evolution (Figures 1, 3). It also hosts, together with the NKR, several of the most important sites for hominin remnants (Bishop, 1978; Hill et al., 1986; Andrews and Banham, 1999; Senut et al., 2001; Prat et al., 2005). The CKR has not been intensely studied in terms of its hydrocarbon potential. It is composed of a set of two major parallel north–south-trending half-grabens, the Kerio Basin to the west and the Baringo-Bogoria Basin to the east, separated by the Tugen Hills structural block (Chapman et al., 1978; Tiercelin and Vincens, 1987; Hill, 1999) (Figure 3). The National Oil Corporation of Kenya (NOCK) acquired preliminary gravity and seismic reflection data in the Kerio Basin in 1990 (Pope, 1992; Ngenoh, 1993; Mugisha et al., 1997). This study was later complimented by a magnetotelluric survey in the Baringo-Bogoria Basin (Marigat-Loboi Plain) under contract with Elf Petroleum Norge AS (1996–1997) (Hautot et al., 2000). Together with a reinterpretation of the gravity and seismic data from the Kerio Basin, this study resulted in a new interpretation of the age and development history of the CKR (Hautot et al., 2000). The Kerio and Baringo-Bogoria half grabens were interpreted to be 7 to 8 km (4.4–5 mi) deep and filled by

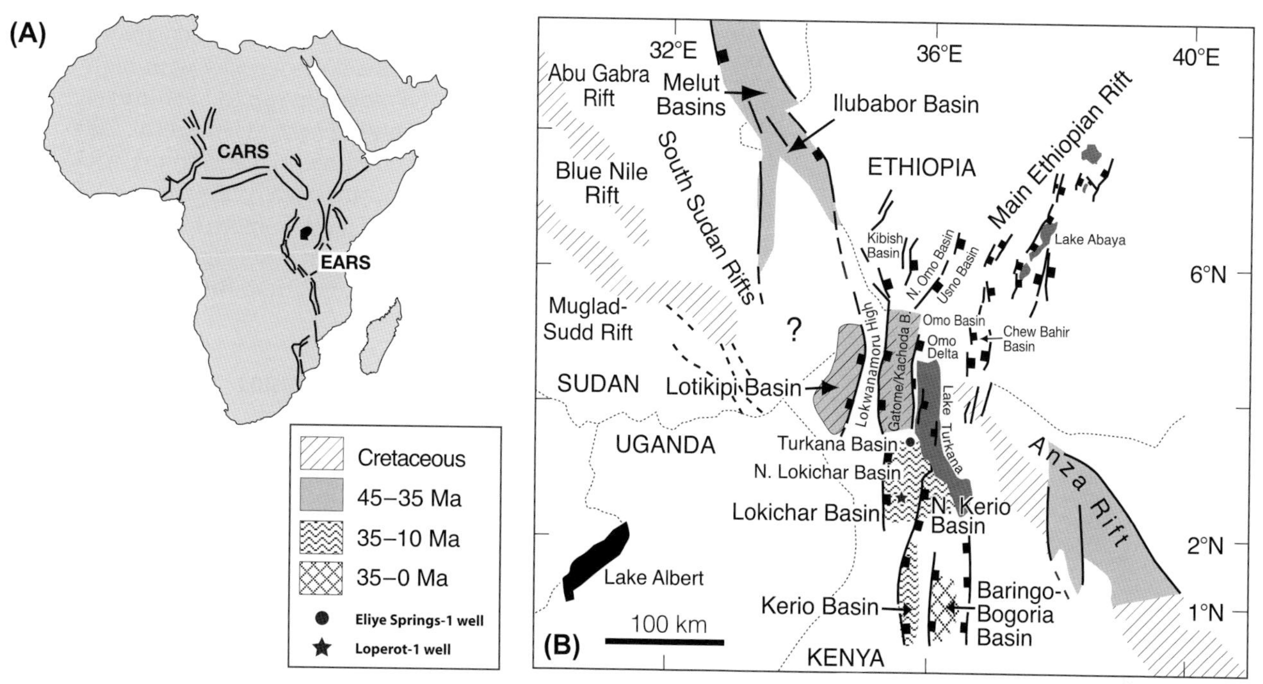

Figure 1. (A) The Central African (CARS) and East African rift systems (EARS). (B) A map showing the distribution of Cretaceous? to Paleogene–Neogene rift basins in south Sudan and northern-central Kenya. In northeastern Kenya, the Anza Rift is a well-developed northwest–southeast-trending rift system of Cretaceous–Paleogene age. To the northwest, the Lotikipi and Gatome basins are described as faulted synformal basins (Wescott et al., 1999) of Cretaceous?–Paleogene age, in trend with the south Sudan rifts. To the immediate east of the Gatome Basin is a small, elongated, 1.5 km (0.9 mi) deep, north–south-oriented half graben named the Kachoda Basin (Abdelfettah, 2009) that possibly extends north into the Kibish, North Omo, and Usno basins in southern Ethiopia. To the northeast is the Chew Bahir Rift, which is interpreted from geophysical data as containing between 2.5 and 5 km (1.6 and 3.1 mi) of basin fill. To the south are the Lokichar, North Kerio, North Lokichar, and Turkana basins that are both of Paleogene–Neogene age. The Loperot-1 well (black star), drilled in the Lokichar Basin, and the Eliye Springs-1 well (black dot), drilled in the Turkana/Lothidok Basin, are indicated (modified from Morley et al., 1992; Tiercelin et al., 2004; Thuo, 2009). 100 km (62 mi).

alternating packages of sediment and lavas, the oldest sediments being possibly of Paleogene age or even older. Up to now, no exploration wells have been drilled in these two basins.

In early 2008, a resurgence of interest by oil companies to acquire exploration acreage in the Kenya Rift occurred. In northern Kenya, the exploration activity for oil and gas is high, with several companies at various stages of their exploration programs in various exploration blocks. As a consequence of this renewal of interest for oil exploration in this region of the EARS, new stratigraphic and sedimentologic/petrographical studies have been directed toward source rock quality and reservoir properties within the most important sedimentary basins identified in the NKR and the CKR: (1) in the NKR, with the Lotikipi, Gatome, and Lokichar basins; and (2) in the CKR, with the Kerio and Baringo-Bogoria basins (Figure 3). These respective sedimentary basins have demonstrated a large number of structural and sedimentary aspects of hydrocarbon exploration in rifts (Lambiase and Morley, 1999; Morley, 1999b), with an extreme variability between individual half grabens, particularly in regions where both volcanics and sediments are present. Finally, each studied basin has been tentatively interpreted and classified in terms of hydrocarbon prospectivity.

THE NORTHERN KENYA RIFT BASINS

All of the buried basins identified in the NKR (Figure 1) are interpreted to be filled by 5 to 8 km (3.1 to 5 mi) thick fluvial or lacustrine sediments of Cretaceous?–Paleogene to Neogene age (Morley et al., 1999a; Wescott et al., 1999; Tiercelin et al., 2004; Thuo, 2009) locally associated with thick volcanics of upper Eocene–Neogene age known as the Turkana Volcanics (Bellieni et al., 1981, 1987; Zanettin et al., 1983; Thuo, 2009). Sediments of Neogene age outcrop widely on the eastern and western shores of the present-day Lake Turkana and have been intensely investigated by multiple paleontological and archeological projects under the auspices of the

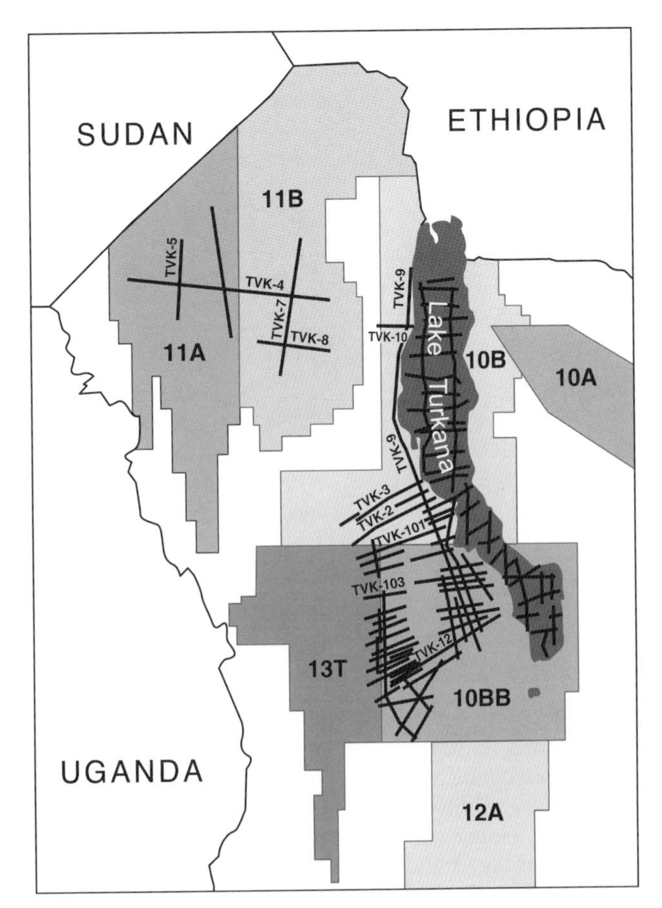

Figure 2. A Kenya exploration acreage map showing exploration blocks in the western Turkana region and location of seismic lines acquired by Amoco Kenya Petroleum Company between 1985 and 1988. The offshore Lake Turkana Project PROBE seismic reflection survey is also indicated (seismic lines in black). Blocks 11A and B: Lotikipi and Gatome basins. Block 10B: Turkana Basin, northern half. Blocks 13T and 10BB: Lokichar and North Kerio basins (from NOCK, 2009).

National Museums of Kenya. Cretaceous?–Paleogene sediments are only suspected from subsurface data with the exception of some outcropping arkosic formations originally described as the "Turkana Grits," a term created by Murray-Hughes (1933). These refer to packages of immature siliciclastic sediments directly resting on basement rocks in the northern and central parts of the Kenya Rift (Wescott et al., 1993; Tiercelin et al., 2004, 2009; Thuo, 2009).

The proximity of these NKR basins to the southern Sudanese oil-producing fields (Melut and Muglad rift basins) (Figure 4) and the similarity in the structural trends of the two regions (Schull, 1988; Salama, 1997; Wescott et al., 1999) make these basins an important target for oil exploration. In addition, the possibility for positive hydrocarbon potential in these basins was confirmed by the Loperot-1 exploration well drilled in the Lokichar Basin (Figures 1, 3), which encountered well-established lacustrine source rocks with high total organic carbon (TOC) content (up to 17%) and significant oil shows (Morley et al., 1999a; Maende et al., 2000; Talbot et al., 2004). Reservoir potential and source rock quality have been examined for each of the three NKR basins, the Lotikipi, Gatome, and Lokichar basins, then compared with other types of reservoir and source rocks encountered in the other Kenya Rift basins.

The Lotikipi and Gatome Basins

The Lotikipi and Gatome basins have been shown from gravity and seismic data to lie below the 120 km (75 mi) wide Lotikipi Plain, which is located 80 km (50 mi) northwest of Lake Turkana (Figures 1, 3). Because of the poor quality of the seismic data occasioned by the interpreted presence at depth of a thick pile of volcanics, as confirmed by ties between the seismic data and outcrops, the deep geometry of the Lotikipi and Gatome basins, as well as depth to basement, is not clearly defined. The Lotikipi Basin was interpreted either as a thermal sag basin or a rift basin, with a better preference for the rift interpretation (Wescott et al., 1999). The deep stratigraphy of the basin has only been deduced from seismic facies data and outcrop geology (Figure 5). No exploration well has been drilled in this basin. The upper part of the basin fill is formed by approximately 1 km (~0.6 mi) thick Neogene sediments (undetermined facies; possibly fluvial to shallow lacustrine) that overlie a thick pile, up to 3.5 km (2.2 mi), of basaltic and rhyolitic rocks belonging to the Turkana Volcanics (Wescott et al., 1999) that have been dated upper Eocene (35–37 Ma) at base (McDougall and Brown, 2009; Thuo, 2009) and early to mid-Miocene at top (Morley et al., 1992; Zanettin et al., 1983). Subvolcanic strata seen on the TVK-4 seismic line (Figure 5) are assumed to be arkosic (fluvial) sandstones equivalent to the Turkana Grits and lacustrine deposits forming one single sedimentary unit possibly several hundreds of meters thick (Wescott et al., 1999). To the immediate east (Figure 5), the Gatome Basin shows a similar stratigraphic succession with a greater infill thickness at its northern part. A revised interpretation of the TVK-7 seismic line (initially interpreted by Wescott et al., 1999) (Thuo, 2009) (Figure 5) suggests the presence below the Turkana Volcanics of a sedimentary unit with a thickness of 700 (±100) m (2996 ft). The lowest part of this unit is described as unconformably overlying an eroded tilted block paleotopography resulting from normal faulting affecting a stratified basement that cannot be clearly identified (sedimentary formation? or Precambrian basement?) (Figure 6).

Fifty kilometers (31 mi) to the east of the Gatome Basin, in the Lapur Range (Figures 3, 5), is an approximately 500 m (~1640 ft) thick pile of sandstones outcrops above the Precambrian basement in an impressive cliff and range and is overlain by the Turkana Volcanics (Figures 5, 7, 8). This sedimentary formation, previously described as the Lubur Series by Arambourg (1935) and recently renamed the Lapur Sandstone (LS) by Thuo (2009), is dated from possible post-Cenomanian near its base (on the basis of poorly preserved dinosaur bones) (Arambourg and Wolff, 1969; Sertich et al., 2005, 2006) to upper Eocene at its top; the uppermost beds of the LS being locally interbedded within the lowest lava flows of the Turkana Volcanics dated 37–35 Ma (McDougall and Brown, 2009; Thuo, 2009) (Figure 8). Considering its stratigraphic position below the Turkana Volcanics, the 700 m (2996 ft) thick sedimentary unit identified on the reinterpreted TVK-7 seismic line in the nearby Gatome Basin (Figure 6) can be tentatively correlated to the middle Cretaceous–upper Eocene LS. The tilted block paleotopography underlying this formation (Figure 6) could be interpreted as relating to one of the Lower Cretaceous rifting events identified in the southern Sudan and northern Kenya regions (Giedt, 1990; Genik, 1993; Salama, 1997; Wescott et al., 1999). This topography possibly controlled the deposition of the lower beds of the LS during a middle Cretaceous postrift episode (Winn et al., 1993; Bosworth and Morley, 1994; Morley et al., 1999a; Thuo, 2009) (Figure 6). Thus, the presence of the LS in the deepest part of the Gatome Basin is envisaged, and extension of this formation toward the west, in the Lotikipi Basin, can be suspected. Nevertheless, a major change in thickness along the south–north trend is observed in the LS, from 500 m (1640 ft) in the Lokitaung Gorge type section in the south up to less than 50 m (<164 ft) at the north end of the Lapur Range (Thuo, 2009) (Figure 7). Such important lateral variation in thickness possibly suggests that other sedimentary packages in a similar stratigraphic position may exist in the Gatome Basin and, most likely also, in the Lotikipi Basin. This possibility is illustrated by the presence at the south end of the Gatome Basin of the arkosic sandstone formation known as the Muruanachok Sandstone (MS) (Walsh and Dodson, 1969; Morley et al.,

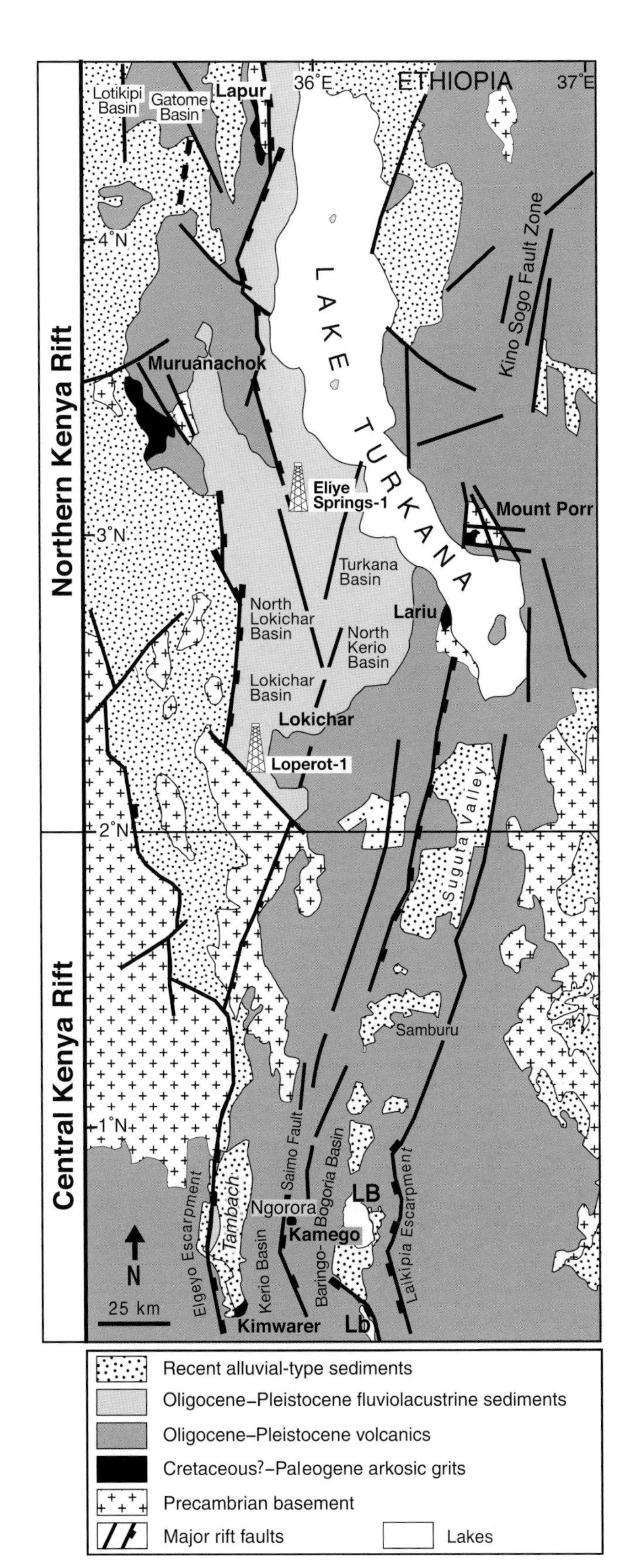

Figure 3. A geologic sketch map of the Northern Kenya (NKR) (Turkana depression) and the Central Kenya rifts (CKR) showing the different Cretaceous?–Paleogene to Neogene rift basins cited in this work. The different Turkana Grits formations (Cretaceous?–Paleogene) from north to south are indicated (in bold): Lapur Sandstone, Muruanachok Sandstone, Lokone and Auwerwer sandstones, Lariu Sandstone, and Mount Porr Sandstone in the NKR; Kimwarer and Kamego formations in the CKR. LB = Lake Baringo; Lb = Lake Bogoria.

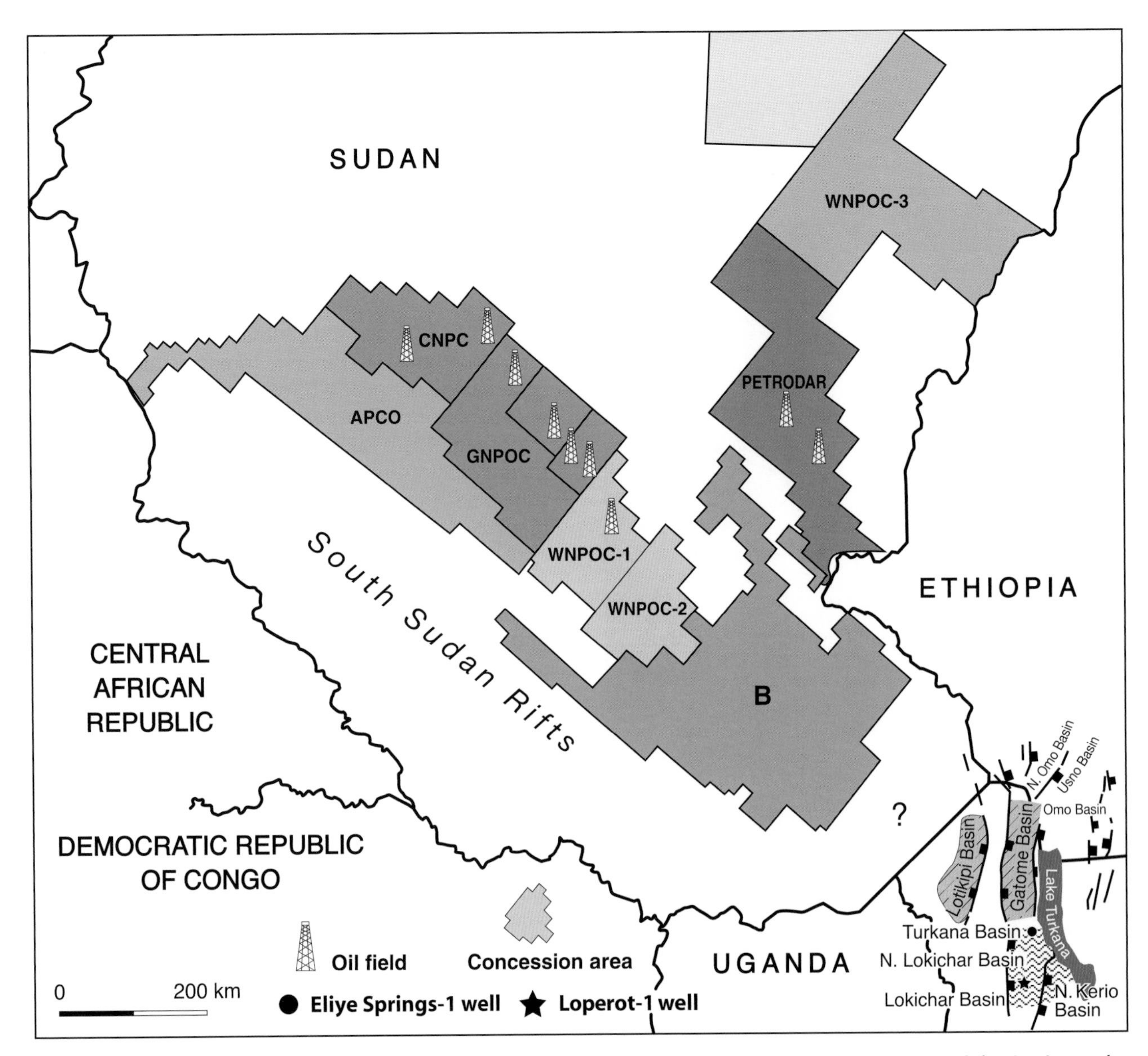

Figure 4. Map of the south Sudan rift oil fields showing their proximity to the Cretaceous?–Paleogene rift basins in northern Kenya (from European Coalition on Oil in Sudan [ECOS], 2007). CNPC = China National Petroleum Corporation; APCO = Advanced Petroleum Company; GNPOC = Greater Nile Petroleum Operating Company; WNPOC-1 = White Nile Petroleum Operating Company-1; WNPOC-2 = White Nile Petroleum Operating Company-2. 200 km (124 mi).

1992, 1999a; Wescott et al., 1993; Thuo, 2009) (Figure 3). These sandstones crop out in several isolated small hills (Figures 3, 9), where they are locally seen to overlie Precambrian basement and are at some places unconformably overlain by volcanics of late Oligocene to Miocene age (Morley et al., 1999a). No evidence to precisely determine the age of the MS exists (no fossil evidence to the exception of few fossil wood and poorly preserved prints of plant leaves) (Thuo, 2009) (Figure 10). Consequently, they can be of late Oligocene age or older.

Despite this uncertainty concerning the lateral distribution of the LS that can only be resolved by drilling, the LS, because of its significant thickness, may offer interesting reservoir potential. This formation has an original detrital composition made predominantly of polycrystalline quartz grains and K-feldspars sourced from metamorphic basement (Thuo, 2009). Arkosic sandstones such as the ones in the LS tend to undergo a marked reduction in permeability following the diagenetic breakdown of feldspars (Morley, 1999b) (Figures 11; 12A, B). In the LS, the main control on reservoir quality is cementation by calcite (in the lower 85 m [279 ft] in the type section; Zone 1), hematite (middle 125 m [410 ft]; Zone II), and kaolin (upper 275 m [902 ft]; Zone III)

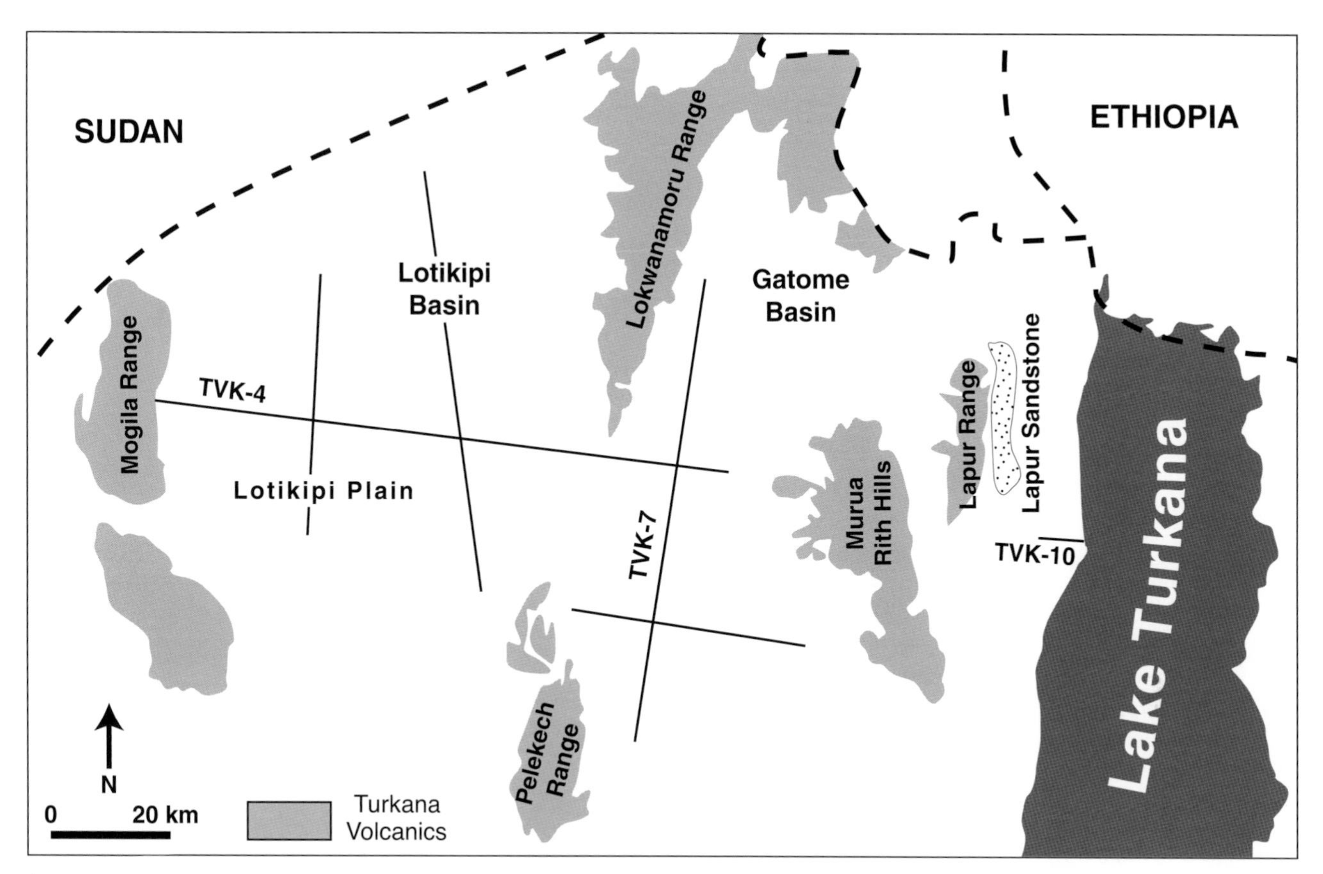

Figure 5. Regional location map showing the Lotikipi Plain, the Lotikipi and Gatome basins, the Lapur Range, the Lapur Sandstone, and the TVK-4 and TVK-7 seismic lines acquired by the Amoco Kenya Petroleum Company (AKPC) (redrawn from Wescott et al., 1999). 20 km (12.4 mi).

(Thuo, 2009) (Figure 8). Multiple generation cementation occurs in the upper parts of the LS (Figure 13A).

The initial depositional porosity of the LS may have been as high as 40% as is common with many sandstones during deposition (Selley, 1978; McBride et al., 1996), which has since been reduced to values ranging from 3 to 25% (Thuo, 2009) (Figure 8). In the lower part of the formation, cementation mainly by poikilotopic calcite has reduced porosity to between 7 and 11%. The lowest porosity measured in the LS is reported from the top of the formation (Figure 8; upper part of Zone III) that is also cemented by poikilotopic calcite (Figure 13B). In terms of hydrocarbon reservoir potential, the low-porosity values and the corresponding low-permeability values expected to be associated with the lower part of the LS make it a less prospective horizon. This would hold true unless, at depth, porosity-enhancing processes exist, such as dissolution of calcite or dolomitization, which can improve reservoir potential. The middle 100 m (328 ft) of the LS (Figure 8; Zone II) is characterized by the abundance of hematite cement associated with locally abundant ferric crusts measuring up to 20 mm in thickness that occur at the top of various beds (Figure 11B, C). The effects of abundant hematite have reduced the porosity of this section of LS to between 11 and 25% (Figure 8). Kaolin cement is also present within the section where it occurs as later cement after hematite. Although the measured porosity values are fair to good, the presence of iron crusts in this part of the section, if these are laterally continuous in reservoir scale, could create possible reservoir heterogeneities, resulting in compartmentalization and hence production difficulties. In the top zone of the LS type section (Figure 8; Zone III), the most prominent cement is kaolin (Figure 13C), with subordinate amounts of hematite (except for the topmost calcite-cemented beds). Porosity in this zone has been relatively well preserved, with values ranging from 12 to 22%. Secondary dissolution porosity from the dissolution of feldspars and secondary intragranular porosity from partial dissolution of feldspars are important contributors to the total porosity.

The MS (Figures 3, 9A, B) is composed of conglomerates and sandstones and is interpreted as having been deposited within alluvial plains and fluvial channels. At the type locality, the thickness of the measured section is 104 m (341 ft), and this therefore represents a

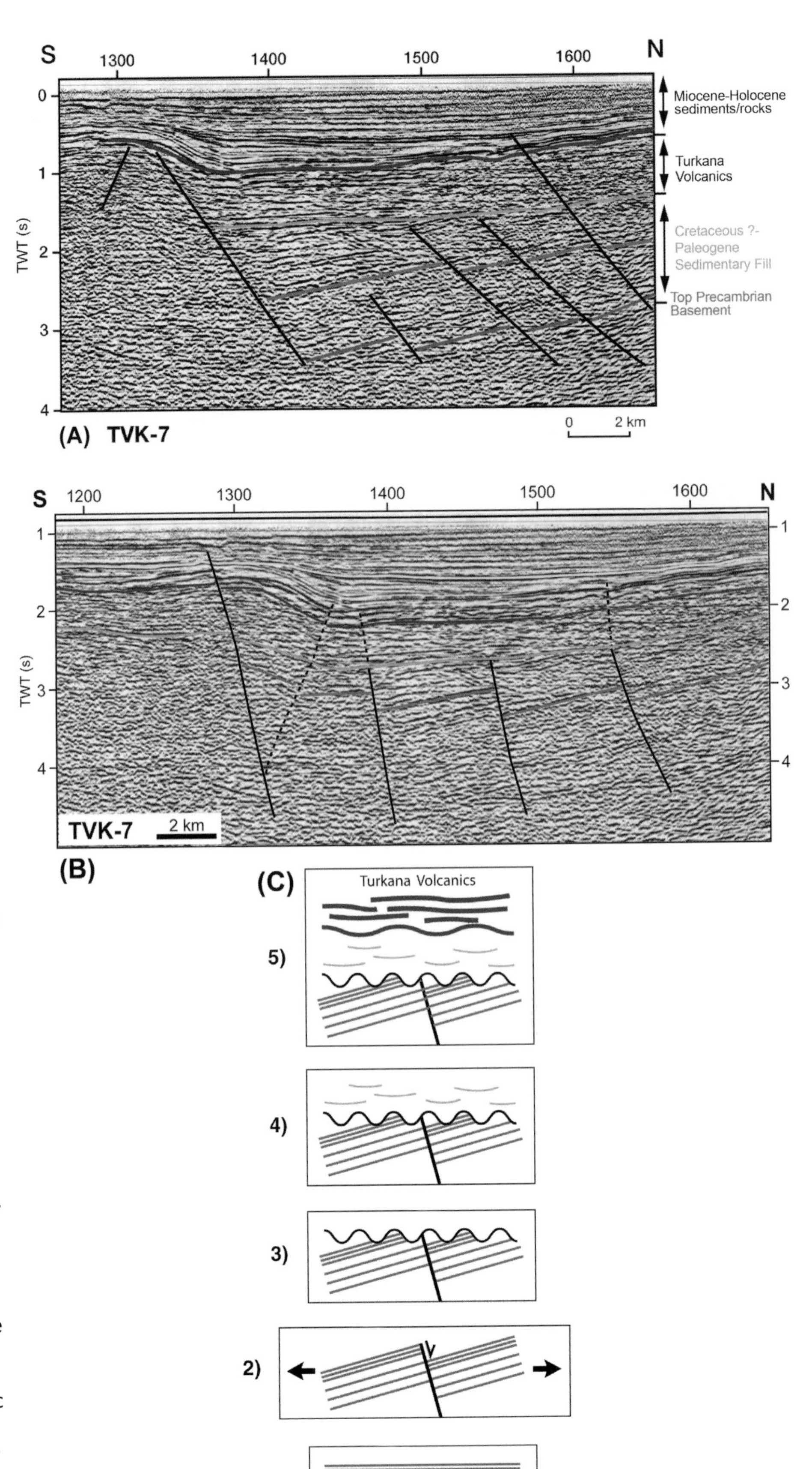

Figure 6. Reinterpretation of the north–south-oriented TVK-7 seismic line through the Gatome Basin (see location Figure 5). (A) Earlier interpretation by Wescott et al., 1999 (figure 6, page 59). (B) Our interpretation. (C) Successive phases of evolution of the Gatome Basin. (1) A stratified basement (in light brown) is visible at the base of the seismic line and can be interpreted as a stratified sedimentary series or metamorphic basement of Precambrian age?. (2) A first phase of deformation associated with a normal faulting episode creates a series of small north-dipping tilted blocks similar to the geometries that can be observed in the south Sudan rifts (Bosworth, 1992). This tilted block topography possibly developed during Early–middle Cretaceous. (3) A phase of erosion affects the crests of the tilted blocks. (4) On top of this erosive unconformity, an approximately 700 m (~2296 ft) thick sedimentary unit (orange color) is deposited. This unit can be interpreted as the middle Cretaceous–upper Eocene Lapur Sandstone (LS) or as a lateral sedimentary equivalent. (5) Basaltic flows belonging to the Turkana Volcanics (violet color) unconformably overlying the sedimentary formation interpreted as the LS. 2 km (1.2 mi).

Figure 7. The Lapur Sandstone (LS). (A) General view of the northern half of the LS from the east (Todenyang Plain) at lat. 04°25′N. The LS extends from lat. 04°14′N toward the north over more than 50 km (31 mi), forming the prominent Lapur Range (Figures 3, 5). The LS sediments appear resting on a surface of basement rocks (gneiss, migmatites, and amphibolites of Precambrian age; Walsh and Dodson, 1969). (B) At the Lapur Peak, the LS outcrops as an impressive 120 m (394 ft) high cliff that lies on Precambrian basement and culminates at an elevation of 1481 m (4859 ft). View from helicopter. (C) In the Lokitaung Gorge, the LS (white arrow) shows a general tilt of 10 to 15° toward the west or southwest and disappears rapidly below a thick (500 m [1640 ft]) accumulation of basaltic lavas forming the lower part of the Turkana Volcanics (see black arrow). White box: Land Rover for scale.

significant potential reservoir unit (Figure 9C). The MS is not generally a well-cemented formation to the exception of the top of the formation where quartzitic facies can be observed. Total cement ranges in abundance between trace amounts to 30%. Three main cements have been identified, in the case of the LS: kaolin, calcite, and hematite (Figure 13D, E). Iron ooids are concentrated near the middle part of the MS containing abundant hematite cement (Figure 13). Porosity values based on visual estimation in thin sections studied throughout the vertical section of the MS range from 5 to 20%, mainly in the form of primary intergranular porosity (2–20%) as well as secondary intragranular and dissolution porosity resulting from the partial and occasionally total dissolution of feldspar grains. The formation contains fractures that are partially healed by the deposition of calcite cement, and these could significantly enhance permeability, thereby increasing the reservoir potential of the MS.

In terms of source rocks, the potential occurrence of oil-prone sediments is also a major question within the Gatome and Lotikipi basins. Good-quality source rocks in continental rift basins are typically shales of lacustrine facies (Talbot, 1988), as it is for the Lokichar Basin (Morley et al., 1999a; Talbot et al., 2004). In the Gatome and Lotikipi basins, no seismic evidence for thick accumulations of lacustrine shales exists. Only lenses of dark-gray or black silty mudstones a few meters to a few tens of meters in thickness can be seen at the outcrop scale in the LS, where they may represent postrift flood-plain or shallow lake deposits (Thuo, 2009) (Figure 11D). Nevertheless, these mudstones, sampled at the outcrop level, do not demonstrate organic content. Only few wood stems are found in the sandstone facies of LS. No source rock evidence from outcrops of the MS exists. Only 1 m (3.3 ft) thick layer of gray and purple silty mudstone containing small dark organic debris and prints of plant leaves has been found in the lower MS section (Figure 10), possibly representing a flood-plain depositional environment (Khalifa and Catuneanu, 2008). Because it is reasonable to expect that, in a conformity with common trends in continental rift basin characteristics, thick lacustrine shales having a high source rock potential may exist in the deeper parts of the Gatome Basin (e.g., Morley, 1999b), the environments of deposition of formations such as the LS and MS appear to be postrift and are not expected, as confirmed on outcrops, to offer good source rock potential.

The Lokichar Basin

Located close to the southwest end of Lake Turkana (Figures 1, 3), the Lokichar Basin is a north–south-trending, 60 km (37 mi) long and 30 km (18.6 mi) wide,

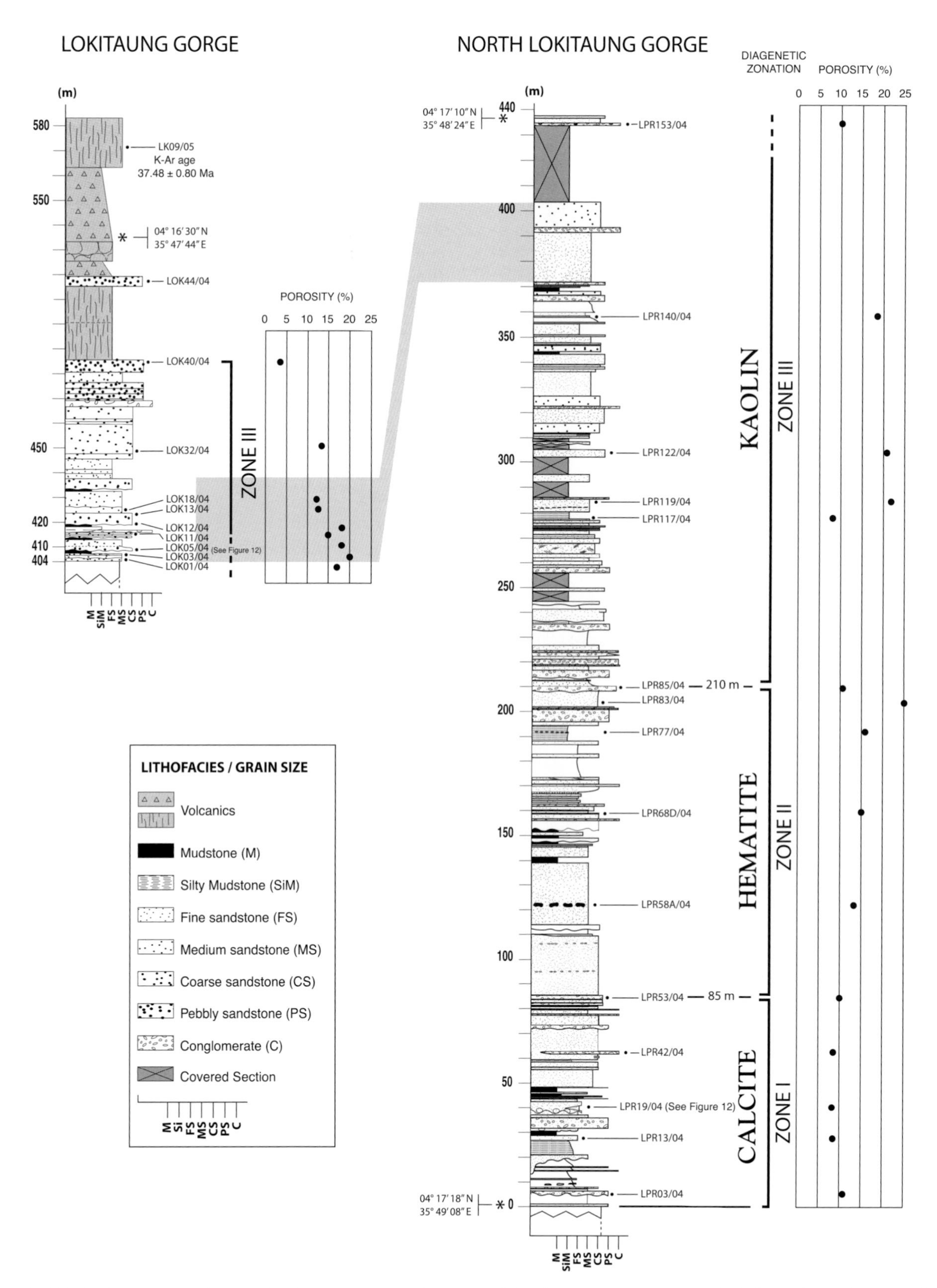

Figure 8. Lithologic log of the Lapur Sandstone, type section, established at the northwest end of the Lake Turkana Basin (Lapur Range, immediately north of Lokitaung Gorge), with diagenetic zonation and porosity plot, showing measured porosities ranging from 3 to 25% (modified from Thuo, 2009). Low porosities are generally associated with cementation by calcite (Zone I), whereas the kaolin-cemented zone (Zone III) retains higher porosity values. LPR = Lapur; LK/LOK = Lokitauna.

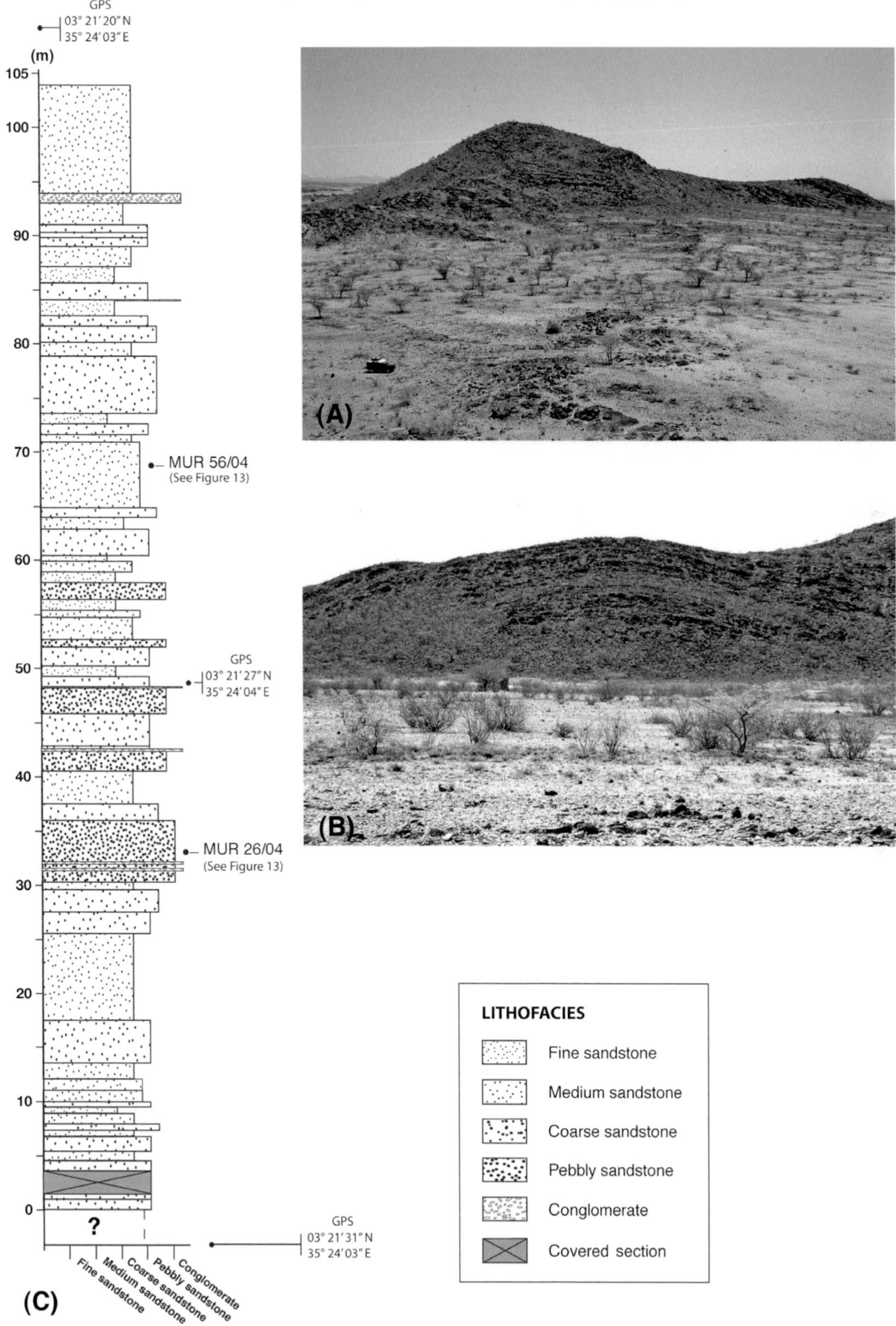

Figure 9. The Muruanachok Sandstone (MS). (A) View of the main Muruanachok Hill on the left side of the main road from Lodwar to Lokichoggio, 45 km (28 mi) northwest of Lodwar. Land Rover for scale. (B) Anticlinal attitude of the MS beds from northeast side of the main hill. (C) Stratigraphic section measured on the main hill exposure whose base is at lat. 03°21′31″N and long. 35°24′03″E, with a total thickness of 104 m (341 ft).

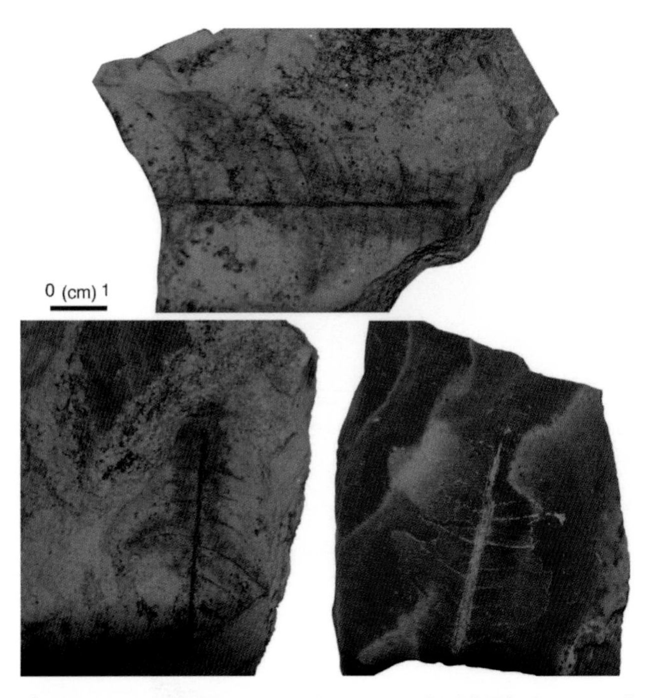

Figure 10. Fossil leaves found in a 1 m-thick bed of gray and purple silty mudstones near the base of the Muruanachok Sandstone on the left side of the main road from Lodwar to south Sudan. A preliminary investigation by J. Broutin, National Museum of Natural History, Paris, suggests that the leaves could belong to *Lauriacae*, which in terms of chronostratigraphy, are known from the Cretaceous to the Present.

east-facing half graben floored by Precambrian crystalline basement. This basin has been described in detail by Morley et al. (1992, 1999a). Its western boundary is a prominent east-dipping listric fault, the Lokichar Fault, which has no present-day topographic expression. As shown by reflection seismic data (TVK-12 and TVK-13) (Morley et al., 1999a), the basin fill, which is 7 km (4.4 mi) thick, consists of interbedded Paleogene to middle Miocene lacustrine and fluviodeltaic sediments (Figure 14A, B). Capping the basin fill is a 300 m (984 ft) thick basaltic sequence, the Auwerwer Basalts, dated from 12.5 to 10.7 Ma (Figure 14C). The 2.9 km (1.8 mi) deep Loperot-1 exploration well (Figures 14B, 15) was drilled close to the Lokone Horst, a basement high that separates the Lokichar Basin from the North Kerio Basin, another parallel half graben located to the immediate east (Figure 1). Alternating packages of coarse to fine sandstones, tens to hundreds of meters thick, were encountered in the well. These packages were divided into two main sedimentary units, from bottom to top: the Lokone Sandstone and the Auwerwer Sandstone (Morley et al., 1999a) (Figure 14B). The Lokone Sandstone includes two major lacustrine shale intervals, the Lokone Shale Member and the Loperot Shale Member, which are both several hundreds of meters thick in the well but may be as thick as 1.3 km (0.8 mi) to the west toward the Lokichar boundary fault (Figure 14B). Palynological data from the Loperot-1 well indicate a Paleogene to middle Miocene age for the whole formation. Mammal fauna discovered in sandstones overlapping the western side of the Lokone Horst indicates a late Oligocene age for this part of the basin infill (Ducrocq et al., 2010) (Figure 14C).

Because of low relief, sediment exposures are poor in this area and only represent the flexural margin sediments. Coarse-grained facies, mainly pebbly sandstones and minor conglomerates, onlap the gneissic basement of the Lokone Horst (Figures 14C, 16A–C). Quartz, K-feldspar, biotite, and muscovite are the dominant minerals in these rocks and demonstrate a sediment source from basement outcrops located south of the basin. These sediments illustrate a fluviodeltaic environment. The upper part of the Lokone Sandstone as well as the Auwerwer Sandstone show basement-sourced minerals but with significant amounts (more than 50% of the detrital material) of fresh volcaniclastic material such as amphibole, pyroxene, and plagioclase. The presence of volcanic components indicate a major change in the sediment source, from basement outcrops for the lower part of the Lokone Sandstone to an alkaline volcanic source, such as basaltic flows occurring south of the Lokichar Basin, possibly linked to the widespread lower Miocene Samburu Basalts in the CKR (Chapman and Brook, 1978; Tiercelin et al., 2004) (Figure 3). Carbonate beds are locally present in the Lokone Sandstone, mainly mollusk (gastropods and bivalves) shells accumulations (Figure 16D), indicating a nearby lacustrine shoreline or shallow pond environment (Figure 16E, F). These facies could present an interesting potential reservoir. However, only minimal thicknesses have been observed at the outcrop scale (Figure 17).

Diagenetic changes in the upper Lokone and Auwerwer sandstones are marked by calcite (Figure 16G) or by analcite/calcite precipitation (Figure 16H), resulting in a rather low porosity, with values of 1 to 15% (Tiercelin et al., 2004). Analcite and calcite precipitated from Na/Ca-rich fluids related to the dissolution of the volcanic material. Data from the Loperot-1 well (Visser, 1993) indicate that the dominant grain type in the main reservoir horizons is made up of fine- to medium-grained sandstone, poorly to very poorly sorted, with a few moderately sorted exceptions. The main detrital components are quartz (predominantly monocrystalline) and feldspars (mainly plagioclase with lesser amounts of alkali feldspars), with significant amounts of lithic fragments and clay matrix. The lithic fragments are composed of mainly argillaceous sedimentary fragments with lesser amounts of altered quartzo-feldspathic metamorphic fragments and rare volcanic fragments. Authigenic

Figure 11. A selection of sedimentary facies of the Lapur Sandstone (LS) in terms of reservoirs and possible source rocks. (A) Trough cross-laminated sandstone (facies St), middle part of the LS. At the base of the unit, planar cross-stratified sandstone (facies Sp), very coarse-grained sandstone, with individual cross beds measuring 4 to 6 cm (1.6–2.4 in.) in thickness. (B, C) Iron crusts, 2 to 3 cm (0.8–1.2 in.) thick, made of hematite, are abundant in the middle part of the LS (see Figure 8; Zone II, hematite cementation). They commonly act as a boundary between metric-size beds of sandstone (C; white arrow) and may reflect periods of emersion and/or nonsedimentation. At the scale of a reservoir unit, such iron crusts may generate significant reservoir discontinuities. (D) Dark-gray laminated silty mudstone (facies Fl) forming a 3 m (9.8 ft) thick bed conformably overlying massive medium-grained sandstones and unconformably overlain by a massive sandstone bed. Lapur Range, left bank of the Lokitaung Gorge. This facies possibly reflects a floodplain, marsh, or shallow lake environment. It shows a poor organic content, minimal thicknesses, and limited lateral continuity, thus demonstrating a poor source rock potential at outcrop level. Scale bar is 1 m (3.3 ft) long.

minerals within the Loperot reservoirs include quartz and feldspar overgrowths, carbonate cements, zeolites, authigenic clays, and opaque minerals. The quartz overgrowths, which generally form a minor component in the sandstones, occur as small syntaxial overgrowths that partially occlude adjacent pore spaces. The authigenic feldspars occur as overgrowths on detrital feldspar grains as well as albitization of plagioclase and neomorphism of albite. Carbonate cements are represented mainly by coarsely crystalline to locally poikilotopic pore-filling ferroan calcite, with minor amounts of nonferroan calcite and siderite. Zeolites present in the sandstones include analcite crystals intermixed with pore-filling clay and laumontite, similar to the sandstones found in the Chalbi-1 well drilled in the Anza Rift, where porosity was severely degraded because of cementation by zeolites, particularly laumontite (Winn et al., 1993; Morley et al., 1999b). Authigenic clays occur as grain-coating rims, pore-filling clays, and as pore-filling and grain-replacive patches of kaolinite. The reservoir properties for the studied reservoir section are poor to very poor, with the point-counted porosity never exceeding approximately 5%. The main agent responsible for the reduction in porosity is a combination of sediment compaction and cementation by laumontite and quartz overgrowths. There appears, however, to be a

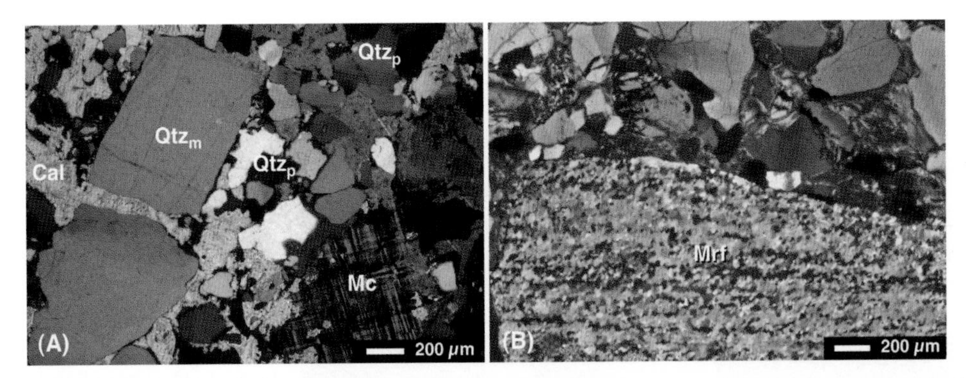

Figure 12. Petrographic micrographs of Lapur Sandstone (LS). (A) Detrital grains in the LS include polycrystalline quartz, monocrystalline quartz, and feldspars. Well-preserved feldspar grains (such as the microcline showing characteristic cross-hatched twinning) are not common in the LS. The cement in this sample is poikilotopic calcite (sample LPR 19/04, LPA). (B) Metamorphic rock fragments (Mrf) contribute a significant amount of framework grains in the LS (sample LOK 05/04, LPA). LPA = cross-polarized light; Qtzp = polycrystalline quartz; Qtzm = monocrystalline quartz; Mc = microcline; Cal = calcite.

discrepancy between point-count porosity and log-calculated porosity, which indicates values of 10 to 20% for the same section. Porosities calculated from wireline logs are significantly higher because they represent total porosity, including microporosity (Visser, 1993).

Possible equivalents of the lower part of the Lokone Sandstone (arkosic sandstones and conglomerates) are known to outcrop in the Lariu Range at the southeastern end of the North Kerio Basin, located immediately east of the Lokone Horst (Figures 1, 3). This half-graben basin, imaged by the TVK-12 seismic line of AKPC (Figures 14A, 16), is north–south oriented, parallel to the Lokichar Basin but is less well studied compared with the Lokichar Basin. In the Lariu Range, Wescott et al. (1993) described coarse-grained arkosic pebbly sandstones with thin, discontinuous, red and gray siltstone-mudstone units up to 610 m (2000 ft) thick—named the Lariu Sandstone—overlain by a thick sequence of lava flows that are, in places, heavily intruded by volcanic dykes. The lowest lava flow immediately overlying the sandstones was dated 15.7 ± 0.7 Ma. An age of 14.7 ± 0.17 Ma for one of the intruding dikes was obtained (Wescott et al., 1993). Dike intrusions of similar age (16.9–12.6 Ma) have been described in the Lokichar Basin (Tiercelin et al., 2004).

McGuire et al. (1985) described two types of strata exposed at the Lariu Range: (1) well-indurated cross-bedded sandstones and conglomerates, similar in appearance to the LS, at the lower part of the Lariu section; and (2) finer grained sandstones with colors that alternate from red to white at the upper part of the section. These basement-derived sediments are characterized by quartz pebbles and subangular metamorphic rock fragments near the base of the section. The conglomerates are composed predominantly of polycrystalline quartz with lesser amounts of microcline and orthoclase as the framework grains, whereas the cement is a combination of quartz overgrowths, authigenic kaolinite, calcite, and hematite. The arkosic arenites consist predominantly of quartz and feldspar, with minor amounts of altered volcanic rock fragments. Plagioclase grains occur as relatively unaltered to highly altered grains, the alteration mainly involving replacement by calcite with minor amounts of sericitization. All framework grains show secondary dissolution. Calcite-replaced plagioclase grains are almost completely dissolved, forming moldic and intragranular secondary porosity, whereas the relatively unaltered plagioclase commonly shows intragranular porosity along twin lamellae. Potassium feldspars (microcline and orthoclase) are abundant and are relatively unaltered, whereas volcanic rock fragments, which occur in trace amounts, have been mostly replaced by clays and hematite. Cement material consists mostly of authigenic kaolinite, with minor amounts of clays and grain-replacement calcite. Porosity, which is estimated to be a maximum of 20%, is mainly made up of primary intergranular with sufficient secondary intragranular to form an interconnected pore network capable of producing an effective permeable potential reservoir unit (Dunn, 1985).

Sandstones similar to the Lariu and Lokone sandstones can also be found at Mount Porr on the southeastern side of Lake Turkana (Williamson and Savage, 1986; Wescott et al., 1993; Tiercelin et al., 2004) (Figure 3). As in the Lokone and Lariu sandstones, the Mount Porr Sandstone (also known as the Sera Iltomia Formation of Savage and Williamson, 1978) overlies the Precambrian basement and is in turn overlain by volcanics dated 18.5 ± 0.45 Ma (Williamson and Savage, 1986). No precise age (because of a lack of fossil evidence) can be given for the Mount Porr Sandstone. It could be Cretaceous in age (as is the case of the base of the LS) or Paleogene as the sandstones of the lower Lokone Sandstone or the upper

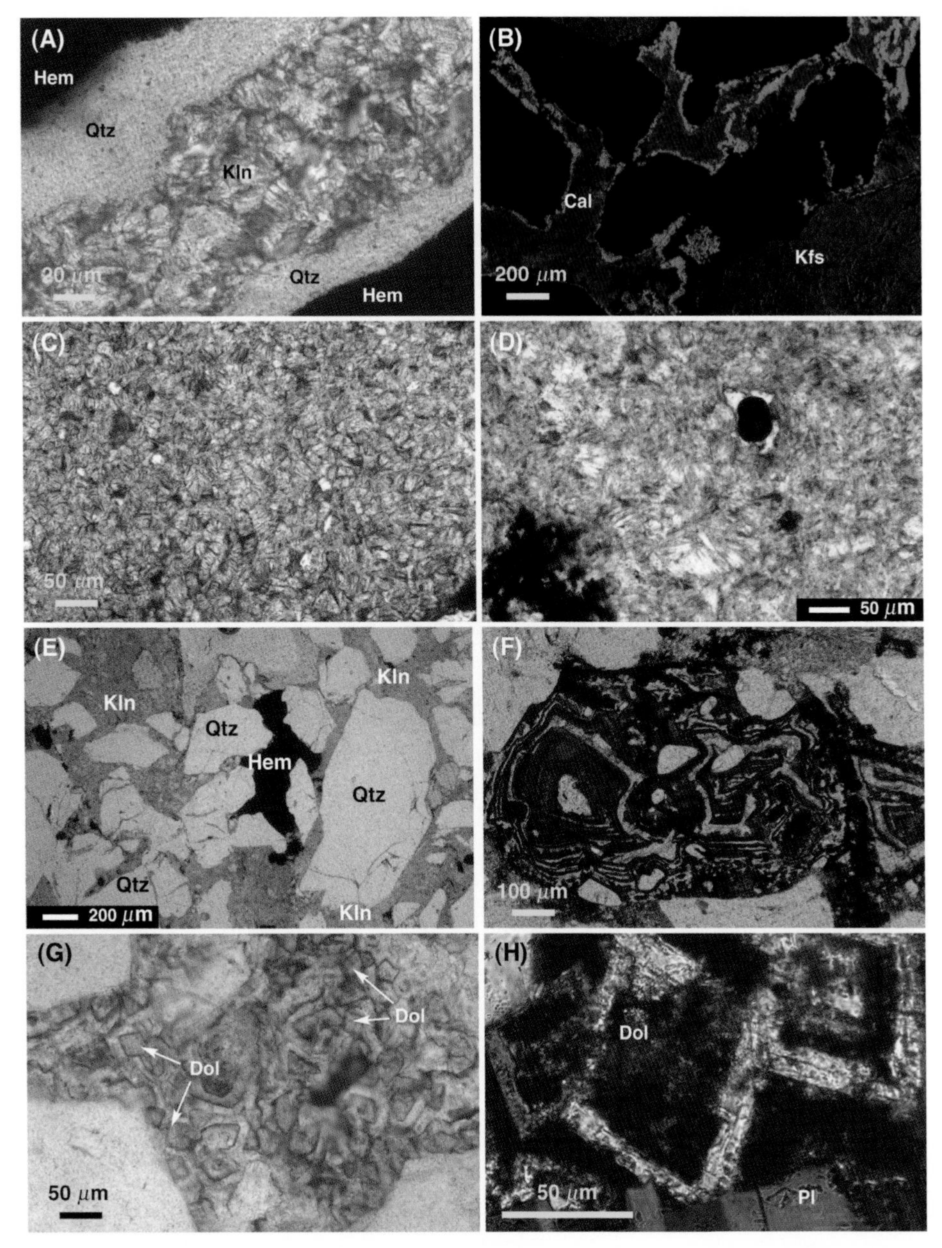

Figure 13. Cementation models in the sandstones of the basins of the Northern Kenya Rift: Lapur Sandstone (LS), Muruanachok Sandstone (MS), and Mount Porr Sandstone. (A) Multiple-generation cementation in the upper parts of the Lapur Sandstone. The diagenetic sequence is such that the hematite cement is the earliest phase of cementation, followed later by a phase of cryptocrystalline quartz cement, and finally by pore-filling authigenic kaolin. (B) Two phases of calcite cementation as seen from cathodoluminescence petrography of the sandstones from the lower parts of the LS. An earlier bright orange luminescent phase is followed by a later dull orange luminescent phase. Calcite cementation is pervasive and results in total obliteration of porosity in the lower LS with severe reduction in reservoir potential. (C) Authigenic kaolin cement within the upper parts of the LS is associated with abundant microporosity (in blue). This microporosity, although contributing to the overall bulk porosity of the sandstones, may not be effective because of the small size of the pores. (D) Vermicular kaolin commonly occurs as a pore-filling cement (sample MUR 56/04 [MS]; LPA photograph). (E) Hematite is another common pore-lining and sometimes pore-filling cement in the MS (sample MUR 26/04; LPNA). (F) Two phases of cementation within hematitic ooids within the MS. The pseudoconcentric rings are composed of alternating brown hematite and white calcite rings. Some porosity has been retained within the oolitic structures. (G) Ghosts of dolomite rhombs resulting from dedolomitization of calcite cements in the Mount Porr sandstones. The dolomitization of calcite is normally accompanied by enhancement of porosity in the sandstones, but these have suffered later dedolomitization, although overall, porosity retention occurs. (H) Mount Porr Sandstone. Calcite-iron oxide pseudomorphs after dolomite (Dol). The initial crystal shape of dolomite is still visible and is now occupied by calcite and iron oxides. Pl = plagioclase; LPA = polarized analyzed light; LPNA = polarized nonanalyzed light; Qtz = quartz; Kln = kaolin; Hem = hematite; Cal = calcite; Kfs = K-feldspar.

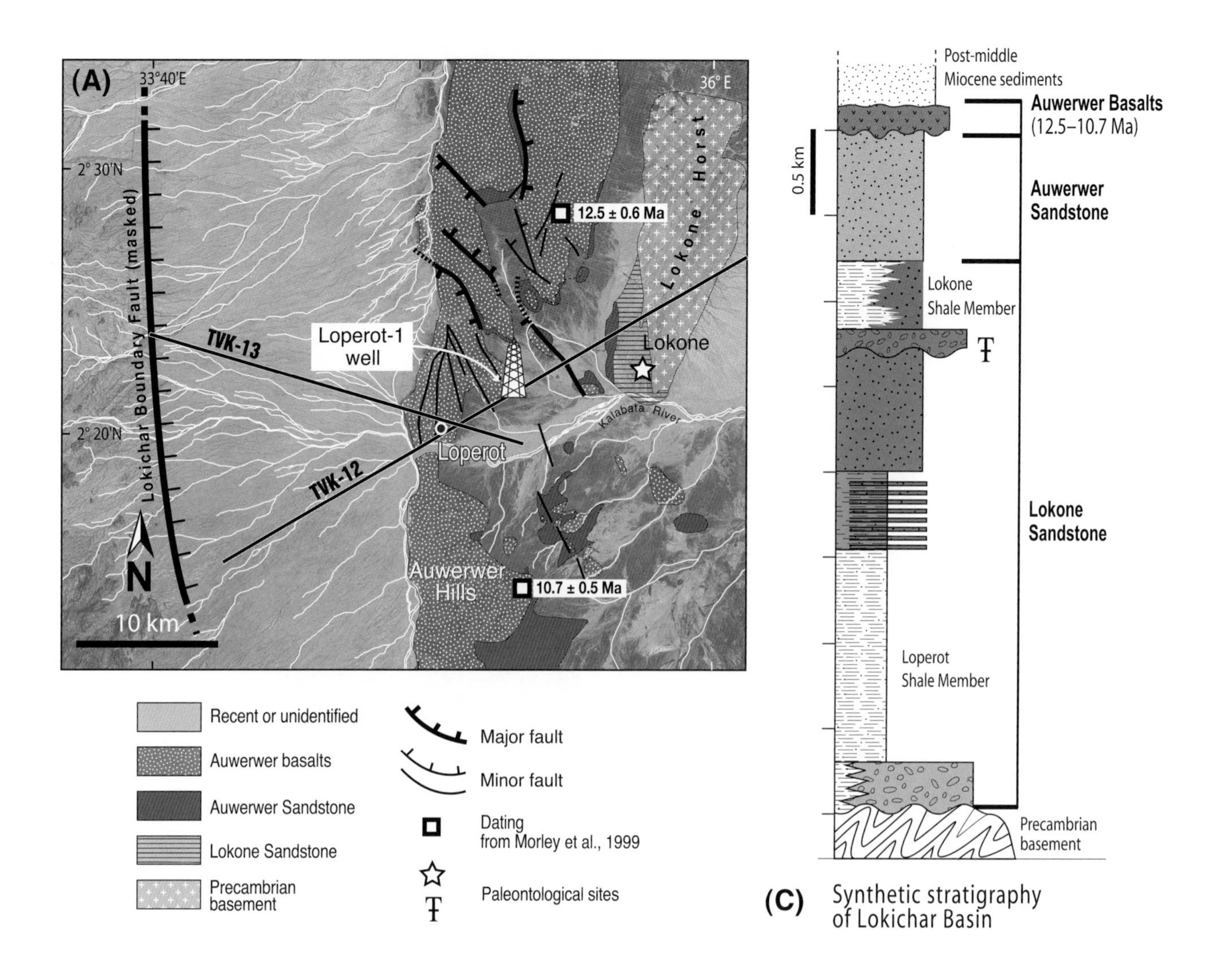

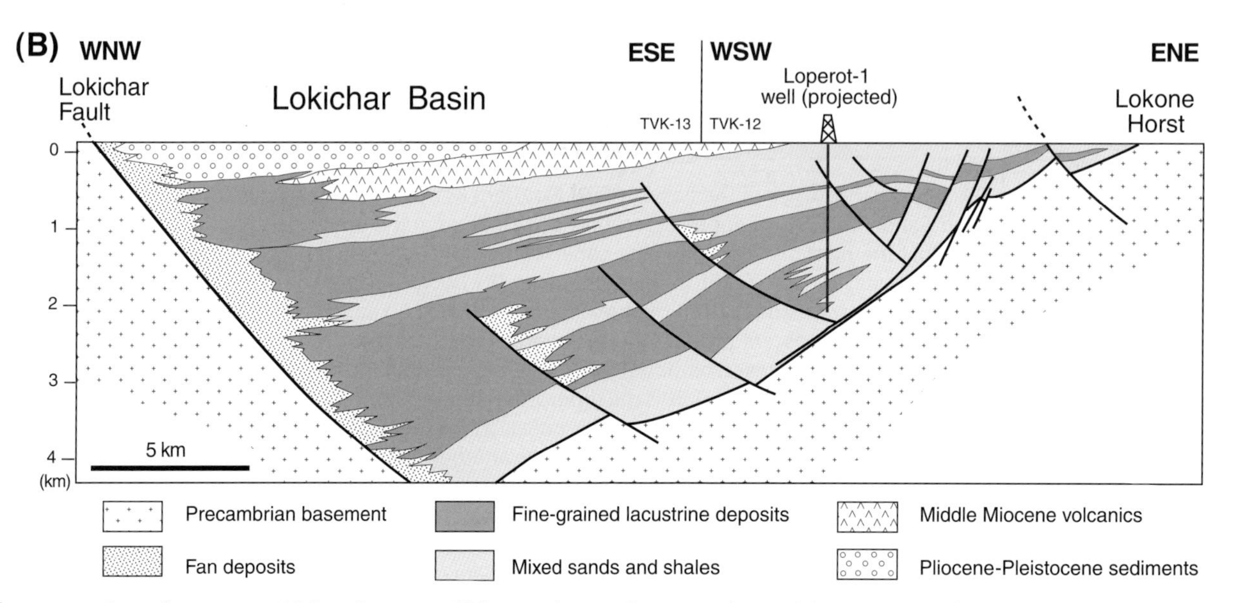

Figure 14. The Paleogene–middle Miocene Lokichar Basin, southwest Turkana. (A) Location of the Loperot-1 well and seismic lines TVK-12 and TVK-13. The white star indicates the Lokone vertebrate locality, of late Oligocene age (Ducrocq et al., 2010). (B) Basin structure and stratigraphy based on reflection seismic data (TVK-12 and 13 lines) (modified from Morley et al., 1999a). (C) Synthetic stratigraphy of the Lokichar Basin infill showing the vertical distribution of the potential source rocks (Loperot and Lokone shale members) and reservoirs (Lokone and Auwerwer sandstones).

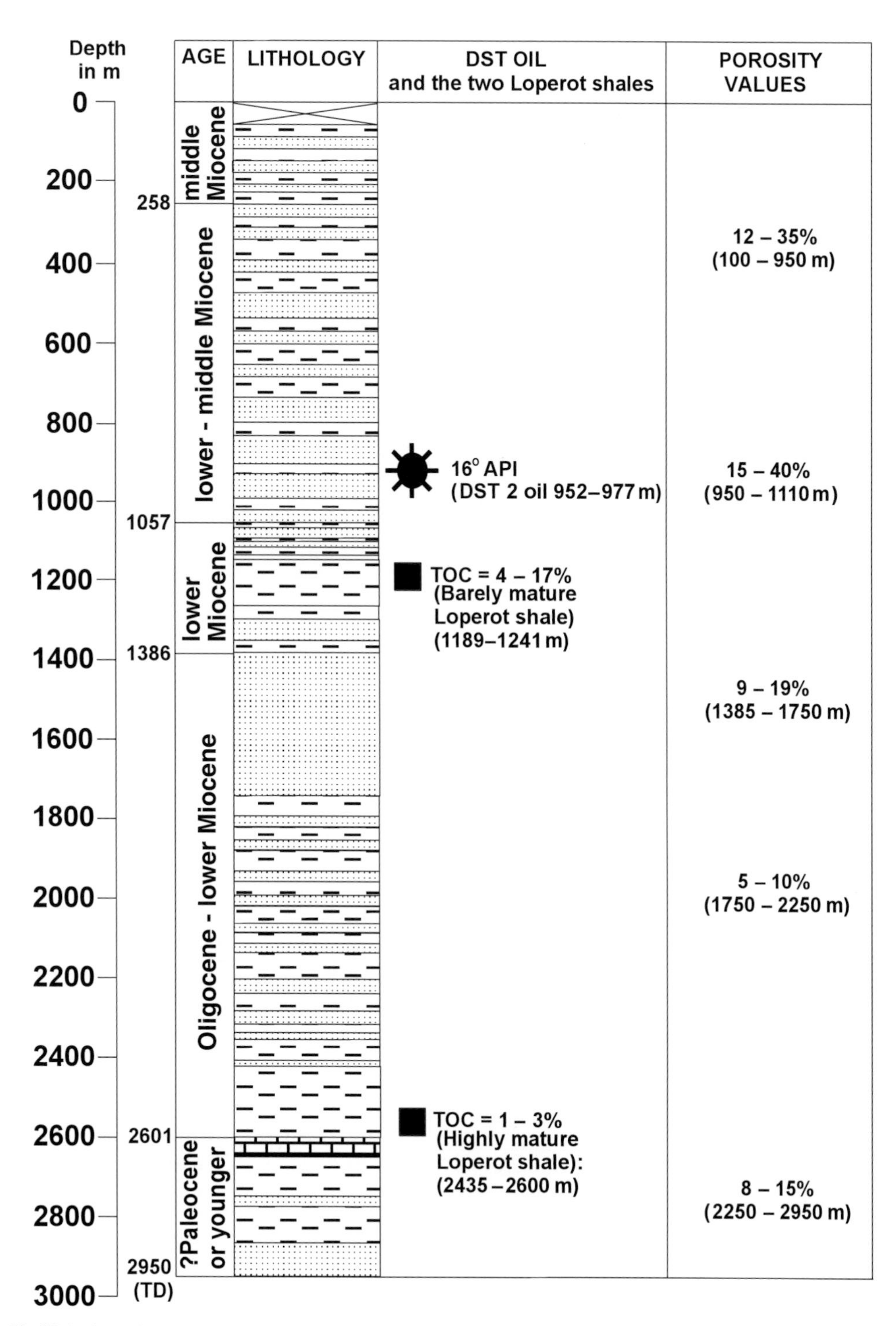

Figure 15. Detailled lithologic log of the Loperot-1 well (total depth [TD] = 2950 m) drilled in the Lokichar Basin, with indications of total organic carbon (TOC) values, oil show, and evolution of porosity values with depth (redrawn from Maende et al., 2000).

part of LS. They possibly represent part of the infill of a basin developed east of the North Kerio Basin, a part of it being intersected and downwarped by the recent (Pleistocene?) southern part of the Lake Turkana Basin. The Mount Porr Sandstone is composed mainly of quartz and feldspars (microcline, albite, and oligoclase), indicating a regional metamorphic source (Tiercelin et al., 2004). Diagenetic events are recorded by calcite, quartz, or kaolin cements. Some of these events may relate to migration of fluids originating from igneous intrusions

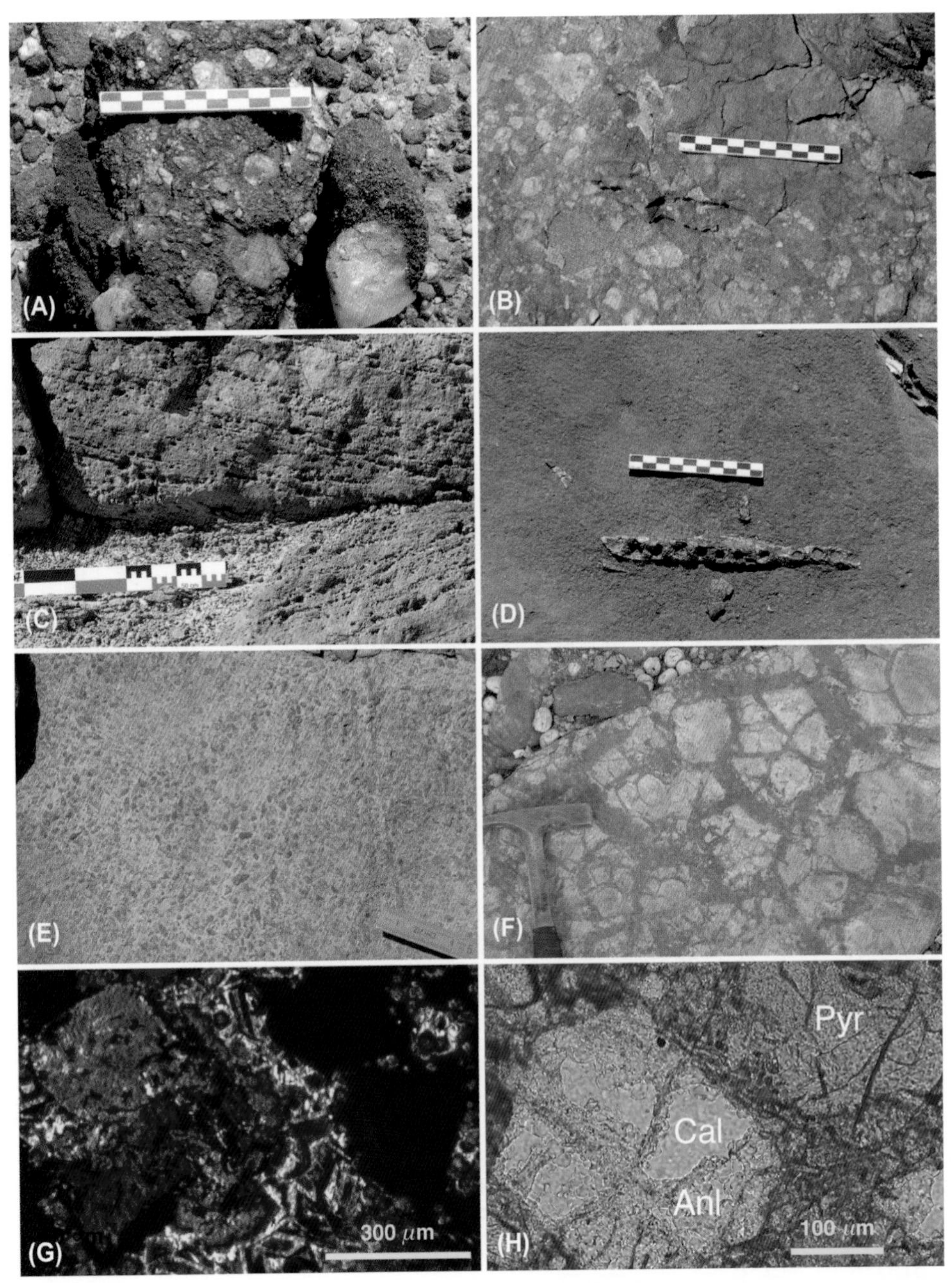

Figure 16. A selection of sedimentary facies and cements of the Lokone Sandstone, Lokichar Basin, in terms of reservoirs. (A, B) Arkosic pebbly sandstones onlapping the Lokone Horst, forming the upper part of the Lokone Sandstone. These basement-derived strongly cemented sediments offer a poor reservoir potential because of low porosity. (C) Coarse-grained, cross-bedded sandstone of fluvial or fluviodeltaic type. Ghosts of clay pebbles are abundant in this coarse facies that locally overlie mudstones, perhaps as rip-up clasts, indicating erosion of flood-plain deposits accompanied by sediment bypass. (D) Fine-grained sandstone containing abundant debris of fish, tortoise, and crocodile, mainly represented by fragmented mandibles. (E) Carbonate beds made of mollusk shell (gastropods and few bivalves) accumulations in shallow fresh water, lacustrine shoreline, or shallow pond environment, upper part of the Lokone Sandstone. These carbonate facies are closely associated with the organic-rich shales (Lokone Shale Member) that accumulated in the deepest part of the basin. (F) Mud crack facies associated with the carbonate facies, indicating abrupt drying episodes. (G) Carbonate cement. The sandstones in the Lokone Sandstone show homogeneous orange-luminescent cement. (H) Calcite-analcite (Anl) pseudomorphs after pyroxene (Pyr). Analcite forms microcrystalline ribbons that preserve the initial shape of the detrital minerals. Calcite (Cal) fills the pores remaining between the Anl ribbons. The Cal-Anl cement is characteristic of sandstones containing significant amounts of fresh volcaniclastic materials (amphibole, pyroxene).

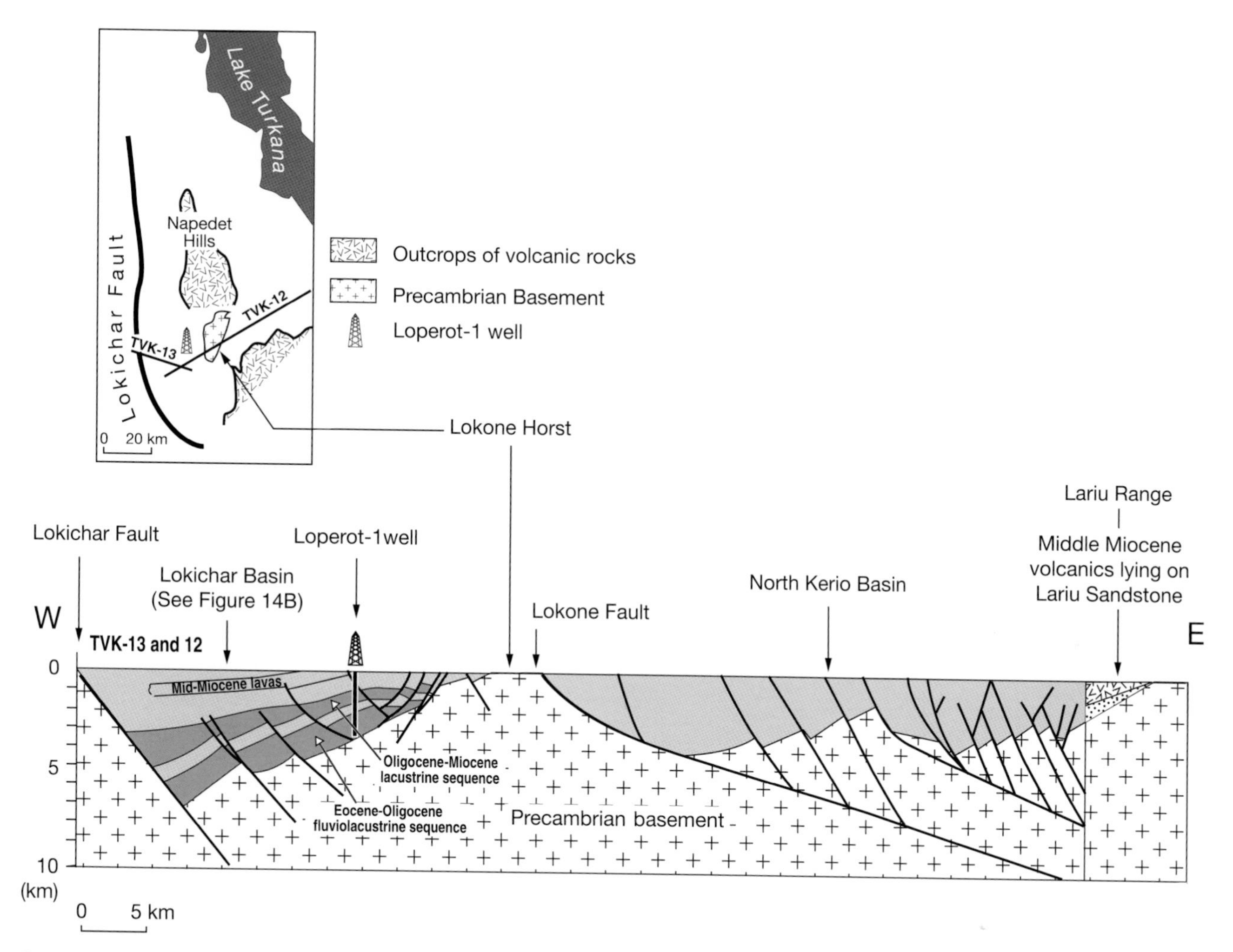

Figure 17. A geologic cross section based on the TVK-13 and TVK-12 seismic lines, showing the Lokichar Basin (of Paleogene-middle Miocene age) to the west and the North Kerio Basin (of Paleogene-Miocene and Pliocene age), separated by the Lokone Horst, which represents the footwall uplifted area to the Lokone Fault, which bounds the North Kerio Basin to the west (redrawn from Morley, 1999a, and Morley et al., 1999a). The presence of arkosic grits (Lariu Sandstone) below volcanics dated middle Miocene suggests for the North Kerio Basin a stratigraphy similar to the Lokichar Basin. Stratigraphical equivalents of the Eocene–Oligocene and Oligocene–middle Miocene shale sequences observed in the Lokichar Basin may have been deposited in the deepest parts of the North Kerio Basin during periods of high lake levels associated with high rainfall climatic episodes. These suspected deposits can offer a very high hydrocarbon potential as it is for the Lokichar Basin.

of Miocene age. A high porosity (up to 33%) was nevertheless preserved, probably associated with late diagenetic dolomitization (Figure 13G, H).

In terms of source rock potential, the presence of a source rock in the Lokichar Basin is evident. The two shale intervals encountered by the Loperot-1 well have been described as good-quality source rocks (Morley, 1999a; Talbot et al., 2004) (Figures 14C, 15). The upper shale interval (Lokone Shale Member) is dated late Oligocene–early Miocene, whereas the lower interval (Loperot Shale Member) is of possible Eocene to early Oligocene age (Morley et al., 1999a). Organic matter studies demonstrated a good source rock potential with high TOC values (1–17%; average, 2.4%) (Morley et al., 1999a) and proved algal lacustrine origin for this organic matter. Multicellular algae, *Botryococcus* spp., and *Pediastrum* spp. indicated freshwater conditions. The hydrogen index (HI) values in the Loperot-1 well suggest a type I/II composition for the productive kerogen (Talbot et al., 2004). Similar source rocks characterized by the presence of botryococcanes derived from *Botryococcus brauni* and kerogen of type I and type II have been identified in the Paraa and Kibiro oil seeps in the Albertine Graben of Uganda (Lirong et al., 2004).

Wide lateral extension of the Lokone shales into the North Kerio Basin is suspected from interpretation of

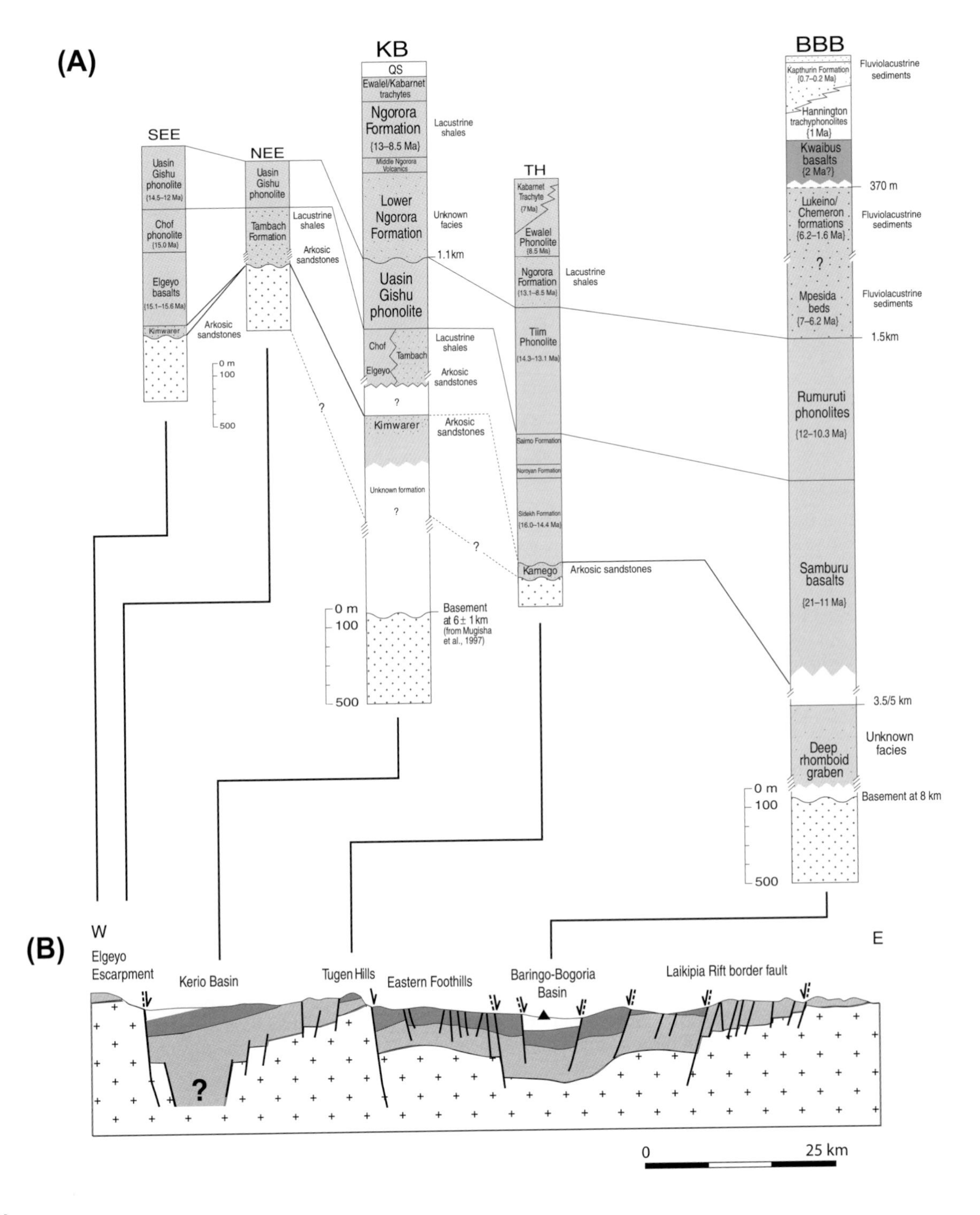
(A)
SEE
Uasin Gishu phonolite
(14.5–12 Ma)
Chof phonolite
(15.0 Ma)
Elgeyo basalts
(15.1–15.6 Ma)
Kimwarer
Arkosic sandstones
0 m
100
500
NEE
Uasin Gishu phonolite
Tambach Formation
Lacustrine shales
Arkosic sandstones
KB
QS
Ewalel/Kabarnet trachytes
Ngorora Formation
(13–8.5 Ma)
Lacustrine shales
Middle Ngorora Volcanics
Lower Ngorora Formation
Unknown facies
1.1 km
Uasin Gishu phonolite
Chof
Tambach
Elgeyo
Lacustrine shales
Arkosic sandstones
Kimwarer
Arkosic sandstones
Unknown formation
0 m
100
500
Basement at 6 ± 1 km (from Mugisha et al., 1997)
TH
Kabarnet Trachyte
(7 Ma)
Ewalel Phonolite
(8.5 Ma)
Ngorora Formation
(13.1–8.5 Ma)
Lacustrine shales
Tiim Phonolite
(14.3–13.1 Ma)
Saimo Formation
Noroyan Formation
Sidekh Formation
(16.0–14.4 Ma)
Kamego
Arkosic sandstones
BBB
Kapthurin Formation
(0.7–0.2 Ma)
Fluviolacustrine sediments
Hannington trachyphonolites
(1 Ma)
Kwaibus basalts
(2 Ma?)
370 m
Lukeino/ Chemeron formations
(6.2–1.6 Ma)
Fluviolacustrine sediments
Mpesida beds
(7–6.2 Ma)
Fluviolacustrine sediments
1.5km
Rumuruti phonolites
(12–10.3 Ma)
Samburu basalts
(21–11 Ma)
3.5/5 km
Deep rhomboid graben
Unknown facies
0 m
100
500
Basement at 8 km
(B)
W
E
Elgeyo Escarpment
Kerio Basin
Tugen Hills
Eastern Foothills
Baringo-Bogoria Basin
Laikipia Rift border fault
?
0
25 km

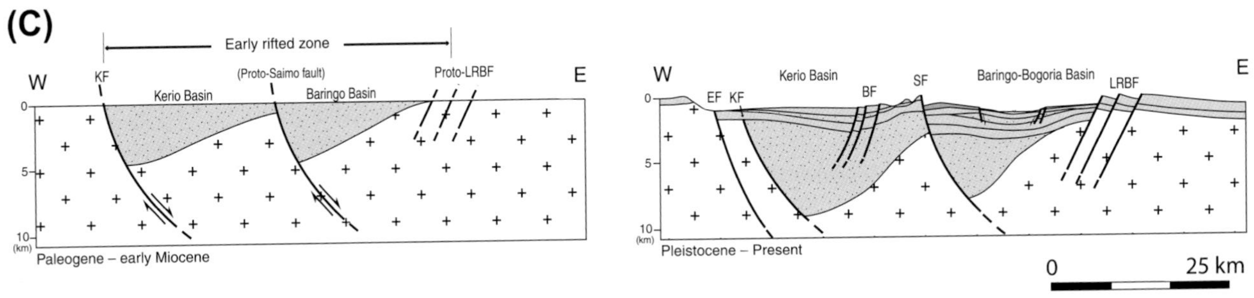
(C)
Early rifted zone
W
E
KF
(Proto-Saimo fault)
Proto-LRBF
Kerio Basin
Baringo Basin
0
5
10
(km)
Paleogene – early Miocene
W
E
EF
KF
Kerio Basin
BF
SF
Baringo-Bogoria Basin
LRBF
0
5
10
(km)
Pleistocene – Present
0
25 km

seismic data across the Lokone Horst (Figure 16). These shales were deposited during a time when displacement on the main Lokichar boundary fault system was greater and adjacent basins (the North Lokichar Basin to the north, and the North Kerio Basin to the east) had been established (Morley et al., 1999a). Consequently, lacustrine conditions offering good source rock potential may have extended into these adjacent basins, thus allowing accumulation of high concentrations of organic matter in the basins. Pollen analyses conducted on the Lokone shale interval from the Loperot-1 well (Vincens et al., 2006) demonstrate the presence in the Lokichar region of a mosaic environment of semideciduous forest and humid woodland presenting strong affinities with vegetation known today in the Guinea-Congolia/Zambezia phytogeographical zone. Rainfall of more than 1000 mm/yr characterized the Lokichar region during late Oligocene–lower/middle Miocene. The existence of a quite large, possibly deep, freshwater lake is supported by paleoclimatic data and by the discovery of wide populations of fish, crocodiles, and turtles in arkosic sandstones of late Oligocene age (27–28 Ma) onlapping the basement at Lokone Horst (Ducrocq et al., 2010). The Eocene–middle Miocene Lokichar Lake could be interpreted as similar to some of the modern large lakes of the western branch of the EARS, such as the modern Lake Albert or Lake Edward. Such climatic conditions also support the possibility of high lake level conditions that, combined with tectonic displacements along the main boundary fault, induced the development of a large (possibly 100 km [62 mi] wide and more than 150 km [93 mi] long), highly productive freshwater lake covering the Lokichar, North Lokichar, and North Kerio basins (Figures 1, 3).

Evidence of oil generation from the Loperot shales was provided by good oil shows encountered at several levels in the Loperot-1 well (Maende et al., 2000; NOCK, 2001) (Figure 15). The upper Lokone Shale Member still has relatively high HI. It is thus probable that significant volumes of liquid hydrocarbons have already been generated from this unit, which still has considerable generative potential (Talbot et al., 2004).

THE CENTRAL KENYA RIFT BASINS

The Kerio and Baringo-Bogoria Basins

The Kerio and Baringo-Bogoria basins are located in the CKR between 0°15′ and 0°45′N and are separated by the uplifted Tugen Hills fault block (bounded to the east by the major Saimo and Kito Pass faults) (Figures 1; 3; 18A, B). The present-day Kerio Basin is occupied by the Kerio River, a seasonal river draining into Lake Turkana. This half-graben basin is considered as tectonically quiescent, whereas the Baringo-Bogoria Basin, which today is occupied to the north by the freshwater Lake Baringo and to the south by the saline alkaline Lake Bogoria, is considered as tectonically and volcanically active, as demonstrated by important seismicity, intense hydrothermal activity, and recent (Holocene) volcanic activity (Tiercelin and Vincens, 1987; Renaut and Tiercelin, 1994; Tongue et al., 1994) (Figure 18B). Gravity and seismic investigations were conducted by the NOCK in the Kerio Valley in 1989. A magnetotelluric survey was conducted in 1996 in the Baringo-Bogoria Basin (Marigat-Loboi Plain), associated with a reinterpretation of the Kerio seismic and gravity data (Hautot et al., 2000). This work demonstrated that the modern Kerio and Baringo-Bogoria basins are superimposed on two deep sedimentary basins that possibly initiated during the Paleogene. The Kerio Basin is described as a more than 8 km (>5 mi) deep typical half graben, whereas the Baringo-Bogoria Basin is estimated to be 7 km (4.4 mi) deep (Hautot et al., 2000; Tiercelin and Lezzar, 2002). Both basins are suspected to be filled by alternating fluvial and lacustrine sediments and thick piles of volcanics (Figure 18A, B). In the case of the Baringo-Bogoria Basin, the upper part

Figure 18. (A) Interpreted stratigraphic series of the Central Kenya Rift from the Elgeyo Fault Escarpment (South Elgeyo Escarpment [SEE] and North Elgeyo Escarpment [NEE]) to the Kerio Basin (KB) to the west, the Tugen Hills uplifted block (TH) to the Baringo-Bogoria Basin (BBB) to the east (redrawn from Hautot et al., 2000). These lithostratigraphic columns (SEE, NEE, KB, and TH) show the stratigraphic position and sedimentary facies of the fluvial Paleogene? Kimwarer and Kamego formations and the fluviolacustrine Miocene Tambach and Ngorora formations. (B) A geologic cross section through the KB, the TH, and the BBB, showing the suspected deep structure of the KB and BBB, respectively. To the east, the BBB log shows the deep structure of the BBB below the 2 km (1.2 mi) thick Samburu Basalts (21–11 Ma). The deep part of the BBB could contain potential source rocks and reservoirs of Paleogene age. The small black triangle in the central part of the BBB (Lake Baringo) indicates the location of Holocene volcanism and present-day hydrothermal activity in the axial rift (Tiercelin and Vincens, 1987). (C) Simplified diagrams of the rifting evolution of the Central Kenya Rift (KB and BBB) during the Paleogene–Early Miocene, and during the Pleistocene to Present (black triangle: hydrothermal activity and volcanism in the axial rift). The early stage of rifting is possibly contemporaneous with the Lokichar-North Kerio rifting events in the Northern Kenya Rift. But contrary to the case of the Lokichar Basin, evolution in the KB has continued until the upper Miocene and up to the Present in the BBB (modified from Hautot et al., 2000; Tiercelin and Lezzar, 2002). 25 km (15.5 mi).

of the basin infill, mainly fluviolacustrine deposits of Miocene to Pleistocene age (Mpesida, Lukeino, and Chemeron formations), outcrops widely on the hanging wall of the Saimo Fault (Figure 18A, B).

The lowest part of the Kerio Basin infill is thought to be sedimentary in origin and possibly of Paleogene age. At the southern end of the Kerio Basin, few tens of meters thick arkosic sandstones known as the Kimwarer Formation (Figures 3; 18A; 19A, B) outcrop along the footwall of the Elgeyo border fault—the western faulted margin of the CKR—where they overlie the Precambrian basement (Figures 3; 19A, B). Renaut et al. (1999) described green, waxy, laminated tuffaceous shales associated with these sandstones. To the east, in the Baringo-Bogoria Basin, the oldest sediments, described as quartzites, sandstones, and siltstones resting on an irregular surface of Precambrian basement are poorly exposed along the western faulted side of the basin (Tugen Hills fault block). These sediments are known as the Kamego Formation (Williams and Chapman, 1986) (Figures 3, 18A). The Kimwarer and Kamego sediments are mainly arkosic fluviatile sandstones sourced from Precambrian rocks (Chapman et al., 1978; Renaut et al., 1999). These sediments may represent the earliest sections of the Kerio and Baringo-Bogoria basin infills, possibly indicating fluvial and lacustrine environments similar to the ones found north of the Lokichar and North Kerio basins. Fluvial networks of Paleogene age draining the basement floor of the Kerio and Baringo-Bogoria basins are interpreted as flowing northward, feeding the Lokichar and North Kerio basins (Morley and Wescott, 1999; Tiercelin et al., 2004). The Paleogene history of the Kerio and Baringo-Bogoria basins is unfortunately poorly known. This is because outcrops are scarce and poorly exposed and because of the absence of sufficient geophysical data and exploration wells. Nevertheless, a possible similarity between the Kimwarer and Kamego formations with the Lokichar-North Kerio sediments should encourage oil exploration in this area.

The Miocene history of the Kerio and Baringo-Bogoria half grabens is much better documented by sediments and lavas outcropping along the Elgeyo border fault and the western flank of the Tugen Hills fault block. Thick sedimentary deposits are found interbedded between the main volcanic units (Figure 18A) and relate to the existence of at least two wide lacustrine domains known as Lake Tambach and Lake Ngorora, both dated middle Miocene. In the Kerio Basin, more than 2 km (>1.2 mi) of volcanic rocks have been deposited between 23 and 15 Ma, then between 14 and 7 Ma (Chapman et al., 1978) (Figure 18A). The oldest sequence of Neogene age is the Tambach Formation, consisting of up to 400 m (1312 ft) of colluvial, fluvial, and lacustrine sediments, which are exposed discontinuously along the Elgeyo border fault, and lie on an irregular topography of basement rocks (Ego, 1994; Renaut et al., 1999). The upper limit of the Tambach Formation is provided by the overlying phonolites of the Uasin Gishu Formation (Lippard, 1973) that is dated from 14.5 to 12 Ma (Figure 18A). The overall Tambach depositional sequence is formed by alluvial fan and gravelly braided stream deposits near its base, followed by sandy braided deposits, to shallow deltaic and lacustrine sediments at the top (Renaut et al., 1999) (Figure 19C, D).

The Ngorora Formation also is of Miocene age but younger than the Tambach Formation (Bishop and Chapman, 1970; Bishop and Pickford, 1975) (Figure 18A). The 400 m (1312 ft) thick Ngorora sediments lie on lavas of the Tiim Phonolite dated 13.1 Ma and are unconformably overlain by the Ewalel Phonolite dated 8.5 Ma (Hill, 1999). Clastic sediments in the Ngorora Formation are mainly fluvial sediments and lahar deposits. The main component of the Ngorora Formation consists of dark-green or gray to black paper shales of lacustrine facies. Petroliferous calcarenites are reported from the Ngorora Formation (Martyn, 1969; Chapman, 1971). Thus, the Tambach and Ngorora formations were considered as attractive for hydrocarbon exploration. In terms of reservoir rocks, arkosic sandstones near the base of the Tambach Formation are quartz rich, well sorted, and reflect dominant basement provenance (Figure 19C). Cements are calcite and iron oxides, as with the LS in the NKR. Quartz and albite overgrowths are locally present. Near the top of the formation, the sandstone composition is characterized by an increase in volcanic grains as a consequence of the occurrence of regional volcanic activity (Ego, 1994). The fluvial sediments and lahar deposits of the Ngorora Formation have been poorly studied but are not suspected to be good potential reservoir rocks.

In terms of source rocks, the lacustrine facies in the Tambach Formation are represented by pale green, gray, and white shales. Some are waxy in texture and appearance or show a paper shale facies (Figure 19C). They are interpreted as having accumulated in fresh to slightly brackish waters (Renaut et al., 1999). For the Ngorora Formation, of the two lacustrine members (C and E) described by Pickford (1978), only Member E, which forms the upper part of the formation, represents full lacustrine conditions, with a large freshwater lake domain extending from the Elgeyo border fault escarpment to the west up to the axial part of the Baringo-Bogoria Basin. This episode may show strong interest in terms of source rock and/or hydrocarbon potential. Petroliferous shales have been described at Poi locality, situated along the Elgeyo border fault, as well as on the western flank of the Tugen Hills. Renaut et al. (1999) cited a TOC value of up to 4.3%. These paper shales are

Figure 19. (A, B) Kimwarer Formation: massive sandstones (see white arrow) unconformably overlying Precambrian basement at the southern end of the Elgeyo Border Fault Escarpment, Kerio Basin, near the Fluorspar Mine site (light brown area, center left of photograph [A]). These sandstones are a few tens of meters thick outcrop on top of a small faulted block that is part of the Elgeyo Border Fault. On the far horizon of the photograph (A) is the crest of the Tugen Hills fault block separating the Kerio and Baringo-Bogoria basins (Figure 18). (C) Tambach Formation (Figure 18): outcrops along the Elgeyo fault scarp along the road from Kabarnet to Eldoret, south of the Tambach village. Pale green, gray, and white lacustrine shales form the upper part of the formation. Some shales are waxy; some show a paper shale facies. (D) Fluvial sandstones in the Tambach Formation. At the base of the formation are coarse-grained deposits mainly arkosic paraconglomerates. Photograph (D) shows a small deltaic-type sand body with northeast-dipping foresets, interbedded with lacustrine shales. These sandstones are quartz rich but show an increase of volcanic minerals toward the top of the section. Cements are predominantly calcite and iron oxides.

dark green or dark gray to black in color and contain very well preserved fossil fish and leaf fragments. The Poi shales can tentatively be compared, in terms of depositional environment, with the Oligocene–Miocene shales found in the Lokichar Basin. A very great thickness is suspected for the Ngorora Formation and possibly older sediments, as suggested by the strong negative gravity anomaly identified in that part of the Kerio Basin (Swain et al., 1981; Williams and Chapman, 1986) and the reinterpretation of NOCK gravity and seismic data by Hautot et al. (2000) (Figure 18A). The Ngorora Member C represents an association in small downwarped basins of different lake environments (Pickford, 1978), from shallow freshwater lakes resembling the present-day Lake Baringo in the CKR (Figure 3) or to small saline alkaline lakes, as confirmed by the presence in the sediments of abundant zeolites such as analcite and clinoptilolite associated with other authigenic minerals (Renaut et al., 1999). These lakes resemble the modern saline lakes of the CKR, Lake Nakuru, Lake Elmenteita, or Lake Bogoria, a small meromictic lake with an anoxic monimolimnion preserving the accumulation of organic-rich zeolitic muds in the deeper parts of the lake (Tiercelin and Vincens, 1987; Renaut and Tiercelin, 1994).

DISCUSSIONS AND CONCLUSIONS

Although the basins of the NKR and the CKR have not attracted a lot of interest from oil exploration companies in the past, recent oil discoveries in the Albertine Graben in Uganda have shifted the focus of many oil companies to such rift basins. The presence of hydrocarbons in rift basins is strongly controlled by the style of postrift tectonics (Morley, 1999a). In the EARS, postrift tectonic structures that may cause major loss in hydrocarbon potential are not frequent, with the exception of inversion structures that were identified thanks to the dense seismic coverage in the Anza, Rukwa, and Turkana rift basins (Morley et al., 1999c). In these long to very long lived rifts, inversion tectonics appear to be most effective at the end of the rifting. In the case of the Turkana area, Morley et al. (1999c) demonstrate that inversion structures are weak or absent in the Lokichar Basin and the southern part of the North Kerio Basin. In this region, lateral migration of rifting is one of the major characteristics of the early–middle Cenozoic rifting episodes, as demonstrated by the string of Paleogene–early Miocene rift basins. This confirms that different areas in this part of the NKR will have considerable variation in potential plays. Hydrocarbon potential is also mostly controlled by the nature of the basin-fill stratigraphy and facies. Sedimentologic, petrological, and geochemical studies conducted on the reservoir and source rock potential in the major basins belonging to the NKR and CKR have permitted the establishment of a provisional classification of these basins that can be used as a predictive tool for evaluating every half-graben for source rock quality and quantity and for reservoir potential:

1) In terms of source rocks: Lacustrine source rocks are widespread in the EARS (Tiercelin et al., 1992; Wolela, 2006) and have been particularly encountered in areas with well-developed natural oil seeps (Tiercelin et al., 1993a; Simoneit et al., 2000). The best example in the EARS (western branch) is the Albertine Graben where oil seeps are numerous and where major oil discoveries have recently been made (PEPD, 2007). Other natural oil seeps are known in the northern basin of Lake Tanganyika, at Cape Banza and Cape Kalamba (Tiercelin et al., 1993a, b), where asphaltic oil (radiocarbon dated at ~25,000 yr BP) is generated from late Pleistocene organic-rich lacustrine muds associated with hydrothermal systems (Simoneit et al., 2000). In the NKR, asphaltic inclusions are known from fluorite veins crossing the Precambrian basement in the hanging wall of the Saimo Fault, western flank of the Baringo-Bogoria Basin. Oil seeps are also suspected on the northwestern shores of Lake Turkana, from synthetic aperture radar surveys (Diggens and Hall, 2010). The best known source rocks (qualified as good- to high-quality source rocks) have been identified in the Lokichar Basin, particularly in the Lokone Shale Member. According to calculations by Morley (1999b), source rocks 500 to 1000 m (1640–3280 ft) in thickness can be expected for the Lokone shales near the major Lokichar boundary fault, and huge volumes (about 10–20 billion of bbl) of hydrocarbons could be generated from the Lokichar Basin. The Poi source rocks found in the Ngorora Formation of the Kerio-Baringo basins, CKR, can be also classified as good-quality source rocks and comparable to the Lokone source rocks. In terms of oil potential, the main questions for the Ngorora Formation are (1) the quantity of source rock that is dependent on the basin size and the depositional setting of the shales. The minimum thickness is evaluated at outcrop scale, but a very great thickness is suspected from geophysical investigations (Swain et al., 1981; Hautot et al., 2000). Paleogeographical reconstructions proposed by Pickford (1978) indicate small-size lake basins with the exception of Member E period in the course of which a large lake extended almost more than the whole width of the CKR. Nevertheless, such data have been deduced from two-dimensional observations along the Elgeyo and Saimo fault escarpments. Dense seismic reflection coverage of the Kerio Basin will probably help to obtain a better description of the Ngorora Basin paleogeography. No other source rocks of good quality have been identified in the central or northern Kenya basins. The dark gray to black silty mudstones described at outcrop scale within the LS (Thuo, 2009) show no organic content; minimal thicknesses and limited lateral continuity characterize these sediments. In addition, interpreted depositional environments of floodplains or shallow lakes relate to postrift episodes and, thus, suggest minimal source rock potential.

Nevertheless, thicker accumulations of similar dark mudstones can be expected in the deeper parts of Gatome Basin. But only the acquisition of new good-quality seismic coverage and subsequent deep drilling will permit to solve this question.

2) In terms of reservoir rocks: The different basins observed in the NKR and CKR—Lotikipi/Gatome, Lokichar/North Kerio, Kerio and Baringo—offer good to very good reservoir potential. Sandstone quality in this part of the EARS is highly dependent on the source area and the transport distance (Morley, 1999b). Early stages of development for the older basins in northern Kenya (middle Cretaceous–Paleogene; Lotikipi, Gatome, and Lokichar basins) are characterized by alluvial, fluvial, or fluviodeltaic environments. Sandstones associated with these early stages of basin development and specific depositional environments are most commonly basement-derived arkoses (mainly quartz and feldspar rich). The better examples are provided by the LS and MS, as well as the lowest parts of the Lokichar Basin infill (Lokone Sandstone). Such arkosic sandstones are characterized by a marked reduction of porosity mainly because of diagenetic evolution of feldspars. In the Cretaceous–Paleogene Anza Rift of northeastern Kenya, the quality of reservoirs is locally significantly modified by diagenetic growth of zeolites (Winn et al., 1993). No zeolitic cementation has been encountered in the arkosic sandstones belonging to the LS, MS, and Mount Porr Sandstone, but this has been observed in the Lokone Sandstone. The major cements are calcite, kaolinite, and hematite; some of these cements being strongly associated with climatic changes prevailing in central-eastern Africa from the middle Cretaceous. Hydrothermal fluid circulation during episodes of volcanic dike intrusions, in the LS as well as in the Lokichar Basin, might also have been the source of such cementations (Thuo, 2009). Results of porosity measurements in the LS yielded values ranging from a minimum of 3% to a maximum of 25% (Thuo, 2009). These values compare well with other observed porosity values in sandstones of similar depositional environments in the southern Sudanese basins where porosity values ranging from 8 to 38% have been reported (Schull, 1988; Giedt, 1990). In the Lokichar Basin, arkosic sandstones show calcitic and zeolitic cements and porosities between 1 and 15%. In the Loperot-1 well, reservoir horizons show quartz and feldspar overgrowths, carbonate cements, zeolites, authigenic clays, and opaque minerals, particularly in levels (Auwerwer Sandstone) where sandstones contain large quantities of volcanic-derived material, which are prone to diagenetic alteration. Good porosity values of 12 to 30% have been obtained at different intervals in the Loperot-1 well (Morley, 1999b).

In the CKR, potential reservoirs may be expected from the Kimwarer and Kamego arkosic sandstones, which are comparable to the LS, MS, Lariu, or Mount Porr sandstones. Unfortunately, very few detailed sedimentologic and/or petrological studies have been conducted on these formations. Nevertheless, they may represent deep targets within the context of the Paleogene Kerio and Baringo-Bogoria basins. Sandstone units associated with the Miocene Tambach and Ngorora formations show similar characteristics with the upper part of the Lokichar Basin infill (Auwerwer Sandstone), as large quantities of volcanic-derived minerals are directly derived from a volcanically delineated basin watershed. As a consequence, low porosities can be expected from these formations.

To conclude, in terms of hydrocarbon prospectivity classification for the NKR and CKR, the Lokichar/North Kerio Basin (NKR) contain source rocks showing enormous petroleum potential. These basins can be ranked 1, that is, having the highest potential for hydrocarbons based on their source rock quality and quantity and reservoir potential. In addition, basin evolution (lateral shift of rift activity and inversion tectonics) during synrift and postrift episodes does not represent a major risk factor. In the Lokichar Basin, only trap size and seal are the largest risk factors. Rank 2 (medium to high potential) can be attributed to the Gatome Basin (NKR) because of its thick infill and the proximity of good potential reservoirs represented by the LS and MS. No good source rocks have been identified in this basin at the outcrop scale. Nevertheless, a strong possibility to encounter such source rock at depth may exist. Finally, rank 3 (medium potential) can be attributed to the Kerio and Baringo-Bogoria basins. Both basins offer good source rock but low-quality potential reservoir rocks. Nevertheless, in similar conditions, reservoir quality sandstones can be encountered in wells at depths 3 km or more (≥1.9 mi). With the increased oil exploration activity within Kenya rift basins, it will only be a matter of time before deep exploration wells are drilled in these basins, providing a better understanding of the evolution of basin architecture and sedimentary fill.

ACKNOWLEDGMENTS

Permission to conduct field geologic research in the West Turkana Basin was provided by the Office of the President and the Ministry of Education, Science and

Technology of the Republic of Kenya (research permits to Jean-Jacques Tiercelin, no. OP/13/001/23C 290 and MOEST 13/001/23C 290). This study was supported by grants from the French Ministry of Foreign Affairs. We are grateful to the National Oil Corporation of Kenya (NOCK) for administrative and logistic support in the field. The authors thank Chris Morley and Ian Hutchinson for their careful reviews and useful suggestions on ways to improve this chapter, as well as the Editor Dengliang Gao for valuable comments at various stages of manuscript preparation. Peter Thuo and Jean-Jacques Tiercelin dedicate this chapter to the memory of F. M. Mbatau, former NOCK exploration manager, who died in a road accident in July 2010.

REFERENCES CITED

Abdelfettah, Y., 2009, Inversion conjointe des données magnétotelluriques et gravimétriques: Application à l'imagerie géophysique crustale et mantellique: Thèse Université de Bretagne Occidentale, Brest, France, 141 p.: http://tel.archives-ouvertes.fr/tel-00424413/fr/ (accessed October 20, 2009).

Andrews, P., and P. Banham, eds., 1999, Late Cenozoic environments and hominid evolution: A tribute to Bill Bishop: Geological Society (London), 276 p.

Arambourg, C., 1935, Esquisse géologique de la bordure occidentale du Lac Rodolphe. Mission scientifique de l'Omo, 1932–1933: Bulletin Muséum National d'Histoire Naturelle, Paris, vol. 1, 55 p.

Arambourg, C., and R. G. Wolff, 1969, Nouvelles données paléontologiques sur l'âge des grès du Lubur (Turkana Grits) à l'Ouest du lac Rodolphe: Comptes Rendus Société géologique de France, v. 6, p. 190–192.

Bellieni, G., E. Justin Visentin, B. Zanettin, E. M. Piccirillo, F. Radicati di Brozolo, and F. Rita, 1981, Oligocene transitional-tholeiitic magmatism in northern Turkana (Kenya): Comparison with the coeval Ethiopian volcanism: Bulletin of Volcanology, v. 44, p. 411–427, doi:10.1007/BF02600573.

Bellieni, G., E. Justin Visentin, E. M. Piccirillo, and B. Zanettin, 1987, Volcanic cycles and magmatic evolution in northern Turkana (Kenya): Tectonophysics, v. 143, p. 161–168, doi:10.1016/0040-1951(87)90085-0.

Bishop, W. W., 1978, Geological background to fossil man: Geological Society (London) Special Publication 6, 585 p., doi:10.1144/GSL.SP.1978.006.01.16.

Bishop, W. W., and G. R. Chapman, 1970, Early Pliocene sediments and fossils from the northern Kenya Rift Valley: Nature, v. 226, p. 914–918, doi:10.1038/226914a0.

Bishop, W. W., and M. Pickford, 1975, Geology, fauna and paleoenvironments of the Ngorora Formation, Kenya Rift Valley: Nature, v. 254, p. 185–192, doi:10.1038/254185a0.

Boschetto, H. B., F. H. Brown, and I. M. McDougall, 1992, Stratigraphy of the Lothidok Range, Northern Kenya, and K-Ar ages of its Miocene primates: Journal of Human Evolution, v. 22, p. 47–71.

Bosworth, W., 1992, Mesozoic and early Tertiary rift tectonics in East Africa: Tectonophysics, v. 209, p. 115–137.

Bosworth, W., and C. K. Morley, 1994, Structural and stratigraphic evolution of the Anza Rift, Kenya: Tectonophysics, v. 236, p. 93–115, doi:10.1016/0040-1951(94)90171-6.

Chapman, G. R., 1971, The geological evolution of the northern Kamasia Hills, Baringo District, Kenya: Ph.D. dissertation, University of London, London, United Kingdom.

Chapman, G. R., and M. Brook, 1978, Chronostratigraphy of the Baringo Basin, Kenya: Geological Society (London) Special Publication 6, p. 207–223, doi:10.1144/GSL.SP.1978.006.01.16.

Chapman, G. R., S. J. Lippard, and J. E. Martyn, 1978, The stratigraphy and structure of the Kamasia Range, Kenya Rift Valley: Journal of the Geological Society (London), v. 135, p. 265–281, doi:10.1144/gsjgs.135.3.0265.

Desprès, A., 2008, Evolution tectono-sédimentaire des bassins de rift Crétacé-Paléogène du Nord du Kenya: Master 2, Université de Rennes 1, France, 37 p.

Diggens, J., and M. Hall, 2010, East African rift system regional geological interpretation study, Utilizing earth observation data: Petroleum Exploration 2010 Conference, London, November 23–25, 2010.

Ducrocq, S., et al., 2010, New Oligocene vertebrate localities from northern Kenya (Turkana Basin): Journal of Vertebrate Paleontology, v. 30, p. 293–299, doi:10.1080/02724630903413065.

Dunkelman, T. J., B. R. Rosendahl, and J. A. Karson, 1988, Structural style of the Turkana Rift, Kenya: Geology, v. 16, p. 258–261.

Dunkelman, T. J.,. B. R. Rosendahl, and J. A. Karson, 1989, Structure and stratigraphy of the Turkana Rift from seismic reflection data: Journal of African Earth Sciences, v. 8, p. 489–510, doi:10.1016/S0899-5362(89)80041-7.

Dunn, J. L., 1985, Petrography of selected samples from the Lake Turkana area, northwestern Kenya: Unpublished report prepared by National Lead Erco/National Lead Industries, Inc., for Amoco Production Company, 23 p.

Genik, G. J., 1993, Petroleum geology of the Cretaceous–Tertiary rift basins in Niger, Chad and Central African Republic: AAPG Bulletin, v. 77, p. 1405–1443.

ECOS (European Coalition on Oil in Sudan), 2007, ECOS oil map: http://www.ecosonline.org.

Ego, J. K., 1994, Sedimentology and diagenesis of Neogene sediments in the central Kenya Rift Valley: M.Sc. thesis, University of Saskatchewan, Saskatoon, Saskatchewan, Canada, 148 p.

Giedt, N. R., 1990, Unity Field-Sudan, Muglad Rift Basin, Upper Nile Province: Structural traps III—Atlas of oil and gas fields: AAPG Treatise of Petroleum Geology, p. 177–197.

Hautot, S., P. Tarits, K. Whaler, B. Le Gall, J.-J. Tiercelin, and C. Le Turdu, 2000, Deep structure of the Baringo Rift Basin (central Kenya) from three-dimensional magnetotelluric imaging: Implications for rift evolution: Journal of Geophysical Research, v. 105, no. B10, p. 23,493–23,518, doi:10.1029/2000JB900213.

Hill, A., 1999, The Baringo Basin, Kenya: From Bill Bishop to

BPRP, *in* P. Andrews and P. Banham, eds., Late Cenozoic environments and hominid evolution: A tribute to Bill Bishop: Geological Society (London), p. 85–97.

Hill, A., G. Curtis, and R. Drake, 1986, Sedimentary stratigraphy of the Tugen Hills, Baringo District, Kenya, *in* L. E. Frostick, R. W. Renaut, I. Reid, and J.-J. Tiercelin, eds., Sedimentation in the African rifts: Geological Society (London) Special Publication 25, p. 285–295, doi:10.1144/GSL.SP.1986.025.01.23.

Khalifa, M. A., and O. Catuneanu, 2008, Sedimentology of the fluvial and fluviomarine facies of the Bahariya Formation (early Cenomanian), Bahariya Oasis, Western Desert, Egypt: Journal of African Earth Sciences, v. 51, p. 89–103, doi:10.1016/j.jafrearsci.2007.12.004.

Lambiase, J. J., and C. K. Morley, 1999, Hydrocarbons in rift basins: The role of stratigraphy: Philosophical Transactions of the Royal Society of London, v. 357, p. 877–900, doi:10.1098/rsta.1999.0356.

Lippard, S. J., 1973, The petrology of phonolites form the Kenya Rift: Lithos, v. 6, p. 217–234, doi:10.1016/0024-4937(73)90083-2.

Lirong, D., C. Dingsheng, W. Jianjung, E. N. T. Rubondo, R. Kasande, A. Byakagaba, and F. Mugisha, 2004, Geochemical significance of seepage oils and bituminous sandstones in the Albertine Graben, Uganda: Journal of Petroleum Geology, v. 27, p. 299–312.

Maende, A., F. M. Mbatau, and P. K. Thuo, 2000, Source rock potential of two Loperot shales of Lodwar South subbasin, Tertiary Rift Valley, Kenya: Petroleum geochemistry and exploration in the Afro-Asian region: 5th International Conference and Exhibition, New Delhi, India, November 25–27, 2000, p. 147–154.

Martyn, J. E., 1969, The geological history of the country between Lake Baringo and the Kerio River, Baringo District, Kenya: Ph.D. thesis, University of London, London, United Kingdom.

McBride, E. F., A. Abdel-Wahab, and A. M. K. Salem, 1996, The influence of diagenesis on the reservoir quality of Cambrian and Carboniferous sandstones, southwest Sinai, Egypt: Journal of African Earth Sciences, v. 22, p. 285–300.

McDougall, I., and F. H. Brown, 2009, Timing of volcanism and evolution of the northern Kenya rift: Geological Magazine, v. 146, p. 34–47, doi:10.1017/S0016756808005347.

McGuire, M., M. Serra, and R. Day, 1985, Surface geological evaluation Block 10 concession, NW Kenya: Denver, Amoco Production Company.

Mohamed, A. Y., M. J. Pearson, W. A. Ashcroft, J. E. Iliffe, and A. J. Whiteman, 1999, Modeling petroleum generation in the southern Muglad Rift Basin, Sudan: AAPG Bulletin, v. 83, p. 1943–1964.

Mohamed, A. Y., M. J. Pearson, W. A. Ashcroft, J. E. Iliffe, and A. J. Whiteman, 2002, Petroleum maturation modeling, Abu Gabra-Sharaf area, Muglad Basin, Sudan: Journal of African Earth Sciences, v. 35, p. 331–344.

Morley, C. K., 1999a, Basin evolution trends in East Africa: *in* C. K. Morley, ed., Geoscience of rift systems: Evolution of East Africa: AAPG Studies in Geology 44, p. 131–150.

Morley, C. K., 1999b, Comparison of hydrocarbon prospectivity in rift systems: *in* C. K. Morley, ed., Geoscience of rift systems: Evolution of East Africa: AAPG Studies in Geology 44, p. 233–242.

Morley, C. K., and W. A. Wescott, 1999, Sedimentary environments and geometry of sedimentary bodies determined from subsurface studies in East Africa, *in* C. K. Morley, ed., Geoscience of rift systems: Evolution of East Africa: AAPG Studies in Geology 44, p. 211–231.

Morley, C. K., W. A. Wescott, D. M. Stone, R. M. Harper, S. T. Wigger, and F. M. Karanja, 1992, Tectonic evolution of the northern Kenya Rift: Journal of the Geological Society (London), v. 149, p. 333–348.

Morley, C. K., W. A. Wescott, D. M. Stone, R. M. Harper, S. T. Wigger, R. A. Day, and F. M. Karanja, 1999a, Geology and geophysics of the western Turkana basins, Kenya, *in* C. K. Morley, ed., Geoscience of rift systems: Evolution of East Africa: AAPG Studies in Geology 44, p. 19–54.

Morley, C. K., W. Bosworth, R. A. Day, R. Lauck, R. Bosher, D. M. Stone, S. T. Wigger, W. A. Wescott, D. Haun, and N. Bassett, 1999b, Geology and geophysics of the Anza Basin, *in* C. K. Morley, ed., Geoscience of rift systems: Evolution of East Africa: AAPG Studies in Geology 44, p. 67–90.

Morley, C. K., R. M. Harper, and S. T. Wigger, 1999c, Tectonic inversion in East Africa, *in* C. K. Morley, ed., Geoscience of rift systems: Evolution of East Africa: AAPG Studies in Geology 44, p. 193–210.

Mugisha, F., C. J. Ebinger, M. Strecker, and D. Pope, 1997, Two-stage rifting in the Kenya rift: Implications for half-graben models: Tectonophysics, v. 278, p. 63–81.

Murray-Hughes, R., 1933, Notes on the geological succession, tectonics and economic geology of the western half of the Kenya colony: Report Geological Survey of Kenya, no. 3, 8 p.

Ngenoh, D. K. A., 1993, Hydrocarbon potential of South Kerio Trough basin (Kenya) from seismic reflection data: M.Sc. thesis, University of Western Ontario, London, Canada, 126 p.

NOCK (National Oil Corporation of Kenya), 2001, Petroleum potential of Lake Turkana area, Kenya: Nairobi, Kenya, National Oil Corporation of Kenya, 12 p.

NOCK (National Oil Corporation of Kenya), 2009, Petroleum exploration opportunities in Kenya: Promotional brochure: Nairobi, Kenya, National Oil Corporation of Kenya.

Obaje, N. G., H. Wehner, H. Hamza, and G. Scheeder, 2004, New geochemical data from the Nigerian sector of the Chad basin: Implications on the hydrocarbon prospectivity: Journal of African Earth Sciences, v. 38, p. 477–487.

PEPD (Petroleum Exploration and Production Department), 2007, Petroleum potential of the Albertine Graben, Uganda: Petroleum Exploration and Production Department, Ministry of Energy and Mineral Development, Republic of Uganda, internal report, 17 p.

Peterson, J. A., 1986, Geology and petroleum resources of central and east-central Africa: Modern Geology, v. 10, p. 329–364.

Pickford, M. H., 1978, Geology, paleoenvironment and vertebrate faunas of the mid-Miocene Ngorora Formation, Kenya, *in* W. W. Bishop, ed., Geological background to fossil man: Geological Society (London) Special Publication 6, p. 237–262.

Pope, D. A., 1992, Analyses and interpretation of seismic reflection profiles from the Kerio Valley, Kenya Rift: M.Sc. thesis, University of Leeds, England, United Kingdom, 133 p.

Prat, S., J. P. Brugal, J.-J. Tiercelin, J.-A. Barrat, M., Bohn, A. Delagnes, S. Harmand, K. Kimeu, M. Kibunjia, J. P. Texier, and H. Roche, 2005, First occurrence of early Homo in the Nachukui Formation (west Turkana, Kenya) at 2.3–2.4 Myr: Journal of Human Evolution, v. 49, p. 230–240.

Renaut, R. W., and J.-J. Tiercelin, 1994, Lake Bogoria, Kenya Rift Valley: A sedimentological overview, *in* R. W. Renaut and W. M. Last, eds., Sedimentology and geochemistry of modern and ancient saline lakes: SEPM, v. 50, p. 101–123.

Renaut, R. W., J. K. Ego, J.-J. Tiercelin, C. Le Turdu, and R. B. Owen, 1999, Saline, alkaline paleolakes of the Tugen Hills-Kerio Valley region, Kenya Rift Valley, *in* P. Andrews and P. Banham, eds., Late Cenozoic environments and hominid evolution: A tribute to Bill Bishop: Geological Society (London), p. 41–58.

Rosendahl, B. R., 1987, Architecture of continental rifts with special reference to East Africa: Annual Review of Earth Planetary Sciences, v. 15, p. 445–503.

Rosendahl, B. R., D. J. Reynolds, P. M. Lorber, C. F. Burgess, J. McGill, D. Scott, J.-J. Lambiase, and S. J. Derksen, 1986, Structural expressions of rifting: Lessons from Lake Tanganyika, Africa, *in* L. E. Frostick, R. W. Renaut, I. Reid, and J.-J. Tiercelin, eds., Sedimentation in the African rifts: Geological Society (London) Special Publication 25, p. 29–43.

Rosendahl, B. R., J. W. Versfelt, C. A. Scholz, J. E. Buck, and L. D. Woods, 1988, Seismic atlas of Lake Tanganyika, East Africa: Project PROBE Geophysical Atlas Series, folio 1: Durham, North Carolina, Duke University, 82 p.

Rosendahl, B. R., E. Kilembe, and K. Kaczmarick, 1992, Comparison of the Tanganyika, Malawi, Rukwa and Turkana Rift zones from analyses of seismic reflection data, *in* P. A. Ziegler, ed., Geodynamics of rifting, volume II: Case history studies on rifts—North and South America and Africa: Tectonophysics, v. 213, p. 235–256.

Salama, R. B., 1997, Rift basins in the Sudan, *in* R. C. Selley, ed., African basins: Sedimentary basins of the world: Amsterdam, Netherlands, Elsevier, v. 3, p. 105–149.

Savage, R. J. G., and P. G., Williamson, 1978, The early history of the Turkana Depression, *in* W. W. Bishop, ed., Geological background to fossil man: Geological Society (London) Special Publication 6, p. 375–394.

Schull, T. J., 1988, Rift basins of interior Sudan: Petroleum exploration and discovery: AAPG Bulletin, v. 72, p. 1128–1142.

Selley, R. C., 1978, Porosity gradients in the North Sea oil-bearing sandstones: Journal of the Geological Society (London), v. 135, p. 119–132.

Senut, B., M. Pickford, D. Gommery, P. Mein, K. Cheboi, and Y. Coppens, 2001, First hominid from the Miocene (Lukeino Formation, Kenya): Comptes Rendus de l'Académie de Sciences, v. 332, p. 137–144.

Sertich, J. J. W., S. D. Sampson, M. A. Loewen, P. N. Gathogo, and F. H. Brown, 2005, Dinosaurs of Kenya's rift: Fossil preservation in the Turkana Grits of northern Kenya: Journal of Vertebrate Paleontology, v. 25, p. 114A.

Sertich, J. J. W., F. K. Manthi, S. D. Sampson, M. A. Loewen, and M. Getty, 2006, Rift valley dinosaurs: A new Late Cretaceous vertebrate fauna from Kenya: Journal of Vertebrate Paleontology, v. 26, p. 124A.

Simoneit, B. R. T., T. A. T. Aboul-Kassim, and J.-J. Tiercelin, 2000, Hydrothermal petroleum from lacustrine sedimentary organic matter in the East African rift: Applied Geochemistry, v. 15, p. 355–368.

Swain, C. J., M. A. Khan, T. J. Wilton, P. K. H. Maguire, and D. H. Griffiths, 1981, Seismic and gravity surveys in the Lake Baringo-Tugen Hills area, Kenya Rift Valley: Journal of the Geological Society (London), v. 138, p. 93–102.

Talbot, M. R., 1988, The origins of lacustrine oil source rocks: evidence from the lakes of tropical Africa, *in* A. J. Fleet, K. Kelts, and M. R. Talbot, eds., Lacustrine petroleum source rocks: Geological Society (London) Special Publication 40, p. 29–43.

Talbot, M. R., C. K. Morley, J.-J. Tiercelin, A. Le Hérissé, J.-L. Potdevin, and B. Le Gall, 2004, Hydrocarbon potential of the Meso-Cenozoic Turkana Depression, northern Kenya: II. Source rocks—Quality, maturation, depositional environments and structural control: Marine and Petroleum Geology, v. 21, p. 63–78.

Thuo, P., 2009, Stratigraphic, petrographic and diagenetic evaluation of Cretaceous/Paleogene reservoir sandstones of western Turkana, Kenya: Implications on the petroleum potential of northwestern Kenya: Thèse Université de Bretagne Occidentale, Brest, 139 p.: http://tel.archives-ouvertes.fr/tel-00534181/fr/ (accessed November 9, 2010).

Tiercelin, J.-J., and K. E. Lezzar, 2002, A 300-million-year history of rift lakes in Central and East Africa: An updated broad review, *in* E. Odada and D. Olago, eds., The East African Great Lakes: Limnology, paleolimnology and biodiversity—Advances in global change research: Amsterdam, Netherlands, Kluwer Publishers, v. 12, p. 3–60.

Tiercelin, J.-J., and A. Vincens, eds., 1987, Le demi-graben de Baringo-Bogoria, Rift Gregory, Kenya: 30.000 ans d'histoire hydrologique et sédimentaire: Bulletin des Centres de Recherches Exploration-Production Elf-Aquitaine, v. 11, p. 249–540.

Tiercelin, J.-J., M. Soreghan, A. S. Cohen, K.-E. Lezzar, and J.-L. Bouroullec, 1992, Sedimentation in large rift lakes: Examples from the middle Pleistocene–Modern deposits of the Tanganyika trough, East African rift system: Bulletin des Centres de Recherches Exploration-Production Elf-Aquitaine, v. 16, p. 83–111.

Tiercelin, J.-J., et al., 1993a, Hydrothermal vents in Lake Tanganyika, East African rift system: Geology, v. 21, p. 499–502.

Tiercelin, J.-J., J. Boulègue, and B. R. T. Simoneit, 1993b, Hydrocarbons, sulfides, and carbonate deposits related to sublacustrine hydrothermal seeps in the North Tanganyika Trough, East African Rift, *in* J. Parnell, H. Kucha, and P. Landais, eds., Bitumen in ore deposits: Berlin, Germany, Springer-Verlag, p. 96–113.

Tiercelin, J.-J., J. L. Potdevin, C. K. Morley, M. R. Talbot, H. Bellon, A. Rio, B. Le Gall, and W. Vétel, 2004, Hydrocarbon potential of the Meso–Cenozoic Turkana Depression, northern Kenya: I. Reservoirs—Depositional environments, diagenetic characteristics, and source rock-reservoir relationships: Marine and Petroleum Geology, v. 21, p. 41–62.

Tiercelin, J.-J., P. Thuo, T. Nalpas, and J. L. Potdevin, 2009, Hydrocarbon prospectivity in Mesozoic and early Cenozoic rift basins in central/northern Kenya: Search and Discovery article 10188: http://www.searchanddiscovery.net/documents/2009/10188tiercelin/images/tiercelin_ppt.pdf (accessed April 23, 2009).

Tongue, J., P. Maguire, and P. Burton, 1994, An earthquake study in the Lake Baringo basin of the central Kenya Rift: Tectonophysics, v. 236, p. 151–164.

Vincens, A., J.-J. Tiercelin, and G. Buchet, 2006, New Oligocene-early Miocene microflora from the southwestern Turkana Basin. Paleoenvironmental implications in the northern Kenya Rift: Palaeogeography, Palaeoclimatology, Palaeoecology, v. 239, p. 470–486.

Visser, C., 1993, Petrographic analysis of thirteen percussion sidewall samples from well Loperot-1, Kenya: Unpublished report prepared by Core Laboratories for Shell Internationale Petroleum Maatschappij, Besloten Vennootschap.

Walsh, J., and R. G. Dodson, 1969, Geology of northern Turkana: Report Geological Survey of Kenya, no. 82, 42 p.

Wescott, W. A., C. K. Morley, and F. M. Karanja, 1993, Geology of the Turkana Grits in the Lariu range and Mt. Porr areas, southern Lake Turkana, northwestern Kenya: Journal of African Earth Sciences, v. 16, p. 425–435.

Wescott, W. A., S. T. Wigger, D. M. Stone, and C. K. Morley, 1999, Geology and geophysics of the Lotikipi Plain, *in* C. K. Morley, ed., Geoscience of rift systems: Evolution of East Africa: AAPG Studies in Geology 44, p. 55–65.

Williams, L. A. J., and G. R. Chapman, 1986, Relationships between major structures, salic volcanism and sedimentation in the Kenya Rift from the equator northward to Lake Turkana, *in* L. E. Frostick, R. W. Renaut, I. Reid, and J.-J. Tiercelin, eds., Sedimentation in the African rifts: Geological Society (London) Special Publication 25, p. 59–74.

Williamson, P. G., and R. J. G. Savage, 1986, Early sedimentation in the Turkana Basin, northern Kenya, *in* L. E. Frostick, R. W. Renaut, I. Reid, and J.-J. Tiercelin, eds., Sedimentation in the African rifts: Geological Society (London) Special Publication 25, p. 267–283.

Winn, R. D., J. C. Steinmetz, and W. L. Kerekgyarto, 1993, Stratigraphy and rifting history of the Anza Rift: AAPG Bulletin, v. 77, p. 1989–2005.

Wolela, A., 2006, Fossil fuel energy resources of Ethiopia: Oil shale deposits: Journal of African Earth Sciences, v. 46, p. 263–280, doi:10.1016/j.jafrearsci.2006.06.005.

Zanettin, B., E. Justin Visentin, G. Bellieni, E. M. Piccirillo, and F. Rita, 1983, Le volcanisme du Bassin du Nord-Turkana (Kenya): Age, succession et évolution structurale, *in* M. Popoff and J.-J. Tiercelin, eds., Rifts et Fossés anciens: Bulletin des Centres de Recherches Exploration-Production Elf-Aquitaine, v. 7, p. 249–255.

9

Abeinomugisha, Dozith, and Robert Kasande, 2012, Tectonic control on hydrocarbon accumulation in the intracontinental Albertine Graben of the East African rift system, *in* D. Gao ed., Tectonics and sedimentation: Implications for petroleum systems: AAPG Memoir 100, p. 209–228.

Tectonic Control on Hydrocarbon Accumulation in the Intracontinental Albertine Graben of the East African Rift System

Dozith Abeinomugisha and Robert Kasande
Petroleum Exploration and Production Department, PO Box 9, Entebbe, Uganda (e-mails: d.abeinomugisha@petroleum.go.ug; r.kasande@petroleum.go.ug)

ABSTRACT

The Albertine Graben is a Tertiary intracontinental rift that developed on the Precambrian orogenic belt of the African craton. It forms the northern termination of the western arm of the East African rift system (EARS). The western branch of the EARS consists of three sectors, the Rukwa rift in the south, the Tanganyika rift in the central, and the Albertine rift in the north. The Albertine rift stretches from the border between Uganda and Sudan in the north and to Lake Edward in the south. It is commonly referred to as the Albertine Graben and is composed of the four lakes of Albert, George, Edward, and Kivu. These lakes overlie discrete depocenters offset by northwest–southeast or east–west-trending transfer zones.

Although fundamentally, the Albertine Graben has been considered an extensional province, undeniable evidence of transpression exists, suggesting a component of lateral shear stress at a late stage in basin evolution. The available geologic and geophysical data indicate that the graben has gone through extension and transpressional episodes, resulting in structures that are typical to both settings. The prevalence of deformation documented by flower structures in the shallow sedimentary sections in some basins in the graben indicates that the neotectonic processes are transpressional.

The Albertine Graben has undergone substantial tectonic movements that created accommodation space for thick sediments (~6 km [~3.7 mi]) that were deposited in lacustrine and fluviodeltaic environments. The sedimentary layers dip gently toward the depocenter on the western margin of the rift. Rapid tectonic subsidence coupled with limited sediment input led to deep stratified lakes with the accompanying deposition of source rocks. Significant oil discoveries during the past few years on the eastern margins of Lake Albert have fueled considerable interest in the hydrocarbon prospectivity within the Albertine Graben. The hydrocarbon exploration wells drilled

DOI:10.1306/13351554M1003539

in the Albertine Graben have proved deposition of source, reservoir, and cap rocks and development of structural and stratigraphic traps.

INTRODUCTION

Among the branches of the East African rift system (EARS), the northern section of the western arm (Albertine Graben) has been the least studied. Limited amounts of literature using mainly gravity and surface geology data document the tectonic setting of the area. However, the discovery of commercial quantities of hydrocarbons in the Albertine Graben has generated interest in the area. Four oil exploration companies are active in five out of the nine exploration areas in the graben and have acquired substantial amounts of geophysical and geologic data to guide in the identification of exploration targets. Researchers have also been attracted into the area, and several research projects are ongoing.

From surface observation, the major structure consists mainly of steep escarpments on both sides of the rift floor, showing that the major tectonic elements are extensional (Ollier, 1990). Deformed young volcanic rocks exposed along the rift flanks indicate that the neotectonic stress regimes are compressional. An imposing feature that stands out prominently right in the center of the rift, the Rwenzori Mountains, reaches an altitude of 4500 m (14,764 ft) above the surrounding Semliki flats, an altitude not seen in any other extensional setting. The cause of this uplift is yet to be established.

Subsurface geophysical data interpretation indicates that the structure of the Albertine Graben is more complex than originally thought. Structures typical of extension have been interpreted. Positive flower structures that are typical of strike slip regimes have also been interpreted, indicating transpression in the area. Hydrocarbon exploration in the graben has targeted these and other structural traps. Although stratigraphic traps have also been interpreted, only structural traps have been evaluated by exploration wells. All of the wells that have been drilled in the Albertine Graben have been on either positive flower (Palm tree) structures or on rotated fault blocks. Fault closures against basin bounding faults or even intrabasinal faults have trapped significant amounts of hydrocarbons. Not only was tectonics a crucial factor in deposition of source, reservoir, and cap rocks and formation of structural traps, but also provided migration pathways for the hydrocarbons. Complex fault patterns as interpreted from geophysical data have provided conduits for hydrocarbon migration, sometimes long-distance migration.

The objective of this study is to analyze the key structural geometries that have been observed in the Albertine Graben and deduce the tectonic regimes that could have created these structures. With the discovery of hydrocarbons in a young sedimentary basin, probably the youngest in the world, it becomes important to understand the development of a petroleum system in such a tectonic setting. At the moment, no geologic model has been established to explain the greater prospectivity of the graben. Despite the now known excellent prospectivity of the graben, several issues remain poorly resolved like the tectonic evolution, the source rock depositional environment, the migration pathways, and the trapping mechanism within the Albertine Graben. This chapter will therefore use the data that have been acquired and interpreted for oil exploration to analyze the tectonic regimes at play in the graben and the control they have had on the development of a petroleum system in the Albertine Graben.

GEOLOGIC SETTING

The EARS is an Oligocene to Holocene series of rift basins extending more than 4000 km (2486 mi) from the Red Sea to Mozambique. South of Sudan, the EARS is composed of two branches, east and west, which appear to have developed independently (Chrowicz, 2005). Initial rifting began in the Gulf of Aden approximately 30 Ma (Kurz et al., 2006), propagating southward, along the preexisting basement lineaments. The Albertine Graben is part of the EARS that is a classic example of the processes of continental breakup. It is characterized by initial phases (incipient) of rifting in the southwest Botswana Rift and by the beginning of sea-floor spreading in the Afar depression (Ebinger, 1989). The Albertine Graben is at the intermediary stage of rifting and forms the northernmost part of the western arm of the EARS (Figure 1). It stretches from the border with Sudan in the north to Lake Edward in the south. It trends in a northeast to southwest direction in the central part, turning to an almost north to south trend in its northern and southern parts. It is about 45 km (28 mi) wide and 500 km (311 mi) long at the Uganda Democratic Republic of Congo (former Zaire) border, between lat. 4°N and 1°S and long. 29° and 32°E. The Albertine Graben is a Cenozoic rift basin, superimposed on Precambrian continental crust. The available geologic data indicate that rifting was initiated during the late Oligocene or early Miocene. Precambrian and Mesozoic lineaments strongly influenced the evolution of the graben.

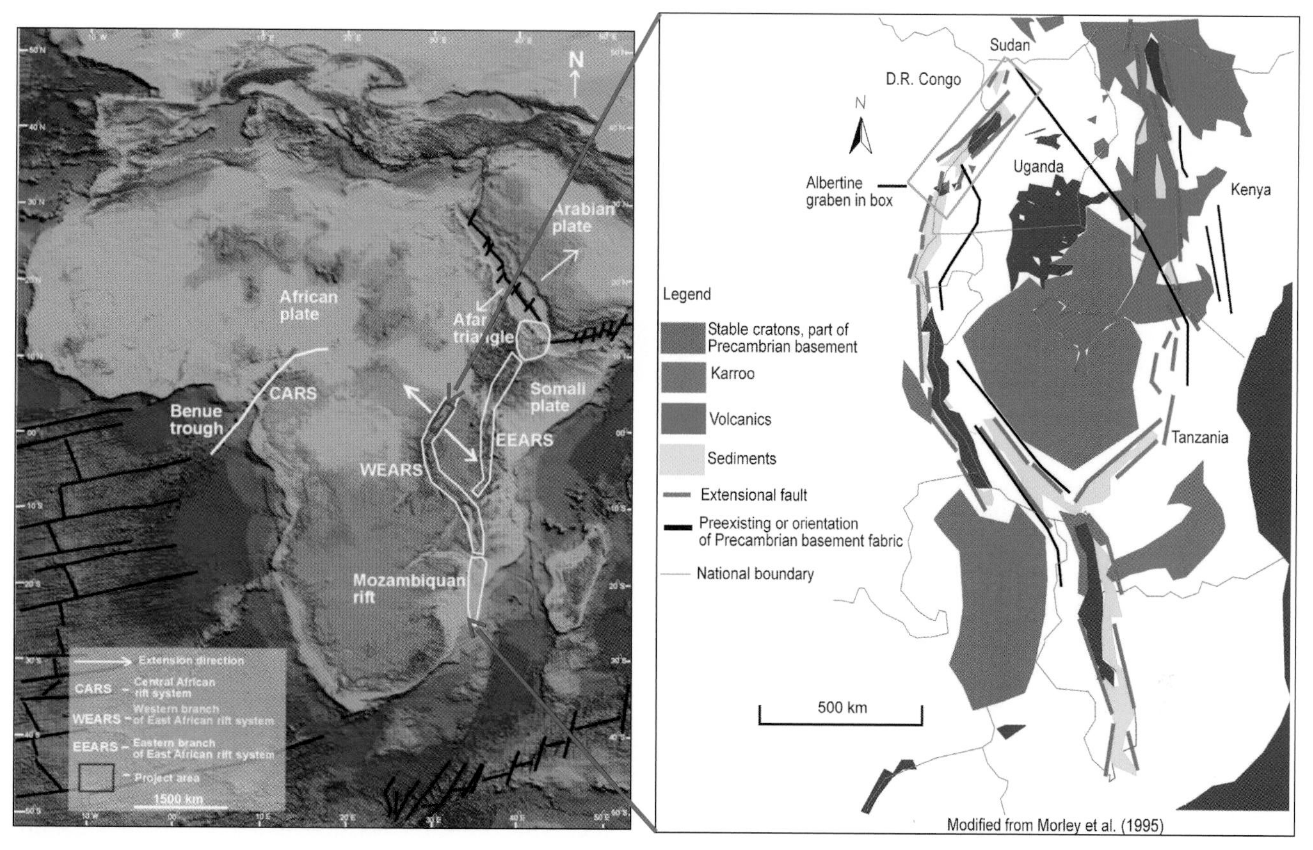

Figure 1. The tectonic setting of the East African rift system showing the location of the Albertine Graben. EEARS = eastern arm of the East African rift system; CARS = Central African rift system; WEARS = western arm of the Weat African rift system. 1500 km (932 mi); 500 km (310 mi).

Structural History

The structure of the Albertine Graben consists mainly of northeast–southwest and north–south-trending extensional faults that formed because of multiple rifting events during the Tertiary. Geophysical and geologic data acquired in the graben suggest at least two episodes of rifting in the Miocene/Pliocene and some episodes of compression, probably during Pliocene/Pleistocene. Each of the rift basins in the graben is bounded by steep border faults and broad uplifted flanks that are predominantly Precambrian basement composed of metamorphic rocks such as gneisses, quartzite, schist, and varying amounts of mafic intrusions. Inversion in the EARS is documented to have occurred at least two times during the Tertiary (Morley et al., 1999a).

The Lake Albert area does not obey the traditional models of rifting, with one faulted basin-controlling margin and a flexural ramp side like the Tanganyika and the Malawi rifts (Kiram et al., 2002; Scholz et al., 2004). The east–west geologic section across Lake Albert indicates faulting on both margins that create an almost symmetrical structure (Figure 2). The north–south geologic section along strike on Lake Albert indicates two subbasins separated by a basement high, the Kaiso-Tonya accommodation zone that is a transpressional structure (Figure 3). The two basins probably evolved separately but have now been linked and are operating as a single depositional model.

The African plate is surrounded on three sides by constructive plate margins: to the west, the mid-Atlantic Ridge; to the south and southeast, the southwest Indian Ridge; and to the east, the central Indian Ridge and the Carlsberg Ridge. In addition, the Sheba Ridge in the Gulf of Aden and the Red Sea floor spreading center are located to the northeast of the African plate (Figure 1). The northern margin of the African plate is the boundary with the European plate formed by the Azores Hinge zone, which progresses eastward into the better defined subduction zones of the Calabrian, Hellenic, and Cypriot arcs (Rose and Curd, 2005).

According to Richardson (1992), ridge push forces are the main driving mechanism for plate movements. He goes on to show that active spreading centers such as those surrounding the African plate generate compressional forces that correlate positively with the direction

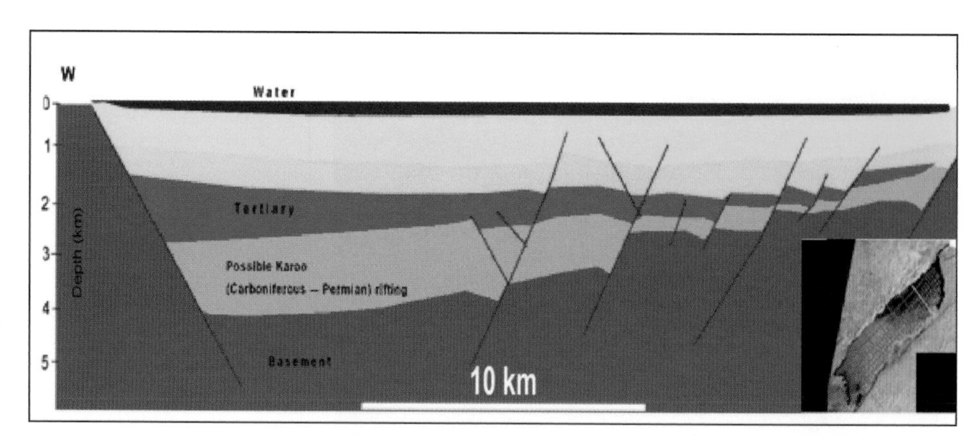

Figure 2. East–west geologic section across Lake Albert in the Albertine Graben, showing a gentle tilt of the basin toward the west. 10 km (6.2 mi).

of the maximum horizontal stress. The African plate should therefore be in a state of compression. The existence of the EARS could be interpreted as a result of increased influence of a dome beneath this area. Much of the rifted area of eastern Africa is coincident with the general area covered by the East African dome. Hence, to have extension in East Africa, Zoback (1992) argued that the buoyancy forces of the upper mantle beneath the rift system must be more dominant than the ridge push compression, and this produces the present-day northwest–southeast direction of extension. The EARS is connected to the worldwide system of oceanic rifts via the Afar Triangle to the Red Sea (Figure 1).

The EARS has evolved through a complex history of rifting characterized by a series of en echelon linked border faults (Figure 1). Individual grabens and half grabens within the rift zone are separated by accommodation zones along which there may be a significant strike-slip component of movement (Morley, 1999). The length of individual faults ranges from 70 to 160 km (44–99 mi) (Morley, 1999). Both soft and hard linkages occur along transfer zones. Studies of the structural interaction between rift basins and ages of rift basin fill indicate that the rifting within the EARS has not been a continuous process but rather has occurred as a series of discrete short-lived pulses (Lambiase and Bosworth, 1995).

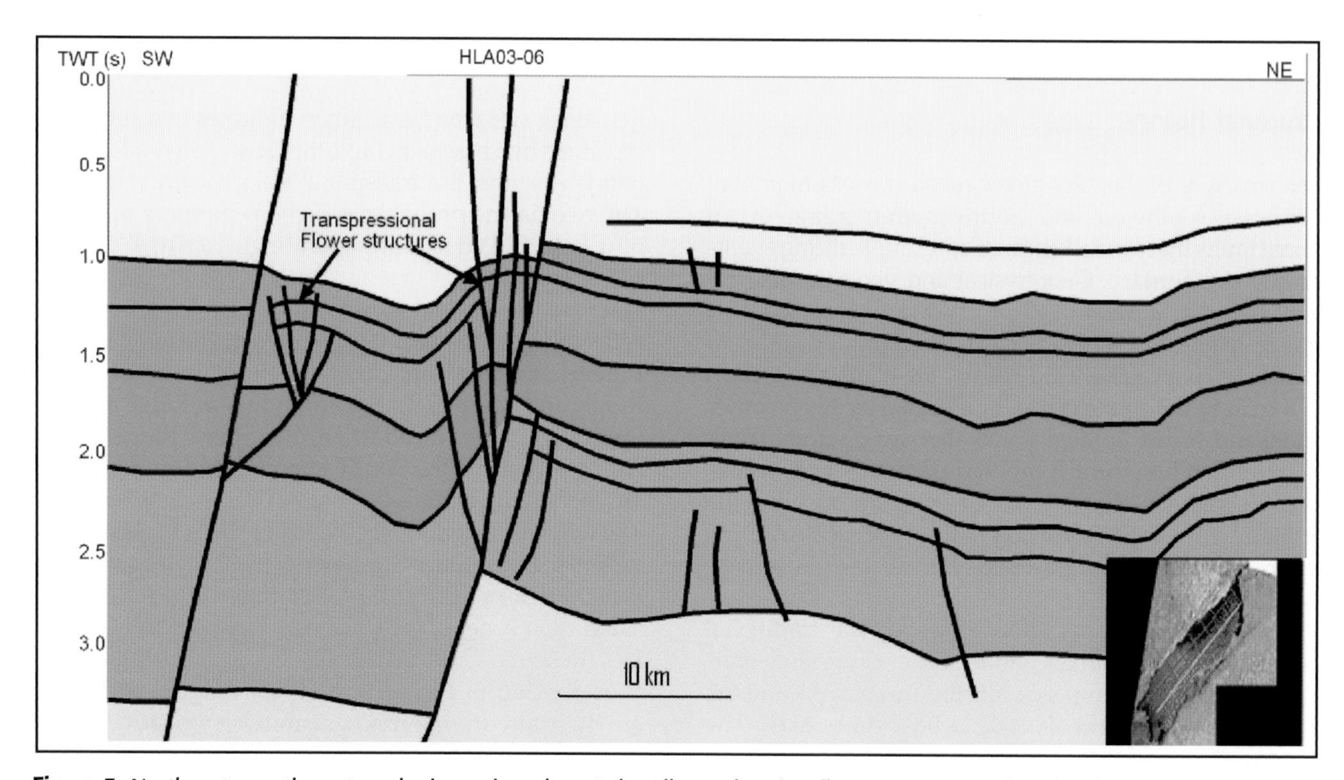

Figure 3. Northeast–southwest geologic section along Lake Albert, showing flower structures that developed in the hanging wall of a major fault in the Kaiso-Tonya accommodation zone. TWT = two-way traveltime. 10 km (6.2 mi).

Two models have so far been advanced for the evolution of the western arm of the EARS. The proponents of the first model argue that the EARS is a classic example of extensional tectonics with associated horsts and grabens bounded by normal faults. Current proponents of this view (Ebinger et al., 1989; Upcott et al., 1996) appeal to plastic deformation of the lithosphere caused by mantle upwelling or hotspot activity. Deformation of the brittle crust brings about brittle extension, which produces the horst and graben structures that are visible at the surface. The larger structures such as the Rwenzori Mountains are accounted for by extensive footwall uplift and sometimes underplating or even isostatic compensation.

The proponents of the second model argue that the western arm of the EARS has evolved through a history of extension and compression, resulting into strike slip and inversion. This second model was proposed by Morley (1999) and maintained by Delvaux (2005). In the Albertine Graben, this model has been advanced by Rose and Curd (2005) and Smith (2005). They argue that the east–west ridge push induced by compression of the African plate may have started during the early Eocene (55 Ma). It is therefore likely that the compression is involved in not only the inversion of the Karoo basins during pre-Miocene but also in the transpression that produces strike-slip structures (Figures 3, 4) during the Miocene and later time. Their argument is derived from seismic data interpretation, acquired for hydrocarbon exploration in the Albertine Graben.

The Albertine Graben is still an active rift system. The most important indicator of neotectonics is the record of the earthquake activity in the Albertine Graben and the deformed young volcanic tuffs of the Katwe Kikorongo volcanic field. Figure 5 shows the locations of earthquakes recorded by the U.S. Geological Survey at their National Earthquake Information Centre. The data cover the years 1973–2002. The eastern arm of the rift is comparatively aseismic, showing only one or two patches of seismicity. The region within the Albertine Graben is moderately seismically active (Girdler and McConnell, 1994). In addition, the eastern arm of the rift is more volcanically active than the western arm. Focal mechanism solutions indicate normal slip, with a few especially around the Rwenzori Mountain Ranges, indicating strike slip (Tugume and Nyblade, 2009) on north–south and northwest–southeast-trending fault planes. Microearthquake epicenters are reported to correlate with the location of steep border faults (Maasha, 1975).

Volcanism

Two Quaternary to Holocene volcanic fields exist in the southern part of the Albertine Graben: the Katwe-

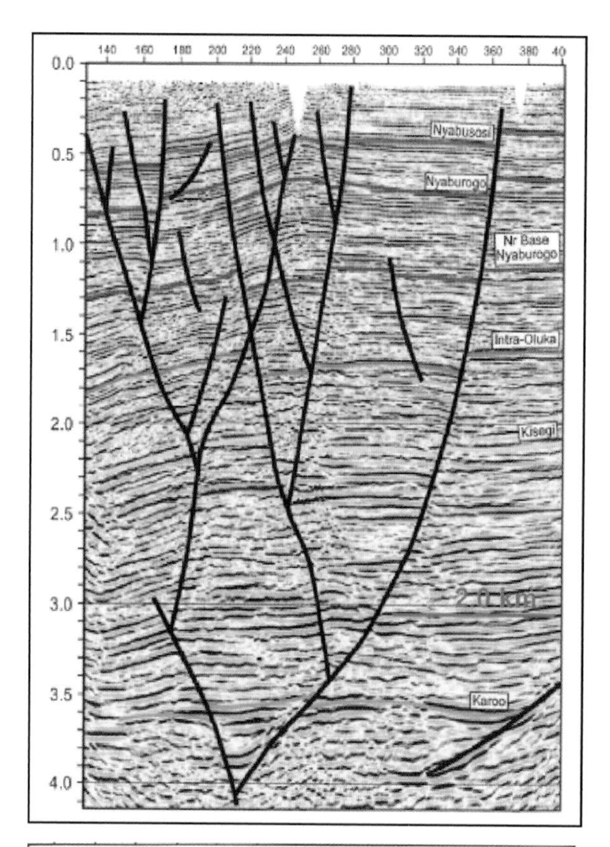

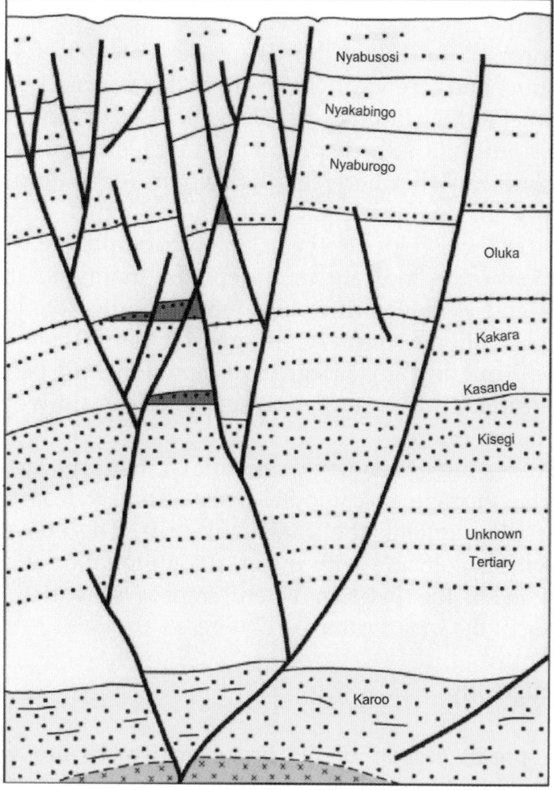

Figure 4. Line HOG 01 06 showing a positive flower structure interpreted from seismic data.

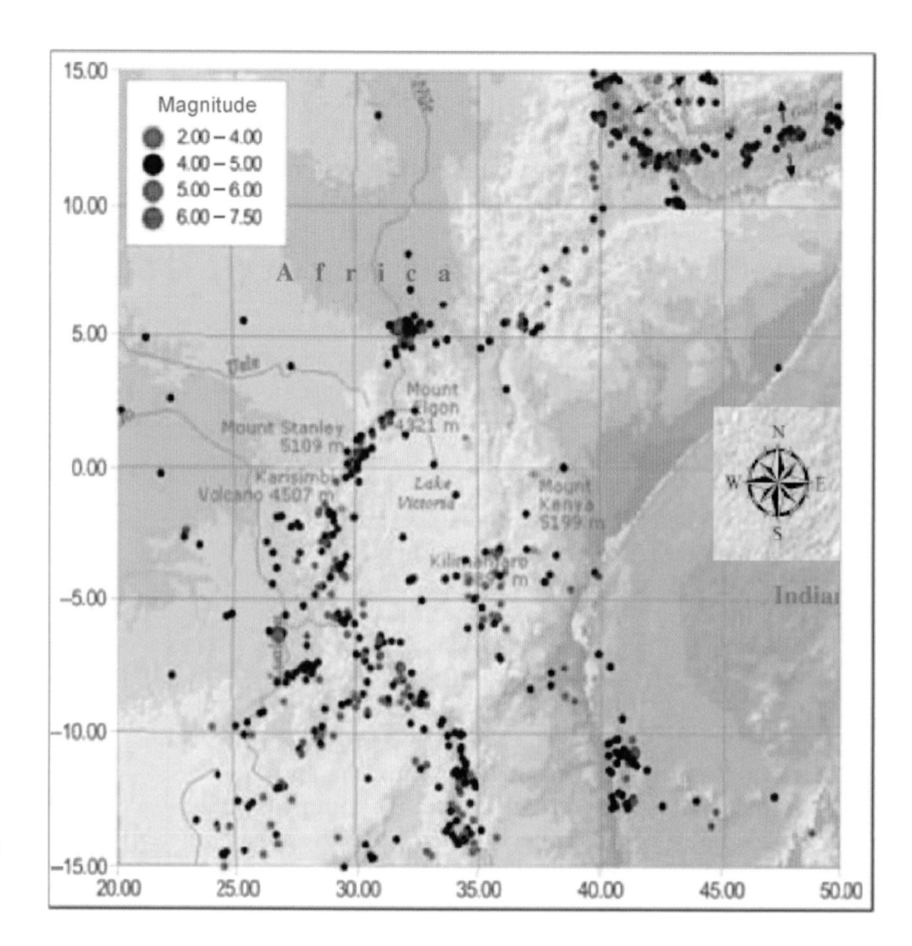

Figure 5. Earthquake distribution in the East African rift system showing that the western arm of the rift is more active compared to the eastern arm.

Kikorongo and the Bunyaruguru volcanic fields (Bishop and Trendall, 1966). Both fields trend near parallel to the rift margins. These fields erupted compositionally distinct products, suggesting that a discontinuity lies between the two at depth along a strong anomalous zone at the lower crust–upper mantle level (Upcott et al., 1996). Incompatible trace element ratios indicate an undepleted mantle source. This and evidence from ultramafic mantle xenoliths suggest magma source depths greater than 50 to 60 km (31–37 mi) and small degrees of partial melting before ascension. Thus, extensive extrusive volcanism would not be expected.

At its southern end, the Albertine Graben is bordered by the Virunga volcanic field, separating it from the next rift segment, the Tanganyika rift. The volcanic provinces coincide with major accommodation zones that mark the location where transverse structures crosscut the major rift axis (Ebinger et al., 1989).

Stratigraphy

The prerift section of the Proterozoic rocks are well exposed on the rift flanks and shoulders of the Albertine Graben. Like the rest of the country, it is composed predominantly of high-grade metamorphosed and igneous rocks of the Precambrian (Bishop, 1969). Although the thickest sedimentary section in the graben is Tertiary to Holocene in age, reflection seismic data interpretation indicates the possible presence of Mesozoic sediments (Jurassic or Cretaceous) at the base of the section. Lower Tertiary sediments may overlie the crystalline basement in many parts of the graben, but a strong possibility that it overlies a Mesozoic section (Karoo) in the western parts of the graben exists, as seismic data indicate.

In the eastern arm of the rift system, Morley et al. (1999b) describe an Eocene–Oligocene shale section with similar characteristics to that postulated to lie beneath the Kisegi Formation in the Albertine Graben. Pre-Tertiary basins have been interpreted from seismic data in the southeastern part of Lake Albert. These basins have been inverted by the early Tertiary compressional episode. Analysis on some of the oil samples indicates two types of sourcing that include marine source. This, therefore, suggests the possibility of Karoo sedimentary units beneath the Albertine Graben.

The synrift section of the Albertine Graben was recorded in the Waki-B1 well, which was drilled in 1938 close to the basin margin fault in the Butiaba region at the northeastern shores of Lake Albert. A section of this well is shown in Figure 6. The well penetrated both the

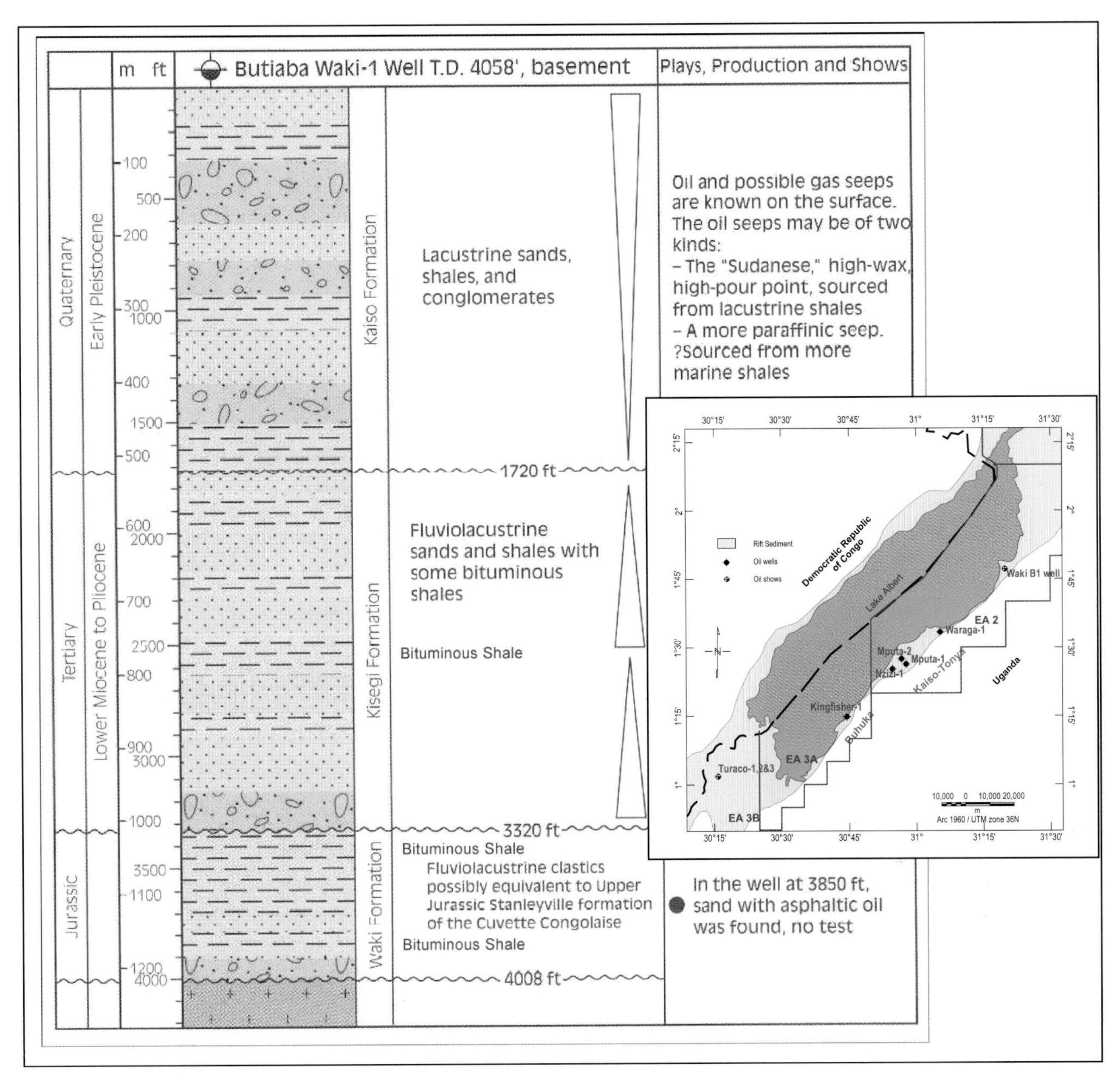

Figure 6. A stratigraphic section of the Waki-1B well drilled 1.3 km (0.8 mi) west of the eastern basin-bounding fault around the Butiaba area.

lower and upper Tertiary synrift section and went through the Mesozoic prerift section before reaching the basement.

The available geologic and geophysical data suggest that the Albertine Graben has undergone substantial tectonic movements and thick sediments (~6 km [~3.7 mi]) have been deposited in fluviodeltaic and lacustrine environments. In the Semliki area, the first phase of rifting commenced probably early Miocene, creating grabens and half grabens in which the first sediments were deposited. From the geologic and geophysical data, the Kisegi, Kasande, Kakara, and Oluka formations (Figures 4, 7) were deposited during the first phase of rifting. The second phase of rifting occurred during the Pliocene to the Pleistocene, when deposition of the Nyaburogo and the Nyakabingo formations occurred. These formations have been roughly described by the Petroleum Exploration and Production Department (PEPD) and the oil companies as they have been encountered in the Turaco wells (Figure 7) drilled in the Semliki Basin.

The oldest section at outcrop is the Kisegi Formation, a predominantly sandy unit, which is made up primarily of stacked channel-fill sands. The Kisegi Formation has

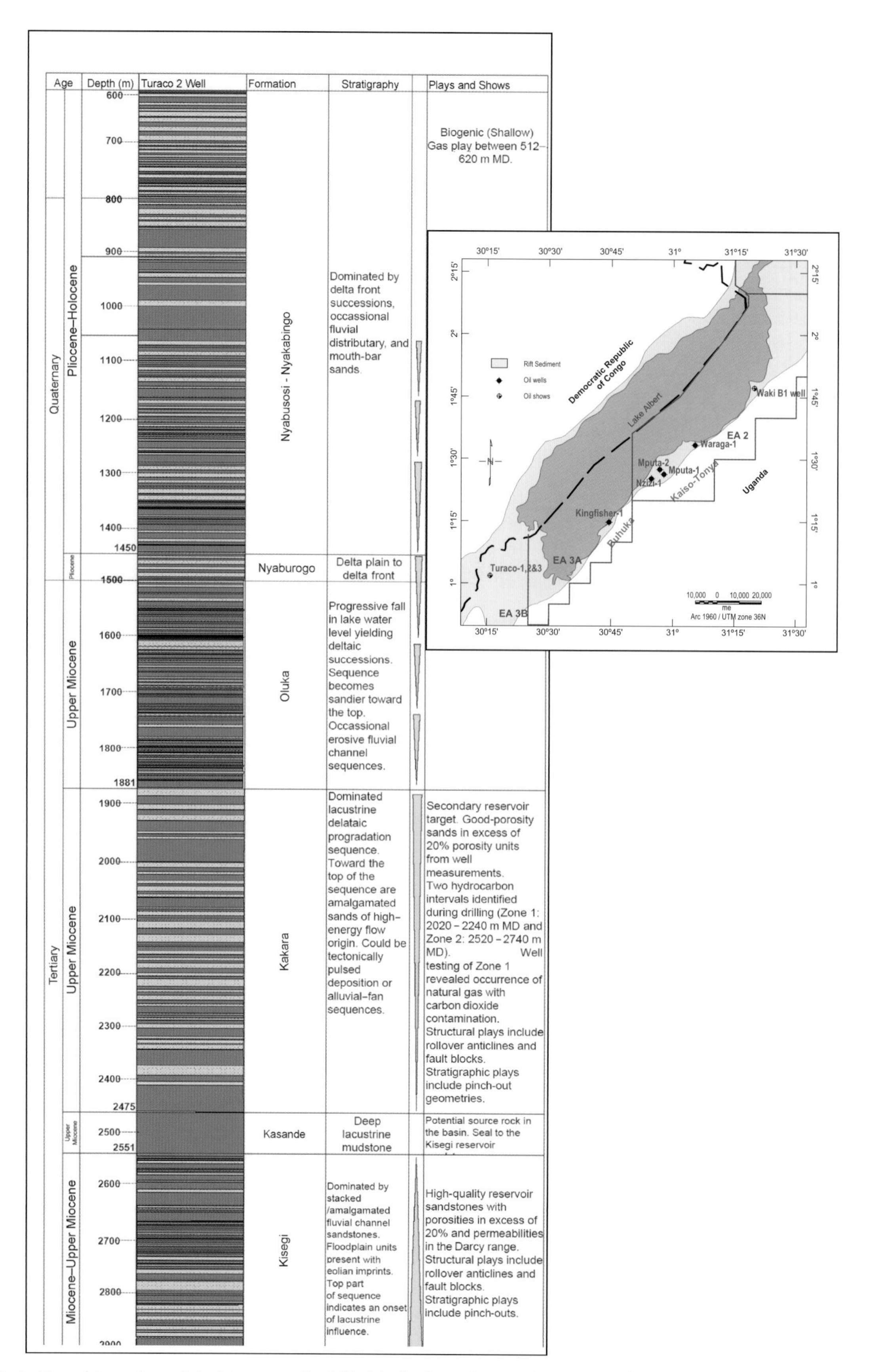

Figure 7. A stratigraphic section of the Turaco wells drilled for hydrocarbon exploration in the Semliki area. MD = measured depth.

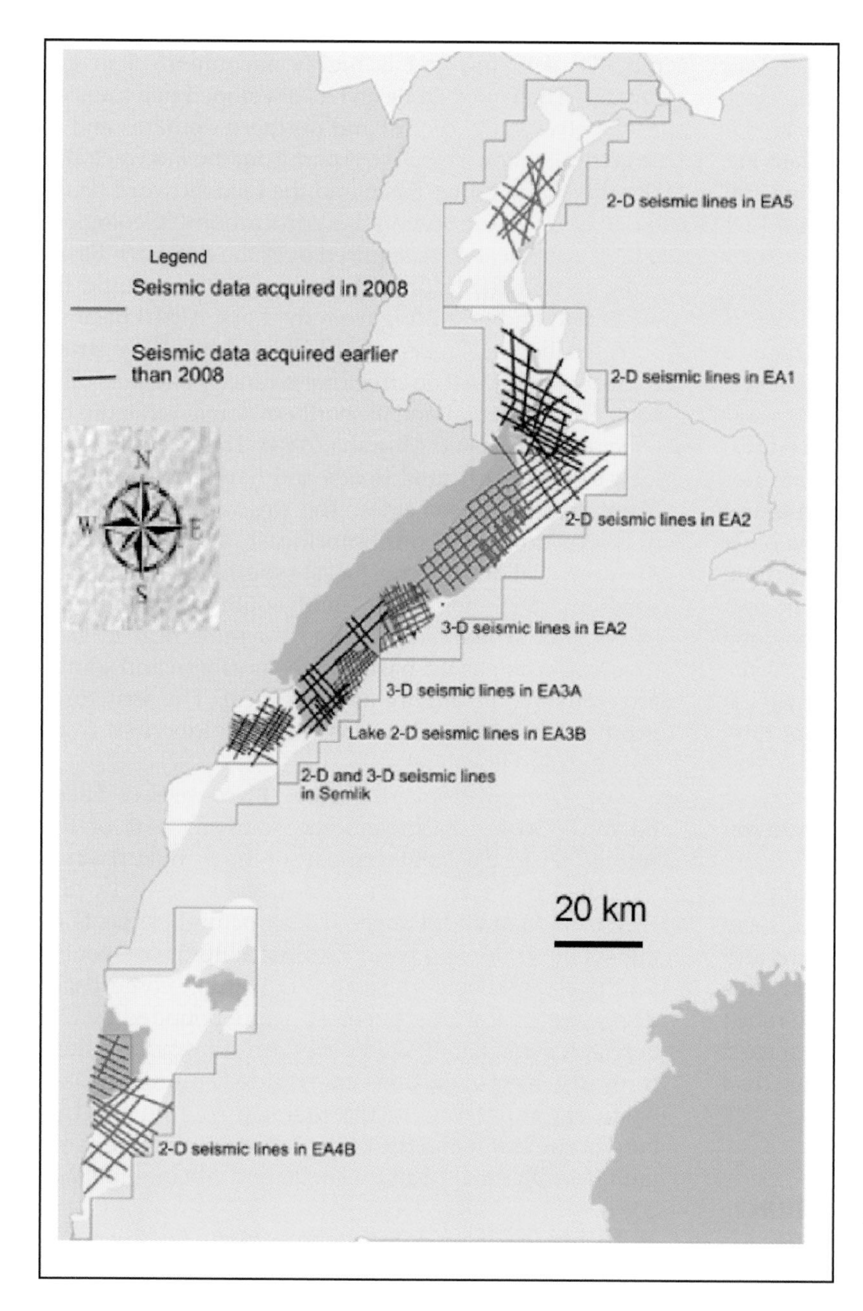

Figure 8. A map of Uganda showing seismic data coverage in the Albertine Graben. 20 km (12.4 mi).

an onlapping relationship with the basement schists and quartzites of the Rwenzori Mountains; therefore, the thickness described at outcrop may grossly underestimate the true total thickness of the formation. Turaco wells penetrated the upper parts of the formation, although its base was not reached. Seismic data may suggest an additional 1000 m (3281 ft) of sedimentary section between the total depth of the well and basement.

The Kasande Formation is a predominantly clay-prone unit that unconformably overlies the Kisegi Formation. Exposed sediments comprise brown to yellowish-brown, highly weathered, and fractured claystone/mudstone. The formation's mudstones and claystones represent deep-water lacustrine sequences.

Above the Kasande Formation lies a predominant clay-prone sequence with subordinate sands. These have been subdivided into the Kakara, Oluka, Nyaburogo, Nyakabingo, and Nyabusosi formations based on subtle changes in sand-to-shale ratio and by microfossil content.

In general, the depositional environments of the Semliki Basin are fluviolacustrine, showing significant variations in gamma-ray character that reflect water level changes and river-lake interactions throughout the depositional period and the influence of rifting tectonics on sediment deposition through time.

DATA SET

The Albertine Graben is the least studied area in the EARS. The available geologic and geophysical data are therefore limited. This chapter has been based on the two-dimensional (2-D) and three-dimensional (3-D) reflection seismic data in the Albertine Graben. The data sets were acquired and provided by the oil companies exploring for hydrocarbons in the graben. Geological and geophysical data acquired by PEPD in Uganda Government's Ministry of Energy and Mineral Development have also been used in this study.

Seismic data coverage in the Albertine Graben, although line spacing is sparse and some areas are not covered (Figure 8), can give a preliminary structural understanding of the basin. However, these data have been collected by different companies over separate exploration areas. Correlation of the different data sets is difficult. Interpretation of this seismic data was done based on standard seismic sequence and stratigraphic methods/procedures using a landmark suite at the PEPD. The interpretation led to the generation of time structure and surface attribute maps and isochron (isopach) or seismic depositional sequence maps to evaluate the structure and its controls on the evolution of the graben synrift sequences.

At the moment, a total of 34 wells have been drilled in the Albert Graben (Figure 9). These wells however, have been drilled on the eastern margin of the basin, and none of them has encountered the source rock. This indicates that the deeper section of the Albert Graben has not been penetrated. Well data have also been used in this study mainly for correlation during seismic data interpretation.

STRUCTURAL OBSERVATION AND INTERPRETATION

The Lake Albert basins of the Albertine Graben does not obey traditional models of continental rift basins, with one major basin-bounding fault and a ramping continental margin like the Tanganyika and Malawi rifts (Specht and Rosendahl, 1989; Scholz, 1995). Geological and geophysical data indicate that this basin, which is in the central domain, is an almost symmetrical structure with almost equal displacement on the two basin-bounding faults. This full graben geometry is characterized by steep border faults with highly uplifted rift flanks that has diverted drainage on either sides, limiting clastic sediment input into the basin and enhancing organic matter deposition in the Lake Albert basins.

The Rhino Camp Basin, together with the Pakwach Basin and the Lake Edward Basin in the northern and southern domains, respectively, conforms to traditional models of rifting and is highly asymmetrical in geometry. The Pakwach Basin has developed in a transfer zone between the central and northern domains and is structurally more complex. Apart from the Pakwach Basin, the Rhino Camp Basin and the Lake Edward Basin have not yet encountered hydrocarbons. Geological and geophysical data acquired over the Pakwach Basin indicate that the hydrocarbons encountered could be generated and migrating from the Lake Albert basins.

The Albertine Graben can be divided into three structural domains based on structural geometry and trend: the southern, the central, and the northern domains (Figure 10) (Abeinomugisha and Mugisha, 2004). These domains exhibit different structural trends and have developed different structural geometries. The structural trend in the northern domain is north-northeast–south-southwest in the central domain, northeast–southwest, and in the southern, north-northwest–south-southeast. The basins in the central domain are almost symmetrical in geometrical, whereas the basins in the northern and southern domains are highly asymmetrical. The structural geometries in these domains could be inherited from Precambrian lineaments.

The southern domain (Figure 10) consists of Lakes Edward–George Basin and the southern part of the Semliki Basin. The southern part of the Semliki Basin is separated from Lakes Edward–George Basin by the Rwenzori Mountain Ranges. The structural elements in the southern domain trend in a north-northeast–south-southwest direction. The Lakes Edward–George Basin displays a typical half-graben structure bounded by Lubero extensional fault on the western side and a faulted ramping margin on the eastern side. This suggests a basin asymmetry with the major throw on the Lubero border fault and the basin depocenter on the same fault (McGlue et al., 2006; Nicholas and Abeinomugisha, 2009).

The basin is terminated to the north by the Rwenzori Mountain Ranges. The rift here branches into two segments, which lie on either side of the Rwenzori: the Lake George rift and the Semliki rift. The Rwenzori Mountain Ranges reach an altitude of 5109 m (16,762 ft) at its highest peak. This is substantially higher than the normal rift-flanking mountains such as the Blue Mountains that reach heights of only 2420 m (7940 ft) (Rose and Curd, 2005). The flanks of Rwenzori Mountains are very steep (>80°) and strongly suggest very steep fault control. Compression structures have been observed on the southeastern and northwestern sides of the mountain ranges, implying that the Rwenzori Range could be rotating counterclockwise.

The southern domain is a geologically challenging area referred to as a frontier basin in terms of petroleum exploration. Unlike the central domain to the north, no

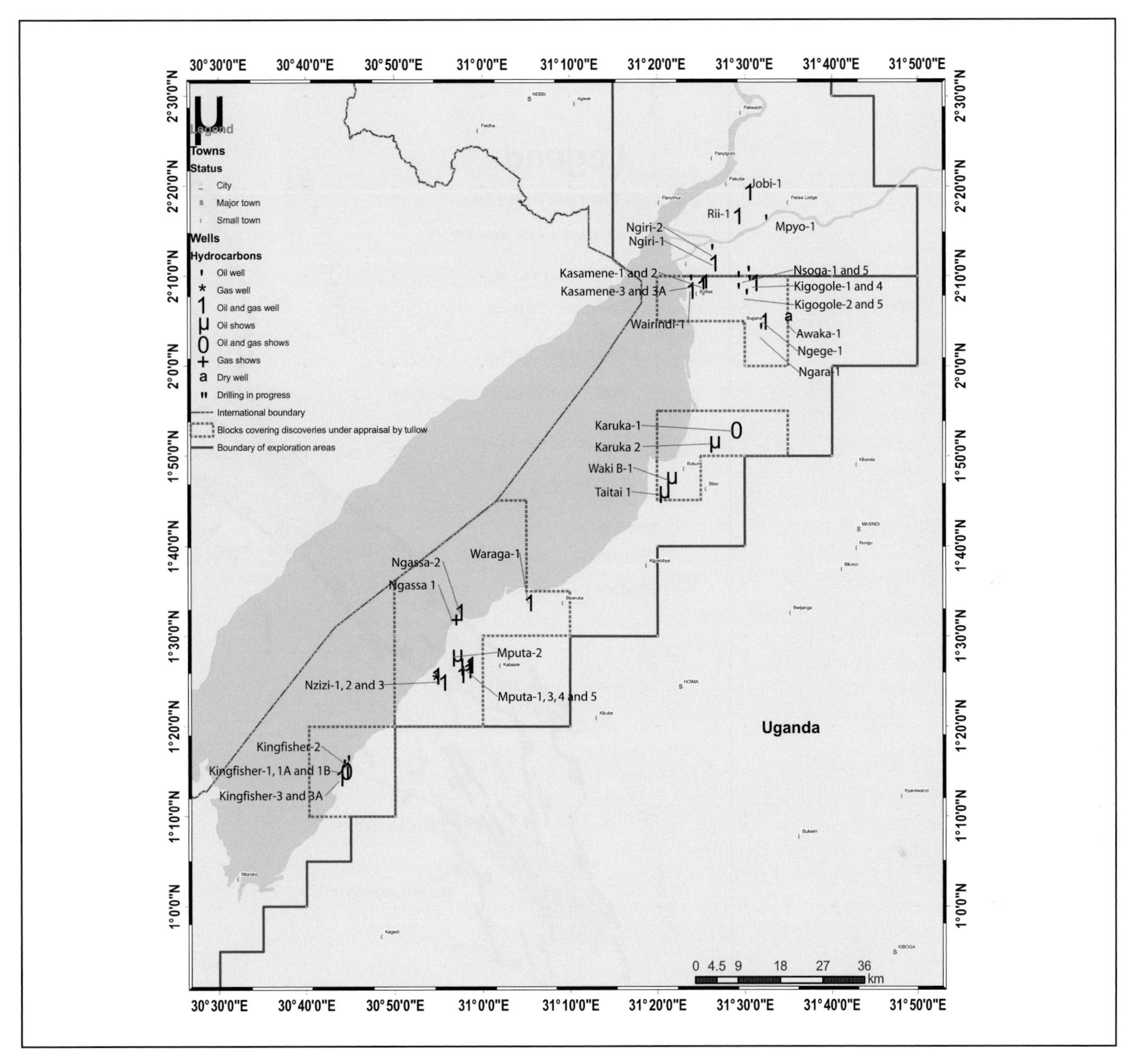

Figure 9. A map of the Albertine Graben showing the locations of some of the wells. 9 km (5.6 mi).

wells have been drilled in this area, and this chapter has been based on sparse grid reflection seismic data over the area and surface geologic data. From interpretation of these data, half-graben structures, characterized by rotated fault blocks (Figure 11), are the major structural elements. Apart from the basin-bounding fault, the faults are segmented, implying slightly oblique extension.

The central domain (Figure 10) consists of two Lake Albert basins: the northern Lake Albert Basin and the southern Lake Albert Basin. These basins are separated by accommodation zones. The central domain is within a low-lying sediment-filled basin floor and flanked by a high topography, including the Rwenzori Mountain to the south. The raised topography controlled the deposition of sediments by allowing fluviodeltaic systems to prograde into the area, bringing weathering products of the metamorphic basement that formed the rift scarp and most of the basement complex of Uganda (Karner et al., 2000; Laerdal et al., 2002). The Rwenzori Range has been both a sediment barrier and also sediment source to this area. The structural trend in the central domain is mainly northeast–southwest. The faults are long and linear, implying that the extension has been normal, as opposed to oblique. Parallel anticline syncline pairs, indicating a series of compressive regimes, have

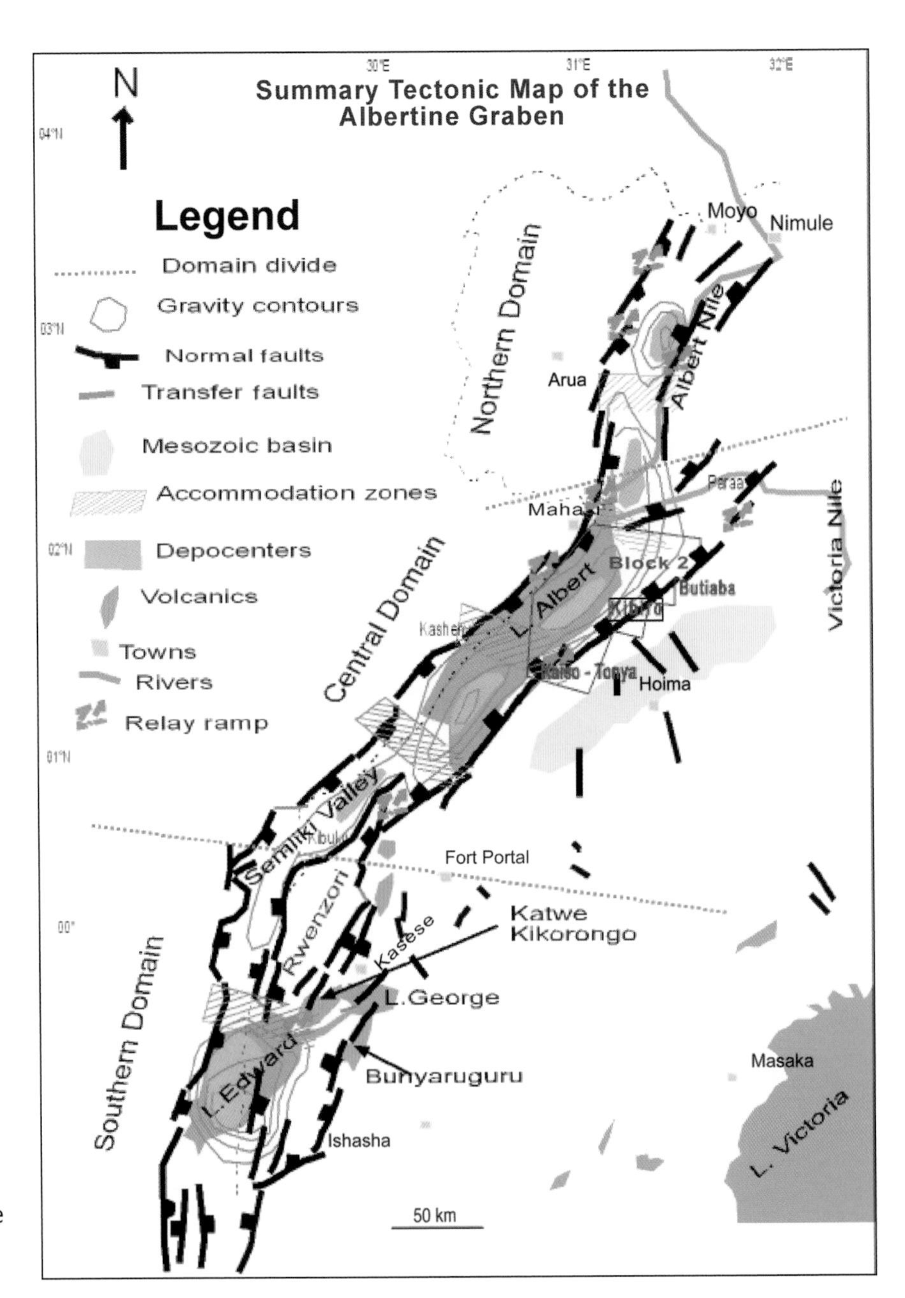

Figure 10. Tectonic map of the Albertine Graben showing the three interpreted structural domains. 50 km (31 mi).

been interpreted along the major boundary faults, plunging to the northwest and verging to the southeast, implying that the tectonic transport is from the northwest to the southeast.

The Lake Albert basins are bounded on the western flank by the Bunia fault and on the eastern side by two faults: the Tooro-Bunyoro border fault and the Butiaba platform fault. The Butiaba platform fault runs parallel to the border fault and is synthetic to it. The trend of these faults is northeast to southwest. The Lake Albert basins do not conform to traditional models of rifted continental basins, defined by discrete basin-bounding faults opposed by a low-gradient flexural margin. Almost similar amounts of displacement on the Butiaba fault and the Bunia border fault produce close to a full-graben structure that gently dips toward the west (Figure 2), in contrast to the typical half-graben structures of the Tanganyika and Malawi rifts.

In the northern domain (Figure 10) toward the northern end of Lake Albert, the Bouguer anomalies increase to the background value, the northeast to southwest rift segment narrows and terminates against north-northeast–south-southwest-trending normal faults. The north-northeast–south-southwest-trending structural patterns then characterize the northern domain of the Albertine Graben. Here, the basin is narrow and asymmetrical. The northern domain is formed by two half grabens: Pakwach Basin in the

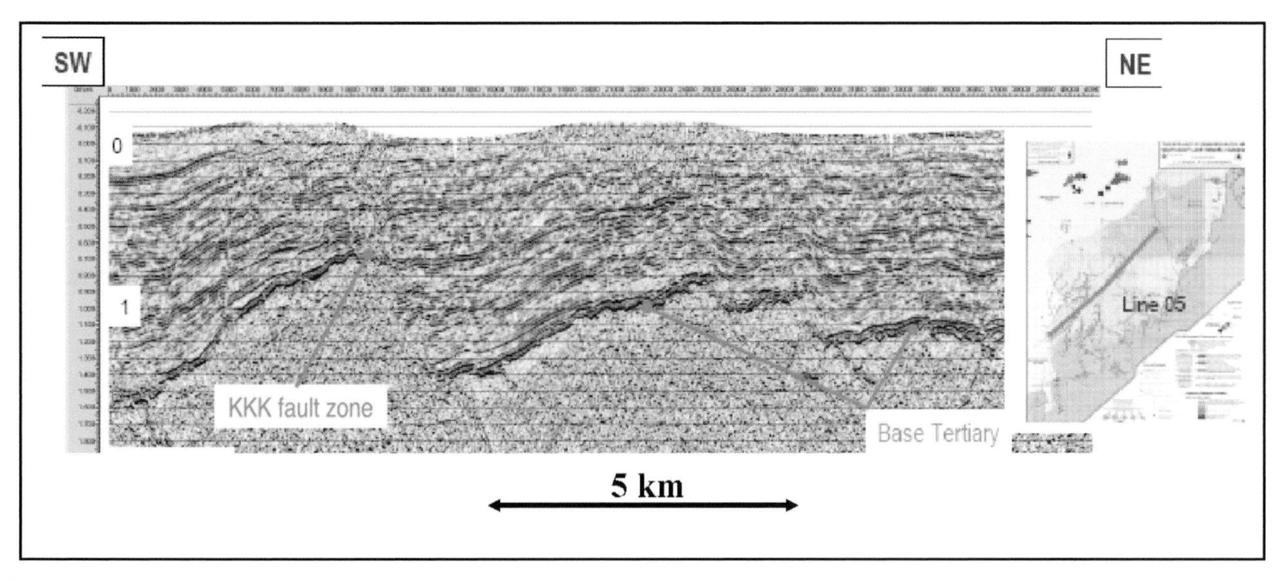

Figure 11. An east–west seismic line in the southern domain, showing half-graben structures that conform to traditional models of rifting in the Lake Edward Basin. 5 km (3.1 mi).

south and Rhino Camp Basin in the north. The two basins are separated by an accommodation zone, the Wadelai accommodation zone. The Pakwach Basin is controlled by Panyimur fault on its western side, and the Rhino Camp Basin is controlled by Rhino Camp Fault on its eastern side.

Like the southern domain, the northern domain is an equally geologically challenging area. Only one well has been drilled in this area, and this chapter has been based on that well and sparse grid reflection seismic data over the area. From interpretation of the data, half-graben structures, characterized by rotated fault blocks (Figure 12), are the major structural elements.

The interpretation of the data acquired shows that the three domains of the Albertine Graben are essentially different. The southern and northern domains display purely extensional tectonics with one major basin-bounding fault and a ramping opposing flexural margin (Figures 11, 12). Fault systems are highly segmented along strike. Internally, the major extensional structures are rotated fault blocks and horst and graben structures (Figure 13). The central domain is in between the southern and northern domains and is separated by either highly faulted structural highs or accommodation zones (Figure 10). Studies of the structural interaction between rift basins and ages of

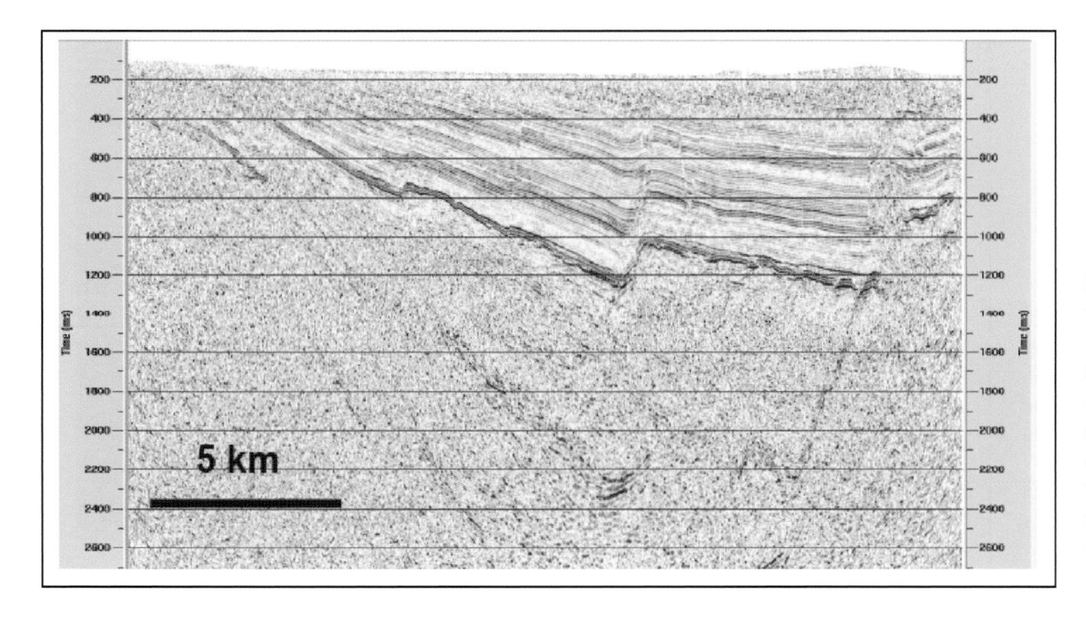

Figure 12. An east–west poststack time migrated (PSTM) stack seismic line in the northern domain, showing half-graben structures that conform to traditional models of rifting in the Rhino Camp Basin. 5 km (3.1 mi).

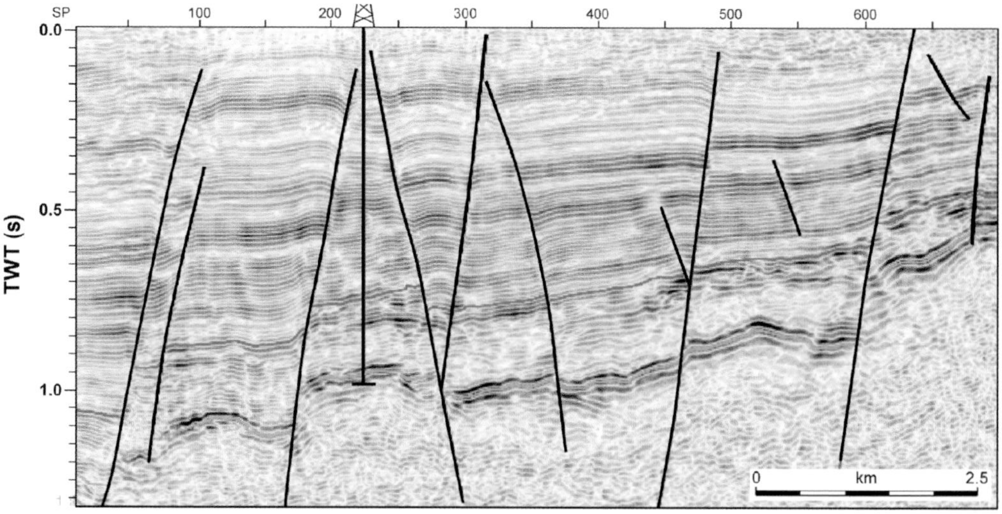

Figure 13. Interpretation of three dip lines from the northern end of the central domain showing rotated and horst blocks. Some of these structures have been drilled and found to contain oil (courtesy of Tullow Oil). TWT = two-way traveltime. 2.5 km (1.5 mi).

rift basin fill indicate that the rifting has not been a continuous process, but rather has occurred as a series of discrete short-lived pulses.

Although the Albertine Graben has for a long time been interpreted as an extensional province, undeniably overwhelming evidence for a compressive regime exists. The evidences for this compressive regime are more pronounced around the Rwenzori Mountains and reduce progressively to the north and south away from this feature. Note that the Rwenzori Mountains reach an altitude of 5109 m (16,762 ft) above sea level and about 4500 m (14,764 ft) above the surrounding Semliki Rift floor, an altitude not seen in any other extensional setting for a mountain that is right in the middle of the rift. The causes for its uplift have not been resolved, and compression tectonics cannot be ruled out.

Positive flower structures have been interpreted from both 2-D and 3-D seismic data (Figures 3, 4, 14, 15). The interpretation of these structures is that they are transpressional flower structures formed by sinistral strike slip. They are formed by restraining fault bends formed by fault offsets or by change in strike of fault systems. The growth of the contained anticline structures sometimes reaches up to the top of the seismic sections, suggesting that this strike-slip regime continued from the Miocene to Holocene (J. F. Rose and B. Smith, 2001, personal communication). This interpretation suggests that the Rwenzori Uplift is related to this

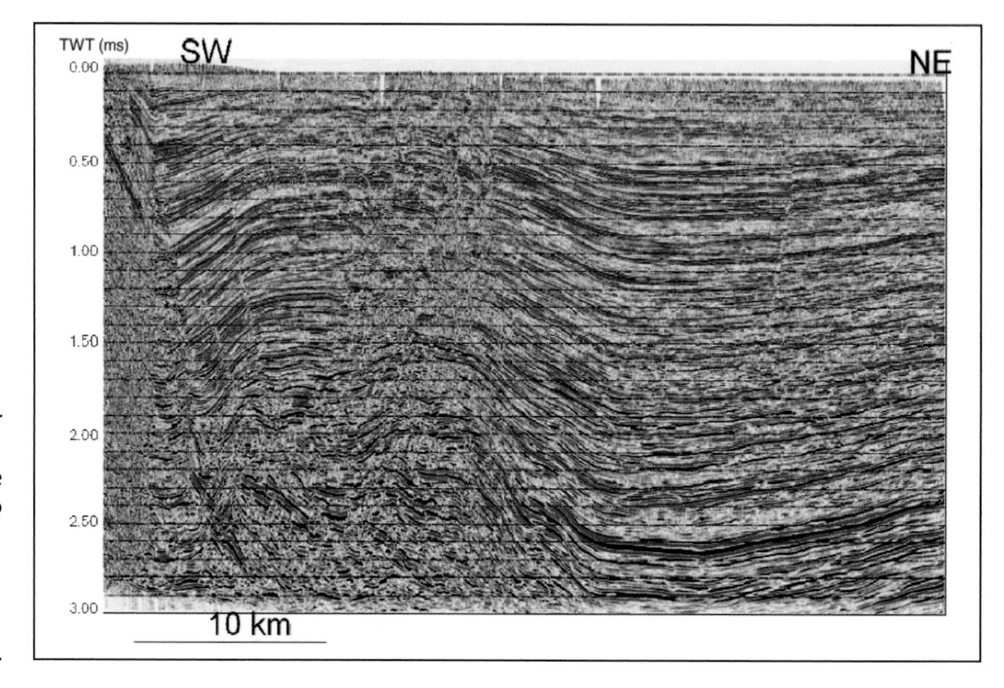

Figure 14. Line HOG-98-01, showing a compression anticline with thrusting in the lower part of the section and normal faulting in the upper part in the Semliki Basin at the northern tip of the Rwenzori Mountain ranges that forms the transfer zone from the central to the southern domain. TWT = two-way traveltime. 10 km (6.2 mi).

transpression and also implies that the Albertine Graben formed as a pull-apart structure by releasing a fault bend. The Rwenzori Mountains according to this interpretation are pop-up structures formed by restraining a fault bend along fault strike.

The seismic data in the Albertine Graben show hanging-wall anticlines typical of compression (Figures 14, 15). The anticlines extend to the top of the seismic sections. This indicates that the compression events persist up to the Holocene. The compressional episode inverted some of the faults at the base of the sections, contributing to the development of anticlines (Figure 14). The top basement reflector is faulted in a reverse manner, indicating that the compressional episode went beyond inversion to thrusting. The whole structure is interpreted as a product of tectonic inversion and reactivation of older normal faults because of compression (Cooper and Williams, 1989). Most of the inverted faults show reverse offset only at depth and retain a dominant normal separation upsection (Figures 4, 14). Similarly, the faults are semilistric at depth but steepen sharply at shallower levels.

Compression is characterized by asymmetrical hanging-wall anticline-syncline pairs, with anticlines displaying well-developed flower structures especially in the transfer zones. The ages of these compressional episodes are difficult to determine but could be attributed to the rifting and eventual opening up of the Red Sea during the Tertiary because structural vergence indicates tectonic transport from the northwest. Note that dextral shear structures have been interpreted in the northwest–southeast-trending Rukwa Rift to the south of the Albertine Graben (Delvaux, 1991; Morley et al., 1999a).

Seismic data interpretation shows that the base Miocene strata can be characterized by extensive reverse faulting (Figure 16). Reverse faulting, however, is more intense in the Semliki Basin than in the Lake Albert basins. The larger reverse faults in Semliki are related to the northward-plunging nose of the Rwenzori Mountains, where it becomes buried under Tertiary cover and could be rotating counterclockwise. Smaller faults die out shortly after this base Miocene section. In the Miocene section, these reverse faults sometimes have small rollovers corresponding to their propagating tips. The orientation of these faults is not known but could be a reactivation of Karoo fault systems.

At least three phases of faulting can be observed from the seismic data in the Albertine Graben. The first phase consists of deep-seated pre-Tertiary fault systems. Synthetic and antithetic faults create grabens and half grabens in which sediments were deposited. These fault patterns are inverted by a later compressional event, which thrusts the top basement reflector (Figure 16).

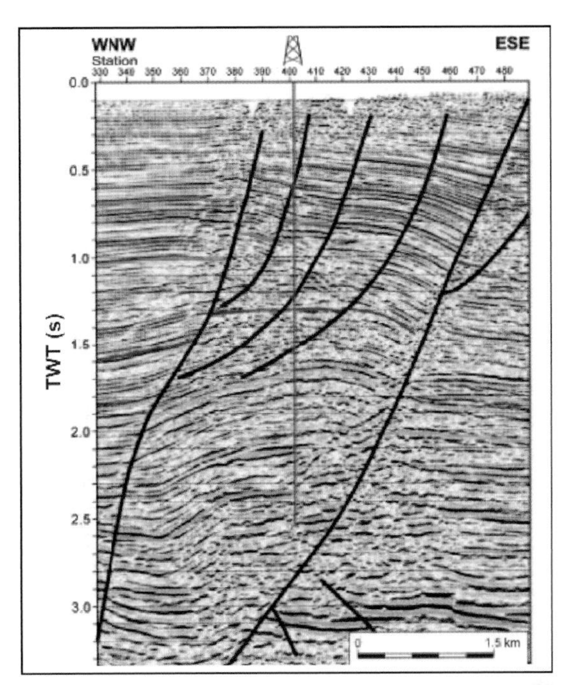

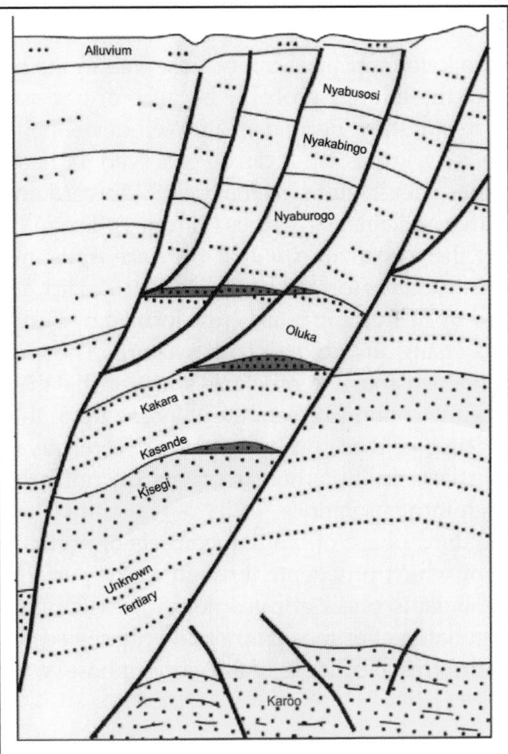

Figure 15. Line HOG-01-04 from the Semliki Basin close to the southern end of the southern domain, showing a hanging-wall anticline as a result of transpression. TWT = two-way traveltime. 1.5 km (0.9 mi).

Later faulting has also rotated these fault systems to a near-vertical position.

The second phase of faulting consists of almost vertical faults. The faults are planar to semilistric in geometry.

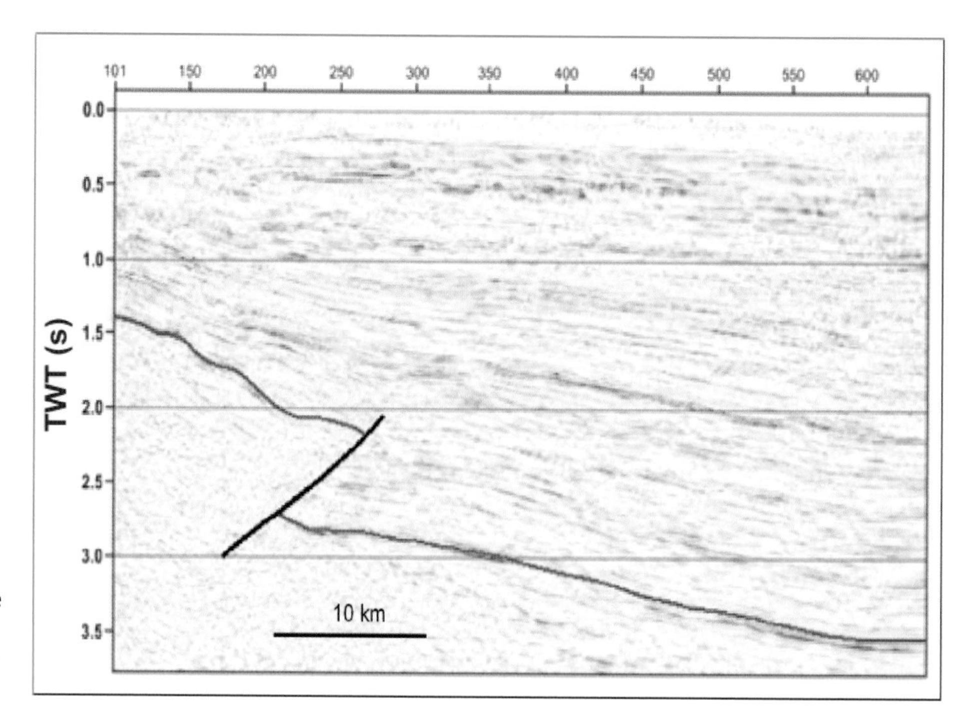

Figure 16. Seismic line from the Semliki Basin, showing reverse faulting at the base of the Cenozoic section. TWT = two-way traveltime. 10 km (6.2 mi).

Tight rollover anticlines can be observed in the hanging walls of these faults probably because of transpression, but they still show normal offset. Well-developed flower structures (Figures 4, 14) can be observed in the rollover anticlines. Here, fault systems branch upward and form palm tree anticlinal structures (Figure 4). The interpretation of these structures is that they are transpressional flower structures formed by sinistral strike slip. They are formed by restraining fault bends formed by fault offsets or by a change in strike of fault systems. The growth of the contained anticline structures suggests that this strike-slip regime continued from the Miocene up to the Holocene. Some of these fault systems are a reactivation of pre-Tertiary faults, whereas others do not propagate through into basement.

The third phase of faulting consists of normal faults, some of which propagate through to the surface. They show planar to semilistric geometry. These fault systems were initiated after the compression phase because they sit on the crestal collapse of the earlier phase of rifting.

TECTONIC CONTROL ON HYDROCARBON ACCUMULATION

Tectonic control on hydrocarbon generation, migration, and accumulation in different basins worldwide cannot be overemphasized. Hydrocarbons have accumulated in basins of different tectonic settings. According to Horn (2004), continental passive margins account for 35%, continental rifts and overlying sag basins account for 31%, terminal collision belts between two continents account for 20%, and other settings including foreland basins, shear margins, and subduction margins account for 14%.

In the Albertine Graben, deep steep-border faults, coupled with less clastic sediment input into the basin, created deep and stratified lakes with anoxic conditions that allowed accumulation of organic-rich source rocks. Thick sediments (6 km [3.7 mi]) accumulated along the basin-bounding faults together with high geothermal gradients as has been interpreted from well data, created conditions for maturation of the source rocks and generation of hydrocarbons. Migration of hydrocarbons has been aided by complex fault patterns as evidenced by numerous oil seeps along the basin margin fault (Figure 17).

The hydrocarbon-generating and -trapping potential of the Albertine Graben has now been proved. Eighteen oil and/or gas discoveries have been made to date, and two of these discoveries are at a field development level, indicating that commercially viable hydrocarbon accumulations are present in the graben. Like any other frontier basin, only structural traps have been tested in the Albertine Graben. Three major types of play have been drilled: the Pliocene Basin Margin Fault Closure Play, the Pliocene Intrabasin Play, and the Miocene Basin Margin Fault Closure Play.

In the Pliocene, discoveries have been made on structures created in the hanging wall of major faults by a compression episode, especially along fault offsets (Figure 18). The trapping mechanism in this play is hanging-wall anticline closures on the basin margin fault. In the Miocene Intrabasin Play, discoveries have been made in the accommodation zone that separates

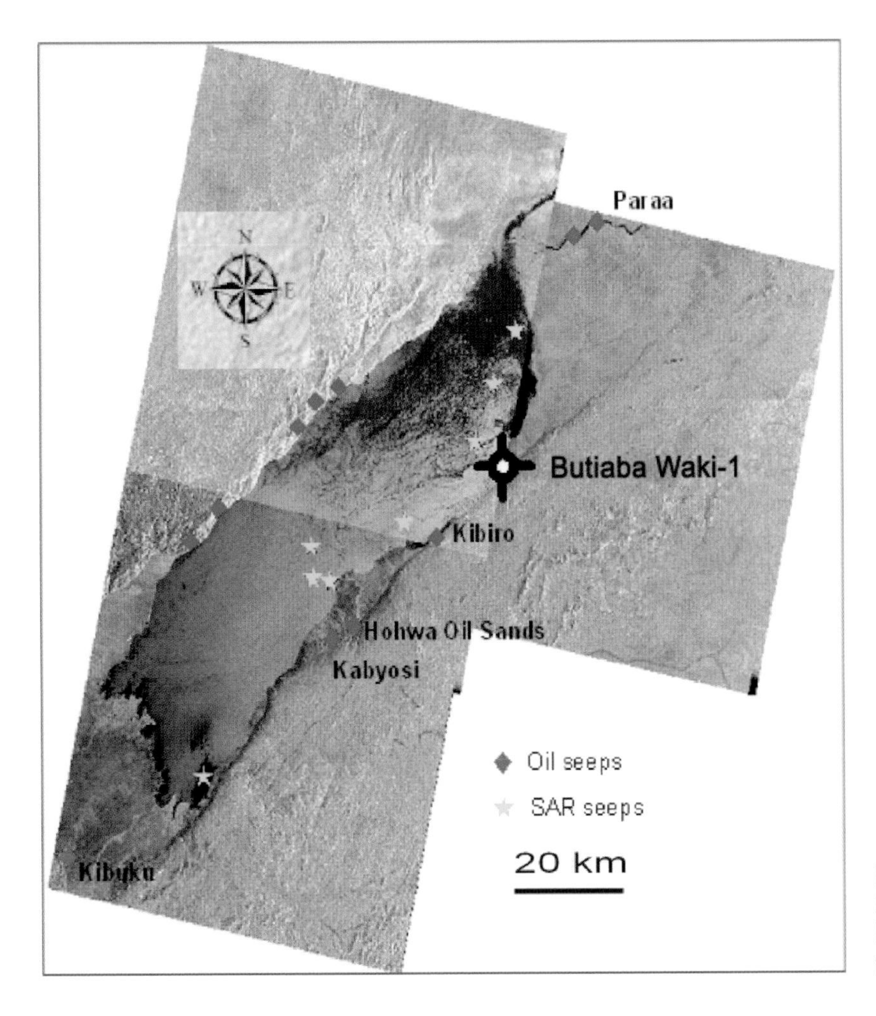

Figure 17. Numerous oil seeps along the basin-bounding faults indicate that oil has been generated and has migrated along the fault systems. 20 km (12.4 mi).

the central domain from the northern domain, and the trapping mechanisms are horst and rotated fault blocks. Significant discoveries have been made on rotated fault blocks created by synthetic faults and on horst blocks created by antithetic faults in an extensional regime. The trapping mechanism in the Miocene Basin Margin Fault Closure Play is similar to the trapping mechanism in the Pliocene Basin Margin Fault Closure Play, only that these sands are deeper in the hanging walls of major extensional faults.

The Lake Albert region of the Albertine Graben has deeply inclined basin margin faults consisting of a network of predominantly normal slip faults that form steep escarpments. The rift flanks have been uplifted, causing drainage diversion into the basin. The drainage on the western side has been diverted away from the Albertine Graben to the west. The major drainage on the eastern side has been diverted to the north, with smaller streams along the escarpment entering the graben along relay ramps (Gawthorpe and Hurst, 1993). This drainage divide combined with the lack of a pronounced ramping margin could have limited clastic sediments input into the basin and allowed deposition of source rocks.

The Lake Albert basins are gently tilted toward the west, with some intrabasin faults in some places penetrating into basement. Both the basement and the overlying synrift strata dip toward the western part of the basin. The dip angle of the synrift strata and basement to the west favors migration to the east and accumulation of hydrocarbons in sandstone and fractured basement reservoirs in the eastern side. Like in any other rift systems of Gulf of Suez in Egypt and the Aden rift in Yemen, the hydrocarbon prospectivity of the Lake Albert basins lies on the ramping side of the basin, where good-quality alluvial/fluvial sands were deposited. This has led to hydrocarbon discoveries on the eastern side of the basin.

Similarly, discoveries have been made north of the Lake Albert basins in the Butiaba-Wanseko area, which lies in the transfer zone between the central domain and the northern domain of the graben. The area is the site of recent oil and gas discoveries in a new play being described as the Miocene intrabasin play. This area is structurally controlled by the Bunia Fault in the west and the Toro-Bunyoro Fault in the east, running parallel in an almost northeast–southwest direction. The fault is hard

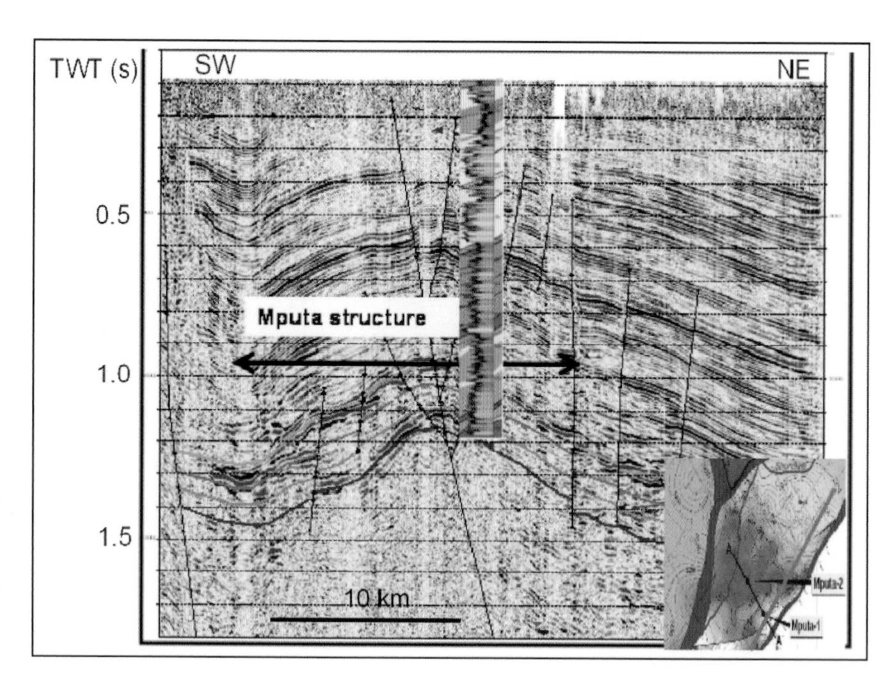

Figure 18. The Mputa anticlinal structure created by compression against the basin margin fault. This structure has been drilled and found to contain significant amounts of hydrocarbons. The pay zones are green on the log. TWT = two-way traveltime. 10 km (6.2 mi).

linked with few soft linkages like at River Sonso. Two relay ramps are prominent, one around Butiaba and the other at River Sonso. Both relay ramps are formed by synthetic overlapping transfer faults. Field studies have shown that the two relay ramps have acted to focus sand input into this area and also could have provided migration pathways for hydrocarbons.

Rift basins have been and continue to be one of the world's leading petroleum producers. Different rift systems evolve through different tectonic regimes, and their petroleum systems are therefore different (Anders and Schlische, 1994). The understanding of key tectonic regimes controlling rift development is key to understanding the petroleum system of a given basin (Morley et al., 1990). Successful hydrocarbon exploration in the Albertine Graben, therefore, requires an understanding of the tectonic regimes at play in the graben that have controlled basin evolution.

CONCLUSIONS

The Albertine Graben has evolved through multiple-phase tectonic deformation involving extension and compression. The extension regime created accommodation space for the accumulation of thick sediments, necessary for the generation of hydrocarbons. Complex faulting during extensional and compressional regimes allowed migration and created structures that have trapped hydrocarbons. All of the discoveries made to date are structurally controlled and are mainly contraction anticlines developed along basin margin faults and rotated fault blocks created by deep-seated synthetic and antithetic faults. The high potential of the petroleum system is spatially associated with the transfer faults that control the along-axis variation in graben geometry and sedimentation.

Geologic and geophysical data interpretation suggests that the Albertine Graben can be divided into three structural domains based on structural trend and internal basin geometries. The central domain in the Lake Albert region trends in a northwest–southeast direction and internally shows an almost full-graben structure dipping gently to the west. The northern and southern domains trend in an almost north–south direction and display typical half-graben structures that are asymmetrical, with one basin-controlling fault and a ramping flexural margin.

As observed from outcrop, seismic, and well data, the basin fill is strongly influenced by tectonics as is evidenced by the cyclicity of the sediments. However, the cyclicity of the sediments may also reflect changes in base level, brought about by climatic fluctuations. Outrop studies in Butiaba-Wanseko and Kaiso-Tonya areas show that the lithofacies types formed are influenced by their positions within the graben. Lithofacies associations formed include alluvial fans (where fault scarps are steep), fluvial channels and deltas, and complex associations of these. With steep scarps, a small rise in lake level leads to a big shift in the shoreline, leading to deposition of fine clastic sediments. Lake level fall, however, leads to progradation of fluvial systems into the basin. Depositing coarse sediments farther into the basin leads to coarsening-upward sequences. The structural configuration being described above and its sediment constituent has been targeted for hydrocarbons and yielded positive results.

ACKNOWLEDGMENTS

This chapter is dedicated to the field staff of the Petroleum Exploration and Production Department in the Ministry of Energy and Mineral Development of the Uganda Government for their contribution to the development of the oil and gas sector in Uganda. The technical discussions both in the field and boardrooms are highly appreciated. Acknowledgement is also given to the international oil companies operating in Uganda for venturing into a frontier basin and for allowing the use of seismic and well data in this chapter. William Hill and Dengliang Gao provided peer reviews for this chapter.

REFERENCES CITED

Abeinomugisha, D., and F. Mugisha, 2004, Structural analysis of the Albertine Graben (abs.): Proceedings of the 2nd East African Rift System Conference, Addis Ababa, Ethiopia, June 20–24, 2004, p. 4–6.

Anders, M. H., and R. W. Schlische, 1994, Overlapping faults, intrabasin highs and the growth of normal faults: The Journal of Geology, v. 102, p. 165–180, doi:10.1086/629661.

Bishop, W. W., 1969, Pleistocene stratigraphy of Uganda: Geological Survey of Uganda Memoir X, 128 p.

Bishop, W. W., and A. F. Trendall, 1966, Erosion-surfaces, tectonics and volcanic activity in Uganda: Quarterly Journal of Geological Society, v. 122, p. 385–420, doi:10.1144/gsjgs.122.1.0385.

Chorowicz, J., 2005, The East African rift system: Journal of African Earth Sciences, v. 43, p. 379–410, doi:10.1016/j.jafrearsci.2005.07.019.

Cooper, M. A., and G. D. Williams, eds., 1989, Inversion tectonics: Geological Society of (London) Special Publication 44, p. 375.

Delvaux, D., 2005, Evidence for a prototransform evolution of the Mbeya rift: Rift–rift triple junction, SW Tanzania: Proceedings of the 3rd East African Rift System Conference, Mbeya, Tanzania, August 16–18, 2005, p. 2–4.

Delvaux, D., 1991, The Karoo to recent rifting in the western branch of the East African rift system: A bibliographical synthesis: Royal Museum for Central Africa, Department of Geology and Mineralogy, Belgium, Annual Report 1989–1990, p. 63–83.

Ebinger, C. J., 1989, Tectonic development of the western branch of the East African rift system: Geological Society of America Bulletin, v. 101, p. 885–90, doi:10.1130/0016-7606(1989)101<0885:TDOTWB>2.3.CO;2.

Ebinger, C. J., T. D. Bechtel, D. W. Forsyth, C. O. Bowin, 1989, Effective elastic plate thickness beneath the east African and Afar plateaus and dynamic compensation of the uplifts: Journal of Geophysical Research, v. 94, p. 2883–2901, doi:10.1029/JB094iB03p02883.

Gawthorpe, R. L., and J. M. Hurst, 1993, Transfer zones in extensional basins: Their structural style and influence on drainage development and stratigraphy: Journal of Geological Society (London), v. 150, p. 1137–1152, doi:10.1144/gsjgs.150.6.1137.

Girdler, R. W., and D. A. McConnell, 1994, The 1990 to 1991 Sudan earthquake sequence and the extent of the East African rift system: Science, v. 264, no. 5155, p. 67–70.

Horn, M. K., 2004, Giant fields 1868–2004: AAPG/Datapages Miscellaneous Data Series, version 1.2 (revision of 2003), CD-ROM.

Karner, G. D., et al., 2000, Distribution of crustal extension and regional basin architecture of the Albertine rift system, East Africa: Marine and Petroleum Geology, v. 17, p. 1131–1150.

Kiram, E. L., J. J. Tiercelin, C. Le Turdu, S. A. Cohen, D. J. Reynolds, B. Le Gall, and C. A. Scholz, 2002, Control of normal fault interaction on the distribution of major Neogene sedimentary depocenters, Lake Tanganyika, East African rift: AAPG Bulletin, v. 86, no. 6, p. 1027–1059, doi:10.1306/61EEDC1A-173E-11D7-8645000102C1865D.

Kurz, T., R. Gloaguen, C. J. Ebinger, M. Casey, and B. Abebe, 2006, Deformation distribution and type in the Main Ethiopian Rift (MER): A remote sensing study: Journal of African Earth Sciences, v. 48 (2007), p. 100–114, doi:10.1016/j.jafrearsci.2006.10.008.

Laerdal, T., and M. R. Talbot, 2002, Basin neotectonics of lakes Edward and George, East African rift. Paleogeography, Paleoclimatology, Paleoecology, v. 187, p. 213–232.

Lambiase, J. J., and W. Bosworth, 1995, Structural controls on sedimentation in continental rifts, *in* J. J. Lambiase, ed., Hydrocarbon habitat in rift basins: Geologic Society (London) Special Publication 80, p. 117–144.

Maasha, N., 1975, The seismicity of the Ruwenzori region in Uganda: Journal of Geophysical Research, v. 80, p. 1485–1496.

McGlue, M. M., C. A. Scholz, K. Tobias, B. Ongodia, and K. E. Lezzar, 2006, Facies architecture of flexural margin lowstand delta deposits in Lake Edward, East African rift: Constraints from seismic reflection imaging: Journal of Sedimentary Research, v. 76, no. 6, p. 942–958, doi:10.2110/jsr.2006.068.

Morley, C. K., 1999, Patterns of displacement along large normal faults: Implications for basin evolution and fault propagation based on examples from East Africa: AAPG Bulletin, v. 83, p. 614–634.

Morley, C. K., R. A. Nelson, T. L. Patton, and S. G. Munn, 1990, Transfer zones in the East African rift system and their relevance to the hydrocarbon exploration in rifts. AAPG Bulletin, v. 74, no. 8, p. 1234–1253.

Morley, C. K., R. M. Harper, and S. T. Wigger, 1999a, Tectonic Inversion in East Africa, *in* C. K. Morley, ed., Geoscience of rift systems: Evolution of East African rift system: AAPG Studies in Geology 44, p. 193–210.

Morley, C. K., D. K. Ngenoh, and J. K. Ego, 1999b, Introduction to the East African rift system, *in* C. K. Morley, ed.,

Geoscience of rift systems: Evolution of East African rift system: AAPG Studies in Geology 44, p. 1–8.

Nicholas, C., and D. Abeinomugisha, 2009, Petroleum exploration in the Lake Edward–Lake George basins, Albertine Graben, Uganda: Fourth East African Petroleum Conference, Mombasa, Kenya, p. 280–289.

Ollier, C. D., 1990, Morphotectonics of the Lake Albert rift valley and its significance for continental margins: Journal of Geodynamics, v. 11, p. 343–355, doi:10.1016/0264-3707(90)90016-N.

Richardson, R. M., 1992, Ridge forces, absolute plate motions and the intraplate stress field: Journal of Geophysical Research, v. 97, p. 739–748, doi:10.1029/91JB00475.

Rose, J. F., 2001, Geological sampling and reevaluation of the geology of the Tertiary outcrop in the Semliki Basin: Heritage Oil and Gas Ltd., Report, 18 p.

Rose, J. F., and S. R. Curd, 2005, Rift development and new play concepts in the Albertine Graben: 10th Paper of the East African Petroleum Conference, Entebbe, Uganda, March 3, 2005, 15 p.

Rose, J. F., and B. Smith, 2001, Seismic interpretation report: Incorporating the HOG-98 and HOG-2001 and including regional and local geological models: Uganda lisence 1/97, block 3, Albertine graben, Heritage Oil and Gas Limited Report, p. 1–77, unpublished.

Scholz, C. A., 1995, Deltas of the Lake Malawi Rift, East Africa: Seismic expression and exploration implications: AAPG Bulletin, v. 79, no. 11, p. 1679–1697.

Scholz, C. A., T. Karp, M. McGlue, K. E. Lezzar, R. Kasande, and F. Mugisha, 2004, The subsurface structure of Lake Albert: Constraints from seismic reflection data (abs.): 2nd East African Rift System Conference, Addis Ababa, Ethiopia, June 20–24, 2004, p. 91.

Specht, T. D., and R. B. Rosendahl, 1989, Architecture of the Lake Malawi Rift, East Africa: Journal of African Earth Sciences, v. 8, no. 2/3/4, p. 355–382.

Smith, B. G., 2005, Exploration strategy in a virgin basin, the Albertine Graben: 9th Paper of the East African Petroleum Conference, Entebbe, Uganda, March 3, 2005, 10 p.

Tugume, F., and A. Nyblade, 2009, The depth distribution of seismicity at the northern Rwenzori Mountains: Implication for heat flow in the western rift: 5th Paper of the 2nd African Rift Geothermal Conference, Entebbe, Uganda, November 24–28, 2008, 30 p.

Upcott, R. K. Mukasa, C. J. Ebinger, G. D. Karner, 1996, Along-axis segmentation and isostasy in the western rift, East Africa: Journal of Geophysical Research, v. 101, p. 3247–3268, doi:10.1029/95JB01480.

Zoback, M. L., 1992, First and second order patterns of stress in the lithosphere: The world stress map project: Journal of Geophysical Research, v. 97, p. 11,703–11,728, doi:10.1029/92JB00132.

10

Gao, Dengliang and Jeff Milliken, 2012, Cross-regional intraslope lineaments on the Lower Congo Basin Slope, offshore Angola (West Africa): Implications for tectonics and petroleum systems at passive continental margins, *in* D. Gao, ed., Tectonics and sedimentation: Implications for petroleum systems: AAPG Memoir 100, p. 229–248.

Cross-regional Intraslope Lineaments on the Lower Congo Basin Slope, Offshore Angola (West Africa): Implications for Tectonics and Petroleum Systems at Passive Continental Margins

Dengliang Gao

Department of Geology and Geography, West Virginia University, 98 Beechurst Ave., Morgantown, West Virginia, 26506, U.S.A. (e-mail: dengliang.gao@mail.wvu.edu)

Jeff Milliken

Marathon Oil Corporation, Houston, 5555 San Felipe St., Texas, 77056, U.S.A. (e-mail: jvmilliken@marathonoil.com)

ABSTRACT

Approximately 20,000 km^2 (7900 mi^2) of three-dimensional (3-D) seismic data, along with the Bouguer gravity and bathymetry, show a series of intraslope lineaments that extend more than tens of kilometers (miles) in the Lower Congo Basin, offshore Angola (west Africa). Most of these lineaments trend to the northeast at approximately 45°, crossing the regional northwest-trending folds and thrusts. Geometric relationships and distribution patterns of folds and faults shown in the postsalt Tertiary section suggest that many of the lineaments might have a significant strike-slip component. Seismic structures, facies, and prospects indicate that the lineaments have been associated with the allochthonous salt bodies, turbidite sands, and oil and gas fields. We interpret that the lineaments are primarily related to the postsalt regional gravitational sliding of the Tertiary sediments. They could provide fairways for turbidite flow and pathways for salt emplacement and hydrocarbon migration. We infer that many cross-regional lineaments are the expression of the presalt basement transfer faults formed during the rifting phase of the continental crust in the Early Cretaceous have influenced the locus and orientation of the lineaments in the postsalt sediments. The obliquity of the continental lineaments (45°) to the oceanic fracture zones (80°) is consistent with and supportive of the previously recognized plate tectonic model depicting the counterclockwise rotation of the west African continent associated with the southward opening of the South Atlantic Ocean. These findings provide important insights to the nature of the cross-regional lineaments and their implications for tectonics, sedimentation, and petroleum systems at the west African passive continental margin.

DOI:10.1306/13351555M1003528

INTRODUCTION

The Lower Congo Basin in the deep-water offshore Angola is located in an important tectonic setting at the west African passive continental margin. It is one of the major petroliferous sedimentary basins globally. The tectonic history of the Lower Congo Basin features early continental rifting, thermal subsidence (sag), and passive margin gravitational progradation. The tectonic processes have been spatially and temporally associated with the development of a working petroleum system (Nombo-Makaya and Han, 2009; Guiraud et al., 2010). The analysis of the Lower Congo Basin structure has important implications for understanding tectonic processes and for assessing hydrocarbon potential at the west African passive continental margin.

Because of its importance to plate tectonics and hydrocarbon systems, the Lower Congo Basin has been one of the focus areas of research in both academia and the energy industry. Many authors have discussed the basin-slope structures with particular interest in the regional northwest-trending folds and thrusts and salt tectonics (e.g., Hempton et al., 1990; Lundin, 1992; Jackson et al., 1998; Anderson et al., 2000; Cramez and Jackson, 2000; Marton et al., 2000; Hudec and Jackson, 2002; Tari et al., 2003; Hudec et al., 2004; Jackson et al., 2004; Kilby et al., 2004; Rowan et al., 2004; Jackson et al., 2008; Jackson and Hudec, 2009; Nombo-Makaya and Han, 2009). These previous research efforts have contributed significantly to the understanding of the structural style and hydrocarbon potential of the basin (Nombo-Makaya and Han, 2009); however, compared with the northwest-trending regional structures, not much has been documented on the cross-regional lineaments that extend perpendicular or oblique to the regional folds and thrusts. Even less is documented regarding the economic implications of the cross-regional lineaments for hydrocarbon exploration at the passive continental margin. Although prospect generators have been mapping reservoir-scale faults and attempting to evaluate migration potential for individual prospects, little is known about the nature of the cross-regional lineaments and their relations to fluid migration pathways and reservoir sand fairways. A more focused investigation on the cross-regional lineaments could lead to a better understanding of the nature of the lineaments and their implications for hydrocarbon exploration at the passive continental margin.

An active and successful exploration program in the Lower Congo Basin in the deep-water offshore Angola (west Africa) has led to an increased interest in investigating the basin slope structures, facies, and play fairways. New investment from the energy industry in the hydrocarbon exploration program has resulted in large amounts of high-quality three-dimensional (3-D) seismic data that cover the major part of the basin. These 3-D seismic data help shed new light on the geologic complexities of the basin. In this chapter, we present the seismic expression of different structural styles in the postsalt Tertiary sedimentary section of the basin, with particular reference to the cross-regional lineaments and their tectonic and economic implications. We first make 3-D seismic observations and descriptions of critical structural features and their spatial relationships to the turbidite flow, salt evacuation, and hydrocarbon accumulation in the postsalt section. We then present our interpretations regarding the nature of the cross-regional lineaments and their kinematic relationships to the regional postsalt folds and thrusts in the Tertiary section. Given the limited depth range of the high-quality 3-D seismic data, we make speculations regarding the possible link of the postsalt seismic lineaments to the presalt basement transfer faults (Versfelt, 2010). Finally, we discuss the implications of the cross-regional lineaments for hydrocarbon exploration at the west African passive continental margin and in other similar geologic settings around the world.

GEOLOGIC SETTING

The Lower Congo Basin is one of the major petroliferous sedimentary basins along the west African passive continental margin (Figure 1). The basin was associated with the Early Cretaceous opening of the South Atlantic Ocean (Rabinowitz and LaBrecque, 1979; Da Costa et al., 2001; Eagles, 2007). The Early Cretaceous rifting of continental crust formed lacustrine basins filled with continental clastics and rich organic material (Brice et al., 1982; Harris et al., 2004). In addition to clastics, carbonates deposited in a lacustrine environment (112–131.8 Ma) are also important reservoir rocks as part of the presalt Barremian–Aptian carbonate play (Harris, 2000; Jameson et al., 2011; Beglinger et al., 2012). At the cessation of continental rifting in the Aptian, a marine incursion and restricted circulation created an evaporitic condition for deposition of the Loeme salt. During the Cenomanian and Turonian, anoxic conditions led to deposition of organic-rich shales of the Lower Iabe Formation, the principal source rock for the deep-water Angola petroleum system (Harris et al., 2004). After a period of very slow deposition in the early Tertiary, sedimentation accelerated during the Oligocene, coincident with a proposed regional tectonic uplift of the Atlantic hinge zone (Cramez and Jackson, 2000).

A)

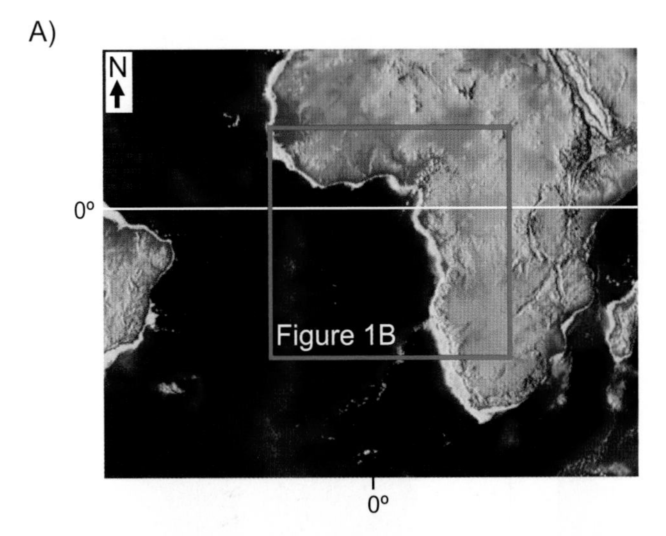

B)

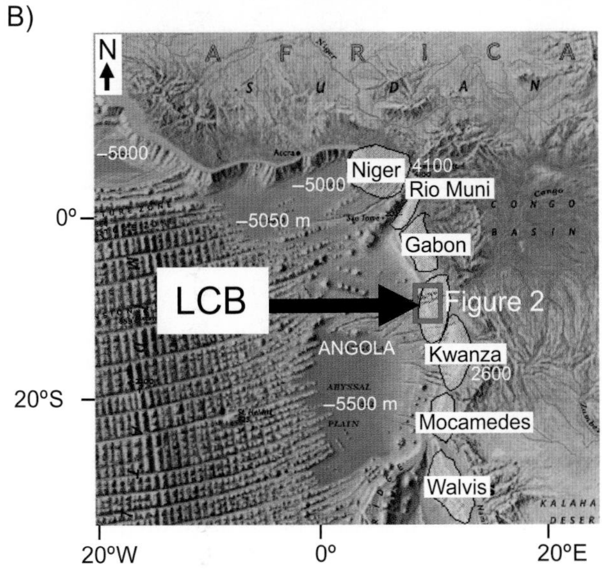

Figure 1. (A) Global plate tectonotopographic image (brown and green indicate area above the sea level, and blue indicates area below the sea water), demonstrating the present-day passive continental margins of the south Atlantic Ocean. (B) Regional tectonotopographic image showing the location of the Lower Congo Basin (LCB) in the deep-water offshore Angola at the west African passive continental margin. The Atlantic sea floor is dominated by the north–south-trending Mid-Atlantic Ridge and the northeast–east-trending (~80°) fracture zones (bathymetry data source: http://topex.ucsd.edu/marine_grav/mar_grav.html).

Deltaic and fluvial systems associated with the ancestral Congo River supplied sediment to the deep-water Congo Fan in the form of sandy turbidites of the Malembo Formation (Hempton et al., 1990; Kolla et al., 2001). These canyon-fed sandy turbidites were deposited in mostly slope environments, causing active degradation of the basin slope and deposition of reservoir sand in the deep-marine setting at the passive continental margin (Gardner and Borer, 2000; Kolla et al., 2001; Posamentier and Kolla, 2003).

The west African passive continental margin consists of three primary tectonostratigraphic units: the presalt rifted basement and its associated sedimentary fill, the Aptian salt, and the postsalt clastic wedge (Da Costa et al., 2001; Valle et al., 2001; Nombo-Makaya and Han, 2009; Versfelt, 2010). These three rheologically distinct units feature independent presalt and postsalt structural systems, with the intervening salt layer serving as a major geologic horizon of regional extent (Marton et al., 2000). The presalt, relatively rigid basement rocks contrast strongly with the overlying, mechanically weak sheet salt and postsalt clastic wedge (Marton et al., 2000). The updip part of the system consists of extensional faults infilled with syntectonic sediments (Vendeville and Jackson, 1992; Anderson et al., 2000; Rowan et al., 2004). As the thermal subsidence rate increased during the Neogene and as the cratonic interior of Africa was uplifted, erosion and sediment supply were accelerated, leading to progradation of clastics into the deep water of the basin (Hempton et al., 1990). The progradation of sedimentary load and the resulting gravity flow caused downdip contraction, effectively balancing the extension on the upper slope (Anderson et al., 2000; Rowan et al., 2004). Beneath this thin-skinned deformational progression lies the underpinning of the presalt section of rifted blocks that were formed in the rifting phase during the Early Cretaceous and draped by a thick sequence of rift and passive margin sediments (Brice et al., 1982). The deeply buried basement transfer faults developed during the early continental rifting might have cryptically influenced the growing sea-floor topography at the subsequent continental margin, thereby influencing the deposition, degradation, and deformation of the overlying Tertiary prograding sediment wedge. Description of the postsalt structural elements, particularly the cross-regional intraslope lineaments in the contractional domain and their implications for tectonics, sedimentation, and hydrocarbon accumulation, is the focus of this chapter.

THREE-DIMENSIONAL SEISMIC DATA AND METHODOLOGIES

Unlike field geologic investigation that is based on direct observation and mapping of outcrops, subsurface geologic investigation in a frontier deep-marine setting relies primarily on visualizing, mapping, and interpreting reflection seismic data. The 3-D seismic survey in this study extends for approximately 20,000 km^2 (7900 mi^2) and reaches more than 4 km (3 mi) below the sea floor in the deep-water offshore Angola (west Africa). Although seismic data do not provide such a high resolution as

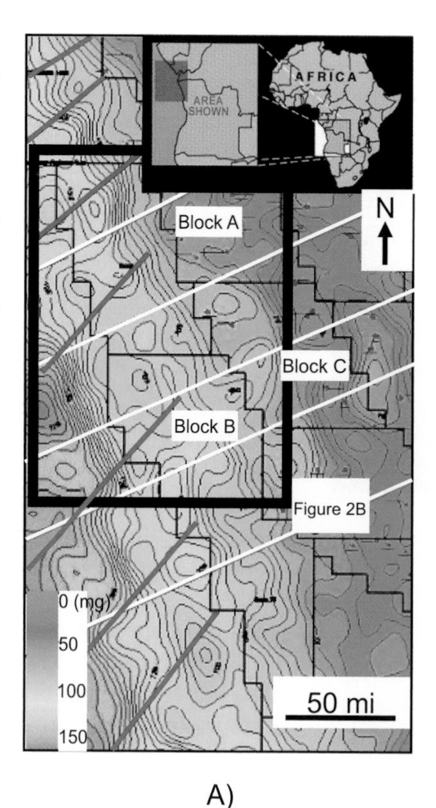

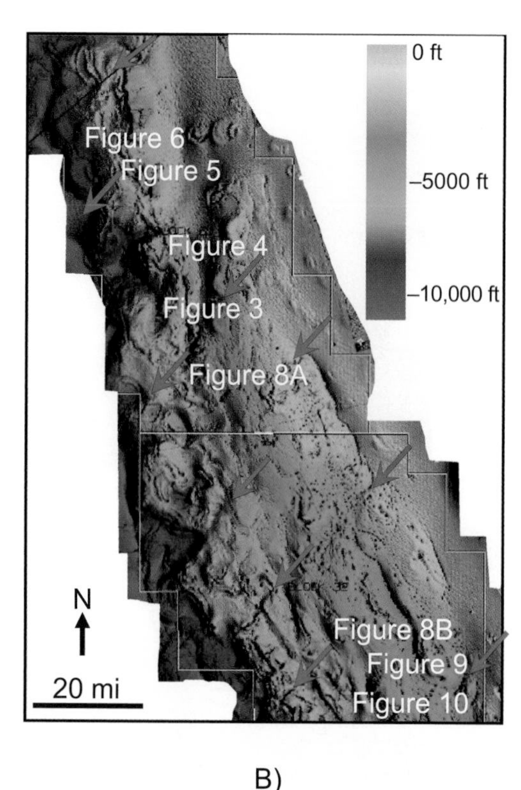

Figure 2. (A) Bouguer satellite gravity map of the Lower Congo Basin slope. The regional (northwest-trending) intensity contours are interrupted by a series of cross-regional (northeast-trending) kinks or deflections. The white solid lines show the approximate projection of the oceanic fracture zones (see Figure 1B) to the continental margin as a reference, whereas the red solid lines are interpreted kinks crossing the regional gravity contours. (B) Sea-floor bathymetric image showing the regional (northwest-trending) and the cross-regional (northeast-trending) lineaments. Notice the regional curvilinear and cuspate character of the escarpments of the sea floor (modified from Gao and Milliken, 2007) (bathymetry and gravity data source: http://topex.ucsd.edu/marine_grav/mar_grav.html). 20 mi (32 km); 50 mi (80 km); 1000 ft (304.8 m).

outcrops, they provide a continuous, volumetric seismic coverage of the subsurface geology, making it possible to investigate deeply buried structures, facies, and petroleum systems from a 3-D perspective. Furthermore, digital 3-D seismic data makes computer-aided volume observation in an interactive manner feasible, eventually leading to robust interpretation of subsurface structures, facies, and petroleum systems.

Seismic attribute analysis has been helpful for imaging seismic structures, facies, and reservoir systems. Among the many attributes, we compute the three most effective and relevant ones that help visualize the complexities of subsurface geology in the basin. First, we calculate a coherence attribute (Bahorich and Farmer, 1995) to highlight the major faults and fractures. The coherence attribute denotes the seismic waveform similarity between neighboring traces along structural horizons. A low similarity indicates incoherent seismic signal associated with discontinuities, whereas a high similarity indicates coherent seismic signal associated with continuous structures. The coherence attribute is particularly effective at highlighting critical structural and stratigraphic features such as high-angle faults, fractures, salt bodies, and channels in the map view. Second, we generate a structure attribute from the regular amplitude data for an improved structure visualization and interpretation (Gao, 2004, 2006a). The structure attribute is superior to the regular wiggle-trace imagery in imaging and mapping folds and faults. Unlike the coherence attribute, the structure attribute is particularly useful for investigating folds and faults and their relationships that are otherwise not easily discernible from the regular seismic image in the map and cross sectional views. Third, we derive a seismic facies attribute (Gao, 2004, 2006b). Unlike the structure attribute, the facies attribute removes the effect of phase of trace to minimize the structure interference in seismic facies visualization. The facies attribute helps identify and distinguish among different facies elements that are otherwise not easily recognizable from the regular seismic data.

We use volume visualization technology to interrogate subsurface structures and facies from seismic attribute data. The technology enables us to make seismic observations from a 3-D perspective interactively. For instance, by continuously slicing through the coherence and structure attribute volumes, we are able to unravel structural geometry and stratigraphic relationship in 3-D space. By continuously slicing through the facies attribute volumes using a series of stratigraphic surfaces, we are able to define depositional facies and play fairways at different geologic times, thereby unraveling the evolution of depositional facies and play fairways. These can significantly reduce the nonuniqueness and minimize potential pitfalls in subsurface seismic exploration.

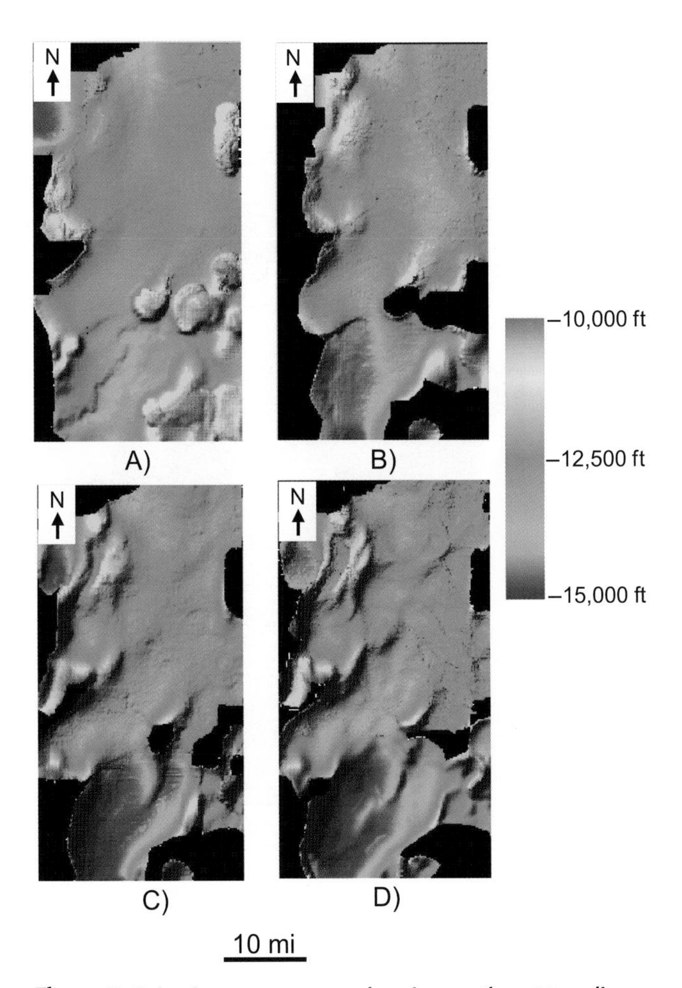

Figure 3. Seismic structure map showing northeast-trending (45°) faults and associated folds and salt at different stratigraphic levels. The en-echelon pattern of the faults and folds suggests a possible lateral sense of shear along the faults. (A) Lower Pliocene. (B) Upper Miocene. (C) Mid-Miocene. (D) Top Eocene. 10 mi (16 km); 1000 ft (304.8 m).

Extensive high-quality 3-D seismic data, new seismic attributes, and state-of-the-art volume visualization technologies have led to many new geophysical observations of the subsurface geology in the Lower Congo Basin. These new observations help unravel the spatial and temporal relationships among deformation, deposition, and hydrocarbon accumulation. These are fundamental to the success of hydrocarbon exploration on the Lower Congo Basin slope in the deep-water offshore Angola at the west African passive continental margin.

GEOPHYSICAL OBSERVATIONS

The Bouguer satellite gravity map on the Lower Congo Basin slope in the deep-water offshore Angola (west Africa) (Figure 2A) is dominated by northwest-trending intensity contours. This general trend of gravity-intensity contours is consistent with the regional structural trend at the passive continental margin. The regional northwest-trending contours of the gravity intensity are interrupted by a series of cross-regional deflections that can be easily overlooked. The nature of deflections is unknown, but they indicate the complexity of regional structural grains on the basin slope.

Sea-floor bathymetric imagery (Figure 2B) can be the surface expression of the underlying subsurface structural trends. The color-coded bathymetric contours curve gradually along the slope, featuring a seaward cuspate bulge. In association with this bathymetric bulge, a series of bathymetric lineaments running across the regional bathymetric contours exists. These bathymetric lineaments at the continental margin trend to the northeast at approximately 45° and deflect significantly from the oceanic fracture zones that trend to the east northeast at approximately 80° (Figure 1) in the South Atlantic Ocean (Rabinowitz and LaBrecque, 1979; Eagles, 2007).

In the subsurface down to 2 km (1 mi) below the sea floor, high-resolution reflection seismic data indicate that the cross-regional lineaments extend perpendicular or oblique to the trend of regional folds and thrusts in the Tertiary section. Seismic structure maps indicate en-echelon, northeast-trending discontinuities at the Eocene, Miocene, and Pliocene levels (Figure 3). Structure contour maps show that the fold axes curve gradually toward the lineaments (Figure 4). A continuum of depth slices through the structure attribute volume indicates that the cross-regional lineaments are major vertical or subvertical faults that penetrate the Tertiary sections from the Eocene, Miocene, to Pliocene structural levels.

Although salt bodies are primarily associated with the regional, northwest-trending folds and thrusts, salt canopies that are aligned preferentially along the cross-regional lineaments also exist (Figures 5A, B; 6A, B). A continuum of depth slices through the structure attribute volume demonstrates that both regional and cross-regional trends of salt bodies exist in the Tertiary section. Most salt bodies follow the regional trend of the folds and thrusts, whereas other salt bodies become segregated and deviated from the regional trend to follow the cross-regional lineaments. The anticlines associated with the cross-regional lineaments are symmetrical with a flower-shaped geometry as seen in the cross sectional view (Figure 5C). Similarly, salt bodies associated with the cross-regional lineaments are generally symmetrical, tall, and rooted to the autochthonous basal salt (Figure 7). These contrast with the asymmetric rootless salt canopies associated with the regional asymmetric folds and low-angle thrusts.

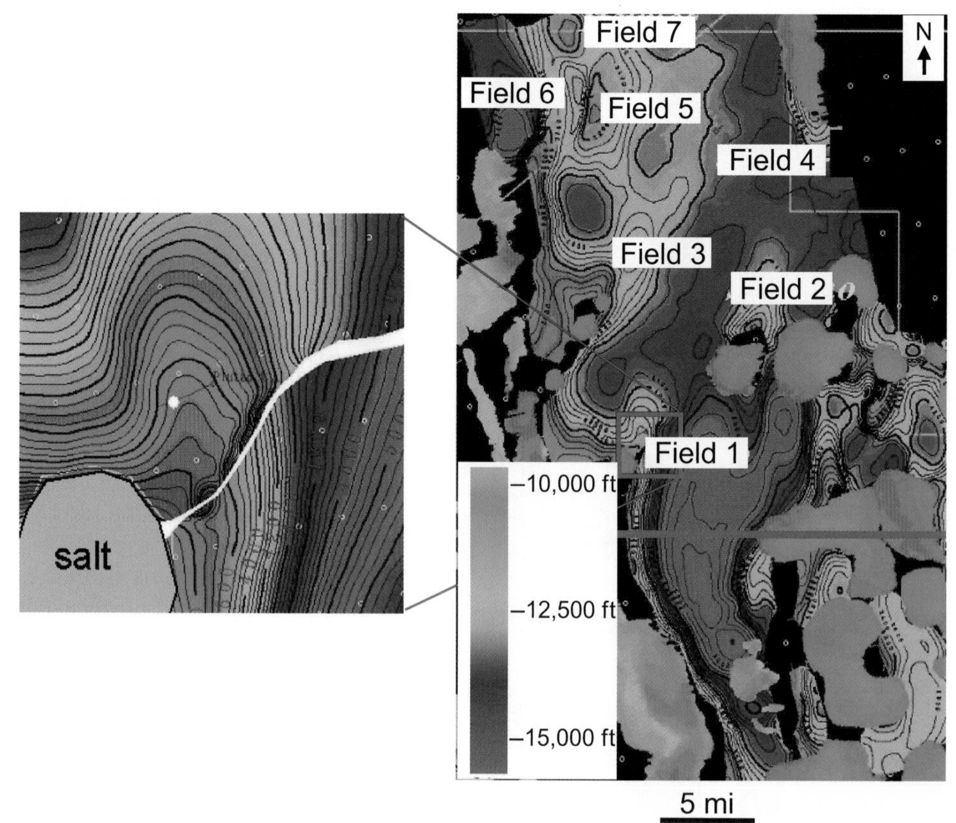

Figure 4. Detailed structure contour maps (lower Miocene) showing the fold trend change at the intersection with northeast-trending, cross-regional lineaments, suggesting a possible right-lateral sense of shear along the lineaments (modified from Gao and Milliken, 2007). 5 mi (8 km); 1000 ft (304.8 m).

One of the most intriguing observations is the spatial association of the cross-regional lineaments with oil and gas field discoveries. Regional well operation maps (Gawenda et al., 2004) indicate that the exploration success rate in this part of the basin is significantly both higher than in the neighboring regions. A map overlay of field discoveries (J. Helmich and E. Zhurina, 2007, personal communications) with major structural features shows a spatial correlation of hydrocarbon accumulation to the cross-regional lineaments. Similarly, newly generated prospects and leads based on direct hydrocarbon indicators (J. Helmich and E. Zhurina, 2007, personal communications) suggest an increased potential to find oil and gas along the cross-regional lineaments. At the shallow structural levels less than 330 m (1082 ft) below the sea floor, clusters of amplitude anomalies related to gas hydrate deposits are also spatially associated with the lineaments.

A detailed investigation of seismic structure attribute indicates that antiforms and minibasins associated with the cross-regional lineaments are different from those associated with the regional folds and thrusts in scale and geometry. The folds associated with the cross-regional lineaments are relatively small, symmetrical, and variable in trend with an en-echelon pattern, which contrast with those associated with the regional thrusts. Furthermore, the cross-regional lineaments are typically characterized by a major vertical fault at depth, that diverges upward to become a symmetrical antiform at the shallow structure level (Figure 5C). These might represent a possible flower structure that has been widely recognized as being indicative of transpressional strike-slip faults with both strike-slip and dip-slip components (Harding, 1990).

Secondary fractures were imaged in the vicinity of the primary lineaments, showing an en-echelon pattern and possibly representing synthetic faults and fractures that are induced by shearing along the primary lineaments. The relationship and pattern shown in the map view are similar to those observed in clay-model experiments (Wilcox et al., 1973) and at outcrops (Fleming and Johnson, 1989). These suggest that the primary lineaments might have a significant strike-slip component, and the secondary fractures could be indicative of the sense of shear along the primary lineaments. For example, the angular relationship between the secondary fractures (synthetic) and the primary lineaments shown in Figures 6 and 8A suggests that the primary lineaments might have a significant strike-slip component with a right-lateral sense of shear, whereas that shown in Figure 8B suggests that the primary lineament might have a significant strike-slip component with a left-lateral sense of shear.

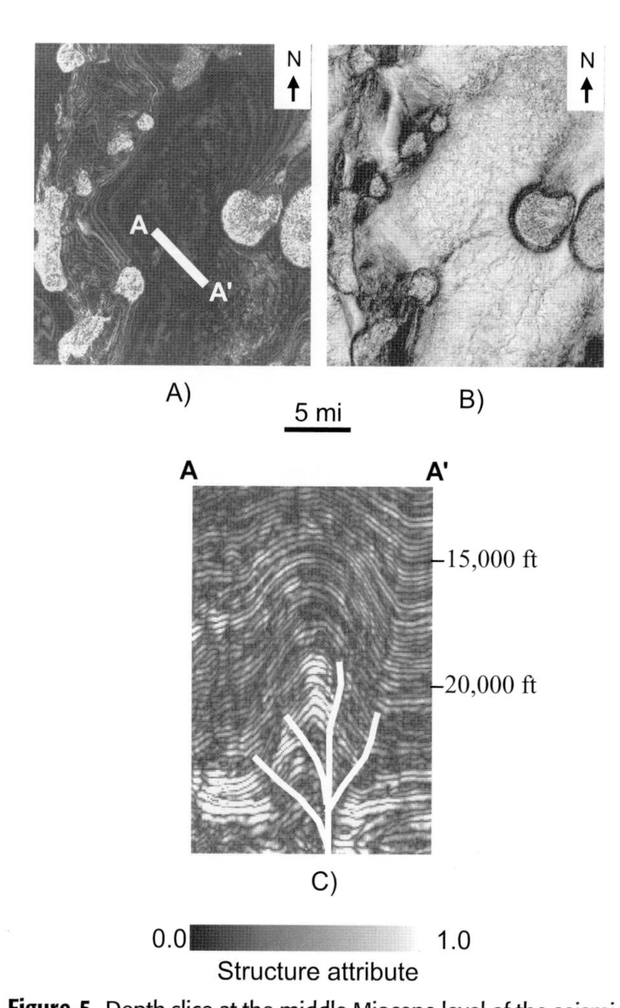

Figure 5. Depth slice at the middle Miocene level of the seismic structure attribute volume (A) and the coherence attribute volume (B) along with a cross sectional view (C), showing a linear distribution of allochthonous salt bodies along northeast-trending (45°) faults. Note that a possible flower structure as shown in the cross sectional view occurs at the bend of the lineament as shown in the map view. 5 mi (8 km); 1000 ft (304.8 m).

Other possible seismic kinematic indicators are the local drag folds and systematic curving of the fold axes toward the lineaments as shown in the map view. In the northern part of the 3-D seismic survey, the curving of anticlinal axes is suggestive of a right-lateral sense of shear along the cross-regional lineaments (Figures 3, 4), whereas in the southern part of the 3-D seismic survey, the curving of the anticlinal axes is suggestive of a left-lateral sense of shear along the cross-regional lineaments (Figures 9, 10). A preexisting channel complex was left-laterally offset several kilometers (miles) by a cross-regional lineament (Figure 11C).

Turbidite channel flow directions are spatially and temporally associated with the regional folds and thrusts, the cross-regional lineaments, and the salt-withdrawal minibasins at different scales. From a basinwide perspective, trends of channels and sand fairways are parallel or subparallel to the present-day bathymetric gradient. As the structural trend changes along the basin slope, the flow direction of channels changes accordingly (Figure 11A, B, C), indicating the tectonic influence to topographic gradient and, in turn, to syntectonic turbidite flow fairways on the basin slope. From a local perspective, the intraslope lineaments are coeval with channels (Figure 12A), and thick sequences of sediments are associated with salt-withdrawal minibasins along the lineaments. These suggest that the cross-regional lineaments could have influenced the turbidite flow direction (Figure 12B, C) and sediment thickness.

Because of the decrease in seismic image resolution and quality with depth, it is more difficult to map all the lineaments and turbidite fairways on a regional basis below the salt than above the salt. Seismically, it is not quite evident that all these postsalt Tertiary lineaments penetrate below the Aptian salt and directly connect to the basement faults (Versfelt, 2010) formed in the

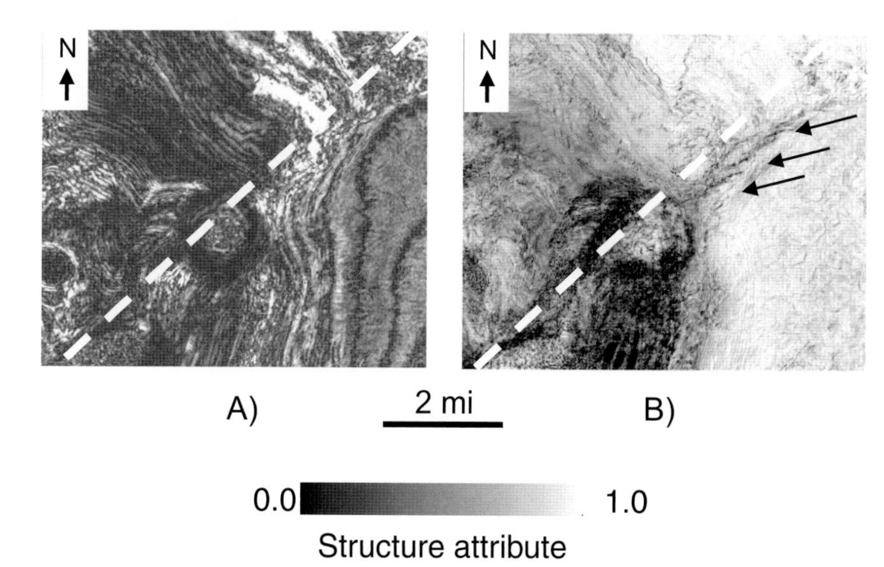

Figure 6. Depth slice at the lower Miocene level of the seismic structure attribute volume (A) and the coherence attribute volume (B), showing a primary northeast-trending (45°) fault and secondary faults suggesting a right-lateral sense of shear along the primary fault. 2 mi (3.2 km).

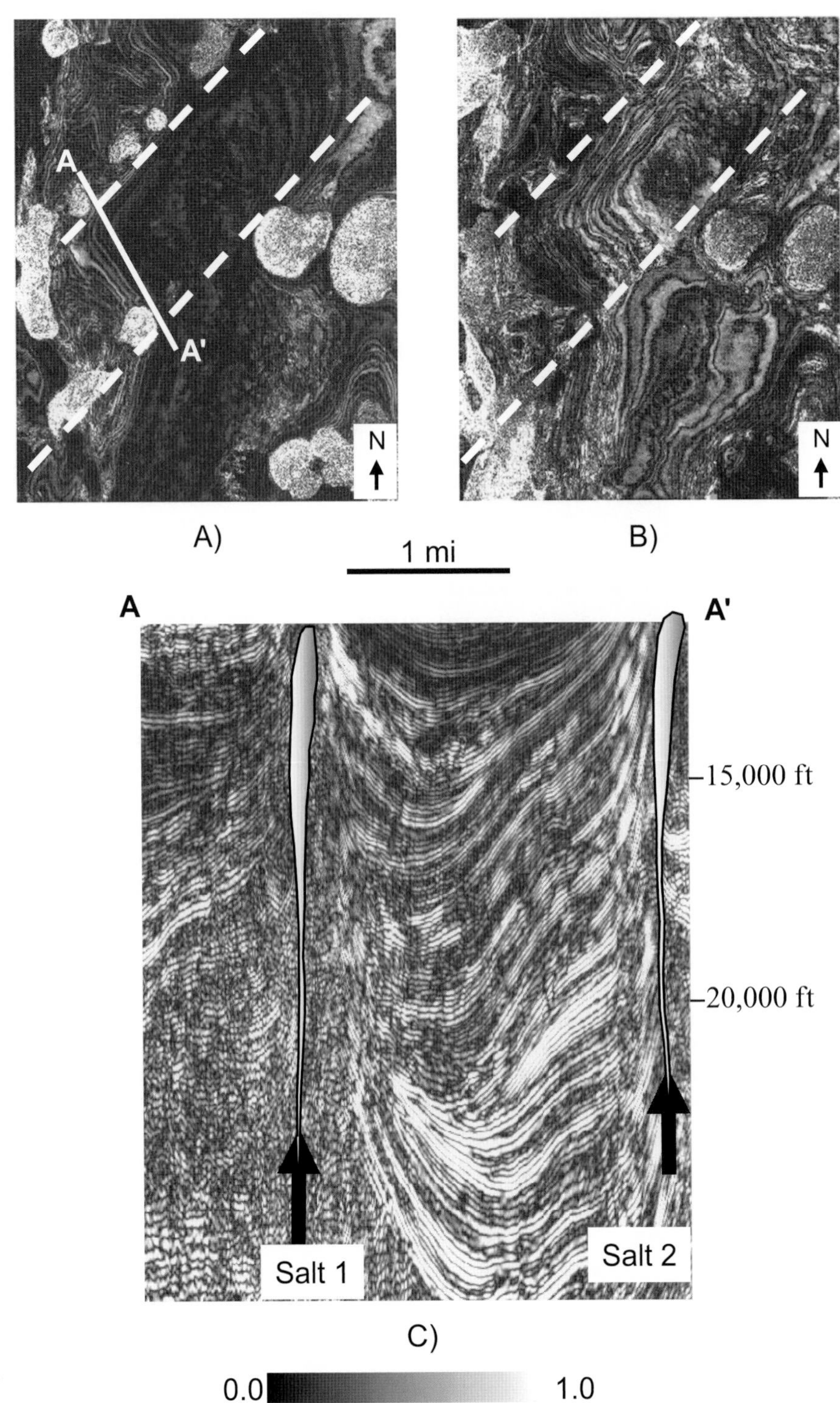

Figure 7. Depth slices at the Pliocene level (A) and the Eocene level (B) of the seismic structure attribute volume along with a cross sectional view (C), showing primary northeast-trending (45°) faults and salt bodies along the faults. The salt bodies that emplaced vertically along the cross-regional lineaments are tall and thin and symmetrical as shown in the cross sectional view, which are distinct from those associated with regional thrusts. The cross sectional view shows the interplay among two vertical, cross-regional faults, salt canopies, and salt-withdrawal minibasins. Note that the depocenter reversal of the salt-withdrawal minibasins may suggest relative timing of salt emplacement along the active lineaments.

continental rifting phase in the Early Cretaceous (Brice et al., 1982; Cramez and Jackson, 2000). However, a limited number of seismic lines with high quality do indicate that the deeply rooted presalt basement faults are connected to the postsalt lineaments (C. Cramez, 2005, personal communication). This suggests that the basement faults might have been influential to the development of the Tertiary lineaments, in which case the Tertiary lineaments are the expression of the underlying, deeply buried basement faults.

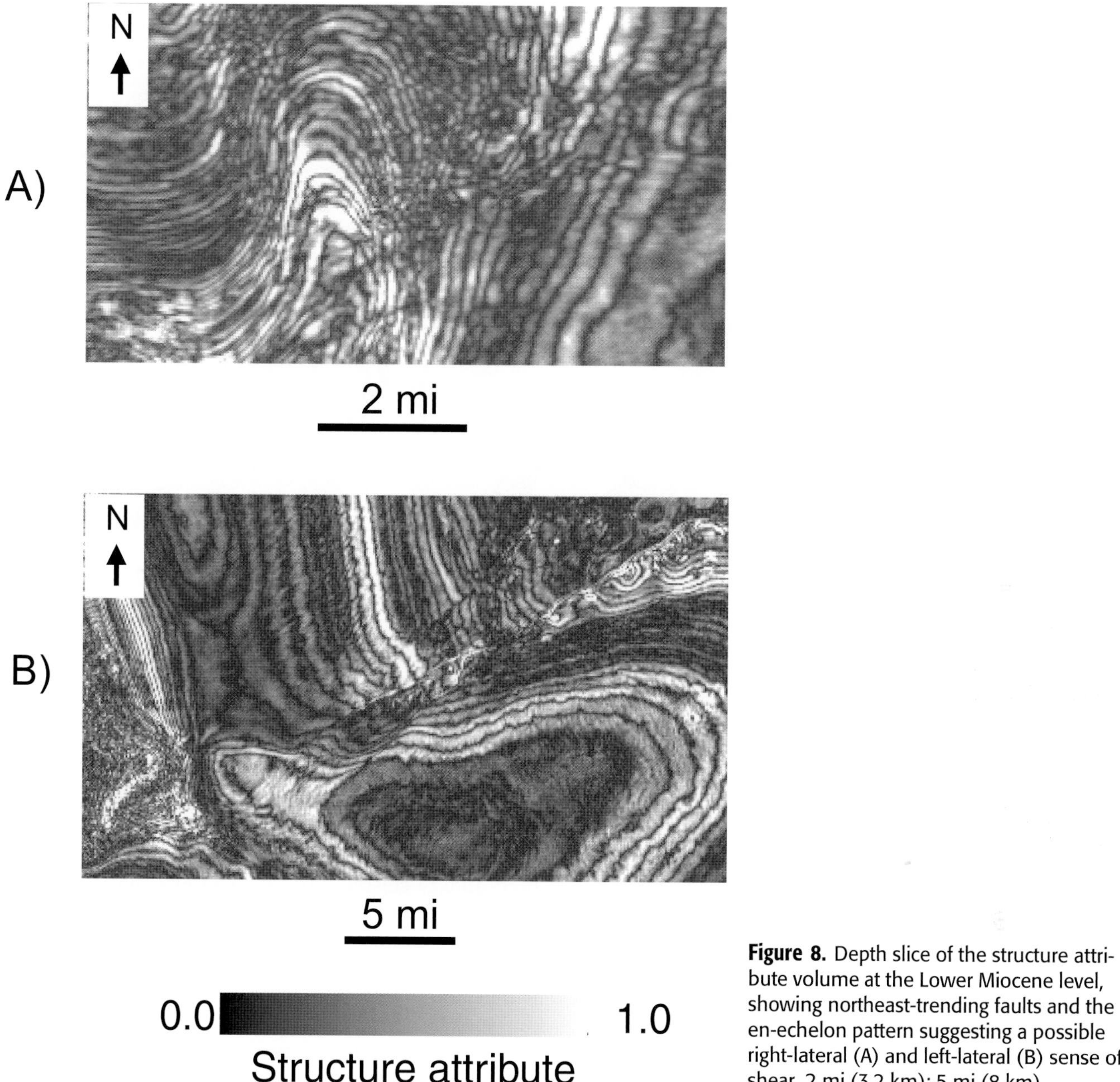

Figure 8. Depth slice of the structure attribute volume at the Lower Miocene level, showing northeast-trending faults and the en-echelon pattern suggesting a possible right-lateral (A) and left-lateral (B) sense of shear. 2 mi (3.2 km); 5 mi (8 km).

Finally, it is evident that the cross-regional lineaments are oblique to the South Atlantic oceanic fracture zones. Many lineaments at the continental margin trend to the northeast at approximately 45°, whereas fracture zones in the South Atlantic Ocean trend to the east northeast at approximately 80°. A 35° obliquity between the continental lineaments and the oceanic transforms is strikingly similar, but with an opposite polarity, to that observed from the 3-D seismic data in the Campos Basin at the eastern Brazil passive continental margin on the other side of the South Atlantic Ocean (Gao et al., 2009). This comparative observation, regarding the obliquity of the continental cross-regional lineaments to the oceanic fracture zones, indicates tectonic similarity between the west African and eastern Brazil passive continental margins across the South Atlantic Ocean.

GEOLOGIC INTERPRETATIONS

Located at the west African passive continental margin, the Lower Congo Basin experienced an Early Cretaceous rifting and salt deposition, a Late Cretaceous thermal subsidence and progradation, followed by an episodic Tertiary uplift and clastic influx (Guiraud et al., 2010). In

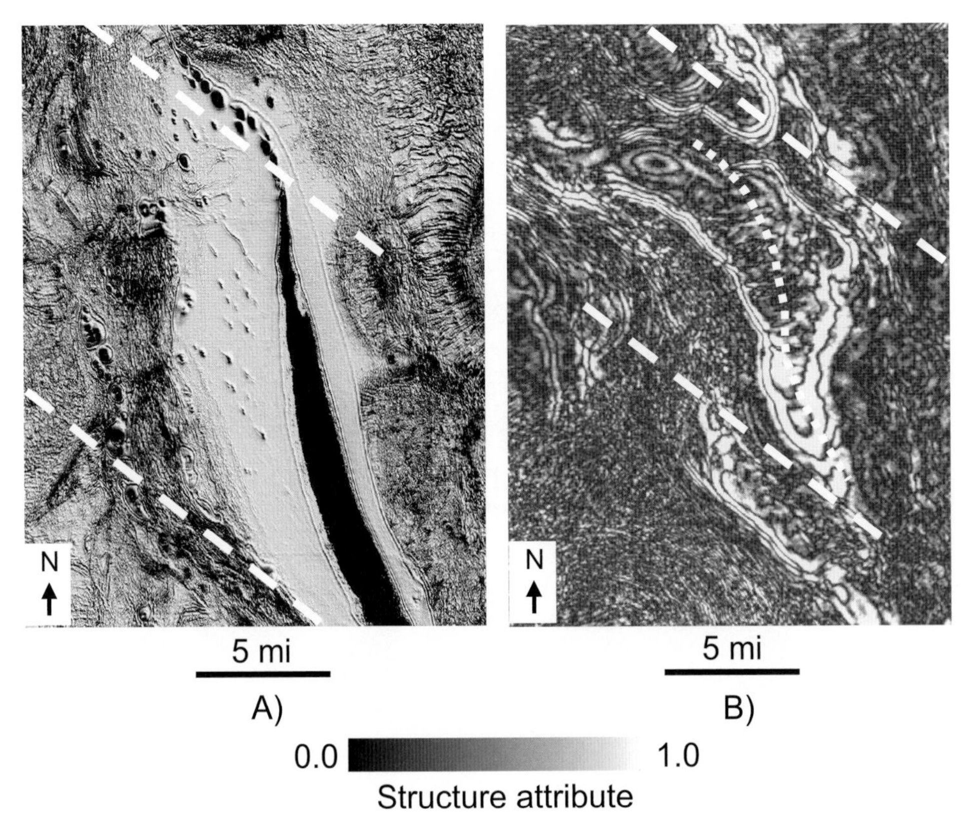

Figure 9. Depth slice of the structure attribute volume close to the sea floor (A) and at the Paleocene level (B), showing southeast-trending (120°) faults and associated folds and footprints of fluid flow (pockmarks). 5 mi (8 km).

the Early Cretaceous, continental rifting created regional basement rifts segmented by cross-regional transfer faults (e.g., Hudec and Jackson, 2002; Guiraud et al., 2010; Versfelt, 2010). In the Late Cretaceous, thermal subsidence caused an incipient upslope extension and a mild downslope deformation of basin-slope sediments (Hempton et al., 1990). In the Tertiary, an increased clastic input and modification of the slope gradient (Hempton et al., 1990) eventually triggered massive gravitational sliding, leading to intensified downslope contraction that is manifested as folding, thrusting, and allochthonous salt emplacement (Cramez and Jackson, 2000; Rowan et al., 2004; Jackson and Hudec, 2009).

The cross-regional intraslope lineaments seismically imaged in the Tertiary section as reported in this study extend over that part of the gravitational sliding system within the contractional province on the lower basin slope. High-quality seismic data in the postsalt Tertiary section indicate that the lineaments were developed above the major detachment horizon of the Aptian salt (Figure 13). In this regional setting, we interpret that the cross-regional lineaments are an integral component of a gravitational sliding system above the Aptian salt. Coeval with the northwest-trending regional folds and thrusts at the toe region of the system, the cross-regional lineaments could have served as tear (transfer) faults to accommodate the differential basinward movement of the Tertiary slope-forming sediments.

Because of the degraded image quality below the Tertiary section, our 3-D seismic data do not provide sufficient evidence that the cross-regional Tertiary lineaments are linked to the basement transfer faults associated with the Early Cretaceous rift (Versfelt, 2010). Based on high-quality seismic lines (C. Cramez, 2005, personal communication), we infer that most of the northeast-trending lineaments in the postsalt Tertiary section could be the expression of presalt, rift-related transfer faults. If that is the case, the obliquity between the present-day, northeast-trending (45°) lineaments at the continental margin and the northeast–east-trending (80°) oceanic transforms in the South Atlantic Ocean is suggestive of approximately 35° of counterclockwise rotation of the African continent during the drift-spreading phase since the initial rifting in the Early Cretaceous. Similar obliquity but with an opposite polarity was observed in the Campos Basin at the eastern Brazilian passive continental margin (Gao et al., 2009). It then follows that the present-day trend difference between the older continental lineaments and the younger oceanic fracture zones is supportive of the conjugate rotation of the two continents on both sides of the South Atlantic, which was coeval with the northward propagation of continental rift and the southward opening of the South Atlantic Ocean since the Early Cretaceous breakup of Gondwana (Rabinowitz and LaBrecque, 1979; Brice et al., 1982; Eagles, 2007).

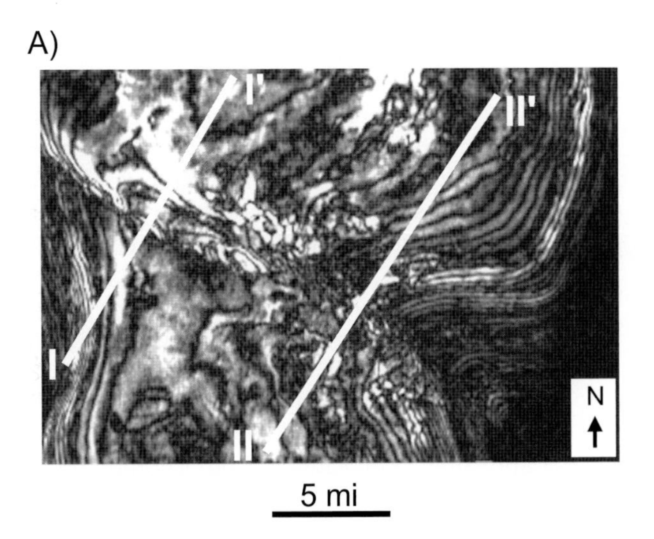

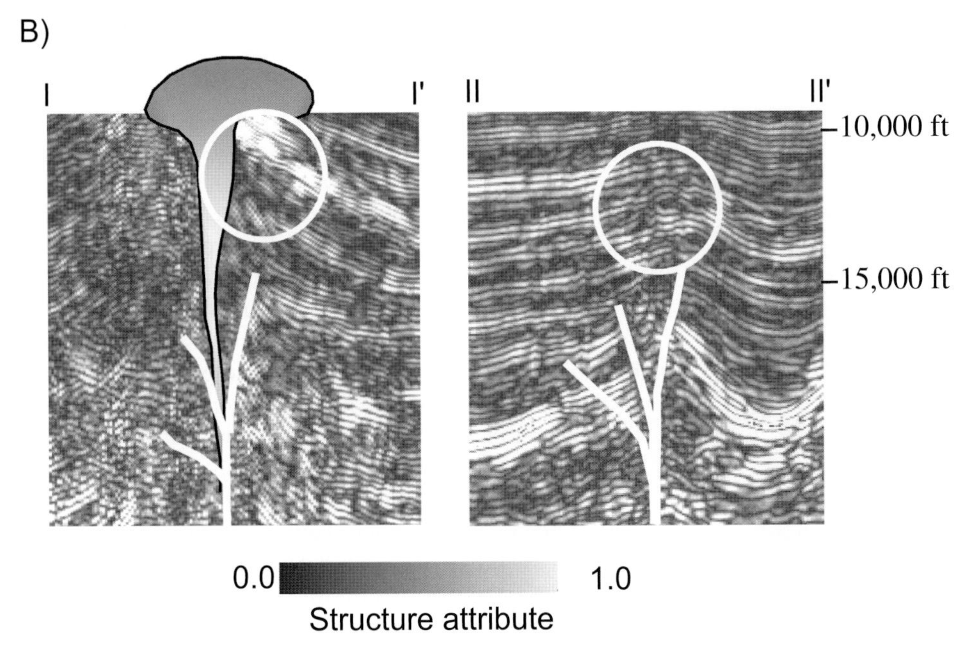

Figure 10. (A) Depth slice at the upper Miocene level of the structure attribute volume showing a southeast-trending fault. (B) Vertical slices across the southeast-trending fault showing two different potential traps, with the sub-salt trap (left) having a higher sealing capacity than the faulted anticline (right). 5 mi (8 km); 1000 ft (304.8 m).

Observations of seismic structures, seismic facies, and their spatial relationships suggest a strong tectonic influence to sedimentation both regionally and locally. At basin scale, gravitational sliding and associated sediment influx and progradation have been dynamically modifying the paleobathymetric gradient. That could have controlled the regional flow direction of the turbidite system as indicated by the alongslope variation in channel trends and play fairways. Locally, the cross-regional lineaments and the associated salt-withdrawal minibasins appear to have created local accommodation space that was responsible for the intrabasinal variability in flow direction and sediment thickness of the turbidite system.

Based on observations from the high-quality 3-D seismic data in the Tertiary section along with gravity and bathymetry, we propose a conceptual model (Figure 14) highlighting the postsalt intraslope lineaments in the context of a regional gravitational sliding system that extends more than thousands of square kilometers (miles). Two- and three-dimensional computer modeling demonstrates the concept of cross-regional shearing associated with regional differential contraction in a gravitational sliding system (Figures 15, 16). Both the conceptual and computational models focus on the structural level above the basal salt detachment and relate the cross-regional lineaments to the differential movement of the slope-forming sediments in a gravitational sliding system. These lineaments have influenced the salt evacuation and hydrocarbon accumulation, thereby adding to the complexity in salt tectonics and petroleum systems. The model, however, does not explicitly link the postsalt lineaments to the presalt basement transfer faults because the degraded quality of seismic data in the presalt section makes it difficult to substantiate such a relationship. Nevertheless, we speculate that the northeast-trending lineaments are the expression of the underlying basement transfer faults that were formed during the continental rifting phase in the Early Cretaceous. The present-day obliquity of the continental lineaments to the oceanic fracture zones can be explained by the existing plate tectonic model depicting the conjugate rotation of the continents (Figure 17) associated with the southward widening of the South Atlantic Ocean (Rabinowitz and LaBrecque, 1979; Klitgord and Schouten, 1986; Janssen et al., 1995; Eagles, 2007).

IMPLICATIONS FOR HYDROCABON EXPLORATION: DISCUSSION

Although the cross-regional lineaments are not as well recognized and as extensively documented as the regional folds, thrusts, and salt tectonics, they are spatially associated with major oil and gas discoveries in the Lower Congo Basin. In addition to the Lower Congo Basin, oil and gas fields (Mann et al., 2003) in other sedimentary basins at passive continental margins show a similar relationship to the cross-regional lineaments. For instance, in the Campos Basin offshore Brazil, the cross-regional northwest-trending lineaments, together with the regional northeast-trending folds and thrusts, are spatially associated with clusters of major hydrocarbon discoveries in the Upper Cretaceous and Tertiary sections (Gao et al., 2009). In the offshore Equatorial Guinea (west Africa), the cross-regional north–south-trending lineaments, along with the regional northeast-trending folds and thrusts, controlled the deposition of reservoir sands and the development of hydrocarbon migration pathways and traps as shown in the Alba field (Lawrence et al., 2002; Wolak and Gardner, 2008). In offshore Gabon (west Africa) (Kilby et al., 2004), the cross-regional lineaments are associated with migration pathways and structural traps as seen in the Tchitamba field that is located at a

A)

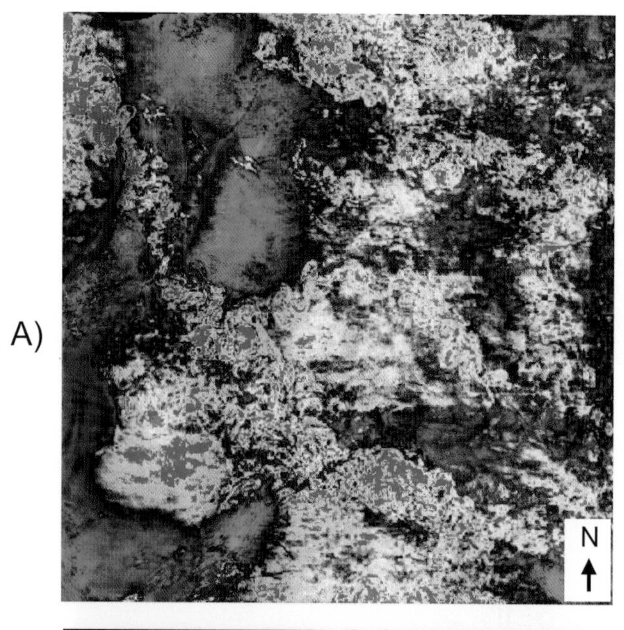

B)

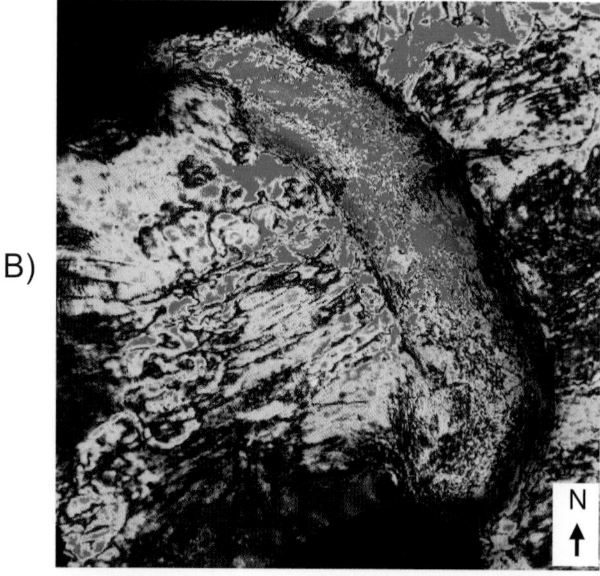

C)

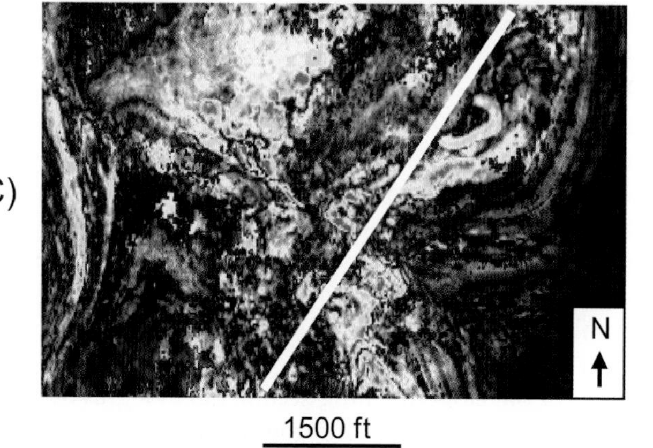

Figure 11. Seismic facies maps showing changes in turbidite flow direction along the slope. (A) In the northern part of the slope, Miocene channels flow toward the northwest that is perpendicular to the northeast-trending lineaments. (B) In the middle of the slope, Miocene channels flow toward the southwest that is perpendicular to the northwest-trending folds and thrusts. (C) In the southern part of the slope, Miocene channels flow to the south-southwest that is perpendicular to the northwest-trending lineaments. The channel was left laterally offset by the lineament. 1500 ft (457 m).

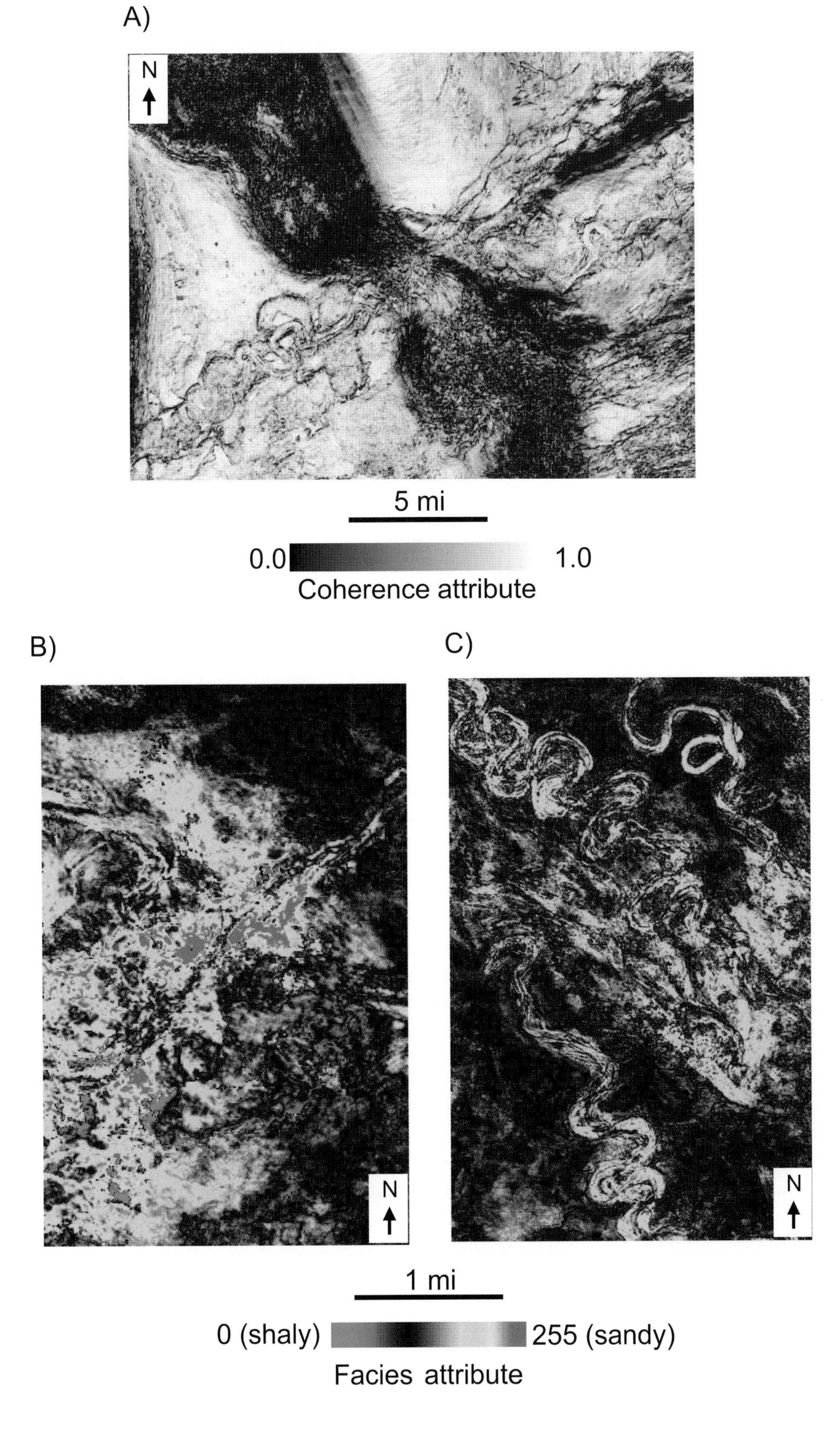

Figure 12. (A) Depth slice of the coherence attribute at the Miocene level showing the northeast-trending lineament and meandering turbidite channels that run along the lineament. The lineament provided accommodation space to influence the locus and direction of the turbidite channel flow. (B) Horizon slice showing a cross-regional southwest-flowing channel-fan system that is parallel to the cross-regional lineaments on the basin slope at the upper Miocene level. (C) Horizon slice showing a regional northwest-flowing channel system that is parallel to the slope contours at the upper Pliocene level. 5 mi (8 km); 1 mi (1.6 km).

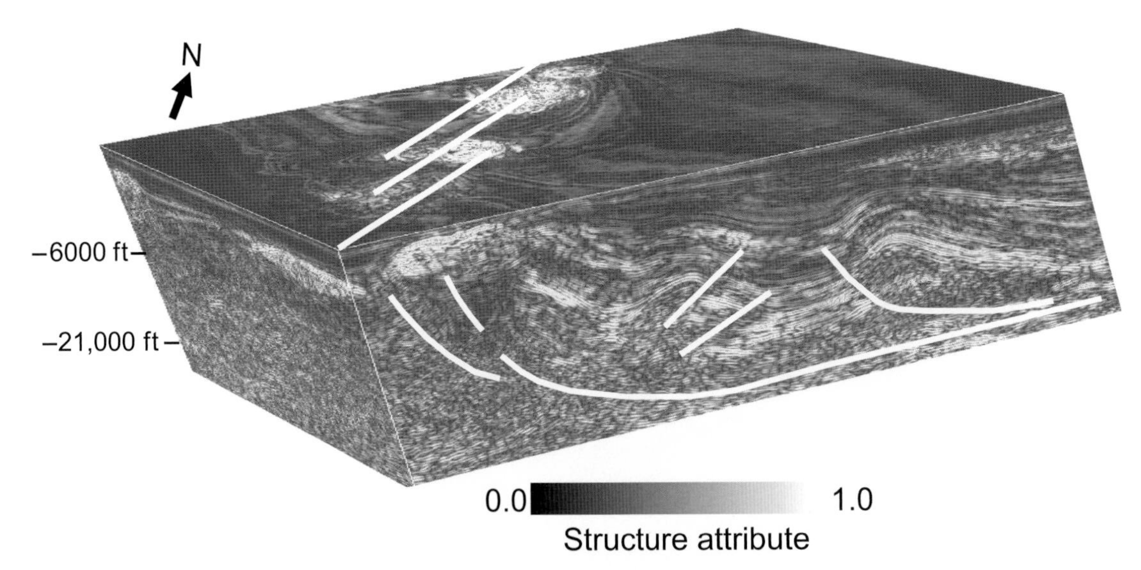

Figure 13. Northern part of the three-dimensional (3-D) seismic survey showing the cross-regional (northeast-trending) lineaments and the regional (northwest-trending) folds and thrusts. The 3-D seismic extends approximately 30 km (20 mi) in the northwestern direction and approximately 60 km (40 mi) in the northeastern direction. 6000 ft (1828.8 m); 21,000 ft (6400.8 m).

transpressional structural bend along the lineaments. In the deep-water Gulf of Mexico, the cross-regional lineaments correlate with oil and gas fields at their intersections with contractional structures such as the Mississippi Fan fold and thrust belt (Weimer and Buffler, 1992; Rowan, 1997). In all these and many other cases, cross-regional lineaments could be a critical component of petroleum systems and play an important function in the accumulation of hydrocarbons at passive continental margins worldwide.

The cross-regional intraslope lineaments in the Lower Congo Basin at the west African passive continental margin have two fundamental characteristics that are typical of strike-slip faults and make the lineaments effective migration pathways in the petroleum system. First, the lineaments feature vertical or subvertical fault planes. Steeply dipping faults and associated fractures are most effective and straightforward pathways for hydrocarbons to migrate from the deeply buried source rocks to the shallow reservoirs. Second, the lineaments have a major lateral component of shear. A major lateral component of shearing along the primary fault could generate extensive secondary fractures in response to the induced shear stress along the primary lineament (Wilcox et al., 1973). These could generate vertically well-connected fault networks, further enhancing the charging potential of oil and gas from sources to reservoirs.

Differentiating the cross-regional lineaments from the regional thrust faults can be instructive for evaluating the potential and risk of petroleum systems at the passive continental margin. The mode and slip partitioning of the cross-regional lineaments could affect the openness and migration potential of the lineaments at the margin. A simple-shear strike-slip lineament, which is typically orthogonal to the regional structural trend and parallel to the primary tectonic transport direction, features little contractional or extensional component. Increasing contractional component (transpressional) could lead to reduced porosity and permeability by closing down open pores and fractures along the lineament. Increasing extensional component (transtensional) could lead to enhanced porosity and permeability by opening up tight pores and fractures along the lineament (Wilcox et al., 1973).

Differentiating postsalt lineaments developed above the salt horizon (detachment structure) from those penetrating the salt all the way down to the basement transfer faults (basement structure) can be important in evaluating the hydrocarbon potential in the presalt and postsalt petroleum systems. Vertical penetration of the lineaments into the basement could jeopardize the top seal integrity for the presalt reservoirs but enhance the charging potential for the postsalt reservoirs by opening salt windows as migration pathways. In contrast, in cases where lineaments disappear at the salt horizon, they contribute little to the charging potential of the postsalt reservoirs from the presalt source rock. Thus, relating the detachment structural style to the basement structural style by investigating the relationships between postsalt and presalt lineaments can be instrumental in evaluating hydrocarbon potential of both presalt and postsalt petroleum systems at the South Atlantic passive continental margins.

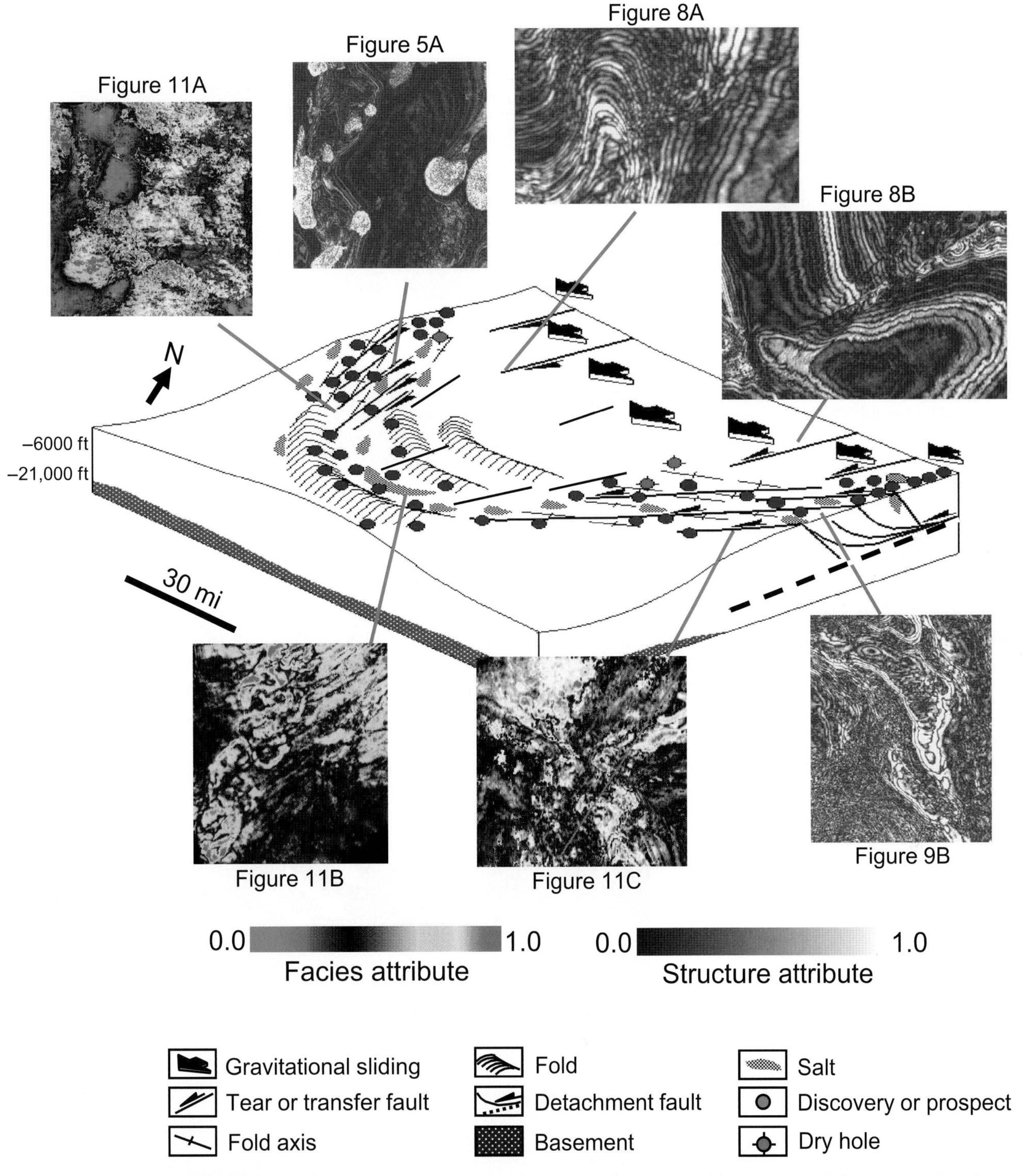

Figure 14. Schematic block diagram (Miocene subcrop) generalizing the geometry and kinematics of the cross-regional lineaments and regional folds and thrusts in a gravitational sliding system in the postsalt section on the Lower Congo Basin slope in the deep-water offshore Angola (west Africa). The inset images are extracted from the seismic structure and facies attributes of the three-dimensional seismic survey, showcasing the structural styles and depositional facies at some locations in the system. The solid circles indicate the schematic locations of hydrocarbon discoveries and prospects based on data from exploration operations and seismic direct hydrocarbon indicator (DHI) analysis (J. Helmich and E. Zhurina, 2007, personal communications; Gawenda et al., 2004). The diagram highlights the regional, curvilinear or cuspate fold-and-thrust belt and the cross-regional transfer faults. It also highlights the significance of gravitational sliding structures to the assessment of hydrocarbon potential on the basin slope (modified from Gao and Milliken, 2007). 30 mi (48 km); 1000 ft (304.8 m).

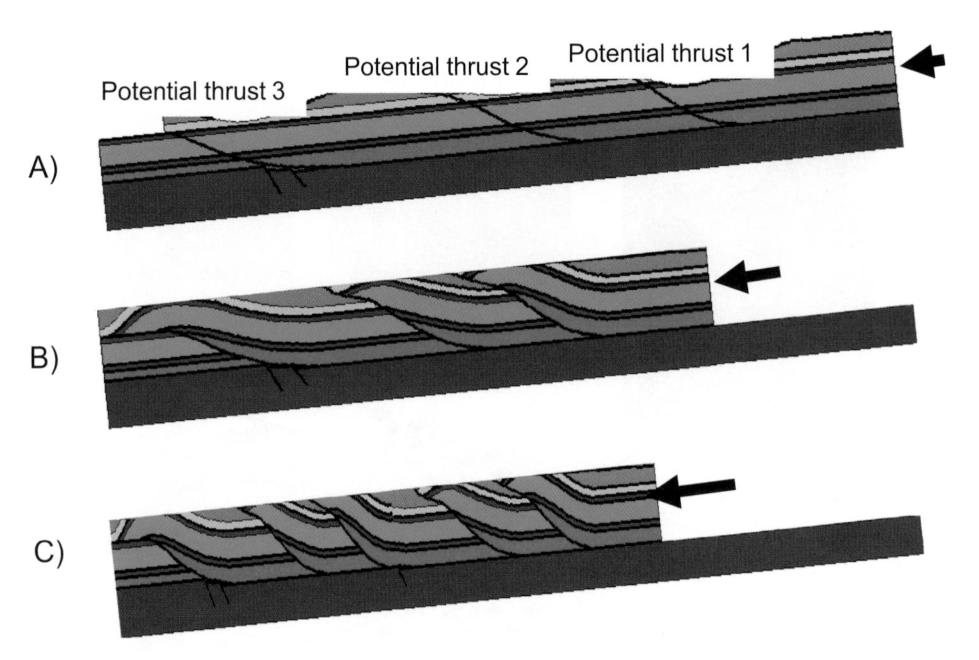

Figure 15. Two-dimensional computational modeling (not to scale) for the gravitational sliding above a salt detachment horizon, demonstrating the alongstrike differential contraction. (A) Little or no contraction outside of the gravitational sliding system. (B) Mild contraction within the gravitational sliding system. (C) Intensive contraction in the central front of the gravitational sliding system. The alongslope differential contraction is interpreted here to be accommodated by the cross-regional lineaments as possible tear faults.

CONCLUSIONS

The Bouguer gravity and bathymetric data show potential northeast-trending intraslope lineaments in the Lower Congo Basin, offshore Angola (west Africa). In concert with the Bouguer gravity and bathymetry, 3-D reflection seismic data and seismic attributes reveal many intraslope lineaments in the Tertiary section, which trend to the northeast at approximately 45° and extend tens of kilometers (miles) across the northwest-trending regional folds and thrusts. Seismic structure analysis suggests that these cross-regional lineaments might have a significant strike-slip component and are spatially associated with but kinematically different from the regional folds and thrusts. We speculate that the postsalt lineaments seen in the seismic and the sea-floor bathymetric data are the expression of the presalt basement transfer faults associated with rifting in the Early Cretaceous. In that case, the obliquity of the continental lineaments relative to the oceanic fracture zones observed in this study

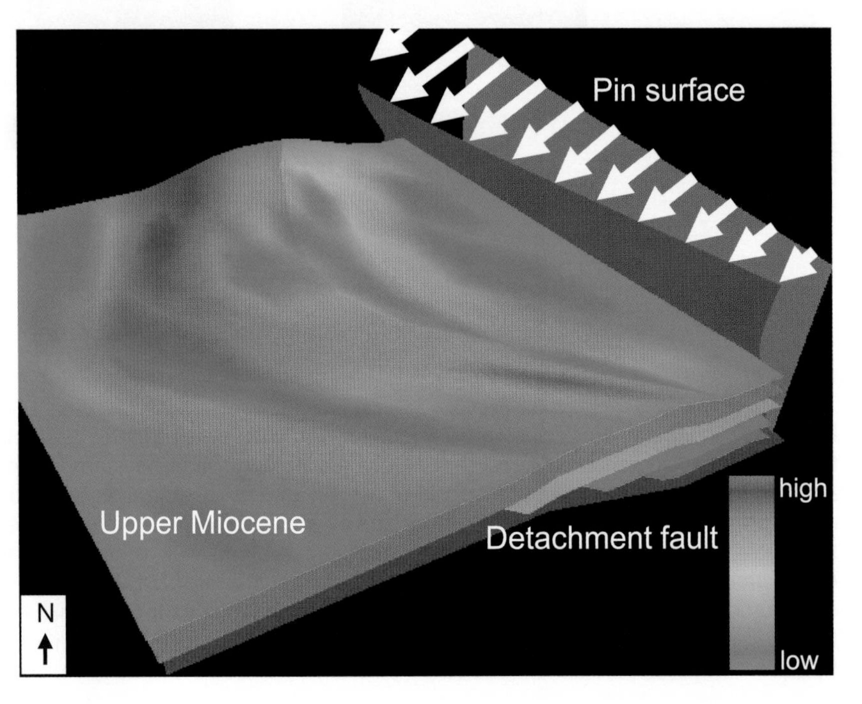

Figure 16. Three-dimensional computational modeling (not to scale) for the gravitational sliding system above a salt detachment horizon, showing the curvilinear fold belt along with cross-regional lineaments caused by the alongstrike differential movement (indicated by arrows and pin surface) of slope-forming sediments above the basal detachment horizon (red). The color on the top surface represents the local strain of the deformed surface.

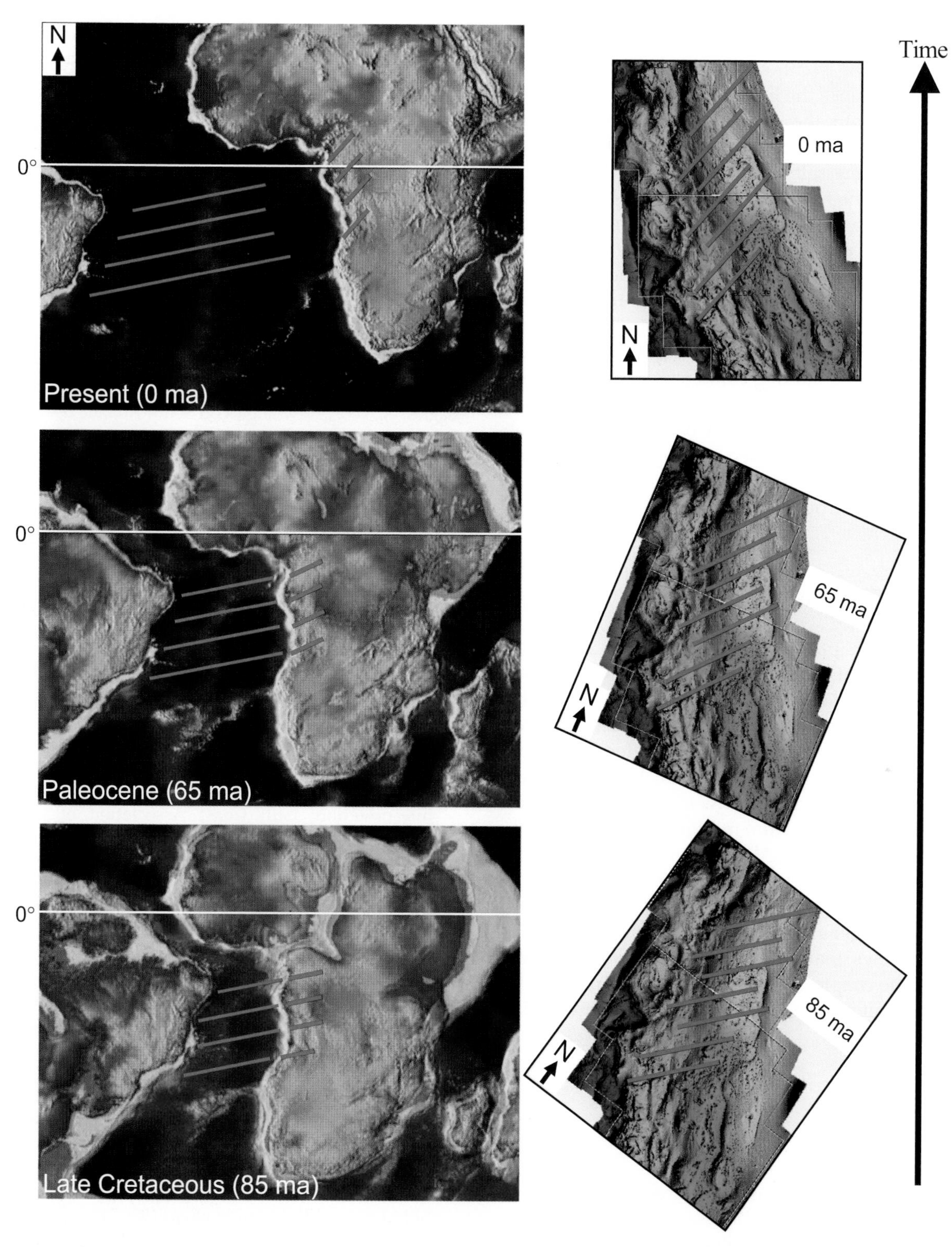

Figure 17. A schematic explanation for the present-day obliquity of the west African continental lineaments (short red lines) relative to the Atlantic oceanic fracture zones (long red lines). This explanation is consistent with and supportive of the existing plate tectonic model depicting the counterclockwise rotation of the west African continent associated with the southward opening of the South Atlantic Ocean (Rabinowitz and LaBrecque, 1979; Klitgord and Schouten, 1986; Janssen et al., 1995; Eagles, 2007) (bathymetry data source: http://topex.ucsd.edu/marine_grav/mar_grav.html).

is consistent with and supportive of the previously recognized plate tectonic model depicting the counterclockwise rotation of the west African continent associated with the southward opening of the South Atlantic Ocean since the Early Cretaceous.

Associated with the cross-regional lineaments are allochthonous salt bodies that differ from the regional northwest-trending salt ridges. Also aligned along the lineaments are the hydrocarbon discoveries, prospects, and gas hydrate deposits. These spatial relationships suggest that the cross-regional lineaments might have provided pathways for salt evacuation and hydrocarbon migration. We argue that the cross-regional lineaments, which could have a major strike-slip component, have relatively high fluid migration potential, whereas continued faulting could jeopardize the fluid retention permanency. Investigating the cross-regional lineaments in the Lower Congo Basin has important implications for a better understanding of the interplay among tectonics, sedimentation, and petroleum systems at the west African passive continental margin.

ACKNOWLEDGMENTS

A part of this chapter was presented at the 2007 Gulf Coast Associate of Geologic Societies (GCAGS) annual conference. We thank Marathon Oil Corporation for permission to publish the chapter. Marathon technology services generated 3-D depth-migrated data used in this study. Paradigm Geophysical Inc., and Midland Valley Inc., provided software for 3-D and two-dimensional (2-D) structure modeling. We also thank Shankar Mitra, Dozith Abeinomugisha, and Maria José R. Oliveira for their peer reviews. Colin P. North served as the guest editor for this chapter, and Suzanne Beglinger and Mike Sweet provided additional peer reviews that helped to further improve the quality of the chapter. This chapter is a contribution to the West Virginia University Advanced Energy Initiatives program.

REFERENCES CITED

Anderson, J. E., J. Cartwright, S. J. Drysdall, and N. Vivian, 2000, Controls on turbidite sand deposition during gravity-driven extension of a passive margin: Examples from Miocene sediments in Block 4, Angola: Marine and Petroleum Geology, v. 17, p. 1165–1203.

Bahorich, M., and S. Farmer, 1995, 3-D seismic discontinuity for faults and stratigraphic features: The coherence cube: The Leading Edge, v. 14, p. 1053–1058.

Beglinger, S. E., H. Doust, and S. Cloetingh, 2012, Relating petroleum system and play development to basin evolution: West African South Atlantic Basins: Marine and Petroleum Geology, v. 30, p. 1–25, doi:10.1016/j.marpetgeo.2011.08.008.

Brice, S. E., M. D. Cochran, G. Pardo, and A. D. Edwards, 1982, Tectonics and sedimentation of the South Atlantic rift sequence, Cabinda, Angola, *in* J. S. Watkins and C. L. Drake, eds., Studies in continental margin geology: AAPG Memoir 34, p. 5–18.

Cramez, C., and M. P. A. Jackson, 2000, Superposed deformation straddling the continental-oceanic transition in deep-water Angola: Marine and Petroleum Geology, v. 17, p. 1095–1109, doi:10.1016/S0264-8172(00)00053-2.

Da Costa, J. L., T. W. Schirmer, and B. R. Laws, 2001, Lower Congo Basin, deep-water exploration province, offshore west Africa: AAPG Memoir 74, p. 517–530.

Eagles, G., 2007, New angles on South Atlantic opening: Geophysical Journal International, v. 166, p. 353–361, doi:10.1111/j.1365-246X.2006.03206.x.

Fleming, R. W., and A. M. Johnson, 1989, Structures associated with strike-slip faults that bound landslide elements: Engineering Geology, v. 27, p. 39–114, doi:10.1016/0013-7952(89)90031-8.

Gao, D., 2004, Texture model regression for effective feature discrimination: Application to seismic facies visualization and interpretation: Geophysics, v. 69, p. 958–967, doi:10.1190/1.1778239.

Gao, D., 2006a, Structure-oriented texture model regression for seismic structure visualization and interpretation: Gulf Coast Association of Geological Societies Transactions, v. 56, p. 207–216.

Gao, D., 2006b, Theory and methodology for seismic texture analysis: Implications for seismic facies visualization and interpretation: Gulf Coast Association of Geological Societies Transactions, v. 56, p. 217–226.

Gao, D., and J. Milliken, 2007, Gravity-induced transfer faults on the Lower Congo Basin slope, offshore Angola (west Africa): Implications for deep-marine hydrocarbon exploration on passive continental margins: Gulf Coast Associate of Geologic Societies Transactions, v. 57, p. 271–289.

Gao, D., L. Seidler, D. Quirk, M. Bissada, M. Farrell, and D. Hsu, 2009, Intraslope northwest-trending lineaments and geologic implications in the central Campos Basin, offshore Brazil: AAPG Search and Discovery article 90100, http://www.searchanddiscovery.net/documents/2009/90100gao/ndx_gao.pdf (accessed August 29, 2012).

Gardner, M. H., and J. M. Borer, 2000, Submarine channel architecture along a slope to basin profile, Brushy Canyon Formation, west Texas, *in* A. H. Bouma and C. G. Stone, eds., Fine-grained turbidite systems: AAPG Memoir 72/SEPM Special Publication 68, p. 195–214.

Gawenda, P., J.-M. Conne, A. Hayman, and M. Marchat, 2004, Offshore west Africa offers exceptional opportunities: Offshore, February, p. 42–44.

Guiraud, M., A. Buta-Neto, and D. Quesne, 2010, Segmentation and differential postrift uplift at the Angola margin as recorded by the transform-rifted Benguela and oblique-to-orthogonal–rifted Kwanza basins: Marine and Petroleum Geology, v. 27, p. 1040–1068, doi:10.1016/j.marpetgeo.2010.01.017.

Harding, T. P., 1990, Identification of wrench faults using subsurface structural data: Criteria and pitfalls: AAPG Bulletin, v. 74, p. 1590–1609.

Harris, N. B., 2000, Toca carbonate, Congo Basin: Response to an evolving rift lake, *in* M. R. Mello and B. J. Katz, eds., Petroleum systems of South Atlantic margins: AAPG Memoir 73, p. 341–360.

Harris, N. B., K. H. Freeman, R. D. Pancost, T. S. White, and G. D. Mitchell, 2004, The character and origin of lacustrine source rocks in the Lower Cretaceous synrift section, Congo Basin, west Africa: AAPG Bulletin, v. 88, p. 1163–1184, doi:10.1306/02260403069.

Hempton, M. R., M. A. Rosen, R. M. Coughlin, and A. D. Scardina, 1990, The geology of west Africa: A regional review (abs.): AAPG Bulletin, v. 75, p. 583.

Hudec, M. R., and M. P. A. Jackson, 2002, Structural segmentation, inversion, and salt tectonics on a passive margin: Evolution of the inner Kwanza Basin, Angola: Geological Society of America Bulletin, v. 114, p. 1222–1244, doi:10.1130/0016-7606(2002)114<1222:SSIAST>2.0.CO;2.

Hudec, M. R., M. P. A. Jackson, and D. Jennette, 2004, Influence of precursor salt structures on thrust faulting, deep-water Lower Congo Basin, Gabon (abs.): AAPG Annual Meeting Program 13, p. A67.

Jackson, M. P. A., and M. R. Hudec, 2009, Interplay of basement tectonics, salt tectonics, and sedimentation in the Kwanza Basin, Angola: AAPG Search and Discovery article 30091, http://www.searchanddiscovery.net/documents/2009/30091jackson/ndx_jackson.pdf (accessed August 29, 2012).

Jackson, M. P. A., C. Cramez, and W. U. Mohriak, 1998, Salt-tectonic provinces across the continental-oceanic boundary in the lower Congo and Campos Basins on the South Atlantic margins (abs.): AAPG International Conference and Exhibition Extended Abstracts, p. 40–41.

Jackson, M. P. A., M. R. Hudec, and D. C. Jennette, 2004, Insights from a gravity-driven linked system in deep-water Lower Congo Basin, Gabon, *in* P. J. Post, D. L. Olson, K. T. Lyons, S. L. Palmes, P. F. Harrison, and N. C. Rosen, eds., Salt-sediment interactions and hydrocarbon prospectivity: Concepts, applications, and case studies for the 21st century: 24th Annual Gulf Coast Section SEPM Foundation Bob F. Perkins Research Conference, Houston, Texas, December 5–8, 2004, p. 735–752.

Jackson, M. P. A., M. R. Hudec, D. C. Jennette, and R. E. Kilby, 2008, Evolution of the Cretaceous Astrid thrust belt in the ultradeep-water Lower Congo Basin, Gabon: AAPG Bulletin, v. 92, p. 487–511, doi:10.1306/12030707074.

Jameson, M., S. Wells, J. Greenhalgh, and R. Borsato, 2011, Prospectivity and seismic expression of pre- and postsalt plays along the conjugate margins of Brazil, Angola and Gabon: Society of Exploration Geophysicists Annual Meeting, San Antonio, Texas, September 18–23, 2011: http://www.onepetro.org/mslib/servlet/onepetropreview?id=SEG-2011-1062 (accessed October 2, 2012).

Janssen, M. E., R. A. Stephenson, and S. Cloetingh, 1995, Temporal and spatial correlations between changes in plate motions and the evolution of rifted basins in Africa: Geological Society of America Bulletin, v. 11, p. 1317–1332, doi:10.1130/0016-7606(1995)107<1317:TASCBC>2.3.CO;2.

Kilby, R. E., M. P. A. Jackson, and M. R. Hudec, 2004, Preliminary analysis of thrust kinematics in the Lower Congo Basin, deep-water southern Gabon (abs.): Geological Society of America Abstracts with Programs, v. 36, p. 505.

Klitgord, K. D., and H. Schouten, 1986, Plate kinematics of the central Atlantic, *in* P. R. Vogt and B. E. Tucholke, eds., The geology of North America: The western North Atlantic region: The Geological Society of America Bulletin, v. M, p. 351–378.

Kolla, V., P. Bourges, J.-M. Urruty, and P. Safa, 2001, Evolution of deep-water Tertiary sinuous channels offshore Angola (west Africa) and implications for reservoir architecture: AAPG Bulletin, v. 85, p. 1373–1405.

Lawrence, S. R., S. Munday, and R. Bray, 2002, Regional geology and geophysics of the eastern Gulf of Guinea (Niger Delta to Rio Muni): The Leading Edge, v. 21, p. 1112–1117, doi:10.1190/1.1523752.

Lundin, E. R., 1992, Thin-skinned extensional tectonics on a salt detachment, northern Kwanza Basin, Angola: Marine and Petroleum Geology, v. 9, p. 405–411, doi:10.1016/0264-8172(92)90051-F.

Mann, P., L. M. Gahagan, and M. B. Gordon, 2003, Tectonic setting of the world's giant oil fields, *in* M. Halbouty and M. Horn, eds., Giant oil and gas fields of the decade, 1990–2000: AAPG Memoir 78, p. 15–105.

Marton, L. G., G. C. Tari, and C. T. Lehmann, 2000, Evolution of the Angolan passive margin, west Africa, with emphasis on postsalt structural styles, *in* W. Mohriak and M. Talwani, eds., Atlantic rifts and continental margins: American Geophysical Union Geophysical Monograph 115, p. 129–149.

Nombo-Makaya, N. L., and C. H. Han, 2009, Presalt petroleum system of Vandji-Conkouati structure (Lower Congo Basin), Republic of Congo: Research Journal of Applied Sciences, v. 4, p. 101–107.

Posamentier, H. W., and V. Kolla, 2003, Seismic geomorphology and stratigraphy of depositional elements in deep-water settings: Journal of Sedimentary Research, v. 73, p. 367–388, doi:10.1306/111302730367.

Rabinowitz, P. D., and J. LaBrecque, 1979, The Mesozoic South Atlantic Ocean and evolution of its continental margins: Journal of Geophysical Research, v. 84, p. 5973–6002, doi:10.1029/JB084iB11p05973.

Rowan, M. G., 1997, Three-dimensional geometry and evolution of a segmented detachment fold, Mississippi Fan fold belt, Gulf of Mexico: Journal of Structural Geology, v. 19, p. 463–480, doi:10.1016/S0191-8141(96)00098-3.

Rowan, M. G., F. J. Peel, and B. C. Vendeville, 2004, Gravity-driven fold belts on passive margins, *in* K. R. McClay, ed., Thrust tectonics and hydrocarbon systems: AAPG Memoir 82, p. 157–182.

Tari, G., J. Molnar, and P. Ashton, 2003, Examples of salt

tectonics from west Africa: A comparative approach: Geological Society (London) Special Publication 207, p. 85–104.

Valle, P. J., J. G. Gjelberg, and W. Helland-Hansen, 2001, Tectonostratigraphic development in the eastern Lower Congo Basin, offshore Angola, west Africa: Marine and Petroleum Geology, v. 18, p. 909–927, doi:10.1016/S0264-8172(01)00036-8.

Vendeville, B. C., and M. P. A. Jackson, 1992, The fall of diapirs during thin-skinned extension: Marine and Petroleum Geology, v. 9, p. 354–371, doi:10.1016/0264-8172(92)90048-J.

Versfelt, J. W., 2010, South Atlantic margin rift basin asymmetry and implications for presalt exploration: AAPG Search and Discovery article 30112, http://www.searchanddiscovery.net/documents/2010/30112versfelt/ndx_versfelt.pdf (accessed August 29, 2012).

Weimer, P., and R. Buffler, 1992, Structural geology and evolution of the Mississippi Fan fold belt, deep Gulf of Mexico: AAPG Bulletin, v. 76, p. 225–251.

Wilcox, R. E., T. P., Harding, and D. R. Seely, 1973, Basic wrench tectonics: AAPG Bulletin, v. 57, p. 74–96.

Wolak, J. M., and M. H. Gardner, 2008, Synsedimentary structural growth in a deep-water reservoir, Alba field, Equatorial Guinea: AAPG International Conference and Exhibition, Cape Town, South Africa October 26–29, 2008: AAPG Search and Discovery article 90082, http://www.searchanddiscovery.com/abstracts/html/2008/intl_capetown/abstracts/495780.htm (accessed October 2, 2012).

11

Linzer, Hans-Gert, and Gabor C. Tari, 2012, Structural correlation between the Northern Calcareous Alps (Austria) and the Transdanubian Central Range (Hungary), *in* D. Gao, ed., Tectonics and sedimentation: Implications for petroleum systems: AAPG Memoir 100, p. 249–266.

Structural Correlation between the Northern Calcareous Alps (Austria) and the Transdanubian Central Range (Hungary)

Hans-Gert Linzer

Exploration and Production Oil RAG Rohöl-Aufsuchungs Aktiengesellschaft Schwarzenbergplatz 16, A-1015 Vienna, Austria (e-mail: hans-gert.linzer@rag-austria.at)

Gabor C. Tari

OMV Exploration and Production GmbH Trabrennstrasse 6-8, A-1020 Vienna, Austria (e-mail: gabor.tari@omv.com)

ABSTRACT

The classical Alpine folded belt of the Northern Calcareous Alps (NCA) of Austria is correlated with the Transdanubian Central Range (TCR) of Hungary using structural and stratigraphic relationships to restore the system. The semiquantitative map-view restoration of several consecutive Alpine deformational periods reveals unexpected similarities between the NCA and TCR. In fact, some west–northwest-trending right lateral strike-slip faults in the TCR (e.g., Telegdi-Roth, Padrag, and Vargesztes faults) are interpreted here for the first time to be analogous to those described from the NCA (e.g., Lammertal, Wolfgangsee-Windischgarsten, and Hochwart faults). These middle to late Miocene transpressional faults are reactivated in the Late Cretaceous tear faults, as can be documented by reflection seismic data in the subsurface of the southeastern Danube Basin. The structural correlation between the NCA and TCR provides further evidence for the much debated interpretation of the TCR in terms of a large Eo-Alpine (Cretaceous) nappe-system in an Uppermost Austroalpine structural position. Furthermore, recognition of a once continuous, regional-scale, right lateral strike-slip fault system in the NCA-TCR areas has a significant impact on the pre-Tertiary kinematic reconstructions of the broader Eastern Alps and Pannonian Basin region.

INTRODUCTION

The Northern Calcareous Alps (NCA) and the Transdanubian Central Range (TCR), although located some 200 km (~124 mi) apart in their present-day position (Figure 1), have very similar stratigraphic successions. However, the level of understanding of their respective tectonic styles is quite different. It is widely

DOI:10.1306/13351556M1003533

accepted that the NCA of Austria and Germany represents a thin-skinned fold and thrust belt along the northern margin of the Eastern Alps, thrust onto the Penninic series of the Rhenodanubian flysch, and that the Helvetic nappes are derived from the distal European passive margin and were thrust onto the Molasse foreland basin (Figure 1). In fact, the concept of the nappe system of the NCA was established almost a century ago (Ampferer, 1912, 1932, 1939; Tollmann, 1976). The thin-skinned folded belt nature of the NCA was also confirmed and documented by industry reflection seismic lines and deep wells (Wachtel and Wessely, 1981; Kröll et al., 1981; Bachmann et al., 1982). In contrast, the TCR of Hungary is traditionally considered a simple autochthonous unit without any internal deformation (e.g., Haas et al., 1995). An alternative interpretation was presented by Tari (1994, 1996) based on the analysis of reflection seismic and well data across the Danube Basin, describing the TCR as a stack of Alpine nappes with very similar deformational styles to that of the NCA. The allochthonous versus autochthonous nature of the TCR is still a subject under debate (see discussion in Tari and Horvath, 1995).

Presently, the NCA and TCR are separated by the Danube and Styrian basins that formed because of the large-scale crustal extension in the Pannonian Basin system during the Middle Miocene to Pliocene (Tari, 1995). As to the present-day topography of these major tectonic units, the NCA has an average elevation of about 2000 m (~6562 ft), whereas the TCR has a very subtle topography barely emerging from underneath the Pannonian Basin. The average elevation of the TCR is about 150 m (~492 ft) above sea level, with the highest point just a little more than 700 m (>2297 ft).

Structural analysis and interpretation of reflection profiles in both areas facilitated a quantitative reconstruction of early Alpine shortening (Linzer et al., 1995; Tari, 1995) and late Alpine extension (Tari and Horvath, 1995). The late Alpine (Miocene) extension between the NCA and the TCR was superposed on the Early Cretaceous nappe stacking of the TCR (Tari, 1995).

The structural evolution of the Alpine-Pannonian transition area presently separating the NCA and the TCR (Figure 1) is very complex with several deformational episodes recorded in the broader Danube and Styrian basins and their flanks. The Early Cretaceous nappes were sealed by Late Cretaceous sequences, which were deposited in a foredeep basin, typical for flexural basins. After a major break in the sedimentation between the Senonian and the middle Eocene, the Senonian basins were replaced by a set of Paleogene flexural basins in retro-arc setting (Tari et al., 1993). The Early Miocene extensional collapse and the formation of metamorphic core complexes at the western margin of the Pannonian Basin resulted in a minimum crustal extension of 80 km (50 mi) in an east–northeast-west–southwest direction. This early core complex style extension occurred during the Ottnangian-Karpatian (17.5–16.5 Ma). After the middle Miocene (Badenian, 16.5–13.8 Ma), the northwest–southeast-directed rift-style extension was followed by the late Miocene (Pannonian, 12–1.65 Ma) postrift subsidence phase forming the Danube Basin. Quarternary inversion because of east–west compression (Horváth, 1995; Peresson and Decker, 1997) began to slightly invert the Danube Basin and the preexisting Alpine fault systems in the NCA.

The aim of this chapter is to show that the proper palinspastic restoration of the postnappe deformations in the transitional zone between the NCA and TCR reveals the close Austro-Hungarian relationship of these major Alpine units despite their present-day separation.

REGIONAL SETTING AND STRATIGRAPHIC CORRELATION OF THE NORTHERN CALCAREOUS ALPS AND THE TRANSDANUBIAN CENTRAL RANGE

The following brief description of the NCA and TCR units is intended to summarize the elements that are critical to the understanding of the correlation approach used in this chapter.

In the Eastern Alps, the Austroalpine accretionary wedge consists (from north to south) of Molasse imbricates, the Helvetic nappes, Rhenodanubian Flysch nappes, the NCA, the Grauwacken zone, the Austroalpine basement nappes (AA), and the Penninic windows (Figure 1).

The Tauern Window (TW) is composed of metasedimentary rocks and granitoids subdivided into three units: the Zentralgneiss core, the Paleozoic/Permian–Mesozoic Lower Schieferhülle (distal European crust), and the Mesozoic Upper Schieferhülle that represents transitional and oceanic sequences of the Penninic realm (Frisch, 1974, 1977, 1980; Morteani, 1974; Frisch et al., 1987). The eclogite zone at the base of the Upper Schieferhülle shows ^{39}Ar-^{40}Ar ages between 36 and 32 Ma, indicating the metamorphic peak in the early Oligocene. Retrograde metamorphism started in the late Oligocene because of slab detachment (~32 Ma) and continental collision (Selverstone, 1988; Blanckenburg et al., 1989; Blanckenburg and Kagami, 1998; Ratschbacher et al., 1991; Selverstone, 1993; Zimmermann et al., 1994; Froitzheim et al., 1996; Schmid et al., 1996).

The Rechnitz and Bernstein windows (RW and BW, respectively) at the eastern end of the Alps are composed of oceanic-type sediments and ophiolites similar to the TW (Koller and Pahr, 1980; Ratschbacher et al., 1990). Deformation structures in the RW and BW suggest

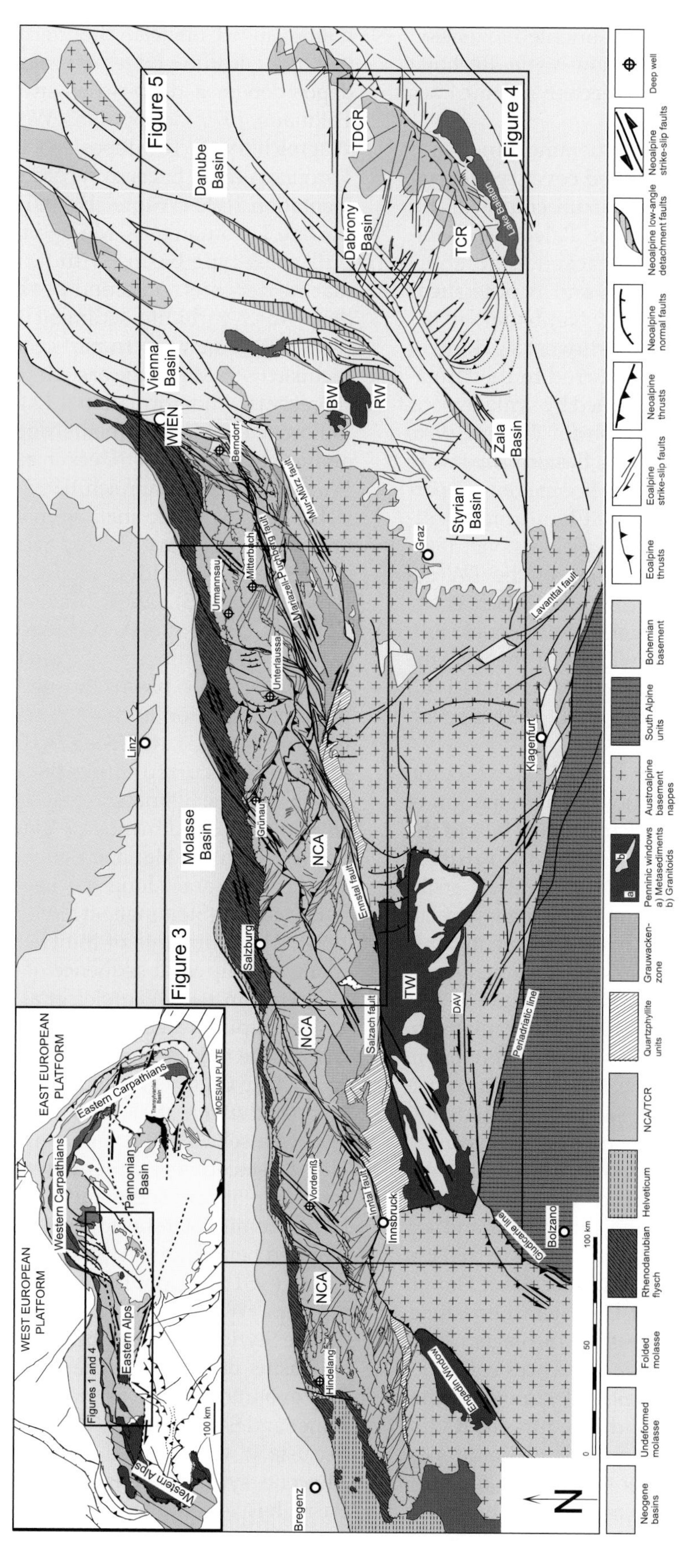

Figure 1. Alpine structural elements of the Eastern Alps and the Western Pannonian Basin. NCR = Northern Calcareous Alps; TCR = Transdanubian Central Range.

east–west to northeast–southwest-oriented extension (Ratschbacher et al., 1990). The low-angle Rechnitz detachment fault was traced by reflection seismic lines in the Danube Basin (Tari, 1996).

The AA are composed of polymetamorphic basement complexes and their associated cover sequences. The basement complexes (e.g., Ötztal complex and Innsbruck Quartzphyllite) and the Paleozoic metasedimentary rocks of the Grauwacken zone show Variscan penetrative deformation and Alpine shear zones (Satir and Morteani, 1978; Neubauer et al., 1995). East–southeast to west–northwest stacking of the basement nappe complex occurred in the Cretaceous (Ratschbacher, 1986), as indicated by synkinematic temperature-dominated metamorphism (Thöni, 1986; Schmid and Haas, 1989). The central Eastern Alps were strongly reactivated by Oligocene–Miocene orogen parallel extension and low-angle ductile normal faulting and lateral extrusion of the AA (Selverstone, 1988; Ratschbacher et al., 1991). Updoming of the TW occurred in the early to middle Miocene (Frisch et al., 1998), synchronous with the low-angle detachment faulting on both sides of the window. East of the TW, the Austroalpine units are detached at the contact with the Penninic units (Ratschbacher et al., 1991; Becker, 1993).

The Northern Calcareous Alps form a 500 km (311 mi) long fold and thrust belt at the northeastern margin of the Alps (Figures 1, 2). The fold and thrust belt of the NCA consists of a 3- to 5-km (1.9- to 3.1-mi) thick sequence of Permian–Mesozoic sedimentary rocks (Tollmann, 1976; Lein, 1987). Competent Triassic platform carbonates are separated by incompetent marls and evaporitic series, forming major detachment horizons and well-defined seismic reflectors (Figure 3). The lower carbonate series is composed of reef platform carbonates (up to 1500 m [4921 ft] thick Wetterstein limestone and dolomite), with a lateral facies transition into basinal shales and limestones (Bechstädt and Mostler, 1976). The intrashelf basins were filled during the Carnian with clastic terrigenous input and evaporites caused by rapid subsidence of the carbonate platforms (Schlager and Schöllnberger, 1974; Brandner, 1978). The upper carbonate complex (up to 2000 m [6562 ft] thick) is composed of the ultrabackreef Hauptdolomite, the backreef and reef Dachstein Limestone, and the pelagic Hallstatt Limestone (Schlager, 1967; Schöllnberger, 1973). The Jurassic sequences consist of Early Jurassic breccia complexes and crinoidal limestones, Middle Jurassic marls and Late Jurassic Radiolarites interbedded by pelagic marls and limestones. Locally preserved erosional relics of syntectonic sediments such as the late Valanginian to the early Aptian Roßfeld beds, the Albian Losenstein beds, and the Cenomanian to Coniacian Branderfleck beds were deposited along leading edges of major thrusts and indicate deposition in a deep-water environment (Faupl and Tollmann, 1979; Gaupp, 1983; Weidich, 1984; Faupl and Wagreich, 1992). The deposits of the Gosau Group (Late Turonian–early Eocene) record a change in the sedimentation style (Wagreich, 1991). High subsidence in the late Turonian (~90 Ma) led to the development of pull-apart-type basins with terrestrial and shallow-marine deposits. A second peak in subsidence led to deep-water turbidite sedimentation in the Santonian (~85 Ma; in the western part of the NCA) to the Maastrichtian (~66 Ma; in the eastern part). The Gosau Group represents syndeformational Late Cretaceous (90 Ma) to Eocene (50 Ma) sediments, deposited on the Austroalpine nappe system (Decker et al., 1987; Faupl and Wagreich, 1992; Wagreich, 1995). The lower part consists of freshwater to shallow-marine coarse sediments, and the upper part of deep-water sediments, which are related to a sudden deepening of the whole NCA (Wagreich, 1995). The Gosau sediments cover some traces of thrust sheets, but the faults were generally reactivated by Miocene deformation.

Intraorogenic basins formed because of orogen-parallel extension and are markers for the dating of tectonic events. Late Eocene to Oligocene (37–27 Ma) sediments occurred along the Inn fault system and lower Miocene (Ottnangian–Karpatian–Badenian; 18–16 Ma) clastic sediments are situated along the SEMP (Salzach-Ennstal-Mariazell-Puchberg) fault (Enns Valley, Hieflau) and along the Lavanttal and Mur-Mürz fault systems (Steininger et al., 1988; Ratschbacher et al., 1991). The sediments of the Hieflau Basin represent a coarsening-upward sequence of clastics of Ottnangian to Karpatian age (Wagreich et al., 1997).

The Alpine foredeep, the Molasse Basin, contains late Eocene to Miocene sediments unconformably overlaying Late Cretaceous sediments on the European foreland (Bachmann et al., 1987; Nachtmann and Wagner, 1987; Wessely, 1987; Bachmann and Müller, 1991). The Late Cretaceous sediments were deposited on Late Jurassic platform sediments. The Mesozoic sediments form prominent reflectors on seismic data and were mapped beneath the Alpine nappes to the south, some 50 km (~31 mi) from the present-day edge of the Eastern Alps (Wessely, 1987).

The Styrian and Danube basins are the westernmost subbasins of the Pannonian Basin system (Figure 1). The evolution of the Styrian Basin (Kröll et al., 1988; Ebner and Sachsenhofer, 1991) is fairly well understood because it was formed in a single stage of Middle Miocene synrift extension. In contrast, the Danube Basin has a complex multistage history (Tari et al., 1995). These deformational episodes left behind several

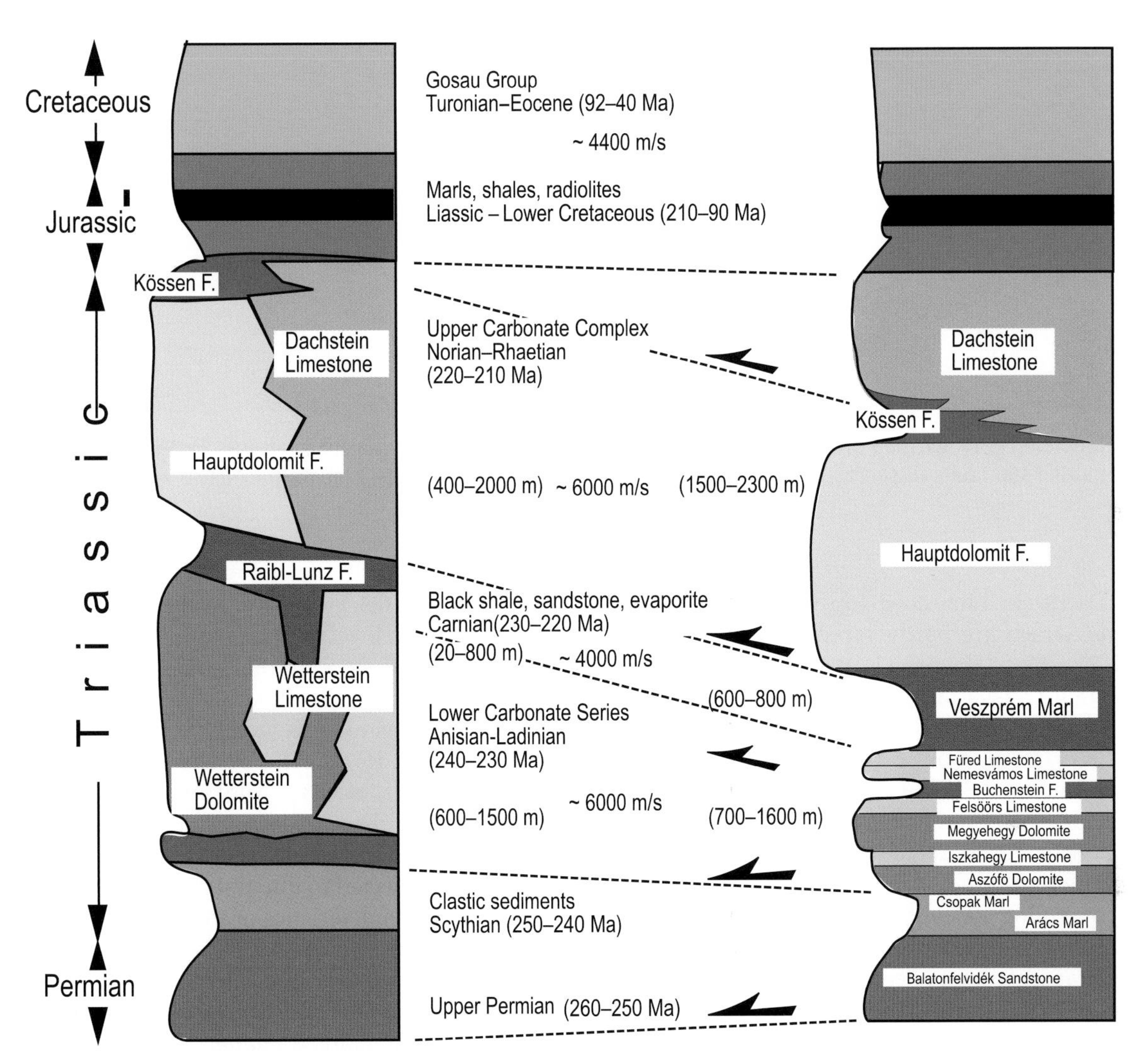

Figure 2. Simplified stratigraphy and correlation of documented detachment surfaces within the Mesozoic sequences of the Northern Calcareous Alps (NCA) and the Transdanubian Central Range (TCR). Compiled from various sources listed in the text.

structural elements including (1) Eo-Alpine (i.e., Mesozoic) nappes and flexural basins (Tari, 1994) commonly simplified in the literature as "basement"; (2) Paleogene basin fragments (upper Eocene–lower Miocene) as part of a larger retro-arc foredeep basin (Tari et al., 1993); (3) middle Miocene synrift extensional structures; (4) upper Miocene–Pliocene postrift sequences; and (5) Quaternary uplift and erosion, locally developed flower structures related to the regional neotectonic inversion of the Pannonian Basin system (Tari, 1994; Horváth, 1995; Decker and Peresson, 1996).

The TCR (Figure 4) consists of Austroalpine basement units of slightly metamorphosed Paleozoic strata and Permian sandstones and evaporites. The Triassic sequences are composed of a predominantly carbonate platform succession with thicknesses up to 3000 m (9843 ft), alternating with dolomites and limestones (Figure 3). Some of the more important formations include the Carnian Veszprém Marl (500–800 m [1640–2625 ft]), which is composed of interbedded sandstone and limestone, representing fine-grained deep-water terrigenous rocks. The overlying carbonate complex is the Hauptdolomite (up to 1500 m [4921 ft])

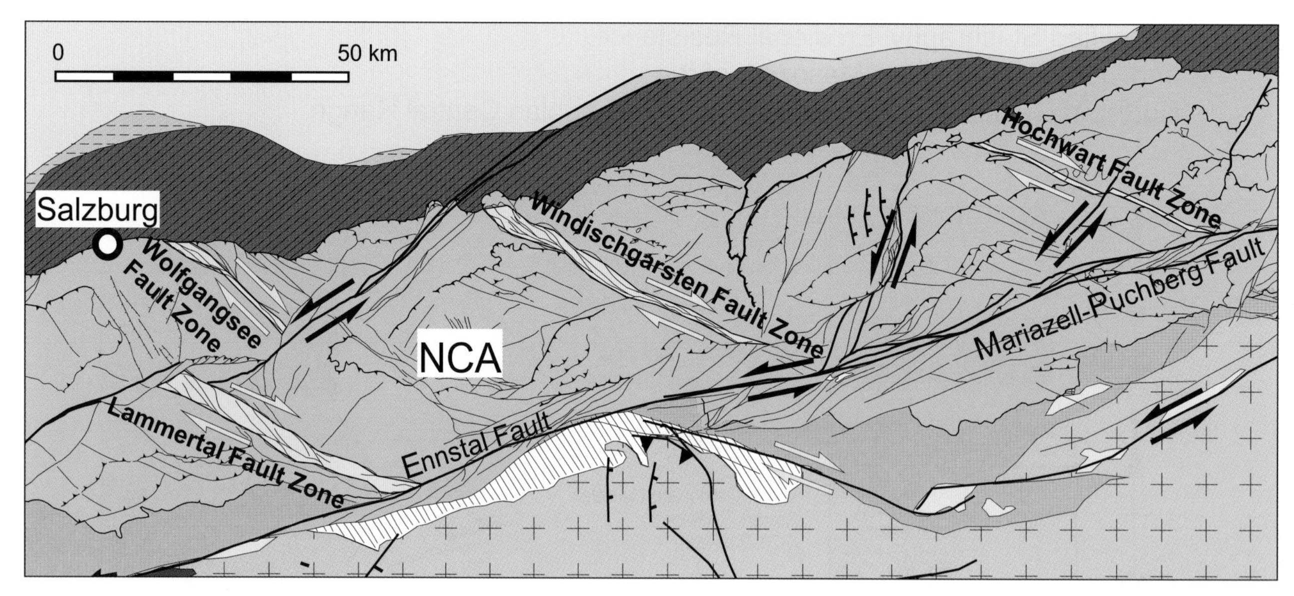

Figure 3. Simplified geologic map of the eastern Northern Calcareous Alps (NCA) (Linzer et al., 1997) (Figure 1) Note the large number of strike-slip faults dissecting the Alpine thrusts trending west-northwest–east-southeast. 50 km (31 mi).

and the Dachstein Limestone (up to 800 m [2625 ft]) formations, which are partly separated by the marly Kössen Formation (Haas et al., 1995). Compared with the very thick Triassic strata, the predominantly deep-water Jurassic of the TCR is fairly condensed with a typical thickness of about 300 m (~984 ft). The overlying Cretaceous sediments (up to 1000 m [3281 ft] thick) developed in a siliciclastic facies with significant unconformities related to several Eo-Alpine deformational periods. After a major stratigraphic gap, the overlying Paleogene basin fragments have a typical foredeep sequence, that is, neritic carbonates at the onset, transitioning to deep-water marls and shales (Tari et al., 1993). Miocene and younger sediments are poorly preserved in the TCR because of the significant Quaternary uplift and erosion (Figure 4).

CHARACTERISTIC REFLECTION SEISMIC PROFILES OF THE NORTHERN CALCAREOUS ALPS AND THE TRANSDANUBIAN CENTRAL RANGE

Despite their first-order lithologic and structural similarities, the NCA and the TCR differ fundamentally in their exhumation history. Although large parts of the fold and thrust belt of the NCA were uplifted and eroded in the postcollisional Miocene (13–10 Ma) (Frisch et al., 1998), the TCR was under extension and subsidence during most of the Miocene. The exhumation of the NCA truncated most of the anticlines, and younger strata were preserved only in the cores of prominent synclines. The level of preservation of the overall structure, including the synkinematic sequences is better in the TCR. Another important difference is the amount of industry seismic and well data available in the NCA and the TCR (Figure 5). Whereas the TCR has seen several hydrocarbon exploration campaigns on its northwestern flank (Körössy, 1987), the NCA remains practically unexplored. The following seismic sections were selected to illustrate the similarities between the Alpine characteristics of the NCA and the TCR despite differences in the seismic data quality.

As in many other fold and thrust belts throughout the world, the seismic imaging of the NCA imbricates and nappes remains a challenge, even with carefully designed, acquired, and processed regional seismic profiles (Figure 6). The subsurface geometry of the individual imbricates of the NCA is poorly constrained by the reflection seismic data, although the surface geometry is very well constrained by the excellent outcrop conditions.

Another seismic example from the NCA is located in the central part, in the Weyer Arc region (Figure 7). The Weyer Arc structure in the eastern part of the NCA (Figure 4) is composed of the east–west-striking footwall thrust structure, which are covered by synorogenic Gosau sediments and the bent hanging-wall thrust sheets (Linzer et al., 2002). Erosion of the Gosau beds was prevented by the load of the large-scale post-Gosau thrust. These beds extend in a north–south direction across nappes and internal thrusts of the

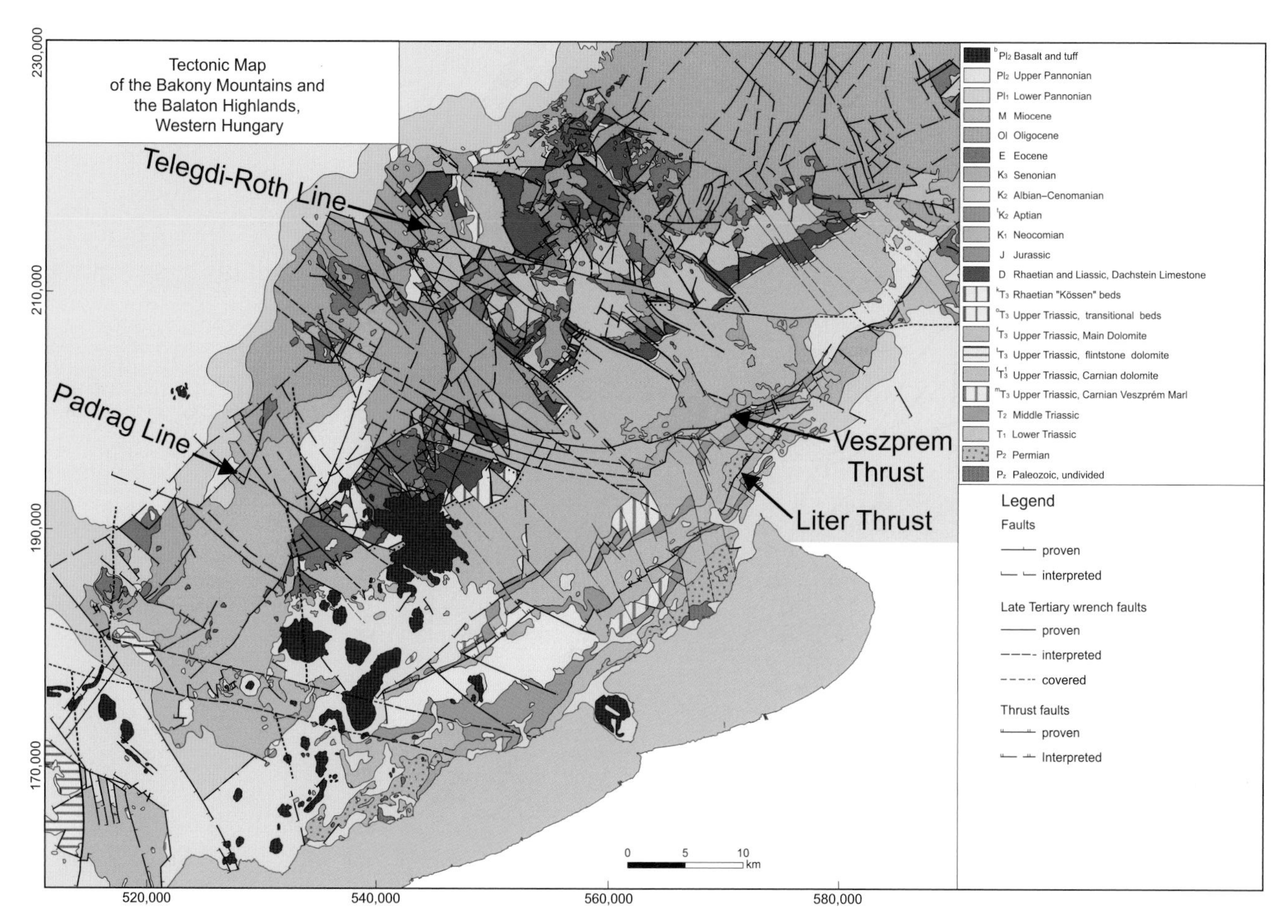

Figure 4. Simplified pre-Quaternary geologic map of the Bakony Mountains and the Balaton Highlands of the westernmost Transdanubian Central Range (Mészáros, 1983). For location, see Figure 1. Note the overall northeast–southwest-trending synformal structure with several thrusts on the southeast and the large number of strike-slip faults dissecting it trending west–northwest-east–southeast.

NCA. The overthrust Gosau beds were drilled by the Unterlaussa-1 well. The reflection seismic profile in front of the complex Weyer Arc structure offers a view of the subsurface continuation of pre- and post-Gosau structures (Figure 7).

The north–south-oriented reflection seismic profile (Figure 8) shows the east–west-striking thrust sheets below the Gosau beds. Early Cretaceous thrust sheets were covered by the Late Cretaceous to Eocene Gosau sedimentation. The eastern subsurface continuation of the Wenger thrust, an internal thrust sheet within the Reichraming nappe, is traced below the Gosau beds. The overturned southern limb of an east-plunging syncline is projected from surface map data. The late Triassic to Jurassic sequences in the hanging wall of the post-Gosau Wenger thrust were cut by a west–northwest-east–southeast-striking wrench fault, which is terminated at depth along the decollement of the Wenger thrust (Figure 8; dashed line 1). Two other strike-slip faults (Figure 8; dashed lines 2 and 3) in the north of the Wenger thrust show similar compressional structures and are also terminated at depth at the detachment of internal thrust decollement. Clearly, these strike-slip faults do not cross the basal decollement of the NCA.

On the northwestern flank of the TCR, numerous industry seismic reflection profiles were analyzed by Tari (1994). Two of these sections are reproduced here to show the typical seismic signature of the Alpine structures (Figures 9, 10). The dip-oriented line (Figure 9) illustrates the thick Miocene to Pliocene basin fill of the Danube Basin thinning and onlapping against the Mesozoic strata of the TCR. The Senonian basin fill, with its gently east- to southeast-dipping reflectors below the Neogene strata, suggests imbricated older strata that are interpreted here as northwest-verging imbricates of the Eo-Alpine fold and thrust belt. Based on the systematic seismic mapping of these subsurface

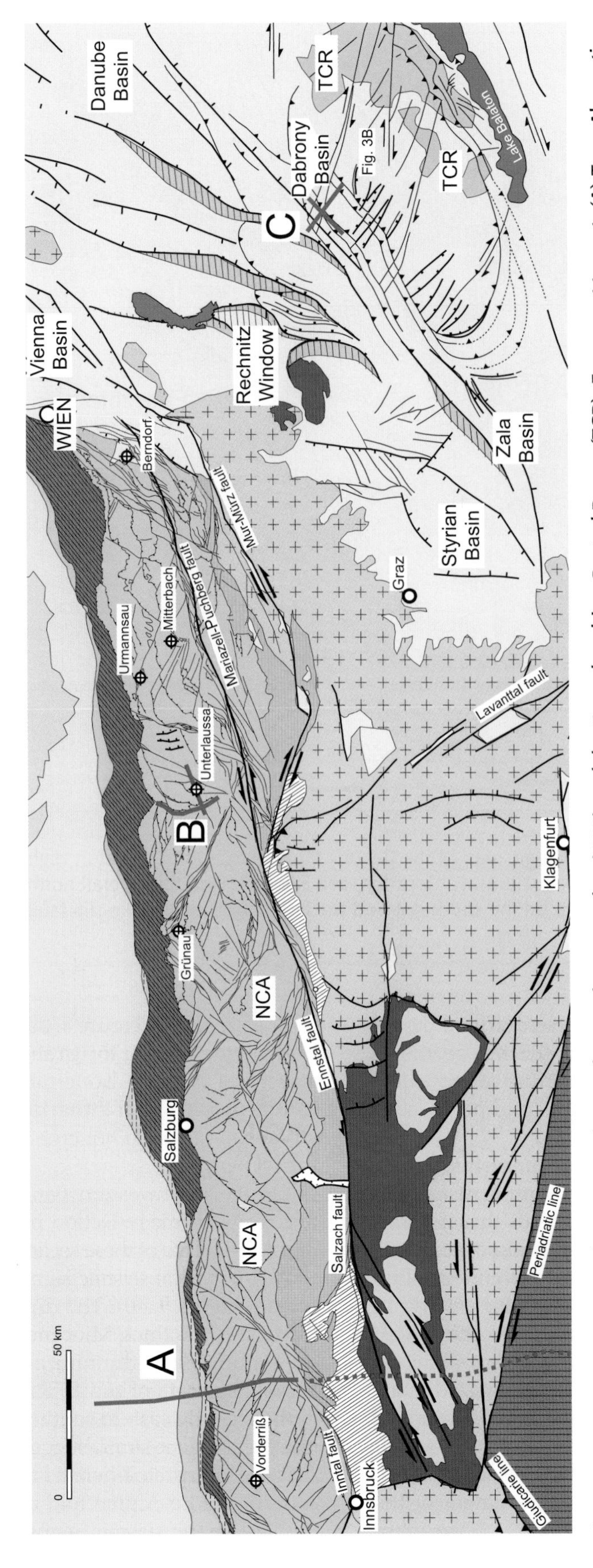

Figure 5. Index map of seismic lines across the Northern Calcareous Alps (NCA) and the Transdanubian Central Range (TCR). From west to east: (A) TransAlp section, (B) Weyer Arc sections, and (C) Dabrony Basin sections.

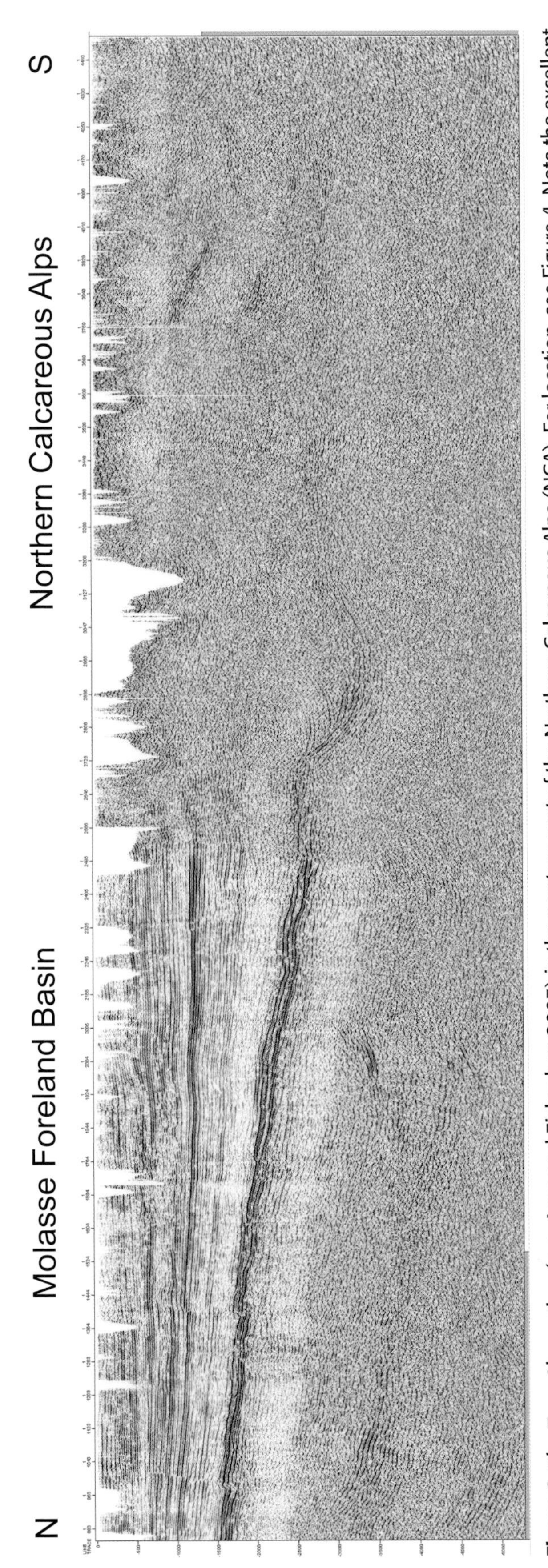

Figure 6. The TransAlp section (e.g., Auer and Eisbacher, 2003) in the western part of the Northern Calcareous Alps (NCA). For location, see Figure 4. Note the excellent seismic image of the European foreland and the overlying foreland basin in the north as opposed to the very poor expression of the NCA nappe systems in the south.

structures (Tari, 1996), these 3 to 4 km (1.9 to 2.5 mi) thick imbricate units are composed of slightly metamorphosed Paleozoic successions and an overlying nonmetamorphosed Permian–Triassic sequence (Figure 3). Some of the major thrust contacts were tentatively correlated with similar thrusts outcropping on the southeastern flank of the TCR (Figure 4).

A strike-parallel profile in the same area (Figure 10) shows that a major strike-slip fault of the TCR can be studied on the seismic data. Interestingly, a multistage structural history can be outlined for the approximately 5 km (~3.1 mi) wide zone of the Telegdi-Roth Line (cf. Figure 4). This right lateral fault zone was definitely active during the middle Miocene, and the transpressional inversion along previous normal faults is quite clear. A total of 4.7 km (2.9 mi) dextral offset was suggested by Mészáros (1983) for this fault based on map relations within the Bakony Mountains (Figure 2). The strain partitioning caused by Cretaceous and Miocene strike-slip movements (cf. Telegdi-Roth, 1935; Mészáros, 1983; Tari, 1991) still cannot be properly determined. At any rate, the Telegdi-Roth fault zone shown in Figure 10 cannot be traced much farther to the northwest. This lends credibility to an earlier speculation of Tari (1991) that this and other similar strike-slip faults in the Bakony Mountains should be detached at depth on an Eo-Alpine thrust plane (cf. Figure 8). The same explanation holds true for the still unexplained termination of strike-slip faults at major thrust faults in the Balaton Highland (Figure 4) already reported by Mészáros (1983). Also, extensional reactivation of Eo-Alpine thrust contacts can be seen on several seismic profiles on the northeastern flank of the TCR, underlining the importance of the preexisting Alpine nappe fabric (Tari, 1996).

ALPINE STRUCTURAL PATTERN OF THE NORTHERN CALCAREOUS ALPS AND THE TRANSDANUBIAN CENTRAL RANGE

The NCA are made up of 17 thrust sheets with large lateral continuity (Ampferer, 1932; Tollmann, 1976). The geometry of thrust sheets is controlled by lateral facies variations, for example, the transition of the mid-Triassic competent reef platform of the Inntal nappe in the western NCA to the incompetent basin facies, acting as a detachment horizon. The tectonically lower nappes are detached along higher stratigraphic levels (base of Norian Hauptdolomit unit). The tectonically higher nappes in the south are detached at the Upper Permian–Schythian evaporitic level. Only the southernmost nappes, south of the SEMP line, rest on the

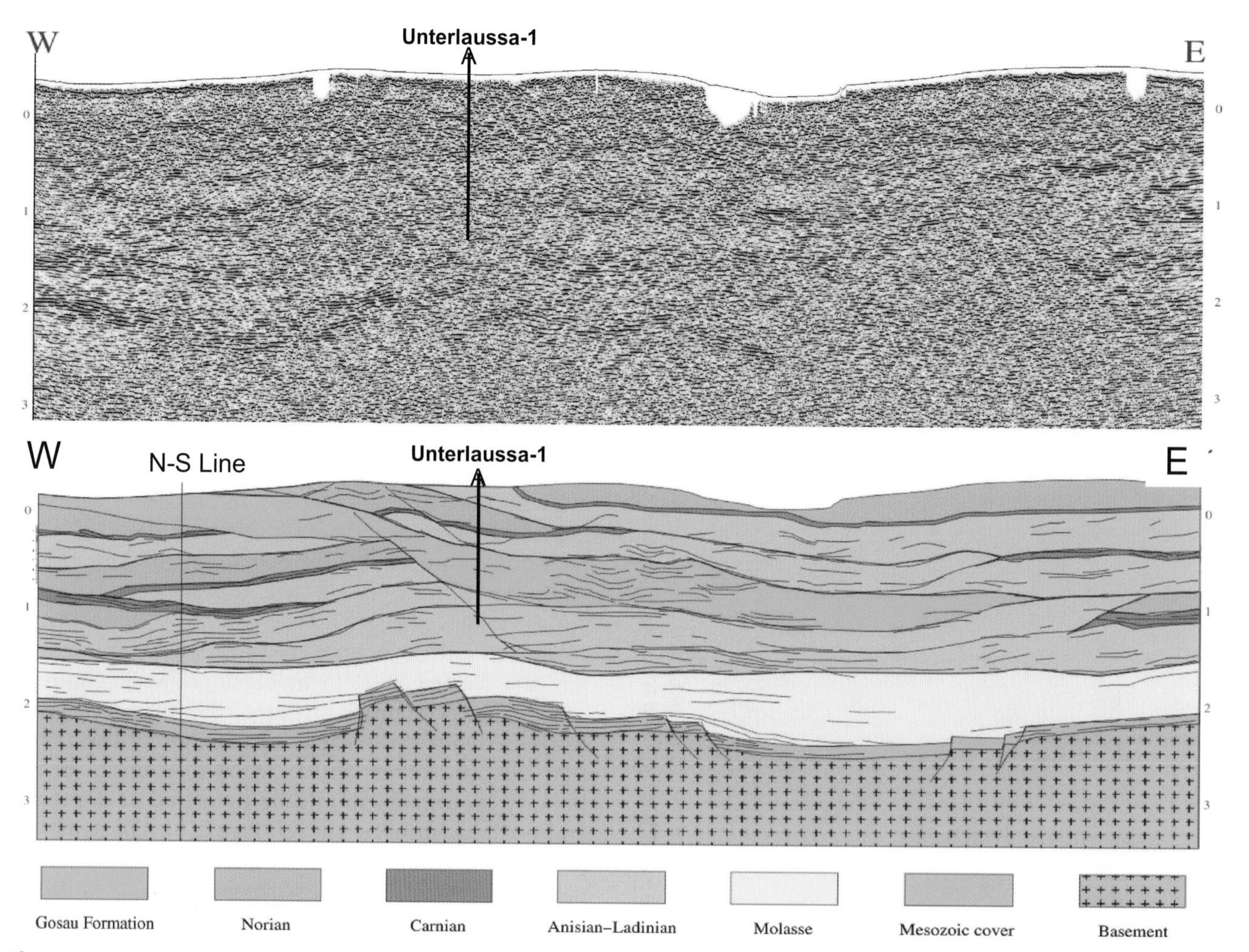

Figure 7. Industry reflection seismic profile in the eastern part of the Northern Calcareous Alps (NCA) across the Hieselberg Gosau Basin west of the Weyer Arc structure (Figure 5).

Grauwackenzone basement in the eastern NCA and on the basement rocks in the western segment of the NCA (Figure 2). The basal decollement of the NCA climbs up on ramp-flat systems from the basement to upper detachment levels (Figure 3B). Total shortening in the NCA (calculation is based on balanced cross sections) was estimated to range from 55 to 65% (Eisbacher et al., 1990; Linzer et al., 1995). The internal structures of the NCA (nappe boundaries, general strike of beds, and major fold axis) show a general east–northeast trend. These structures were displaced by a set of dextral strike-slip faults interpreted as tear faults or transfer faults caused by a right lateral oblique convergence (Linzer et al., 1995). The synsedimentary character of the faults is documented in syndeformational clastic basins in the western NCA (Brandenberg and Muttekopf Gosau basins), where deep-water sediments were separated from shallow-water sediments by these transfer faults (Eisbacher and Brandner, 1995). Major dextral fault sets (Figure 1; thin lines) show a spacing on the order of 10 to 30 km (6.2–18.6 mi), with horizontal displacements of up to 15 km (9.3 mi), and they dismember the NCA into large blocks. The dextral displacement on these wrench faults is indicated by the offset of nappe and facies boundaries. The area between the major faults is covered with numerous secondary faults with well-defined stratigraphic offsets on the order of a few meters to 1 km (0.6 mi). The dextral fault set was overprinted and reactivated by north–northeast to northeast-oriented sinistral strike-slip faults (Figure 1; thick lines) because of a reorientation of the strain field in response to the collision with the European foreland during the Eocene–Oligocene. The contractional direction rotated from the Cretaceous northwest to a north orientation (Decker et al., 1993; Linzer et al., 1995, 1997). The NCA were further dismembered during this period because of lateral extrusion of the central Austroalpine nappes in the Miocene (Linzer et al., 2002). The central Austroalpine extrusional wedge was bordered in the north by the SEMP line. The rhomboid blocks

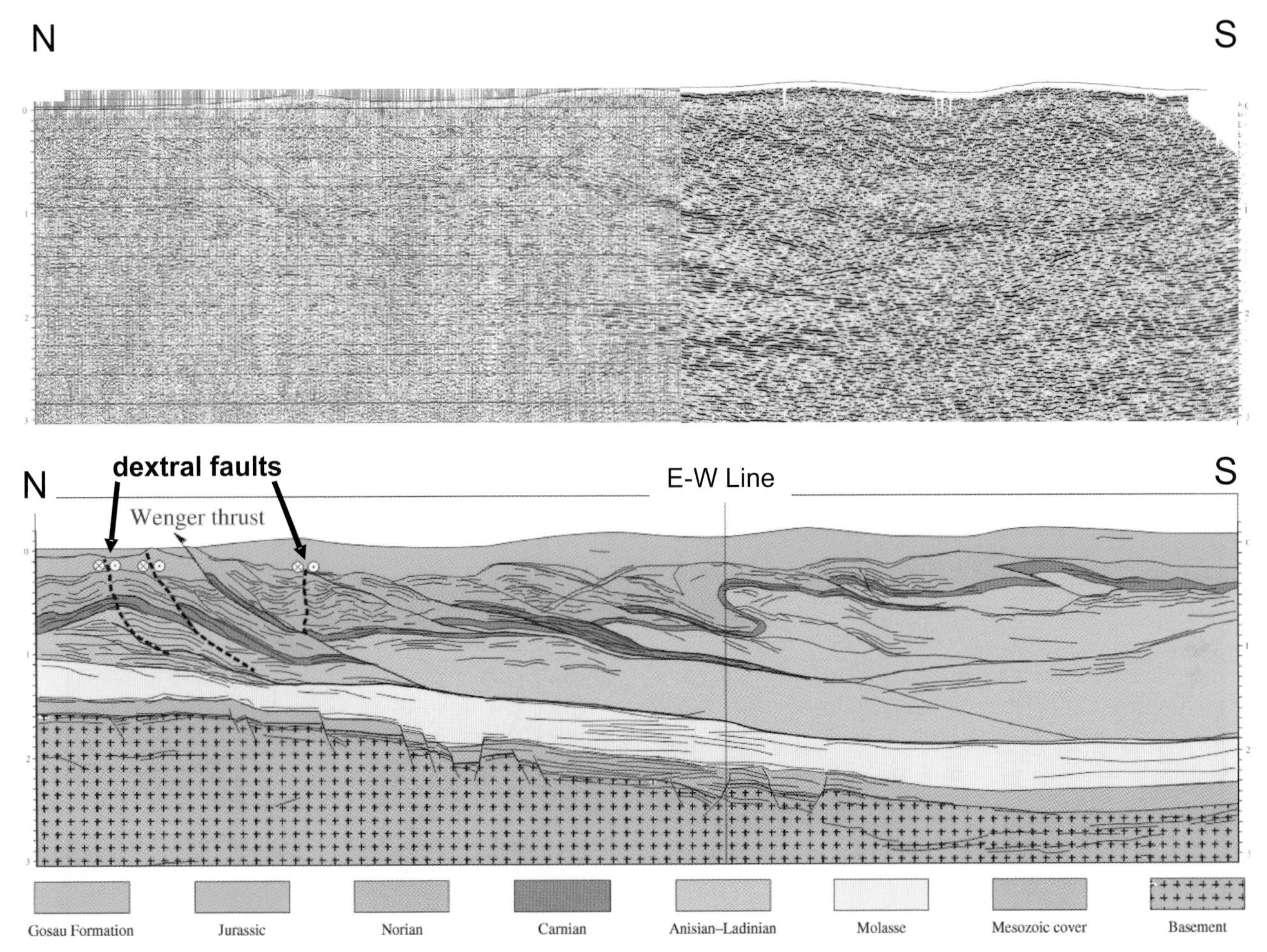

Figure 8. Industry reflection seismic profile in the eastern part of the Northern Calcareous Alps (NCA) across the Hieselberg Gosau Basin, west of the Weyer Arc structure (Figure 4).

of the NCA north of the SEMP line were moved to the northeast and formed contractional and extensional structures because of releasing and restraining geometries at the block boundaries. These block movements transferred part of the left lateral displacement along the SEMP line to the north and caused northeast-directed stacking in the central and eastern segment of the NCA, for example, the Weyer Arc thrust with a minimum offset of about 15 km (~9.3 mi) (Linzer et al., 2002).

In comparison with the NCA, the Eo-Alpine thrust systems of the TCR are much less known, mostly because of the poor outcrop conditions. However, subsurface data sets are more abundant and generally of better quality in the TCR than in the NCA. Therefore, subsurface Eo-Alpine thrust contacts could be mapped by the interpretation of reflection seismic lines on the northwestern and western flanks of the TCR. These overthrust surfaces were correlated with the outcropping southeast-vergent thrusts in the Balaton Highland (Tari, 1994, 1995). Based on this interpretation, the TCR has been subdivided into the Balaton nappe and Bakony nappe systems with at least five internal thrust sheets. The structurally lower Balaton nappe forms a large-scale syncline and is composed of very low grade Paleozoic rocks similar to the Grauwacken Zone of the Eastern Alps and an overlying Triassic sedimentary sequence. The structurally higher Bakony nappe is generally detached at the base of Middle Triassic strata, and it is made up mainly of Upper Triassic to Jurassic rocks. The TCR was, just as the NCA, dismembered by west–northwest-east–southeast-oriented dextral strike-slip faults during the Cretaceous. Some of them, such as the Telegdi-Roth Line, have been known for a long time (i.e., Telegdi-Roth, 1935). However, most of them were systematically described for the first time by Mészáros (1983) based on data acquired during the extensive bauxite and coal exploration in the broader

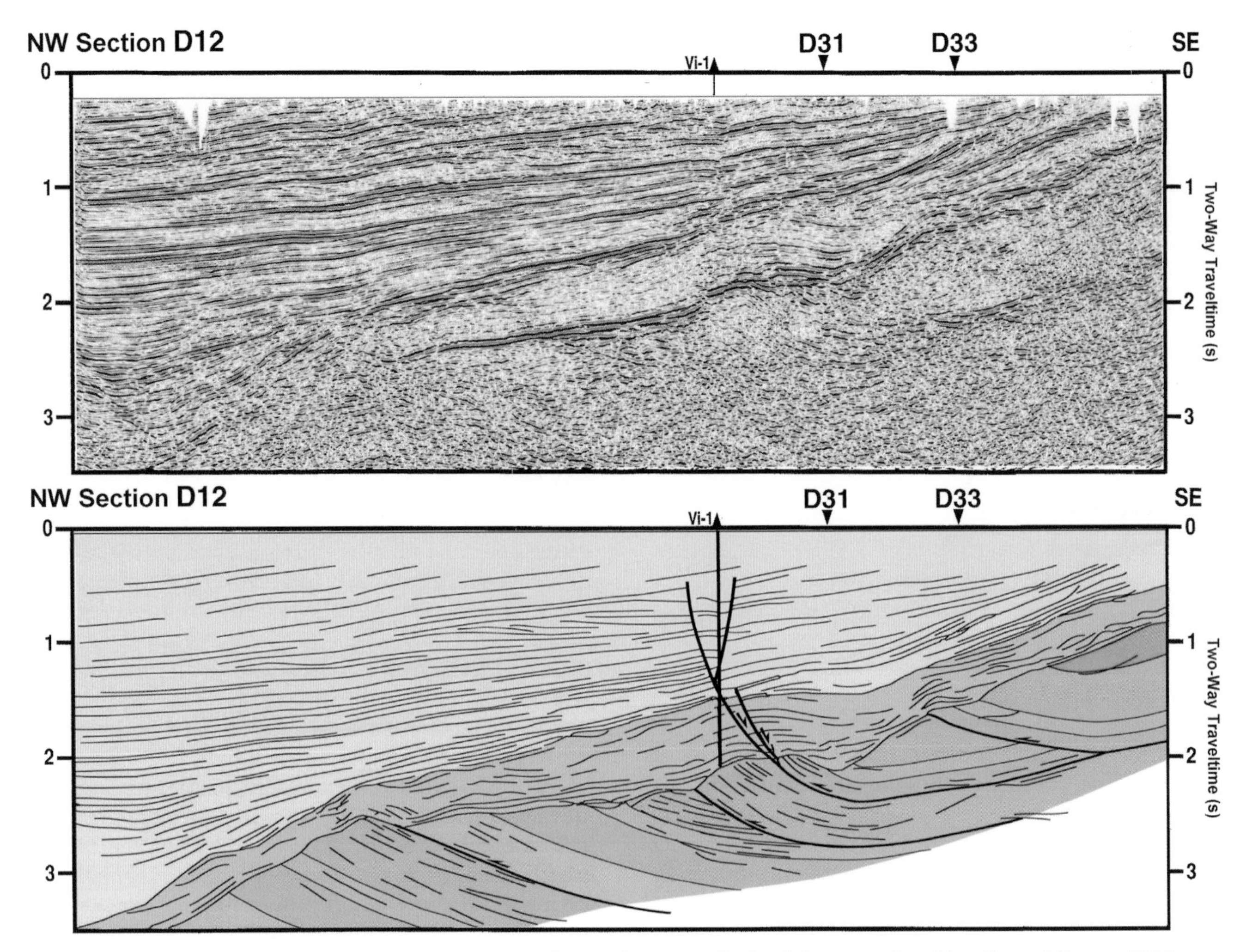

Figure 9. Dip-oriented industry seismic section across the northwestern flank of the Transdanubian Central Range (TCR) (Tari, 1994), illustrating the Eo-Alpine nappe structure beneath the Neogene to Senonian basin fill. For location, see Figure 4. An intersecting strike profile is shown in Figure 10.

area (Figure 4). In the TCR, the major strike-slip faults are spaced about 30 km (~18.6 mi) apart and the typical reported strike-slip offsets are on the order of 1 to 5 km (0.6–3.1 mi). However, in contrast to the NCA, some of these dextral faults were partly reactivated during the Miocene.

STRUCTURAL CORRELATION BETWEEN THE NORTHERN CALCAREOUS ALPS AND THE TRANSDANUBIAN CENTRAL RANGE

The important contribution of this chapter is the map-view correlation of these major Alpine units by semi-quantitative palinspastic restoration of the deformation that displaced them and caused their present-day separation (Figures 11–14). The lower right corner of these figures illustrates the kinematics model inferred for the given period, whereas the map shows the actual retrodeformed geometry during the given deformation phase. The kinematic constraints for the map-view restoration were provided by (1) the present-day position of Upper Triassic facies distribution in the NCA (e.g., Tollmann, 1976; Haas and Budai, 1995), (2) structural data on major fault systems in the NCA (Linzer et al., 1995, 1997, 2002), (3) structural maps compiled in the TCR (Mészáros, 1983; Budai and Fodor, 2008), and (4) interpretation of an extensive grid of 2-D reflection seismic lines in the Danube Basin and along the western flank of the TCR (Tari, 1994, 1995, 1996). Also, paleomagnetic data were also taken into account as declination anomalies suggest different block rotations for these units: clockwise in the NCA (Mauritsch and Frisch, 1978; Channell et al., 1992) and counterclockwise in the TCR (Marton and Marton, 1981).

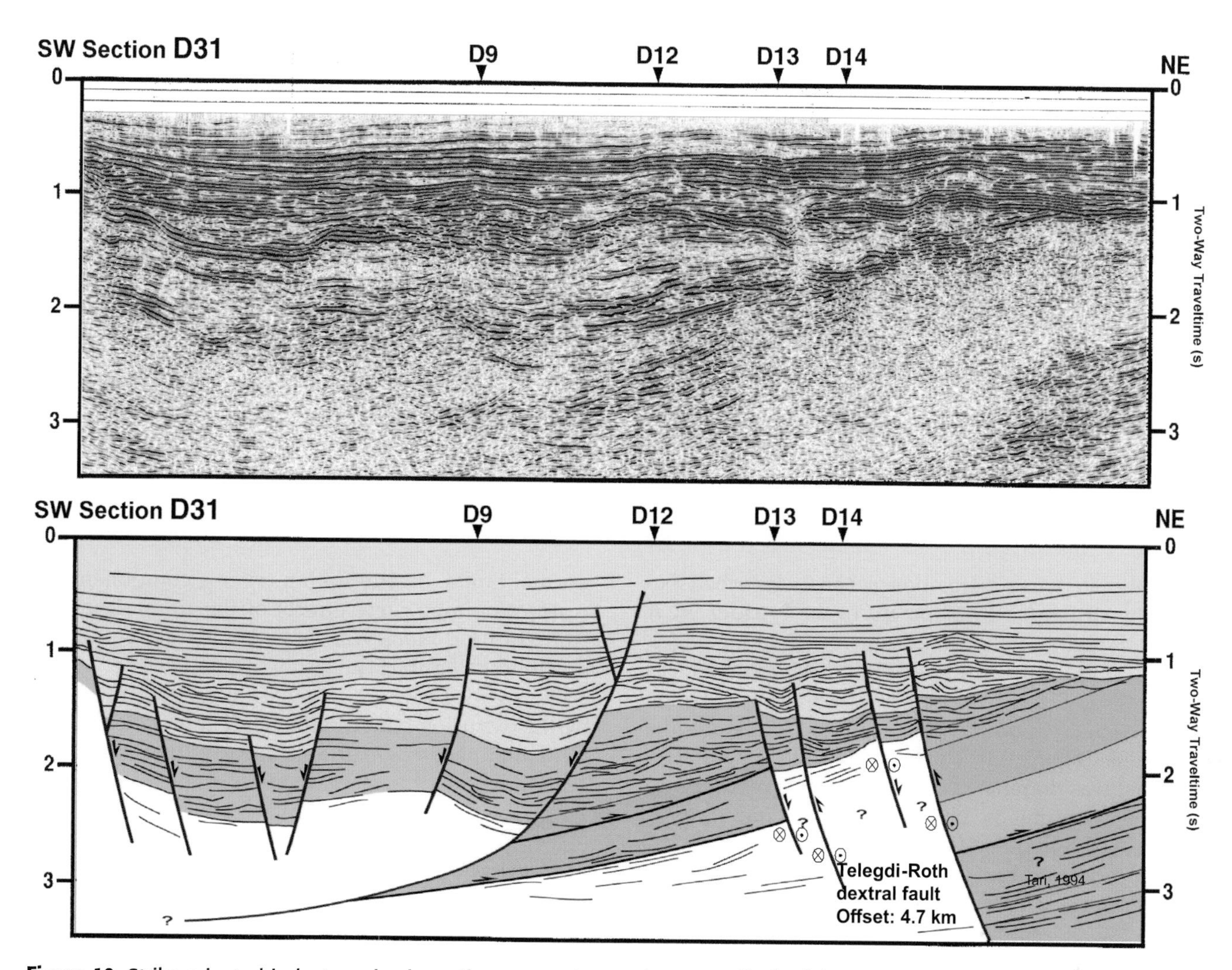

Figure 10. Strike-oriented industry seismic section across the northwestern flank of the Transdanubian Central Range (TCR) (Tari, 1994), illustrating the Eo-Alpine nappe structure beneath the Neogene to Senonian basin fill. For location, see Figure 4. An intersecting dip profile is shown in Figure 9.

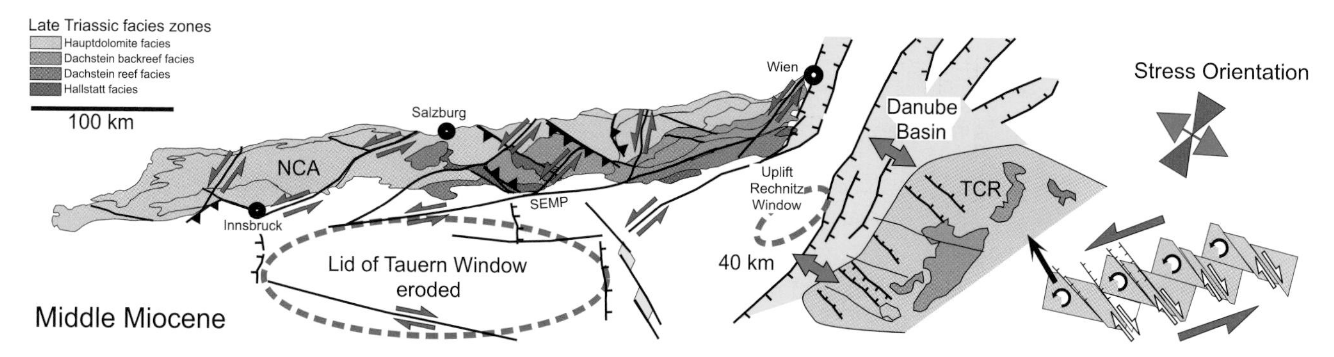

Figure 11. Middle Miocene structural activity and relative position of the Northern Calcareous Alps (NCA) and the Transdanubian Central Range (TCR). Note left lateral strike-slip faulting within the NCA and northwest–southeast-oriented extension in the Danube Basin to the northwest of the TCR. SEMP Line = Salzach-Ennstal-Mariazell-Puchberg Line.

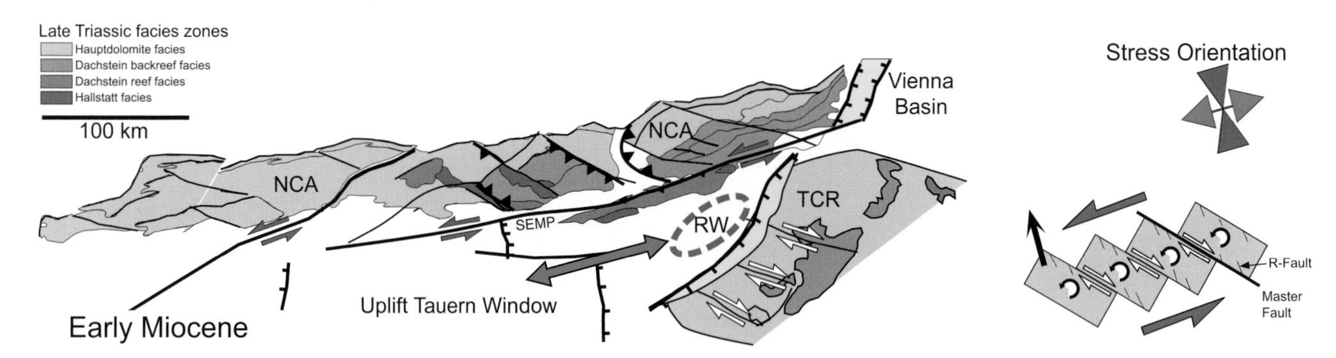

Figure 12. Early Miocene structural activity and relative position of the Northern Calcareous Alps (NCA) and the Transdanubian Central Range (TCR). Note the large left lateral strike-slip faulting along the SEMP (Salzach-Ennstal-Mariazell-Puchberg) Line to the south of the NCA. The area south of the NCA has been extended in an east–northeast-west–southwest direction because of the extrusion of the Central Alpine area (sensu Ratschbacher et al., 1991) toward the east. RW = Rechnitz Window metamorphic core complex.

In the stepwise restoration scheme described below, the structural development of the broader NCA and TCR regions has been subdivided into four stages. Two of these are major Neo-Alpine (Miocene) stages (Figures 11, 12), whereas the end of the Meso-Alpine contractional period has been captured as the middle Oligocene (Figure 13). The Eo-Alpine stage (Late Cretaceous) was dominated by contraction and strike-slip faulting dissecting the NCA and TCR (Figure 14).

During the middle Miocene (Badenian), rift-style extension occurred on a set of northeast-trending low-angle detachment faults in the Danube Basin (Tari, 1996), with an estimated total extension of about 40 km (~25 mi) (Figure 11). Within the TCR, Late Cretaceous strike-slip faults were reactivated as steeply dipping normal faults, locally with a reverse component. To restore the map-view effects of the rift-style extension, the TCR block was moved 40 km (25 mi) to the northwest (Figure 11).

Figure 12 shows our map-view extension caused by the extrusion between the NCA and TCR. This deformation occurred during the early to middle Miocene (Karpatian), in a relatively short period of about 1 m.y. During this brief time interval, up to 20 km (12.4 mi) of exhumation occurred in the TW (2 mm/a uplift rate) and 100 km (62 mi) of the extension occurred in the Austroalpine lid. The TCR block, as a part of the central Austroalpine nappe pile, was moved out of the Austroalpine that caused 80 ± 10 km (50 ± 6.2 mi) of east–northeast-west–southwest-oriented

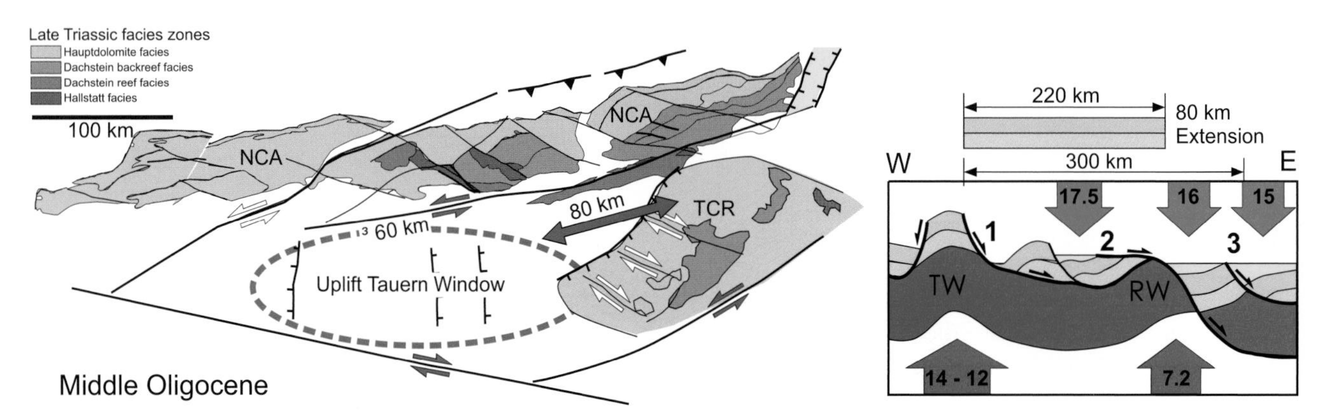

Figure 13. Middle Oligocene structural activity and relative position of the Northern Calcareous Alps (NCA) and the Transdanubian Central Range (TCR). Note left lateral strike-slip faulting along the SEMP (Salzach-Ennstal-Mariazell-Puchberg) Line to the south of the NCA and the unroofing of both the Tauern Window (TW) and Rechnitz Window (RW). The magnitude of the early Miocene metamorphic core complex-style extension of the RW is estimated to be 80 km (50 mi) in a west–southwest-east–northeast direction, to the west of the TCR. In the cartoon, red and blue arrows stand for apatite and zircon FT ages, respectively. FT = fission track.

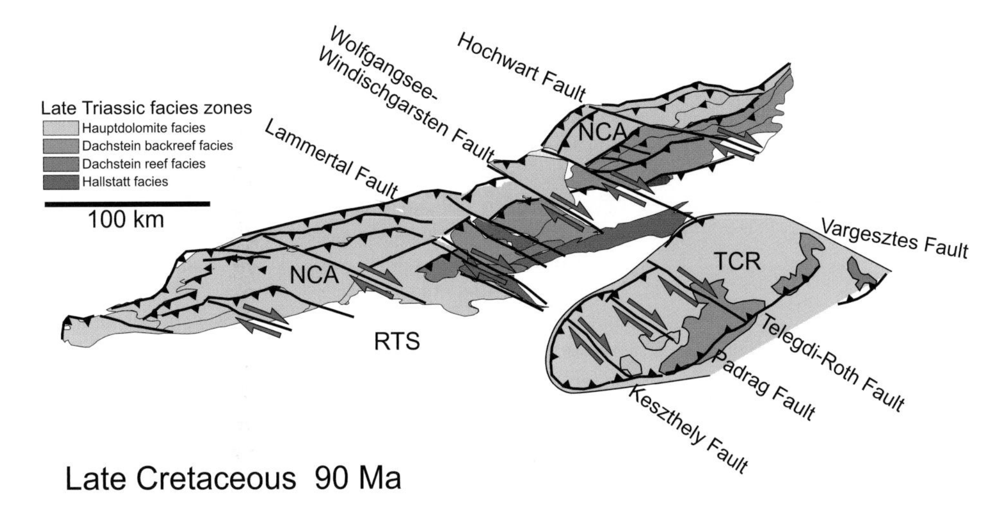

Figure 14. Late Cretaceous structural activity and relative position of the Northern Calcareous Alps (NCA) and the Transdanubian Central Range (TCR). This map-view restoration shows that the right lateral strike-slip faults of the NCA can be correlated with those of the TCR. For example, the Wolfgangsee-Windischgarsten fault face with the Telegdi-Roth fault. RTS = Radstadt Thrust System.

extension (Linzer et al., 2002). Extension in the central Austroalpine was synchronous with strike-slip faulting at the boundaries of the extruding wedge. The northeast-striking thrust sheets of the NCA, which form the northern border of the extruding wedge, were reactivated as left lateral strike-slip faults. The northwest-striking strike-slip faults were reactivated as thrust faults, which partly show large-scale displacements.

As the next step, the metamorphic core complex–style extension of the TW in the west and the RW in the east needs to be restored (Figure 13). This extension occurred during the late Oligocene to early Miocene interval, and thus Figure 13 shows the middle Oligocene restoration. The simplified tectonic sketch (Figure 13) shows the core complex-style extension in the Austroalpine nappes. Uplift and unroofing of the Penninic units in the TW occurred slightly earlier than in the RW based on fission track data.

As the final restoration step (Figure 14), the northwest–southeast-trending right lateral strike-slip faults of the NCA (cf. Figure 2) face the right lateral strike-slip faults of the TCR (cf. Figure 4). The restored Late Cretaceous position of the TCR block is shown directly south of the central NCA and east of the Radstadt Thrust System. The structural pattern of the northeast-trending thrust boundaries and fold axes and the west–northwest to northwest orientation of the dextral strike-slip faults appears to be identical with the structural pattern of the NCA. Whereas it might be still too early to claim direct correlations between some of the well-known fault zones within the NCA and the TCR (e.g., the Wolfgangsee-Windischgarsten Fault seems to be the continuation of the Telegdi-Roth Line), the overall style of these faults, such as geometry and timing, suggests their close relationship in a coherently deforming Eo-Alpine nappe complex.

CONCLUSIONS

Palinspastic restoration of several Alpine structural elements in the broader Alpine-Pannonian transitional area indicates the close relationship between the NCA and the TCR despite their present-day geographic separation. Both consist of similar sedimentary and basement sequences and strongly resemble each other in their tectonic style, although the level of structural understanding of these Alpine nappe systems is markedly different. However, a compilation of critical surface geologic data integrated with the interpretation of subsurface geophysical data such as seismic reflection profiles and exploration wells facilitated a semiquantitative map-view reconstruction of Early Alpine convergence and superimposed Late Alpine extrusion and extension. This restoration brings the NCA and the TCR into close proximity. As a corollary, some west-northwest–trending right lateral strike-slip faults in the TCR (e.g., Telegdi-Roth, Padrag, and Vargesztes faults) are interpreted here to be at least analogous, or perhaps even identical with those described from the NCA (e.g., Lammertal, Wolfgangsee-Windischgarsten, and Hochwart fault systems).

ACKNOWLEDGMENTS

Hans-Gert Linzer appreciates the support of the German Science Foundation (grant Li-575) for a research fellowship at Rice University, Houston. Thanks are due to OMV for making some of the reflection seismic lines available for this publication. Many discussions with A. W. Bally, Rice University, Houston, are gratefully acknowledged. We also acknowledge the help of many colleagues from Austria and Hungary such as Tamas Budai, Lászlo Fodor, Wolfgang Frisch, Frank Horváth, Wolfgang Nachtmann, Franz Neubauer, Afred Pahr, Lothar Ratschacher, and Godfrid Wessely. Peer reviews by Terry Pavlis, Hermann Lebit, and Dengliang Gao are greatly appreciated.

REFERENCES CITED

Ampferer, O., 1912, Gedanken über die Tektonik des Wettersteingebirges: Verhandlungen der Geologischen Reichsanstalt, Vienna, Austria, v. 1912, p. 197–212.

Ampferer, O., 1932, Erläuterungen zu den geologischen Karten der Lechtaler Alpen: Geologischen Bundes-Anstalt, Vienna, Austria, 125 p.

Ampferer, O., 1939, Im Kampfe für Reliefüberschiebung und O-W-Bewegung: Verhandlungen der Zweigstelle Wien der Reichsstelle für Bodenforschung, Vienna, Austria, v. 1939, p. 196–205.

Auer, M., and G. H. Eisbacher, 2003, Deep structure and kinematics of the Northern Calcareous Alps (TRANSALP Profile): International Journal of Earth Sciences, v. 92, p. 210–227.

Bachmann, G. H., and M. Müller, 1991, The Molasse basin, Germany: evolution of a classic petroliferous foreland basin, *in* A. M. Spencer, ed., Generation, accumulation and production of Europe's hydrocarbons: European Association of Petroleum Geologists Special Publication 1, p. 263–276.

Bachmann, G., G. Dohr, and M. Müller, 1982, Exploration in a classic thrust belt and its foreland: Bavarian Alps, Germany: AAPG Bulletin, v. 66, p. 2529–2542.

Bachmann, G. H., M. Müller, and K. Weggen, 1987, Evolution of the Molasse Basin: Tectonophysics, v. 137, p. 77–92, doi:10.1016/0040-1951(87)90315-5.

Bechstädt, T., and H. Mostler, 1976, Riff-Beckenentwicklung in der Mitteltrias der westlichen Nördlichen Kalkalpen: Zeitschrift der Deutschen Geologischen Gesellschaft, Berlin, Germany, v. 127, p. 271–289.

Becker, B., 1993, The structural evolution of the Radstadt thrust system, Eastern Alps, Austria: Kinematics, thrust geometries, strain analysis: Tübinger Geowissenschaftliche Arbeiten, Reihe A, Geologie, Paleontologie, Stratigraphie, Tübingen, Germany, v. 14, p. 92.

Brandner, R., 1978, Tektonisch kontrollierter Sedimentationsablauf im Ladin und Unterkarn der westlichen Nördlichen Kalkalpen: Geologie Paläontologie Mitteilungen Innsbruck, Innsbruck, Austria, v. 8, p. 317–354.

Budai, T., and L. Fodor, eds., 2008, Explanatory book to the geological map of the Vertes Hills, Innova: Budapest, Hungary, scale 1:50,000, p. 368.

Channell, J. E. T., R. Brandner, A. Spieler, and J. S. Stoner, 1992, Paleomagnetism and paleogeography of the NCA: Tectonics, v. 11, p. 792–810, doi:10.1029/91TC03089.

Decker, K., and H. Peresson, 1996, Tertiary kinematics in the Alpine-Carpathian-Pannonian system: Links between thrusting, transform faulting and crustal extension, *in* G. Wessely and W. Liebl, eds., Oil and gas in Alpidic thrust belts and basins of Central and Eastern Europe: European Association of Geoscientists and Engineers Special Publication 5, p. 96–77.

Decker, K., P. Faupl, and A. Müller, 1987, Synorogenic sedimentation on the Northern Calcareous Alps during the Early Cretaceous, *in* H. W. Flügel and P. Faupl, eds., Geodynamics of the Eastern Alps, Wien, Vienna, Austria, Deuticke, p. 126–141.

Decker, K., M. Meschede, and U. Ring, 1993, Fault slip analysis along the northern margin of the Eastern Alps (Molasse, Helvetic nappes, North and South Penninic Flysch, and the Northern Calcareous Alps): Tectonophysics, v. 223, no. 3–4, p. 291–312, doi:10.1016/0040-1951(93)90142-7.

Ebner, F., and F. Sachsenhofer, 1991, Die Entwicklungsgesichte des Steirischen Tertiärbeckens: Mitteilungen der Abteilung Geologie und Paläontologie am Landesmuseum Joanneum, Graz, Austria, v. 49, p. 1–96.

Eisbacher, G. H., and R. Brandner, 1995, Role of high-angle faults during heteroaxial contraction, Inntal Thrust Sheet, Northern Calcareous Alps, Western Austria: Geologie Paläontologie Mitteilungen Innsbruck, Innsbruck, Austria, v. 20, p. 389–406.

Eisbacher, G. H., H. G. Linzer, L. Meier, and R. Polinski, 1990, A depth-extrapolated structural transect across the Northern Calcareous Alps of western Tirol: Eclogae Geologica Helvetica, v. 83, p. 711–725.

Faupl, P., and A. Tollmann, 1979, Die Roßfeldschichten: Ein Beispiel für Sedimentation im Bereich einer tektonisch aktiven Tiefseerinne aus der Kalkalpinen Unterkreide: Geologische Rundschau, Berlin, Germany, v. 68, p. 93–120.

Faupl, P., and M. Wagreich, 1992, Cretaceous flysch and pelagic sequences of the Eastern Alps: Correlations, heavy minerals, and paleogeographic implications: Cretaceous Research, v. 13, p. 387–403.

Frisch, W., 1974, Die stratigraphisch tektonische Gliederung der Schieferhülle und die Entwicklung des penninischen Raumes im westlichen Tauernfenster (Gebiet Brenner-Gerlosspass): Mitteilungen der Österreichischen Geologischen Gesellschaft, Vienna, Austria, v. 66/67, p. 9–20.

Frisch, W., 1977, Die Alpen im westmediterranen Orogeneine Plattentektonische Rekonstruktion: Mitteilungen der Gesellschaft der Geologie-und Bergbaustudenten, Vienna, Austria, v. 24, p. 263–275.

Frisch, W., 1980, Plate motions in the Alpine region and their correlation to the opening of the Atlantic ocean: Mitteilungen der Österreichischen Geologischen Gesellschaft, Vienna, Austria, v. 71/72, p. 45–48.

Frisch, W., K. Gommeringer, U. Kelm, and F. Popp, 1987, The upper Bündner Schiefer of the Tauern Window: A key to understanding Eo-Alpine orogenic processes in the Eastern Alps, *in* H. W. Flügel and P. Faupl, eds., Geodynamics of the Eastern Alps: Wien, Deuticke, Vienna, Austria, p. 55–69.

Frisch, W., J. Kuhlemann, I. Dunkl, and A. Brügel, 1998, Palinspastic reconstruction and topographic evolution of the Eastern Alps during late Tertiary tectonic extrusion: Tectonophysics, v. 297, p. 1–15.

Froitzheim, N., S. M. Schmid, and M. Frey, 1996, Mesozoic paleogeography and the timing of eclogite-facies metamorphism in the Alps: A working hypothesis: Eclogae Geologicae Helveticae, v. 89, no. 1, p. 81–110.

Gaupp, R. H., 1983, Die paläogeographische Bedeutung der Konglomerate der Losensteiner Schichten (Alb, Nördliche Kalkalpen): Zitteliana, Munich, Germany, v. 8, p. 33–72

Haas, J., and T. Budai, 1995, Upper Permian–Triassic facies zones in the Transdanubian Range: Rivista Italiana di Paleontologia e Stratigrafia, Milano, Italy, v. 101, no. 3, p. 249–266.

Haas, J., S. Kovacs, L. Krystyn, and R. Lein, 1995, Significance of Late Permian-Triassic facies zones in terrane reconstructions in the Alpine–North Pannonian domain: Tectonophysics, v. 242, p. 19–40, doi:10.1016/0040-1951(94) 00157-5.

Horváth, F., 1995, Phases of compression during the evolution of the Pannonian Basin and its bearing on hydrocarbon exploration: Marine and Petroleum Geology, v. 12, p. 837–844, doi:10.1016/0264-8172(95)98851-U.

Koller, F., and A. Pahr, 1980, The Penninic ophiolites on the eastern end of the Alps: Ofioliti, v. 5, p. 65–72.

Körössy, L., 1987, Hydrocarbon geology of the Little Plain in Hungary. In Hungarian with English summary: General Geological Review, v. 22, p. 99–174.

Kröll, A., K. Schimunek, and G. Wessely, 1981, Ergebnisse und Erfahrungen bei der Exploration in der Kalkalpenzone in Ostösterreich: Erdöl-Erdgas-Z., Hamburg, Germany, v. 97, p. 134–148.

Kröll, A., H. W. Flügel, W. Seiberl, F. Weber, G. Walach, and D. Zych, 1988, Erläuterungen zu den Karten über den steirischen Beckens und der südburgenlandischen Schwelle: Geologische Bundesanstalt, Vienna, Austria, 47 p.

Lein, R., 1987, Evolution of the Northern Calcareous Alps during Triassic times, *in* W. Flügel and P. Faupl, eds., Geodynamics of the Eastern Alps: Vienna, Austria, Deuticke, p. 85–102.

Linzer, H. G., L. Ratschbacher, and W. Frisch, 1995, Transpressional collision structures in the upper crust: The fold-thrust belt of the Northern Calcareous Alps: Tectonophysics, v. 242, p. 41–61, doi:10.1016/0040-1951(94)00152-Y.

Linzer, H. G., F. Moser, F. Nemes, L. Ratschbacher, and B. Sperner, 1997, Build-up and dismembering of a classical fold-thrust belt: From non-cylindircal stacking to lateral extrusion in the eastern Northern Calcareous Alps: Tectonophysics, v. 272, p. 97–124.

Linzer, H. G., K. Decker, H. Peresson, R. Dell'Mour, and W. Frisch, 2002, Balancing lateral orogenic float of the Eastern Alps: Tectonophysics, v. 354, no. 3-4, p. 211–237, doi:10.1016/S0040-1951(02)00337-2.

Márton, P., and E. Márton, 1981, Mesozoic paleomagnetism of the Transdanubian Central Mountains: Tectonophysics, v. 72, p. 129–140.

Mauritsch, H. J., and W. Frisch, 1978, Paleomagnetic data from the central part of the Northern Calcareous Alps, Austria: Journal of Geophysics, v. 44, p. 623–637.

Mészáros, J., 1983, Structural and economic-geological significance of strike-slip faults in the Bakony Mountains (In Hungarian with English summary): Annual Report of the Hungarian Geological Survey, v. 1981, p. 485–502.

Morteani, G., 1974, Petrology of the Tauern Window, Austrian Alps: Fortschritte in der Mineralogie, v. 52, p. 195–220.

Nachtmann, W., and L. Wagner, 1987, Mesozoic and early Tertiary evolution of the Alpine foreland in Upper Austria and Salzburg, Austria: Tectonophysics, v. 137, p. 61–76, doi:10.1016/0040-1951(87)90314-3.

Neubauer, F., R. D. Dallmeyer, I. Dunkl, and D. Schirnik, 1995, Late Cretaceous exhumation of the metamorphic Gleinalm dome, Eastern Alps: Kinematics, cooling history and sedimentary response in a sinistral wrench corridor: Tectonophysics, v. 242, p. 79–98, doi:10.1016/0040-1951(94)00154-2.

Peresson, H., and K. Decker, 1997, The Tertiary dynamics of the Northern Eastern Alps (Austria): Changing paleostresses in a collisional plate boundary: Tectonophysics, v. 272, p. 125–157, doi:10.1016/S0040-1951(96)00255-7.

Ratschbacher, L., 1986, Kinematics of Austro-Alpine cover nappes: changing translation path due to transpression: Tectonophysics, v. 125, p. 335–356, doi:10.1016/0040-1951 (86)90170-8.

Ratschbacher, L., J. H. Behrmann, and A. Pahr, 1990, Penninic windows at the eastern end of the Alps and their relation to the intra-Carpathian basins: Tectonophysics, v. 172, p. 91–105, doi:10.1016/0040-1951(90)90061-C.

Ratschbacher, L., W. Frisch, H. G. Linzer, and O. Merle, 1991, Lateral extrusion in the Eastern Alps: Part 2. Structural analysis: Tectonics, v. 10, p. 257–271, doi:10.1029 /90TC02623.

Satir, M., and G. Morteani, 1978, P-T-conditions of the high-pressure Hercynian event in the Alps as deduced from petrological, Rb-Sr and O 18/0 16 data on Phengites from the Schwazer Augengneise (Eastern Alps, Austria): Schweizerische Mineralogische und Petrographische Mitteilungen, Zurich, Switzerland, v. 58, p. 289–302.

Schlager, W., 1967, Hallstätter-und Dachsteinkalkfazies am Gosaukamm und die Vorstellung parautochthoner Hallstätter Zonen in den Ostalpen: Verhandlungen der Geologischen Bundesanstalt, Vienna, Austria, 1967, v. 1/2, p. 50–70.

Schlager, W., and W. Schöllenberger, 1974, Das Prinzip der stratigraphischen Wenden in der Schichtfolge der Nördlichen Kalkalpen: Mitteilungen del Geologischen Gesellschaft in Wien, v. 66/67, p. 165–193.

Schmid, S. M., and R. Haas, 1989, Transition from near-surface thrusting to intrabasement decollement, Schlinig Thrust, eastern Alps: Tectonics, v. 8, no. 4, p. 697–718, doi:10.1029/TC008i004p00697.

Schmid, S. M., O. A. Pfiffner, N. Froitzheim, G. Schönborn, and E. Kissling, 1996, Geophysical-geological transect and tectonic evolution of the Swiss-Italian Alps: Tectonics, v. 15, no. 5, p. 1036–1064, doi:10.1029/96TC00433.

Schöllnberger, W. E., 1973, Zur Verzahnung von Dachsteinkalk-Fazies und Hallstätter Fazies am Südrand des Toten Gebirges (Nördliche Kalkalpen, Österreich): Mitteilungen

der Gesellschaft der Geologie-und Bergbaustudenten in Österreich, Vienna, Austria, v. 22, p. 95–153.
Selverstone, J., 1988, Evidence for east-west crustal extension in the Eastern Alps: Implications for the unroofing history of the Tauern Window: Tectonics, v. 7, no. 1, p. 87–105, doi:10.1029/TC007i001p00087.
Selverstone, J., 1993, Micro-to macroscale interactions between deformational and metamorphic processes, Tauern Window, Eastern Alps: Schweizerische Mineralogische und Petrographische Mitteilungen, v. 73, p. 229–239.
Steininger, F. F., C. Müller, and F. Rögl, 1988, Correlation of Central Paratethys, Eastern Paratethys, and Mediterranean Neogene Stages, *in* L. H. Royden and F. Horváth, eds., The Pannonian Basin: A study in basin evolution: AAPG Memoir 45, p. 79–87.
Tari, G., 1991, Multiple Miocene block rotation in the Bakony Mts. (Transdanubian Central Range, Hungary.): Tectonophysics, v. 199, p. 93–108, doi:10.1016/0040-1951(91)90120-H.
Tari, G., 1994, Alpine tectonics of the Pannonian Basin: Ph.D. thesis, Rice University, Houston, Texas, 501 p.
Tari, G., 1995, Eo-Alpine (Cretaceous) tectonics in the Alpine-Pannonian transition zone, *in* F. Horváth, G. Tari, and C.s. Bokor, eds., Extensional collapse of the Alpine orogene and hydrocarbon prospects in the basement and basement fill of the western Pannonian Basin: AAPG International Conference and Exhibition, Nice, France, Guidebook to Fieldtrip 6, p. 133–156.
Tari, G., 1996, Extreme crustal extension in the Rába River extensional corridor (Austria/Hungary): Mitteilungen der Gesellschaft der Geologie-und Bergbaustudenten Österreich, Vienna, Austria, v. 41, p. 1–17.
Tari, G., and F. Horváth, 1995, Overview of the Alpine evolution of the Pannonian Basin, *in* F. Horváth, G. Tari, and Cs. Bokor, eds., Extensional collapse of the Alpine orogene and hydrocarbon prospects in the basement and basement fill of the western Pannonian Basin: AAPG International Conference and Exhibition, Nice, France, Guidebook to Fieldtrip 6, p. 7–19.
Tari, G., T. Báldi, and M. Báldi-Beke, 1993, Paleogene retroarc flexural basin beneath the Neogene Pannonian Basin: A geodynamic model: Tectonophysics, v. 226, p. 433–455.
Tari, G., F. Horváth, and G. Weir, 1995, Palinspastic reconstruction of the Alpine/Carpathian/Pannonian system, *in* F. Horváth, G. Tari, and C.s. Bokor, eds., Extensional collapse of the Alpine orogene and hydrocarbon prospects in the basement and basement fill of the western Pannonian Basin: AAPG International Conference and Exhibition, Nice, France, Guidebook to Fieldtrip 6, p. 119–131.
Telegdi-Roth, K., 1935, Daten aus dem Nördlichen Bakony Gebirge zur Jungmesozoischen Entwicklungsgesichte der "Ungarischen Zwischenmasse" (In Hungarian with German summary): Matematikai és Természettudományi Értesítö, Budapest, Hungary, v. 52, p. 205–252.
Thöni, M., 1986, The Rb-Sr thin slab isochron method: An unreliable geochronologic method for dating geologic events in polymetamorphic terrains?: Memorie di Scienze Geologiche, Padova, Italy, v. 36, p. 283–352.
Tollmann, A., 1976, Analyse des klassischen nordalpinen Mesozoikums-Monographie der Nördlichen Kalkalpen, Teil 2: Wien, Verlag Franz Deuticke, v. 580, p.
von Blanckenburg, F., I. M. Villa, H. Baur, G. Morteani, and R. H. Steiger, 1989, Time calibration of a PT-path from the Western Tauern Window, Eastern Alps: The problem of closure temperatures: Contributions to Mineralogy and Petrology, v. 1401, p. 1–11.
von Blanckenburg, F., and H, Kagami, 1998, The origin of alpine plutons along the Periadriatic Lineament: Schweizerische Mineralogische und Petrographische Mitteilungen, Zurich, Switzerland, v. 78, p. 55–66.
Wachtel, G., and G. Wessely, 1981, Die Tiefbohrung Berndorf 1 in den stlichen Kalkalpen und ihr geologischer Rahmen: Mitteilungen der Österreichischen Geologischen Gesellschaft, Vienna, Austria, v. 74/75, p. 137–165.
Wagreich, M., 1991, Subsidenzanalyse an kalkalpinen Oberkreideserien der Gosau-Gruppe (Österreich): Zentralblatt für Geologie und Paläontologie, Teil I, Geologie, Stuttgart, Germany, v. 1990, p. 1645–1657.
Wagreich, M., 1995, Subduction tectonic erosion and Late Cretaceous subsidence along the northern Austroalpine margin (Eastern Alps, Austria): Tectonophysics, v. 242, p. 63–78, doi:10.1016/0040-1951(94)00151-X.
Wagreich, M., R. Zetter, G. Bryda, and H. Peresson, 1997, Das Tertiär von Hieflau (Steiermark): Untermiozäne Sedimentation in den östlichen Kalkalpen. Zentralblatt für Geologie und Paläontologie, Teil I., Stuttgart, Germany, v. 1996, p. 633–645.
Weidich, K. F., 1984, Feinstratigraphie, Taxonomie planktonischer Foraminiferen und Palökologie der Foraminiferengesamtfauna der kalkalpinen tieferen Oberkreide (Untercenoman-Untercampan) der Bayerischen Alpen: Abh. Bayer. Akad. Wiss., math.-naturwiss. KI., N. F., Munich, Germany, v. 162, p. 1–151.
Wessely, G., 1987, Mesozoic and Tertiary evolution of the Alpine-Carpathian foreland in eastern Austria: Tectonophysics, v. 137, p. 45–59, doi:10.1016/0040-1951(87)90313-1.
Zimmermann, R., K. Hammerschmidt, and G. Franz, 1994, Eocene high pressure metamorphism in the Penninic units of the Tauern Window (Eastern Alps): Evidence from 40Ar-39Ar dating and petrological investigations: Contributions to Mineralogy and Petrology, v. 117, p. 175–186, doi:10.1007/BF00286841.

12

Verzhbitsky, Vladimir E., Sergey D. Sokolov, Erling M. Frantzen, Alice Little, Marianna I. Tuchkova, and Leopold I. Lobkovsky, 2012, The South Chukchi Sedimentary Basin (Chukchi Sea, Russian Arctic): Age, structural pattern, and hydrocarbon potential, *in* D. Gao, ed., Tectonics and sedimentation: Implications for petroleum systems: AAPG Memoir 100, p. 267–290.

The South Chukchi Sedimentary Basin (Chukchi Sea, Russian Arctic): Age, Structural Pattern, and Hydrocarbon Potential

Vladimir E. Verzhbitsky[1]

Gazpromneft Science and Technology Center, 5A Galernaya St., Saint-Petersburg, 190000 Russia (e-mail: torsek1@mail.ru)

Sergey D. Sokolov and Marianna I. Tuchkova

Geological Institute, Russian Academy of Sciences, 7 Pyzhevsky Ln., Moscow, 119017 Russia (e-mails: sokolov@ginras.ru; tuchkova@ginras.ru)

Erling M. Frantzen

TGS-NOPEC Geophysical Company ASA, Hagaløkkveien 13, N-1383 Asker, Norway (e-mail: erling.frantzen@tgsnopec.no)

Alice Little[2]

Aker Solutions, Lagerveien 30, NO-4033 Stavanger, Norway (e-mail: alice.little@akersolutions.com)

Leopold I. Lobkovsky

Shirshov Institute of Oceanology, Russian Academy of Sciences, 36 Nahimovskiy Ave., Moscow, 117997 Russia (e-mail: llobkovsky@ocean.ru)

ABSTRACT

The South Chukchi Basin separates the late Mesozoic Chukotka Fold Belt from the Wrangel Arch and represents the northwestern continuation of the Hope Basin of the United States sector of the Chukchi Sea, which is filled with middle Eocene–Quaternary nonmarine, marine, and lacustrine rocks. The main stages of South Chukchi Basin development in the Cenozoic are comparable to those of the Hope Basin, although the analysis of onshore data from Chukotka and Wrangel Island points to the beginning of sedimentation during the Aptian–Albian–Late Cretaceous. In the South Chukchi Basin, the sediment thickness seldom exceeds 3 to 4 km (1.9–2.5 mi) but can locally reach 5 to 6 km (3.1–3.7 mi). The geometry of the faults indicates an extensional and/or transtensional setting for the South Chukchi Basin, although folds, reverse

[1] *Previous address*: TGS-NOPEC Geophysical Company Moscow, Donskaya St. 4, Bldg. 3, Moscow, Russia.
[2] *Previous address*: TGS Geological Products and Services, Professor Olav Hanssensvei 7A, PO Box 8034, M-4068 Stavanger, Norway.

DOI:10.1306/13351557M1003534

and thrust faults, pop-up and positive flower structures also occur, pointing to the local development of compressional and transpressional stress. Low-angle thrust faults predating the Aptian(?)–Paleogene extension (most likely of Late Jurassic–Neocomian age) are recognized at the base of the South Chukchi Basin. This could support the idea that the extension in the basin was driven by gravitational collapse of the Wrangel-Herald-Lisburne fold and thrust belt in the post-Neocomian. Based on the interpretation of new seismic data and analysis of published material, we believe that the hydrocarbon potential of the South Chukchi Basin may be significantly higher than what has been previously suggested.

INTRODUCTION

The South Chukchi Basin is a sedimentary basin to the south of the Wrangel-Herald-Lisburne Arch and to the north of the Chukotka Fold Belt (Figures 1–4). It extends for approximately 1000 km (~600 mi) in a northwest–southeast direction from the eastern East Siberian Sea (east of Chauna Bay) through the Longa Strait (south of Wrangel Island), along the northern coast of Chukotka and, farther to the east, between the Lisburne and Seward peninsulas (Alaska) toward the Kotzebue Sound area. The basin is split longitudinally between the Russian (west) and the United States (east) sectors, the latter part is commonly being referred to as the Hope Basin. The South Chukchi Basin contains many subbasins, with the Hope Subbasin being the deepest.

In previous publications, the South Chukchi Basin was also referred to as the South Chukchi Sea Basin (Mazarovich and Sokolov, 2003) and the Longa-Chukchi Basin (Petrovskaya et al., 2008). The thickness of sedimentary rocks above acoustic (folded) basement seldom exceeds 4 km (2.5 mi) but can locally reach 8 km (5 mi).

The structural style, age of sediments, and hydrocarbon potential of the South Chukchi Basin are still poorly understood. This is caused by the limited seismic coverage, the absence of offshore wells, and an incomplete understanding of the geologic structure of the coastal areas of the Chukotka Peninsula and nearby islands. It is commonly believed that a low hydrocarbon potential could be expected in the South Chukchi Basin, thus geologic and geophysical exploration in the South Chukchi Basin has not been a high priority within the Chukchi Sea region. However, Orudzheva et al. (1999) reported that the South Chukchi Basin has potential for the future energy in the northern regions of Russia.

Several stages of geophysical data acquisition between 1976 and 1990 resulted in many fine publications on the geology of the South Chukchi Basin (Kogan, 1981; Pol'kin, 1984; Shipilov et al., 1989; Otochkin and Ivanov, 1989; Vladimirtseva et al., 2001; Kos'ko et al., 1993; Burlin and Shipel'kevich, 2006; Petrovskaya et al., 2008; and others). These studies divided the basin fill in several different ways and dated the basin fill as ranging from Late Jurassic (Petrovskaya et al., 2008) or even Late Triassic (Shipilov et al., 1989) at the base to Holocene at the top.

The only direct evidence for the age of sedimentary fill comes from two wells drilled in the United States sector by the Standard Oil of California (SOCAL). These recovered sedimentary rocks down to the Eocene before encountering strongly deformed and metamorphosed pre–Upper Cretaceous (Paleozoic?) rocks of the basement (Tolson 1987). These wells, however, are located on the southeastern periphery of the Hope Basin and may not represent the full section contained in the basin. No wells have been drilled in the Russian sector of the South Chukchi Basin.

We base our study primarily on new two-dimensional (2-D) seismic data acquired in the Russian part of the Chukchi Sea in 2006 by TGS and LLC Integrator of Geophysical Solutions and some results of onshore observations on the Chukotka Peninsula and Wrangel Island. The main goal of this chapter is to describe the available geologic and geophysical data to shed some light on the structural style, tectonic development, and hydrocarbon potential of the South Chukchi Basin.

GEOLOGIC SETTING

The South Chukchi Basin is surrounded by Late Mesozoic fold and thrust belts (Figures 2, 3). In the Russian sector, it is bounded in the south by the Chukotka north to northeast-vergent fold and thrust belt, which resulted from the Late Jurassic–Early Cretaceous (pre-Aptian) collision between Eurasia and the Chukotka microcontinent along the South Anyui Suture Zone (Zonenshain et al., 1990; Parfenov et al., 1993; Sokolov et al., 2002, 2009; Bondarenko et al., 2003; Miller et al., 2008, 2009; and many others). To the north, the South Chukchi Basin is bounded by the Wrangel-Herald Arch (Figures 2–4), the frontal thrust zone of the Chukotka

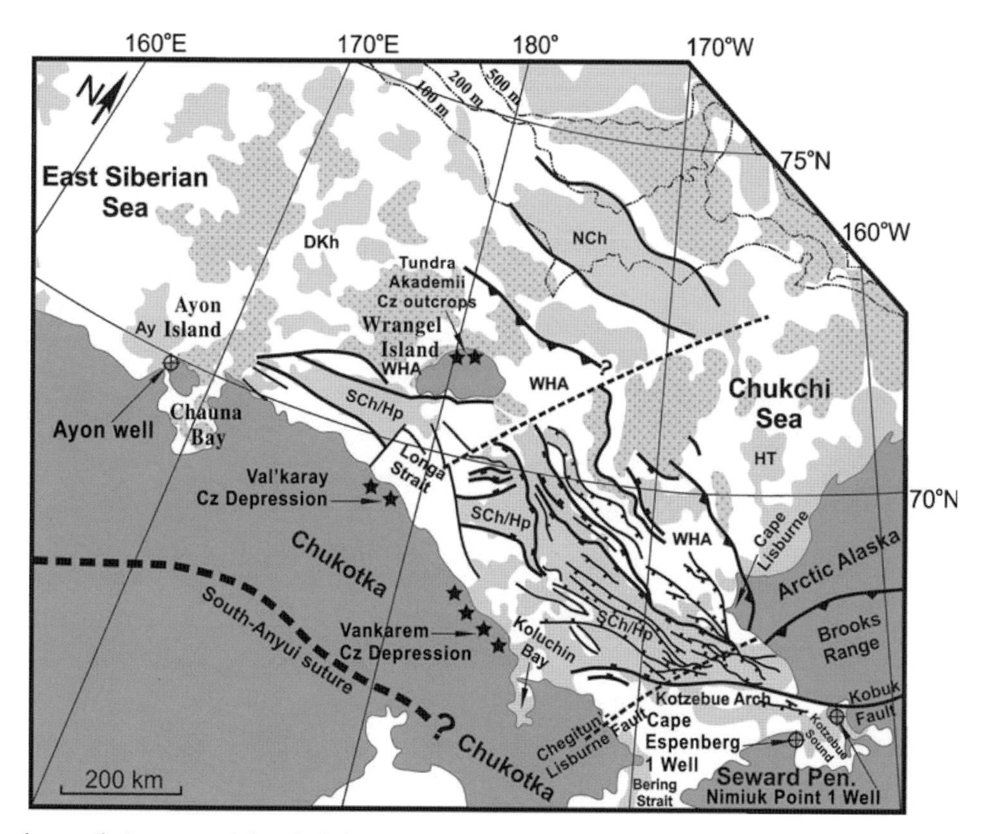

Figure 1. Regional compilation map of the Chukchi Sea, East Siberian Sea, and adjoining regions based on previously published maps, seismic sections, and satellite-free air gravity data. Compiled from many sources including Kogan (1981), Tolson (1987), Shipilov et al. (1989), Grantz et al. (1990a, b; 2002), Kos'ko et al. (1993, 2003), Drachev et al. (1998, 2001), Natal'in (1999), Sokolov et al. (2002; 2009), and Mazarovich and Sokolov (2003). Modified from Miller and Verzhbitsky (2009). Faults are denoted by black lines, normal faults having ticks on the downthrown side, thrusts having triangles on the upthrown side. Shaded areas denote gravity lows (light gray) and gravity highs (stippled gray). Black stars indicate the locations of onshore Cenozoic outcrops. Circles with crosses correspond to well locations. The South Anyui Suture Zone is also shown. Ay = Ayon depression; DKh = Drem-Khed Basin; NCh = North Chukchi Basin; SCh/Hp = South Chukchi (Hope) Basin; WHA = Wrangel Herald Arch; HT = Hanna Trough. 200 km (124 mi).

fold and thrust belt. Thus, it could be inferred that the entire basin is underlain by late Mesozoic folded basement (e.g., Drachev et al., 2010).

The Chukotka Belt fold and thrust belt is mostly composed of deformed Triassic turbidites and a Late Jurassic–Neocomian synorogenic clastic sequence outcropping in central Chukotka, known as the Myrgovaam/Rauchua Basin (Figure 2). Lower Jurassic sedimentary rocks are known only locally in the Myrgovaam/Rauchua Basin, and Middle Jurassic rocks have not yet been found in the region. Outcrops of Ordovician to Permian rocks are very rare and thus still poorly understood.

The onset of deformation in central and northern Chukotka is dated by the deposition of synorogenic terrigenous rocks in the Late Jurassic–Early Cretaceous in the Rauchua-Myrgovaam Basin and in the South Anyui Suture Zone (Baranov, 1995; Sokolov et al., 2002, 2009; Bondarenko et al., 2003; Miller et al., 2008) (Figure 2). The youngest terrigeneous rocks involved in the collisional deformation on Chukotka are dated from different localities as Valanginian (e.g., Belik and Sosunov, 1969), approximately 140 to 136 Ma (numbers are from Gradstein et al., 2004); late Hauterivian, approximately 131 Ma (Bondarenko et al., 2003); and Barremian, 130 to 125 Ma (Miller et al., 2009; Sokolov et al., 2009). The final stage of Chukotkian orogeny occurred in the Early Cretaceous (Neocomian) (Zonenshain et al., 1990; Sokolov et al., 2002, 2009; Katkov et al., 2007; Miller et al., 2008, 2009). Triassic rocks display normal fault-related deformation that predates the Chukotkian orogeny. The age of this deformation is approximately 200 Ma (Tuchkova et al., 2007).

The U-Pb ages of zircons from postcollisional granite plutons in Central Chukotka range from approximately 117 Ma (Aptian) to approximately 108 Ma

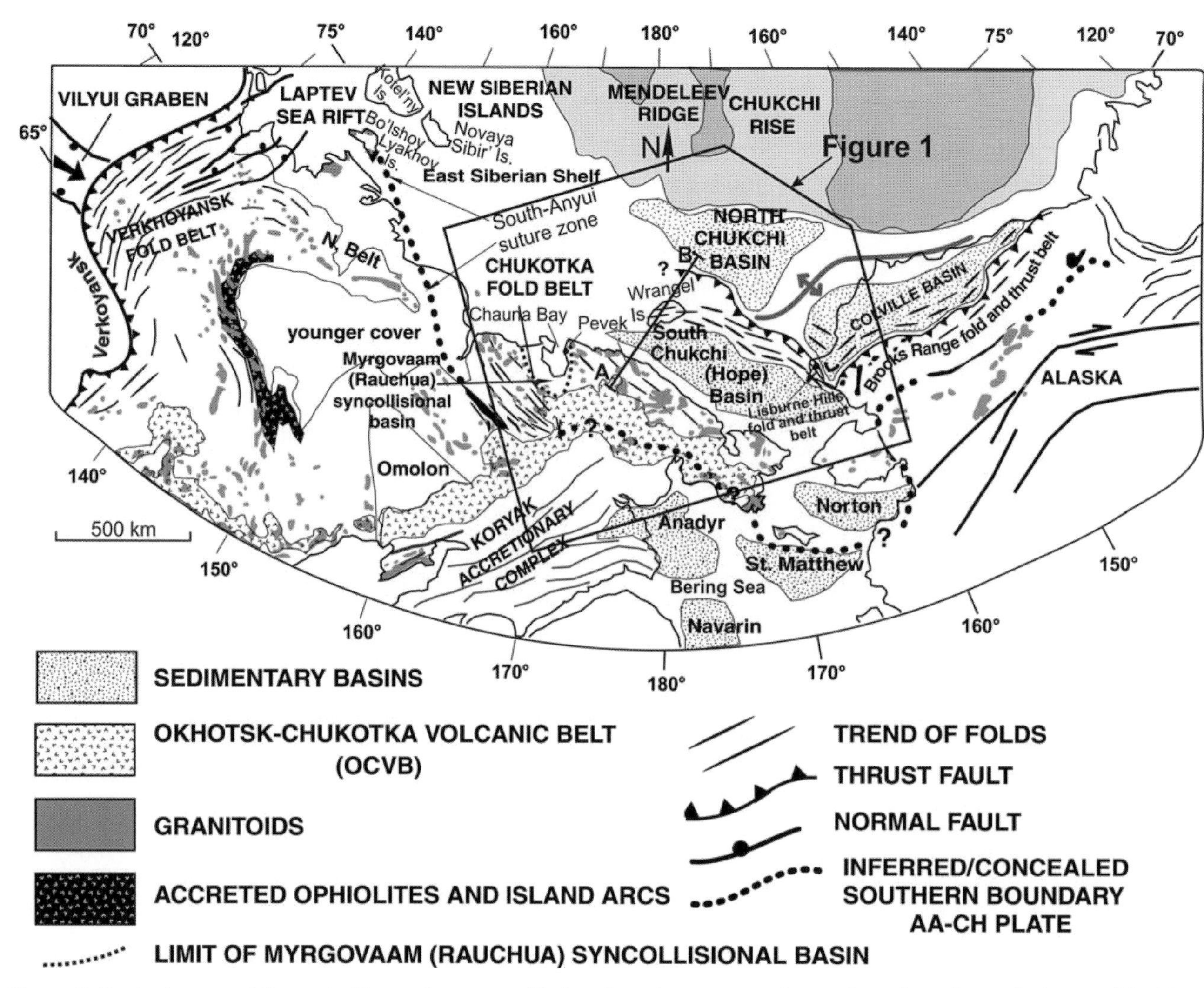

Figure 2. Tectonic map of the main Mesozoic structural belts of northeastern Arctic Russia and northern Alaska. Modified from Miller et al. (2006). See Figure 3 for cross section AB. 500 km (310.7 mi).

(Albian) (Katkov et al., 2007; Miller et al., 2009) (Figures 2, 3). The emplacement of the plutons on Chukotka occurred in an extensional setting and thus marks a drastic change in the regional tectonic regime (Miller and Verzhbitsky, 2009). Dating (U-Pb, K-Ar, and Ar-Ar) of postorogenic granitoids on Bol'shoy Lyakhov Island, at the northwest of the South Anyui Suture Zone (Drachev and Savostin, 1993; Kuzmichev et al., 2005) (Figure 2), also revealed an Aptian–Albian age (~122–106 Ma) (Layer et al., 2001; Kos'ko and Trufanov, 2002; Kuzmichev, 2009). In the South Anyui Suture Zone, a series of small depressions are filled by slightly deformed sedimentary and volcanic deposits of Aptian–Albian age. The latter overlay all major contractional structures of the suture zone, with a sharp angular unconformity (Sokolov et al., 2009). Sokolov et al. (2009) also showed that the Barremian–Aptian (possibly early Albian) late collisional stage of the South Anyui Suture Zone development was characterized by dextral strike-slip faulting.

During the Middle–Late Cretaceous, the region was subjected to postcollisional extension, which resulted in exhumation of amphibolite and granulite complexes, forming a metamorphic core complex both in eastern Chukotka and northern Alaska (Miller and Hudson, 1991; Dumitru et al. (1995); Natal'in, 1999; Akinin and Calvert, 2002; Amato and Miller, 2004; Akinin et al., 2009; Miller et al., 2009; Miller and Verzhbitsky, 2009).

The Wrangel-Herald Arch is mostly composed of intensively deformed Neoproterozoic metamorphic rocks (Wrangel Complex), unconformably overlain by the Silurian–Triassic sedimentary sequences that are involved in the late Mesozoic contractional deformation (Figures 5–7). The timing of the contractional deformation on Wrangel Island is not precisely dated

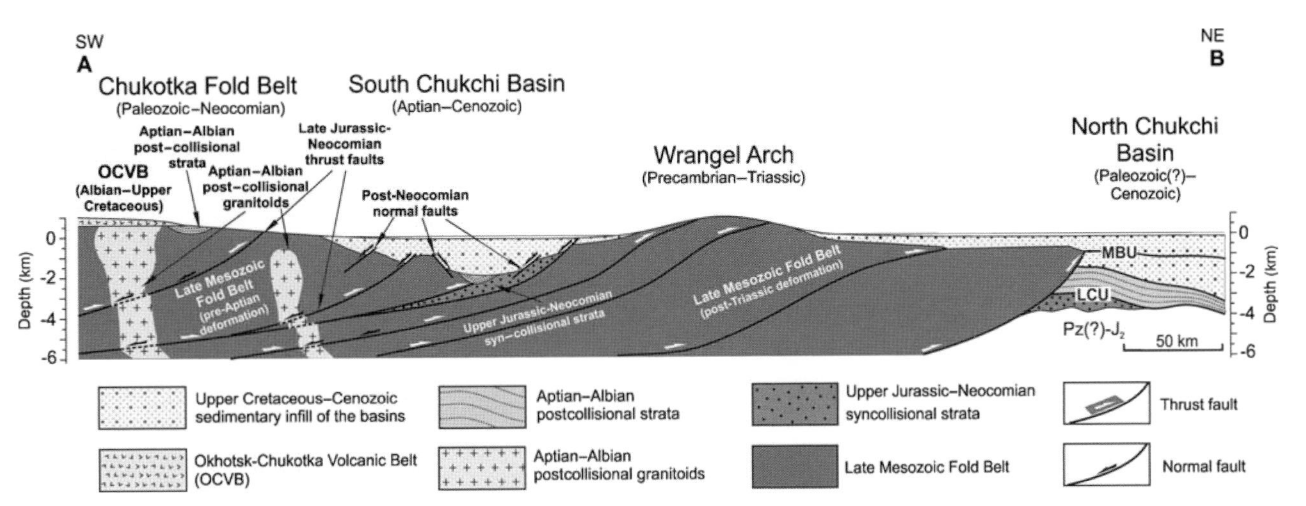

Figure 3. Schematic structural cross section through northern Chukotka Fold Belt, South Chukchi Basin, Wrangel Island, and southern North Chukchi Basin. LCU = Lower Cretaceous Unconformity (pre-Aptian); MBU = Mid-Brookian Unconformity (Cretaceous-Cenozoic boundary); OCVB = Okhotsk-Chukotka Volcanic Belt. Compiled from many sources including Grantz et al. (1990a), Kos'ko et al. (1993, 2003), Orudzheva et al. (1999), Sokolov et al. (2002, 2009), Burlin and Shipel'kevich (2006), Katkov et al. (2007), Verzhbitsky et al. (2008); Miller et al. (2009), Miller and Verzhbitsky (2009), and Drachev et al. (2010). See Figure 2 for the location of the cross section. See text for explanation. 50 km (31 mi).

mainly because the island complexes lack postcollisional plutons, but it must have occurred in the post-Triassic (the youngest units involved in deformation) and before the deposition of the Late Cretaceous–Tertiary strata that remain undeformed (Kos'ko et al., 1993, 2003). It is likely that the contractional deformation

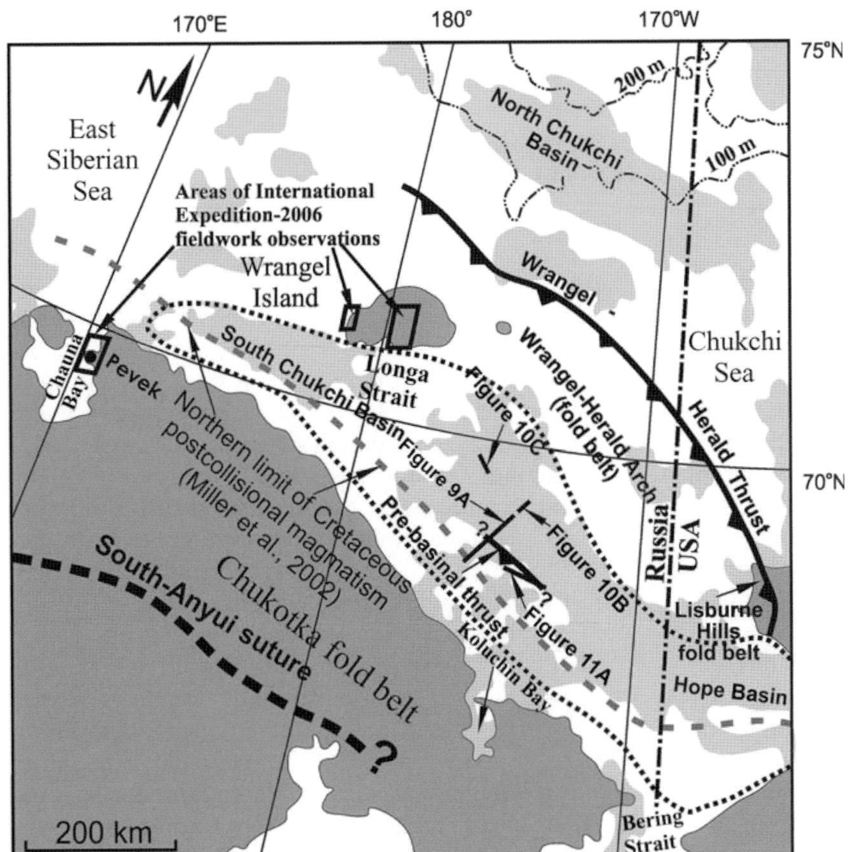

Figure 4. Map showing the main structural elements of the Russian sector of the Chukchi Sea region. Light-gray shaded areas denote gravity lows, dark-gray land. Solid black lines show the location of the TGS/Integrator seismic profiles. Legend and references as for Figure 1.

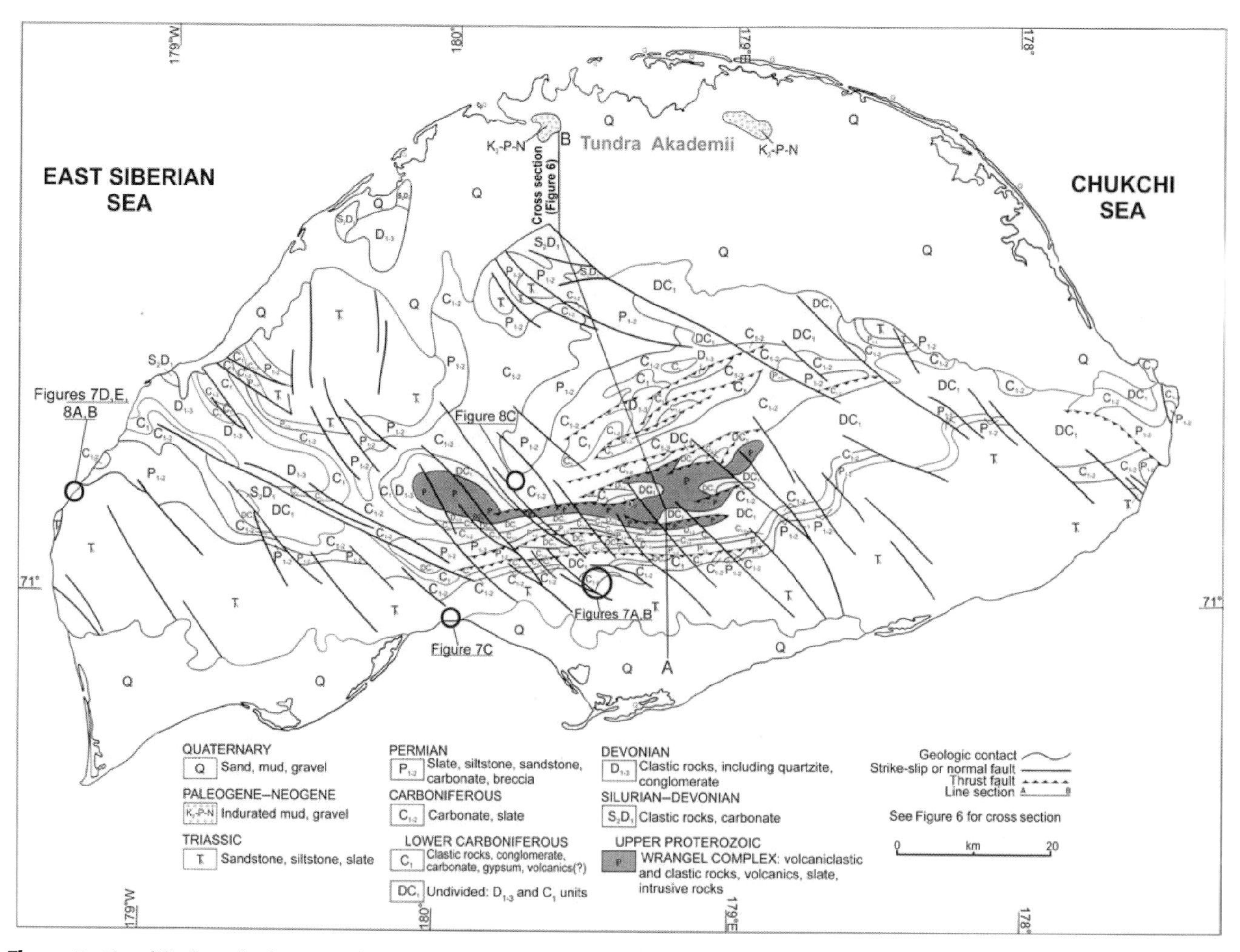

Figure 5. Simplified geologic map of Wrangel Island (Kos'ko et al., 1993). Modified from Kos'ko et al. (2003). The locations of photographs (circles), for Figures 7 and 8, are shown. 20 km (12.4 mi).

along the Wrangel-Herald Arch was roughly synchronous with the Chukotkian orogeny (Kos'ko et al., 1993).

During field work in 2006 of the central and western parts of Wrangel Island, we investigated Precambrian and Paleozoic–Triassic rocks (Miller et al., 2010). In general, our field observations confirmed a rather uniform, north-vergent, low-angle thrust fault and fold structural pattern (Figures 3, 5–7), complicated by northwest–southeast-trending dextral strike-slip faults (Til'man et al., 1970; Kos'ko et al., 1993, 2003). We also recognized that the latest stage of structural evolution on the island is expressed as normal displacement along south-dipping cleavage and formerly reverse faults and extensional quartz veins superimposed on the preexisting contractional fabric (Figure 8), which we relate to the formation of the South Chukchi Basin. It should be noted that Kos'ko et al. (1993) also pointed to the limited evidence for Tertiary extensional tectonism on Wrangel Island. The prevalence of south-dipping normal faults with right lateral component indicates a roughly northeast–southwest extensional setting, which is supported in places by measurements of other types of extensional mesostructures (Figure 7E).

The origin of the South Chukchi Basin is still a major issue that has not been resolved. Kos'ko et al. (1993) noticed that the basin is located well south of the Chukotkian-Brookian deformation front (Wrangel-Herald Arch) and thus is in an intermontane (piggy-back?) position. Most researchers considered the basin as extensional and/or transtensional in origin (Tolson, 1987; Shipilov et al., 1989; Natal'in, 1999; Klemperer et al., 2002; Elswick, 2003; Elswick and Toro, 2003; Filatova and Khain, 2007; Khain et al., 2009). Some geologists relate the basin's formation to Cenozoic dextral strike-slip displacement along the east–west-trending Kobuk Fault in western Alaska and related fault zones in the Russian sector (Tolson, 1987; Filatova and Khain, 2007; Khain et al., 2009; and others) (Figure 1).

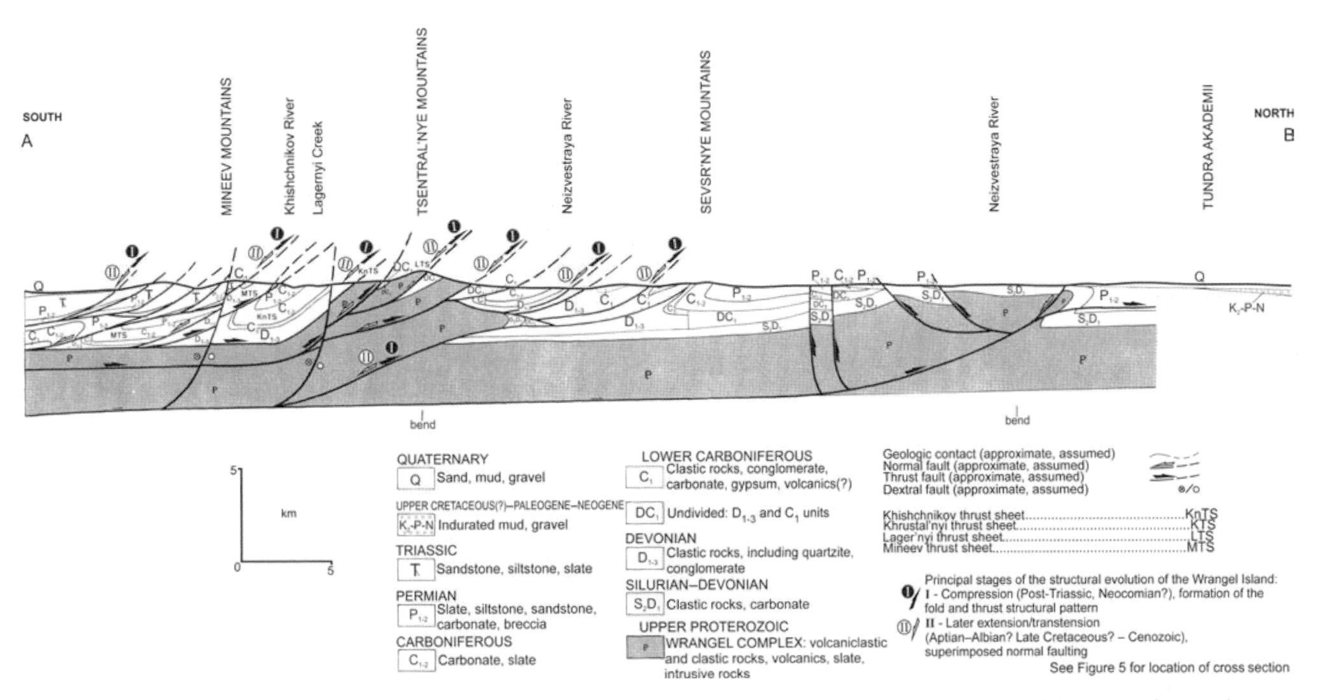

Figure 6. Schematic structural cross section through central Wrangel Island. Simplified and modified from Kos'ko et al. (1993, 2003). 5 km (3.1 mi).

Natal'in (1999) attributed the formation of the basin to the latest stage (Late Cretaceous–Eocene–Oligocene) of the orogenic collapse of the thickened crust (late Mesozoic fold belts) in the Bering Strait region. This collapse was initiated in the Albian (~108 Ma) and was accompanied by the formation of metamorphic core

Figure 7. Evidence for the contractional deformation of the Late Triassic sequence at several localities on Wrangel Island. (A) South-dipping Mineev Thrust Fault Zone (red dashed line), with Carboniferous carbonates overthrust by Triassic turbidites (exposure, ~3 km [~1.9 mi]). (B) North-vergent tight folds and axial planar cleavage in the Triassic sequence close to the Mineev Thrust Fault Zone. (C) Subvertical Triassic beds dipping to the south. (D) North-vergent open fold with axial plane cleavage in Triassic turbidites. (E) South-vergent thrust sheet of Triassic turbidites complicated by a series of duplexes in the hanging wall. Footwall consists of a slaty unit (exposure, ~100 m [~328 ft]). See Figure 5 for the locations.

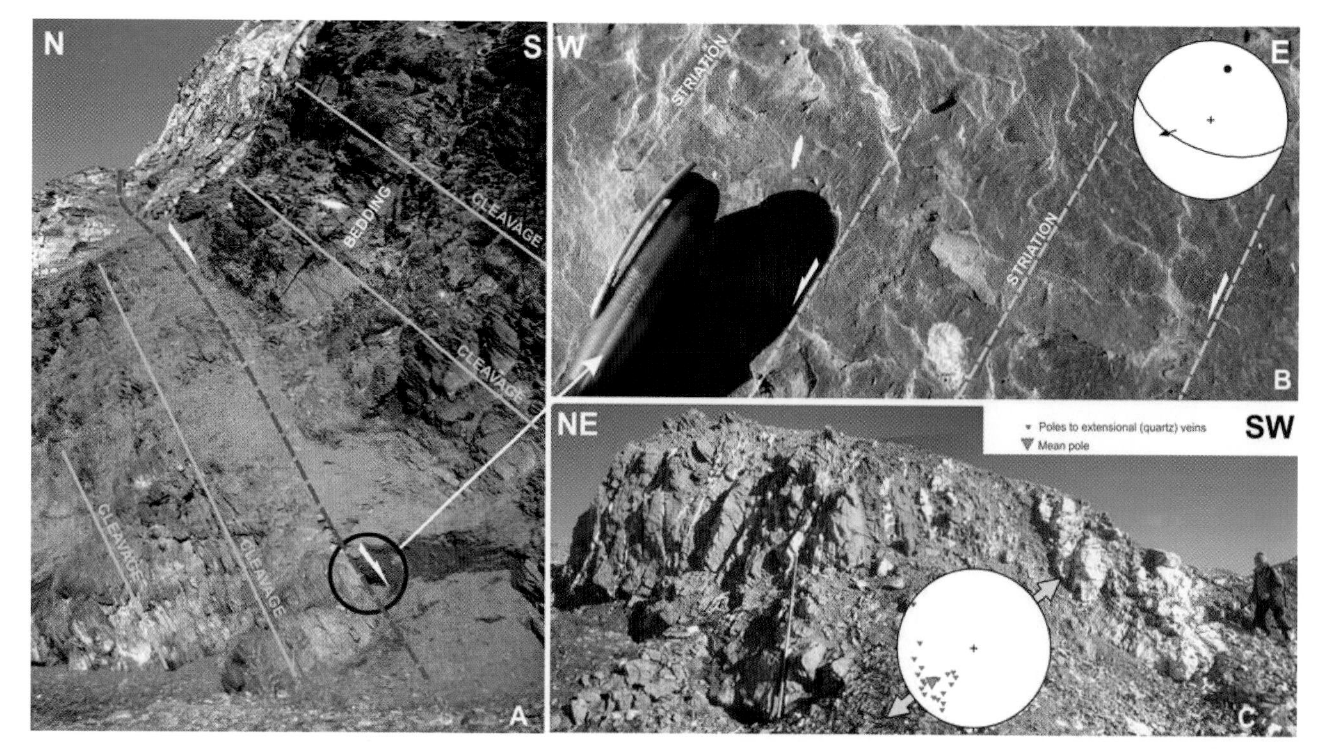

Figure 8. Evidence for postcontractional extensional tectonics on Wrangel Island. (A) Normal fault superimposed onto the south-dipping cleavage plane and displacing preexisting contractional structural pattern. (B) Example of striations with a normal dextral sense of displacement on the fault surface. (C) Extension-related veins in Carboniferous carbonates filled by hydrothermal quartz. Veins strike northwest–southeast and are superimposed on the earlier contractional fabric. See Figure 5 for the locations.

complexes on the Chukotka and Seward peninsulas (Natal'in, 1999). He also pointed out that the development of the Hope Basin was accompanied by the formation of northwest–southeast-trending normal faults on Chukotka and the associated prominent northeast-southwest-trending right lateral Chegitun'-Lisburne Transfer Fault (Figure 1). The formation of the Herald-Lisburne Hills fold and thrust belt (Figure 2) was attributed to the compensation of the extension and displacement to the northeast (from Chukotka to Alaska) along normal faults and orthogonal dextral strike-slip faults, respectively (Natal'in, 1999).

Elswick (2003) and Elswick and Toro (2003) showed that little evidence exists for strike-slip or transtensional features in the seismic data of the Hope Basin and argued that the extension in the basin was driven by gravitational collapse of the Wrangel-Herald-Lisburne fold and thrust belt and thus postdates basin formation. Nevertheless, they added that there might be some dextral movement that influenced the creation of the Hope Basin as well but was not a major factor.

Recently, geologists from the Scientific Technical Center of Rosneft Oil Company (Malyshev et al., 2010a) concluded that three main stages of development for the Russian South Chukchi Basin could be distinguished. This was based on a large seismic data set combined with geologic data on both Russian and United States parts of the Chukchi Sea and adjoining onshore. During the first stage (Albian–Late Cretaceous), a series of extensional depressions were formed, related to the Chukotka-Brooks Range orogenic collapse. During the Paleocene–Miocene, the extensional tectonic setting continued but was characterized by a significant right lateral component during the late Oligocene–early Miocene. The third stage, a relatively stable general subsidence, occurred in the Pliocene–Quaternary (Malyshev et al., 2010a).

Based on thermal maturity, fission-track, and structural data, Moore et al. (2002) concluded that the Lisburne Hills fold and thrust belt was formed mainly in the Early Cretaceous (Neocomian). Thus, the age of folded basement for the Russian and United States parts of the South Chukchi Basin appears to be roughly coeval.

AGE OF SEDIMENTARY INFILL

Complexes of the Upper Part of Sedimentary Cover

Two SOCAL onshore wells were drilled in the Cotzebue Sound area: Cape Espenberg well 1 (2450 m [~8000 ft]) and Nimiuk Point well 1 (1817 m [~6000 ft]) (Tolson, 1987) (Figure 1). These data, combined with U.S. Geological Survey seismic data, were used by Tolson (1987) for describing the age and main stages of the development of the Hope Basin. In the wells, Eocene lacustrine and nearshore sediments overlie Paleozoic(?) crystalline carbonate rocks of the folded basement. According to Tolson (1987), the Hope Basin was developed mainly in three stages, corresponding to three seismic sequences (units). During the Paleocene(?)–Eocene–Oligocene (unit 1), the northwest–southeast-trending half-grabens were formed and filled by continental to possibly lacustrine and shallow-marine sediments. The sedimentation was accompanied by basaltic eruptions (40–42 Ma) and deposition of tuffs. During the early Miocene (unit 2), the entire basin thermally subsided, leading to deposition of nonmarine sediments over the entire basin. The middle to late Miocene stage (unit 3) was characterized by the extension and reactivation of the Paleogene normal faults and deposition of marine and nonmarine sediments in the basin lows (Tolson, 1987).

A common observation from studies of outcrops and wells on Chukotka and Ayon Island is that a weathered crust (Grinenko et al., 1989; Slobodin et al., 1990; Avdyunichev et al., 2004) of a broad range of age (Late Cretaceous–Paleocene) (Biske, 1975; Kos'ko et al., 2003) was developed at the interface between folded basement (Paleozoic–Mesozoic) and Tertiary strata. A hiatus in sedimentation is noted, the extent of which varies regionally. The onset of sedimentation in the Vankarem and Ayon depressions is dated as Paleocene, whereas in the Val'karay depression, it corresponds to the Miocene (Grinenko et al., 1989; Slobodin et al., 1990; Avdyunichev et al., 2004). The Ayon depression was a predominantly marine basin in the early Oligocene, Miocene, and Pliocene (Slobodin et al., 1990), whereas the Vankarem and Val'karay depressions (Grinenko et al., 1989) show transitions from terrestrial to nearshore to marine depositional environments during the middle Miocene to Pliocene (Figure 1).

These onshore studies also served to confine the younger limit for the onset of sedimentation in the South Chukchi Basin. Tolson (1987) noted the occurrence of Late Cretaceous and Tertiary outcrops in the Kotzebue Sound area (Figure 1), and it has been speculated by Elswick and Toro (2003) that, in the part of the Hope Basin that has undergone the most subsidence (Hope Subbasin), sediments of latest Cretaceous age may be present.

It is clear that the lower limit of onset of the sedimentation in the South Chukchi Basin is controlled by the age of collision of the Chukotka microcontinent and Eurasia (South Anyui collision) and the regional peneplain created during collisional and early postcollisional processes that were completed in the Late Neocomian, pre-Aptian.

Folded Basement and Lower Part of the Sedimentary Cover

The Late Jurassic–Neocomian synorogenic clastic sequence was deposited during the South Anyui collisional event and was involved in the late stages of contractional deformation. It represents part of the folded basement. The U-Pb age of the oldest postorogenic granite plutons is Aptian, approximately 117 Ma (Katkov et al., 2007; Miller et al., 2009), marking the lower limit for completion of the collision process. This age is in good agreement with the regional unconformity at the base of Aptian sequence across the East Arctic region (Zonenshain et al., 1990; Kos'ko and Trufanov, 2002; Kos'ko and Korago, 2009; Sokolov et al., 2009).

According to Natal'in (1999), the orogenic collapse both on Chukotka and Alaska started at approximately 108 Ma, in the Albian. Dumitru et al. (1995) pointed to an earlier regional episode of north–south extension during approximately 120 to 70 Ma (Aptian–Late Cretaceous), which predated Paleogene–Eocene extension in the Norton and Hope basins (Figure 2). Similarities in age of the regional gravitational collapse of the crust on the Seward Peninsula, Alaska (105–90 Ma, Albian–Turonian), and eastern Chukotka (104–94 Ma, Albian–Cenomanian) were presented by the Bering Strait Field Party (1997). The final cooling and exhumation of one of the gneiss core complexes on eastern Chukotka was coeval with motion along a low-angle extensional fault between 88 and 84 Ma, during the Coniacian–Santonian (Akinin and Calvert, 2002). The northeast-trending right lateral strike-slip faults and northeast-dipping normal faults of the north of eastern Chukotka (likely related to the South Chukchi rift development) are superimposed on the earlier foliation that was dated by $^{40}Ar/^{39}Ar$ as young as approximately 92 to 95 Ma (Cenomanian–Turonian) (Natal'in, 1999).

The oldest rocks of the Okhotsk-Chukotka Volcanic Belt (OCVB) (Figure 2) that overlie the regional peneplain surface are dated as Albian (~106 Ma) (Tikhomirov et al., 2008). Thus, the postcollisional and postmagmatic period of general uplift, erosion, and leveling might have occurred during the Albian, between 108 (the age of the youngest U-Pb-dated postcollisional plutons) and 106 Ma (the age of the oldest volcanic rocks for the northern segment of OCVB). It was proposed that this peneplain surface extends farther north offshore Chukotka toward the East Siberian shelf, where it constitutes the top of basement onto which overlying sedimentary sequences were deposited (Miller and Verzhbitsky,

2009). Thus, the sedimentary sequences that could overlay the folded basement of East Siberia and the Chukchi Sea should correspond to the late(?) Aptian–Albian age or younger. It is notable that a similar Aptian–Cenozoic age was proposed by Kos'ko and Trufanov (2002) for the sedimentary cover of the Laptev and East Siberian seas based on stratigraphic data from the New Siberian Islands (Figure 1) and their comparison with offshore seismic data (Franke et al., 2001). An Albian–Cenozoic age was proposed for the sedimentary cover of the East Siberian Sea by Drachev et al. (2001) (Figure 2).

Information on the geologic structure of the South Chukchi Basin can also be obtained from Wrangel Island (Figures 1, 5, 6). Using a seismic profile, Shipilov et al. (1989) proposed that Triassic strata on Wrangel Island overlies intensively folded Carboniferous and is not involved in contractional deformation. Here, we present an alternative interpretation based on our field observations on Wrangel Island and on Chukotka and previous studies. The Carnian–Norian (Late Triassic) turbidites, both in the southern and western parts of the island, are involved in folding and low-angle thrusting (Figure 7) and thus should represent folded basement onto which the younger sedimentary fill was deposited. The folded Triassic strata (together with the older sequences) are also involved in the latest extensional deformation (Figure 8), which we relate to the formation and development of the South Chukchi Basin. Therefore, the age of the basin's sedimentary infill should be younger than Triassic (Figure 4).

Undeformed post-Triassic (Late Cretaceous–Cenozoic) rocks, which could represent the equivalents of the South Chukchi Basin sedimentary infill, are described mostly in the northern part of Wrangel Island (Tundra Akademii) (Kos'ko et al., 1993; 2003) (Figure 5). The weathered crust on the southeastern part of the island, represented by clays approximately 1.5 to 2 m (~5–6.5 ft) thick with fragments of siltstones from the underlying Late Triassic sequence, has been referred to as Late Cretaceous–Paleocene in age (Kos'ko et al., 2003). Two main outcrops of claystone and siltstone in the north of Wrangel Island with a total thickness a few tens of meters revealed (mostly by palynological data and foraminifer analysis) a rather wide (Late Cretaceous–Miocene or Paleogene–Miocene) age range (Kos'ko et al., 1993, 2003). The Pliocene is represented by sands with gravel, pebbles, and boulders (Kos'ko et al., 2003). Thus, Late Cretaceous–Paleocene sediments could be present at the base of the South Chukchi Basin as well. Late Cretaceous clastic deposits of 300 m (~1000 ft) with brown coal and plant remnants are identified on the New Siberian Islands. They were dated, on the basis of leaf prints and palynological data, as Cenomanian–Turonian (Kos'ko and Trufanov, 2002; Kos'ko and Korago, 2009).

Shipilov et al. (1989) pointed to the existence on the published seismic profile of relatively short and intensive seismic reflectors near the base of sedimentary cover in the South Chukchi Basin. These were interpreted by them as volcanic units probably related to the northern segment of the voluminous OCVB. The magmatic activity was initiated in the Albian (~106 Ma) but was most intense during the Late Cretaceous until the Campanian (~78 Ma) (Tikhomirov et al., 2008, and others). If that is the case, the lower sections of the basin's sedimentary cover, below these anomalies, could include Aptian–Albian–Late Cretaceous sediments. It is also possible that these proposed volcanic rocks could be coeval with middle Eocene basalts and tuffs, encountered by the SOCAL wells in the Hope Basin (Tolson, 1987).

All these data and field investigations in the coastal area of the South Chukchi Basin show that it is likely that rocks older than middle Eocene in age, but not older than Aptian–Albian, could occur in the lower part of the basin's sedimentary fill. Pre-Aptian sequences are shown in outcrop to be involved in the regional collisional deformation and thus should constitute the folded and thrust-faulted basement of the basin.

STRUCTURAL GEOLOGY OF THE SOUTH CHUKCHI BASIN

The seismic data acquired in 2006 provide new information on both the structure of the sedimentary cover and underlying folded basement (Figures 9–11). Combined with onshore geologic investigations, these data provide a unique opportunity to build a more realistic model of the present-day structure of the South Chukchi Basin and a model of its geologic evolution from the Late Jurassic–Neocomian to the Pliocene–Quaternary.

The Russian sector of the Chukchi Sea is composed of several regional tectonic subdivisions (from south to north), including the Chukotka Fold-Thrust Belt, the South Chukchi Basin, the Wrangel (Wrangel-Herald) Arch, and the North Chukchi Basin (Figures 1–4). Although the TGS-Integrator seismic survey covered all of these structural elements, in this chapter, we focus on the data covering the South Chukchi Basin and adjoining Chukotka offshore area.

In our interpretation, units labeled SCB-1, SCB-2, SCB-3, and SCB-4 represent the sedimentary fill of the South Chukchi Basin (Figure 9). Seismic units labeled SCB-FB-1 and SCB-FB-2 are interpreted as basement

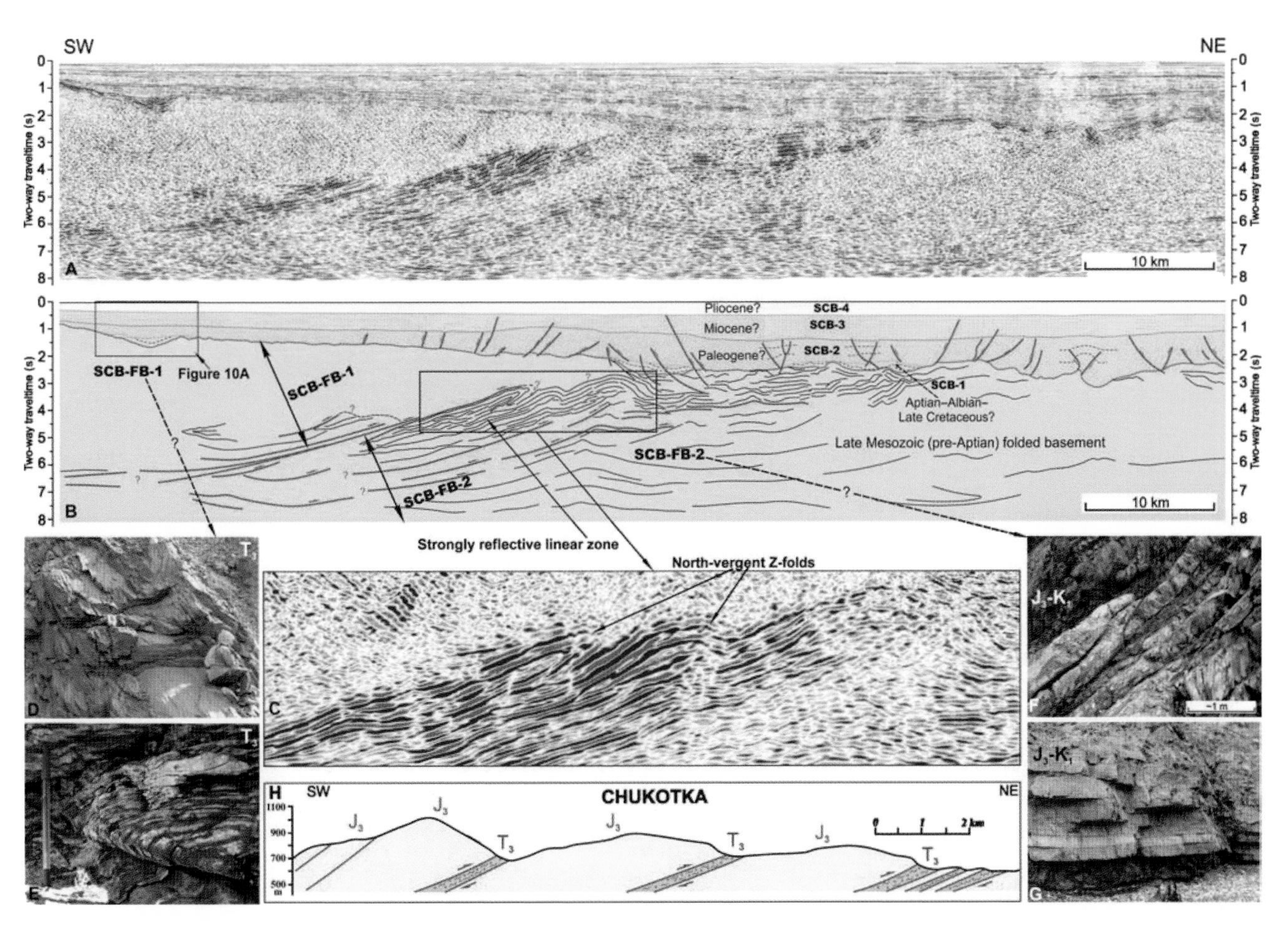

Figure 9. Interpreted seismic section across the South Chukchi Basin and examples of probable analogs for the seismic units from Chukotka and Wrangel Island. The locations of the seismic line and photographs are shown in Figure 4. (A) Seismic section (data courtesy of TGS). (B) Schematic interpretation of seismic section A. (C) Enlarged north-vergent asymmetric Z-folds along the proposed southwest-dipping low-angle thrust fault of pre-Aptian age. (D) North-vergent tight to isoclinal folds of Late Triassic age, complicated by axial plane cleavage, an analog for SCB-FB-1. (E) Two generations of cleavage, south-vergent folds of earlier cleavage (slaty, subparallel, close to flat lying), and later north-dipping crenulation cleavage. Note the south-vergent thrust displacements along later generation of cleavage. (F) Northeast-dipping monocline in the Late Jurassic–Neocomian strata complicated by a southwest-vergent mesoscale thrust fault (duplex). (G) Section with subhorizontal bedding in the Late Jurassic–Neocomian. (H) Schematic cross section of the Myrgovaam Nappe Zone: thrust sheets of synorogenic Late Jurassic clastic sequence and preorogenic Late Triassic clastic rocks. Modified from Baranov (1995). Red lines correspond to the main faults developed in the sedimentary cover and upper parts of the basement. Black solid lines correspond to the interpretation of reflectors, dashed where inferred. Arrows show the sense of displacement. 2 km (1.2 mi).

units (Figure 9). A prominent strongly reflective linear zone divides the two basement units. This zone dips gently southward.

The SCB-FB-1 unit (South Chukchi Basin Folded Basement 1) is generally acoustically transparent in character. It extends in a tapering wedge shape into the basinal area (thinning toward the north) and occurs above the linear dipping zone. The more northerly unit SCB-FB-2 (South Chukchi Basin Folded Basement 2) is internally reflective, suggesting that it has some elements of stratification. SCB-FB-2 is identifiable below the sedimentary fill and in the south below the linear dipping zone. The SCB-FB-2 unit is predicted (from seismic) to be present beneath the entire basin and may subcrop the sea floor along the Wrangel-Herald Arch to the north. We believe that the reflective SCB-FB-2 unit could be equivalents of Late Jurassic–Neocomian syncollisional deposits of the Myrgovaam (Rauchua) area (see Figure 2 and text below).

Within the linear dipping zone, a series of asymmetric Z-folds are interpreted on the seismic (Figure 9). A secondary (intraformational) zone with a flat and ramp geometry is traceable directly to the northeast (completely within the SCB-FB-2 unit) and is roughly

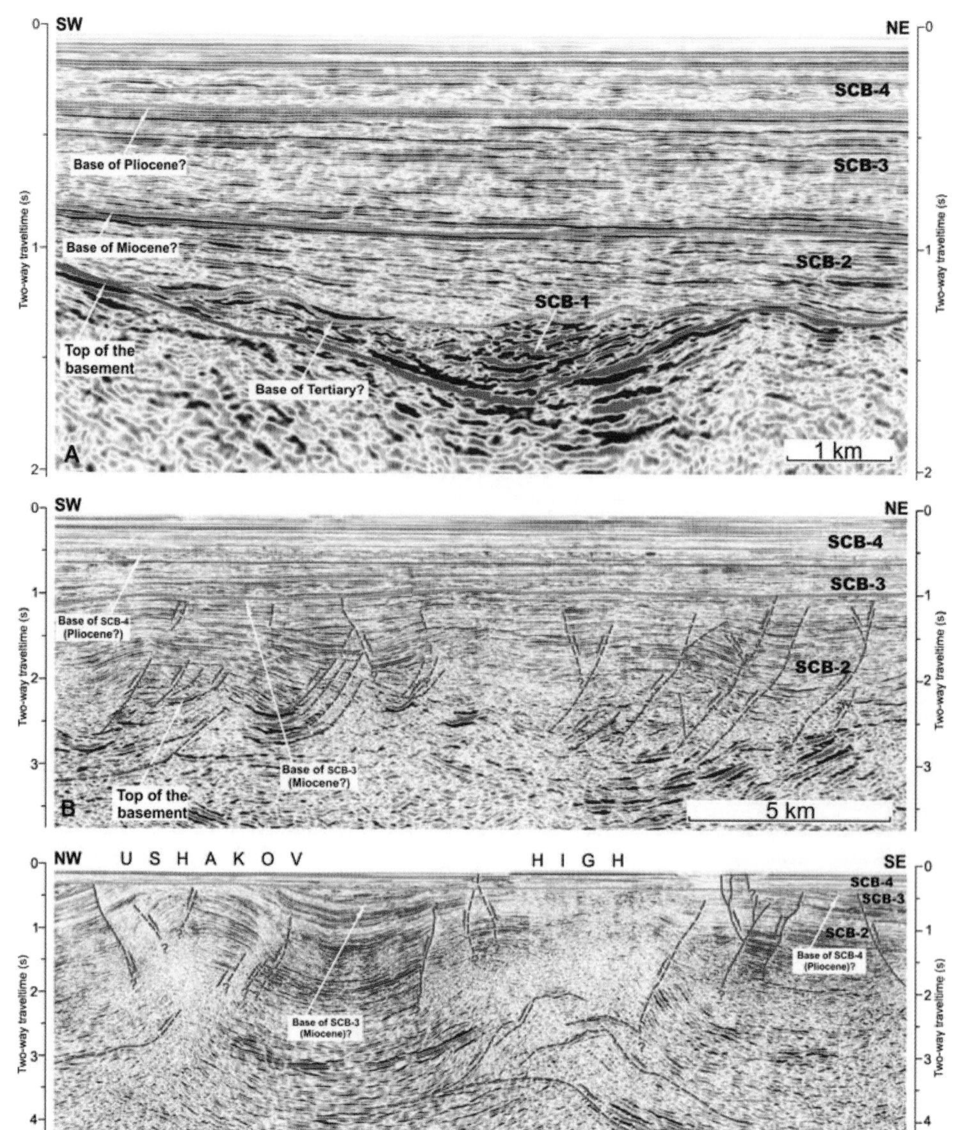

Figure 10. Seismic sections from various parts of the South Chukchi Basin (data courtesy of TGS). (A) Southwest–northeast section illustrating the pre-Tertiary(?) SCB-1 unit, interpreted to be of Aptian–Albian–Late Cretaceous age, filling depressions in the SCB-FB-1 basement unit. (B) Southwest–northeast section illustrating general extensional and/or transtensional structural pattern of the basin. Most of the normal faults are dipping to the south, which is in agreement with the inherited preexisting basement trend. (C) Northwest–southeast section through the east–west-trending Ushakov High, a transpressional pop-up (positive flower) structure formed within the sedimentary cover. See Figure 4 for the line locations.

subparallel to the linear dipping zone. The folded and thrusted units in the linear dipping zone are truncated at their intersection with the overlying sedimentary units (SCB-1 and 2), forming a major unconformity onto which the sedimentary fill onlaps (Figure 9A, B).

It is proposed that the boundary between SCB-FB-1 and SCB-FB-2 represents a basement fault dividing two different geologic blocks involved in contractional deformation (Figure 9B). Because of its generally north-vergent thrust sense (Figure 9C), we named this structure the South Chukchi Sea Thrust Fault Zone. The SCB-FB-2 basement unit is obviously folded and involved in the thrusting, underscored by the flat and ramp geometry of the reflectors and by the existence of the footwall and hanging-wall cutoffs. Most likely, the north-vergent Z-folds are developed immediately below the main fault plane, representing the subthrust sheet deformation of the SCB-FB-2 unit.

The obvious differences in the style and degree of deformation between the Late Triassic and the Late Jurassic–Neocomian sequences are evident in northern Chukotka in the vicinity of Pevek (Figure 4). The Triassic rocks were strongly deformed, forming tight to isoclinal folds and axial planar cleavage (Figure 9D). In some places, two generations of cleavage were observed, whereas the original bedding was not recognizable at all the localities (Figure 9E). Although the Late Jurassic–Neocomian rocks were obviously involved in the contractional deformation, their structural pattern is much simpler compared with that of the Triassic-aged

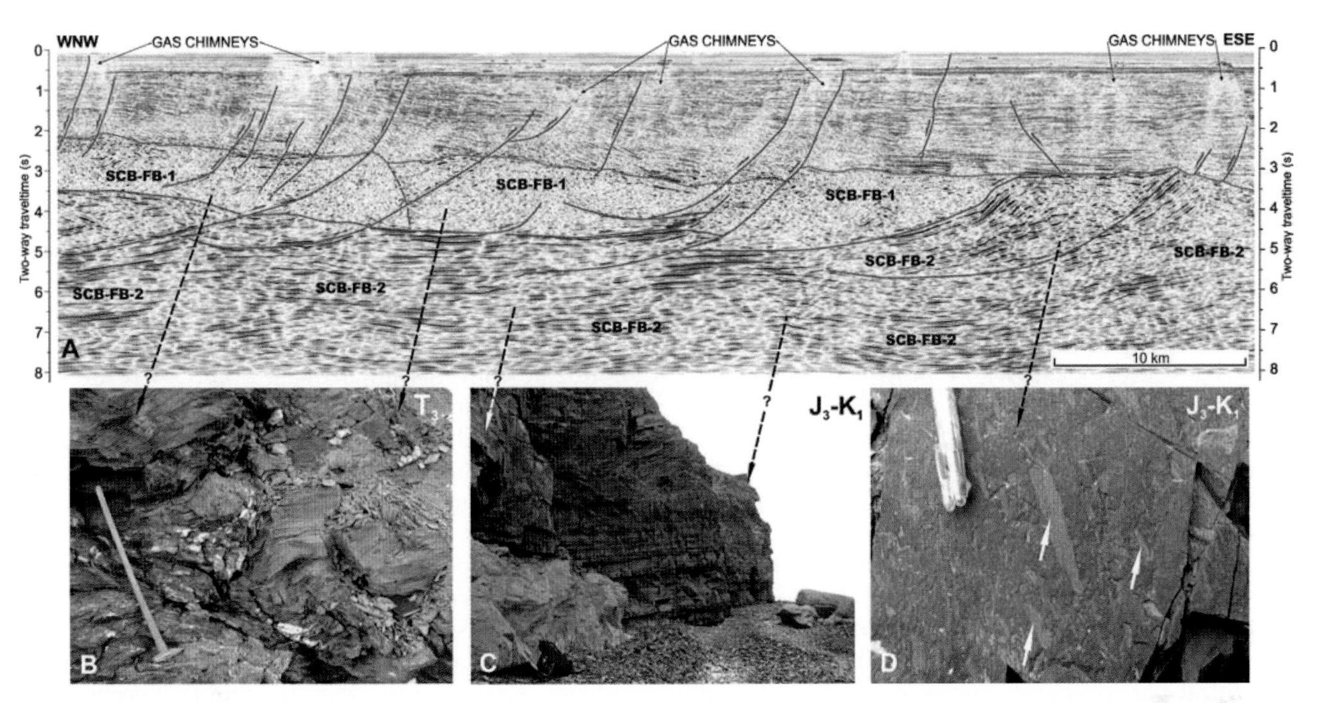

Figure 11. Seismic strike section (A) with outcrop analogs (B–D) for the interpreted seismic basement units (see Figure 4 for approximate locations). (A) West–northwest-east–southeast seismic section, an approximate strike line along the South Chukchi Basin (data courtesy of TGS), see Figure 4 for the line location. (B) Intensively folded Late Triassic analog for SCB-FB-1. (C) Flat-lying section of the Upper Jurassic–Neocomian sequence, analog for the flat-lying SCB-FB-2. (D) Outcrop of Upper Jurassic–Neocomian strata with plant remnants (indicated by white arrows).

rocks. The strata were deformed in open to tight folds, which were commonly tilted and complicated by reverse faults (Figure 9F) and younger normal faults (Miller and Verzhbitsky, 2009). The original bedding is recognizable being tilted or subhorizontal (Figures 9G, 11C). Cleavage is not visually identifiable at all the sites but, in places, is quite strongly developed (Miller and Verzhbitsky, 2009).

The South Chukchi Sea Thrust Fault Zone (Figure 9B) may represent overthrusting of the intensively deformed Chukotkian Paleozoic–Triassic rocks (SCB-FB-1) onto the moderately deformed, early syncollisional, Upper Jurassic–Lower Cretaceous sediments (SCB-FB-2). Similar relationships exist between the Late Triassic and Lower Jurassic strata observed and mapped in central northern Chukotka (Baranov, 1995) (Figure 9H). Based on our interpretation, it is likely that during the Late Neocomian before the post-Neocomian extensional stage, the area of the future South Chukchi Basin (at least in part) was an intermountain depression filled with syncollisional sediments and then overthrusted by Triassic and older sequences during the Late Neocomian (pre-Aptian). This is in good agreement with the interpretation of Kos'ko et al. (1993) except that, in our model, the intermontane basin development predates the formation of the extensional South Chukchi Basin.

The paleotectonic and/or paleogeographic relations between Chukotka and Wrangel Island in the Paleozoic and the Triassic have not been explored in detail. Hypothetically, the described South Chukchi Sea fault zone could correspond to the overthrust of the Wrangel Island terrane by the Chukotka microcontinent.

The South Chukchi Sea Thrust Fault Zone has a northwest–southeast strike and is roughly subparallel to the regional trend defined by the South Anyui Suture, the Wrangel-Herald Arch (and thrust zone), and also to the proposed northern limit of the Cretaceous postcollisional plutonic magmatism of the Chukotka Fold Belt (Miller et al., 2002) (Figure 4).

According to our interpretation, the magnitude of horizontal displacement of the thrust zone may exceed 20 to 30 km (~12–19 mi). The latter value is approximately 10 times greater than the known horizontal displacement of thrusts observed on Wrangel Island (3 km [1.9 mi]), but comparable to the general horizontal shortening (~29 km [~18 mi]) along the compiled cross section (Figure 6) (Kos'ko et al., 1993; 2003). This comparison potentially shows the decrease in the magnitude of the Late Mesozoic contractional deformation toward the north (from Chukotka to Wrangel Island), away from the South Anyui Suture Zone.

Mazarovich and Sokolov (2003) proposed that the main thrust front of the Brooks Range is traceable below the South Chukchi Basin, although it contradicts other results on the similarities in age and structural style of the Lisburne Hills and Brooks Range fold belts (Moore et al., 2002). However, the existence of high-amplitude north-vergent thrust sheets below the sedimentary cover of the South Chukchi Basin could be considered a supporting fact for the proposal by Mazarovich and Sokolov (2003).

The late Mesozoic thrust and/or reverse faults were reactivated in the post-Neocomian as extensional and/or transtensional features. Thus, it appears that extensional deformation in the South Chukchi Basin is superimposed on north-vergent basement fault zone(s). Most of the normal faults dip to the south, in accordance with the prevailing dip directions of the preexisting thrust or reverse faults, cleavage planes, and fold axial planes of the Chukotka and Wrangel Island fold belts. Several antithetic (north-dipping) normal faults identified in the central part of the seismic profile (Figure 9A, B) could be explained by preexisting pre-Aptian backthrusts that developed in the vicinity of the tip line of the main thrust fault (Figure 9A, B). These observations lead to the suggestion that the South Chukchi Basin inherited pre–late Mesozoic structural fabrics and the basin itself was formed during the Aptian–Albian–Late Cretaceous–Cenozoic as the structural style changed from contractional to extensional.

Interpretation of the seismic data available for the South Chukchi Basin (Figure 9A, B) led us to interpret similar stages of development in the post-Neocomian with those defined for the Hope Basin (Tolson, 1987). The principal difference is that we define an additional seismic package (SCB-1) lowermost in the basin (Figures 9B, 10A) that we infer to be pre-Cenozoic (most likely Aptian–Albian). We have noted that several isolated small depressions occur at the base of the sedimentary fill of the basin. The maximum thickness of the lowermost sedimentary unit on the interpreted profile (Figures 9, 10A) reaches approximately 0.4 s two-way traveltime, roughly corresponding to 500 m (~1600 ft). The strata unconformably overlie the folded basement and in turn are separated from the above sequences by an angular unconformity with obvious signs of erosion (Figure 10A). Although the isolated patches of SCB-1 cannot be directly correlated with each other on the seismic sections, their similarities suggest that they are of the same age.

The Aptian–Albian unit on the New Siberian Islands (Kotel'ny Island; Figure 2) is gently folded (Kos'ko and Trufanov, 2002; Kos'ko and Korago, 2009). Sokolov (2008, personal communication) noted that the Aptian–Albian strata in Central Chukotka (South Anyui Suture Zone) was also slightly folded. This folding occurred before the formation of the Albian–Late Cretaceous volcanic and/or volcaniclastic rocks of OCVB (before 106 Ma). The Late Cretaceous–early Paleocene (mid-Brookian) contractional event is well known for the adjoining North Chukchi Basin (Burlin and Shipel'kevich, 2006; Petrovskaya et al., 2008; Verzhbitsky et al., 2008), A second(?) leveling event, in the early Paleocene, led to a widespread peneplain over the entire northeast Asian continent (e.g., Biske, 1975). The pre-Cenozoic compressional phase could correspond to the intra-Albian or the early Paleocene (or both) events. Thus, although we are not able to determine the age of the SCB-1 unit, it is very likely that it is pre-Tertiary (Aptian–Albian or Aptian–Albian–Late Cretaceous) in age.

Taking into consideration the Late Cretaceous–Paleocene rocks known on Wrangel Island and Paleocene sediments in the Vankarem depression (Figure 1), we propose that the main stage of sedimentation (subsidence) started earlier than the middle Eocene, most probably in the Paleocene, just after the completion of the regional Late Cretaceous–early Paleocene peneplain formation (Biske, 1975; Kos'ko et al., 2003). The SCB-2 unit progressively onlaps the unconformity, overstepping the basement (SCB-FB-1 and SCB-FB-2) and the SCB-1 units to infill the basin. Small faults with normal offset are common in this unit, suggesting syndepositional accommodation. Commonly, these faults link into inferred faults in the basement.

The Paleocene–Oligocene extension and/or transtension was followed by general thermal subsidence of the basin, resulting in deposition of the SCB-3 unit, probably in the early Miocene (Figure 8A, B). The SCB-3 unit has less abundant faulting. It onlaps the SCB-2 unit with significant unconformity, but with no significant erosion or truncation. At the base of SCB-4, a significant unconformity exists in the Ushakov High area, where erosional truncation of inversion anticlines occurred (Figure 10C). Subsequent subsidence led to deposition of the Pliocene–Quaternary SCB-4 unit (up to ~500 m [~1766 ft] thick) (Figures 9A, B; 10). The most recent unit, SCB-4, is present across the whole basin, with continuous planar events of alternating low and moderate amplitudes.

The extensional and/or transtensional reactivation of the Paleogene faults occurred before deposition of SCB-4, probably during the middle to the late Miocene, as proposed for the Hope Basin by Tolson (1987). Drachev et al. (2010) proposed that during the Oligocene–Miocene, the South Chukchi Basin was affected by compression, which was caused by northeast–southwest convergence between North American and Eurasian lithospheric plates.

Although we found little evidence for the deformation of the SCB-4 complex on the seismic lines (Figures 9A, B; 10A, B), transpressional and transtensional faults locally penetrate all the Pliocene–Quaternary sequences, reaching the sea floor in places (e.g., Ushakov High area, Figure 10C). More significant indications of recent seismotectonic activity are expected well to the southeast, closer to the Bering Strait region, where a diffuse plate boundary between the North American plate and Bering Block is proposed by many researchers (Lander et al., 1996; Mackey et al., 1997, 2009; Fujita et al., 2002). It is also notable that the early opening of the Bering Strait near the Miocene–Pliocene boundary (~4.8–5.5 Ma) (Marincovich and Gladenkov, 1999), led to the connection of North Atlantic and Pacific waters, roughly coincides with the beginning of deposition of the SCB-4 unit.

Although most of the rift-related faults in the South Chukchi Basin are listric normal faults, obvious signs of the transtensional and/or strike-slip tectonics are also recognizable on the seismic lines (e.g., Ikhsanov, 2010; Malyshev et al., 2010b). Upward-bifurcating faults or normal faults (similar to the negative flower structures) are observed on the seismic records (e.g., Ikhsanov, 2010; Malyshev et al., 2010b) (Figure 10B). Pop-up and positive flower structures are also widespread in the sedimentary cover of the basin, showing both the syndepositional and postdepositional deformation (Figures 9A, B; 10B; 11C). According to the analysis provided by geoscientists from the Russian state oil company Rosneft, the Ushakov High, an inversion anticline (Figure 10C), was formed in the Paleocene–late Miocene as a result of dextral strike-slip displacements along a restraining bend of a northwest–southeast-trending fault zone (Ikhsanov, 2010; Malyshev et al., 2010b). A right lateral component of displacement of the youngest south-dipping normal faults was also detected on Wrangel Island (Figure 8). Thus, a significant function of dextral strike-slip movements in the development of the South Chukchi Basin (Tolson, 1987) is also confirmed by both seismic and onshore data.

It is commonly accepted that right lateral strike-slip movements along the series of large Alaskan faults occurred during the Campanian and the Paleocene, accommodating the change in the direction of movement of the Kula paleoplate in the northern Pacific Ocean region and the rotation of western Alaska (Plafker and Berg, 1994). Offshore sedimentary basins in the Bering Sea (such as Norton; Figure 2) were formed near the western termination of these large dextral strike-slip faults and filled by post–middle Eocene strata (Worrall, 1991). Following this model, the formation of the Hope (South Chukchi) Basin could be the result of dextral transtensional displacements along the Kobuk Fault Zone (Tolson, 1987) (Figure 1).

Our observations support previously published ideas that the opening of the South Chukchi Basin is related to gravitational orogenic collapse (Natal'in, 1999; Elswick, 2003; Elswick and Toro, 2003) and also to dextral strike-slip displacement along the Kobuk Fault Zone (Tolson, 1987). Based on the seismic data available and our knowledge of the onshore geology, Aptian–Albian–Late Cretaceous strata could occur at the base of the sedimentary infill of the South Chukchi Basin. Thus, we are in agreement with Malyshev et al. (2010a), who proposed that the Albian–Late Cretaceous gravitational collapse stage predated the Paleocene–Miocene extension and/or transtension. We believe that the Aptian–Albian and possibly the Late Cretaceous strata (SCB-1 unit) are not ubiquitous across the basin but are localized in isolated depressions and were slightly involved in the contractional deformation during the Albian and/or Early Tertiary (mid-Brookian event). It is likely that the overlying Cenozoic sediments are separated from the older sequences by the Late Cretaceous–early Paleocene regional planar peneplain, thus most probably representing the unconformity between SCB-1 and SCB-2 seismic units.

In summary, one could conclude that although the South Chukchi Basin in general represents an extensional offshore depression, it experienced several episodes of extension, dextral transtension, strike slip, contraction and/or transpression, and thermal subsidence during its development from the Aptian to the Cenozoic (Figure 12). The multiphase deformation and overprinting of different structural styles make the basin structure complicated in spatial variation and temporary evolution. Unrevealing the geohistory of the South Chukchi Basin can be instrumental in evaluating its hydrocarbon potential in the Chukchi Sea, Russian Arctic.

IMPLICATIONS FOR HYDROCARBON POTENTIAL

The hydrocarbon potential of the South Chukchi Basin has been considered to be low in comparison with the North Chukchi Basin, which is believed to be one of the most promising among the East Arctic offshore sedimentary basins (Orudzheva et al., 1999; Avdyunichev et al., 2004; Burlin and Shipel'kevich, 2006; Khain and Polyakova, 2007; Burlin and Stupakova, 2008; Khain et al., 2009; Malyshev et al., 2010a, b). Tolson (1987), after analyzing seismic reflection data from the Hope Basin together with drilling results from the two SOCAL wells (Cape Espenberg 1 and Nimiuk Point 1), concluded that the hydrocarbon potential in the U.S. sector of the South Chukchi Basin was quite low. The nonmarine sediments in the wells appear to be gas prone, thermally immature (even for oil generation), and of little hydrocarbon potential in general. Nevertheless, he proposed that, in the most subsided parts of the basin, thermally mature sedimentary rocks of marine or lacustrine origin could occur (Tolson, 1987). Taking into consideration an estimated thermal gradient of 50°C/km obtained from the two SOCAL wells, it was expected that maximum generation of oil would occur in the 2.5 to 5.0 km (1.6 to 3.1 mi) depth range. Because the seismic data indicate a variety of possible structural and stratigraphic traps, significant hydrocarbon accumulations could exist in the depocenters of the basin, assuming the presence of an organic-rich source and appropriate reservoir rocks in these areas (Tolson, 1987). The possible occurrence of volcaniclastic rocks in the deeper parts of the basin (as in Cape Espenberg 1) could potentially lead to a reduction in porosity and permeability of potential reservoir rocks (Tolson, 1987).

Orudzheva et al. (1999) pointed out that the South Chukchi Basin is filled by Cretaceous–Cenozoic sediments with no deltaic sequences, so it is unlikely to expect reservoirs of good quality. The insufficient thickness of sedimentary cover of 2 to 4 km (1.2–2.5 mi) is

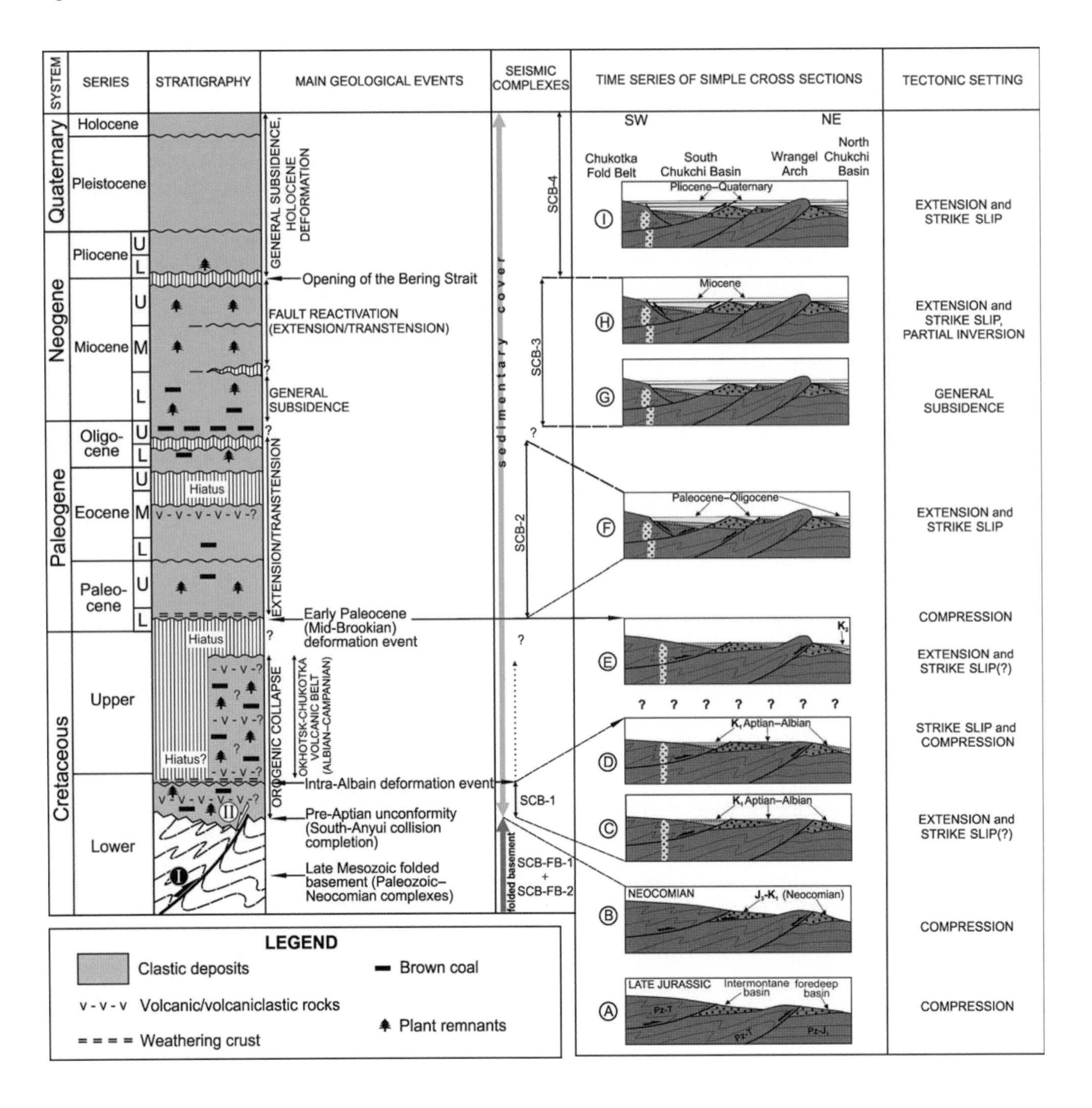

unlikely to generate significant volumes of hydrocarbons. However, the absence of deltaic sequences does not exclude that other reservoir facies types are present.

As it was shown, sedimentary rocks older than the middle Eocene could be present in the lower parts of the sedimentary cover of the South Chukchi Basin. These could be lower Eocene–Late Cretaceous, or even Albian–Aptian, in age. This may lead to significant widening of the range of potential source and reservoir rocks in the South Chukchi Basin. Additional hydrocarbon prospectivity could also be related to the Late Jurassic–Neocomian-aged source rock proposed in the basement complexes.

Kim et al. (2007) proposed that the clay-rich sediments of the Late Cretaceous and the Paleogene age could enhance gas potential in the basin. Petrovskaya et al. (2008) related the assumed hydrocarbon prospectivity of the basin mainly with the lower Brookian (Aptian–Albian–Late Cretaceous) terrigenous sequence. It is notable that Aptian–Albian and Late Cretaceous sequences on the New Siberian Islands also contain a significant amount of coal beds (Kos'ko and Trufanov, 2002; Kos'ko and Korago, 2009).

Because we proposed that the SCB-FB-2 basement unit corresponds to the synorogenic Late Jurassic–Early Cretaceous clastic sequence, it is possible that this moderately deformed sequence could act as a gas-generating source. Khain and Polyakova (2007) pointed to the occurrence of gas shows onshore in the vicinity of Chauna Bay (Chauna Lowland), where Upper Jurassic–Neocomian rocks of the Rauchua-Myrgovaam Basin (Figures 1, 2, 4) are overlain by a thin cover of Cenozoic sediments. Klubov and Semenov (1973) showed that among all the Lower Cretaceous sedimentary rocks, the Utuveem Formation in the Rauchua-Myrgovaam Basin is the most promising source rock. The age of this formation is interpreted by Baranov (1995) as Berriasian.

The Utuveem Formation contains a significant amount of black shale with organic matter corresponding to sapropelic and, to a lesser extent, humic-sapropelic type. In some sandy argillaceous units of the sequence that are enriched by coalified plant detritus, the total organic content of the rock can reach 3.36% (Klubov and Semenov, 1973). During fieldwork in the vicinity of the city of Pevek, we also observed that synorogenic sediments of the Utuveem Formation in places contained visible plant remnants.

The catagenetic maturity of the rocks of the Utuveem Formation is less than the underlying Paleozoic and Mesozoic sequences. Vitrinite reflectance measurements indicate that the organic matter in the Utuveem Formation is semianthracite (Klubov and Semenov, 1973). Offshore analogs of this sedimentary unit within the folded basement of South Chukchi Basin could be less deformed and metamorphosed because they are more remote from the South Anyui Suture Zone and outside of the northern limit of the Cretaceous postcollisional magmatism (Figures 2, 4). Thus, the syncollisional Upper Jurassic–Neocomian sequence may contain regional gas source rocks.

Figure 12. Inferred stratigraphy and main stages of tectonic development of the South Chukchi Basin based on onshore geology and offshore seismic data (more detailed with respect to the seismic interpretation). Compiled from many sources including Tolson (1987), Grinenko et al. (1989), Slobodin et al. (1990), Kos'ko et al. (1993; 2003), Dumitru et al. (1995), Marincovich and Gladenkov (1999), Natal'in (1999), Kos'ko and Trufanov (2002), Sokolov et al. (2002, 2009), Elswick and Toro (2003), Avdyunichev et al. (2004), Burlin and Shipel'kevich (2006), Miller et al. (2008, 2009), Tikhomirov et al. (2008), Kos'ko and Korago (2009), Miller and Verzhbitsky (2009), Drachev et al. (2010), and Malyshev et al. (2010a, b). Main stages of tectonic development: (A) Late Jurassic, beginning of collision and deposition of synorogenic rocks (SCB-FB-2 unit); (B) Neocomian, collision development, synorogenic rocks deposition (SCB-FB-2 unit) and their subsequent completion prior to the Aptian. Formation of the folded basement of the South Chukchi Basin; (C) Aptian–early Albian, postcollisional general extension related to postorogenic collapse, accompanied by dextral transpressional displacements (Sokolov et al., 2009); (D) Gentle folding occurred before the Late Albian (106 Ma). Deposition of SCB-1 unit and its subsequent deformation; (E) Late Albian–Late Cretaceous, Okhotsk-Chukotka Volcanic Belt (OCVB) activity, general extension and sinistral strike-slip fault development. North–northeast-south–southwest-trending sinistral strike-slip faults offset the Albian–Cenomanian rocks of the OCVB on Chukotka and are accompanied by Late Cretaceous dikes (Sokolov et al., 2009). During this stage, deposition of SCB-1 unit also could occur. The end of the stage was marked by Mid-Brookian compression event in the Late Cretaceous–early Cenozoic, related to the North American and Eurasian plate convergence (Harbert et al., 1990), accompanied by the widespread regional peneplain formation (e.g., Biske, 1975); (F) Paleocene–Oligocene, main subsidence of the basin (transtensional setting), controlled by dextral motion between the North American and Eurasian plates; (G) (early Miocene) General subsidence, followed by middle–late Miocene fault reactivation in extensional and/or transtensional setting (H), partial inversion (e.g., Ushakov High). Deposition and subsequent deformation of the SCB-3 unit; (I) The recent or present-day northeast–southwest dextral strike slip faulting under north–south extension in the easternmost Chukotka and South Chukchi Basin region, related to the diffuse plate boundary between the North American and Beringia plate (Bering Block) development (Mackey et al., 2009).

It is known that the Middle Cretaceous (~120–80 Ma) was a time of deposition of carbon-rich sediments (e.g., Larson, 1991); however, because the South Chukchi Basin region was uplifted during this time (Natal'in, 1999), it is unlikely that the deposition of organic-rich marine black shale occurs here.

The recently obtained well data for the sedimentary cover over the Lomonosov Ridge (Central Arctic Ocean) by the Integrated Ocean Drilling Program Arctic Coring Expedition (IODP ACEX, August 2004) brought new insights to the hydrocarbon potential of the circum-Arctic region. It gave rise to the proposal that the lowermost middle Eocene (~48.6 Ma) organic-rich Azolla sediments may constitute possible Arctic-wide source rock (Brinkhuis et al., 2006; Moran et al., 2006; Sluijs et al., 2006; Durham, 2007; Bujac and Houseknecht, 2009; Mann et al., 2009; etc.). Additional source rocks may also correspond to the strata deposited during the time interval of Paleocene–Eocene Thermal Maximum (~55 Ma), when generally good conditions for preservation of organic matter existed in the central Arctic (Sluijs et al., 2006; Mann et al., 2009). In theory, these two stratigraphic intervals could have a significant function in hydrocarbon generation in the South Chukchi Basin, but connection to the Arctic Ocean is problematic because no strong evidence for either of these events in the South Chukchi Basin exists (Tolson, 1987; Grinenko et al., 1989; Slobodin et al., 1990; Avdyunichev et al., 2004).

Numerous coal interbeds were also reported from the upper Oligocene in the Ayon well (Slobodin et al., 1990). The lower Miocene unit II in the Cape Espenberg 1 well contains woody coal interbeds (Tolson, 1987). Plant remnants (coalified) were described in almost all Cenozoic units of the Valkarem depression, from the Paleocene to the lower Pliocene (Grinenko et al., 1989). Given favorable burial conditions, it could be expected that the Paleogene sediments have good potential for gas generation in the South Chukchi Basin.

All of the seismic lines indicate that the basin is populated with tilted block, horst and graben, inversion anticlines, and pop-up structures. Pinch-outs and unconformities are also quite common. These features provide ample potential for structural and stratigraphic traps. The changes in phase or polarity of reflection seismic waveform in the upper parts of the sedimentary cover (bright spots) and areas of reduced seismic reflectivity in the upper sediments (gas chimneys) may point to a working hydrocarbon system. On the seismic sections that are roughly parallel to the east–southeast-west–northwest structural trend of the South Chukchi Basin, we have noticed that the principal normal faults are characterized by listric geometry, and that gas chimneys are spatially associated with them and have similar trends (Figure 11A). It is visible in places that they also penetrate the acoustic basement (SCB-FB-1) and the prerift reflective sequences (SCB-FB-2). It is likely that generation of gas could have been within the SCB-FB-2 unit in the folded basement and migrated to the higher levels of the sedimentary cover using the widespread fault plane network as the flow pathway.

Independent evidence for the prospectivity of the South Chukchi Basin comes from the results of bottom-sediment sampling to study the concentration and origin of hydrocarbon seeps (Kim et al., 2007; Yashin and Kim, 2007) (Figure 13). The authors proposed that anomalous concentrations of migrated hydrocarbon gases within the sea bottom sediments could be considered as direct evidence for the presence of oil and/or gas in the underlying sedimentary units (Kim et al., 2007; Yashin and Kim, 2007).

Some 200 samples were taken from the South Chukchi Basin and the Wrangel-Herald Arch areas (Yashin and Kim, 2007). The most convincing geochemical evidence for the presence of oil and gas was obtained for the southeastern part of the South Chukchi Basin (Hope Subbasin, Russian sector), representing the main local depocenter in the South Chukchi Basin. For methane, anomalously high values ranging from 0.1 to 1 cm^3/kg are quite common. Values of more than 10 cm^3/kg (up to 57 cm^3/kg) were discovered at five stations. Heavy gases ranging from ethane to butane were present, with average values of approximately 0.005 cm^3/kg, with background values of 0.001 cm^3/kg, with a prevalence of ethane (Yashin and Kim, 2007). It was also noted that the Hope Subbasin is characterized by a thick (up to 6–8 km [3.7–5 mi]) sedimentary cover, which is complicated by a series of faults. In general, this makes for favorable conditions for the generation and upward migration of hydrocarbon gases (Yashin and Kim, 2007).

The zones of high concentrations of hydrocarbon gases in the central part of the South Chukchi Basin trend approximately east–west, which is oblique to the general northwest–southeast trend of the main structural elements (Figure 13). This could hypothetically be explained by the influence of poorly studied east–west-trending prerift late Mesozoic thrust and reverse fault zones, widely mapped on Wrangel Island (e.g., Kos'ko et al., 1993, 2003), whereas the compressional structural pattern of Chukotka is characterized by a general west–northwest strike, similar to those of the South Chukchi Basin.

Taking into consideration all the data on the Aptian–Late Cretaceous–Cenozoic age of sedimentary infill, the variety of potential structural and stratigraphic traps, the possible occurrence in the entire Arctic of the lower–middle Eocene source rocks and gas geochemical analysis of the bottom sediments, it is very likely that the

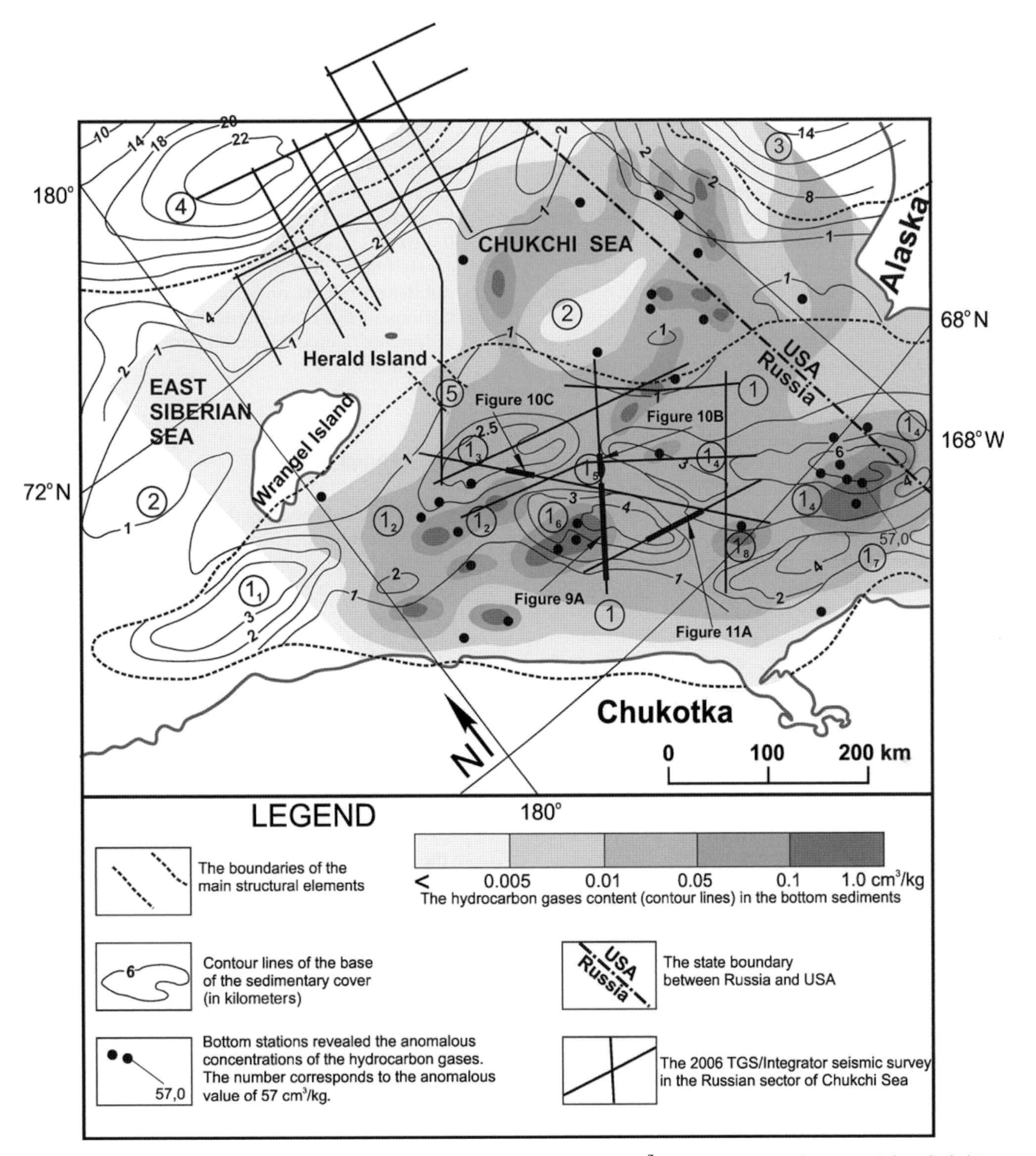

Figure 13. Map of anomalous concentrations of hydrocarbon gases (>0.05 cm^3/kg) in bottom sediments of the Chukchi Sea shelf. Modified from Yashin and Kim (2007). Circled numbers correspond to the main structural elements: 1 = South Chukchi/Hope Basin (including 1_1 = Longa Subbasin; 1_2 = Ushakov High; 1_3 = North-Schmidt Subbasin; 1_4 = Hope Subbasin; 1_5 = Onman High; 1_6 = South-Schmidt Subbasin; 1_7 = Koluchin Subbasin; 1_8 = Inkigur High); 2 = Wrangel-Herald Arch; 3 = Colville Foredeep Basin; 4 = North Chukchi Basin; 5 = Chukchi (Herald) Channel (canyon). Sediment thickness contour lines are redrawn from Yashin and Kim (2007). 200 km (124 mi).

hydrocarbon potential of the South Chukchi Basin could be significantly higher than what has been acknowledged in previous studies.

CONCLUSIONS

Analysis of newly acquired seismic data and published geologic, geophysical, and geochemical data, combined with results of onshore field observations on northern Chukotka and Wrangel Island, led us to the following conclusions:

1) The South Chukchi Basin is an offshore depression underlain entirely by a late Mesozoic folded basement. The basin was formed during several stages of extension and/or right lateral transtension and subsidence from at least Paleocene to Quaternary. A major low-angle preexisting thrust fault was recognized at the base of the basin. Most probably, it brought the highly deformed Triassic (and older) thrust sheet over much less deformed Upper Jurassic–Lower Cretaceous syncollisional strata. Thus, we propose the existence of two different types of folded basement in the Chukchi and East Siberian Sea regions: acoustically transparent basement (SCB-FB-1) (Paleozoic–Triassic) and reflective basement (SCB-FB-2) (Late Jurassic–Neocomian). The tectonic contact of the two basement units is a late Mesozoic thrust and/or reverse fault that was reactivated in the post-Neocomian as extensional and/or transtensional features.
2) Recognition of the preexisting thrust faults at the base of the basin leads to the suggestion that the South Chukchi Basin inherited the late Mesozoic thrust structure as the result of the change of regional tectonic setting—from compression in the Neocomian to a later extension. This relationship supports the idea that the extension in the basin was driven by gravitational collapse of the Wrangel-Herald-Lisburne fold and thrust belt (Elswick, 2003; Elswick and Toro, 2003). The right lateral component of the youngest south-dipping normal faults on Wrangel Island and transtensional structures identified on seismic sections also confirm the significant function of dextral strike-slip movements in the development of the South Chukchi Basin (Tolson, 1987).
3) Onshore outcrop investigation provides a geologic evidence for the occurrence of Aptian–Albian to middle Eocene strata in the deeper parts of the sedimentary section. The Aptian–Albian–Late Cretaceous are separated from the overlying strata by an angular unconformity interpreted to be an early Paleocene erosion peneplain surface.
4) Seismic data reveal a variety of structural and stratigraphic traps associated with bright spot anomalies and gas chimneys. We speculate that the moderately deformed Late Jurassic–Neocomian synorogenic sequence in the folded basement and organic-rich (mostly coal-bearing) strata in the sedimentary cover could be crucial for hydrocarbon (gas) generation in the South Chukchi Basin.
5) In general, we believe that the hydrocarbon potential of the South Chukchi Basin may be significantly higher than previously acknowledged. The published results on anomalously high concentrations of migrated natural gases in the bottom sediments (Kim et al., 2007) strongly support this speculation.

ACKNOWLEDGMENTS

We thank TGS for providing the opportunity to use the proprietary seismic data in our research. The participants of the International Geological Expedition on northern Chukotka and Wrangel Island in 2006 are grateful to the staff of the Wrangel National Reserve for their help in the fieldwork. We thank our colleagues from the expedition, E. Miller and V. Pease, for their kind help and assistance in the fieldwork. We thank V. Khain, S. Drachev, N. Malyshev, V. Obmetko, A. Borodulin, B. Ikhsanov, A. Khudoley, M. Kopp, and O. Kurakina for their comments and suggestions during preparation of the manuscript. J. Toro, D. Pivnik, and D. Gao provided thorough peer reviews that helped improve the quality of the manuscript. The authors from the Russian Academy of Sciences (Sergey Sokolov, Marianna Tuchkova, and Leopold Lobkovsky) were supported by Program 9 of the Department of Earth Sciences of the Russian Academy of Sciences, Russian Foundation for Basic Research (grants 11-05-00074 and 11-05-00787), the Leading Scientific Schools (NSh-5177.2012.5), and the state contract programs 14.740.11.0190 and Plate-Tectonic Reconstructions and Stress-State Model of the Lithosphere of Arctic region.

REFERENCES CITED

Akinin, V. V., and A. T. Calvert, 2002, Cretaceous midcrustal metamorphism and exhumation of the Koolen gneiss dome, Chukotka Peninsula, NE Russia, *in* E. L. Miller, A. Grantz, and S. L. Klemperer, eds., Tectonic evolution of the Bering Shelf–Chukchi Sea–Arctic margin and adjacent landmasses: Geological Society of America Special Paper 360, p. 147–165.

Akinin, V. V., E. L. Miller, and J. Wooden, 2009, Petrology and geochronology of crustal xenoliths from the Bering Strait region: Linking deep and shallow processes in extending continental crust, *in* R. B. Miller and A. W. Snoke, eds., Crustal cross sections from the western North America Cordillera and elsewhere: Implications for tectonic and petrologic processes: Geological Society of America Special Paper 456, p. 39–68.

Amato, J. M., and E. L. Miller, 2004, Geologic map and summary of the evolution of the Kigluaik Mountains gneiss dome, Seward Peninsula, Alaska, *in* D. L. Whitney, C. Teyssier, and C. S. Siddoway, eds., Gneiss domes in orogeny: Geological Society of America Special Paper 380, p. 295–306.

Avdyunichev, V. V., et al., 2004, Arctic seas (In Russian), *in* I. S. Gramberg, V. L. Ivanov, and Y. E. Pogrebitsky, eds., Geology and mineral resources of Russia, v. 5, book 1: Saint Petersburg, Russia, A. P. Karpinsky Russian Geological Research Institute (VSEGEI), 468 p.

Baranov, M. A., 1995, Nappe tectonics of the Myrgovaam Basin in northwestern central Chukotka (In Russian): Geology of the Pacific Ocean, v. 12, p. 441–448.

Belik, G. Ya., and G. M. Sosunov, 1969, Geologic map of USSR (In Russian): Northeastern Geological Directorate, Anyuisko-Chaunskaya Series R-58-XXIX, XXX, scale 1:200,000, 1 sheet.

Bering Strait Field Party, 1997, Koolen metamorphic complex, NE Russia: Implications for the tectonic evolution of the Bering Strait region: Tectonics, v. 16, p. 713–729.

Biske, S. F., 1975, Paleogene and Neogene of the extreme northeastern USSR (In Russian): Novosibirsk, Nauka, 268 p.

Bondarenko, G. E., A. V. Soloviev, M. I. Tuchkova, J. I. Garver, and I. I. Podgornyi, 2003, Age of detrital zircons from sandstones of the Mesozoic flysch formation in the South Anyui suture zone (western Chukotka): Lithology and Mineral Resources, v. 38, no. 2, p. 162–176.

Brinkhuis, H., et al., 2006, Episodic fresh surface waters in the Eocene Arctic Ocean: Nature, v. 441, p. 606–609.

Bujac, J., and D. Houseknecht, 2009, The Azolla story: Implications for climate change and Arctic petroleum source rocks (abs.): 7th Petroleum Geology Conference, From Mature Basins to New Frontiers, London, United Kingdom, Programme and Abstract Book, p. 34.

Burlin, Yu. K., and A. V. Stupakova, 2008, Geological grounds of oil and gas prospects of offshore Russian sector of northern Arctic Ocean (In Russian): Oil and Gas Geology, v. 4, p. 13–23.

Burlin, Y. K., and Y. V. Shipel'kevich, 2006, Principal features of the tectonic evolution of sedimentary basins in the western Chukchi Shelf and their petroleum resource potential: Geotectonics, v. 40, no. 2, p. 135–149.

Drachev, S. S., and L. A. Savostin, 1993, Ophiolites of Bol'shoi Lyakhovsky Island (New Siberian Islands) (In Russian): Geotektonika, v. 3, p. 98–107.

Drachev, S. S., L. A. Savostin, V. G. Groshev, and I. E. Bruni, 1998, Structure and geology of the continental shelf of the Laptev Sea, Eastern Russian Arctic: Tectonophysics, v. 298, p. 357–393.

Drachev, S. S., A. V. Elistratov, and L. A. Savostin, 2001, Structure and seismostratigraphy of the East Siberian Sea Shelf along the Indigirka Bay–Jannetta Island Seismic Profile: Transactions (Doklady) of the Russian Academy of Sciences/Earth Science Section, v. 377A, no. 3, p. 293–297.

Drachev, S. S., N. A. Malyshev, and A. M. Nikishin, 2010, Tectonic history and petroleum geology of the Russian Arctic shelves: An overview, *in* B. A. Vinning and S. C. Pickering, eds., Petroleum geology: From mature basins to new frontiers: 7th Petroleum Geology Conference, Geological Society London, v. 7, p. 591–619, doi:10.1144/0070591.

Dumitru, T. A., E. L. Miller, P. B. O'Sullivan, J. M. Amato, K. A. Hannula, A. T. Calvert, and P. B. Gans, 1995, Cretaceous to Recent extension in the Bering Strait region, Alaska: Tectonics, v. 14, p. 549–563.

Durham, L. S., 2007, Source ideas boost Arctic promise: AAPG Explorer, v. 28, p. 6, 8, 14.

Elswick, V. L., 2003, Seismic interpretation and structural evaluation of the Hope Basin, Alaska: M.S. thesis, West Virginia University, Morgantown, West Virginia, 27 p.

Elswick, V. L., and J. Toro, 2003, Seismic interpretation and structural evaluation of the Hope Basin, Alaska (abs.): Geological Society of America, Annual Meeting Abstracts, v. 35, no. 6, p. 28.

Filatova, N. I., and V. E. Khain, 2007, Tectonics of the eastern Arctic region: Geotectonics, v. 41, no. 3, p. 171–194.

Franke, D. K., K. Hinz, and, O. Oncken, 2001, The Laptev Sea Rift: Marine and Petroleum Geology, v. 18, p. 1083–1127.

Fujita, K., K. G. Mackey, R. C. McCaleb, L. V. Gunbina, V. N. Kovalev, V. S. Imaev, and V. N. Smirnov, 2002, Seismicity of Chukotka, northeastern Russia, *in* E. L. Miller, A. Grantz, and S. L. Klemperer, eds., Tectonic evolution of the Bering Shelf–Chukchi Sea–Arctic margin and adjacent landmasses: Geological Society of America Special Paper 360, p. 269–272.

Gradstein, F., J. Ogg, and A. Smith, 2004, A geologic time scale: Cambridge, United Kingdom, Cambridge University Press, 589 p.

Grantz, A., S. D. May, and P. E. Hart, 1990a, Geology of Arctic continental margin of Alaska, *in* A. Grantz, L. Johnson, and J. F. Sweeney, eds., The geology of North America: The Arctic Ocean region: Geological Society of America, v. L, p. 257–288.

Grantz, A., S. D. May, P. T. Taylor, and L. A. Lawver, 1990b, Canada Basin, *in* A. Grantz, L. Johnson, and J. F. Sweeney, eds., The geology of North America: The Arctic Ocean region: Geological Society of America, v. L, p. 379–402.

Grantz, A., D. W. Scholl, J. Toro, and S. L. Klemperer, 2002, Plate 1, Geologic structure of Bering and Chukchi shelves adjacent to Bering-Chukchi deep transect and tectonostratigraphic terranes of adjacent landmasses, scale 1:3,000,000, *in* E. L. Miller, A. Grantz, and S. L. Klemperer, eds., Tectonic evolution of the Bering Shelf–Chukchi Sea–Arctic margin and adjacent landmasses: Geological Society of America Special Paper 360, 1 p.

Grinenko, O. V., L. P. Zharikova, and A. F. Fradkina, et al., 1989, Paleogene and Neogene of northeastern Russia (in Russian): YaNTs SO RAN, Yakutsk, 184 p.
Harbert, W., L. Frei, R. Jarrard, S. Halgedahl, and D. Engebretson, 1990, Paleomagnetic and plate-tectonic constraints on the evolution of the Alaskan–eastern Siberian Arctic, *in* A. Grantz, L. Johnson, and J. F. Sweeney, eds., The Arctic Ocean region: The geology of North America: Geological Society of America, v. L, p. 567–592.
Ikhsanov, B. I., 2010, Structure and history of development of Ushakov postsedimentional structural zone in late Mesozoic–Cenozoic time (in Russian), *in* N. B. Kuznetsov, ed., XLIII Tectonic Conference: Tectonics and geodynamics of fold belts and platforms in Phanerozoic: Moscow, Russia, GEOS, v. 1, p. 277–280.
Katkov, S. M., A. Strickland, E. L. Miller, and J. Toro, 2007, Ages of granite batholiths from Anyui-Chukotka Foldbelt: Doklady Earth Sciences, v. 414, no. 4, p. 515–518.
Khain, V. E., and I. D. Polyakova, 2007, Sedimentary basins and prospects of oil and gas deposits on the shelf of the eastern Arctic: Oceanology, v. 47, no. 1, p. 104–115.
Khain, V. E., I. D. Polyakova, and N. I. Filatova, 2009, Tectonics and petroleum potential of the East Arctic province: Russian Geology and Geophysics, v. 50, p. 334–345.
Kim, B. I., N. K. Evdokimova, O. I. Suprunenko, and D. S. Yashin, 2007, Oil geological zoning of offshore areas of the East-arctic seas of Russia and their oil and gas potential prospects (in Russian): Oil and Gas Geology, v. 2, p. 49–59.
Klemperer, S. L., E. L. Miller, and D. W. Scholl, 2002, Crustal structure of the Bering and Chukchi shelves: Deep seismic reflection profiles across the North American continent between Alaska and Russia, *in* E. L. Miller, A. Grantz, and S. L. Klemperer, eds., Tectonic evolution of the Bering Shelf–Chukchi Sea–Arctic margin and adjacent landmasses: Geological Society of America Special Paper 360, p. 1–24.
Klubov, B. A., and G. A. Semenov, 1973, Prospects of oil-and-gas potential of Rauchua Basin (In Russian), *in* A. A. Trofimuk, ed., Problems of oil-and-gas potential of the northeastern USSR: Magadan, Russia, SVKNII (Severo-Vostochnyi Kompleksnyi Nauchno-Issledovatel'skii Institut), p. 40–60.
Kogan, A. L., 1981, Marine seismic survey in the Chukchi Sea (in Russian): Marine geophysical research in the World Ocean: Leningrad, Russia, Vniiokeangeologiya, p. 38–40.
Kos'ko, M. K., and E. A. Korago, 2009, Review of geology of the New Siberian Islands between the Laptev and the East Siberian Seas, northeast Russia, *in* D. B. Stone, K. Fujita, P. W. Layer, E. L. Miller, A. V. Prokopiev, and J. Toro, eds., Geology, geophysics and tectonics of northeastern Russia: A tribute to Leonid Parfenov: European Geosciences Union, Stephan Mueller Publication Series 4, p. 45–64.
Kos'ko, M. K., and G. V. Trufanov, 2002, Middle Cretaceous to Eopleistocene sequences on the New Siberian Islands: An approach to interpret offshore seismic: Marine and Petroleum Geology, v. 19, p. 901–919.
Kos'ko, M. K., M. P. Cecile, J. C. Harrison, V. G. Ganelin, N. V. Khandoshko, and B. G. Lopatin, 1993, Geology of Wrangel Island, between Chukchi and East Siberian seas, northeastern Russia: Geological Survey Canada Bulletin, v. 461, 101 p.
Kos'ko, M. K., et al., 2003, The Wrangel Island: Geological structure, mineragenesis, environmental geology (in Russian): Saint Petersburg, Russia, VNIIOkeangeologia, 137 p.
Kuzmichev, A. B., 2009, Where does the South Anyui suture go in the New Siberian Islands and Laptev Sea?: Implications for the Amerasia basin origin: Tectonophysics, v. 463, p. 86–108.
Kuzmichev, A. B., E. V. Sklyarov, and I. G. Barash, 2005, Pillow basalts and blueschists on Bol'shoi Lyakhovsky Island (the New Siberian Islands): Fragments of the South Anyui oceanic lithosphere: Russian Geology and Geophysics, v. 46, p. 1367–1381.
Kuzmichev, A. B., A. V. Soloviev, V. E. Gonikberg, M. N. Shapiro, and O. V. Zamzhitskii, 2006, Mesozoic syncollision siliciclastic sediments of the Bol'shoi Lyakhov Island (New Siberian Islands): Stratigraphy and Geological Correlation, v. 14, p. 30–48.
Lander, A. V., B. G. Bukchin, A. V. Kiryushin, and D. V. Droznin, 1996, The tectonic environment and source parameters of the Khailino, Koryakia earthquake of March 8, 1991: Does a Beringia plate exist?: Computational Seismology and Geodynamics, v. 3, p. 80–96.
Larson, R. L., 1991, Latest pulse of Earth: Evidence for a mid-Cretaceous superplume: Geology, v. 19, p. 547–550.
Layer, P., R. Newberry, K. Fujita, L. Parfenov, V. Trunilina, and A. Bakharev, 2001, Tectonic setting of the plutonic belts of Yakutia, northeast Russia, based on $^{40}Ar/^{39}Ar$ geochronology and trace element geochemistry: Geology, v. 29, p. 167–170.
Mackey, K. G., K. Fujita, L. V. Gunbina, V. N. Kovalev, V. S. Imaev, B. M. Koz'min, and L. P. Imaeva, 1997, Seismicity of the Bering Strait region: Evidence for a Bering Block: Geology, v. 25, p. 979–982.
Mackey, K. G., K. Fujita, B. M. Sedov, L. V. Gounbina, and S. Kurtkin, 2009, A seismic swarm near Neshkan, Chukotka, northeastern Russia, and implications for the boundary of the Bering plate, *in* D. B. Stone, K. Fujita, P. W. Layer, E. L. Miller, A. V. Prokopiev, and J. Toro, eds., Geology, geophysics and tectonics of northeastern Russia: A tribute to Leonid Parfenov, European Geosciences Union, Stephan Mueller Publication Series 4, p. 261–271.
Malyshev, N. A., V. V. Obmetko, A. A. Borodulin, E. M. Barinova, and B. I. Ikhsanov, 2010a, Tectonics of the sedimentary basins of the Russian Chukchi Sea shelf (in Russian), *in* N. B. Kuznetsov, ed., XLIII Tectonic Conference: Tectonics and geodynamics of fold belts and platforms in Phanerozoic: Moscow, Russia, GEOS, v. 2, p. 23–29.
Malyshev, N. A., V. V. Obmetko, and A. A. Borodulin, 2010b, Hydrocarbon potential of the eastern Arctic sedimentary basins (In Russian): The Rosneft Scientific and Technical Reporter, v. 1, p. 20–28.

Mann, U., J. Knies, S. Chand, W. Jokat, R. Stein, and Z. Janine, 2009, Evaluation and modeling of Tertiary source rocks in the central Arctic Ocean: Marine and Petroleum Geology, v. 26, no. 8, p. 1624–1639.

Marincovich, L., and A. Y. Gladenkov, 1999, Evidence for an early opening of the Bering Strait: Nature, v. 397, p. 149–151.

Mazarovich, À. O., and S. Yu. Sokolov, 2003, Tectonic subdivision of the Chukchi and East Siberian seas: Russian Journal of Earth Sciences, v. 5, no. 3, p. 185–202.

Miller, E. L., and T. L. Hudson, 1991, Mid-Cretaceous extensional fragmentation of a Jurassic–Early Cretaceos compressional orogen, Alaska: Tectonics, v. 10, no. 4, p. 781–796, doi:10.1029/91TC00044.

Miller, E. L., and V. E. Verzhbitsky, 2009, Structural studies near Pevek, Russia: Implications for formation of the East Siberian Shelf and Makarov Basin of the Arctic Ocean, *in* D. B. Stone, K. Fujita, P. W. Layer, E. L. Miller, A. V. Prokopiev, and J. Toro, eds., Geology, geophysics and tectonics of northeastern Russia: A tribute to Leonid Parfenov, European Geosciences Union, Stephan Mueller Publication Series 4, p. 223–241.

Miller, E. L., T. R. Ireland, S. L. Klemperer, K. R. Wirth, V. V. Akinin, and T. M. Brocher, 2002, Constraints on the age of formation of seismically reflective middle and lower crust beneath the Bering Shelf: SHRIMP zircon dating of xenolith from Saint Lawrence Island, *in* E. L. Miller, A. Grantz, and S. L. Klemperer, eds., Tectonic evolution of the Bering Shelf-Chukchi Sea-Arctic Margin and adjacent landmasses: Geological Society of America Special Paper 360, p. 195–208.

Miller, E. L., J. Toro, G. Gehrels, J. M. Amato, A. Prokopiev, M. I. Tuchkova, V. V. Akinin, T. A. Dumitru, T. E. Moore, and M. P. Cecile, 2006, New insights into Arctic paleogeography and tectonics from U-Pb detrital zircon geochronology: Tectonics, v. 25, p. TC3013, doi:10.1029/2005TC001830.

Miller, E. L., A. Soloviev, A. Kuzmichev, G. Gehrels, J. Toro, and M. Tuchkova, 2008, Jura–Cretaceous foreland basin deposits of the Russian Arctic: Separated by birth of Makarov Basin?: Norwegian Journal of Geology, v. 88, p. 227–250.

Miller, E. L., S. M. Katkov, A. Strickland, J. Toro, V. V. Akinin, and T. A. Dumitru, 2009, Geochronology and thermochronology of Cretaceous plutons and metamorphic country rocks, Anyui-Chukotka fold belt, northeastern Arctic Russia, *in* D. B. Stone, K. Fujita, P. W. Layer, E. L. Miller, A. V. Prokopiev, and J. Toro, eds., Geology, geophysics and tectonics of northeastern Russia: A tribute to Leonid Parfenov: European Geosciences Union, Stephan Mueller Publication Series 4, p. 157–175.

Miller, E. L., G. E. Gehrels, V. Pease, and S. Sokolov, 2010, Stratigraphy and U-Pb detrital zircon geochronology of Wrangel Island, Russia: Implications for Arctic paleogeography: AAPG Bulletin, v. 94, no. 5, p. 665–692.

Moore, T. E., T. A. Dumitru, K. E. Adams, S. N. Witebsky, and A. G. Harris, 2002, Origin of the Lisburne Hills-Herald Arch structural belt: Stratigraphy, structural, and fission-track evidence from the Cape Lisburne area, northwestern Alaska, *in* E. L. Miller, A. Grantz, and S. L. Klemperer, eds., Tectonic evolution of the Bering Shelf–Chukchi Sea–Arctic margin and adjacent landmasses: Geological Society of America Special Paper 360, p. 77–109.

Moran, K., et al., 2006, The Cenozoic paleoenvironment of the Arctic Ocean: Nature, v. 441, p. 601–605.

Natal'in, B. A., 1999, Late Cretaceous–Tertiary deformations in the Chukotka Peninsula: Implications for the origin of the Hope Basin and the Heruld Thrust Belt (Chukchi Sea): Geotectonics, v. 33, no. 6, p. 489–504.

Orudzheva, D. S., A. N. Obukhov, and, D. D. Agapitov, 1999, Prospects for oil and gas offshore exploration in Chukchi Sea (In Russian): Oil and Gas Geology, v. 3/4, p. 28–33.

Otochkin, V. V., and, V. A. Ivanov, 1989, Report on the regional seismic works in Chukchi and Bering seas during 1988–1989 (In Russian): Northern Pacific Ocean Geological Prospecting Expedition PGO "Dalmorgeologia," Ministry of Geology of Union of Soviet Socialist Republics, Geological Funds.

Parfenov, L. M., L. M. Natapov, S. D. Sokolov, and N. V. Tsukanov, 1993, Terrane analyses and accretion in northeastern Asia: Island Arc, v. 2, p. 35–54.

Petrovskaya, N. A., S. V. Trishkina, and M. A. Savishkina, 2008, The main features of geological structure of Russian sector of Chukotsk Sea (In Russian): Oil and Gas Geology, v. 6, p. 20–28.

Plafker, G., and, H. C. Berg, 1994, An overview of the geology and tectonic evolution of Alaska, *in* G. Plafker and H. C. Berg, eds., The geology of North America: The geology of Alaska: The Geological Society of America, v. G-1, p. 989–1021.

Pol'kin, Ya. I., 1984, Chukchi Sea (in Russian), *in* I. S. Gramberg and Yu.E.Pogrebitskii, eds., The seas of Soviet Arctic: Leningrad, Russia, Nedra, p. 67–97.

Shipilov, E. V., B. V. Senin, and A. Yu. Yunov, 1989, Sedimentary cover and basement of Chukchi Sea from seismic data: Geotectonics, v. 23, no. 5, p. 456–463.

Slobodin, V. Ya, B. I. Kim, G. V. Stepanova, and F. Ya. Kovalenko, 1990, The stratification of the core of Ayon well based on the new biostratigraphic data: Stratigraphy and paleontology of Meso–Cenozoic of the Soviet Arctic: Leningrad, Russia, Sevmorgeologia, p. 42–58.

Sluijs, A., S. Schouten, and M. Pagani, 2006, Subtropical Arctic Ocean temperatures during the Paleocene/Eocene thermal maximum: Nature, v. 441, p. 610–613.

Sokolov, S. D., G. Y. Bondarenko, O. L. Morozov, V. A. Shekhovtsov, S. P. Glotov, A. V. Ganelin, and I. R. Kravchenko-Berezhnoy, 2002, South Anyui suture, northeast Arctic Russia: Facts and problems, *in* E. L. Miller, A. Grantz, and S. L. Klemperer, eds., Tectonic evolution of the Bering Shelf–Chukchi Sea–Arctic margin and adjacent landmasses: Geological Society of America Special Paper 360, p. 209–224.

Sokolov, S. D., G. Y. Bondarenko, P. W. Layer, and I. R. Kravchenko-Berezhnoy, 2009, South Anyui suture: Tectonostratigraphy, deformations, and principal tectonic events, *in* D. B. Stone, K. Fujita, P. W. Layer, E. L. Miller, A. V. Prokopiev, and J. Toro, eds., Geology, geophysics and tectonics of northeastern Russia: A tribute to Leonid Parfenov, European Geosciences Union, Stephan Mueller Publication Series 4, p. 201–221.

Tikhomirov, P. L., E. L. Kalinina, K. Kobayashi, E. Nakamura, and I. Yu. Cherepanova, 2008, Dynamics of late Mesozoic volcanism of Chukotka (on the data of $^{40}Ar/^{39}Ar$ rock datings) (In Russian), *in* N. B. Kuznetsov, ed., XLIII Tectonic Conference: Tectonics and geodynamics of fold belts and platforms in Phanerozoic: Moscow, Russia, GEOS, v. 2, p. 23–29.

Til'man, S. M., N. A. Bogdanov, S. G. Byalobzhesky, and A. D. Chekhov, 1970, The Wrangel Island (in Russian), *in* B. V. Tkachenko and B. Kh. Egiazarov, eds., The islands of the Soviet Arctic: Leningrad, Russia, Nedra, p. 377–404.

Tolson, R. B., 1987, Structure and stratigraphy of the Hope Basin, southern Chukchi Seas, Alaska, *in* D. W. Scholl, A. Grantz, and J. G. Vedder, eds., Geology and resource potential of the continental margin of western North America and adjacent ocean basins: Beaufort Sea to Baja California: Houston, Texas, Circum-Pacific Council for Energy and Mineral Resources, Earth Science Series 6, p. 59–71.

Toro, J., J. M. Amato, and B. A. Natal'in, 2003, Cretaceous deformation, Chegitun River area, Chukotka Peninsula, Russia: Implications for the tectonic evolution of the Bering Strait region: Tectonics, v. 22, no. 3, p. 1021, doi:10.1029/2001TC001333.

Tuchkova, M. I., G. E. Bondarenko, M. I. Buyakaite, D. I. Golovin, I. O. Galuskina, and E. V. Pokrovskaya, 2007, Deformation of the Chukchi microcontinent: Structural, lithologic, and geochronological evidence: Geotectonics, v. 34, no. 4, p. 294–301.

Verzhbitsky, V., E. Frantzen, T. Savostina, A. Little, S. D. Sokolov, and M. I. Tuchkova, 2008, The Russian Chukchi Sea shelf: GEO ExPro, v. 5, no. 3, p. 36–41: http://www.geo365.no/TGS-Chukchi/ (accessed May 2008).

Vladimirtseva, Y. A., E. A. Dykanuk, A. M. Manukyan, T. S. Stepina, and E. P. Surmilova, 2001, Explanatory notes of the state geological map of Russian Federation, scale 1:1,000,000, sheet Q-2-Uelen (In Russian): St. Petersburg, Russia, A. P. Karpinsky Russian Geological Research Institute (VSEGEI), 139 p.

Worrall, D. M., 1991, Tectonic history of the Bering Sea and the evolution of Tertiary strike-slip basins of the Bering Shelf: Geological Society of America Special Paper 257, 120 p.

Yashin, D. S., and B. I. Kim, 2007, Geochemical features of oil and gas potential of eastern Arctic shelf of Russia (in Russian): Oil and Gas Geology, v. 4, p. 25–29.

Zonenshain, L. P., M. I. Kuzmin, and L. M. Natapov, 1990, Geology of the U.S.S.R.: A plate tectonic synthesis: American Geophysical Union, Geodynamics Series 21, 242 p.

13

Anderson, Arlene V., Donald K. Sickafoose, Tim R. Fahrer, and Richard R. Gottschalk, 2012, Interaction of Oligocene–Miocene deep-water depositional systems with actively evolving structures: The Lower Congo Basin, offshore Angola, *in* D. Gao, ed., Tectonics and sedimentation: Implications for petroleum systems: AAPG Memoir 100, p. 291–313.

Interaction of Oligocene–Miocene Deep-water Depositional Systems with Actively Evolving Structures: The Lower Congo Basin, Offshore Angola

Arlene V. Anderson, Donald K. Sickafoose, Tim R. Fahrer, and Richard R. Gottschalk
ExxonMobil Exploration Company, 233 Benmar Dr., Houston, Texas, 77060, U.S.A.
(e-mails: arlenanderson@aol.com; kim.sickafoose@exxonmobil.com; tim.r.fahrer@exxonmobil.com; richard.r.gottschalk@exxonmobil.com)

ABSTRACT

The Angola continental margin has undergone a complex history of gravity-driven deformation. The interaction of actively evolving structures and depositional systems is a primary control on Oligocene–Miocene reservoir distribution and architecture. Structurally driven changes in sea-floor gradient can lead to rapid lateral changes in reservoir distribution and geometry. Understanding this complexity as the Lower Congo Basin (LCB) evolved through time is an important focus of exploration efforts in the basin.

Deep-water clastic reservoirs in the LCB range from the Rupelian to the Messinian and were deposited on the LCB slope; they are generally organized into confined and weakly confined slope-channel systems. As the Angola margin evolved, sediment gravity flows were diverted through a maze of active structures, forming multicycle stacked channel systems. In the distal reaches of the paleoslope where the paleobathymetry was more subdued, the sands form single-cycle, digitate, anastomosing to distributary bodies. Sheet geometries are mostly absent because of the continued presence of a low-slope regional gradient through the extent of three-dimensional (3-D) coverage.

Considerable lateral variability in structural style exists along the Angola continental margin, where updip extension and translation are accommodated by downslope contraction and extrusion of salt. Because Oligocene–Miocene strata are syntectonic deposits, the active structures strongly influenced deposition and, therefore, the present-day distribution of reservoir sands on the LCB slope. Significant changes occur in the interaction of the depositional system and structures during the early Miocene. From the Rupelian to the early Burdigalian, proximal extensional normal faults had minimal impact on sediment distribution. Downdip, in the contractional domain, the interaction was more subtle and episodic. The style of interaction and depositional system response depend on the rate of local sedimentation compared with the rate of uplift for each structure. This can vary significantly even with structures in close

DOI:10.1306/13351558M1003535

proximity. From the late Burdigalian until the Messinian, active structures in both the extensional and contractional domains controlled the distribution of coarse clastic sediments. In the central LCB, sediment capture in the updip grabens created a coarse clastic sediment shadow downdip. In the contractional domain, the channel systems were deflected between and around the active structures.

INTRODUCTION

The Lower Congo Basin (LCB) is one of the world's major petroleum basins and has been a focus of petroleum exploration since the 1960s. Industry activity accelerated in the late 1990s with the discovery of the Girassol field in 1996, which proved the viability of the Oligocene–Miocene turbidite slope-channel play. Interaction between evolving structures and deep-water depositional systems is a primary control on Oligocene–Miocene reservoir distribution and architecture in the LCB. Understanding the lateral variability and the resulting complex basin evolution has been critically important to focusing exploration efforts. A combination structural-stratigraphic trap is an important trap style in the deep water. Channel systems of the thick Oligocene–Miocene depositional regime respond to local gradient changes around the active structures. This interaction is more complex in the contractional domain because rapid local gradient changes occur as the structures evolve through time.

Because of the proprietary nature of industry activity in the LCB, most published articles and seismic examples from the Angola-Congo region have focused on the less prospective Kwanza Basin to the south (Figure 1) (e.g., Hudec and Jackson, 2004; Jackson and Hudec, 2005). Although the LCB and the Kwanza Basin are both characterized by detached deformation on an Aptian salt substrate, important differences in structural style and depositional history exist between the two basins. The Kwanza Basin is dominated by autochthonous salt structures and has a relatively thin Oligocene–Miocene section (~2 km [~1.2 mi] maximum thickness). In contrast, the LCB is characterized by a thick Oligocene–Miocene clastic section (~8 km [~5 mi] maximum thickness) and contains a large number of allochthonous salt structures that result from a complex basin history. Because much of the deformational activity in the LCB occurred during the Oligocene–Miocene, actively evolving salt and tectonic structures had a significant impact on the distribution and thickness of synkinematic Oligocene–Miocene strata. As a result, lateral variations in structural style and timing have strongly influenced the distribution of reservoir sandstones on the LCB slope.

Several studies in the LCB indicate that active structures create sea-floor topography and local gradients that control the location and architecture of deep-water reservoir systems (Anderson et al., 2000; Mayall and Stewart, 2000; Kolla et al., 2001; Ferry et al., 2005; Gee and Gawthorpe, 2006; Porter et al., 2006). These authors focused on the detailed effect that active structures can have on channel morphology and the internal architecture of deep-water turbidite channel systems. In this chapter, we examine the broader Oligocene–Miocene evolution of the LCB and how variations in structural history and style along the strike length of the basin affect the distribution of reservoir sandstone on a regional scale. We also provide several examples of interaction between active structures and syntectonic sediments and discuss the resulting reservoir distribution and architecture.

DATA AND METHODS

Our structural and stratigraphic analysis is based mainly on an evaluation of 3-D and two-dimensional (2-D) seismic data sets. Although most of the LCB is covered by 3-D seismic data (more than 100 surveys) (Figure 2), 2-D seismic data provide extended coverage in the far updip and downdip reaches of the basin. Major sequence boundaries were identified by analyzing biostratigraphic data from more than 70 exploratory wells and by tying these data to regionally correlative seismic reflectors. Overall, six to seven major second-order sequence boundaries were correlated across the entire LCB. Using this stratigraphic framework, the basin-wide environment of deposition maps for four time intervals were built using a variety of data and techniques including (1) 3-D seismic amplitude extractions, (2) reservoir geometry mapping, (3) well-log suites, (4) biostratigraphic data, and (5) conventional core descriptions. These environment of deposition maps were integrated with paleotectonic interpretations to produce a set of four paleogeographic maps illustrating the interaction of the depositional system with actively evolving structures throughout the Oligocene–Miocene.

We have evaluated the timing and kinematic evolution of structures in the LCB by conducting sequential restorations of several regional structural cross sections. Because the structural style varies significantly along strike, sections were restored to study differences

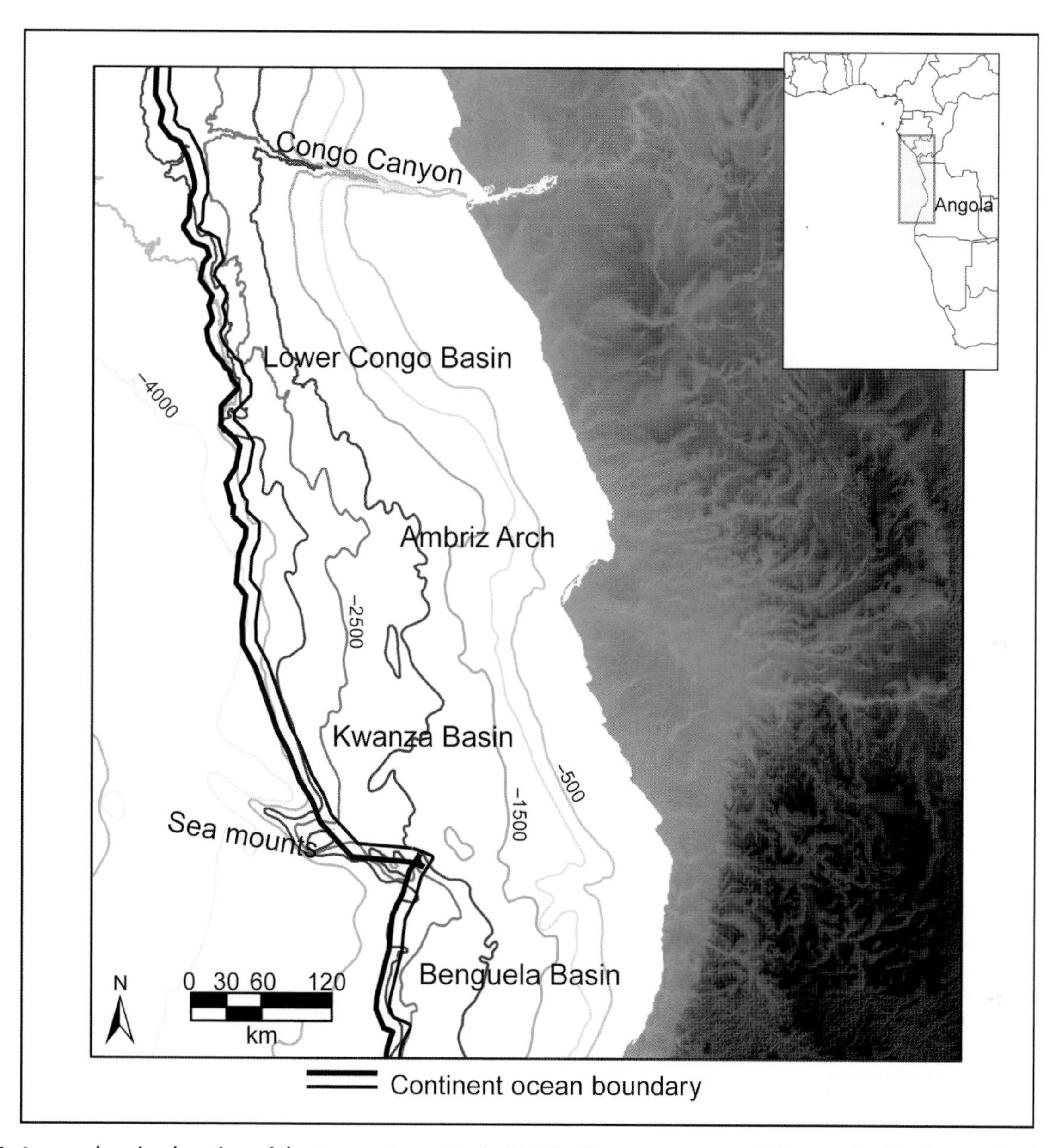

Figure 1. A map showing location of the Lower Congo Basin (LCB), off the west coast of Africa. A double line marks the ocean-continent boundary (OCB). Bathymetric contour lines are 500 m (1640 ft) intervals. The crustal flexure associated with the Ambriz Arch separates the Kwanza Basin and the LCB (Lundin, 1992; Spathopoulos, 1996).

in structural timing and evolution in various segments of the basin. We restored and evaluated critical 2-D and 3-D structures using 2-D and 3-D structure modeling and restoration technologies.

REGIONAL SETTING

The LCB is located offshore Angola and Congo, west Africa (Figure 1). It is one of a series of basins that contain Aptian salt deposited during the tectonic transition from rifting to passive-margin subsidence of the South Atlantic (Burke, 1996; Cramez and Jackson, 2000; Marton et al., 2000).

The Angola margin has a complex history of gravity-driven, salt-involved deformation that spans the Cretaceous–Holocene. The Aptian Loeme salt is the major detachment surface (Duval et al., 1992; Lundin, 1992; Spathopoulos, 1996; Cramez and Jackson, 2000; Marton et al., 2000) above which strata are decoupled from underlying structures. Because of the effectiveness of a salt detachment, deformation can be rapidly

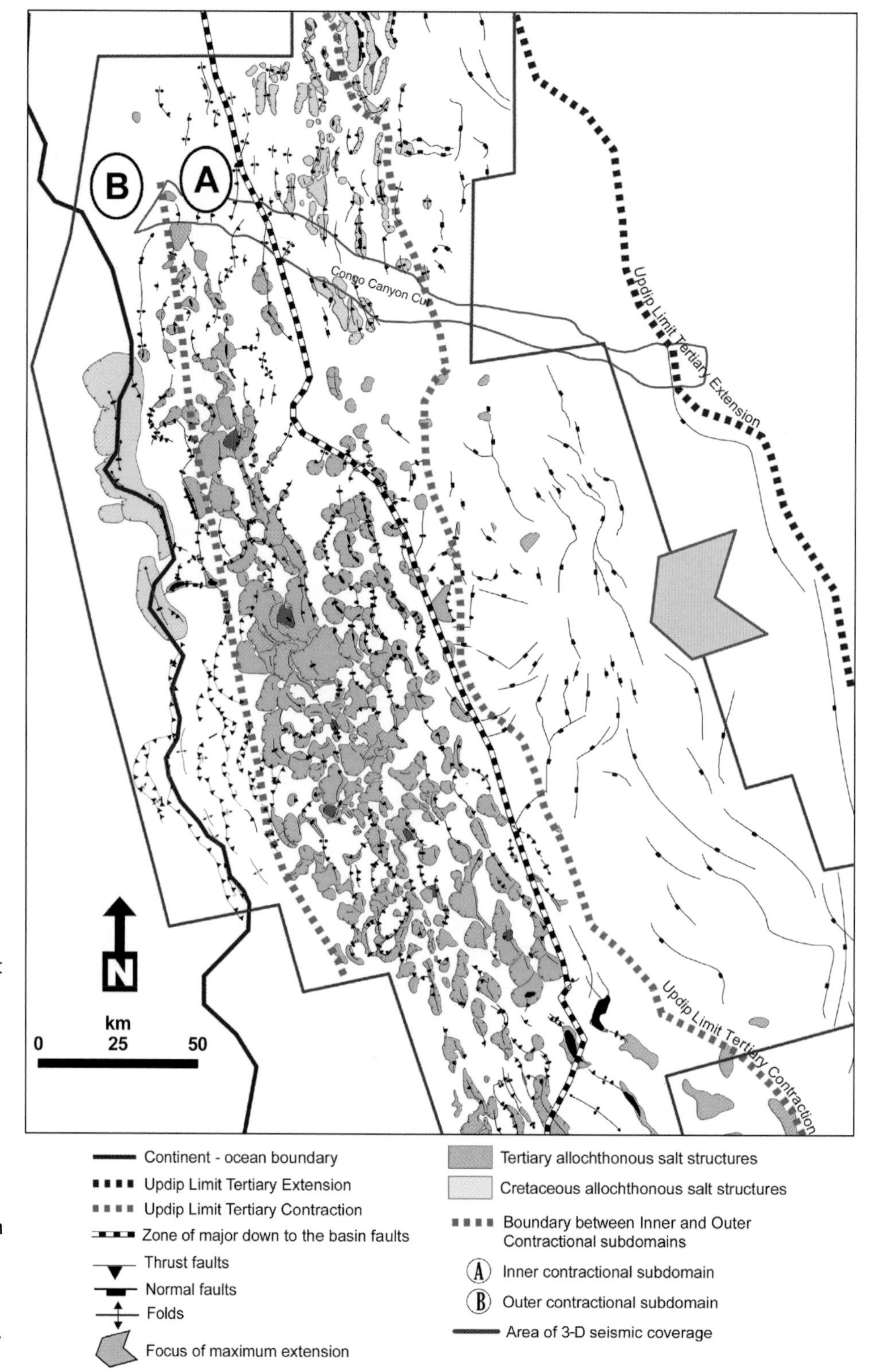

Figure 2. A structural elements map of the Lower Congo Basin. This map shows the updip limit of Tertiary extension and contraction and the ocean-continent boundary. Allochthonous salt structures are shown as well as major faults and folds. The large arrow indicates the area of maximum extension and downdip shortening. The area of three-dimensional (3-D) coverage is outlined. Our mapping of the ocean-continent boundary (OCB) is based on interpretation of gravity and magnetic data in combination with seismic data. The result is consistent with the described results of Marton et al. (2000). 25 km (15.5 mi).

transmitted from updip to the outer edge of the salt basin (Costa and Vendeville, 2002; Rowan et al., 2004). Both Albian and Rupelian contractional structures exist near the edge of salt along the ocean-continent boundary (OCB). This effective salt detachment causes a very low angle depositional slope (estimated to be 0.5–2°) for gravity flows of the deep-water sedimentary system. This is consistent with the physical modeling results of

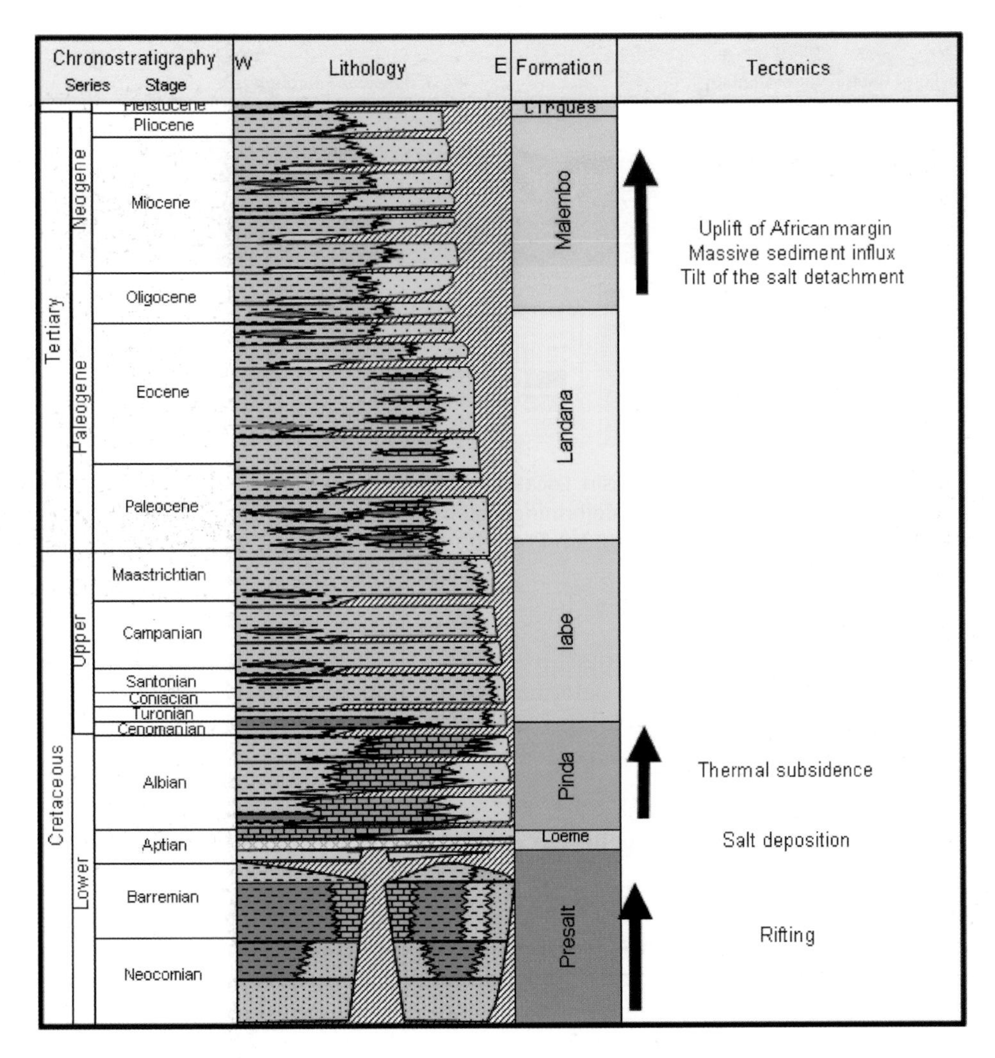

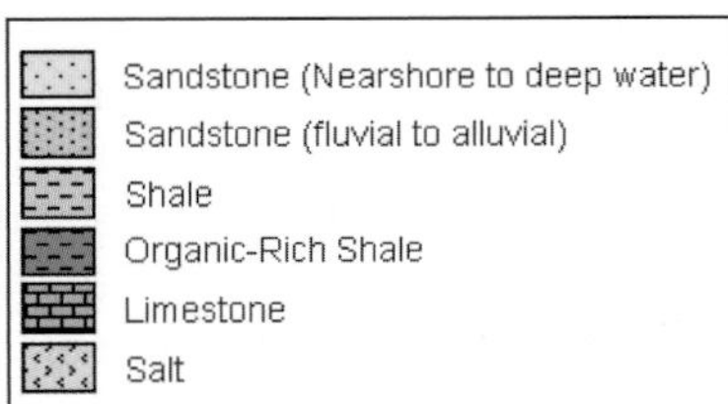

Figure 3. A generalized stratigraphic column of the Lower Congo Basin.

Davis and Engelder (1987), Letouzey et al. (1995), and Costa and Vendeville (2002).

The Aptian Loeme salt is overlain by Cretaceous and Tertiary marine strata (Brice et al., 1982; Burke, 1996) (Figure 3). Cretaceous strata are composed of carbonate shelf and shallow-marine clastic sedimentary rocks and their deep-marine equivalent marls and shales. These rocks include the Upper Cretaceous Iabe Formation, which is the main petroleum source interval in the deep-water parts of the LCB. A thin condensed Paleocene–Eocene section of marl and shale is overlain by the thick coarse clastic sedimentary rocks of the Oligocene–Holocene Malembo Formation (Coward et al., 1999; Ardill et al., 2002). This formation provides the deep-water reservoir sandstones of the prolific LCB hydrocarbon system.

After deposition of the Loeme salt, the LCB experienced two distinct phases of deformation, one of Albian to Cenomanian age and one of Oligocene–Holocene age (Bond, 1978; Sahagian, 1988; Walgenwitz et al., 1990; Lunde et al., 1992; Burke, 1996; Marton et al., 2000). Both phases of deformation are characterized by a gravitationally driven linked system of deformation that detached on the Aptian salt with updip extension compensated by downdip contraction and extrusion of salt (Spathopoulos, 1996; Cramez and Jackson, 2000; Gottschalk et al., 2004; Rowan et al., 2004) (Figure 2).

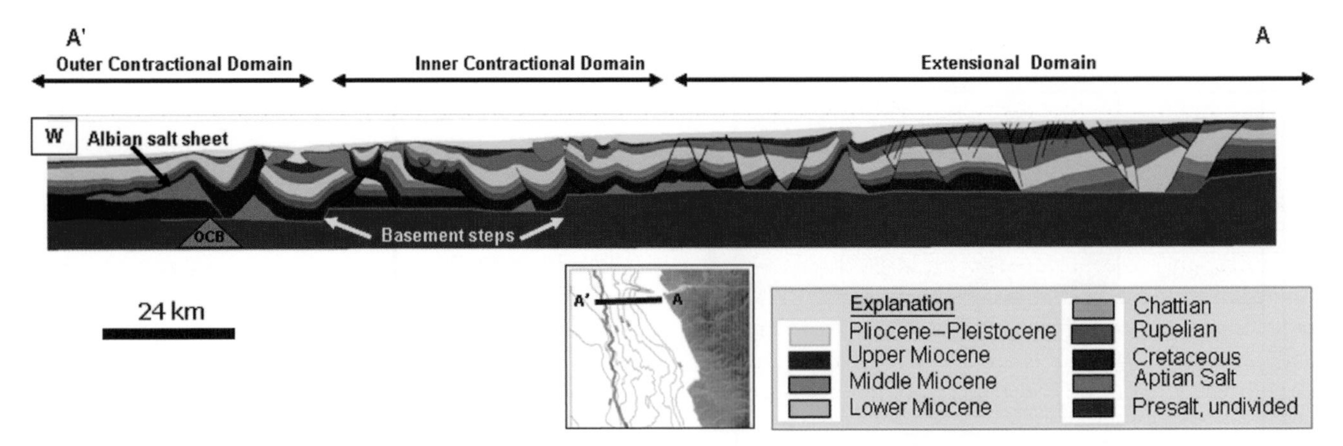

Figure 4. A regional structural cross section across the Lower Congo Basin. Location shown in the insert map. The depth section shows updip extension accommodated by downdip contraction. The deforming system detaches on Aptian salt. The line does not include the more proximal updip grabens. The cross section is 240 km (149 mi) long (2× vertical exaggeration). 24 km (14.9 mi).

Postrift differential thermal subsidence created a west-directed basinward tilt that induced gravity-driven translation of the Cretaceous strata (e.g., Duval et al., 1992; Marton et al., 2000). Cretaceous extensional faults have long been recognized on the shelf (e.g., Dale et al., 1992; Brown, 1993; Eichenseer et al., 1999; Anderson et al., 2000). Basinward Cretaceous contractional structures are more difficult to image and recognize because of overprinting by Oligocene–Miocene deformation. A well-organized imbricate thrust belt of Cretaceous age has been recognized in the northern LCB (Cramez and Jackson, 2000; Gottschalk, 2002). Recently acquired seismic data shows Cretaceous allochthonous salt sheets that flowed laterally during the Albian along the OCB, as illustrated in Figure 4. A Cretaceous age for these salt sheets is indicated by the termination of Albian strata against the base of these distal salt sheets. Locally, folds and thrusts of apparent Albian age are also imaged. Gottschalk et al. (2004) have inferred the presence of salt diapirs downdip during the Cretaceous based on stratal geometries observed in Cretaceous age strata.

A major change in the evolution of the LCB began in the Rupelian. From the Rupelian–Holocene, large volumes of clastic sediments were transported over the shelf edge and deposited on the slope and abyssal plain. Bond (1978), Sahagian (1988), and Burke (1996) have argued that these sediments are derived from the uplifted margin of the African continent. Uplift of the African margin created both a new slope for gravity sliding on the salt detachment and a source of thick Oligocene–Miocene clastic deposits that underwent gravitational collapse and spreading (Spathopoulos, 1996; Cramez and Jackson, 2000; Marton et al., 2000; Hudec and Jackson, 2004; Rowan et al., 2004). These changes triggered a new cycle of deformation, with the deforming wedge detaching on the Aptian salt/salt weld and with proximal extension being accommodated by distal contraction.

OVERVIEW OF BASIN STRUCTURE

The Oligocene–Holocene rocks within the LCB are divided into a proximal extensional domain and a distal contractional domain (Figure 4). A present-day structural elements map of the basin (Figure 2) illustrates the structural domain boundaries and the location of the OCB. Allochthonous salt structures are shown as well as major faults and folds. The proximal extensional domain ranges from 150 to 200 km (62–124 mi) wide, whereas the contractional domain ranges from 90 to 100 km (56–62 mi) in width. Note also that the contractional domain corresponds, in part, to the part of the basin with the highest density of allochthonous salt structures.

The Oligocene–Miocene extensional domain is characterized by a complex system of down-to-the-basin and counterregional normal faults, with the greatest magnitude of extension occurring south of the present-day Congo Canyon (Figure 2). Salt structures are primarily salt rollers in the footwalls of normal faults, but a few isolated salt diapirs are present. Where extensional strain was greatest, deformation resulted in a series of deep salt-detached grabens of Miocene age. These large updip grabens, filled with Miocene sedimentary strata, are prominent structural features in the proximal part of the basin. They occur in the part of the LCB where the Tertiary sediment thickness is the greatest. Some major graben-bounding faults continue to be active.

Within the contractional domain, shortening is accommodated by the complex interaction of folds, thrust faults, and inflation or extrusion of salt from squeezed diapirs (Figure 2). Subsequently, significant lateral variation in structural style exists. Structures vary from areas dominated by detachment folds, to areas dominated by contracted salt diapirs, salt sheets, and salt canopies, and to other areas with large displacement thrust faults. This variability occurs in response to differences in the volume and distribution of original salt, magnitude of Neogene extension-contraction, and structural relief on the basal autochthonous salt surface (Marton et al., 2000; Rowan et al., 2004). We divide the contractional domain into an inner contractional subdomain and an outer contractional subdomain (Figure 2). The inner subdomain is dominated by allochthonous salt structures. Significantly less salt exists in the outer subdomain. In this chapter, we observe the changing response of the Oligocene–Miocene depositional system to active structures as the system evolved through time.

Most Oligocene–Miocene allochthonous salt structures occur to the west of an interpreted system of faults and ramps along which the base of autochthonous salt is down-faulted to the west (Figure 2). With development of the autochthonous salt weld, base-of-salt structural relief had significant impact on developing structures as they translated westward across the fault and ramp system. The impact of structural relief on the structural-stratigraphic evolution of the Kwanza Basin has been discussed by several authors (Peel et al., 1998; Jackson and Hudec, 2005). We see similar structural and stratigraphic patterns in the LCB. In response to the underlying relief, ramp synclines (Jackson and Hudec, 2005) form above the hanging wall of the deep fault and contractional structures, generally thrust faults, form above the footwall. The contractional structures create a bathymetric high that diverts the depositional system, whereas the ramp syncline forms a bathymetric low that fills with synkinematic sediments.

The inner contractional subdomain is characterized by allochthonous salt structures (mainly salt stocks, sheets, and canopies), thrust faults, and to a lesser extent detachment folds. Many of the allochthonous salt structures show significant amounts of lateral flow of salt in response to shortening against salt feeder diapirs. Suturing of allochthonous salt sheets has resulted in several sizable salt canopies. Three-dimensional seismic data reveal early complex amalgamation and suturing of these allochthonous salt structures.

Within the outer contractional subdomain, significantly less salt exists. Structures consist mainly of open to tight detachment folds and thrust faults (Figure 2). The area to the north is dominated by detachment folds. A few cases of allochthonous salt flow from breached detachment folds occur. Along the OCB, structural restorations of seismic profiles show that earlier formed Cretaceous allochthonous salt sheets first were inflated during Oligocene shortening and then folded into the cores of the Miocene detachment folds. In other parts of the outer contractional subdomain, the structural character and general scarceness of salt structures suggest the development of a more frictional detachment. This structural style is characterized by an asymmetric basinward vergent stack of low-angle thrust faults. In physical models, this structural style has been shown to be characteristic of a more frictional detachment or of change from salt to a more frictional detachment (Costa and Vendeville, 2002; Bahroudi and Koyi, 2003). These structural styles contrast to areas farther south in the Outer Kwanza Basin. A northward transition from the inflated salt massifs, seen on Kwanza Basin seismic profiles (Cramez and Jackson, 2000; Hudec and Jackson, 2004), to the region of more limited salt supply to the north exists.

Basinward of the area of maximum extension, both the inner and outer contractional subdomains contain thrust faults with significant amounts of displacement (10–20 km). These faults are mainly Miocene in age and commonly carry earlier formed Oligocene folds and thrusts in the hanging wall. Typically, these older structures are short-wavelength (3–5 km [1.8–3.1 mi]) folds with Rupelian and Chattian strata thinning onto the fold crests. Thrust faults along the OCB frequently do not carry Oligocene folds in the hanging wall but do show thinning of Rupelian and younger strata toward the fault tip, indicating that these faults began to move during the Rupelian. Folds overlying the propagating thrust tip of these outermost thrust faults show significant erosion because of exposure at the sea floor during thrust propagation. Major episodes of thrust movement can be dated by eroded debris flow deposits on the abyssal plain in front of these thrust faults. Abyssal plain strata can be imaged and interpreted beneath the thrust hanging wall within close proximity to the footwall cutoffs. However, the terminations of the footwall cutoffs are not imaged.

Driven by active extension, structures form early across the contractional domain. With continuing translation, additional shortening is accommodated by continued growth of existing structures (Costa and Vendeville, 2002; Rowan et al., 2004). Actively evolving structures (uplift or subsidence) are a major control on syndepositional channel evolution and deposition in the deep water (Ferry et al., 2005; Gee and Gawthorpe, 2006; Clark and Cartwright, 2009; Kane et al., 2010). Deep-water channel systems shift position, seeking new bathymetric lows in response to changing local gradients caused by these active structures.

OVERVIEW OF BASIN STRATIGRAPHY

The beginning of the Rupelian (33 Ma) marks a dramatic change in the stratigraphic history of the LCB. Before the mid-Tertiary uplift of the African margin, sediments were accumulating at very slow rates as pelagic alternations of Turonian through Eocene marl and shale (Lavier et al., 2000; Valle et al., 2001). From the Rupelian, and continuing to present day (Anka et al., 2009), the LCB has received coarse clastic sediments mostly through deep-water slope-channel systems. These coarse clastic sediments are sourced from Proterozoic Angolan shield granites and metaclastics within the very large interior drainage system of the proto-Zaire (Congo) River. The cratonic provenance of these subarkose and quartzose sands (McLaughlin and Hood, 1998) results in excellent porosity and permeability of reservoir sandstones. Rupelian–Messinian sandstones exhibit a progradational stacking pattern within the LCB as the shelf edge moves westward approximately 100–120 km (~62–75 mi) through this period. In response to the overall progradation, slope gradients increase from the Rupelian to the Messinian as the physiographic setting changes up the section from lower slope to upper slope. It is estimated that the regional slope gradient changes from less than 0.1° in the distal reaches of the basin during the Rupelian to 2–3° near the shelf edge in the Messinian. Slope gradients are estimated from comparison with analog basins with similar tectonic settings and from deep-water slope-channel morphologies and architecture. A progression of depositional environments from proximal upper slope leveed-channel systems, to erosionally confined channel systems, to weakly confined channel systems, and ultimately to a basin floor distributary channel system have all been recognized on the Angolan slope. However, this model of a prograding margin with a simple slope profile is complicated by interaction of the channel systems with sea-floor relief created by active structures. These structures created rapid local changes in gradient that lead to rapid lateral changes in reservoir distribution, channel architecture, and morphology.

PALEOGEOGRAPHY OF THE LOWER CONGO BASIN

Oligocene (Rupelian and Chattian)

During the Oligocene, the physiographic setting of the LCB was the middle to lower slope. The Rupelian marks the onset of coarse clastic deposits delivered into the basin through a series of deep-water channel systems emanating south of the present-day Congo Canyon. The Oligocene depositional system and evolving structures are shown on the Oligocene paleogeographic map (Figure 5). Confined channel systems shown on the paleogeographic map may result from erosion into the slope substrate, aggradational levees, or both. Emphasis was placed on representing the coarse clastic parts of the depositional system (reservoirs) instead of accurately representing the genetically related very fine grained overbank deposits (mostly nonreservoir prone). A schematic block diagram (Figure 6) illustrates the interaction of the Oligocene depositional system with active structures.

South of the modern Congo Canyon, the proximal extensional domain is dominated by regional and counter-regional listric normal faults. Many of these faults experienced significant amounts of displacement during the Oligocene. Despite the accommodation space created by Oligocene extensional faulting, minimal sediment capture by these updip faults occurs. Most of the confined channel systems flow unhindered down relay ramps and across fault strands.

The extension was accommodated downslope by Oligocene folds and thrust faults and by inflation of salt structures. Because of the effectiveness of the salt detachment, proximal extension was rapidly transferred to the outer basin edge as well as within the basin. Rupelian shortening occurred at the OCB as well as at the inner edge of the contractional domain. Thrust faults along the OCB show thinning of Rupelian strata in the hanging wall. Restorations involving the Cretaceous allochthonous salt sheets located along the OCB indicate inflation of the allochthonous salt body during the Oligocene, which suggests shortening of the feeder diapir. Squeezing of salt diapirs to accommodate shortening on the Angola margin has been recognized and discussed by several authors (Cramez and Jackson, 2000; Rowan et al., 2004). Gottschalk et al. (2004) show a seismic profile and restoration (Figures 10B, 11) illustrating that salt sheet extrusion is contemporaneous with contraction of the feeder diapir. Their restoration shows Oligocene onset of shortening (Figures 11, 12). Elsewhere, short-wavelength (3–5 km [1.9–3.1 mi]) folds are observed. Rupelian–Aquitanian strata thin onto the crest of these folds. Depth images of many salt-sheet keels reveal an outward widening of the salt body during the Oligocene. This indicates that the rate of salt flow was greater than the rate of local deposition. The rate of salt flow increases with contraction of the feeder diapir (Cramez and Jackson, 2000; Gottschalk et al., 2004; Rowan et al., 2004). Taken all together, abundant evidence can be interpreted as

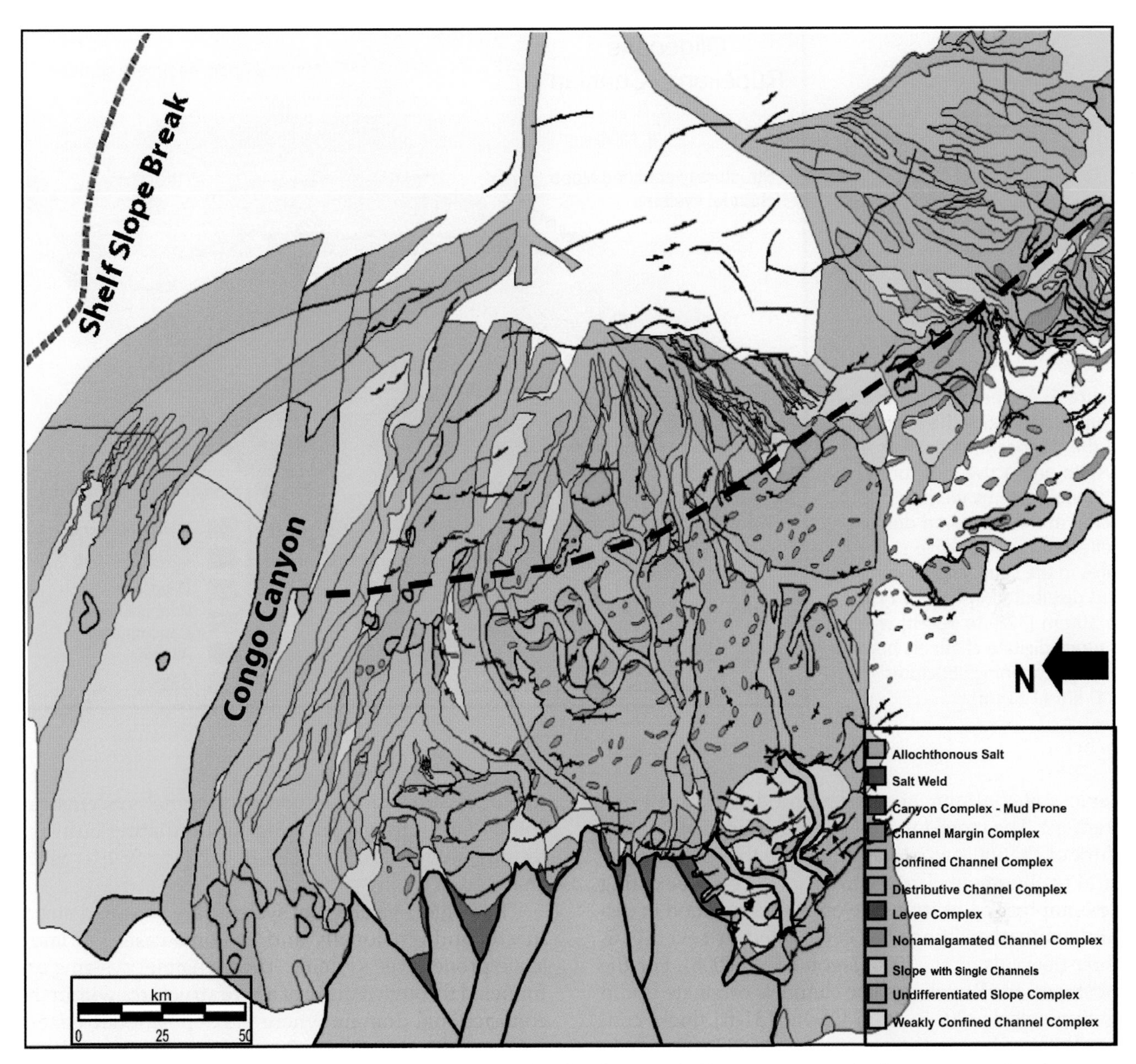

Figure 5. An Oligocene paleogeographic map illustrating the distribution of Oligocene coarse clastic systems and active structures within the Lower Congo Basin. North is to the left. The map area and structure symbols are the same as in Figure 2. The dashed line separates extensional and contractional domains. This figure represents an amalgamation of all reservoir units within the Chattian interval of the Oligocene. The reservoir interpretations are based on three-dimensional and two-dimensional seismic geometries and amplitude extractions calibrated to well penetrations. Channel systems were unhindered in the extension domain, deflected around active structures in the contractional domain, and distributed in narrow (100- to 300-m [328- to 984-ft]-wide) digitate channels (represented by outer distributary channel complexes) beyond the last confining structures.

Rupelian–Aquitanian shortening throughout the contractional domain of the LCB.

For this study, seismic data covering the upper slope was restricted to a few widely spaced 2-D seismic lines. Therefore, little is known about the proximal geometry of reservoir systems in the Oligocene. The following comments are restricted to the behavior of the reservoir systems in the middle to lower slope setting where data coverage is more robust. The Oligocene slope-channel system is dominated by confined channel systems in the middle slope that change to digitate and dispersive distributary channels on the lower slope in response to

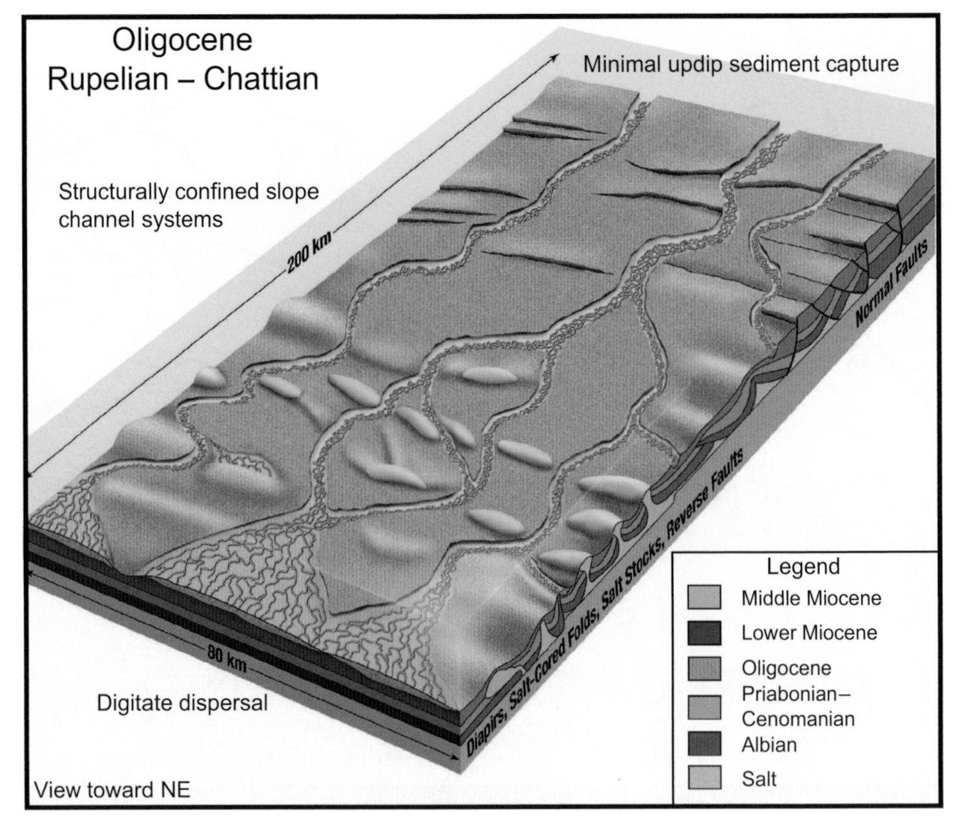

Figure 6. A schematic block diagram summarizing the interaction of the Oligocene depositional system and active structures. Channel systems were unhindered in the extension domain, deflected around active structures in the contractional domain, and distributed in narrow (100- to 300-m [328- to 984-ft] wide), sinuous digitate channels beyond the last confining structures. 200 km (124 mi).

the loss of confining structures and decrease of slope gradient. The largest influence on the distributive network of channels would appear to be the loss of structural confinement. The control exerted by preexisting topography and its influence on lobe deposition in distal deep-water settings has been noted by several authors (Gervais et al., 2006; Jegou et al., 2008). For the most part, the Rupelian slope channels originate updip as single-cycle (20- to 40-m [66- to 131-ft] thick) confined channels about 100 to 300 m [328–984 ft] wide in the extensional updip part of the basin and then stack into multicycle confined channel systems (2–3 km [1.2–1.9 mi] wide) downdip as they pass around active salt diapirs, folds, and thrust faults in the contractional domain. As these Rupelian channel systems emerge from the most outboard confining structures, they disperse back into 100- to 300-m (328- to 984-ft) wide, sinuous, single-cycle confined channels. In contrast to the Rupelian systems, Chattian reservoirs first appear as 2- to 3-km (1.2- to 1.9-mi) wide multicycle (50- to 200-m [164- to 656-ft] thick) confined channel systems in the middle slope extensional domain. They continue to wind through the lower slope contractional domain between the sea-floor topographic highs with little change in reservoir geometry until they emerge from the most outboard confining structures. Here, similar to Rupelian systems, Chattian channel systems become distributive, dispersing, and radiating outward as single-cycle, 100- to 300-m (328- to 984-ft) wide channels.

The Oligocene channel systems are confined structurally and erosionally and, in some cases, by large shale-prone levee systems. These channel systems are funneled through a maze of active structures within the contractional domain where the depositional system forms multicycle (50- to 200-m [164- to 656-ft] thick) stacked channel systems (2–3 km [1.2–1.9 mi] wide). Outboard of the most distal confining structures, the system forms digitate, sinuous, distributary patterns shown as cones at the westernmost edge of the Oligocene paleogeographic map (Figure 5). Each of these cones consists of many (100- to 300-m [328- to 984-ft] wide; 20- to 40-m [66- to 131-ft] thick) dispersive channels. These distal dispersive and distributive sand-filled channels positioned on the lowermost slope are present from the Congo Canyon in the north to points 180 km (112 mi) to the south (shown as distributary channel complexes [DCCs] on Figure 5). For much of the basin, the Oligocene section is topped by a second-order abandonment shale (200–500 m [656–1640 ft] thick), which serves as an excellent regional seal for underlying Chattian reservoirs.

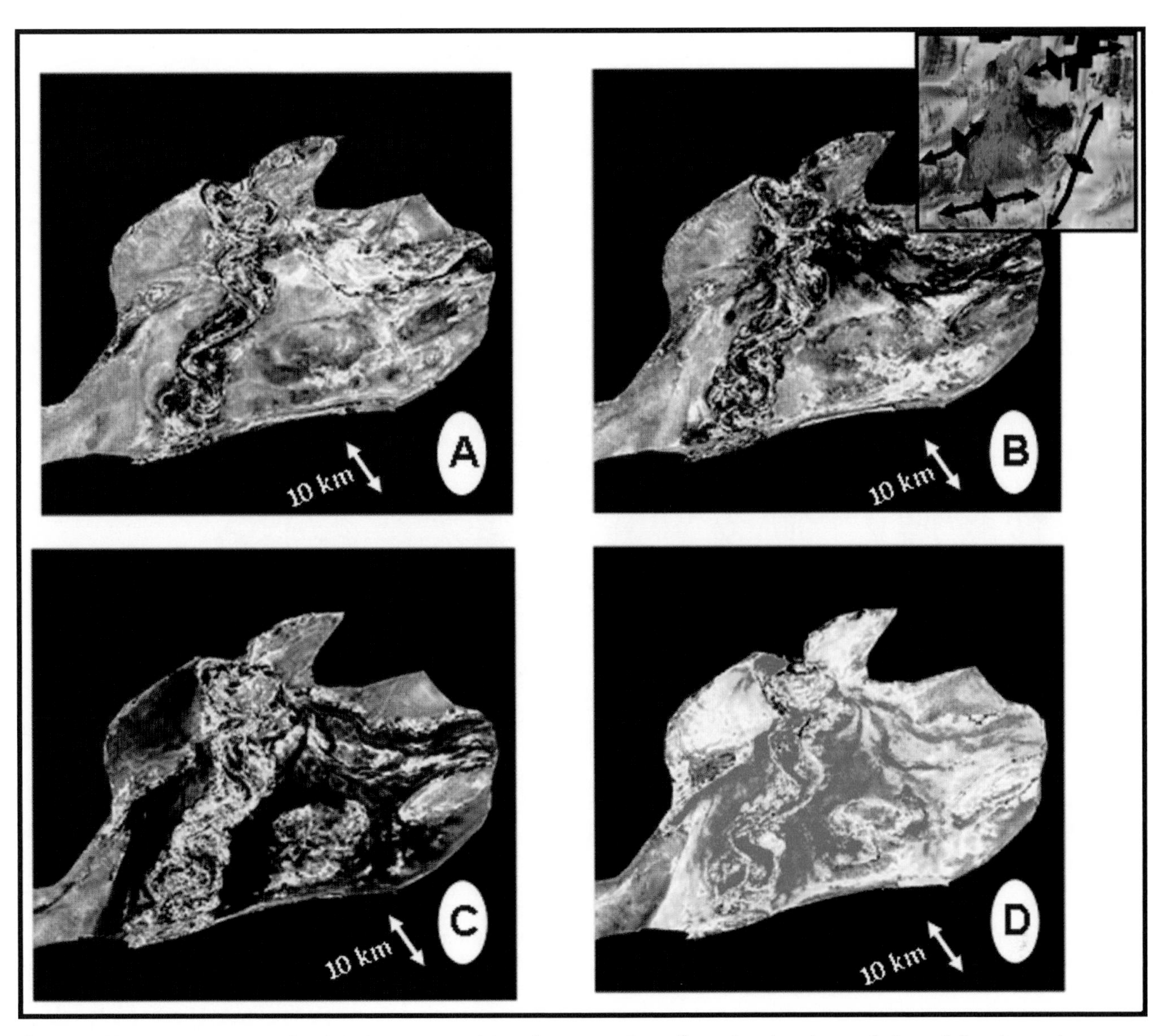

Figure 7. The evolution of a Chattian deep-water channel system. Time slices showing the evolution of the channel system from oldest (A) to youngest (D). (A) Early fairly straight confined channel system with internal sinuous channels. (B, C) Development of increased internal sinuosity. (D) Abrupt change to distributary system in response to a potential change in local gradient caused by continuing growth of the downdip fold now partially blocking the channel systems exit from the basin. The line is 10 km (6.2 mi) long. The inset image shows the location of nearby folds.

The evolution of a Chattian confined channel system to a distributary fan system is shown in Figure 7. Sequential changes in the depositional system are shown in four images derived from a series of sculpted seismic interval amplitude extractions. Initially, the channel system winds between active folds, entering the basin and exiting around the end of an active fold (Figure 7A). Through time, the channel system showed increased sinuosity possibly because of continued flattening of the depositional gradient within the confined channel system (Figure 7B, C). In the youngest phase of deposition, the channel system abruptly changes to a distributary fan (Figure 7D). This ponding may be related to a change in local gradient caused by continual growth of the downdip fold now blocking the channel systems exit from the basin. Alternatively, this basin may have been caught up in the hanging wall of a major thrust sheet, with the change in gradient occurring in response to early uplift and movement of the thrust.

A summary of the interaction of active structures and the Oligocene depositional system is shown in a schematic block diagram (Figure 6). The capture or ponding

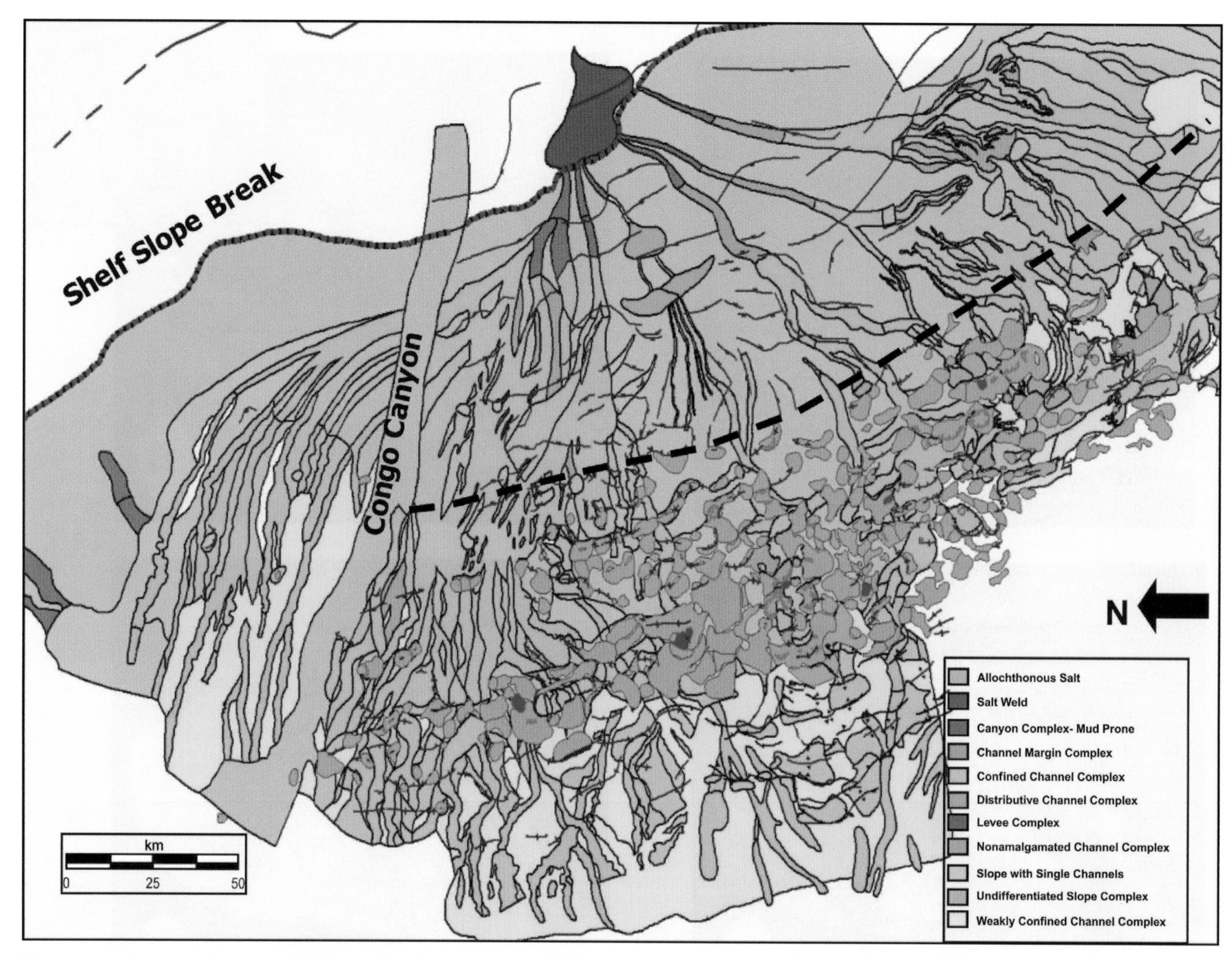

Figure 8. A lower Miocene paleogeography map (Burdigalian) illustrating the distribution of lower Miocene coarse clastic depositional systems and active structures within the Lower Congo Basin. Red polygon is lower Miocene delta at the shelf-slope break. Dashed line separates extensional and contractional domains. North is to the left. The map area and structure symbols are the same as Figure 2. This figure represents an amalgamation of all reservoir units within the lower Miocene interval. The reservoir interpretations are based on three-dimensional and two-dimensional seismic geometries and amplitude extractions calibrated to well penetrations. Note localized catchment of coarse clastic sediments updip in extensional domain (distributary channel complexes) and increased bifurcations outboard of last confining structures in contractional domain (weakly confined channel complexes).

of sediment updip was minimal as the confined channel systems wound their way downslope along relay ramps and across normal faults during times when faults were periodically inactive. In the contractional domain, short-wavelength detachment folds, thrust faults, and inflating salt bodies structurally confined the slope-channel systems. These channel systems shift position as local gradients change around active structures. Outboard of the most distal confining structures, a digitate dispersal system developed. Digitate, sinuous, dispersive 100- to 300-m (328- to 984-ft) wide channels appear to be the terminal mode for sands on the Chattian lowermost slope. Terminal lobes occurring at the end of these dispersive channels were not observed. However, because of limited seismic 3-D coverage, terminal lobes potentially exist farther west. Terminal lobes were observed in the Quaternary Congo River Axial fan described in the ZaiAngo study (Droz et al., 2003).

Lower Miocene (Aquitanian–Burdigalian)

Significant changes in both structural style and deposition occur during the early Miocene (Figure 8). During this time, the shelf edge prograded 50 km (31 mi) westward into the basin. A shelf-edge delta is interpreted in an area where existing wells drilled a thick succession of lower Miocene shallow-water deltaic deposits. Although this shelf-edge delta probably provided coarse clastic sediments to the nearby slope setting (slides and sloughing), the large volume of sand mapped throughout the basin suggests that the shelf was dissected by many incised valleys that fed slope-channel systems to comprise the complete delivery system. As Aquitanian and Burdigalian channel systems traversed the slope, progressive growth of structures caused the channel systems to shift position into new bathymetric lows. This process has been described and documented around an active salt structure by Gee and Gawthorpe (2006) and described by Clark and Cartwright (2009).

A system of proximal salt-detached grabens began to develop in the upper and middle slope over a small area (Figure 8). These active grabens trapped increasing amounts of coarse clastic sediments, allowing a very limited volume of sand to be deposited outboard of these rapidly developing catchments. The potential for better development of fine-grained top seal facies was enhanced downdip of these grabens because of the reduced energy of the depositional systems in the sediment shadow of the accommodating intraslope basin.

Downdip in the contractional domain during the Burdigalian, salt bodies began to flow laterally in response to contraction against the feeder diapir. Contractional pressure against the diapir increased the rate of salt inflation (flow), whereas reduced local sedimentation was insufficient to support a more vertical structure. The laterally flowing salt sheets coalesced and sutured, creating salt canopies. Channel systems that previously funneled around and between these individual salt structures were blocked by the salt canopy. In response, channels were deflected to a new route around the canopy, ponding against the updip side, or ponding until the accommodating space was filled and then spilling around the feature. Downdip of the area of maximum extension, major thrust sheets began to move, carrying earlier formed short-wavelength folds and salt structures in the hanging wall (Rowan et al., 2008).

Both Aquitanian and Burdigalian confined channel systems showed little morphological change across the study area from the updip extensional domain through the downdip contractional domain. Outboard of the most distal confining structure, a tendency for increased channel bifurcations was noted. A good example of one of these Burdigalian confined channel systems is described in detail by Porter et al. (2006). In cross section, these systems are commonly 2 to 3 km (1.2–1.9 mi) wide and 50 to 200 m (164- to 656-ft) thick. Confinement varies along the slope profile in response to gradient changes. Aggradational levees provide the dominant confinement in topographic lows, and erosional confinement is more prevalent in areas of positive relief (Porter et al., 2006). In this latter example, drainage is established and renewed structural growth was slow enough to allow the channel system to erode instead of deflect. The levees were mostly composed of shale, as evidenced by the excellent lateral seals observed for these confined channel systems over a broad variety of structural trap styles.

Some key differences in the responses of the Burdigalian and Aquitanian depositional systems to growing structures were found. In the contractional domain, Burdigalian reservoir systems were commonly deflected by growing structures, indicating that structural growth rates mostly outpaced sedimentation rates (Figures 9, 10). In contrast, Aquitanian depositional systems showed more complex geometries from which we infer that the episodicity of structural growth became the dominant control on depositional geometries. Episodic growth to explain the evolution of some contractional structures has been discussed by Cramez and Jackson (2000).

Our analysis indicates that active structures evolved episodically and that deep-water depositional systems respond to this intermittent movement. An example of an Aquitanian depositional system and its episodically active structure is shown in Figure 9. This image was created by overlaying an Aquitanian stratigraphically sculpted layer showing the depositional system (dark red and black colors) on top of a color-filled and shaded structure map from a slightly deeper horizon. The structure map shows the position of a large fold marked with a fold symbol. The stratigraphic sculpts are 50 m (164 ft) thick slices extracted between two seismic horizons parallel to the stratigraphy. The interval of interest is approximately 150 m (~492 ft) thick. To understand this interaction, we first restored this structure, including the seismic profile. The restoration allowed us to calculate rates of deposition in addition to the structural rate of growth. This enabled correlations between the rate of uplift and the rate of local deposition, with behavior of the depositional system. Restoration of a series of maximum amplitude extractions within the fixed interval of interest across the structure was coupled with restoration of the seismic profile.

In this example, we focused on several successive stratigraphic sculpts each about 50 m (164 ft) thick within the interval of interest, allowing us to study the response of the depositional system to episodic growth of an active structure. The lowest of the subintervals is shown in Figure 9A. Here, Aquitanian east–west-oriented channel systems overwhelm the relatively inactive fold as it is almost completely covered by sinuous channels crossing all but the crest of the fold. Only the

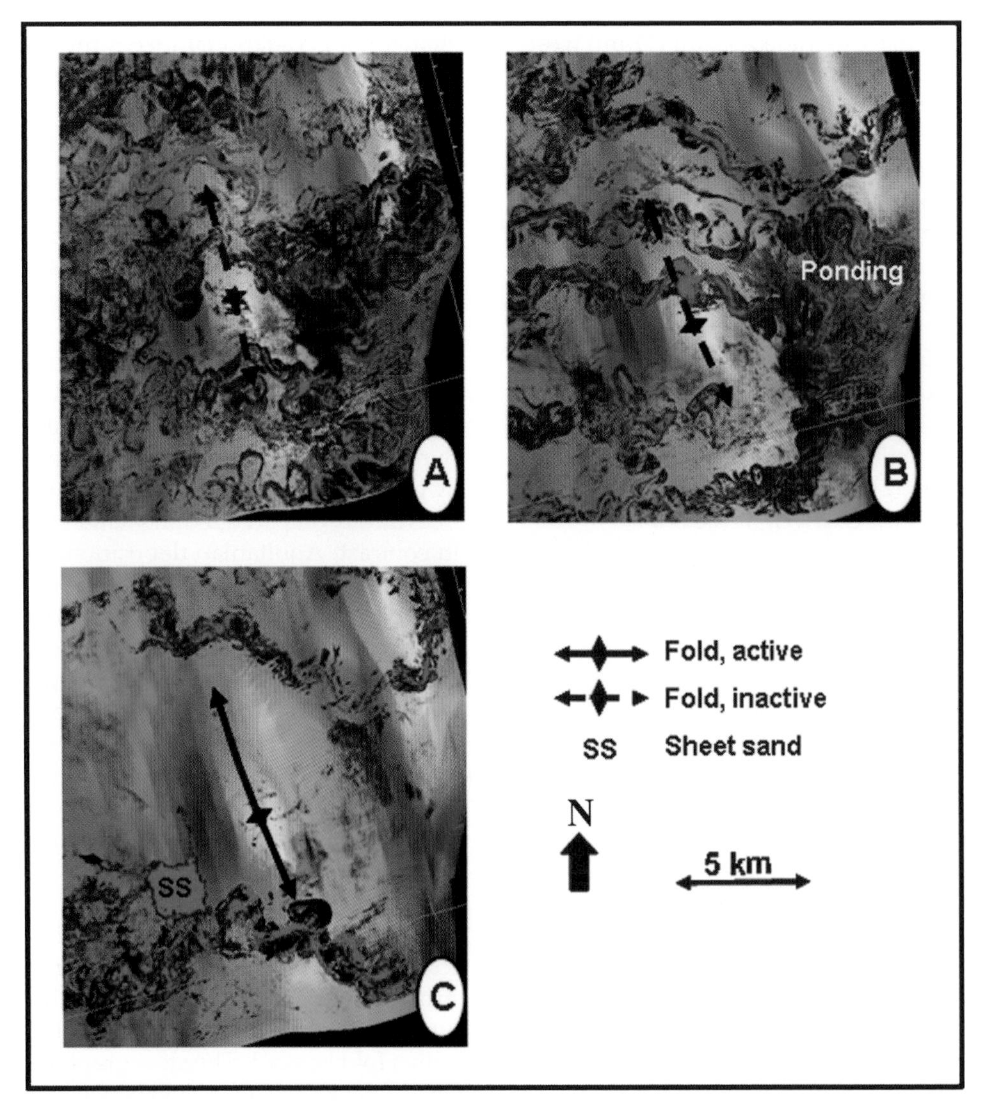

Figure 9. The episodic interaction of a deep-water channel system and an active fold. The image constructed by overlaying an Aquitanian stratigraphically sculpted layer (dark red and black colors) over a color-filled and shaded structure map from a deeper horizon. The structure map shows the position of a large fold marked by a fold symbol. Stratigraphic sculpts (slices about 50 m [164 ft] thick) were extracted from between two seismic horizons to parallel the stratigraphy. These sculpts illustrate an intermittently active fold across the Aquitanian–Burdigalian boundary. Intervals are identified from bottom to top as A, B, and C. (A) Aquitanian east–west channel systems overwhelm the relatively inactive fold. There was very limited bathymetric relief at this time. (B) Ponding updip of the fold axis is more pronounced on the east side of the fold. Channel systems spread laterally in response to subtle bathymetric expressions of the active fold. Later, this ponded complex is traversed by several younger confined channel systems. Note that the center red channel system is now crossing very near the fold crest. (C) This image is located above the interval of interest and shows complete deflection of channel systems around the bathymetric high of the fold crest. Note the small sheet sand (labeled SS) that ponded and onlapped the western side of the fold in response to the now active feature.

small central part of the fold appears to have bathymetric relief at this time. The middle interval is shown in Figure 9B. Ponding updip of the fold is more pronounced on this image, as indicated by the sharp boundary of the large north–south elongated mass (red and black) on the east side of the fold. Here, the channel systems have spread laterally in a possible response to subtle bathymetric expression of the active fold. Later, this ponded complex was traversed by several younger east–west-trending confined channel systems as slope gradient was reestablished. Note that the center channel system is now crossing very near the fold crest. The

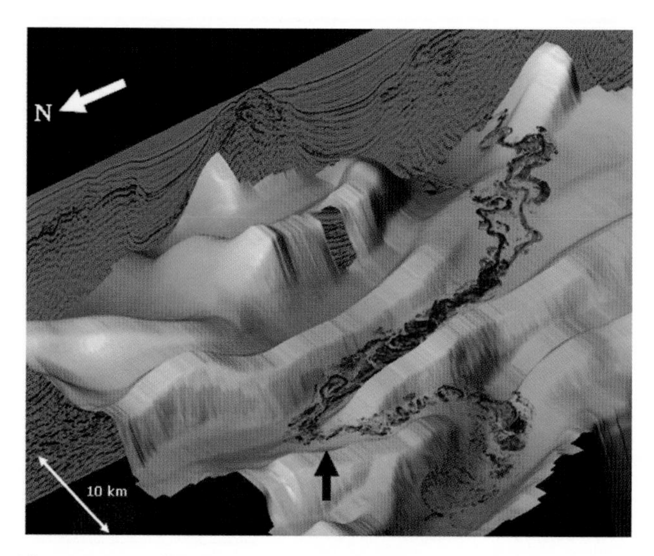

Figure 10. A deep-water confined channel system is shown deviating around and between active folds. This is a stratigraphic sculpt of laterally isolated amplitude anomalies of the Burdigalian interval-confined channel system migrating around structures. Note that once the channel system enters the folds, the width changes in response to the confining structures. Narrowing and straightening in the first constricted (arrow) turn then widening near the exit point at the bottom of the image. Note the well-developed sinuosity. The scale line is 10 km (6.2 mi) long.

episodic movement of the active structure caused this more subtle interaction with the depositional system. During the interval being considered, the rate of local deposition was twice the rate of structural uplift. It appears that the episodic growth of the structure created subtle local changes in gradient. The interval of interest is overlain by a 300 m (984 ft) thick low-amplitude continuous seismic facies interpreted as shale-prone abandonment facies. Figure 9C is from above the interval of interest and represents the stratigraphically younger Burdigalian. This sculpt shows a complete deflection of channel systems around the bathymetric high of the fold crest and a preference to follow the bathymetric lows. Note the small sheet sand (labeled SS) that ponded and onlapped the western side of the fold in response to the now active feature. At this time, the rate of structural growth was equal to or greater than the rate of local deposition.

A similar analysis was conducted on another nearby structure. Comparison of the interaction of depositional system and structural growth for the same interval of time shows that the rate of growth of this structure always equaled or exceeded the rate of local deposition. Therefore, channel systems were always deflected around the crest of the structure and across the flanks.

A second example of strong structural control on lower Miocene deposition is shown in Figure 10. Here a Burdigalian, deep-water, confined channel system was directed around and between active folds as it traveled downslope toward the abyssal plain. The highly sinuous channel patterns suggest that this system was on a very low-gradient slope. The channels were structurally confined as they flowed between and around the folds. The channels in the upper right of this image are two separate systems that merged into one channel system as they were confined between the first folds. The channel system was confined between the folds as the whole system progressed downslope through several tight turns and toward the lower edge of this image. The combined channel system narrows and straightens in the first constricted bend between folds (300 m [984 ft] wide) then returns to the 1 km (0.6 mi) width with a high internal sinuosity before widening to roughly 3 km (1.9 mi) as the structural constraint was decreased near the exit point of the folds. These changes in sinuosity and width occurred in response to structural constraints, which affected the local gradient. Clark and Cartwright (2009) made similar observations for submarine channels flowing through an active fold belt in the Levant Basin.

A summary of the interaction of active structures and the depositional system at the end of the Aquitanian is shown in a schematic block diagram (Figure 11). A proximal graben system had begun to develop in the extensional domain. Most of the Aquitanian channel systems delivered sands through the extensional domain without obstruction. However, in the area undergoing maximum extension, much of the coarse-grained fraction was captured in developing intraslope grabens. Accommodation was slow enough to allow for some channel systems to fill and spill downdip across the active fault. This resulted in a greatly reduced volume of coarse-grained clastic sediments immediately downdip of this graben. An important localized seal interval was created in the central LCB as a result of this sediment capture. In the contractional domain, squeezing of the salt diapirs caused an increased amount of lateral salt flow, leading to the development of salt sheets and salt canopies. Folds and thrust faults continued to be active throughout the contractional domain. Structurally confined aggradational channel systems were funneled around the active structures in response to local gradient changes. However, as structural growth was episodic, channel systems crossed structures during periods of high sedimentation and locally inactive structure. In the outboard region of the basin, the loss of confining structures, along with a

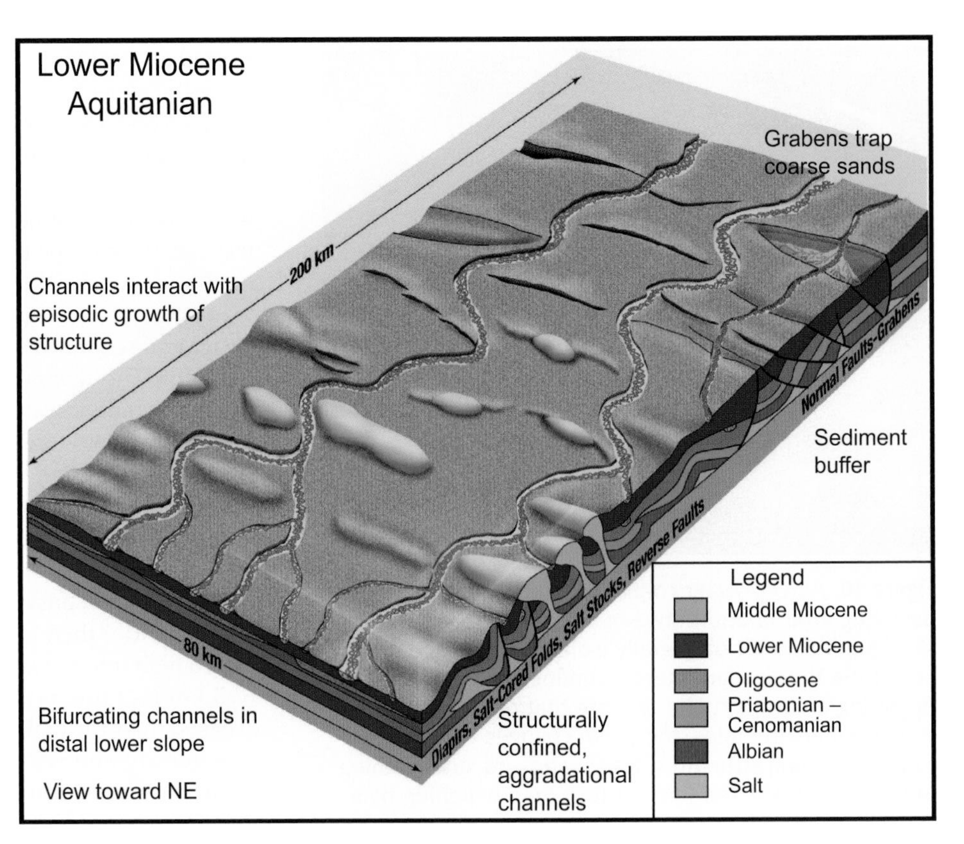

Figure 11. A lower Miocene (Aquitanian) schematic block diagram summarizing the interaction of the depositional system and active structures. Note that most sands were delivered through the extensional domain without resistance. This is the initial occurrence of graben faulting that created a significant intraslope catchment. Accommodation was slow enough to allow for some channel systems to fill and spill downdip. In the contractional domain, lateral salt flow created salt sheets/canopies plus folds and thrust faults continued to be active. Channel systems were funneled around the active structures in response to local gradient changes. Outboard, loss of confining structures and interpreted decrease in bathymetric gradient resulted in increased channel bifurcations. 80 km (49.7 mi); 200 km (124 mi).

potential decrease in bathymetric gradient, was interpreted to cause the observed increase in the bifurcations of channel systems.

Middle Miocene (Langhian–Serravallian)

The middle Miocene marks a time in the LCB evolution where a notable shift of the trunk delivery system to the north was initially recognized along with the first appearance of large second-order style stacking of multiple channel systems within erosion- and levee-confined bounding surfaces. For reasons not clearly understood, sediment feeder systems were becoming fixed for longer periods, and channel systems were following the same pathway downslope in the extensional domain. This is the first appearance of proto-Congo Canyon–style sediment feeder systems. By the middle Miocene, the shelf edge had prograded over the easternmost grabens (Figure 12). Physiographically, the active extensional domain was in an upper to middle slope setting. Middle Miocene sediment thickness patterns suggest that this was the time of maximum accommodation within the proximal grabens. These updip catchments were extremely effective in capturing all the coarse-grained sands in the central part of the basin, allowing only silt and mud to pass through as part of the dilute turbidite suspension load. These Angolan intraslope grabens are described in part by Sikkema and Wojcik (2000). The grabens were controlled by faults; however, capture of coarse clastic sediments was as effective as for salt withdrawal intraslope basins described by authors in the Gulf of Mexico Pliocene–Pleistocene section (Prather et al., 1998; Beaubouef and Friedmann, 2000). The coarse clastic sediment shadow downdip of these grabens is well expressed on the middle Miocene paleogeographic map.

The best-developed channel sands during this time were delivered north of these updip catchment areas. In some cases, these reservoir systems occurred as large, composite, erosion- and levee-confined channel system, some of which were up to 500 m (1640 ft) deep and 5 to 10 km (3.1–6.2 mi) wide. These composite channel systems exhibit an early phase of erosional confinement and a later stage of well-developed levee confinement. The levee systems are composed of very fine grained sand to silt that aggraded adjacent to the channel axes. Levee-confined channel systems remain essentially unchanged in gross channel morphology as they progress downslope, but they were easily deflected by a high relief basin topography created by rapidly growing structures.

In the middle Miocene, shortening was accommodated within the contractional domain by growth of

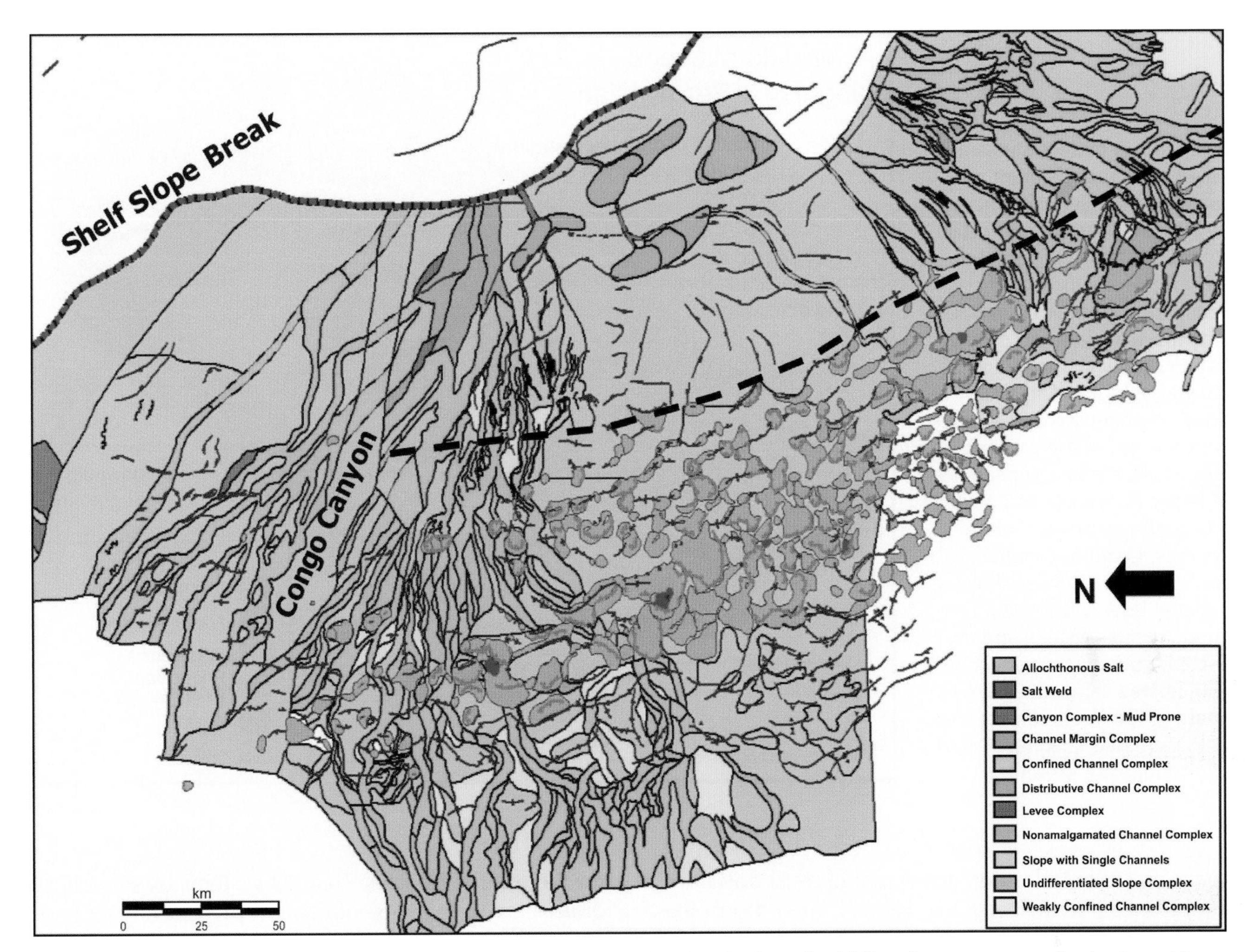

Figure 12. A middle Miocene paleogeographic map illustrating the distribution of middle Miocene coarse clastic deposition systems and active structures within the Lower Congo Basin. North is to the left. The map area and structure symbols are the same as in Figure 2. Dashed line separates extensional and contractional domains. This figure represents an amalgamation of all reservoir units within the middle Miocene interval. The reservoir interpretations are based on three-dimensional and two-dimensional seismic geometries and amplitude extractions calibrated to well penetrations. This is the first recognized shift of the trunk delivery system to the north. The shelf edge prograded to the easternmost grabens that reached maximum accommodation during the middle Miocene (distributary channel complexes [DCCs] on map). In the central part of the basin, a coarse clastic sediment shadow was present downdip of these grabens. The best developed channel sands during this time were delivered north of these DCCs.

existing structures. Within the inner contractional subdomain, as shortening closed the salt feeder diapirs, continued shortening was accommodated by high-angle thrust faults moving up the feeder weld. These steeply dipping thrust faults had synkinematic growth packages that reflect shortening by movement up the thrust-salt weld. Earlier shortening by salt evacuation from the feeder diapir may not be as clearly shown by the stratal geometry. Significant growth or thinning packages may date thrust movement up the feeder weld detachment and not the onset of shortening.

A summary of the interaction of active structures and the middle Miocene depositional system is shown in Figure 13. In the proximal extensional domain, coarse clastic sediments were trapped in the large grabens in the central part of the basin. Only dilute suspended-load silt and mud were deposited downdip of these grabens. The middle Miocene marks the maximum accommodation within these near shelf-edge intraslope basins. Large composite channel systems were observed north of the grabens. This signifies the first appearance of larger (proto-Congo Canyon) composite confined channel systems. In a few local areas, gradient flattening occurred near active extensional faults. In response, weakly confined channel systems spread at the active fault, creating a laterally amalgamated sand system. Once the

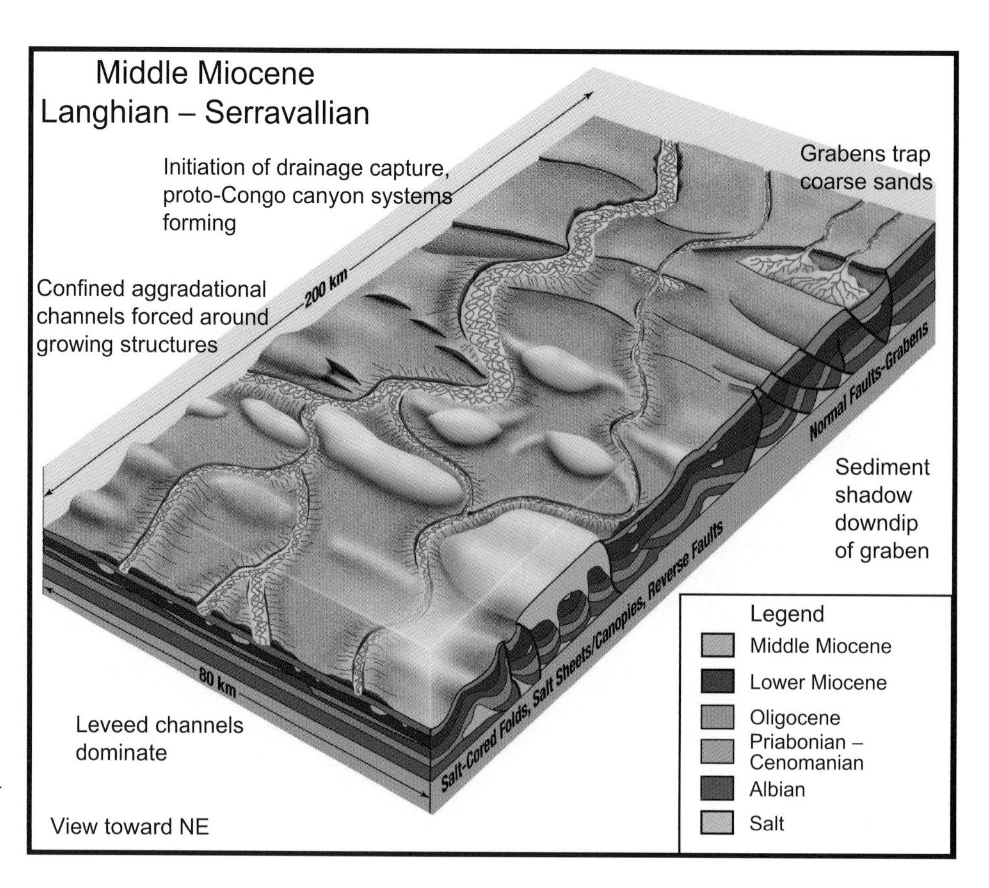

Figure 13. A middle Miocene schematic block diagram summarizing the interaction of the depositional system and active structures. In the proximal extensional domain, coarse clastic sediments are trapped in large grabens. Maximum accommodation occurs in shelf-edge intraslope basins. Large composite channel systems are observed north of the grabens. This marks the first appearance of larger (proto-Congo Canyon) composite confined channel systems. Channel systems are confined by well-imaged aggradational levees. Downdip, confined channel systems are deflected around active structures and inflated salt canopies. Outboard of the last confining structure, increased channel system bifurcations are present.

space was filled, these systems reorganized downdip as confined channel systems. Farther downdip in the contractional domain, middle Miocene confined channel systems were diverted around active structures and the salt canopies. Outboard in the area beyond the last confining structure, increased channel system bifurcations are noted.

Upper Miocene (Tortonian–Messinian)

By the late Miocene, the shelf edge prograded 20 to 30 km (12.4–18.6 mi) to the west, as shown in Figure 14. In the extensional domain, movement was focused on a few major graben-bounding faults. This extension was accommodated downdip by shortening distributed across the entire contractional domain. No newly formed contractional structures existed. However, many of these active structures continued to show topographic relief on the modern sea floor.

The bulk of the volume of sediments being delivered into the LCB was directed northwestward through large composite channel systems north of the present-day Congo Canyon, as described in part by Ferry et al. (2005) (Figure 14). These channel systems express an early phase of erosional confinement and a later stage of well-developed levee confinement similar to that observed for the middle Miocene channel systems. A subsidiary drainage system exists south of the present-day location of the Congo Canyon (Figure 14). These channel systems carried only a small fraction of the coarse clastic sediments compared with the composite channel systems delivering sands to the northwest. Most of these channel systems acted as bypass conduits (steep upper slope gradient) and terminate in salt/fault induced catchments within the contractional domain. The channels were straight and mostly mud filled with occasional interspersed coarse clastic lag deposits. Dilute low-density turbidites overtopping the erosional confinement created levee deposits of silt and occasional very fine grained sand. Subtle local gradient flattening occurred near active extensional faults, and these straight channel systems fed distributary channel systems spreading near the active fault, creating an effective catchment for fine- to medium-grained sands. However, most of the sand collected farther downdip within intraslope basins formed in the lows adjacent to active structures, as partially described in Sikkema and Wojcik (2000). These intraslope basins were sites of sand collection and as a result can contain very high net-to-gross distributary channel systems.

Note that all post-late Miocene deep-water sediments traveled through the paleo-Congo Canyon on

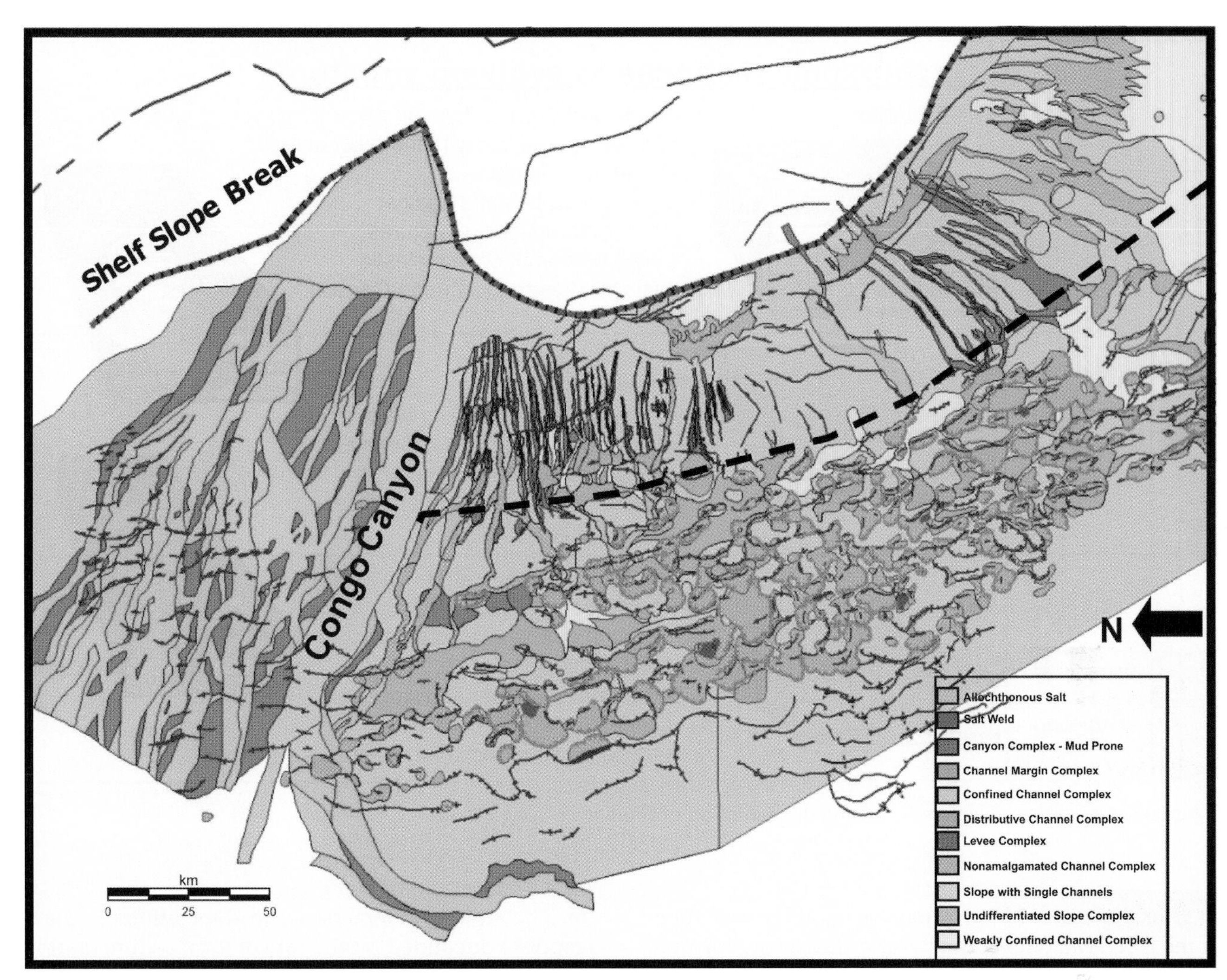

Figure 14. An upper Miocene paleogeographic map illustrating the distribution of Upper Miocene coarse clastic depositional systems and active structures within the Lower Congo Basin. North is to the left. Map area and structure symbols are the same as in Figure 2. Dashed line separates extensional and contractional domains. This figure represents an amalgamation of all reservoir units within the upper Miocene interval. The reservoir interpretations are based on three-dimensional and two-dimensional seismic geometries and amplitude extractions calibrated to well penetrations. The shelf edge prograded 20 to 30 km (12.4–18.6 mi) farther west. The largest volume of sediments was delivered northwest through large composite channel systems with well-developed levees. The subsidiary drainage system was established south of the present-day Congo Canyon. These confined channel systems acted as bypass conduits feeding distributary channel systems downdip within intraslope basins formed in the lows adjacent to active contractional structures. Extension accommodated downdip by shortening distributed across the entire contractional domain.

their way to the abyssal plain through a direct feed from the river mouth into the canyon. This new drainage conduit began about 5.3 Ma and has been described in Anka et al. (2009).

SUMMARY

In the LCB, evolving structures are a primary control on deep-water reservoir distribution and architecture. Large amounts of extension occurred on updip Oligocene listric normal faults. This extension-translation drove early Tertiary shortening. The Oligocene regional and counterregional normal faults evolved into Miocene graben-forming faults that created large coarse clastic traps. This updip extension continued to drive shortening in the contractional domain. We can understand and image the developing graben system. It is the structural complexity within the fold and thrust belt that poses an exploration challenge, especially when targeting 2 to 3 km (1.2 to 1.9 mi) wide channel systems underneath salt canopies and thrust sheets.

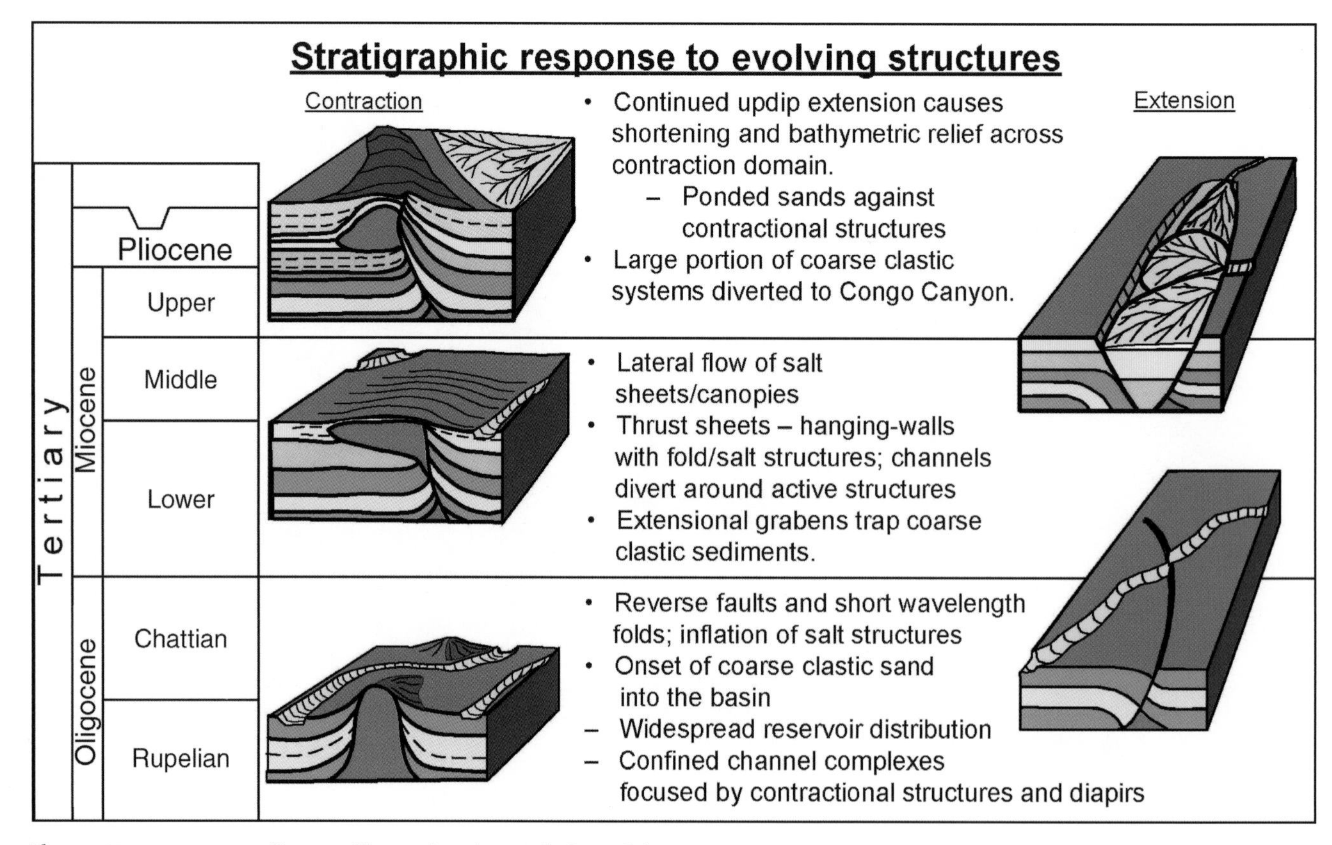

Figure 15. A summary diagram illustrating the evolution of the Lower Congo Basin.

The Oligocene–Miocene paleogeographic evolution of the LCB is an outstanding example of the interaction of deep-water depositional systems, with active structures as a progradational passive margin evolved through time. Slope gradient is one of the primary controls on reservoir distribution and organization. Slope gradients generally get steeper through time as progradation of the shelf edge advances, resulting in systematic changes in large-scale reservoir distribution and geometry. Overprinted on this large-scale slope gradient change are local changes in sea-floor gradient caused by active structures. The local gradient changes result in rapid lateral changes in reservoir distribution and geometry. The interaction between active structures and reservoirs may be subtle and episodic. Subtle gradient changes are detected by observing channel system trend behavior near structures instead of relying entirely on stratal geometries such as onlaps or truncations. Active structures can affect reservoir pathways even when the rate of local deposition is twice the rate of uplift.

The evolution of the LCB through time is summarized in Figure 15. From the Rupelian to the early Burdigalian, proximal extensional normal faults had minimal impact on sediment distribution. Downdip, in the contractional domain, the interaction was more subtle. The style of interaction and depositional system response depended on the rate of local sedimentation compared with the rate of uplift for each individual structure. Active structures interacted with the reservoir system by creating local gradient changes around evolving structures. Because the depositional system is synkinematic with the active structures, these local gradient changes were a major control on the response and resulting architecture of the deep-water depositional system. A progressive displacement of the channel systems occurs as they continuously shift position, seeking newly developing bathymetric lows around the active structures.

From the Late Burdigalian until the Messinian, active structures in both the extensional and contractional domains controlled the distribution of coarse clastic sediments reaching the basin. Sediment capture in the updip grabens created a coarse clastic sediment shadow downdip. By the Messinian, the shelf edge prograded to a position just updip of these grabens. Downdip of the grabens, reduced sedimentation facilitated the lateral flow of salt sheets. The resulting salt canopies further restricted sand fairways. In the contractional domain, structural growth outpaced local sedimentation. The channel systems were deflected between and around the active structures. Post-Burdigalian reservoir systems in the contractional domain can be less attractive exploration targets because of poor positioning on active structures and shallow below-mud-line depths.

ACKNOWLEDGMENTS

We thank ExxonMobil Exploration Company (EMEC) and Sonangol for permission to publish this chapter. Our regional understanding is built with the input of the many EMEC geoscientists who have worked the Lower Congo Basin. Reviews by EMEC geoscientists M. L. Porter, P. D. Snavely III, and T. M. Drexler added clarity to the chapter. Constructive reviews by J. Cartwright (Cardiff University) and M. Hudec (Bureau of Economic Geology, University of Texas) are gratefully acknowledged. Rebecca A. Miller is thanked for careful drafting of the block diagrams. Thanks are also extended to Bright LeMaster, Kim Dinh, and T. M. Drexler for their assistance with several key figures. Midland Valley two-dimensional and 3DMove software was used for the structural restorations.

REFERENCES CITED

Anderson, J. E., J. Cartwright, S. J. Drysdall, and N. Vivian, 2000, Controls on turbidite sand deposition during gravity-driven extension of a passive continental margin: Examples from Miocene sediments in Block 4, Angola: Marine and Petroleum Geology, v. 17, p. 1165–1203, doi:10.1016/S0264-8172(00)00059-3.

Anka, Z., M. Seranne, M. Lopez, M. Scheck-Wenderoth, and B. Savoye, 2009, The long-term evolution of the Congo deep-sea fan: A basinwide view of the interaction between a giant submarine fan and a mature passive margin (ZaiAngo project): Tectonophysics, v. 470, p. 42–56.

Ardill, J., C. Huang-Ting, and O. McLaughlin, 2002, The stratigraphy of the Oligocene to Miocene Malembo Formation of the Lower Congo Basin, offshore Angola (abs.): AAPG Annual Meeting, Expanded Abstracts, p. 9.

Bahroudi, A., and H. A. Koyi, 2003, Effect of spatial distribution of Hormuz salt on deformation style in the Zagos fold and thrust belt: An analog modeling approach: Journal of the Geological Society (London), v. 160, p. 719–733, doi:10.1144/0016-764902-135.

Beaubouef, R. T., and S. J. Friedmann, 2000, High resolution seismic/sequence-stratigraphic framework for the evolution of Pleistocene intraslope basins, western Gulf of Mexico, *in* P. Weimer, R. M. Slatt, J. Coleman, N. C. Rosen, H. Nelson, A. H. Bouma, M. J. Styzen, and D. Lawrence, eds., Deep-water reservoirs of the world: Gulf Coast Section SEPM 20th Annual Research Conference, p. 40–60.

Bond, G., 1978, Evidence for late Tertiary uplift of Africa relative to North America, South America, Australia, and Europe: Journal of Geology, v. 86, p. 47–65.

Brice, S. E., M. D. Cochran, G. Pardo, and A. D. Edwards, 1982, Tectonics and sedimentation of the South Atlantic rift sequence: Cabinda, Angola *in* J. S. Watkins and C. L. Drake, eds., Studies in continental margin geology: AAPG Memoir 34, p. 5–18.

Brown, D., 1993, West Africa challenges explorers (raft tectonics mixes up geology): AAPG Explorer, v. 14, no. 10, p. 12.

Burke, K., 1996, The African Plate: South African Journal of Geology, v. 99, p. 339–409.

Clark, I. R., and J. A. Cartwright, 2009, Interactions between submarine channel systems and deformation in deep-water fold belts: Examples from the Levant Basin, Eastern Mediterranean sea: Marine and Petroleum Geology, v. 26, p. 1465–1482, doi:10.1016/j.marpetgeo.2009.05.004.

Costa, E., and B. C. Vendeville, 2002, Experimental insights on the geometry and kinematics of fold and thrust belts above weak, viscous evaporitic decollement: Journal of Structural Geology, v. 24, no. 11, p. 1729–1739, doi:10.1016/S0191-8141(01)00169-9.

Coward, M. P., E. G. Purdy, A. C. Ries, and D. G. Smith, 1999, The distribution of petroleum reserves in basins of the South Atlantic margins, *in* N. R. Cameron, R. H. Bate, and V. S. Clure, eds., The oil and gas habitats of the South Atlantic: Geological Society (London) Special Publication 153, p. 101–131.

Cramez, C., and M. P. A. Jackson, 2000, Superposed deformation straddling the continental-oceanic transition in deep-water Angola: Marine and Petroleum Geology, v. 17, p. 1095–1109.

Dale, C. T., J. R. Lopes, and S. Abilo, 1992, Takula field and the Greater Takula Area, Cabinda, Angola, *in* M. T. Halbouty, ed., Giant oil and gas fields of the decade 1978–1988: AAPG Memoir 45, p. 197–215.

Davis, D. M., and T. Engelder, 1987, Thin-skinned deformation over salt, *in* I. Lerche and J. J. O'Brian, eds., Dynamical geology of salt and related structures: London, Academic Press, p. 301–337.

Droz, L., T. Marsset, H. Ondreas, M. Lopez, B. Savoye, and F. L. Spy-Anderson, 2003, Architecture of an active mud-rich turbidite system: the Zaire Fan (Congo-Angola margin southeast Atlantic): Results from ZaiAngo 1 and 2 cruises: AAPG Bulletin, v. 87, p. 1145–1168, doi:10.1306/03070300013.

Duval, B. C., C. Cramez, and M. P. A. Jackson, 1992, Raft tectonics in the Kwanza Basin, Angola: Marine and Petroleum Geology, v. 9, p. 389–404, doi:10.1016/0264-8172(92)90050-O.

Eichenseer, H. Th., F. R. Walgenwitz, and P. J. Biondi, 1999, Stratigraphic control on facies and diagenesis of dolomitized oolitic siliciclastic ramp sequences (Pinda Group, Albian, offshore Angola): AAPG Bulletin, v. 83, p. 1729–1758.

Ferry, J. N., T. Mulder, O. Parize, and S. Raillard, 2005, Concept of equilibrium profile in deep-water turbidite systems: Effects of local physiographic changes on the nature of sedimentary processes and the geometries of deposits, *in* D. M. Hodgson and S. S. Flint, eds., Submarine slope systems: Processes and products: Geological Society (London) Special Publication 244, p. 181–193.

Gee, J. J. R., and R. L. Gawthorpe, 2006, Submarine channels controlled by salt tectonics: Examples from 3-D seismic data offshore Angola: Marine and Petroleum Geology, v. 23, p. 443–458.

Gervais, A., B. Savoye, T. Mulder, and D. Gonthier, 2006, Sandy modern turbidite lobes: A new insight from high resolution seismic data: Marine and Petroleum Geology, v. 23, p. 485–502.

Gottschalk, R. R., 2002, The Lower Congo Basin, deep-water Congo and Angola: A kinematically linked extensional/contractional system (abs.): AAPG Abstracts, v. 11, p. A66.

Gottschalk, R. R., A. V. Anderson, J. D. Walker, and J. C. DaSilva, 2004, Modes of contractional salt tectonics in Angola Block 33, Lower Congo Basin, west Africa: 24th Annual Gulf Coast Section SEPM Research Conference, Salt-Sediment Interactions and Hydrocarbon Prospectivity: Concepts, Applications, and Case Studies for the 21st Century, Houston, Texas, December 5–8, 2004, 30 p.

Hudec, M. R., and M. P. A. Jackson, 2004, Regional restoration across the Kwanza Basin, Angola: Salt tectonics triggered by repeated uplift of a metastable passive margin: AAPG Bulletin, v. 88, no. 7, p. 971–990, doi:10.1306/02050403061.

Jackson, M. P. A., and M. R. Hudec, 2005, Stratigraphic record of translation down ramps in a passive-margin salt detachment: Journal of Structural Geology, v. 27, p. 889–911.

Jegou, I., B. Savoye, C. Pirmez, and L. Droz, 2008, Channel-mouth lobe complex of the recent Amazon Fan: The missing piece: Marine Geology, v. 252, p. 62–77, doi:10.1016/j.margeo.2008.03.004.

Kane, I. A., V. Catterall, W. D. McCaffrey, and O. J. Martinsen, 2010, Submarine channel response to intrabasinal tectonics: The influence of lateral tilt: AAPG Bulletin, v. 94, no. 2, p. 189–219, doi:10.1306/08180909059.

Kolla, V., P. Bourges, J. M. Urruty, and P. Safa, 2001, Evolution of deep-water Tertiary sinuous channels off-shore Angola (west Africa) and implications for reservoir architecture: AAPG Bulletin, v. 85, p. 1371–1405, doi:10.1306/8626CAC3-173B-11D7-8645000102C1865D.

Lavier, L., M. Steckler, and F. Brigaud, 2000, An improved method for reconstructing the stratigraphy and bathymetry of continental margins: Application to the Cenozoic tectonic and sedimentary history of the Congo Margin: AAPG Bulletin, v. 84, p. 923–939, doi:10.1306/A9673B6C-1738-11D7-8645000102C1865D.

Letouzey, J., B. Colletta, R. Vially, and J. C. Chermette, 1995, Evolution of salt-related structures in compressional settings, *in* M. P. A. Jackson, D. G. Roberts, and S. Snelson, eds., Salt tectonics: A global perspective: AAPG Memoir 65, p. 41–60.

Lunde, G. K., K. Aubert, O. Lauritzen, and E. Lorange, 1992, Tertiary uplift of the Kwanza Basin in Angola, *in* R. Curnelle, ed., Géologie Africaine: Colloque de Stratigraphie et de Paléogéographie des Bassins Sédimentaires Quest-Africains: Recuile des Communications, Libreville, Gabon, May 6–8, 1991, p. 99–117.

Lundin, E. R., 1992, Thin-skinned extensional tectonics on a salt detachment, northern Kwanza Basin, Angola: Marine and petroleum Geology, v. 9, p. 405–411, doi:10.1016/0264-8172(92)90051-F.

Marton, L. G., G. C. Tari, and C. Lehmann, 2000, Evolution of the Angolan passive margin, west Africa, with emphasis on postsalt structural styles, *in* W. Mohriak and M. Talwani, eds., Atlantic rifts and continental margins: American Geophysical Union Geophysical Monograph 118, p. 129–149.

Mayall, M., and I. Stewart, 2000, The architecture of turbidite slope channels, *in* P. Weimer, R. M. Slatt, J. Coleman, N. C. Rosen, H. Nelson, A. H. Bouma, M. J. Styzen, and D. T. Lawrence, eds., Deep-water reservoirs of the world: Gulf Coast Section SEPM Foundation 20th Annual Research Conference, p. 578–586.

McLaughlin, O., and K. Hood, 1998, Provenance controls on sediment composition and reservoir quality in deep-water depositional systems: A global perspective (abs): AAPG Bulletin, v. 82, p. 1941.

Peel, F. J., M. P. A. Jackson, and D. Ormerod, 1998, Influence of major steps in the base of salt on the structural style of overlying thin-skinned structures in deep-water Angola (abs.): Extended Abstracts Volume, AAPG International Conference and Exhibition, Rio de Janeiro, p. 366–367.

Porter, M. L., et al., 2006, Stratigraphic organization and predictability of mixed coarse- and fine-grained lithofacies successions in a lower Miocene deep-water slope-channel system, Angola Block 15, *in* P. M. Harris and L. J. Weber, eds., Giant hydrocarbon reservoirs of the world: From rocks to reservoir characterization and modeling: AAPG Memoir 88/SEPM Special Publication, p. 281–305.

Prather, B. E., J. R. Booth, G. S. Steffens, and P. A. Craig, 1998, Classification, lithologic calibration and stratigraphic succession of seismic facies of intraslope basins, deep-water Gulf of Mexico: AAPG Bulletin, v. 82, p. 701–728, doi:10.1306/1D9BC5D9-172D-11D7-8645000102C1865D.

Rowan, M. G., F. J. Peel, and B. C. Vendeville, 2004, Gravity-driven fold belts on passive margins, *in* K. R. McClay, ed., Thrust tectonics and hydrocarbon systems: AAPG Memoir 82, p. 157–182.

Rowan, M. G., E. N. Zhurina, M. F. Liebelt, and W. D. Hutchings, 2008, Salt architecture and its impact from deposition, ultra-deep water, Congo Basin, Angola: AAPG International Conference and Exhibition, Cape Town, South Africa: http://www.searchanddiscovery.net/abstracts/html/2008/intl_capetown/abstracts/471212.htm (accessed December 23, 2011).

Sahagian, D., 1988, Epeirogenic motions of Africa as inferred from Cretaceous shoreline deposits: Tectonics, v. 7, p. 125–138, doi:10.1029/TC007i001p00125.

Sikkema, W., and K. Wojcik, 2000, 3-D visualization of turbidite systems, lower Congo Basin, offshore Angola, *in*

P. Weimer, R. M. Slatt, J. Coleman, N. C. Rosen, H. Nelson, A. H. Bouma, M. J. Styzen, and D. T. Lawrence, eds., Deep-water reservoirs of the world: Gulf Coast Section SEPM, 20th Annual Research Conference, p. 928–939.

Spathopoulos, F., 1996, An insight on salt tectonics in the Angola Basin, South Atlantic, *in* G. I. Alsop, D. J. Blundell, and I. Davidson, eds., Salt tectonics: Geological Society (London) Special Publication 100, p. 153–174

Valle, P. J., J. G. Gjelberg, and W. Helland-Hansen, 2001, Tectonostratigraphic development in the eastern Lower Congo Basin, offshore Angola, west Africa: Marine and Petroleum Geology, v. 18, p. 909–927, doi:10.1016/S0264-8172(01)00036-8.

Walgenwitz, F., M. Pagel, A. Meyer, H. Maluski, and P. Monie, 1990, Thermochronological approach to reservoir diagenesis in the offshore Angola Basin: A fluid inclusion, ^{40}Ar-^{39}Ar and K-Ar investigation: AAPG Bulletin, v. 76, p. 547–563.

14

Clark, Ian R., and Joseph A. Cartwright, 2012, A case study of three-dimensional fold and growth sequence development and the link to submarine channel-structure interactions in deep-water fold belts, *in* D. Gao, ed., Tectonics and sedimentation: Implications for petroleum systems: AAPG Memoir 100, p. 315–335.

A Case Study of Three-dimensional Fold and Growth Sequence Development and the Link to Submarine Channel-structure Interactions in Deep-water Fold Belts

Ian R. Clark

Shell International Exploration and Production B.V. PO Box 162, 2501 AN The Hague, The Netherlands (e-mail: ian.clark@shell.com)

Joseph A. Cartwright

3D Lab, Department of Earth and Ocean Sciences, Cardiff University, Main Building, Park Place, Cardiff, United Kingdom (e-mail: joe@ocean.cf.ac.uk)

ABSTRACT

Growth sequences in deep-water fold and thrust belts can preserve a record of the interactions between coeval sedimentation and deformation. These sedimentary sequences can also form hydrocarbon exploration targets as they provide sites where sands can be incorporated into a fold during uplift. This chapter uses three-dimensional (3-D) seismic data to take a combined structural and stratigraphic approach to the analysis of several folds and their adjacent growth sequences from the eastern Nile submarine fan, Eastern Mediterranean Sea. We use along-strike measurements of fold uplift and growth sequence expansion factor to illustrate the irregular spatial and temporal development of sea-floor relief during fold growth. Irregular 3-D fold growth controls growth sequence deposition and affects submarine channel morphology within a specific type of growth sequence (onlapping or overlapping). Submarine channels within these growth sequences can overflow a developing fold or become diverted, depending on the relative rate of uplift and sedimentation. In detail, however, these channel systems show strong variations in sinuosity, which can have important implications for the development of laterally accreted sand packages. This study indicates that variations in folding along strike is a key factor that affects the development of submarine channel systems and provides a case study of how conceptual models of these settings can be improved by fully linking structural and stratigraphic observations.

DOI:10.1306/13351559M1003536

INTRODUCTION

Synkinematic sediments are deposited coevally with growing folds and faults forming geometrically distinctive growth sequences on the forelimbs and backlimbs of folds (Suppe et al., 1992). Growth sequences associated with thrust-related folds have received much interest in the context of structural geology, as they provide useful means of reconstructing the kinematics of fold growth (e.g., Suppe et al., 1992; Poblet et al., 1997; Bernal and Hardy, 2002; Salvini and Storti, 2002). However, there have been surprisingly few detailed studies of growth sequences in submarine fold belts where the preservation potential is high and where the availability of 3-D seismic data allows a fully 3-D approach to be taken (Higgins et al., 2007; Morley and Leong, 2008).

Variations in internal and external geometry of growth sequences such as the positioning and configuration of onlapping and overlapping sedimentary units can be used to constrain the evolution of bathymetric relief during folding. This analytical approach, which focuses on the stratigraphy, is critical for understanding how sedimentary systems respond to growing folds (Puigdefàbregas et al., 1992; Burbank and Verges, 1994; Burbank et al., 1996). Growth sequences are also progressively incorporated into the fold limbs during deformation. Therefore, if the synkinematic sediments include reservoir-prone lithologies, then inclusion of these into the growing fold can result in potential hydrocarbon reservoirs contained within fold closure or on the fold limbs. From a hydrocarbon exploration perspective, understanding how the interactions between sediment deposition and coeval fold growth control the distribution of reservoir (and seal) units is important.

Growth sequences are commonly observed from thin-skinned deep-water fold and thrust belts associated with the gravitational collapse of passive margins. Examples include the Angolan Margin (Brun and Fort, 2004), northwest Borneo (Morley and Leong, 2008; Morley, 2009), the Mississippi Fan Fold Belt (Rowan, 1997), the Niger Delta (Damuth, 1994), and the eastern Nile Delta (Gradmann et al., 2005). In deep-water fold belt settings, growth sequences are commonly composed of deposits derived from gravity currents and mass transport processes, interspersed with background hemipelagic sedimentation (cf. Stow and Mayall, 2000). All of these sedimentary processes may interact with emerging fold relief, which will in turn determine the geometry of the growth sequence (Morley and Leong, 2008; Morley, 2009). Another key factor influencing the sedimentary response to uplift is the along-strike 3-D evolution of folds over time (Higgins et al., 2007). Lateral propagation of folds can exert a strong control on sedimentary pathways and the location of depocenters through time (Burbank et al., 1996; Demyttenaere et al., 2000; Morley and Leong, 2008).

The aim of this study is to link structural and stratigraphic observations of fold growth in three dimensions. These observations are then linked to the detailed evolution of submarine channel systems in response to fold growth over time. An emphasis is placed on describing along-strike variations in fold geometry through time so that the four-dimensional evolution of the system is documented. We attempt to show how these variations can be critical in determining sediment transport routes and ultimately control final growth sequence geometry. Knowledge of the way in which submarine channel systems respond to uplift during growth sequence development can be of wider use in deep-water exploration by helping to predict the presence or absence of potential reservoir units.

GEOLOGIC SETTING AND DATABASE

Pre-Messinian Development of the Levant Basin

The Levant Basin is located in the Eastern Mediterranean Sea and is bounded to the east by the passive continental margin of Israel, Lebanon, and Syria; to the south by the northeastern lobe of the Nile Deep Sea Fan; to the west by the Eratosthenes seamount; and to the north by the subduction zone and transform fault of the Cyprus Arc (Figure 1) (Ben-Avraham et al., 1988, 1995; Vidal et al., 2000). Formation of the Levant Basin and the adjacent margin is related to a sequence of rifting events occurring from the Early Permian to the Middle Jurassic associated with the initial breakup of Pangea (Garfunkel, 1998). Final continental breakup and initiation of ocean spreading occurred at the end of the Middle Jurassic (Garfunkel and Derin, 1984). Compression in the Late Cretaceous and development of the Syrian Arc Fold Belt resulted in a series of northeast–southwest-orientated folds along the Levant Margin (Eyal, 1996; Buchbinder and Zilberman, 1997; Garfunkel, 1998). During the Oligocene, a system of submarine canyons incised along the margin, undergoing headward extension throughout the Miocene because of intermittent uplift and emergence of the Levant Margin (Druckman et al., 1995; Buchbinder and Zilberman, 1997).

Messinian Salinity Crisis and Post-Messinian Basin Development

At the end of the Miocene (5.9 Ma), narrowing of the connection between the Mediterranean Sea and the Atlantic Ocean led to the Messinian Salinity Crisis (Hsu

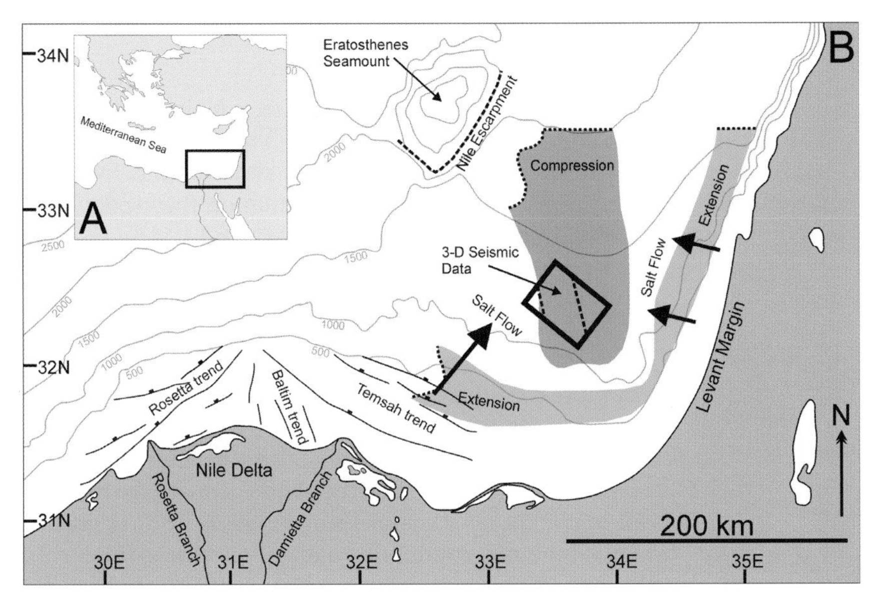

Figure 1. The location of the survey area within the context of the Nile Delta. (A) The inset shows the area of interest in the Eastern Mediterranean Sea. (B) Setting of the Nile Delta and the location of the seismic survey used in this study. The area covers a part of the eastern Nile deep-sea fan that is currently undergoing thin-skinned compression. The zone of compression within the Levant Basin is driven by the gravitational collapse of both the Nile Delta and the Levant Margin. Adapted from Gradmann et al., 2005; Netzeband et al., 2006; and Garziglia et al., 2008. 3-D = three-dimensional. 200 km (124 mi).

et al., 1978). This resulted in a rapid sea level fall estimated at between 800 and 1200 m (2625–3937 ft) below current sea level in the Eastern Mediterranean Sea (Druckman et al., 1995; Bertoni and Cartwright, 2007) and the deposition of a thick evaporitic sequence up to 2 km (1.2 mi) thick in some parts of the basin. Sea level fall was accompanied by erosion along marginal areas of the Levant Basin, resulting in continued incision of a series of prominent canyons around the basin margin (Cita and Ryan, 1978; Garfunkel and Almagor, 1987).

During the Pliocene, the Levant Basin was subjected to increased sedimentation derived primarily from the Nile Delta to the southwest (Mart and Ben Gai, 1982). An increase in sedimentation rate was accompanied by increased basin subsidence due to loading of the Messinian evaporitic sequence (Tibor and Ben-Avraham, 1992; Ben-Gai et al., 2005). In the Levant Basin, sedimentation was predominantly sourced by submarine channels derived from the Nile Delta (Folkman and Mart, 2008).

Database

The 3-D seismic data used in this study covers an area of approximately 1400 km^2 (Figure 2). The survey covers part of the distal northeastern Nile Deep Sea Fan that extends into the Levant Basin, providing a detailed record of post-Messinian sedimentation (Figure 2). Average sea-floor gradients in the study area are between 0.38° in the downslope direction and 0.02° in the cross-slope direction. Water depths range from 1000 to 1350 m (3280–4430 ft) below sea level across the survey area. Submarine channels are ubiquitous throughout the post-Messinian sequence, typically consisting of single channel-levee systems that are rarely erosionally confined (Figure 2). In comparison with the larger slope system, the relatively small scale and lack of complexity of the submarine channels in this study area suggest that they occupy the lower fan region (cf. Babonneau et al., 2002). The most recent of these channels are visible on the present-day sea floor (Figure 2). Submarine channels are common features associated with the Nile Delta, and subsurface examples have been previously described from the deep-water Western Nile (Samuel et al., 2003).

Seismic Stratigraphy

This study focuses on the uppermost sequence of the post-Messinian Pliocene–Quarternary sedimentary section of the Levant Basin. Recent studies from this area have focused on either the Messinian sequence (e.g., Bertoni and Cartwright 2005, 2006, 2007; Gradmann et al., 2005) or along the marginal areas of the basin (e.g., Frey-Martinez et al., 2005). Chronostratigraphic and lithologic data, where given, for the post-Messinian sedimentary cover is based on unpublished well reports (Frey-Martinez et al., 2005). No published wells within the more basinal seismic survey are shown.

Messinian Evaporite Sequence (6.7–5.2 Ma)

The top of the Messinian sequence is marked by Horizon M, a regional high-amplitude positive reflection recognizable throughout the entire Mediterranean Basin (Ryan et al., 1973) (see also Figure 3). Internally, the Messinian sequence shows evidence of deformation with multiple thrust detachment levels observed (Bertoni and Cartwright, 2007; Cartwright and Jackson, 2008). Over more marginal areas of the Levant Basin, Horizon M

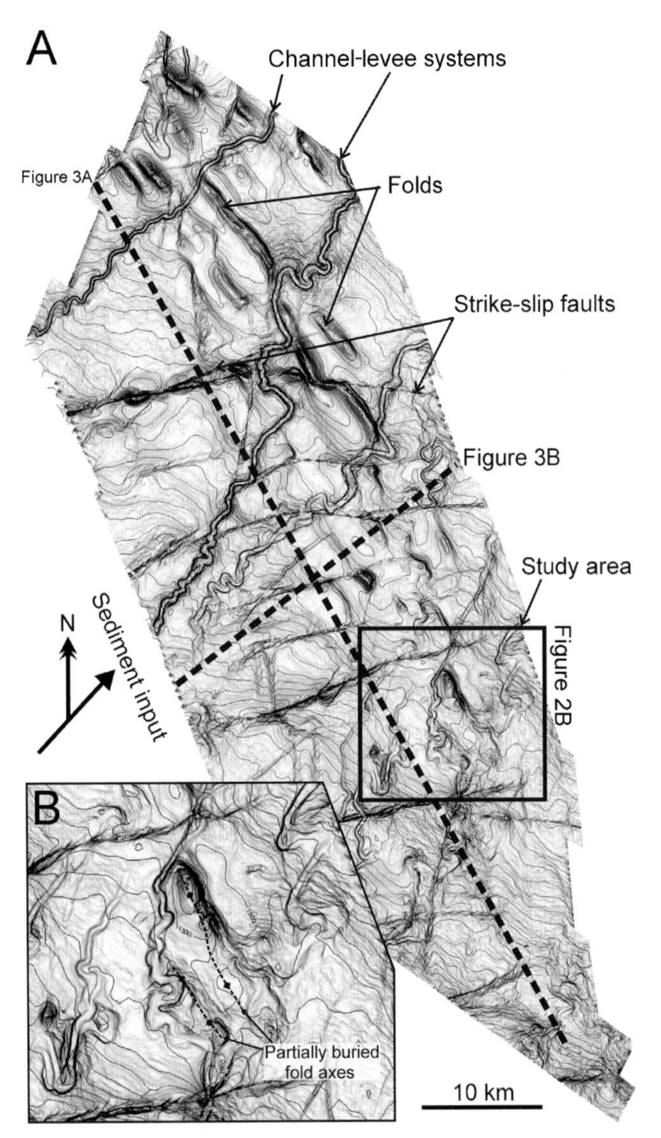

Figure 2. (A) A sea-floor dip (darker shades indicate increases in gradient) attribute map of the seismic survey area. This map is overlain by two-way traveltime contours spaced at 10 ms intervals. Submarine channel systems sourced from the Nile to the southwest are affected by strike-slip fault structures and a series of folds whose strike is perpendicular to the submarine channel flow direction. These folds become progressively buried toward the southeast because of the increasing thickness of the synkinematic interval (unit PM3; see Figure 3). (B) The detailed study area. Note the irregular sea-floor expression of the two folds and the partially buried submarine channel-levee system. 10 km (6.2 mi).

may represent an angular unconformity between the lower, deformed, Messinian evaporites and the post-Messinian sedimentary section (Bertoni and Cartwright, 2007). However, this does not appear to be the case in the Gal C area, where the two sequences are seismically concordant (Figure 3).

Post-Messinian Sequence (5.2 Ma–Present)

The post-Messinian overburden can be subdivided into three intervals (units PM1, PM2, and PM3) based on seismic stratigraphic characteristics and the relationship to thin-skinned deformation throughout the study area (Figure 3). Units PM1 and PM2 comprise the prekinematic section and were deposited before thin-skinned deformation (Figure 3). The lowermost unit PM1 consists of locally continuous high-amplitude reflections whereas the overlying unit PM2 consists of more continuous, but generally lower amplitude, reflections. Both units PM1 and PM2 contain channel-levee systems that cause localized thickness variations (Folkman and Mart, 2008). Only the uppermost interval of unit PM2 lacks channel-levee systems and consists of low-amplitude continuous parallel reflections (Figure 3).

The PM3 interval is the focus of this study because this sequence comprises the synkinematic section recording post-Messinian compression and sedimentation. The base of this unit records the first appearance of onlap caused by folding throughout the study area, with the exception of several folds where the first occurrence of onlap is observed approximately 50 ms below the base of this unit. Over the scale of the survey area, unit PM3 increases in thickness toward the southeast. At the kilometer scale, sedimentation within unit PM3 is strongly controlled by numerous northwest–southeast-trending folds (Figures 2, 3). Unit PM3 shows systematic thinning of sedimentary packages across fold crests, accompanied by thickening into backlimbs and forelimbs of folds (Figure 3). This geometry is typical of growth sequences associated with sedimentation coeval with uplift (e.g., Cartwright, 1989; Suppe et al., 1992; Burbank and Verges, 1994; Masaferro et al., 1999; Morley, 2009). The stratigraphic architecture of unit PM3 consists of a vertical sequence of channel-levee systems, many of which are strongly affected by post-Messinian deformation (Figure 2). These channel-levee systems typically have widths of 500 m (1640 ft) and depths of up to 60 m (197 ft). Many are highly sinuous (average value, 1.51), although variations can occur when channels are confined and diverted around the northwest–southeast-trending folds (Clark and Cartwright, 2009).

Post-Messinian Thin-skinned Deformation

The Gal C survey area is located in the contractional domain of a gravity-driven linked extensional-compressional system in the eastern Mediterranean (Gradmann et al., 2005; Cartwright and Jackson, 2008). Thin-skinned compressional structures result from gravity-induced collapse of the Nile Cone above the ductile Messinian

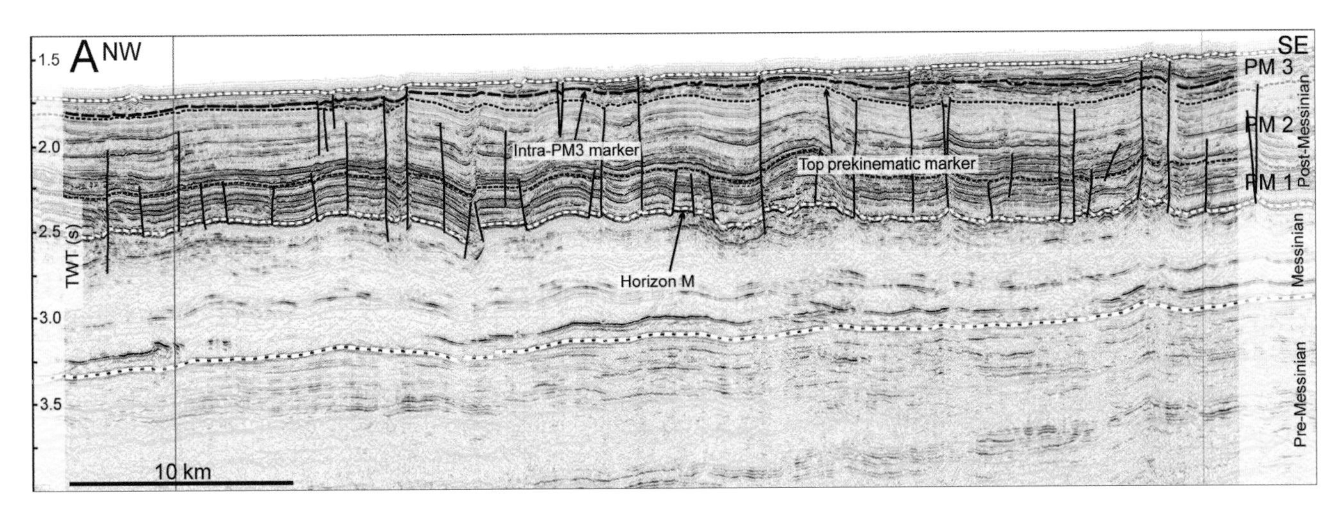

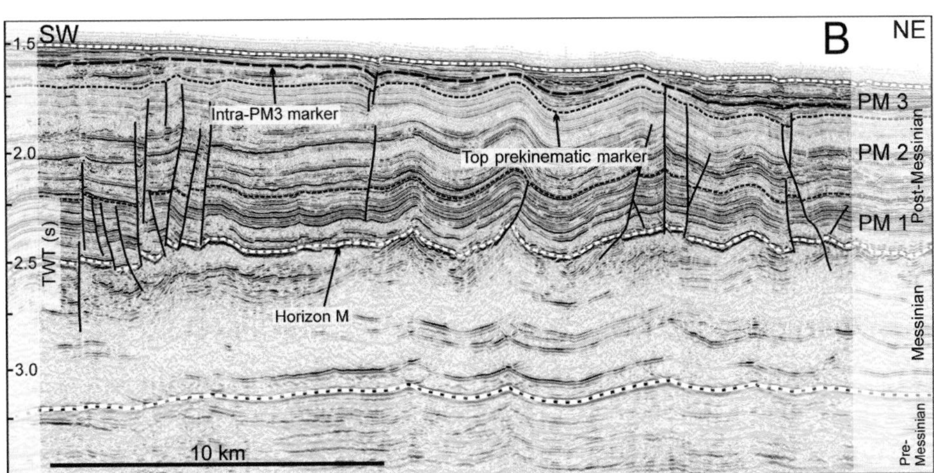

Figure 3. (A) A strike-oriented time seismic profile across the study area. This seismic section shows the principal seismic-stratigraphic units with the interval of interest for this study being unit PM3, which represents the synkinematic interval-associated with thin-skinned compression. This unit thickens toward the southeast of the survey area, resulting in more subdued fold relief. The horizon used to subdivide unit PM3 into upper and lower growth sequences is indicated as the intra-PM3 marker. Clearly visible in this seismic profile are numerous subvertical strike-slip faults that segment the post-Messinian overburden. (B) A dip-oriented time seismic profile on which several thrust faults ramping upward from the uppermost Messinian can be observed. These blind thrusts are associated with the development of overlying folds, the growth of which is recorded by the synkinematic interval unit PM3. TWT = two-way traveltime. 10 km (6.2 mi).

evaporite sequence (Gradmann et al., 2005; Loncke et al., 2006). This is combined with westward tilting and the thin-skinned collapse of the Levant Margin (Cartwright and Jackson, 2008). The variably oriented collapsing margins interact with the buttressing effect of the Eratosthenes seamount, resulting in changes in the orientation of the thrust, fold, and strike-slip structures that characterize this area (Loncke et al., 2006).

Deformation within the Gal C 3-D survey is characterized by northwest–southeast-striking thrust faults that verge northeastward (Figure 4). Thrusts detach within the uppermost Messinian and ramp upward into the overlying section, terminating below the sea floor toward the base of unit PM3 (Figure 3). Northwest of the survey area, a small number (~5) of these thrusts verge in the opposite direction direction towards the southwest (Figure 4). All thrust faults are associated with overlying fold development, which exert a key control on sea-floor bathymetry (Figure 2). Maximum displacement on the thrusts is up to 500 m (~1654 ft), with values of maximum structural relief above the top prekinematic horizon in the range of 300 to 400 m (984–1312 ft). The thrusts generally dip at 30 to 40°, with a downward decrease in dip toward the basal detachment. Detachment levels occur at two or three distinct horizons within the multilayered Messinian evaporite sequence (Cartwright and Jackson, 2008).

Fold structural style typically varies along strike from that of symmetrical detachment folds to thrusted

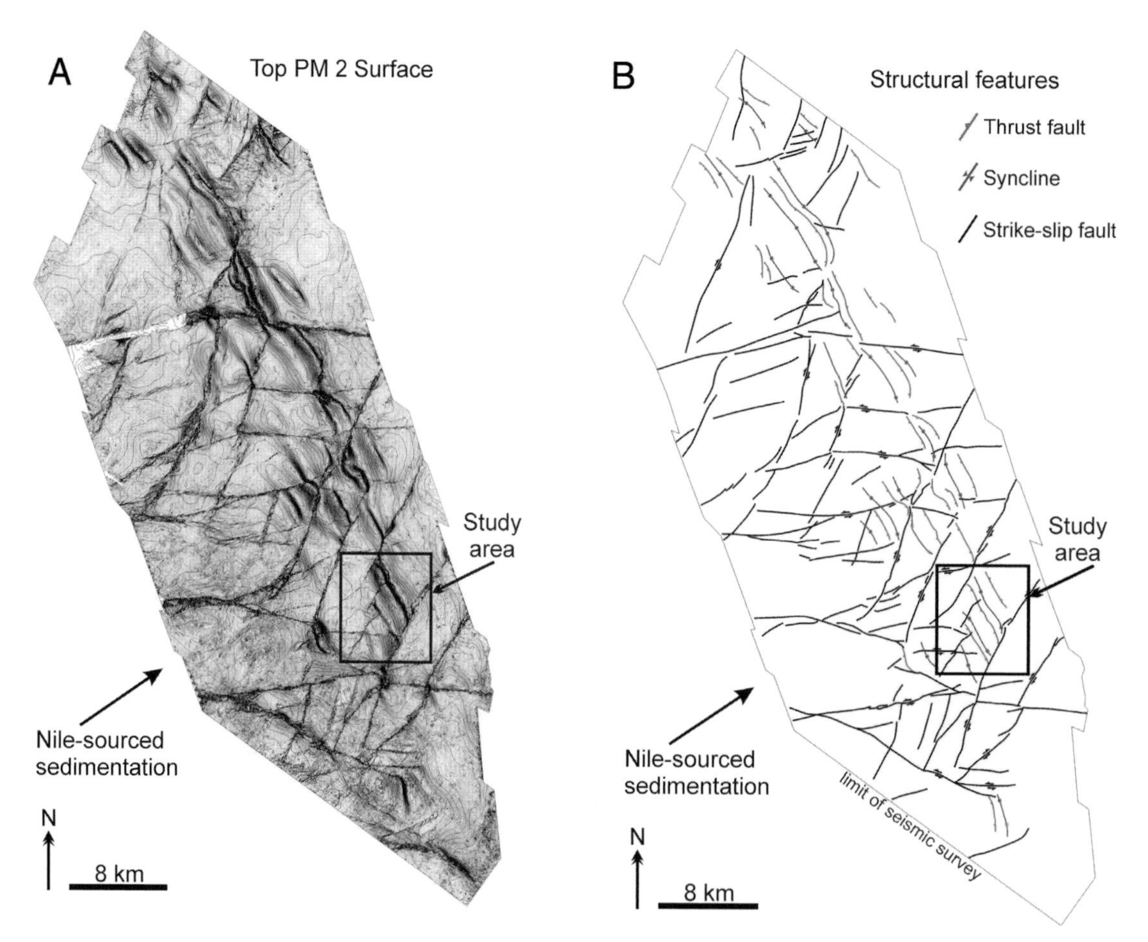

Figure 4. (A) A dip attribute map of the top PM2 horizon (see Figure 3 for the location). This map represents the top of the prekinematic sequence throughout the survey area and more clearly shows the southeast–northwest-oriented fold belt and the conjugate set of strike-slip faults that segment the fold belt. (B) A map showing the distribution of thrusts underlying the folds, as well as associated synclines developed within the hanging walls and footwalls. 8 km (5 mi).

detachment folds and thrust propagation folds (cf. Mitra, 2002). Variations in fold style along strike are linked to the amount of shortening accommodated by each structure (Higgins et al., 2007, 2009). Many of the folds are associated with parallel synclines within the backlimbs and forelimbs (Figures 3B, 4).

A conjugate set of strike-slip faults detaching within the uppermost Messinian sequence is present throughout the survey area (Figures 3, 4). These strike-slip faults are linear in map view and show two dominant trends: an east–west-trending set with a sinistral sense of displacement and a dextral northeast–southwest-trending set (Figure 4). The angle bisecting this conjugate set of strike-slip faults is orthogonal to the orientation of the thrust and fold axes, suggesting a consistent northeastern direction for the maximum compressive stress throughout the deformation (Figure 4). The conjugate strike-slip faults compartmentalize compression and folding in the survey area, most notably in the southern part of the survey (Figure 4).

The displacement direction of the strike-slip faults is measured using the horizontal component of offset measured from channel-levee systems developed at the top Messinian level and within unit PM1. Typical strike-slip displacements are of the order of 200 to 500 m (656–1640 ft). Zones of local transtension and transpression occur where strike-slip fault segments link; these form push-up structures with a sea-floor relief of up to 90 m (295 ft) and also pull-apart structures that are up to 60 m (197 ft) deep (Figure 2).

METHODS

Within the 3-D seismic survey, a detailed study area was selected, showing two folds whose development was coeval with multiple submarine channel systems (Figure 5). To describe the interactions between deformation and sedimentation in this area, basic descriptions of growth sequence geometry linked to mapped isochron patterns were used. This was then combined with measurements of along-strike fold uplift and detailed observations from two channel-levee systems developed within each growth sequence interval. These methods are discussed in more detail below.

The internal geometry of a growth sequence can be described using relationships such as onlap, overlap, and offlap (Burbank and Verges, 1994). When using two-dimensional (2-D) profiles, these relationships provide information about the relative rates of sedimentation versus uplift (Cartwright, 1989; Puigdefàbregas et al., 1992; Suppe et al., 1992; Burbank and Verges, 1994). These observations can be combined with isochron maps that subdivide the growth sequence to aid interpretation of the 3-D growth sequence geometry (cf. Salvini and Storti, 2002). Isochron maps are useful for interpreting 3-D growth sequence geometry as they reveal patterns of thinning over fold crests and changes in accommodation within the fold limbs. Note, however, that these observations could be significantly affected by erosion. In this lower fan setting, however, erosion of the folds and growth strata is limited to the localized effects of incision by channel-levee systems and does not significantly affect the results presented here.

To quantify 3-D variations in folding during sedimentation, along-strike measurements of fold uplift and growth sequence expansion factor were taken at 100 m (328 ft) intervals for each of the two folds within the detailed study area. These measurements were taken for both the lower and upper growth sequence intervals (Figures 2, 5).

Fold uplift measurements were taken using a similar method to that described by Masaferro et al. (1999) and Poblet et al. (2004) for measuring crestal structural relief of folds (Figure 6). This method is based on the assumption that the horizons were initially deposited horizontally (Masaferro et al., 1999). Reference points for measuring stratigraphic thickness where taken between the fold crest and the lowest point of the adjacent forelimb syncline. Where an overlapping growth sequence is observed, the thickness difference of the package between the forelimb syncline and the fold crest is interpreted to represent the cumulative fold uplift that occurred over the time interval of growth sequence deposition (Figure 6) (see also Masaferro et al., 1999; Poblet et al., 2004). Overlapping growth sequence geometries imply that no sea-floor relief developed during folding because of the relative rate of sedimentation outpacing uplift (Burbank and Verges, 1994). For the case of onlapping growth sequences, the minimum value of crestal structural relief is equal to the thickness recorded in the growth sequence adjacent to the fold limbs (Figure 6). The onlap terminations onto the fold limbs imply that the sedimentation rate was not sufficient to bury the growing fold, resulting in a positive sea floor relief during folding.

The expansion factor was measured using the methodology defined for growth sequences associated with normal faults (Thorsen, 1963). For folds, the expansion factor E is defined as $E = Z'/Z$, where Z' is the thickness of the growth sequence within the footwall syncline and Z is the thickness over the fold crest (Figure 6).

The methods presented above assume that no hemipelagic sedimentation occurs during deposition of the growth sequence. The primary mode of deposition within the growth sequence is from gravity currents involved in the formation of channel-levee systems, which are ubiquitous throughout the PM3 synkinematic interval. The abundance of channel-levee systems throughout this interval and the absence of a clear seismic package with the characteristics of a hemipelagic drape deposit (cf. Cartwright, 1989) indicate that the amount of hemipelagic deposition is likely to be low compared with deposition from gravity currents.

RESULTS AND OBSERVATIONS

This section begins by presenting an overview of the key features within the study area, followed by a description of the growth sequence isochron maps and examples of submarine channel systems from the lower and upper growth intervals. Measurements of along-strike fold uplift and growth sequence expansion factor are then described. A summary of key observations is presented at the end of the section in Table 1.

Key Features within the Study Area

Three folds are developed within the detailed study area (Figure 5), with folds 1 (3.8 km [2.4 mi] in length) and 2 (5.6 km [3.5 mi] in length) described in detail. Fold 3 is not expressed as clearly as folds 1 and 2 and is not discussed further. These folds are separated by forelimb and backlimb synclines (synclines 1–3, Figure 5). The folds and adjacent synclines terminate laterally against strike-slip faults, which strongly compartmentalize compression throughout the survey area and are associated with a nonuniform distribution of structural relief along strike (Figure 5; Figure 4). These lateral terminations are associated with zones of increased structural relief observable on the top prekinematic marker horizon (Figure 5). The lateral terminations against the strike-slip faults are recognizable as steep subvertical scarps on the top prekinematic surface, with a total structural relief of up to 120 m (394 ft).

Assuming that the first occurrence of onlap onto the fold limbs marks the onset of fold growth, it can be observed that the lateral terminations of folds 1 and 2 initiated at the base of unit PM3 (Figure 5C, D). However, systematic onlap at the base of the growth sequence is not consistently observed along strike, particularly in the case of fold 1, where the central zone displays overlap. This indicates that the zones of increased structural relief adjacent to the segmenting strike-slip faults record the earliest onset of folding.

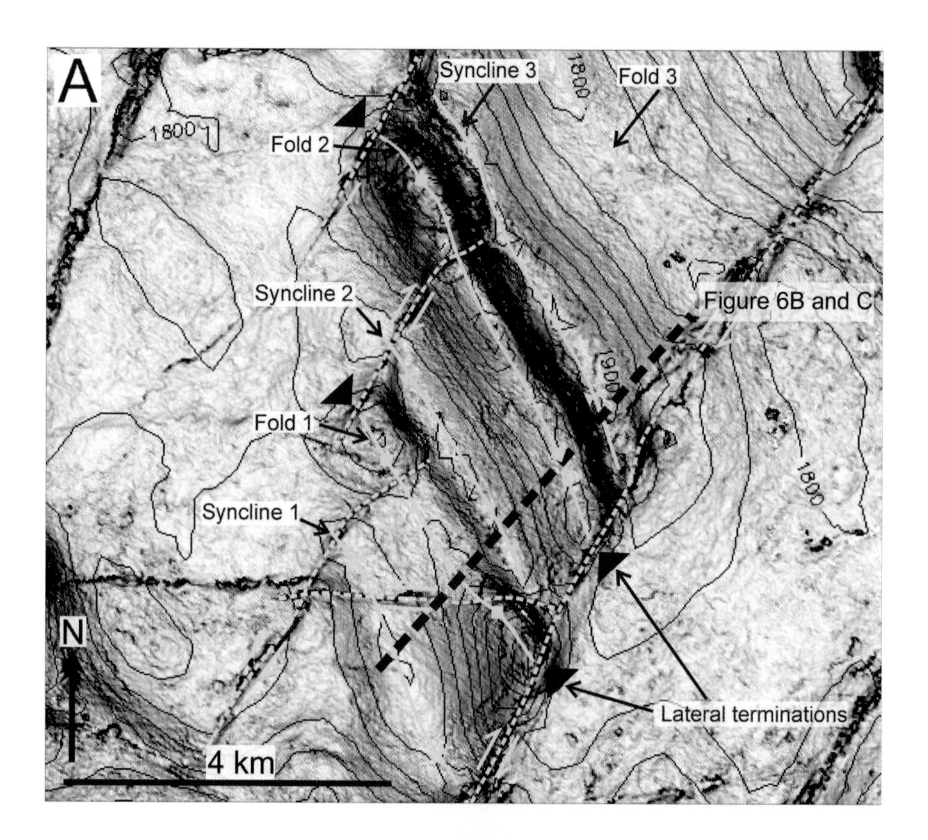

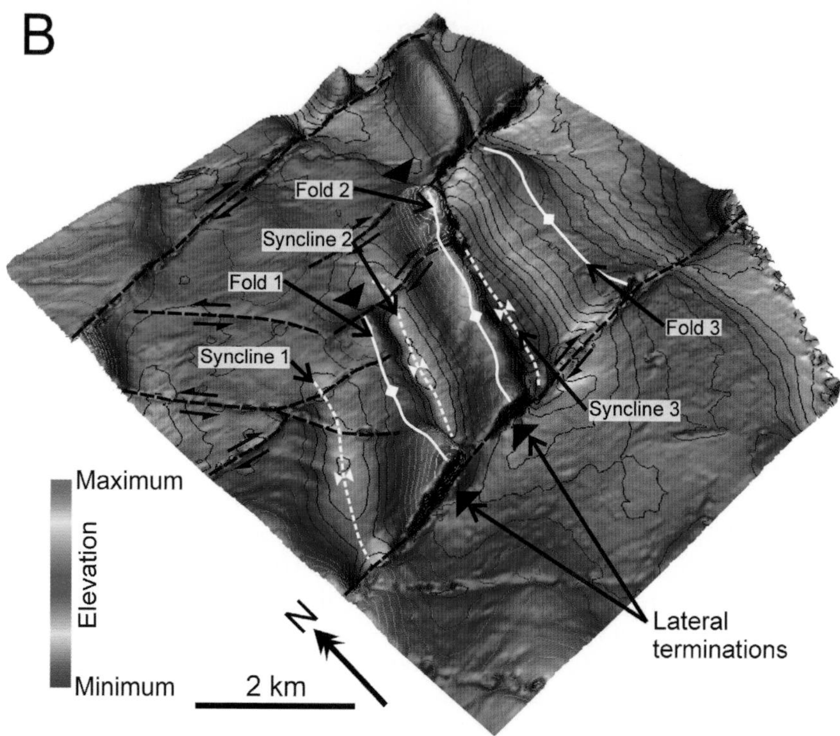

Figure 5. Nomenclature used within the results section of this study. (A) A dip attribute map of the top PM2 surface of the detailed study area. The folds are indicated, as are the synclines adjacent to them. Triangles mark the positions of lateral terminations of the folds against strike-slip faults. (B) A three-dimensional (3-D) surface of the detailed study area generated from the top surface of the pre-growth sequence. This surface illustrates the nonuniform lateral distribution of uplift along strike of each fold. (C) An uninterpreted seismic profile showing folds 1 and 2 and the growth sequence. (D) A line drawing interpretation of the same seismic profile showing the subdivision of the PM3 interval into lower and upper growth sequences. Channel-levee systems described from the lower and upper growth sequences are termed the lower and upper channel-levee systems, respectively. 2 km (1.2 mi).

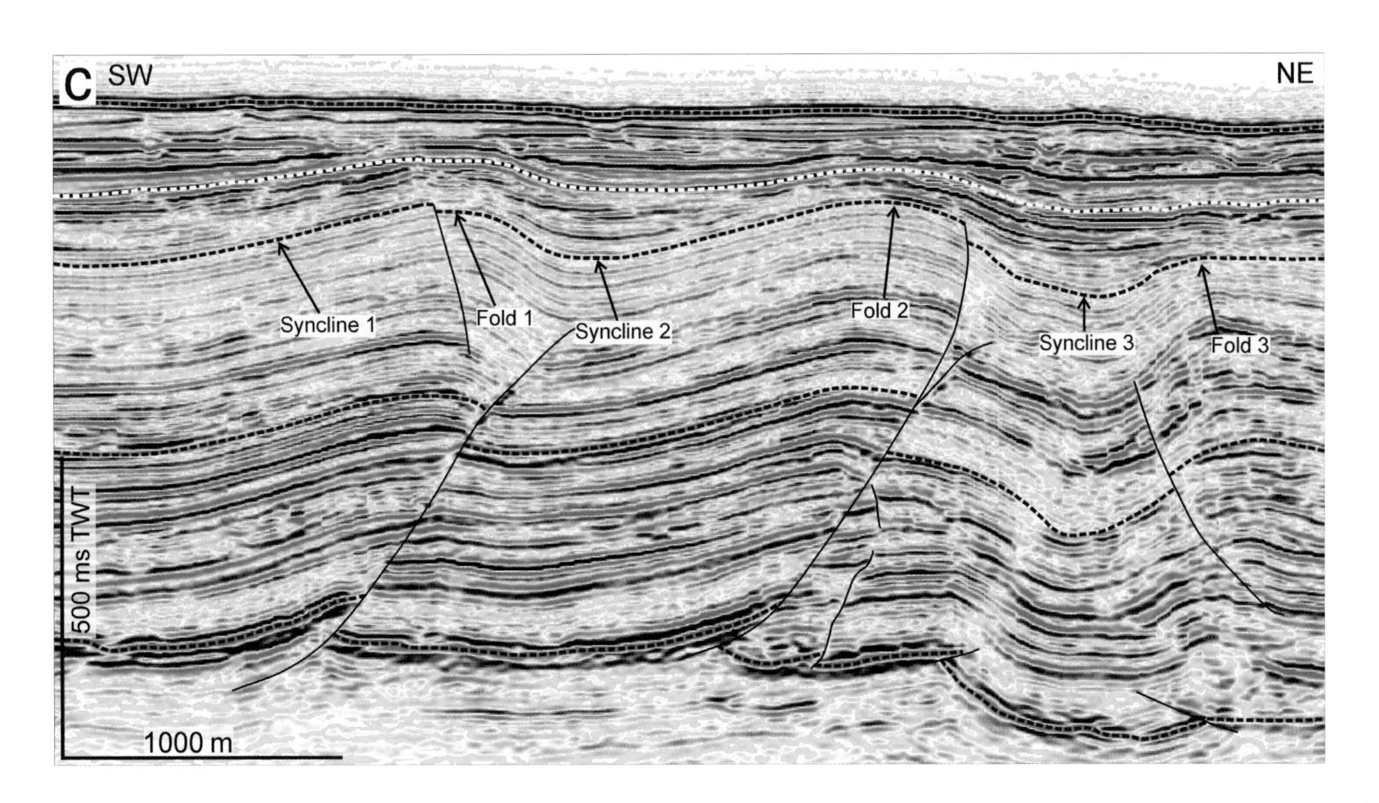

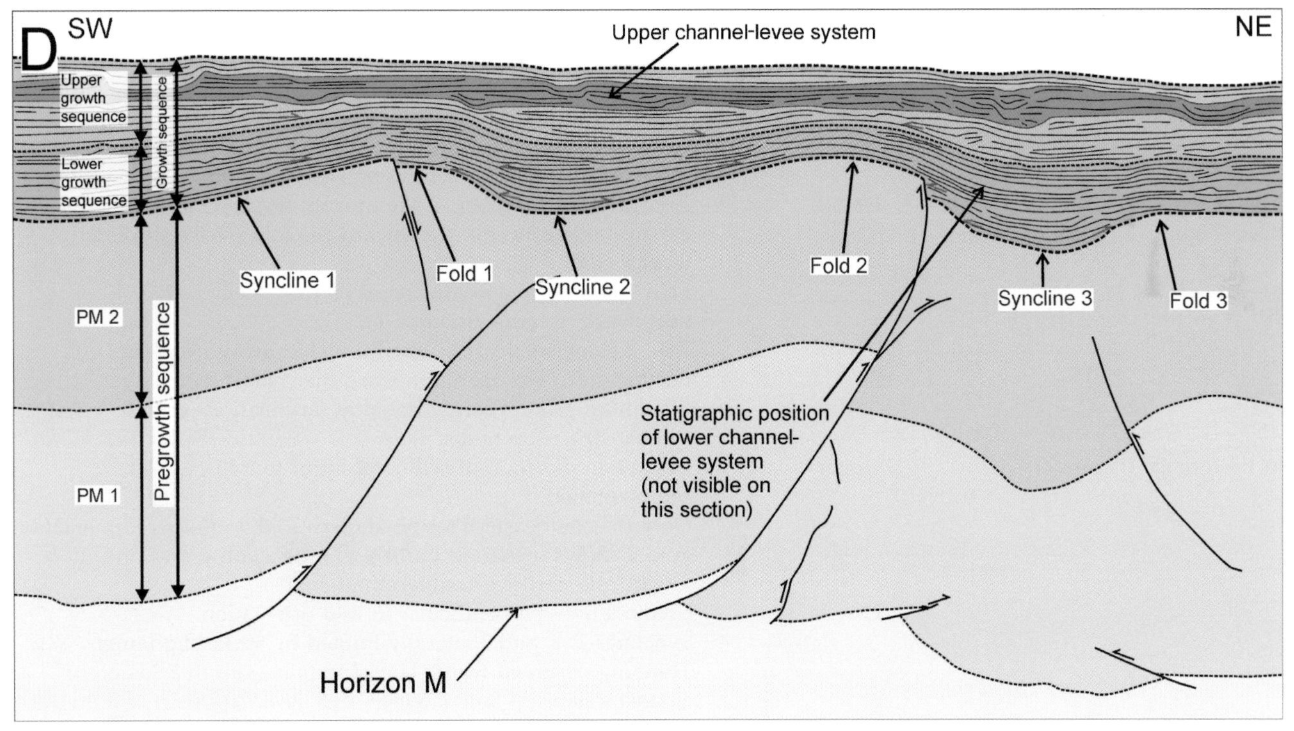

Figure 5. (cont.).

Growth Sequence Isochron Maps

The isochron of the lower growth sequence illustrates the nonuniform development of uplift for folds 1 and 2 (Figure 7A). Maxima in uplift are concentrated toward the lateral terminations against strike-slip faults, resulting in increased stratal thinning at these locations (Figure 7A). This effect is particularly clear for fold 1,

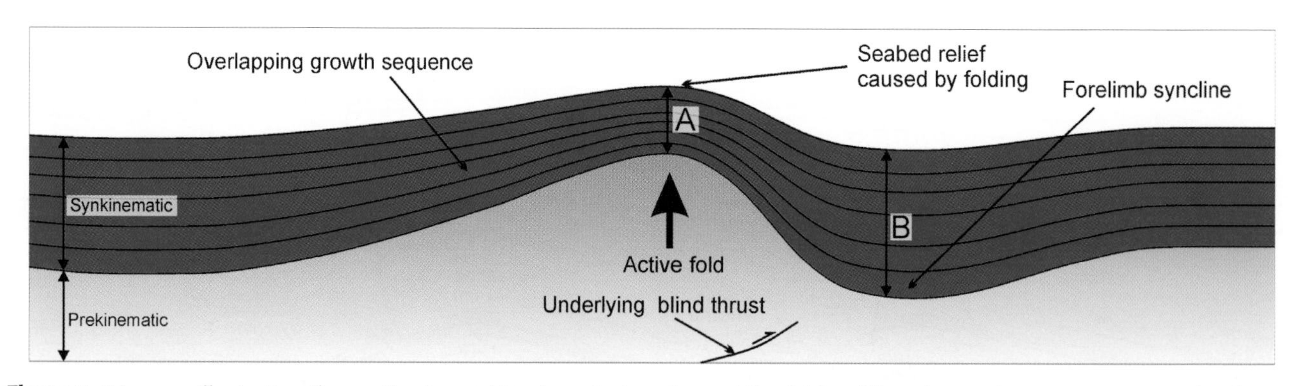

Figure 6. Diagram illustrating the methods used to characterize along-strike fold uplift and growth sequence expansion factor. Measurements at 100 m (328 ft) intervals along strike of both folds 1 and 2 were used to calculate cumulative fold uplift over deposition of both the lower and upper growth sequences (B–A). These measurements were converted to depth using an assumed interval velocity of 2000 ms^{-1}.

Table 1. Summary of structural and stratigraphic observations.

Results	Growth Sequence	Key Observations
Fold uplift measurements (Figure 10)	Upper	Both folds show asymmetric relief development. Fold 1 shows maximum relief recorded towards the southeast, with little additional relief developed at the northwestern termination. Fold 2 shows a shift in the location of maximum relief development toward the central zone of the fold.
	Lower	Folds 1 and 2 show maximum relief developed at the fold–strike-slip fault terminations, with decreased relief relative to these areas across the central part of each fold.
Growth sequence expansion factor (Figure 11)	Upper	Expansion factor along strike mirrors the observations above: expansion is concentrated toward the fold–strike-slip fault terminations. Fold 1 shows little expansion at its previously active northwestern termination. Fold 2 shows increased expansion shift away from the northwestern termination toward the central zone.
	Lower	Maximum growth concentrated at terminations of folds 1 and 2 against strike-slip faults.
Isochron maps (Figure 7)	Upper	Change in direction of sediment input to fold-parallel sedimentation. Only the southeastern termination of fold 1 affects sedimentation. Fold 2 shows increased uplift within its central area and toward the northwestern termination.
	Lower	Channel flow perpendicular to fold orientation. Synclines 1, 2, and 3 laterally limited by strike-slip faults. Thinning observed where lateral terminations of folds occur against strike-slip faults, which also limit deposition within the synclines.
Submarine channel system development (Figures 8, 9)	Upper	The upper channel-levee system is strongly affected by the relief of both folds, resulting in avulsion and spatial variations in sinuosity and lateral migration during diversion.
	Lower	Lower channel-levee system flows perpendicular to fold 2 and crosses the fold crest without diversion. Planform morphology of this channel shows a clear decrease in sinuosity as the fold crest is crossed.

where the lateral terminations act as barriers to sedimentation whereas the central area of the fold allows bypass into syncline 2 (Figure 7A). Synclines 1, 2, and 3 are limited laterally by the effects of the bounding strike-slip faults, particularly toward the southeast of the study area. Stratigraphic thinning across the zones of increased uplift correspond to increased onlap at the base of the growth sequence, typically followed by overlap (Figure 7A). Sedimentation within the lower growth sequence consists of a series of channel-levee systems that flow from southwest to northeast (see next section).

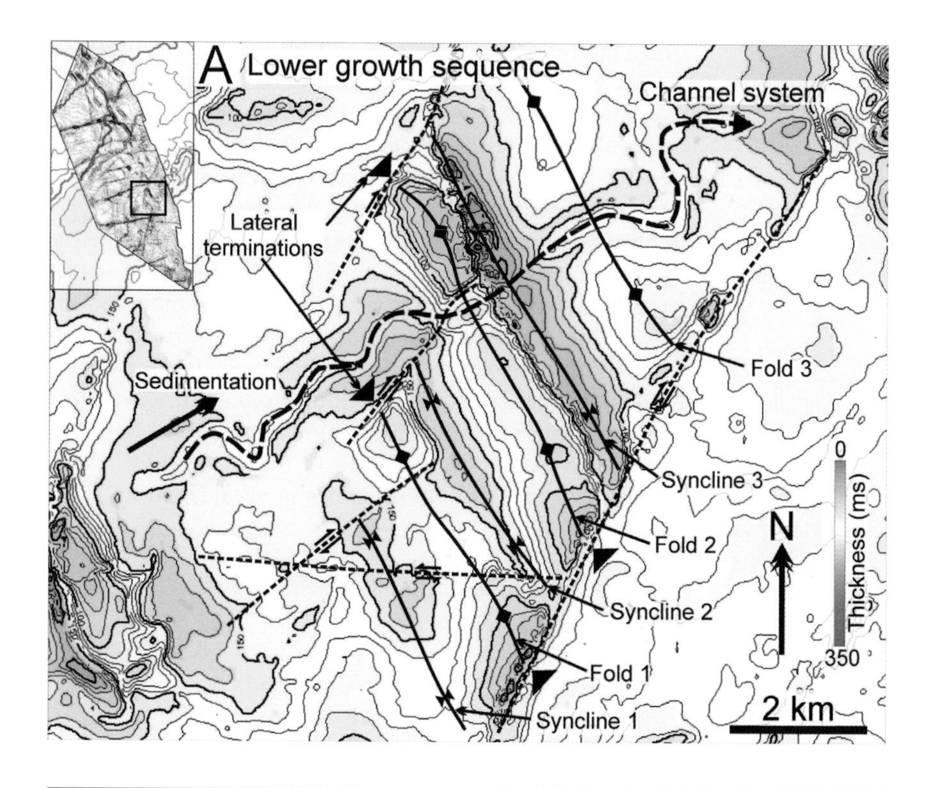

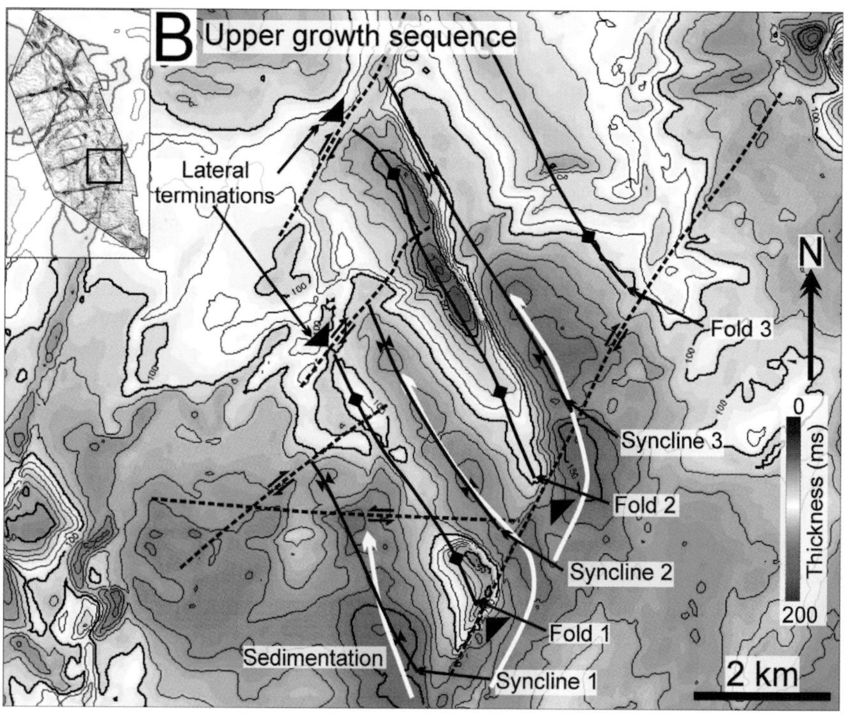

Figure 7. (A) Isochron map of the lower growth sequence, contours spaced at 25 ms intervals. The uplift along folds 1 and 2 is distributed unevenly along strike and exerts a strong control on sedimentation that thins variably across each fold. (B) Isochron map of the upper growth sequence, contours spaced at 25 ms intervals. White arrows indicate the sedimentary input directions for this interval. Clear differences occur in the distribution of uplift along folds 1 and 2 within this interval, most notably the increase in uplift within the central and northwestern areas of fold 2. 2 km (1.2 mi).

The upper growth sequence isochron shows a change in the locations of maximum uplift along folds 1 and 2 (Figure 7B). Fold 1 shows continued uplift at the southeast termination, whereas only minor stratal thinning occurs over the rest of this structure when compared with the lower growth sequence interval. For fold 2, uplift is concentrated toward the central and northwestern parts of this fold, with the southeastern termination becoming inactive (Figure 7B). Synclines 1, 2, and 3 are not limited by the strike-slip faults over this interval and become more open toward the southeast compared with the lower growth sequence (Figure 7B). Sedimentation within the upper growth sequence shows a shift in input direction toward the southeast, again with channel-levee systems being the dominant mechanism that infills the structurally created accommodation (Figure 7B).

Response of Individual Channel-levee Systems to Folding

Significant differences occur in the development of submarine channel systems within the lower and upper growth sequences. Examples of two-channel systems are presented below to show the close relationship between nonuniform fold development and patterns of channel evolution.

Lower Channel-levee System

A small-scale channel-levee system developed at the top of the lower growth sequence is observed to cross the crest of fold 2 without being diverted by any emergent structural relief (Figure 8). This pattern of channel development is consistent with the occurrence of overlap within the lower growth sequence in this area (see Figure 8). Overlap of strata within a growth sequence indicates that little relief caused by folding developed on the depositional surface, such that a growing fold will not present an obstacle to sedimentation (cf. Burbank et al., 1996). However, seismic attributes of this channel show that a straightening in channel course occurs as it crosses the crest of fold 2 and syncline 2 (Figure 8). Patterns of channel development can be strongly affected by variations in slope and accommodation space (cf. Pirmez et al., 2000). In this example, the distinct change in channel planform on crossing the fold crest may indicate a subtle structural control caused by underlying gradient changes, although observations of overlapping growth sequence geometry suggest that little positive relief was expressed on the sea floor at this time.

Upper Channel-levee System

In contrast to the previous example, a channel-levee system developed in the upper growth sequence is significantly affected by the increased uplift that affects fold 2 over this interval (Figure 9). This channel system enters the study area from the southwest and is affected by an avulsion immediately downstream of the southeast lateral termination of fold 1 (Figure 9). The western avulsed channel segment is diverted around fold 2, where it displays obvious lateral migration packages and abandoned meander loops that decrease in scale as the channel is diverted around the northwestern lateral termination of fold 2 (Figure 9). The eastern channel segment displays a uniformly high sinuosity and is not as strongly affected by sea-floor relief as the western segment. The horizon amplitude map shown in Figure 9 also reveals areas of low amplitude on the backlimbs of folds 1 and 2, indicating a lack of deposition of coarse-grained material from channel-related density currents. This observation is consistent with the previously observed increase in uplift for these areas over the upper growth sequence interval (Figure 9).

Along-strike Measurements of Fold Uplift and Expansion Factor

Along-strike uplift profiles for both folds show an irregular distribution where maximum uplift tends to be concentrated toward intersections between folds and strike-slip faults, which limit fold lateral development. This is particularly clear for fold 1, where the central area of the fold experienced relatively uniform uplift during growth sequence deposition (Figure 10A). The northwestern lateral termination of fold 1 shows increased uplift over both the lower and upper growth sequence intervals (Figure 10A). This is to the southeast lateral termination that shows increased uplift over the lower growth sequence interval only.

Fold 2 shows a zone of increased uplift spanning 2 km (1.2 mi) at the northwestern lateral termination during deposition of the lower growth sequence (Figure 10B). The remainder of the fold exhibits relatively uniform values of uplift along strike, although uplift is typically twice that observed for fold 1 over the same interval. The upper growth sequence records a shift in the development of relief away from the northwestern lateral termination toward the central area of the fold (Figure 10B). This is consistent with increased fold relief resulting in channel diversion and increased stratal thinning toward the northwest (Figures 7B, 10B).

Along-strike expansion factor measurements for both folds mirror the observations above (Figure 11).

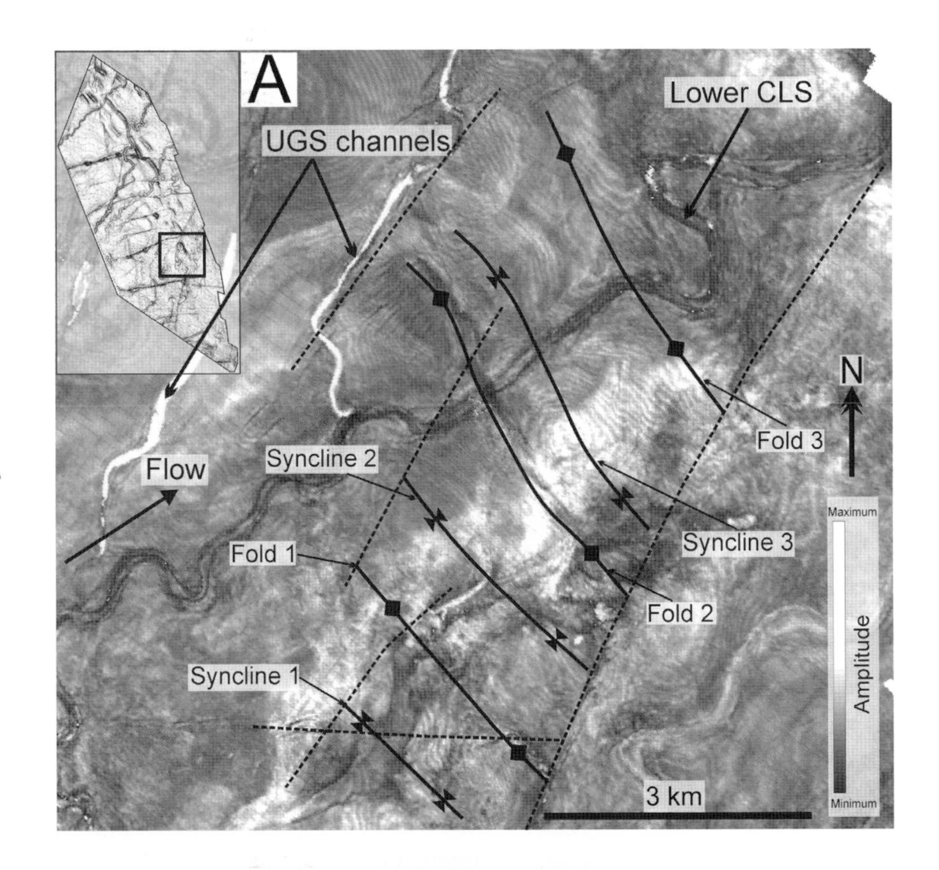

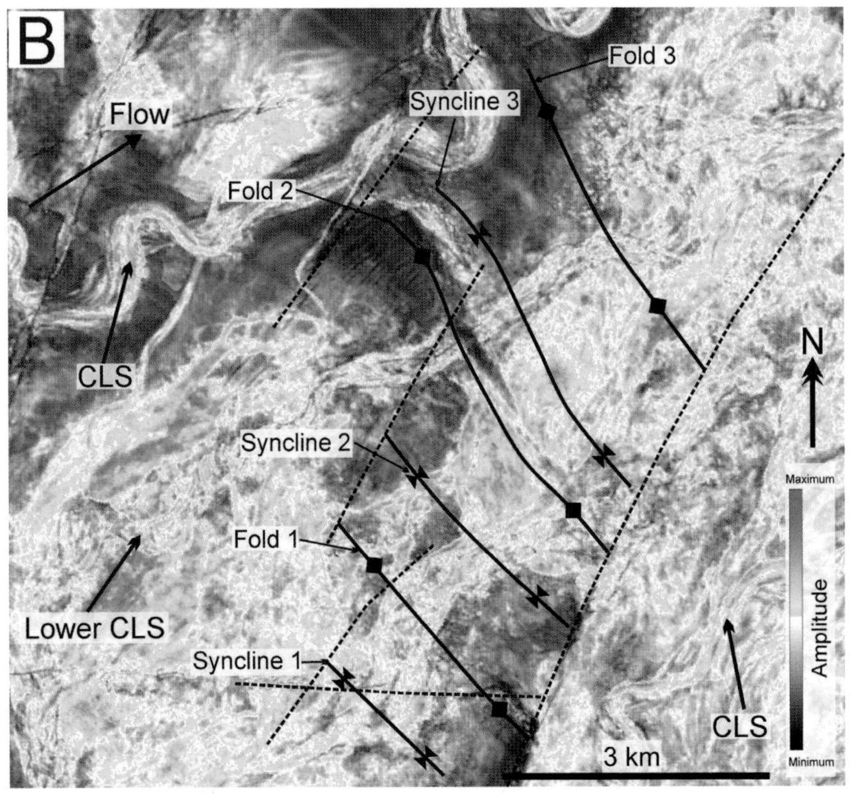

Figure 8. (A) A seismic amplitude map of the top lower growth sequence surface. The upper growth sequence (UGS) channel-levee systems (CLSs) have partially incised down to this stratigraphic level but are not part of the lower growth sequence. The lower CLS crosses the crest of fold 1, where a decrease in channel sinuosity is observed. Otherwise, the course of this channel is unaffected by the development of folds 1 and 2. (B) Root mean square amplitude extracted over a 60 ms window below the top surface of the lower growth sequence. The map shows two CLSs that also flow perpendicular to the strike of folds 1 and 2 but do not cross the fold crests. 3 km (1.8 mi).

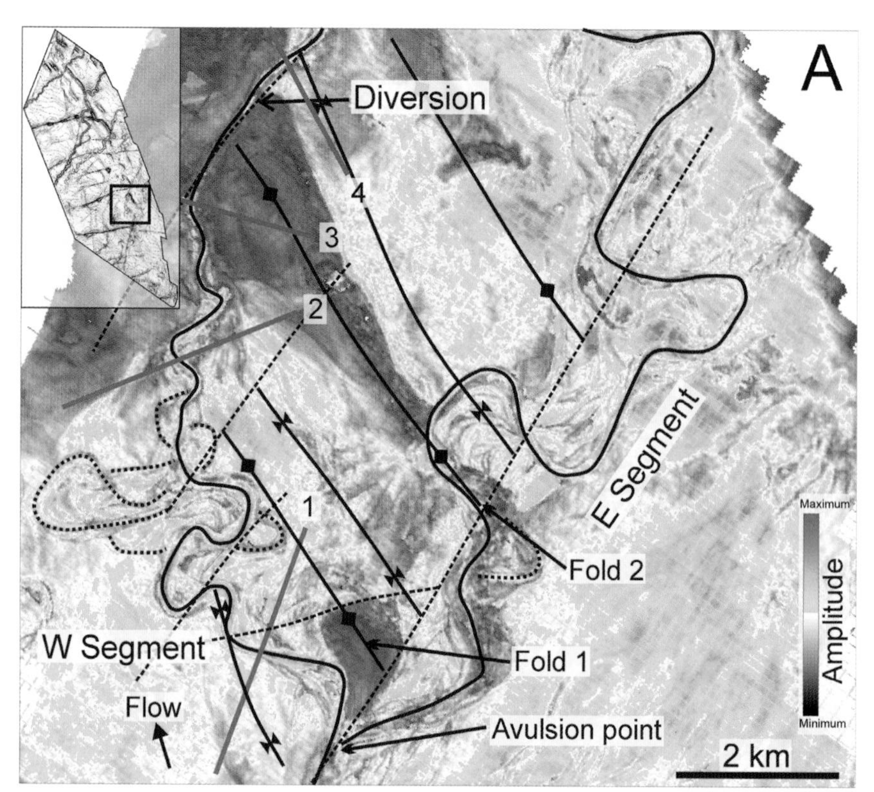

Figure 9. (A) Amplitude map at the base of the upper channel-levee system (CLS). This channel splits at the southeastern lateral termination of fold 1 into western and eastern segments, of which the western segment is younger. The western segment shows spatial variability in channel lateral migration and in the formation of cutoff loops (dashed lines). (B) A series of interpreted seismic profiles showing the variation in channel morphology during diversion around fold 2. 2 km (1.2 mi).

The increase in growth sequence expansion is clearly expressed at the lateral terminations of fold 1, and these measurements more clearly show the decrease in uplift of the northwestern lateral termination of this fold during development of the upper growth sequence (Figure 11A). The shift in the locus of uplift toward the central area of fold 2 during the upper growth sequence is also clearly shown, as is a decrease in uplift toward the southeast over the same stratigraphic interval (Figure 11B).

Summary of Results and Observations

Table 1 summarizes the results and observations presented above.

DISCUSSION

The observations presented above show that 3-D uplift and syncline development through time strongly controls sedimentation in the study area. The structural controls on sedimentation are observed at the scale of growth sequences and also at the scale of individual channel-levee systems preserved within them. The first part of this discussion aims to link observations of growth sequence geometry with patterns of submarine channel development to enable better predictions of channel development and potential reservoir distribution within a growth sequence. The second section then discusses some key factors that affect deep-water growth sequence geometry in three dimensions. We then present some implications of this work for reservoir distribution in deep-water fold belt settings, followed by a conceptual model to link the structural and stratigraphic observations from this study.

Growth Sequence Geometry and Submarine Channel Development

In subaerial fold belts, the link between growth sequence geometry and the pattern of sedimentation within them is well established (Burbank and Verges, 1994; Burbank et al., 1996). Overlapping growth sequences result in fold-perpendicular drainage because of sedimentation outpacing uplift during fold growth. Onlapping growth sequences result in fold-parallel drainage patterns because of an increased relative rate of uplift compared with sedimentation. A similar relationship also applies to deep-water fold and thrust belts where submarine channels and turbidity currents comprise the primary sedimentary components of the growth sequence (Huyghe et al., 2004; Morley and

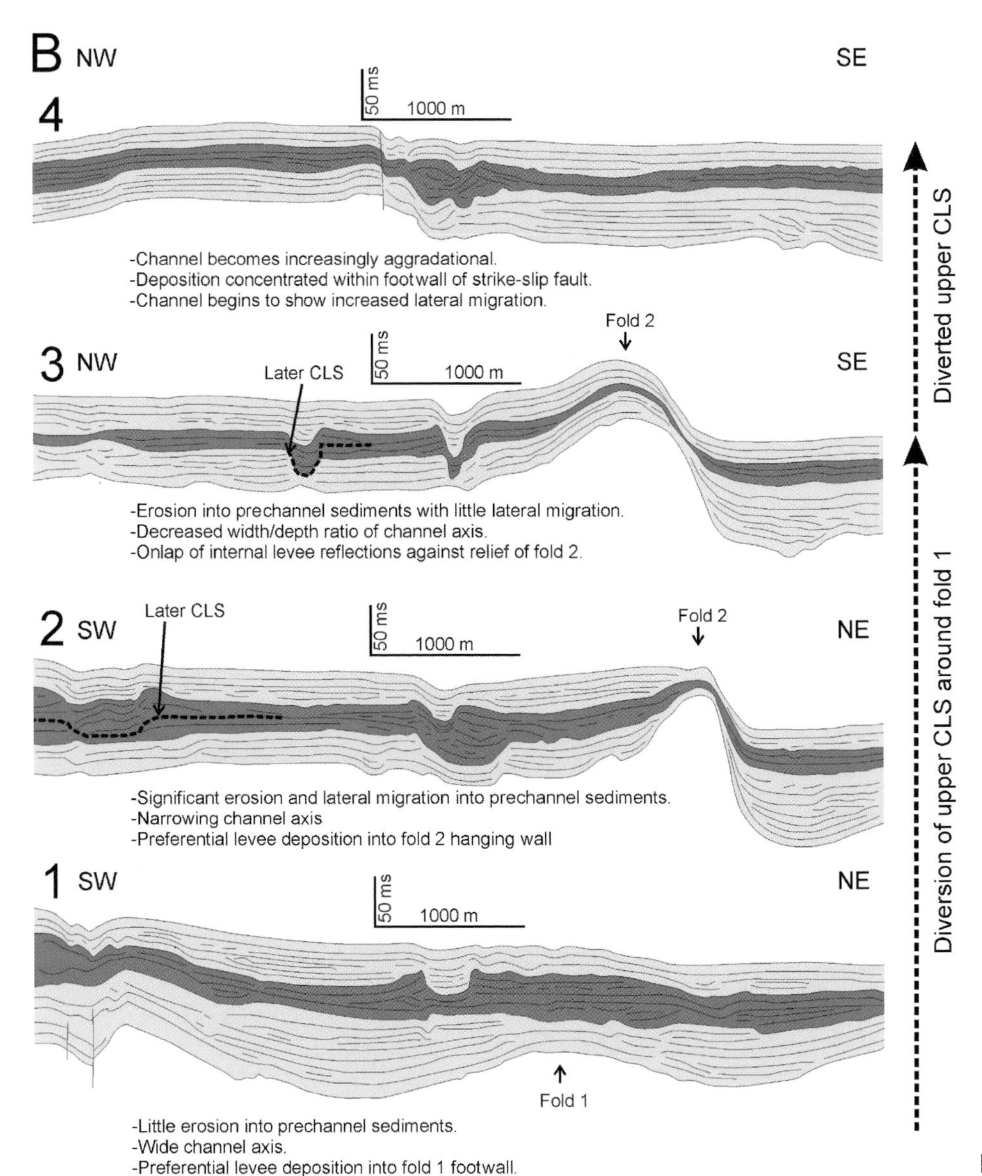

Figure 9. (cont.).

Leong, 2008). Despite these general relationships between growth sequence geometry and drainage patterns, the channel systems described in this study exhibit more subtle variations in morphology within a particular growth sequence style, described in more detail below.

The lower growth sequence shows an overlapping geometry, and the lower channel-levee system contained within it is orientated perpendicular to the strike of fold 2, flowing over the fold crest (Figure 8). The overlapping growth sequence geometry indicates that very little relief (at seismic resolution) would have been present at the sea floor during this depositional interval (Figure 7A). However, the decrease in sinuosity of the lower channel-levee system across the crest of fold 2 implies that folding over time increased the local slope gradient, resulting in a straighter channel course in this area compared with the higher sinuosity reaches upstream and downstream (Figure 8). The structural control on channel sinuosity is well recognized (e.g., Pirmez et al., 2000), and this example shows that channel morphology can be significantly affected even where overlap suggests a minor effect of topography on sedimentation.

The upper growth sequence submarine channel system shows strong onlap onto folds 1 and 2, with the western channel segment diverted around fold 2 (Figure 9). The zone of increased sinuosity and meander loop development is located above the northwestern part of fold 1, which was mostly inactive in this area at this time (Figure 11). Final diversion of the upper channel-levee system around the northwestern lateral

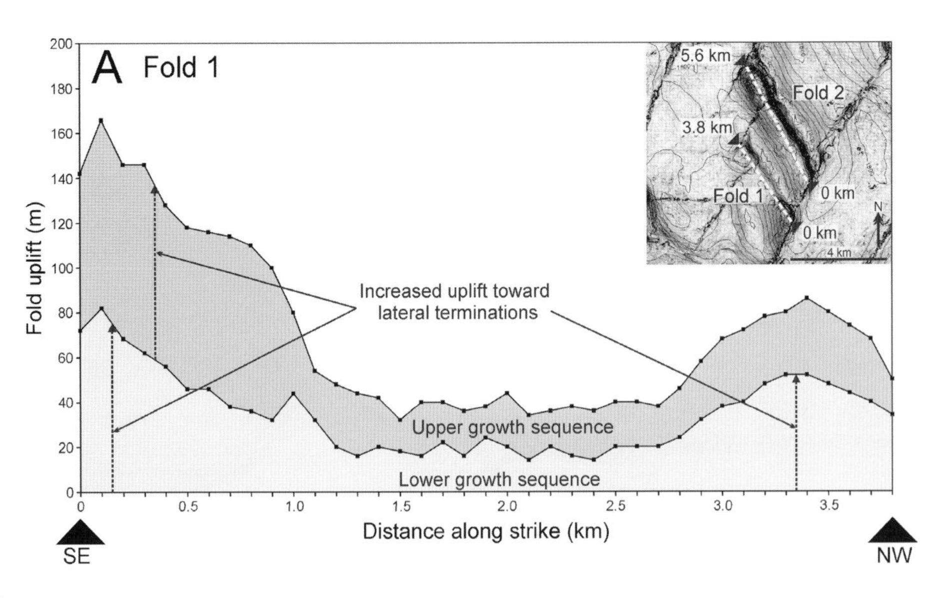

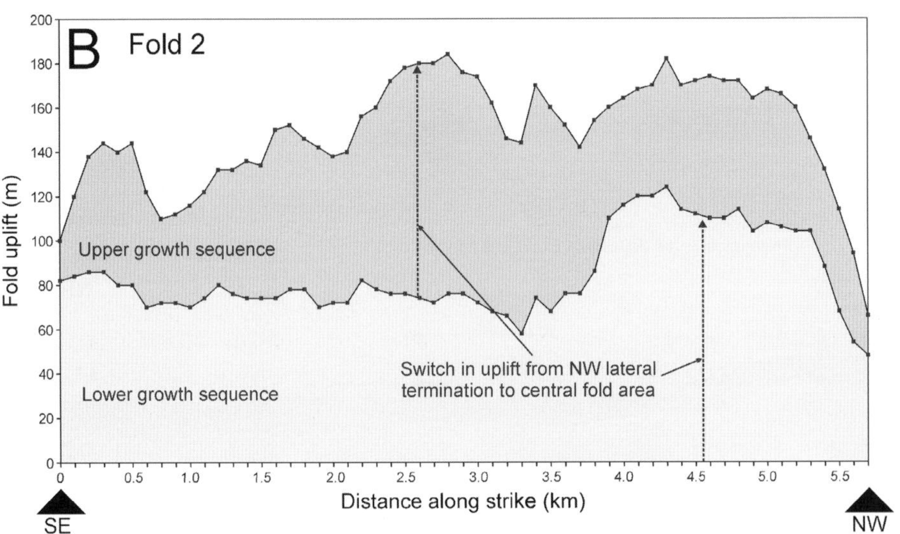

Figure 10. Graphs showing the along-strike distribution of uplift for folds 1 (A) and 2 (B). The inset map shows both folds with lateral terminations marked by black triangles. (A) Both lower and upper growth sequences display areas of increased uplift at the lateral terminations. Zones of increased uplift observed at the lateral terminations are indicated by dashed arrows. (B) A relatively uniform distribution of uplift along-strike over the lower growth sequence for fold 2, with the exception of the northwestern termination that shows a greatly increased uplift over the same interval. The upper growth sequence shows a shift in the area of maximum uplift toward the central part of the fold.

termination of fold 2 results in a straight channel course, with limited levee development. This example shows that channel diversion in an onlapping growth sequence can result in significant spatial variations in channel morphology, particularly where diversion occurs around multiple structures with irregular uplift histories.

Although general relationships apply between the overall growth sequence geometry and patterns of submarine channel development, this study shows that significant 3-D variations in channel morphology can occur, and this can be concealed within simple patterns of onlap and overlap. A 3-D understanding of these systems can therefore be useful when developing conceptual models to understand the evolution of submarine channel systems within growth sequences and to better predict the occurrence (or lack of) hydrocarbon reservoirs.

Factors Affecting Growth Sequence Geometry Along Strike

Along-strike fold amplitude is a key parameter determining 3-D growth sequence geometry (Salvini and Storti, 2002). The folds described above exhibit highly uneven uplift along strike over time, with maximum relief commonly concentrated at the lateral terminations of each fold against strike-slip faults (Figures 10, 11). This mode of fold growth is in contrast to systems where the maximum fold amplitude decreases away from the central area of the structure toward the lateral tip regions (Salvini and Storti, 2002; Higgins et al., 2009; Morley, 2009).

Over the lower growth sequence interval, both folds show uplift concentrated at the lateral terminations; this is clearly expressed on the isochron map and also by measurements of along-strike uplift and

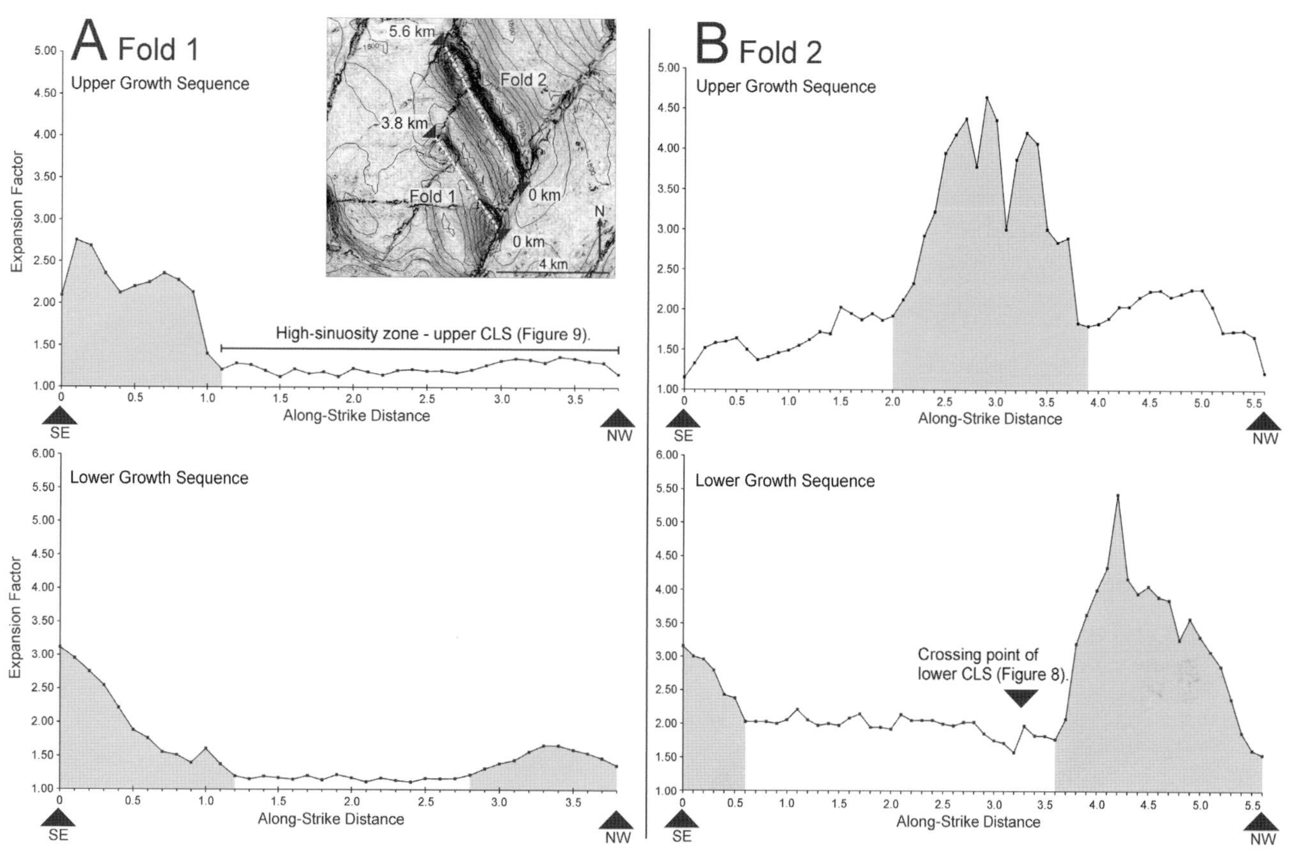

Figure 11. Graphs showing along-strike measurements of expansion factor for folds 1 (A) and 2 (B). Black triangles mark the positions of lateral terminations of the folds against bounding strike-slip faults. Shaded regions indicate zones of increased fold activity. Figure 9A shows graphs for fold 1. See text for details. CLS = channel-levee system.

growth sequence expansion factor (Figures 7, 10, 11). This distribution of uplift along-strike is important as it likely influenced the positioning of the lower channel-levee system where it crosses the crest of fold 2. Although the amount of fold uplift over the lower growth sequence interval was not sufficient to divert the channel path, the irregular along-strike uplift distribution still influences the actual flow path across the fold crest and also channel morphology (see previous section).

The upper growth sequence again shows highly irregular uplift along both structures, with fold 1 becoming mostly inactive apart from the southeastern lateral termination, where continued uplift results in an avulsion of the upper channel-levee system (Figure 9). The inactive part of fold 1 does not affect the planform geometry of the upper channel-levee system, which is highly sinuous in this area with a significant coverage of lateral migration deposits (Figure 9). In contrast, the southeastern lateral termination of fold 2 becomes less active, with maximum uplift switching to the central part of this fold (Figures 10, 11). The localized increase in uplift contributes to the diversion around fold 2, as shown by the western channel segment in Figure 9.

Implications

Previous studies have identified the general relationships between growth sequence geometries and sediment transport paths in fold belts (e.g., Burbank et al., 1996; Morley and Leong, 2008). The results presented here show the dramatic variations in submarine channel morphology that can occur within growth sequences that are controlled by complex 3-D fold evolution. These include spatial variations in channel sinuosity over short distances. This is a key factor for determining reservoir presence within a fold if continued deformation occurs. An example of this effect is shown by the lower channel-levee system (Figure 8). The decrease in sinuosity across the fold crest could potentially lead to a loss in reservoir connectivity because of localized erosion and bypass in this area, producing two separate sand bodies on either fold limb.

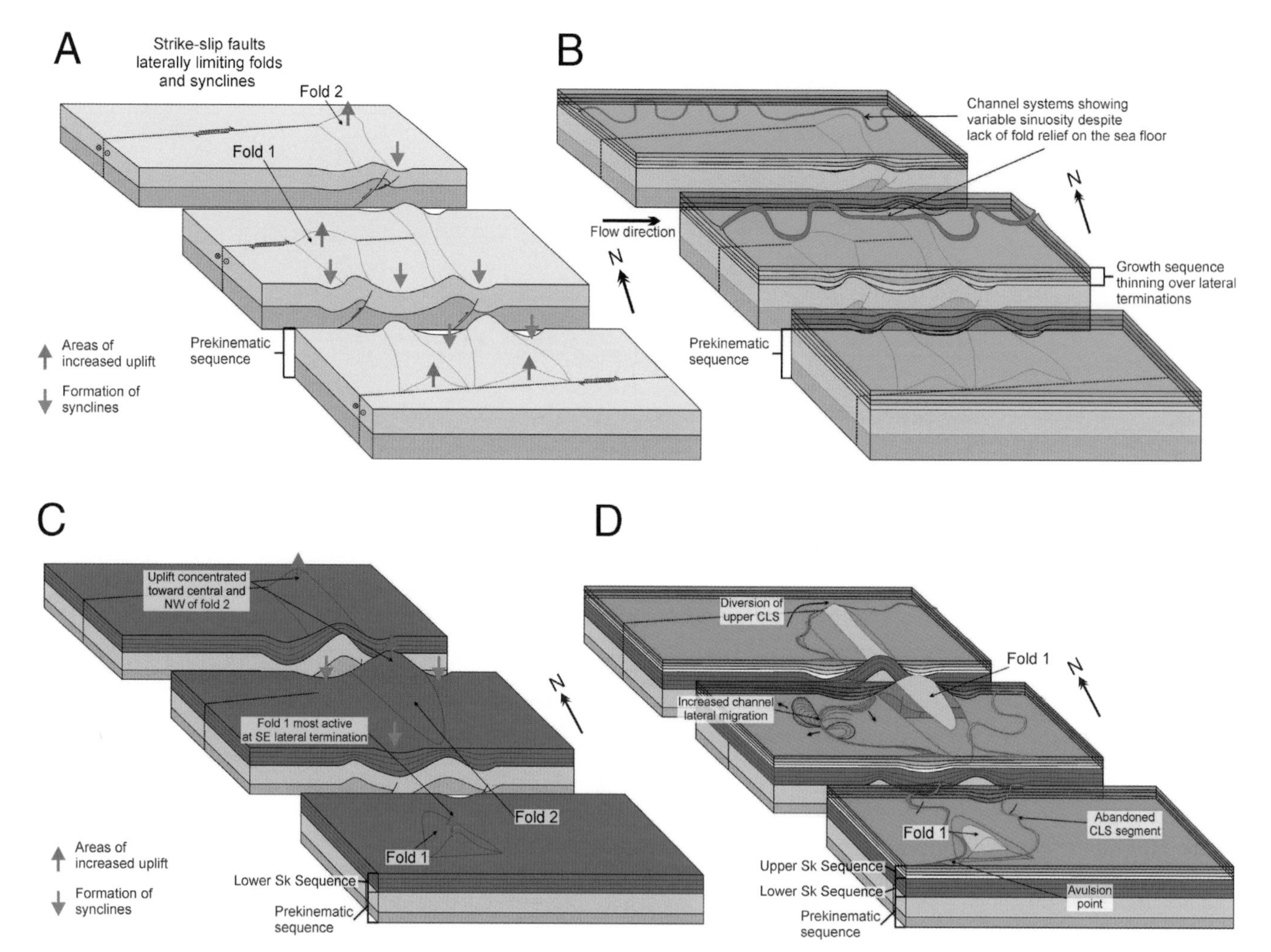

Figure 12. Series of block diagrams illustrating the evolution of folding and growth sequence sedimentation within the study area. Asymmetric development of folds and synclines within the lower growth sequence (A) and the effects on the development of submarine channel systems (B). (C, D) The change in structural development within the upper growth sequence where only the southeastern lateral termination of fold 1 remains active and uplift along fold 2 is concentrated toward the central area of the fold (C). This results in diversion of sedimentation around the emergent relief over this interval (D). See text for details. CLS = channel-levee system; SK = Synkinematic.

In contrast, the diversion of the upper channel-levee system around fold 2 will result in lateral accretion deposits on the upstream fold limb only, although they may be more areally extensive compared to the lower channel-levee system.

Summary: Conceptual Model of Deformation and Sedimentation

Fold uplift during the lower growth sequence was nonuniform and concentrated toward the lateral terminations of folds 1 and 2 (Figure 12A). The magnitude of uplift, from measurements of fold relief, was greater along fold 2 compared with fold 1. Syncline development, as well as folding, was strongly segmented by strike-slip faults over this interval (see also Figure 5).

In response to local gradients, channel-levee systems flowed perpendicular to the fold crest over this interval (Figure 12B). Although channel development is perpendicular to the fold crest, underlying deformation was still sufficient to affect channel sinuosity over the crestal region of fold 2. This was most likely a result of subtle gradient changes, causing the channel to readjust its planform geometry at this location.

The upper growth sequence is characterized by a highly skewed development of structural relief and resultant topography (Figure 12C). Uplift continued at the southeastern termination of fold 1 but discontinued at the northwestern lateral termination (Figure 12C).

Uplift along-strike of fold 2 was concentrated toward the northwestern termination and the central zones but decreased toward the southeast, giving fold 2 a southeastward plunge over this part of the growth sequence (Figure 12C). Growth sequence development over this interval is mostly unaffected by strike-slip faulting and also undergoes a change to fold-parallel sedimentation.

Development of the upper channel-levee system is strongly affected by the nonuniform development of fold relief (Figure 12D). Localized uplift at the southeastern termination of fold 1 results in channel avulsion, separating the eastern and western segments of the upper channel-levee system (Figure 12D). The eastern branch is not affected by fold 2 because of the decreased relief and the southeastward plunge, resulting in unobstructed channel flow. This is in contrast to the western branch, which is strongly affected by diversion around fold 2 and also exhibits marked changes in sinuosity and lateral migration.

CONCLUSIONS

1) This study shows that folding during growth sequence development can be associated with a nonuniform distribution of uplift along strike. Therefore, a fully 3-D approach is needed when characterizing these systems.
2) The focal points of uplift along thrust-related folds can vary both spatially and temporally.
3) Overlapping growth sequences are commonly interpreted to be a response to little or no sea-floor relief over an active structure. Despite the absence of sea-floor relief (at the seismic scale), submarine channel morphology can still be affected, showing a decrease in sinuosity over the growing fold.
4) Onlapping growth sequences imply ponding or diversion of sediments around the emergent fold relief, with the submarine channel systems in this study showing spatial variations in sinuosity and the development of cutoff meander loops.
5) Importance of basin geometry: The geometry of the backlimb and forelimb synclines affects local gradients, the amount of accommodation space, and the ability of these structures to act as depocenters through time.
6) Control of the evolving sea-floor relief on channel sinuosity: Depending on the rate of structural growth and the effects of this on sea-floor gradient, channel-levee system sinuosity can vary greatly over short (hundreds of meters) distances. This can result in preferential deposition of laterally accreted sand bodies, as in this case study.

ACKNOWLEDGMENTS

We thank the Israeli Ministry of National Infrastructures for the permission to use the 3-D seismic volume presented here for the purposes of scientific studies. We also thank the reviewers for their hard work in reviewing the earlier versions of this manuscript.

REFERENCES CITED

Babonneau, N., B. Savoye, M. Cremer, and B. Klein, 2002, Morphology and architecture of the present canyon and channel system of the Zaire deep-sea fan: Marine and Petroleum Geology, v. 19, p. 445–467, doi:10.1016/S0264-8172(02)00009-0.

Ben-Avraham, Z., D. Kempler, and A. Ginzburg, 1988, Plate convergence in the Cyprus Arc: Tectonophysics, v. 146, p. 231–240.

Ben-Avraham, Z., G. Tibor, A. F. Limonov, M. B. Leybov, M. K. Ivanov, M. Y. Tokarev, and J. M. Woodside, 1995, Structure and tectonics of the Eastern Cyprean Arc: Marine and Petroleum Geology, v. 12, p. 263–271, doi:10.1016/0264-8172(95)98379-J.

Ben-Gai, Y., Z. Ben-Avraham, B. Buchbinder, and C. G. St C. Kendall, 2005, Post-Messinian evolution of the southeastern Levant Basin based on two-dimensional stratigraphic simulation: Marine Geology, v. 221, p. 359–379, doi:10.1016/j.margeo.2005.03.003.

Bernal, A., and S. Hardy, 2002, Syntectonic sedimentation associated with three-dimensional fault-bend fold structures: A numerical approach: Journal of Structural Geology, v. 24, p. 609–635.

Bertoni, C., and J. A. Cartwright, 2005, 3-D seismic analysis of circular evaporite dissolution structures, eastern Mediterranean: Journal of the Geological Society (London), v. 162, p. 909–926, doi:10.1144/0016-764904-126.

Bertoni, C., and J. A. Cartwright, 2006, Controls on the basinwide architecture of late Miocene (Messinian) evaporites on the Levant Margin (eastern Mediterranean): Sedimentary Geology, v. 188–189, p. 93–114, doi:10.1016/j.sedgeo.2006.03.019.

Bertoni, C., and J. A. Cartwright, 2007, Major erosion at the end of the Messinian Salinity Crisis: Evidence from the Levant Basin, eastern Mediterranean: Basin Research, v. 19, p. 1–18, doi:10.1111/j.1365-2117.2006.00309.x.

Brun, J.-P., and X. Fort, 2004, Compressional salt tectonics (Angolan Margin): Tectonophysics, v. 382, p. 129–150, doi:10.1016/j.tecto.2003.11.014.

Buchbinder, B., and E. Zilberman, 1997, Sequence stratigraphy of Miocene–Pliocene carbonate siliclastic shelf deposits in the eastern Mediterranean margin (Israel): Effects of eustasy and tectonics: Sedimentary Geology, v. 112, p. 7–32.

Burbank, D. W., and J. Verges, 1994, Reconstruction of topography and related depositional systems during active thrusting: Journal of Geophysical Research, v. 99, p. 281–297, doi:10.1029/94JB00463.

Burbank, D. W., A. Meigs, and N. Brozovic, 1996, Interactions of growing folds and coeval depositional systems: Basin Research, v. 8, p. 199–223, doi:10.1046/j.1365-2117.1996.00181.x.

Cartwright, J. A., 1989, The kinematics of inversion in the Danish Central Graben: Geological Society (London) Special Publication 44, p. 153–175.

Cartwright, J. A., and M. P. A. Jackson, 2008, Initiation of gravitational collapse of an evaporitic basin margin: The Messinian saline giant, Levant Basin, eastern Mediterranean: Geological Society of America Bulletin, v. 120, p. 399–413, doi:10.1130/B26081X.1.

Cita, M. B., and W. B. F. Ryan, eds., 1978, Messinian erosional surfaces in the Mediterranean: Marine Geology, v. 27, 366 p.

Clark, I. R., and J. A. Cartwright, 2009, Interactions between submarine channel systems and deformation in deep-water fold belts: Examples from the Levant Basin, eastern Mediterranean Sea: Marine and Petroleum Geology, v. 26, p. 1465–1482, doi:10.1016/j.marpetgeo.2009.05.004.

Damuth, J. E., 1994, Neogene gravity tectonics and depositional processes on the deep Niger Delta continental margin: Marine and Petroleum Geology, v. 11, no. 3, p. 320–346, doi:10.1016/0264-8172(94)90053-1.

Demyttenaere, R., J. P. Tromp, A. Ibrahim, P. Allman-Ward, and T. Meckel, 2000, Brunei deep-water exploration: From sea-floor images and shallow seismic analogs to depositional models in a slope turbidite setting: Gulf Coast Section SEPM Foundation 20th Annual Research Conference Deep-Water Reservoirs of the World, December 3–6, 2000, p. 304–317.

Druckman, Y., B. Buchbinder, G. M. Martinotti, R. Siman Tov, and P. Aharon, 1995, The buried Afiq Canyon (eastern Mediterranean, Israel): A case study of a Tertiary submarine canyon exposed in late Messinian times: Marine Geology, v. 123, p. 167–185, doi:10.1016/0025-3227(94)00127-7.

Eyal, Y., 1996, Stress field fluctuations along the Dead Sea Rift since the middle Miocene: Tectonics, v. 15, p. 157–170, doi:10.1029/95TC02619.

Folkman, Y., and Y. Mart, 2008, Newly recognized eastern extension of the Nile deep-sea fan: Geological Society of America Bulletin, v. 36, p. 939–942, doi:10.1130/G24995A.1.

Frey-Martinez, J., J. A. Cartwright, and B. Hall, 2005, 3-D seismic interpretation of slump complexes: Examples from the continental margin of Israel: Basin Research, v. 17, p. 83–108, doi:10.1111/j.1365-2117.2005.00255.x.

Garfunkel, Z., 1998, Constraints on the origin and history of the Eastern Mediterranean Basin: Tectonophysics, v. 391, p. 5–35, doi:10.1016/S0040-1951(98)00176-0.

Garfunkel, Z., and G. Almagor, 1987, Active salt dome development in the Levant Basin, southeast Mediterranean, *in* I. Lerche and, J. O'Brien, eds., Dynamical geology of salt and related structures: London, Academic Press, p. 263–300.

Garfunkel, Z., and B. Derin, 1984, Permian–early Mesozoic tectonism and continental margin formation in Israel and its implications for the history of the eastern Mediterranean, *in* J. E. Dixon and A. H. F. Robertson, eds., The geological evolution of the eastern Mediterranean: Geological Society (London) Special Publication 17, p. 187–201.

Garziglia, S., S. Migeon, E. Ducassou, L. Loncke, and J. Mascle, 2008, Mass-transport deposits on the Rosetta province (NW Nile deep-sea turbidite system, Egyptian margin): Characteristics, distribution, and potential causal processes: Marine Geology, v. 250, p. 180–198.

Gradmann, S., C. Hubscher, Z. Ben-Avraham, D. Gajewski, and G. Netzeband, 2005, Salt tectonics off northern Israel: Marine and Petroleum Geology, v. 22, p. 597–611, doi:10.1016/j.marpetgeo.2005.02.001.

Higgins, S., R. J. Davies, and B. Clarke, 2007, Antithetic fault linkages in a deep-water fold and thrust belt: Journal of Structural Geology, v. 29, p. 1900–1914, doi:10.1016/j.jsg.2007.09.004.

Higgins, S., B. Clarke, R. J. Davies, and J. Cartwright, 2009, Internal geometry and growth history of a thrust-related anticline in a deep-water fold belt: Journal of Structural Geology, v. 31, p. 1597–1611, doi:10.1016/j.jsg.2009.07.006.

Hsu, K. J., L. Montadert, D. Bernoulli, M. B. Cita, A. Erickson, R. E. Garrison, R. B. Kidd, F. Melieres, C. Muller, and R. H. Wright, 1978, Initial reports of the Deep Sea Drilling Project, c42 (1): Washington, D. C., U.S. Government Printing Office, 1249 p.

Huyghe, P., M. Foata, E. Deville, and G. Mascle, 2004, Channel profiles through the active thrust front of the southern Barbados prism: Geology, v. 32, p. 429–432, doi:10.1130/G20000.1.

Loncke, L., V. Gaullier, J. Mascle, B. Vendeville, and L. Camera, 2006, The Nile deep-sea fan: An example of interacting sedimentation, salt tectonics, and inherited subsalt paleotopographic features: Marine and Petroleum Geology, v. 23, p. 297–315.

Mart, Y., and Y. Ben Gai, 1982, Some depositional patterns at continental margin of southeastern Mediterranean Sea: AAPG Bulletin, v. 66, p. 460–470.

Masaferro, J., J. Poblet, M. Bulnes, G. P. Eberli, T. Dixon, and K. R. McClay, 1999, Paleogene–Neogene/present-day growth folding in the Bahamian foreland of the Cuban fold and thrust belt: Journal of the Geological Society (London), v. 156, p. 617–631, doi:10.1144/gsjgs.156.3.0617.

Mitra, S., 2002, Fold-accommodation faults: AAPG Bulletin, v. 86, p. 671–693.

Morley, C. K., 2009, Growth of folds in a deep-water setting: Geosphere, v. 5, p. 59–89, doi:10.1130/GES00186.1.

Morley, C. K., and L. C. Leong, 2008, Evolution of deep-water synkinematic sedimentation in a piggy-back basin, determined from 3-D seismic reflection data: Geosphere, v. 4, p. 939–962, doi:10.1130/GES00148.1.

Netzeband, G. L., C. P. Hübscher, and D. Gajewski, 2006, The structural evolution of the Messinian evaporates in the Levantine Basin, Marine Geology, v. 230, p. 249–273.

Pirmez, C., R. T. Beaubouef, S. J. Friedmann, and D. C. Mohrig, 2000, Equilibrium profile and base level in submarine channels: Examples from late Pleistocene systems and implications for the architecture of deep-water reservoirs. Gulf Coast Section SEPM Foundation 20th Annual

Research Conference Deep-Water Reservoirs of the World, December 3–6, 2000, p. 782–805.
Poblet, J., K. McClay, F. Storti, and J. A. Munoz, 1997, Growth strata geometries associated to single-layer detachment fold: Journal of Structural Geology, v. 19, p. 369–381.
Poblet, J., M. Bulnes, K. McClay, and S. Hardy, 2004, Plots of crestal structural relief and fold area versus shortening: A graphical technique to unravel the kinematics of thrust-related folds, *in* McClay, ed., Thrust tectonics and hydrocarbon systems: AAPG Memoir 82, p. 372–399.
Puigdefàbregas, C., J. A. Muñoz, and J. Vergés, 1992, Thrusting and foreland basin evolution in the southern Pyrenees, *in* K. R. McClay, ed., Thrust tectonics: London, Chapman and Hall, p. 247–254.
Rowan, M. G., 1997, Three-dimensional geometry and evolution of a segmented detachment fold, Mississippi Fan fold belt, Gulf of Mexico: Journal of Structural Geology, v. 19, p. 463–480, doi:10.1016/S0191-8141(96)00098-3.
Salvini, F., and F. Storti, 2002, Three-dimensional architecture of growth strata associated to fault-bend, fault-propagation and decollement anticlines in nonerosional environments: Sedimentary Geology, v. 146, p. 57–73, doi:10.1016/S0037-0738(01)00166-X.
Samuel, A., B. Kneller, S. Raslan, A. Sharp, and C. Parsons, 2003, Prolific deep-marine slope channels of the Nile Delta, Egypt: AAPG Bulletin, v. 87, p. 541–560, doi:10.1306/1105021094.
Stow, D. A. V., and M. Mayall, 2000, Deep-water sedimentary systems: New models for the 21st century: Marine and Petroleum Geology, v. 17, p. 125–135, doi:10.1016/S0264-8172(99)00064-1.
Suppe, J., G. T. Chou, and S. C. Hook, 1992, Rates of folding and faulting determined from growth strata, *in* K. McClay, ed., Thrust tectonics: London, Chapman & Hall, p. 105–121.
Thorsen, C. E., 1963, Age of growth faulting in Southeast Louisiana: Transactions, Gulf Coast Association of Geological Societies, v. 13, p. 103–110.
Tibor, G., and Z. Ben-Avraham, 1992, Late Tertiary seismic facies and structures of the Levant passive margin off central Israel, eastern Mediterranean: Marine Geology, v. 105, p. 253–273, doi:10.1016/0025-3227(92)90192-K.
Vidal, N., J. Alvarez-Marron, and D. Klaeschen, 2000, Internal configuration of the Levantine basin from seismic reflection data (eastern Mediterranean): Earth and Planetary Science Letters, v. 180, p. 77–89, doi:10.1016/S0012-821X(00)00146-1.

15

Tari, Gabor, Haddou Jabour, Jim Molnar, David Valasek, and Mahmoud Zizi, 2012, Deep-water exploration in Atlantic Morocco: Where are the reservoirs?, *in* D. Gao, ed., Tectonics and sedimentation: Implications for petroleum systems: AAPG Memoir 100, p. 337–355.

Deep-water Exploration in Atlantic Morocco: Where Are the Reservoirs?

Gabor Tari
OMV Exploration and Production GmbH, Trabennstrasse 6-8, A-1020, Vienna, Austria (e-mail: gabor.tari@omv.com)

Haddou Jabour
Office National des Hydrocarbures et des Mines, 34 Ave. Al Fadila, 10050, Rabat, Morocco (e-mail: jabour@onhym.com)

Jim Molnar
707 Queensmill Ct., Houston, Texas, 77079, U.S.A. (e-mail: jmolnar214@aol.com)

David Valasek
Statoil, 2103 Citywest Blvd., Houston, Texas, 77042, U.S.A. (e-mail: davv@statoil.com)

Mahmoud Zizi
Ziz Geoconsulting, 47 Ave. Fall Oumeir Agdal, Rabat, Morocco (e-mail: mahmoudzizi@menara.ma)

ABSTRACT

The Moroccan salt basin remains one of the least explored of the west African salt basins. Although small producing fields in the onshore Essaouira Basin exist, so far, only subcommercial discoveries on the shelf have been made. During the last decade, three exploration wells were drilled in the deep water between Essaouira and Tarfaya in the central segment of the Atlantic margin of Morocco. These wells documented a general lack of reservoir-facies siliciclastics within the Cenozoic and Upper Cretaceous deep-water sequence.

Compared to the other segments of the Atlantic margin, the Moroccan margin has had a fairly complex structural history since the Middle Jurassic breakup between the North American–African plates involving several well-documented Alpine compressional periods and mountain building in the adjacent Atlas Mountains. In particular, as the Neogene–Holocene inversion, uplift, and erosion of the Atlas system is very well documented onshore, the apparent lack of Upper Cretaceous and Cenozoic reservoirs in the first deep-water wells came as a surprise. Therefore, reservoir presence, as the most critical risk factor in the deep-water exploration of the Moroccan Atlantic margin, needs to be better understood before new exploration wells can be drilled. Based on regional evidence, the Lower Cretaceous and

DOI:10.1306/13351560M1003141

the Jurassic sequences are interpreted to be significantly more sand prone in the deep-water areas than the overlying Upper Cretaceous and Cenozoic strata.

INTRODUCTION

The Atlas mountain belt of Morocco terminates against the present-day coastline of the Atlantic Ocean (Figure 1). Many people have studied the structural evolution of these prominent mountains (Frizon de Lamotte et al., 2009). With the exception of a few notable articles (e.g., Hafid, 2006; Hafid et al., 2006, 2008), much less is known about the westward offshore continuation of the Atlas Mountains.

Whereas numerous wells have been drilled on the shelf of the central segment of the Moroccan margin (Figure 2), to date, only three exploration wells drilled in the deep water have been observed. These include, in chronological order, the Shark B-1, Amber-1, and Rak-1 wells. None of these wells penetrated the Lower Cretaceous sequence.

The Shark B-1 well was drilled to a total depth of 3976 m (13,045 ft, of which 1824 m [5984 ft] was below the mudline) in Upper Albian sediments. The well showed that the predicted reservoir facies, inferred from the detailed interpretation of three-dimensional (3-D) reflection seismic data, were actually chert or redeposited limestone instead of anticipated turbiditic sandstones. The dominance of Mesozoic carbonates outcropping in the present-day Atlas Mountains was acknowledged as a risk factor for the reservoir quality of Cenozoic sequences, but reservoir presence was believed to be a smaller risk factor than, for example, the presence of effective source rocks.

Shortly after the abandonment of Shark B-1, the Amber-1 well was drilled in the neighboring Rimella Block to the south (Figure 2). This well reached a similar stratigraphic horizon (Upper Albian) as the Shark B-1 well, but the post-Albian section was nearly 1000 m (3280 ft) thicker. Like Shark B-1, the Amber-1 well did not have a significant reservoir-facies sequence. Importantly, the third deep-water exploration well, Rak-1 (Figure 2), produced similar results.

In this chapter, the findings from these wells are discussed with particular emphasis on the first deep-water exploration well, Shark B-1, drilled in 2004. The lack of reservoir-quality clastics down to the Albian stratigraphic level in the Shark B-1 and the subsequent Amber-1 and Rak-1 wells could be explained in a broader structural and stratigraphic context. The alternative models suggested here are testable and need to be considered in the next phase of deep-water exploration on offshore Atlantic Morocco.

REGIONAL SETTING

A line-drawing interpretation of a regional composite seismic profile (Figure 3) across the Ras Tafelney Plateau and the adjacent onshore Essaouira Basin (Figures 1, 2) illustrates the overall geometry of the margin. The most striking feature is the presence of salt both offshore and onshore. Indeed, the Tafelney segment of the Moroccan salt basin has the most complex allochthonous salt structures (Tari et al., 2000, 2003; Jabour and Tari, 2007).

Whereas relatively simple salt diapirs exist onshore (Hafid, 2000, 2006), the salt tectonic styles become gradually more advanced farther offshore. The diapirs are replaced by salt tongues that tend to coalesce basinward. Some of the resulting complex salt sheets form large canopies. The allochthonous salt features are dominantly mid-Cenozoic; therefore, most of the subsalt stratigraphy involves pre-Cenozoic strata.

The salt itself is sourced from the latest Triassic to the earliest Jurassic autochthonous salt level that was deposited during the latest synrift period in the evolution of the margin (Tari et al., 2003). For the map-view distribution of salt along the margin and a more detailed discussion of the salt tectonics, the reader is referred to Hafid (2000, 2006), Hafid et al. (2008), Tari et al. (2000, 2003), and Jabour and Tari (2007).

The Moroccan continental margin is the result of the successful opening of the Central Atlantic Basin during the Early Jurassic. Whereas the Triassic rift-related structural features are fairly well mapped onshore (Medina, 1988, 1995; Piqué et al., 1998; Hafid, 2000, 2006; Le Roy and Piqué, 2001; Hafid et al., 2006), the offshore rift architecture is not well understood because of the masking by the overlying salt features and/or by the prominent postrift Jurassic to Lower Cretaceous carbonate platform along the margin (Lancelot and Winterer, 1980a; Hinz et al., 1982; Heyman, 1988; Hafid, 2000; Laville et al., 2004; Tari and Molnar, 2005; Hafid et al., 2008).

At the western end of the High Atlas onshore, in the Argana half-graben system, Medina (1988) documented several synrift fault populations occurring in three directions: north-northeast–south-southwest, east-northeast–west-southwest, and northeast–southwest. The northeast–southwest extensional faults clearly postdate the other fault sets and have the same trend as the ultimate breakup of the Atlantic rift axis. An important synrift feature was documented by Tari and Molnar

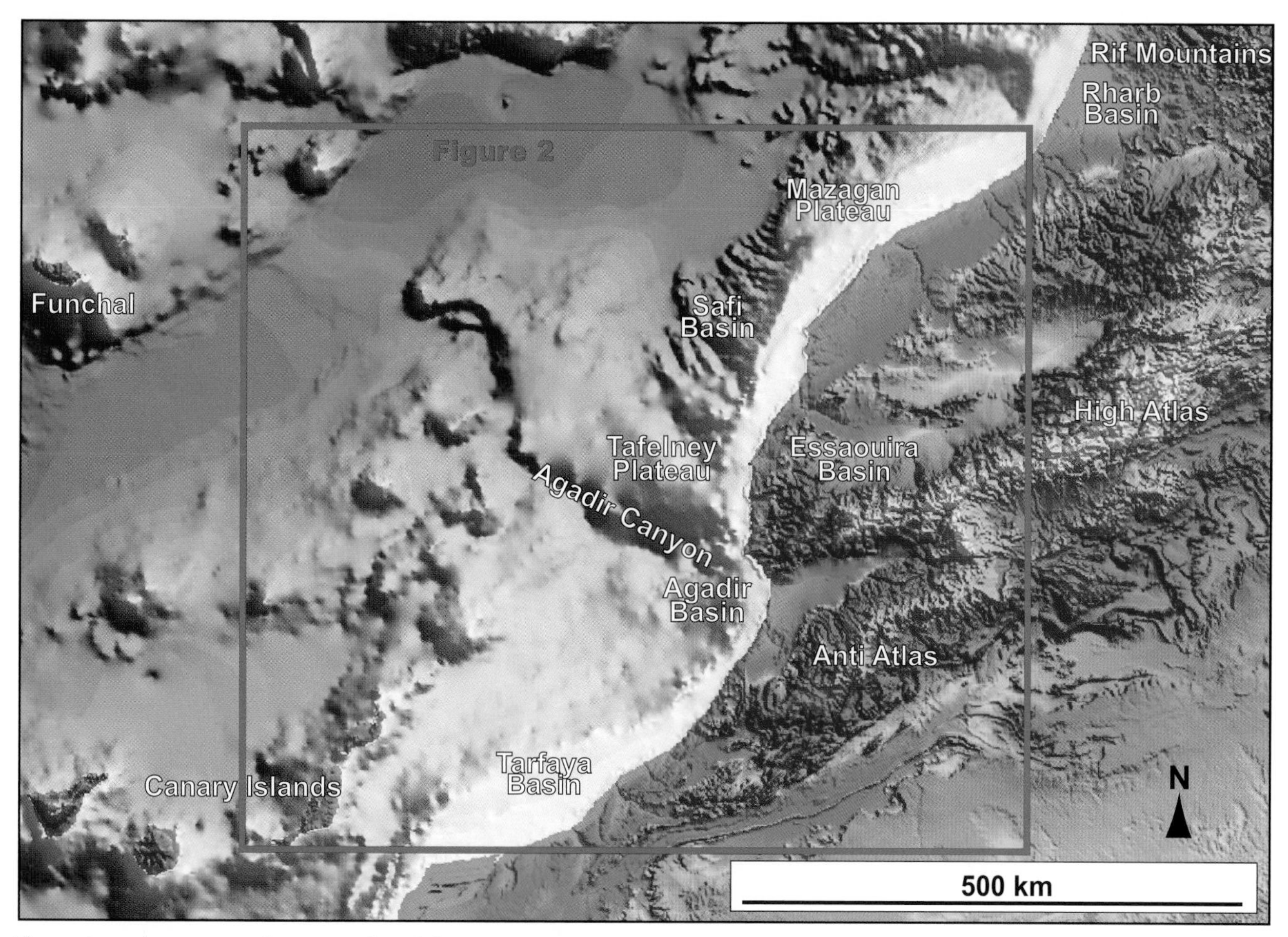

Figure 1. Bathymetry and topography of the Moroccan margin in the Central Atlantic region. The red rectangle shows the location of Figure 2. 500 km (311 mi).

(2005) in the offshore Essaouira Basin. The Tafelney accommodation zone (TAZ) trends northeast-southwest (Figure 2), separating two regional-scale synrift extensional fault systems with opposing polarity on the conjugate margins of Nova Scotia and Morocco. The actual Early to Middle Jurassic breakup occurred obliquely across this prominent synrift feature, leaving most of it on the Moroccan side of the Atlantic. The TAZ might have had an influence on the sedimentary entry points into the deep-water basin during the early postrift period.

The structural evolution of the adjacent Atlas folded belt onshore with several well-documented uplift periods has been described by many (e.g., Beauchamp et al., 1999; Frizon de Lamotte et al., 2000, 2009; Gomez et al., 2000). The High Atlas domain (Figure 2) was also initiated during the Late Triassic as a complex of elongated overall east–west-northeast–southwest-trending synrift basins controlled by major border faults. After a relatively regular postrift thermal subsidence period during the Jurassic and Early Cretaceous (Ellouz et al., 2003), the inversion of the Atlas system started during the Senonian as the consequence of the Africa-Eurasia convergence (Guiraud and Bosworth, 1997). The next major phase of regional inversion and orogenesis of the Atlas system occurred during two distinct episodes, the middle and late Eocene–Oligocene and the late Miocene–Pliocene, respectively. During the intervening period, the Africa-Europe convergence was mainly accommodated in northern Morocco, in the Rif-Tell system (Frizon de Lamotte et al., 2000, 2009).

STRATIGRAPHY

Whereas the stratigraphy of the Atlantic margin of Morocco is very well studied onshore, the offshore basins are much less well known, and the deep-water areas have hardly been documented at all. The following overview reflects the increasingly sparse understanding of the stratigraphy, the onshore areas, and the deep-water parts of the continental margin.

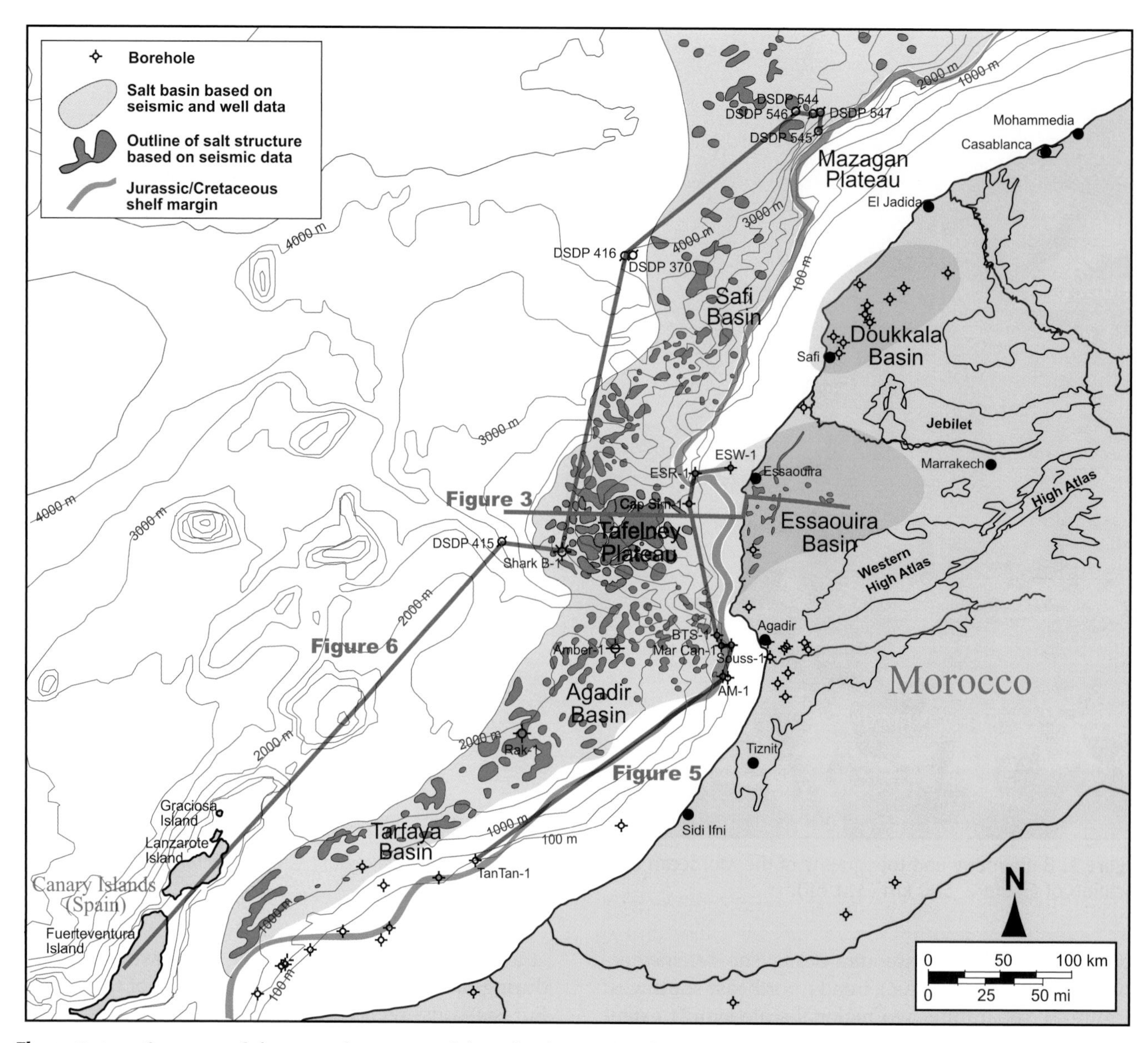

Figure 2. Location map of the central segment of the Atlantic margin of Morocco. The regional dip section shown in Figure 3 is located in the offshore Essaouira Basin segment, just northwest of Agadir. The well correlations shown in Figures 5 and 6 are indicated in red lines. DSDP = Deep Sea Drilling Project. ESW = Essaouira West; ESR = Essaouira; AGM = Agadir Marine.

Onshore Stratigraphy of the Essaouira Basin and Western High Atlas

A composite stratigraphic column illustrates well the syn- and postrift strata in the onshore area of central Atlantic Morocco (Figure 4). This particular compilation by Hafid (2006) is based on several other works, such as Ambroggi (1963), Duffaud et al. (1966), Brown (1980), Le Roy et al. (1998), Hafid (2000), Hafid et al. (2006), Oujidi et al. (2000), Le Roy and Piqué (2001), Ellouz et al. (2003), and Zühlke et al. (2004).

The Triassic synrift half grabens were filled by at least 2000 m (6562 ft) fluviolacustrine continental red beds, locally intercalated with basalt flows. This basin-fill sequence outcrops in the Argana Graben area, at the western end of the High Atlas (Brown, 1980). In the overall Triassic strata, four tectonostratigraphic sequences (TS) have been recognized by Olsen et al. (2003). The lowermost sequence (TS 1) is dated from the Late Permian by its vertebrate fossils; the intermediate sequences (TS 2 and TS 3) are assigned to the Carnian and Norian based on palynology; the sequence with basalt (TS 4) is dated as Rhaetian–Hettangian to lowermost Sinemurian by isotopic results. This stratigraphic record reflects the combined effects of rifting episodes and climatic changes on sedimentation. The overall

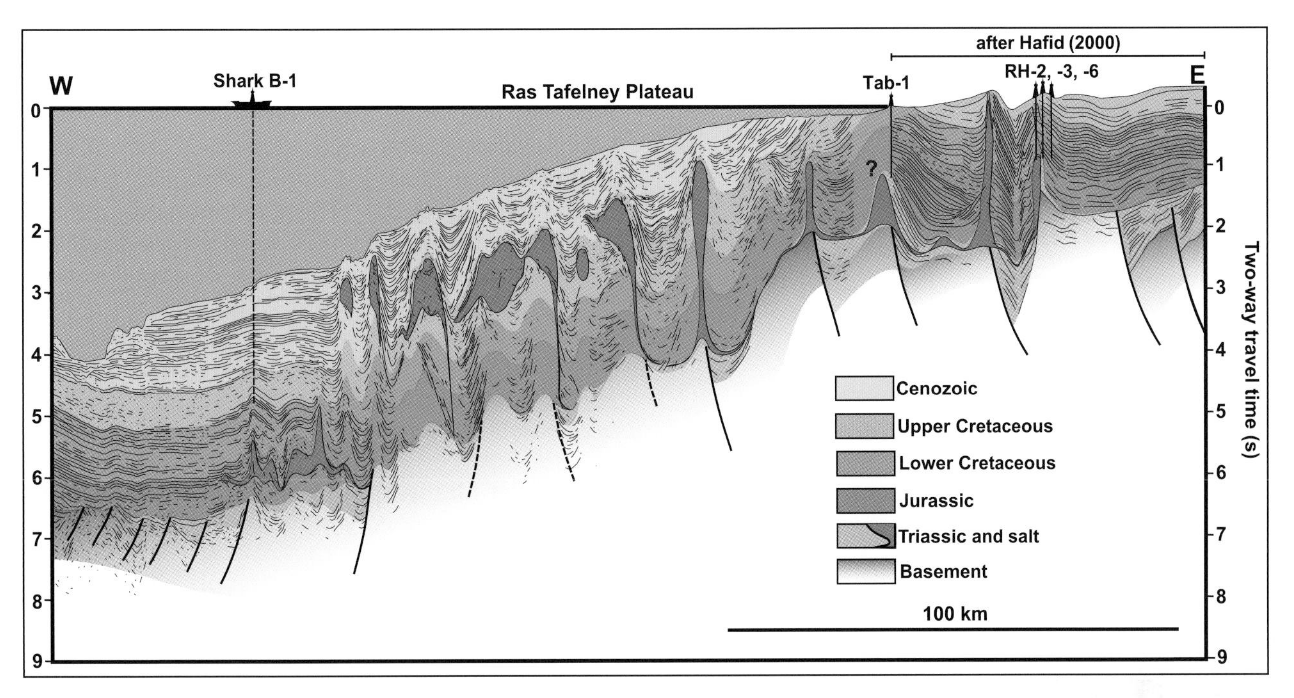

Figure 3. Composite line drawing interpretation of a regional seismic transect across the offshore Essaouira Basin and Tafelney Plateau (Tari et al., 2003) and the onshore Essaouira Basin (Hafid, 2000). Vertical exaggeration is about fivefold at 4 km/s velocity. For location, see Figure 2. The Shark B-1 well was drilled on the toe-thrust anticline at the basinward edge of the salt basin. For a three-dimensional seismic data image, see Figure 7. RH = Sidi Rhalem; Tab-1 = Taboulaouart. 100 km (62 mi).

sequence can be correlated with sediments of the equivalent basins of North America (Olsen et al., 2003).

In the onshore Essaouira Basin, close to the top of the synrift series, thick salt deposits (from a few hundred to more than a thousand meters) developed (e.g., Hafid, 2000). Seismic, surface, and borehole evidence indicates that, toward the end of salt deposition, extrusive volcanic activities occurred in the Moroccan Atlantic basins producing basalt flows that were emplaced at the Triassic–Jurassic boundary. These subaerial lavas and pyroclastics interbedded with shallow-water clastic and evaporitic sequences are part of the Central Atlantic magmatic province (CAMP). The CAMP magmatic events range in age from 197 to 203 Ma but centered around 200 Ma (Knight et al., 2004; Marzoli et al., 2004).

Synrift fault activity gradually declined and marine carbonate deposition was essentially controlled by thermal subsidence in the earliest Liassic time, overlying the synrift sequence (Laville et al., 1995; Medina, 1995; Hafid, 2000). The Lias transgressive marine limestones or alternations of dolomite and anhydrite are overlain by a locally thick sequence of continental clastics. These, in turn, are overlain by Dogger gray limestone deposits of the Middle to Late Callovian. The Malmian saw two main sedimentary cycles. First, during the Oxfordian, more shallow-water carbonates deposited, followed by a transgressive sequence of marls and dolomites. This was followed by the deposition of a thick sequence of Portlandian to Berriasian anhydrites and limestones recording an overall regressive trend. The lower part of the Cretaceous has a more shale-dominated sequence that was the result of the establishment of the Atlas Gulf, a large embayment of the paleoshelf in the area of the present-day Atlas (Hinz et al., 1982; Wiedmann et al., 1982). The Upper Cretaceous–Paleogene sequence is dominated by neritic limestones and shale intercalations with several tectonically enhanced unconformities in between. The Neogene is mostly missing in the onshore area caused by the neotectonic uplift and erosion of the margin (Ellouz et al., 2003).

Shelf Stratigraphy in the Central Segment of Atlantic Morocco

The number of wells drilled on the shelf is much less than onshore (Figure 2); therefore, the shelf stratigraphy is illustrated by a chronostratigraphic chart correlating eight wells (Figure 5). The synrift Triassic is the oldest stratigraphic level encountered in these wells represented by salt and continental redbeds, just like onshore (cf. Figure 4). On top, most of the wells have a massive westward-prograding carbonate platform sequence spanning not only the entire Jurassic, but also the Neocomian (Figure 5). By the end of the Neocomian,

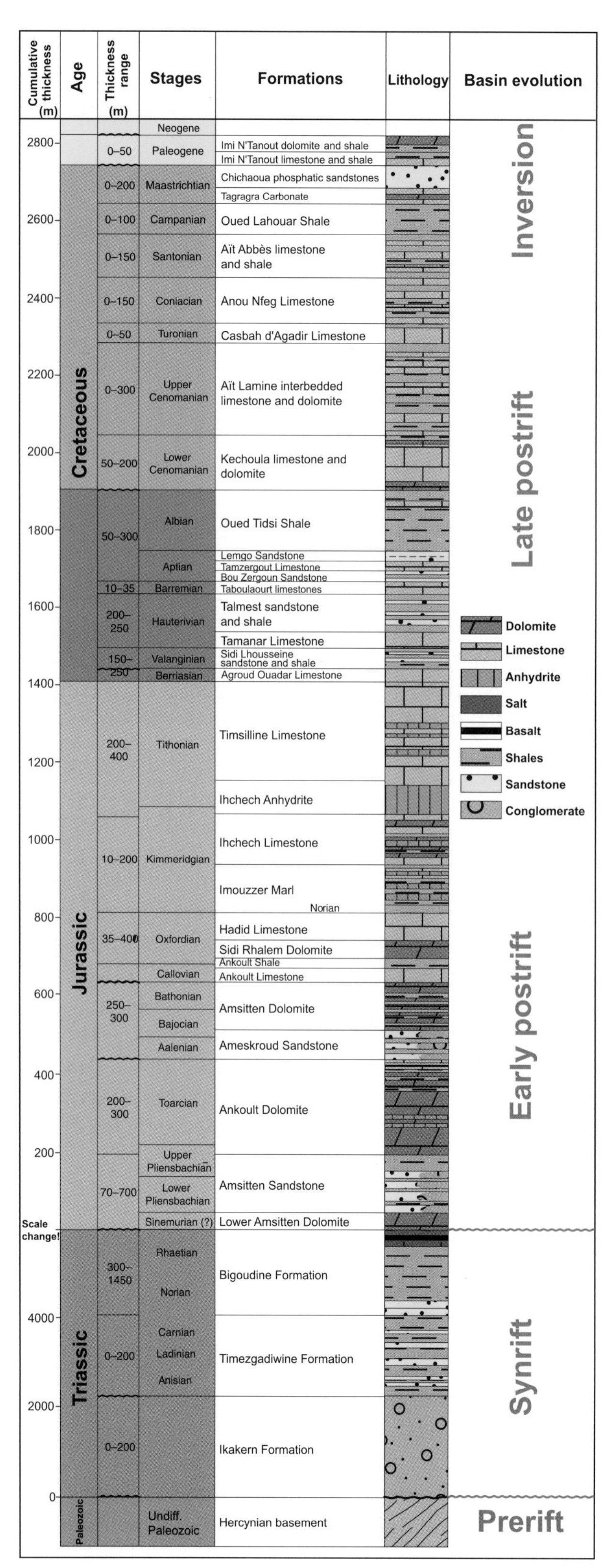

Figure 4. Onshore stratigraphy of the western High Atlas and Essaouira Basin, modified from Hafid (2006).

the vast carbonate platform, which was so widespread in northwest Africa, was drowned (Schlager, 1980). An important exception is the Tarfaya offshore basin farther to the south (Figure 2), where deltaic sedimentation prevailed during the Early Cretaceous (the Tan Tan delta). The predominance of shale deposition throughout most of the Late Cretaceous and Cenozoic indicates the lack of major sediment entry points along the margin. With regards to possible reservoir intervals in wells drilled on the present-day shelf thin, reservoir-quality sands were documented only in a few of them (Figure 5).

Deep-water Stratigraphy in the Central Segment of Atlantic Morocco

The deep-water stratigraphy is poorly documented by just a few exploration and Deep Sea Drilling Project (DSDP) wells (Figure 6). However, as an unusual calibration of the deep-water stratigraphy along the Atlantic margin, the well-studied and stratigraphically fairly complete outcrops of the Fuerteventura Island of the Canaries (Figure 2) also provide important information (Steiner et al., 1998). The Mesozoic deep-water strata on Fuerteventura became uplifted and exposed during the Tertiary magmatism of the Canaries (Davison, 2005).

The Toarcian ultrabasics at the base of the exposed sequence at Fuerteventura represent the oceanic crust of the Central Atlantic Basin and provide a minimum age for the onset of drifting between Nova Scotia and Morocco. The only other penetration of the basement of the deep-water strata occurred in DSDP 544 (Figure 2), where below Triassic continental clastics, crystalline rocks were found (Lancelot and Winterer, 1980b). Well DSDP 544 was drilled basinward from the Mazagan Plateau, on extended continental crust.

The Jurassic part of the postrift stratigraphy (Figure 6) is either characterized by the alternation of pelagic shales, siliciclastic turbidites, and calciturbidites (Fuerteventura) or by a condensed section of pelagic limestones (e.g., DSDP 547) representing a deep-water starved margin outboard of the Mazagan Plateau. The calciturbidites were interpreted to be sourced from the paleoshelf where a carbonate platform rimmed most of the margin during the entire Jurassic (cf. Figure 5). As the carbonate factory on the shelf was drowned by the end of the Neocomian (Schlager, 1980), the calciturbidites were replaced by siliciclastic turbidites (e.g., DSDP 416/370; Figure 2). In Fuerteventura Island, the Neocomian–Barremian interval is dominated by distal turbidites of the Tan Tan delta system. The striking lack of Lower Cretaceous sediments in the Mazagan Slope DSDP wells (Figure 6) is caused by a prominent

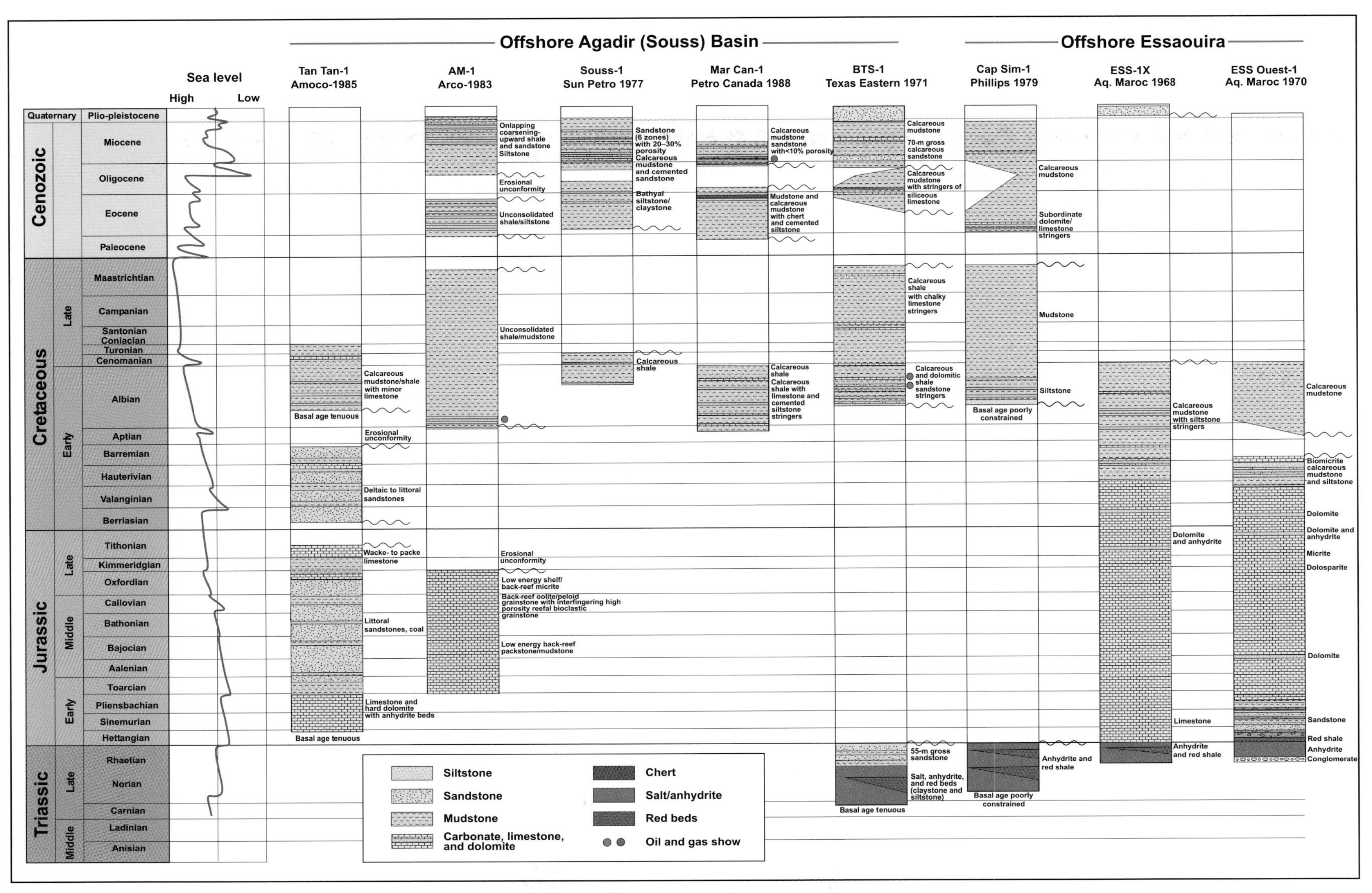

Figure 5. Chronostratigraphic correlation of wells drilled on the shelf of the offshore Agadir and Essaouira basins. For location, see Figure 2. Aq = Aquitaine; AM = Agadir Marine; ESS = Essaouira.

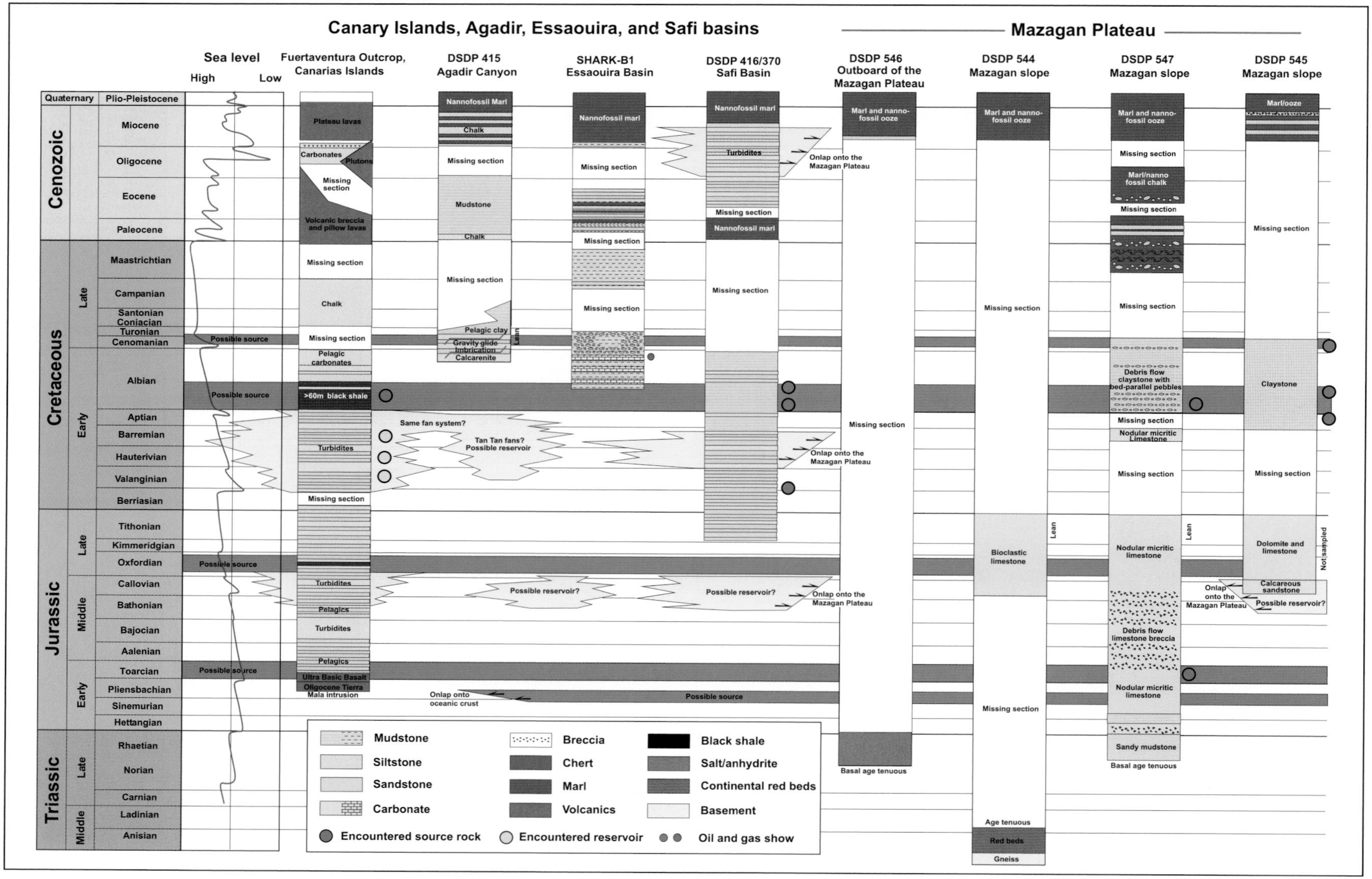

Figure 6. Chronostratigraphic correlation of wells drilled in the deep-water areas of the offshore Agadir, Essaouira, and Mazagan basins. For location, see Figure 2. For a detailed description of the Deep Sea Drilling Project (DSDP) wells, see Lancelot and Winterer (1980a, b); Winterer and Hinz (1984).

basement block on the paleoshelf diverting all the turbidity currents down the continental slopes south of Mazagan (Winterer and Hinz, 1984).

Another important element of the deep-water stratigraphy is a regional-scale, Upper Cretaceous (Cenomanian–Turonian) mass-transport complex that covers a significant part of the central segment of the Moroccan Atlantic margin (Price, 1980). This peculiar unit has been penetrated first in well DSDP 415 and then in the Shark B-1 well (Figure 6).

The presence of multiple mass-transport units within this complex has been documented by Dunlap et al. (2010) using a deep-water 3-D seismic data set. The catastrophic failure of a large segment of the paleocontinental margin is responsible for the emplacement of this complex in the deep-water Essaouira Basin and Tafelney Plateau (Price, 1980).

Based on the well penetrations available to date, most of the Cenozoic strata are dominated by pelagic shales, marls, and even cherts. Siltstones present in some of the wells are observed but, overall, reservoir-quality sands are rare in the deep water (Figure 6).

SHARK B-1 WELL RESULTS

As an example of how well constrained the deep-water stratigraphic scheme described above is (Figure 6), the findings from one of the exploration wells, the Shark B-1, are described in the next section. Because the number of deep-water wells is still very small compared to the entire margin of offshore Morocco, the stratigraphic calibration obtained from individual wells is critical for any future exploration efforts.

Predrill Predictions of Possible Reservoir-quality Stratigraphic Intervals

Based on the detailed interpretation of a high-quality and fairly large (i.e., >3000 km^2 [>1158 mi^2]) 3-D seismic reflection survey, multiple reservoir levels were predicted such as channel and overbank and debris-flow units alternating with sheet sands and sediment wave complexes. These predictions were based primarily on the integrated interpretation of seismic attributes and the geometry seen on 3-D seismic data. Importantly, none of the predicted reservoir units could be directly tied to wells drilled on the shelf (Figure 5) because of the lack of adequate seismic data on the shelf and the obvious differences in the overall stratigraphic architecture. Moreover, seismic correlation of inferred reservoir-facies sequences to the existing DSDP wells in the ultra-deep water (cf. Figures 2, 6) was challenging because the vintage seismic data sets have only poor to moderate data quality. However, the regional seismic correlation provided reasonable age control for the predrill prognosis of reservoir targets.

In particular, the predrill prognosis of the Shark B-1 well had three reservoir targets (Figure 7). From top to bottom, with the associated seismic horizon names included, these were the (1) Upper Cenozoic (H230–H245), (2) Lower Cenozoic (H279–H290) to Upper Cretaceous (H390–H400), and (3) Albian (H490–H505).

Target 1: Upper Cenozoic reservoirs (H230–H245). Based on its seismic signature, this Miocene regional reservoir target was interpreted to be a sand-prone unit. However, its seismic facies is not as well developed at the Shark B location as in other parts of the Ras Tafelney Plateau where the same stratigraphic interval has well-developed channel-levee geometry based on 3-D seismic data.

Target 2: Lower Cenozoic (H279–H290) to Upper Cretaceous reservoirs (H388–H400). The inferred Lower Cenozoic (i.e., Eocene–Paleocene) reservoirs were defined by prominent amplitudes and a map-view seismic attribute pattern that suggested the presence of multiple levels of slope fans. The Upper Cretaceous (i.e., Senonian) target was interpreted to have a combination of basin floor units and also sand-prone sediment waves.

Target 3: Albian reservoirs (H490–H505). The regionally extensive Turonian–Cenomanian debris-flow complex has a special seismic signature labeled as the chaotic zone (cf. Dunlap et al., 2010). As Price (1980) already documented, the mass-transport nature of this package, this Cretaceous interval (H400–H490), was not considered as a reservoir target. However, the underlying Albian sequence with several prominent reflectors showing some conformance to structure (Figure 8) was interpreted to be age equivalent to the turbiditic sandstones seen in outcrop on Fuerteventura Island and in the DSDP 370/416 wells (cf. Figure 6). Moreover, some of the Albian seismic reflectors within the 3-D seismic coverage have a very distinct map-view attribute expression with north–northeast-south–southwest-oriented (essentially coast-parallel) bands with slightly arcuate shape (Figure 9). This amplitude brightening and geometry was interpreted to result from a sand-prone sediment-wave depositional system associated with unconfined turbidite flows.

Postdrill Biostratigraphy and Sequence Stratigraphy of the Shark B-1 Well

A detailed postdrill evaluation of the stratigraphy encountered in the Shark B-1 well based on 120 quantitative analyses of 60 ditch-cutting samples from the

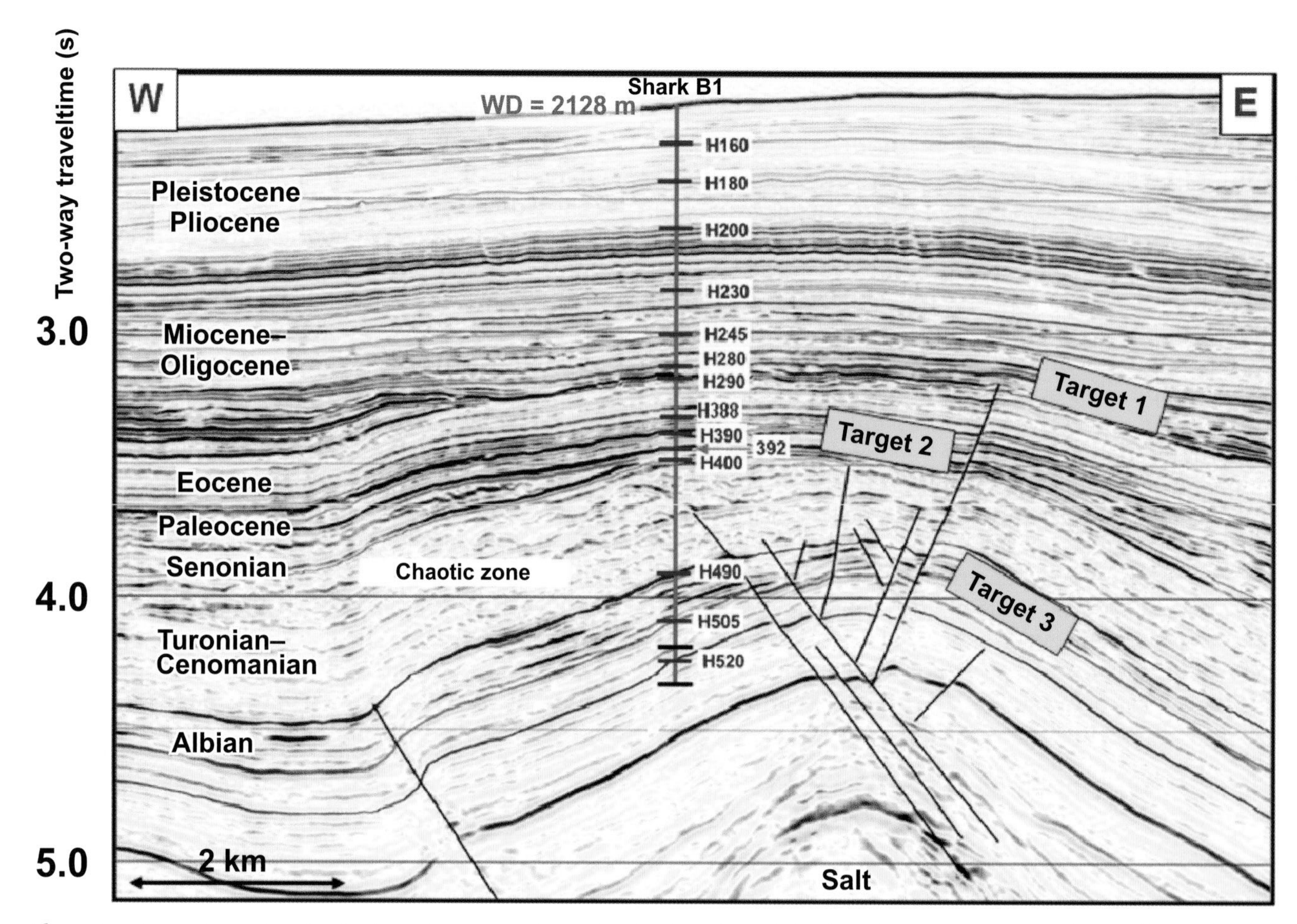

Figure 7. Three-dimensional seismic reflection section, with the location of the first deep-water exploration well (Shark B-1) drilled in the offshore Essaouira Basin. Seismic mapping horizons and predrill stratigraphic age assignments are highlighted in various colors. For approximate location, see Figure 2. WD = water depth. 2 km (1.2 mi).

lower part of the well section (interval, 2820–3975 m [9252–13,041 ft] total depth) have been integrated with lithofacies and wire-line data to provide a well-log sequence-stratigraphic interpretation, which provides the basis for the detailed stratigraphic subdivision of the well section (Figure 10). The middle to early Miocene strata in the Shark B-1 well section unconformably overlies a more or less conformable section of the middle Eocene to late Paleocene, although a significant stratigraphic break is evident within the early Eocene section at 3266 m (10,715 ft) (~2 Ma on the time scales of Haq et al., 1987, or Hardenbol et al., 1998). The late Paleocene strata unconformably overlie a relatively condensed section of the Maastrichtian to Late Campanian, which in turn overlies an expanded interval of Cenomanian to Late Albian strata, in which the Shark B-1 well reached total depth. The late Paleocene unconformity represents a stratigraphic break of approximately 7 Ma on both time scales, and that between Campanian and Cenomanian strata represents a break of approximately 14.5 Ma on the time scale of Haq et al. (1987) and approximately 16.5 Ma on the time scale of Hardenbol et al. (1998).

Within the Cenomanian interval, one depositional sequence is extremely thick (Figure 10), associated with the presence of abundant reworked and allochthonous rock within the lowstand systems tract, that related to the 94.7 Ma maximum flooding surface of the early Middle Cenomanian. Vincent et al. (1980) recorded a comparable sedimentary section in DSDP site 415 (Figure 6) and related this to high rates of gravity-driven mass-transport sedimentation at the Middle Cenomanian as a result of uplift in the western High Atlas Mountains (Figure 2). Based on microfaunal associations, all of the strata were deposited under lower to middle bathyal conditions.

Despite the prediction of several reservoir-facies stratigraphic intervals, no significant reservoir was found in the Shark B-1 well (cf. Figures 7, 10). Although a few sand layers were found in the middle Eocene and Upper Cretaceous, the alternation of pelagic shales and

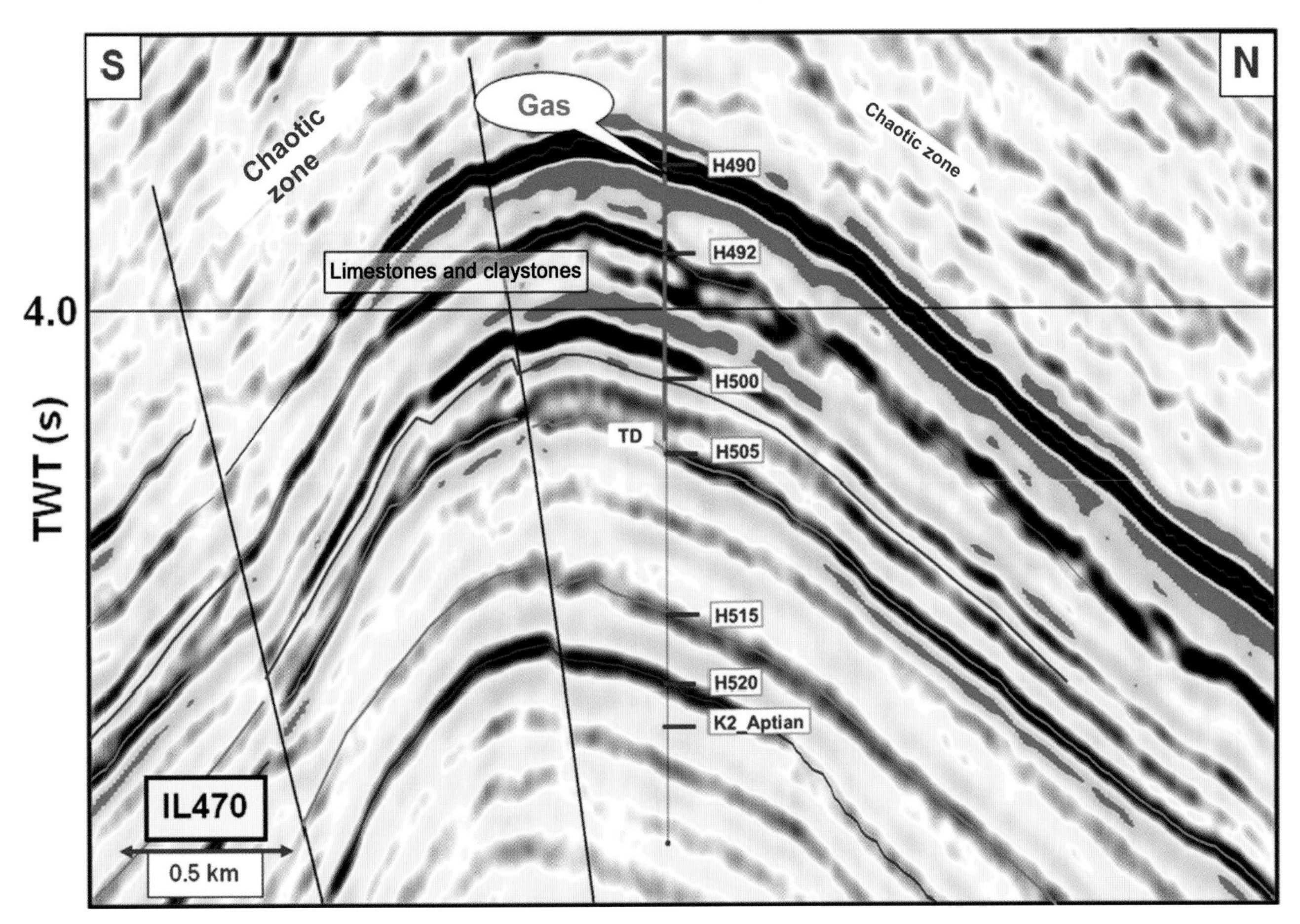

Figure 8. Seismic detail of the deepest part of the Shark B-1 well. See Figure 10 for the sequence and lithostratigraphic results from this drilling interval. TWT = two-way traveltime. 0.5 km (0.3 mi).

fairly thick siltstones dominates the overall stratigraphy in this well. Furthermore, some of the high-amplitude seismic packages within the Cenozoic (Figure 7) turned out to be Eocene chert layers (Figure 10).

The sediment waves identified using 3-D seismic attribute analysis are not sand prone but are composed of reworked shallow-water carbonate material. The Shark B-1 well demonstrated that these features are, for the most part, distinct carbonate layers in the upper part of the Albian sequence (Figure 10). As to reservoir quality, some porosity is present at the top of the uppermost carbonate unit, containing minor dry gas.

Some 60 km (37 mi) to the north-northeast from the Shark B-1 location, Dunlap and Wood (2010) documented Late Albian and Early Cenomanian sediment waves built by along-slope currents on a relatively stable slope showing updip migration. These sediment waves exhibit wave heights of 40 m (131 ft) and wavelengths of approximately 1 km (~0.6 mi). The analysis of the cuttings in the Shark B-1 well showed that the Albian limestone beds are composed of carbonate material reworked from the shelf area into the deep water as calciturbidites. Schlager (1980) reported calciturbidites from the DSDP 416 well (Figure 6), also suggesting a drowning of a carbonate platform during the Valanginian.

The other two deep-water exploration wells, Amber-1 and Rak-1 (Figure 2), were drilled shortly after the Shark B-1 well and documented the same general lack of reservoir units in the entire Cenozoic–Upper Cretaceous succession. The only remaining deep-water well, drilled along the Atlantic margin documenting significant amounts of clastic reservoir units, is the DSDP 416 well (Figure 2) where Lower Cretaceous distal siliciclastic turbidites were found (Figure 6). Interestingly, the same well penetrated Upper Jurassic calciturbidites that are regarded as being analogous to the Albian ones in the Shark B-1 well.

WHERE ARE THE RESERVOIRS?

Three models can explain the apparent lack of reservoirs in the Cenozoic to Upper Cretaceous strata of

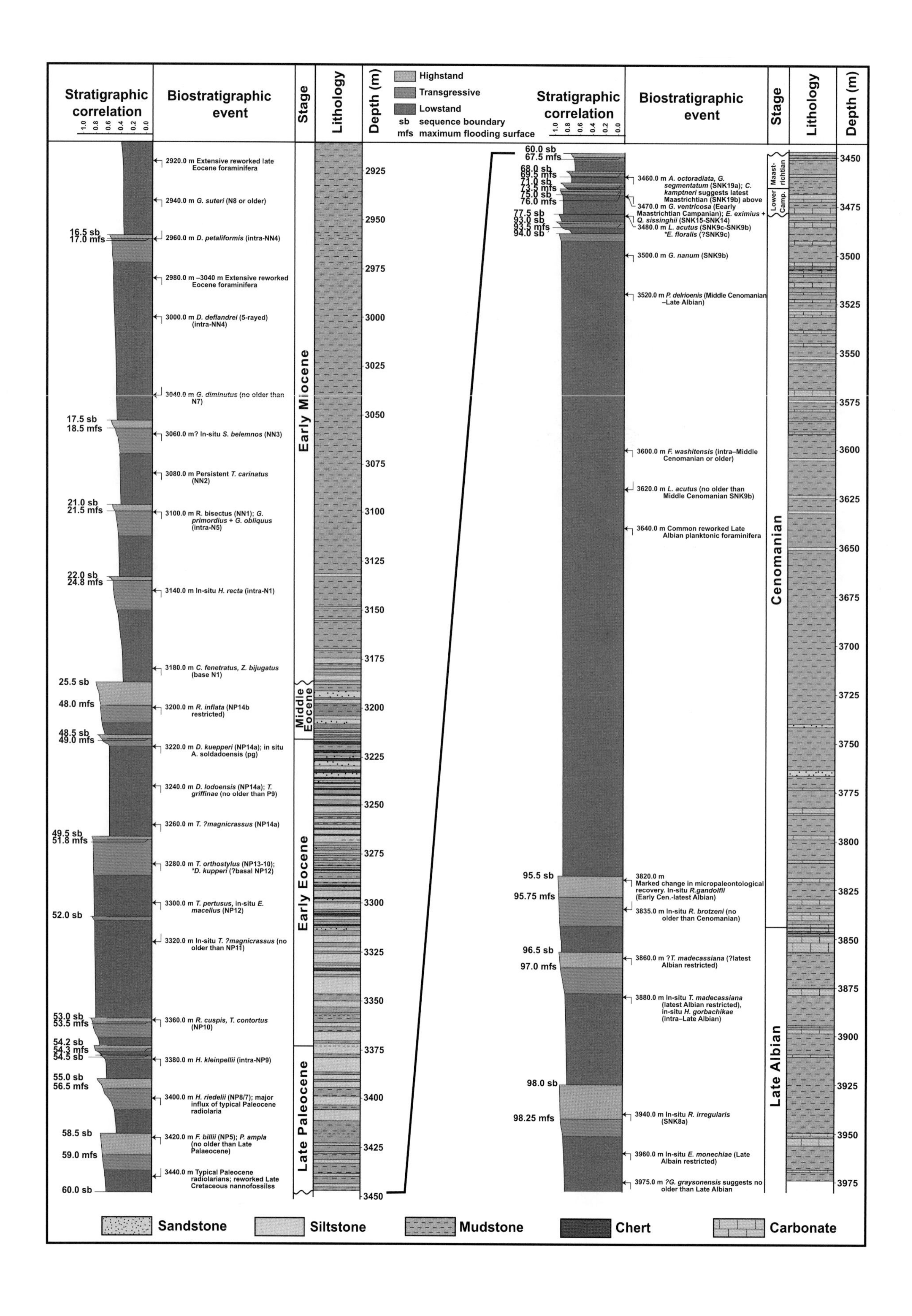

Stratigraphic correlation
1.0 0.8 0.6 0.4 0.2 0.0
Biostratigraphic event
Stage
Lithology
Depth (m)
Highstand
Transgressive
Lowstand
sb sequence boundary
mfs maximum flooding surface
2920.0 m Extensive reworked late Eocene foraminifera
2940.0 m G. suteri (N8 or older)
16.5 sb
17.0 mfs
2960.0 m D. petaliformis (intra-NN4)
2980.0 m –3040 m Extensive reworked Eocene foraminifera
3000.0 m D. deflandrei (5-rayed) (intra-NN4)
3040.0 m G. diminutus (no older than N7)
17.5 sb
18.5 mfs
3060.0 m? In-situ S. belemnos (NN3)
3080.0 m Persistent T. carinatus (NN2)
21.0 sb
21.5 mfs
3100.0 m R. bisectus (NN1); G. primordius + G. obliquus (intra-N5)
22.0 sb
24.8 mfs
3140.0 m In-situ H. recta (intra-N1)
3180.0 m C. fenetratus, Z. bijugatus (base N1)
25.5 sb
48.0 mfs
3200.0 m R. inflata (NP14b restricted)
48.5 sb
49.0 mfs
3220.0 m D. kuepperi (NP14a); in situ A. soldadoensis (pg)
3240.0 m D. lodoensis (NP14a); T. griffinae (no older than P9)
3260.0 m T. ?magnicrassus (NP14a)
49.5 sb
51.8 mfs
3280.0 m T. orthostylus (NP13-10); *D. kupperi (?basal NP12)
3300.0 m T. pertusus, in-situ E. macellus (NP12)
52.0 sb
3320.0 m In-situ T. ?magnicrassus (no older than NP11)
53.0 sb
53.5 mfs
3360.0 m R. cuspis, T. contortus (NP10)
54.2 sb
54.3 mfs
54.5 sb
3380.0 m H. kleinpellii (intra-NP9)
55.0 sb
56.5 mfs
3400.0 m H. riedelii (NP8/7); major influx of typical Paleocene radiolaria
58.5 sb
3420.0 m F. billii (NP5); P. ampla (no older than Late Palaeocene)
59.0 mfs
3440.0 m Typical Paleocene radiolarians; reworked Late Cretaceous nannofossilss
60.0 sb
Early Miocene
Middle Eocene
Early Eocene
Late Paleocene
2925
2950
2975
3000
3025
3050
3075
3100
3125
3150
3175
3200
3225
3250
3275
3300
3325
3350
3375
3400
3425
3450
60.0 sb
67.5 mfs
68.0 sb
69.5 mfs
71.0 sb
73.5 mfs
75.0 sb
76.0 mfs
77.5 sb
93.0 sb
93.5 mfs
94.0 sb
3460.0 m A. octoradiata, G. segmentatum (SNK19a); C. kamptneri suggests latest Maastrichtian (SNK19b) above
3470.0 m G. ventricosa (Eearly Maastrichtian Campanian); E. eximius + Q. sissinghii (SNK15-SNK14)
3480.0 m L. acutus (SNK9c-SNK9b) *E. floralis (?SNK9c)
3500.0 m G. nanum (SNK9b)
3520.0 m P. delrioensis (Middle Cenomanian –Late Albian)
3600.0 m F. washitensis (intra–Middle Cenomanian or older)
3620.0 m L. acutus (no older than Middle Cenomanian SNK9b)
3640.0 m Common reworked Late Albian planktonic foraminifera
95.5 sb
95.75 mfs
3820.0 m Marked change in micropaleontological recovery. In-situ R.gandolfii (Early Cen.-latest Albian)
3835.0 m In-situ R. brotzeni (no older than Cenomanian)
96.5 sb
97.0 mfs
3860.0 m ?T. madecassiana (?latest Albian restricted)
3880.0 m In-situ T. madecassiana (latest Albian restricted), in-situ H. gorbachikae (intra–Late Albian)
98.0 sb
98.25 mfs
3940.0 m In-situ R. irregularis (SNK8a)
3960.0 m In-situ E. monechiae (Late Albain restricted)
3975.0 m ?G. graysonensis suggests no older than Late Albian
Maast-richtian
Lower Camp.
Cenomanian
Late Albian
3450
3475
3500
3525
3550
3575
3600
3625
3650
3675
3700
3725
3750
3775
3800
3825
3850
3875
3900
3925
3950
3975
Sandstone
Siltstone
Mudstone
Chert
Carbonate

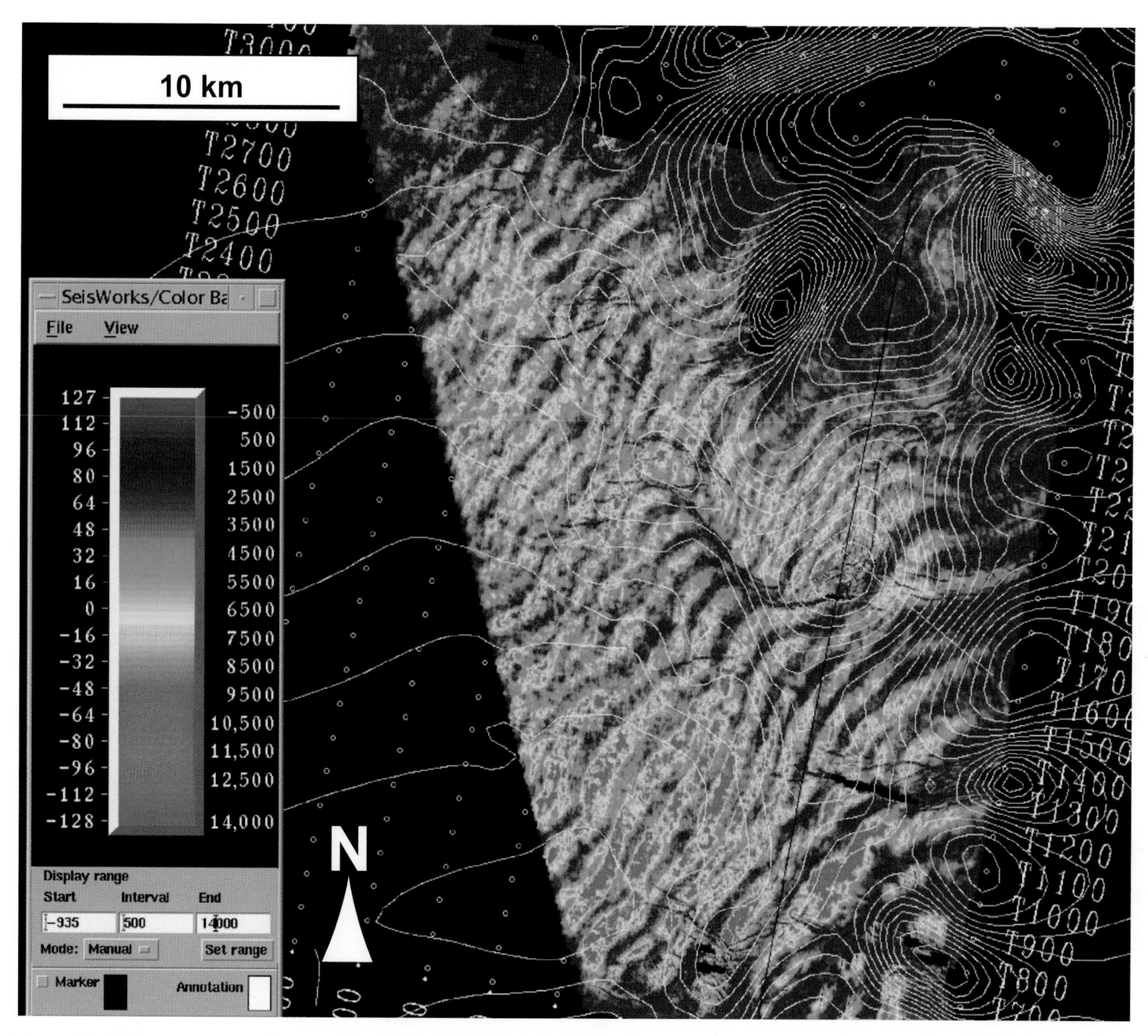

Figure 10. Sediment waves observed in the lower part of the Albian in the broader Shark B-1 area. The root-mean-square amplitude map was generated in a window, 20 m (66 ft) below and above horizon 520 (see Figure 8). The crests of the sediment waves suggest wave migration updip to the east-southeast. These sediment waves are almost identical with the ones reported by Dunlap and Wood (2010) from an interval close to the top Albian, some 60 km (37 mi) to the north from the Shark B-1 location. Similarities include average wavelength, moderate sinuosities, and bifurcations. 10 km (6.2 mi).

deep-water central Atlantic Morocco. However, it remains to be seen which, if any, of these theories can account that no Cenozoic to Upper Cretaceous reservoirs have yet been found in deep-water Morocco.

Present-day Influx of Turbidites into the Deep Water

Because the present is commonly the key to the past, it is worthwhile to look at what happens to the sediments

Figure 9. Detailed lithostratigraphic log of the Shark B-1 well showing the presence of numerous carbonate beds within the Albian sequence. These carbonate layers, typically just a few meters thick, are interpreted to be calciturbidites. For a detailed discussion, see text. Color coding of lithology is identical to that of Figure 5. The depth values shown are sample depths or taken from the well logs in measured depth and are relative to the rotary table elevation on the drill ship of 24 m (79 ft) above sea level. L. Camp. = Late Campanian; Maas. = Maastrichtian.

entering the Moroccan continental margin today. The characteristics of present-day offshore sediment distribution systems, for instance, the Agadir Canyon (Figure 11), suggest a bypass as clastic turbiditic currents deposit much farther outboard, on the order of 200 to 500 km (124–311 mi; Wynn et al., 2002a), than the area of the deep-water exploration wells (cf. Figure 2). These systems were classified as high-efficiency systems where a separation between a channel-levee complex on the slope and detached lobes on the basin floor with a broad channel-lobe transitional zone (CLTZ sensu Wynn et al., 2002b) in between exists. In this context, the siliciclastic reservoirs should be expected outboard from the location of the Shark B-1 well (cf. Figures 2, 11).

Sediment Shadow Zone

The Shark B-1 well targeted a salt-cored toe-thrust anticline (Figures 3, 7). During the growth periods of this structure, it certainly had a bathymetric expression on the slope or the basin floor) and, therefore, turbiditic sediments probably bypassed the apex area of the anticline, making it a bald structure. However, most of the Cretaceous strata do not show growth based on the inspection of the seismic geometry, and salt moved dominantly only during the mid-Late Jurassic and the mid-Cenozoic (Figure 5).

Whereas this could provide a reasonable explanation on a local scale (~1–10 km [~0.6–6.2 mi]) for the Shark B-1 well in particular, the other two wells, Amber-1 and Rak-1 (Figure 2), penetrated the Cenozoic to Upper Cretaceous strata in different traps, for example, in a salt-flank setting. Reservoir accumulation in this particular trap type is less sensitive to a growing structure; yet, no reservoirs were encountered in the Amber-1 well.

On a regional scale (~100 km [~60 mi]), the Shark B-1 area may fall between long-lived sedimentary entry points and corresponding deep-water fan systems, for example, between the Tan Tan and Safi fan systems (cf. Figures 12, 13).

On another regional scale (~10–100 km [~6–60 mi]), a sediment shadow zone could also be considered. The present-day Tafelney Plateau displays neotectonic uplift that manifests itself in the present-day bathymetry (Figure 2). This arching of the Tafelney Plateau, roughly perpendicular to the coastline because of the mountain building in the Atlas Mountains, apparently prevents reservoir-facies clastics from deposition at present (Figure 11). This explanation, however, cannot be applied to the Amber-1 and Rak-1 wells in the Agadir Basin because that area was clearly a depocenter during the entire Cenozoic.

A case for a sediment shadow zone along the Moroccan margin has already been mentioned. Specifically,

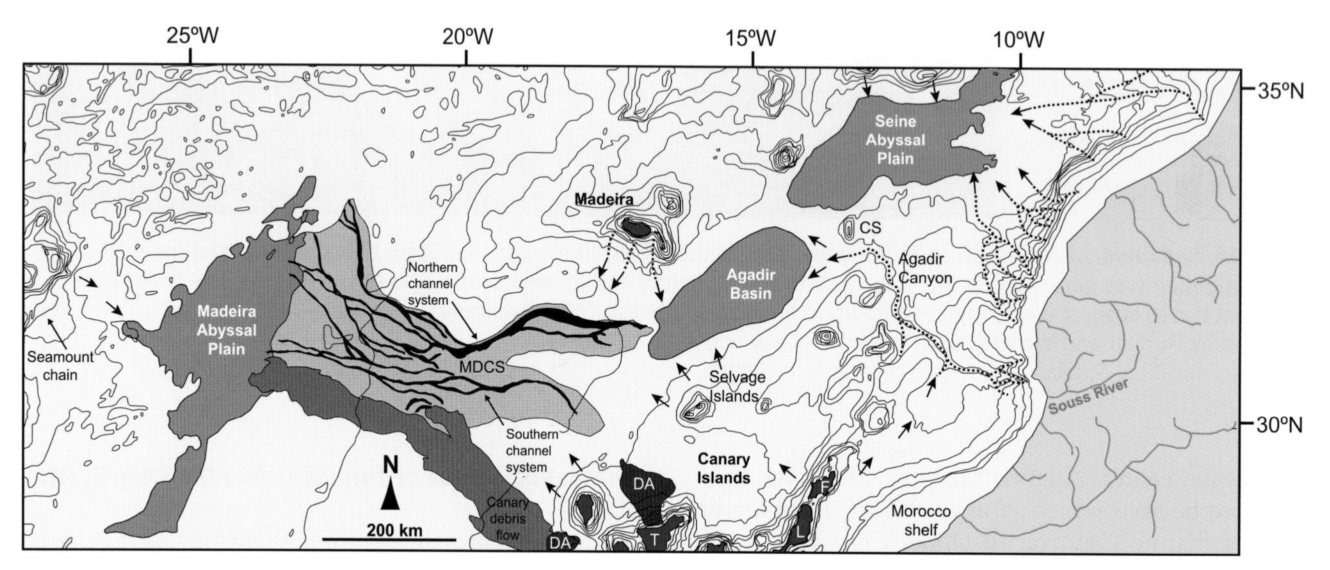

Figure 11. The presently active depocenters are located a few hundred kilometers from the coastline (Wynn et al., 2002a), outboard from the area of the deep-water exploration wells (cf. Figure 2). In fact, the Moroccan turbidite system (MTS) on the northwest African margin extends 1500 km (932 mi) from the head of the Agadir Canyon to the Madeira Abyssal Plain, making it one of the longest turbidite systems in the world (Wynn et al., 2002a). The MTS consists of three interconnected deep-water basins, the Seine Abyssal Plain (SAP), the Agadir Basin, and the Madeira Abyssal Plain (MAP), connected by a network of distributary channels. Principal transport directions for turbidity currents are shown by arrows and dashed lines. CS = Casablanca Seamount; DA = debris avalanche; MDCS = Madeira distributary channel system; T = Tenerife; L = Lanzarote; F = Fuerteventura. Bathymetric contours are spaced at 500 m (1640 ft) intervals. 200 km (124 mi).

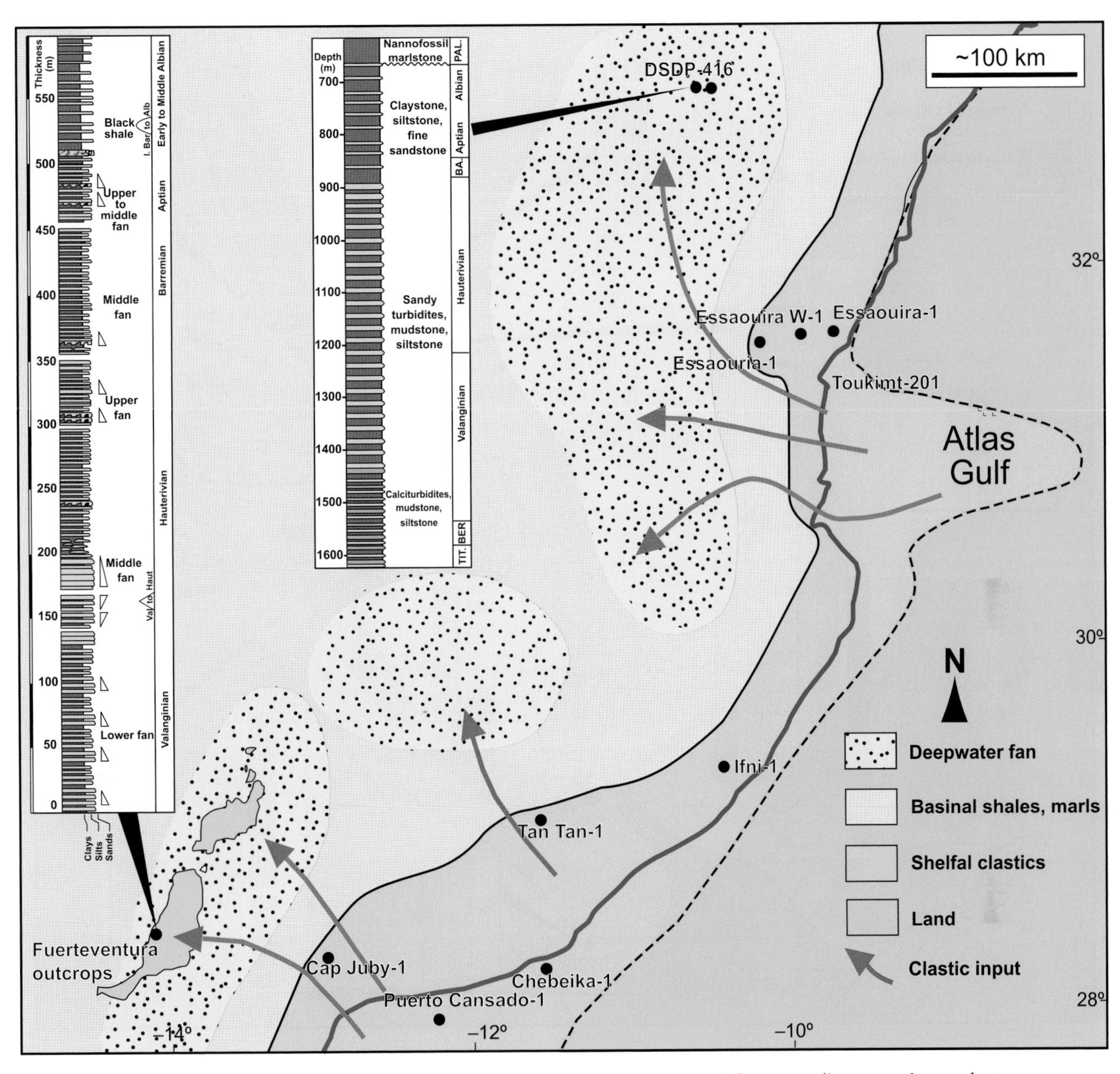

Figure 12. Regional evidence for the presence of Lower Cretaceous clastics in offshore Agadir, Essaouira, and Mazagan basins. See text for a detailed explanation. The outlines of predicted deep-water fans as shown are poorly constrained. TIT = Tithonian; BER = Berriasian; BA = Barremian; PAL = Paleocene. 100 km (62 mi).

the lack of Lower Cretaceous sediments on the flank of the Mazagan Plateau has been interpreted as the result of a subregional fault block diverting all the sediments to the south of the Mazagan area (Winterer and Hinz, 1984).

Limited Reservoir-facies Siliciclastic Supply

Perhaps the simplest reason to explain the lack of siliciclastic reservoirs is the limited sediment supply during the Cenozoic to Upper Cretaceous. The present-day margin does not receive a large volume of sediments because no major river systems entering the Atlantic margin of Morocco (Figure 2) exist, although the Atlas Mountains are actively rising and provide a large catchment area (Figure 1). This is interpreted as a case where climate primarily controls the sediment influx. As a comparison, the conjugate continental margin of Morocco on the Nova Scotia, Canada, side has a much thicker (i.e., about twice as thick on average) postrift sedimentary fill (Tari and Molnar, 2005), despite the lack of mountain ranges in the catchment areas.

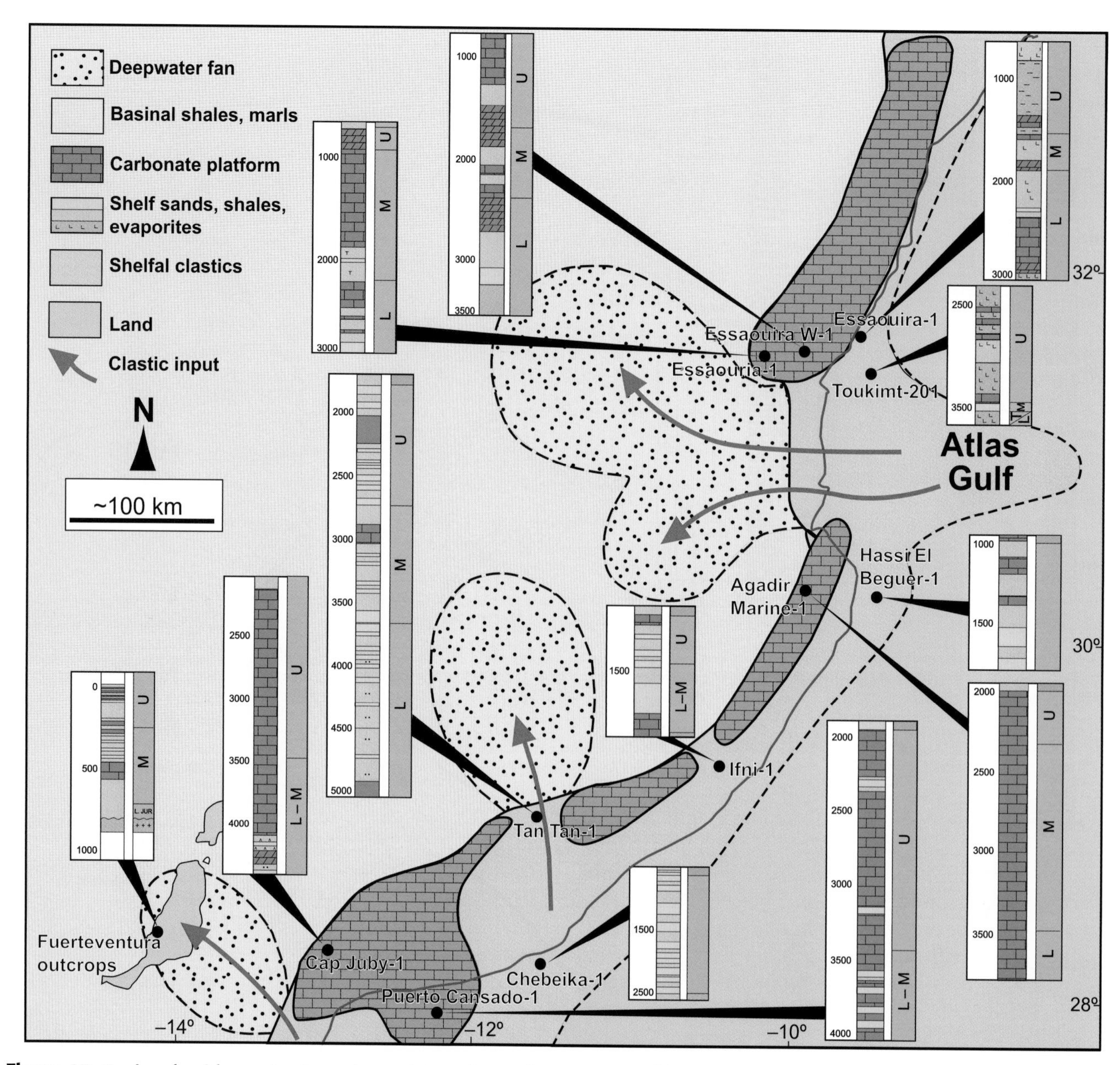

Figure 13. Regional evidence for Jurassic sandstone intervals encountered in various onshore and offshore exploration wells (shown as stick columns, with subsea depth in meters) in the Agadir, Essaouira, and Mazagan Basins. The outlines of predicted deep-water fans as shown are poorly constrained. See text for a detailed explanation. L = Lower; M = Middle; U = Upper; L.JUR. = Lower Jurassic; T = Triassic. 100 km (62 mi).

FUTURE RESERVOIR TARGETS IN DEEP-WATER MOROCCO

Whereas no evidence exists for Upper Cretaceous and Cenozoic reservoir sequences in the three deep-water exploration wells drilled in the central part of Atlantic offshore Morocco to date, the Lower Cretaceous and Jurassic sequence below may have reservoir levels. To build conceptual reservoir models for the pre–Upper Cretaceous strata, the regional paleogeography of the Early Cretaceous and the Jurassic is briefly outlined below.

Early Cretaceous Paleogeography

Regionally, the Neocomian could provide a reservoir facies equivalent to those of the siliciclastic turbidites described from the DSDP 370/416 wells (Lancelot and

Winterer, 1980a, b) and from the outcrops of Fuerteventura Island (Steiner et al., 1998). In the shallow-water to onshore realm, the massive Tan Tan deltaic complex is known in the Tarfaya area to the south (Figure 12), and a Barremian deltaic complex, in the area between Agadir and Essaouira, immediately to the east of the Ras Tafelney Plateau.

The turbiditic sandstones documented in outcrops on Fuerteventura Island, spanning the Neocomian to Aptian interval (Steiner et al., 1998), developed in a lower, middle, and upper fan setting (Figure 12). These outcrops document the presence of Lower Cretaceous deep-water fans located generally basinward from the area of the exploration wells (Figure 2). Similarly, the turbidites described in the DSDP 370/416 wells (Lancelot and Winterer, 1980a, b) prove the existence of clastic turbidites in a distal position in the deep-water areas. Note that none of the deep-water exploration wells reached this stratigraphic interval.

Jurassic Paleogeography

Only the outcrops on Fuerteventura Island provide direct evidence for the presence of deep-water Jurassic siliciclastic turbidites (Figure 6) offshore Morocco (Steiner et al., 1998).

As some of the wells drilled on the shelf show (Figure 5), a widespread Jurassic carbonate platform existed along the margin during this time. However, by looking closely at the map-view distribution of Jurassic lithofacies (Figure 13), certain gaps can be inferred in the carbonate platform pointing to local bypass of the shelf by siliciclastics. The presence of widespread Dogger clastics (Figure 4) just north of Agadir has long been recognized (e.g., Ambroggi, 1963). These continental clastics in between carbonate succession indicate a major basinward shift of the coastline, producing possible deep-water clastic fans in the basin. This major regression is poorly understood at present and cannot be easily interpreted in terms of a eustatic sea level fall. Note that, just like the Lower Cretaceous interval, none of the deep-water exploration wells reached the age-equivalent Middle Jurassic stratigraphic interval.

CONCLUSIONS

The apparent lack of Cenozoic and Upper Cretaceous reservoir-facies siliciclastics in the deep water of central Atlantic Morocco is surprising because the area was (and is) located adjacent to the Atlas Mountains, which had numerous mountain-building episodes during this long period. Several explanations can be considered to explain this anomaly: (1) the wells were drilled inboard of the paleo-CLTZ and, therefore, the reservoirs deposited farther out; (2) the well locations are in a sediment shadow zone, bypassed by the reservoirs; and (3) a limited siliciclastic sediment supply during the entire Cenozoic to Upper Cretaceous caused by the prevailing dry climate and lack of major river systems was observed.

Whereas no evidence exists for Upper Cretaceous and Cenozoic reservoir sequences in the three deep-water exploration wells drilled in the central part of Atlantic offshore Morocco to date, the Lower Cretaceous and Jurassic sequence below is predicted to have various reservoir levels. Reservoir presence, as the key risk factor in the deep-water exploration of the Moroccan Atlantic margin at present, needs to be better understood before new exploration wells can be drilled.

ACKNOWLEDGMENTS

We thank Albert Bally and Mohamad Hafid for the very helpful discussions on the geology of Morocco. We also thank the help and support of our colleagues at Office National des Hydrocarbures et des Mines, Rabat, Morocco. We also thank Katrina Coterill, Marek Kaminski, Charlton Miller, and Eva Moldovanyi, our former colleagues at Vanco Energy Company, Houston, Texas. The biostratigraphic and sequence-stratigraphic subdivision of the Shark B-1 well was done by Time-Trax. The chronostratigraphic framework for the offshore Morocco wells was mostly compiled by Robert Sawyer. We thank Michael Sweet for his helpful and constructive comments on the first draft of this paper and Dengliang Gao for his editorial patience.

REFERENCES CITED

Ambroggi, R., 1963, Etude géologique du Versant Meridional du Haut Atlas Occidental et de la plaine du Souss: Notes et Mémoires du Service Géologique du Maroc, v. 157, p. 321.

Beauchamp, W., R. W. Allmendinger, M. Barazangi, A. Demnati, M. El Alji, and M. Dahmani, 1999, Inversion tectonics and the evolution of the High Atlas Mountains, Morocco, based on a geological-geophysical transect: Tectonics, v. 18, p. 163–184, doi:10.1029/1998TC900015.

Brown, R. H., 1980, Triassic rocks of the Argana Valley, southern Morocco, and their regional structural implications: AAPG Bulletin, v. 64, no. 7, p. 988–1003.

Davison, I., 2005, Central Atlantic margin basins of northwest Africa: Geology and hydrocarbon potential (Morocco to Guinea): Journal of African Earth Sciences, v. 43, p. 254–274, doi:10.1016/j.jafrearsci.2005.07.018.

Duffaud, F., L. Brun, and B. Plaucht, 1966, Le bassin Sud-Ouest

Marocain, *in* D. Reyre, ed., Bassins sedimentaires du littoral Africain: Publications d'Association du Service Geologique Africaine, v. 1, p. 5–26.

Dunlap, D. B., and L. J. Wood, 2010, Seismic architecture and morphology of Mesozoic-age sediment waves, offshore Morocco, northwest Africa: 30th Annual Gulf Coast Section of the SEPM Foundation Research Conference, Houston, Texas, December 5–8, 2010, p. 551–571.

Dunlap, D. B., L. J. Wood, C. Weisenburger, and H. Jabour, 2010, Seismic geomorphology of offshore Morocco's east margin, Safi Haute Mer area: AAPG Bulletin, v. 94, p. 615–642, doi:10.1306/10270909055.

Ellouz, N., M. Patriat, J.-P. Gauier, R. Bouatmani, and S. Sabounji, 2003, From rifting to Alpine inversion: Mesozoic and Cenozoic subsidence history of some Moroccan basins: Sedimentary Geology, v. 156, p. 185–212, doi:10.1016/S0037-0738(02)00288-9.

Frizon de Lamotte, D., B. Saint Bezar, and R. Bracène, 2000, The two main steps of the Atlas building and geodynamics of the western Mediterranean: Tectonics, v. 19, no. 4, p. 740–761, doi:10.1029/2000TC900003.

Frizon de Lamotte, D., P. Leturmy, M. Missenard, S. Khomsi, G. Ruiz, O. Saddiqi, F. Guillocheau, and A. Michard, 2009, Mesozoic and Cenozoic vertical movements in the Atlas system (Algeria, Morocco, Tunisia): An overview: Tectonophysics, v. 475, p. 9–28, doi:10.1016/j.tecto.2008.10.024.

Gomez, F., W. Beauchamp, and M. Barazangi, 2000, Role of Atlas Mountains (northwest Africa) within the African-Eurasian plate-boundary zone: Geology, v. 28, p. 775–778, doi:10.1130/0091-7613(2000)28<775:ROTAMN>2.0.CO;2.

Guiraud, R., and W. Bosworth, 1997, Senonian basin inversion and rejuvenation of rifting in Africa and Arabia: Synthesis and implications to plate-scale tectonics: Tectonophysics, v. 282, p. 39–82, doi:10.1016/S0040-1951(97)00212-6.

Hafid, M., 2000, Triassic–Early Liassic extensional systems and their Cenozoic inversion, Essaouira Basin (Morocco): Marine and Petroleum Geology, v. 17, p. 409–429, doi:10.1016/S0264-8172(98)00081-6.

Hafid, M., 2006, Styles structuraux du Haut Atlas de Cap Tafelney et de la partie septentrionale du Haut Atlas occidental: Tectonique salifère et relation entre l'Atlas et l'Atlantique: Notes et Mémoires du Service Géologique du Maroc, v. 465, p. 172.

Hafid, M., M. Zizi, A. Ait Salem, and A. W. Bally, 2006, Structural styles of the western onshore and offshore termination of the High Atlas, Morocco: Comptes Rendus Geoscience, v. 338, p. 50–64, doi:10.1016/j.crte.2005.10.007.

Hafid, M., G. Tari, D. Bouhadioui, I. El Moussaid, H. Echarfaroui, A. Ait-Salem, M. Nahim, and M. Dakki, 2008, Atlantic basins, *in* A. Michard, ed., Continental evolution: The geology of Morocco: Lecture Notes in Earth Sciences 116: Berlin, Springer-Verlag, p. 303–329.

Haq, B. U., J. Hardenbol, and P. R. Vail, 1987, Mesozoic and Cenozoic chronostratigraphy and cycles of sea level change, *in* C. K. Wilgus et al., eds., Sea level changes: An integrated approach: SEPM Special Publication, v. 42, p. 71–108.

Hardenbol, J., J. Thierry, M. B. Farley, T. Jacquin, P. C. de Graciansky, and P. R. Vail, 1998, Mesozoic and Cenozoic chronostratigraphic framework of the European basins: SEPM Special Publication, v. 60, p. 3–15.

Heyman, M. A., 1988, Tectonic and depositional history of the Moroccan continental margin, *in* A. Tankard and H. Balkwill, eds., Extensional tectonics and stratigraphy of the North Atlantic margin: AAPG Memoir 46, p. 323–340.

Hinz, K., H. Dostmann, and J. Fritsch, 1982, The continental margin of Morocco: Seismic sequences, structural elements and geological development, *in* U. Von Raad, ed., Geology of the northwest African continental margin: Berlin, Springer-Verlag, p. 34–60.

Jabour, H., and G. Tari, 2007, Subsalt exploration potential of the Moroccan salt basin: The Leading Edge, v. 26, p. 1454–1460, doi:10.1190/1.2805765.

Knight K. B., S. Nomade, P. R. Renne, A. Marzoli, H. Bertrand, and N. Youbi, 2004, The Central Atlantic magmatic province at the Triassic–Jurassic boundary: Paleomagnetic and ^{40}Ar/^{39}Ar evidence from Morocco for brief, episodic volcanism: Earth Planetary Science Letters, v. 228, p. 143–160, doi:10.1016/j.epsl.2004.09.022.

Lancelot, Y., and E. L. Winterer, 1980a, Evolution of the Moroccan oceanic basin and adjacent continental margin: A synthesis, *in* Y. Lancelot and E. L. Winterer, eds., Initial reports of the Deep Sea Drilling Project: Washington, U.S. Government Printing Office, v. 50, p. 115–301.

Lancelot, Y., and E. L. Winterer, 1980b, Introduction and summary of results: Deep Sea Drilling Project Leg 50, *in* Y. Lancelot and E. L. Winterer, eds., Initial reports of the Deep Sea Drilling Project: Washington, U.S. Government Printing Office, v. 50, p. 801–821.

Laville, E., A. Charroud, B. Fedan, M. Charroud, and A. Piqué, 1995, Inversion negative et rifting atlasique: L'exemple du bassin triassique de Kerrouchene (Maroc): Bulletin de la Societe geologique de France, v. 166, p. 364–374.

Laville, E., A. Piqué, M. Amrhar, and M. Charroud, 2004, A restatement of the Mesozoic Atlasic Rifting (Morocco): Journal of African Earth Sciences, v. 38, p. 145–153, doi:10.1016/j.jafrearsci.2003.12.003.

Le Roy, P., and A. Piqué, 2001, Triassic–Liassic western Moroccan synrift basins in relation to the Central Atlantic opening: Marine Geology, v. 172, p. 359–381, doi:10.1016/S0025-3227(00)00130-4.

Le Roy, P., F. Guillocheau, A. Piqué, and A. M. Morabet, 1998, Subsidence of the Atlantic Morrocan margin during the Mesozoic: Canadian Journal of Science, v. 35, no. 4, p. 476–493.

Marzoli, A., et al., 2004, Synchrony of the Central Atlantic magmatic province and the Triassic–Jurassic boundary climatic and biotic crisis: Geology, v. 32, p. 973–976, doi:10.1130/G20652.1.

Medina, F., 1988, Tilted-blocks pattern, paleostress orientation and amount of extension, related to Triassic early rifting of the Central Atlantic in the Amzri area (Argana Basin, Morocco): Tectonophysics, v. 148, p. 229–233, doi:10.1016/0040-1951(88)90131-X.

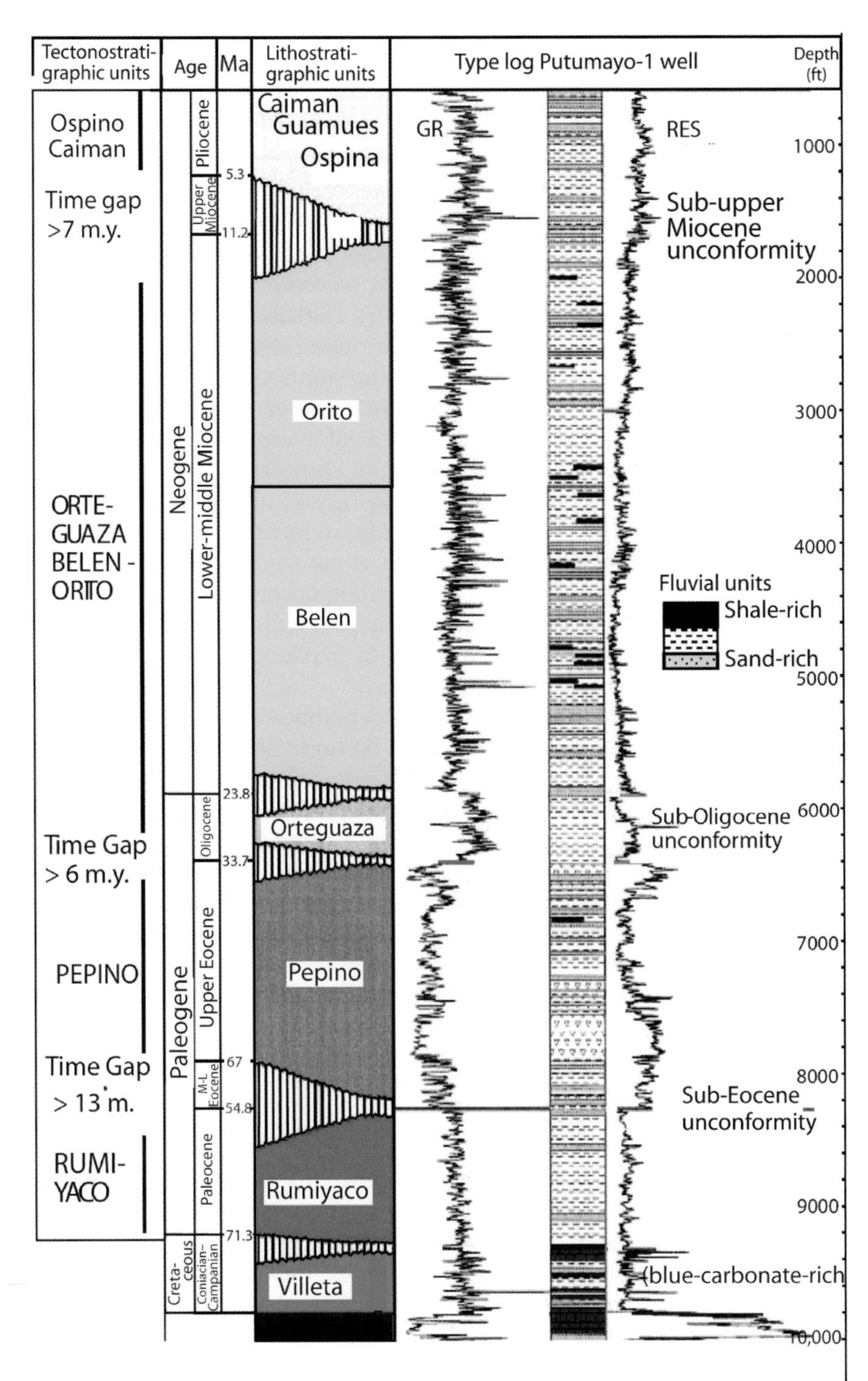

Figure 2. Stratigraphic column and well type log of the Putumayo Basin. The unconformities have been related to main uplift events during the Andes orogeny (Geotec, 1992; Cordoba et al., 1997) and identified in regional seismic lines. GR = gamma ray; RES = resistivity.

thick, adjacent to the Guyana Shield in the east (Dengo and Covey, 1993; Mora et al., 1997; Sarmiento, 2002). The transition from a back-arc basin into a foreland phase occurs during the Late Cretaceous–early Paleogene when the oceanic crust of Western Cordillera today (Figure 1) is accreted to the Central Cordillera (Hendenson, 1979; Aspden et al., 1987) and produces a thick-skin-styled thrust belt and, in the process, tectonically inverted Paleozoic normal faults along the eastern margin of the PRB (Sarmiento, 2002). The mostly continental Rumiyaco Formation (Figure 2) is deposited during this initial collisional period. The sandstone-shale average ratio in this unit has been estimated at 0.21 (Geotec, 1992).

The Tertiary tectonic history of the Putumayo Basin has been divided into three main uplift events related to the Colombian Andean orogeny, which coincide with the dated unconformities, during the middle Eocene, Oligocene, and late Miocene (Geotec, 1992; Cordoba et al., 1997; Figure 2). The first pulse occurred during the late Paleocene–middle Eocene in response to an increase in convergence rate between the Nazca and South American plates (Daly, 1989). It has been interpreted that the continental Pepino Formation is deposited mostly in alluvial fans as a product of the continuous uplift of the Central Cordillera (Mora et al., 1997; Figure 2). The sandstone-shale average ratio in this formation has been estimated at 0.52 (Geotec, 1992). This unit is topped on the west by a regional angular unconformity that cuts the thrusts and back thrusts developed during this period. The second uplift event, caused by the collision between the Andes and the Chocó Terrane during the middle Eocene to the middle Miocene prolonged the exhumation of the Central Cordillera (Sarmiento, 2002) and is responsible for both the continental Oligocene Orteguaza Formation and the Miocene Orito and Belen formations. The sandstone-shale average ratio of this package has been estimated at 0.32 (Geotec, 1992). The third and final tectonic pulse of the foreland stage of the basin was produced by the collision of Panamá with Colombia from the late Miocene to the Holocene (Cordoba et al., 1997; Sarmiento, 2002; Figure 1). This event generated the greatest tectonic shortening and current uplift of the Central Cordillera as well as the development of the doubly vergent thrust belt that constitutes the Eastern Cordillera, in the northern part of the Putumayo Basin. The PRB area remains as an active foredeep depocenter for sediments derived from the Central Cordillera. Fluvial deposits of this period have been informally called Ospina and Caiman formations. The sandstone-shale average ratio of this package has been at 0.47 (Geotec, 1992).

DATA

Over 700 km (438 mi), 2-D, 3 s-long proprietary seismic data from Ecopetrol (Figures 3, 4) form the primary source for interpretations in this study. Composite seismic sections representative of the regional east–west geometry of the PRB (Figure 3) are assembled from more than one single seismic survey for use in flexural analysis (Londono, 2004). Over 50 proprietary well reports from Ecopetrol, containing abundant paleontologic data, are used for the isochore maps of the Rumiyaco, Pepino, and Orteguaza-Belen-Orito intervals (Figure 4; Cordoba et al., 1997). Invariably, all of these units have a characteristic wedge-shaped geometry that thins toward the foreland but shows a little change along the north–south trend, indicating a mostly symmetrical subsidence along strike.

Ecopetrol (Cordoba et al., 1997) subdivides the PRB succession into four tectonostratigraphic units (Rumiyaco, Pepino, Orteguaza-Belen-Orito, and Ospino-Caiman) based on well data that are separated by at least three regional unconformities and/or hiatuses (Figure 2). Time gaps are assessed using paleontologic records (Geotec, 1992; Cordoba et al., 1997). The seismic units between the unconformities are interpreted as seismic-stratigraphic sequences following the methodology of Vail et al. (1977). However, in the composite regional seismic-data sections, the equivalent interpreted seismic unconformable surfaces do not show any angular relationship between the overlying and underlying reflectors. No evidence of structural deformation between these units exists nor do they show any associated erosional feature (Figure 3). Bounding unconformities are not apparent in the isochore maps (Figure 4). Thus, a different mechanism is needed to explain these long-lived regional paraconformities.

We focus our interpretation within unfaulted areas that are most likely to be undisturbed by thrusting at all times and where it appears reasonable to assume that the units have preserved their original reflector geometry and thickness. Among the monotonous, and tabular, seismic facies found within the four sequences, only two types of relevant regional event terminations are recognized: continuous onlapping toward the foreland and as much as eight regional (>50 km [>31 mi]) onlap shifts toward the hinterland (Figure 3).

METHODS AND WORKFLOW

Figure 5 summarizes the basic steps we followed to develop the basin models and analysis. The four sequences are identified in the seismic data, which are depth converted using the proprietary time-to-depth curves of Geotec (1992) and Cordoba et al. (1997) based on interval velocities of more than 30 wells. Interval velocity ranges between 7500 ft/s (2287 m/s) and 18000 ft/s (5488 m/s; Geotec, 1992). The thickness of each unit was calculated and compared against the isochore maps from Ecopetrol (Geotec, 1992; Cordoba et al., 1997) to check for consistency between the two data sets. The estimated difference between the seismic data and the maps is approximately 10% on average. Each sequence, modeled in a composite profile along the seismic lines and extended to the edge of the basin using the isochore maps, is decompacted as a function of its porosity (Watts, 2001; Table 1), assuming a decompacted original porosity of 49% and an exponential constant of 0.27/km (Cordoba et al., 1997). Although

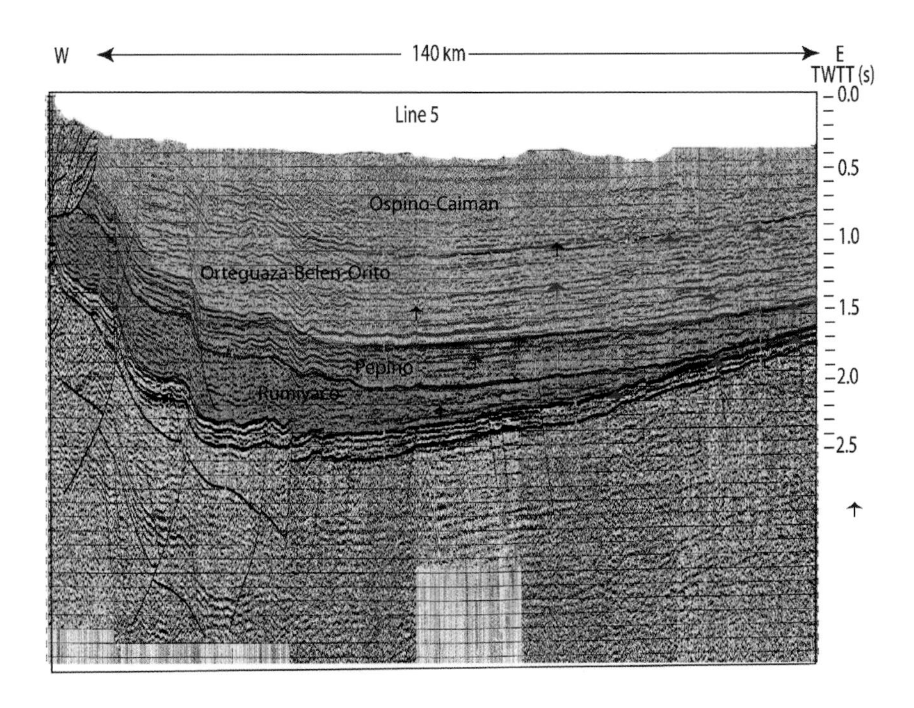

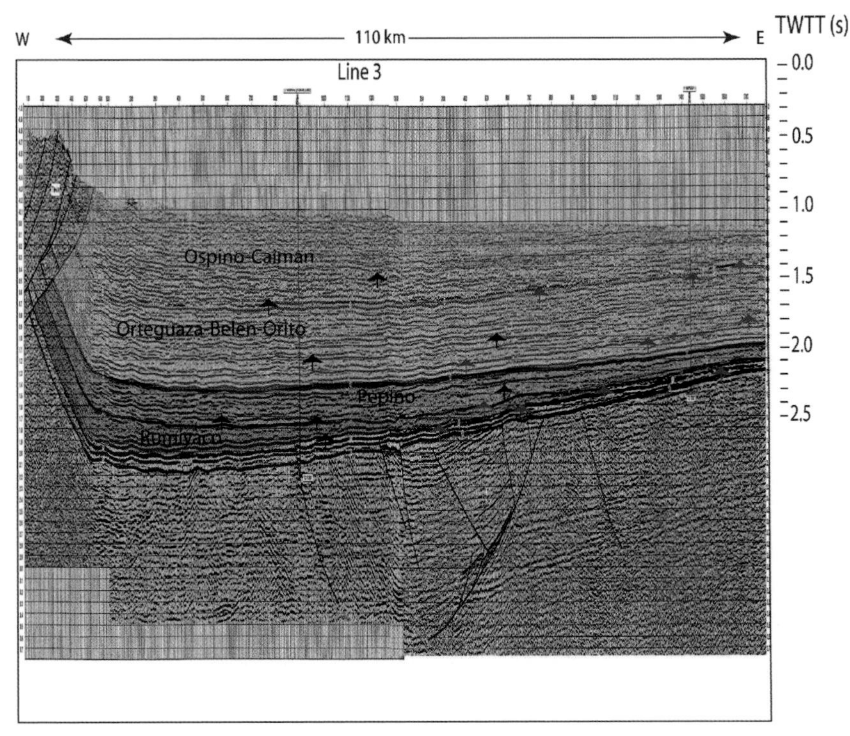

Figure 3. Composite seismic lines 3 (north) and 5 (south; line locations are shown in Figure 4). Color indicates tectonostratigraphic units, as used in this study, bounded by unconformities (thicker lines represent strong events in seismic lines). We interpret foreland onlap toward the east (red arrows) and hinterland onlap shift toward the west (black arrows). The maximum onlap shift within the data boundaries reaches approximately 75 km (~45 mi); as much as eight of these events are found in the foreland sequences (seismic line 3). TWTT = two-way traveltime. 110 km (68.3 mi); 140 km (87 mi).

Jimenez (1997) estimated a 15% tectonic shortening in the Putumayo foothills, thickness values from isochore maps are assumed to be the maximum possible values for each sequence (i.e., maximum flexure when decompacted). The error in thickness when it is not adjusted by tectonic shortening is estimated between 10 and 13% but is considered acceptable, given the unknown nature of the continuation of the thrust belt below the Central Cordillera in the area. Thus, thickness reduction by compaction is estimated in the 15 to 35% range. The parameters used for flexural modeling are summarized in Table 1.

The implementation of flexural modeling assumes that the top of the decompacted continental sequence is originally at base level and that its thickness represents the accommodation of the basin at depositional time, created by the combined effect of tectonic and sedimentary loads deflecting the lithosphere. The amount

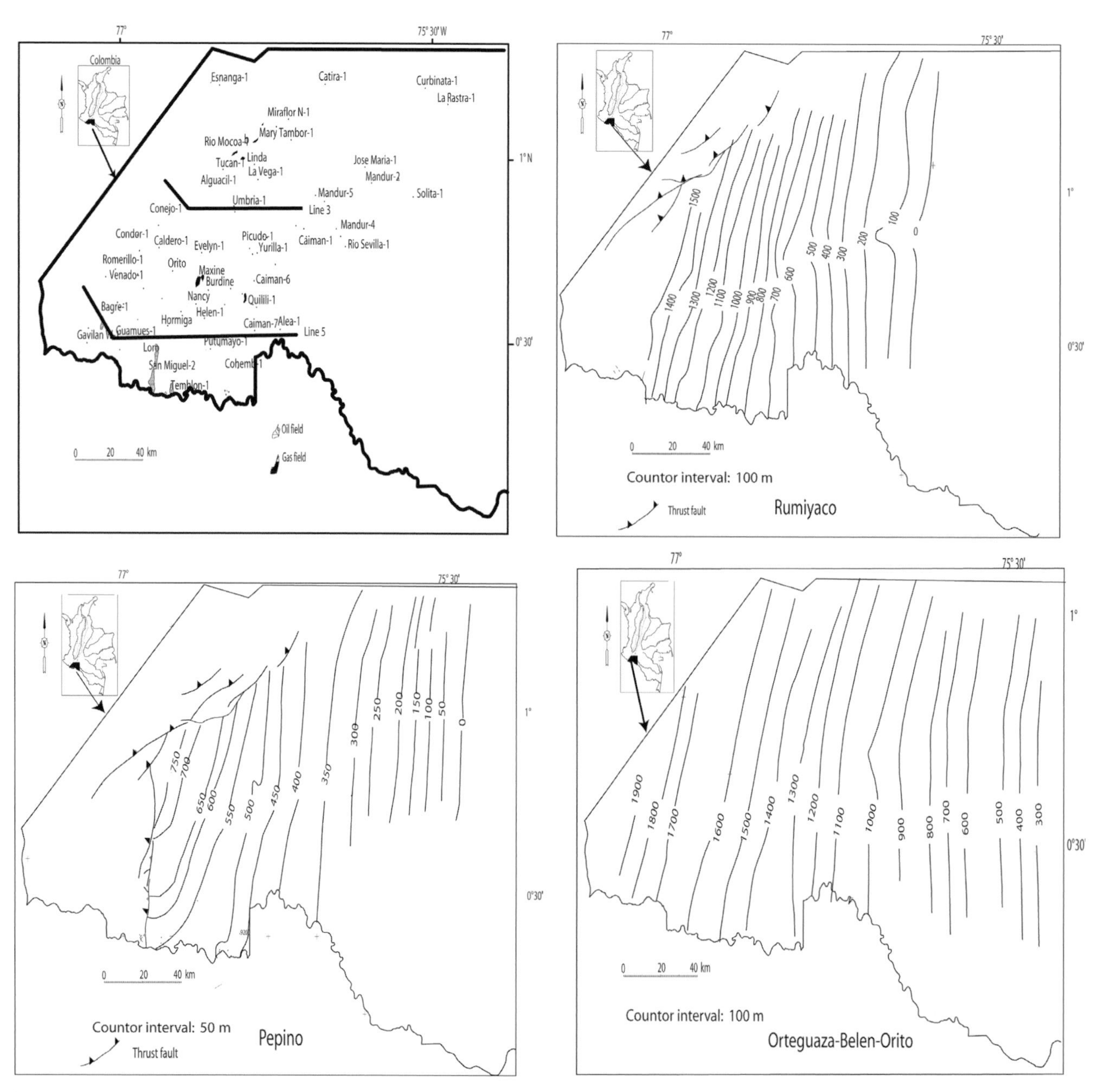

Figure 4. Well location and isochore maps showing wedge-like geometries of stratigraphic units (modified from Cordoba et al., 1997). No erosional scars of regional extent are evident in the maps. Values of these isochore maps were used to constrain flexural modeling. 40 km (24.8 mi).

of flexural subsidence caused by sediment loading is then calculated with a range of effective elastic thickness (10–100 km [6.21–62.1 mi]). The sediment load is discretized into individual linear loads (10-km [6-mi]-wide rectangles) whose heights are a function of the decompacted thickness change along the seismic profiles and isochore maps (Figure 5).The contribution to the overall deflection from each individual rectangle is integrated assuming linear superposition (Hetenyi, 1946). The calculated amount of deflection caused by the decompacted load is subtracted from the original thickness of the decompacted sequence. If the decompacted thickness exceeds the calculated deflection caused by the decompacted sedimentary load, then the residual deflection is taken to represent the minimum accommodation created by thrust-wedge loads at the time just before deposition, this residual flexure is explained by a tectonic load. However, if the decompacted thickness is smaller than the calculated deflection caused by the decompacted load, we assume that no tectonic loading is needed to create the minimum accommodation for the sedimentary sequence.

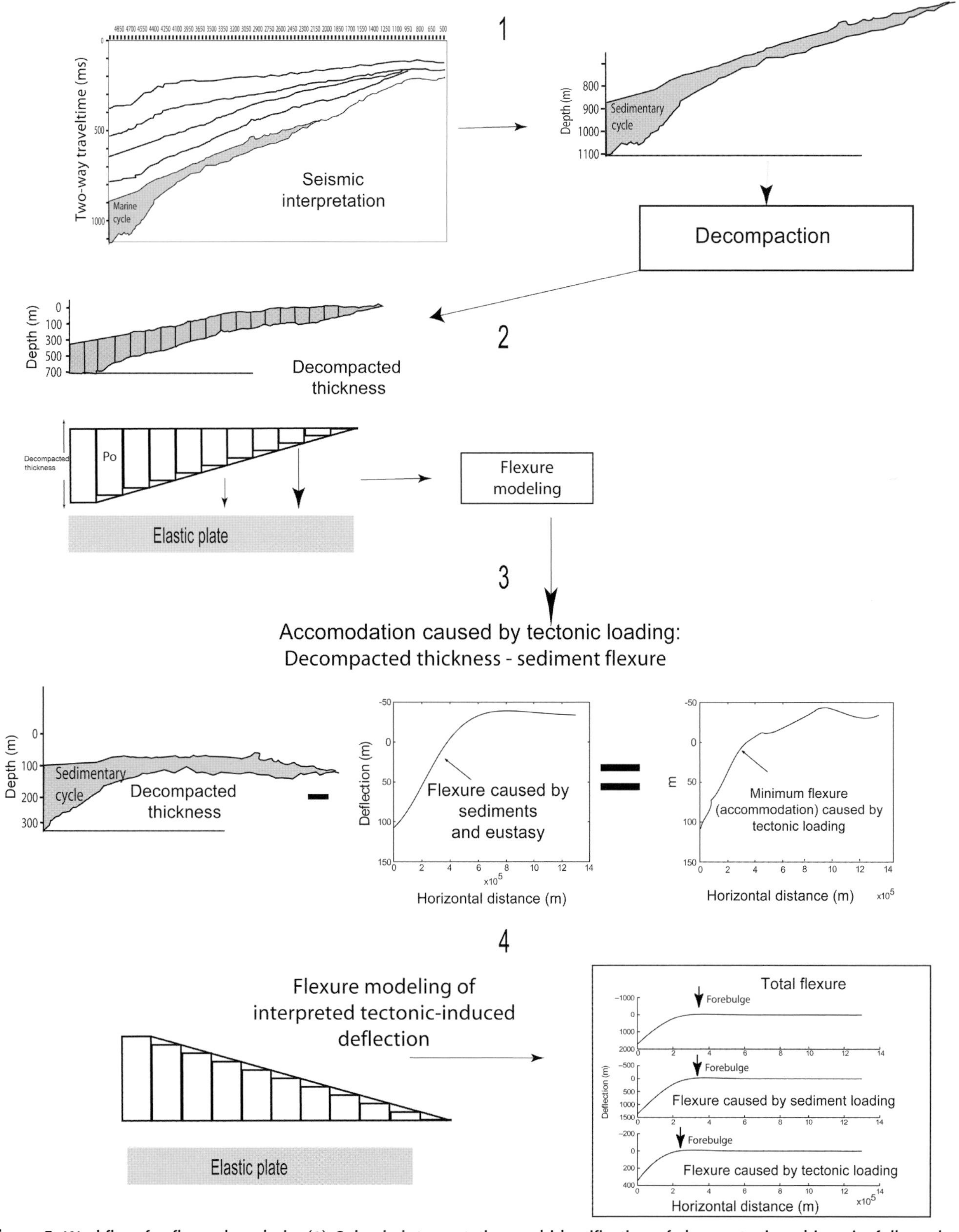

Figure 5. Workflow for flexural analysis. (1) Seismic interpretation and identification of chronostratigraphic units followed by decompaction as a function of porosity. (2) Flexural modeling of sedimentary load. (3) Residual flexure is estimated (represents tectonic-related flexure). (4) Flexural modeling of tectonic wedge. This is an iterative modeling because effective elastic thickness also needs to be modeled. Results show the tectonic, and sedimentary flexure, forebulge location and geometry, the effective elastic thickness, and the estimated width of tectonic wedge. Po = vertical load.

Table 1. Geodynamic constants*

Constant	Symbol	Value	Units
Gravity acceleration	g	9.8	m/s^2
Water density	ρ_w	1035	kg/m^3
Sediment density	ρ_s	1800–1900	kg/m^3
Density of tectonic wedge	ρ_{sw}	2500	kg/m^3
Density mantle	ρ_m	3300	kg/m^3
Young's modulus	E	5×10^{11}	Pa
Poisson's ratio	ν	0.25	
Porosity exponent	c	0.27	1/km

*Porosity exponent taken from Cordoba et al. (1997).

This residual, tectonically induced deflection is then forward modeled iteratively to determine the geometry of the best-fitting tectonic load and an effective elastic thickness comparable to that corresponding to the previously modeled sedimentary package (i.e., we assume that no effective elastic thickness changes during the elapsed time between the onset of tectonic loading and the end of the sedimentary cycle; Figure 5). During modeling, the hinterland end of the tectonic wedge coincides with the edge of the semi-infinite elastic plate in our flexural models, for instance, we assumed that this is the end of the effective load or effective tectonic wedge, although this end does not coincide with the geologic thrust belt hinterland limit that may extend farther toward the arc but has no significant effect on the flexing plate (Watts, 2001). We can estimate the geometry of the tectonic load if we assume that it behaves as a Coulomb wedge, with a critical taper angle that remains constant during thrust belt evolution (Davis et al., 1983). Then, the dimensions of the wedge such as its width and height can be determined. We divided the wedge body along the dip into discrete 1 km (0.6 mi) wide rectangles whose heights are a function of the thrust belt taper angle taken from regional cross sections (Figure 6). The taper angle used during modeling was 7°. It was the average of the cross sections found in published and proprietary articles (Portilla et al., 1993; Geotec, 1992; Dengo and Covey, 1993; Balkwill et al., 1995; Cordoba et al., 1997; Jimenez, 1997).

The computational routine used for forward modeling is developed using MATLAB® (Londono, 2004). The equilibrium equation, $D\frac{d^4w}{dx^4} + (\rho_m - \rho_{if})gw = q(x)$ (Turcotte and Schubert, 1982), is used to calculate the deflection of infinite and semi-infinite elastic beams under vertical loads (where D is flexural rigidity, $w(x)$ is deflection, x is the horizontal distance from plate end, ρ_m is density mantle, ρ_f is the density of basin-filling sedimentary rocks, g is gravity, and $q(x)$ is the vertical load; Table 1). We implement the solution developed by Hetenyi (1946) for semi-infinite beams in which several linear loads load the beam from the end ($x = 0$) inward so that they represent the distributed tectonic and sedimentary loads flexing the plate (Figure 6):

$$
\begin{aligned}
w(x) = {} & \frac{q(x)}{2\alpha(\rho_m - \rho_{if})g}\Big[\{e^{-x/\alpha}(\cos\frac{x}{\alpha} - \sin\frac{x}{\alpha}) \\
& + 2e^{-x/\alpha}(\cos\frac{x}{\alpha})\}\{e^{-x/\alpha}(\cos\frac{x}{\alpha} + \sin\frac{x}{\alpha})\} \\
& - [2\{e^{-x/\alpha}(\cos\frac{x}{\alpha} - \sin\frac{x}{\alpha}) + e^{-x/\alpha}(\cos\frac{x}{\alpha})\} \\
& \times \{e^{-x/\alpha}(\sin\frac{x}{\alpha})\}] + e^{-x/\alpha}(\cos\frac{x}{\alpha} + \sin\frac{x}{\alpha})\Big]
\end{aligned}
\tag{1}
$$

Where α is the flexural parameter (sensu Turcotte and Schubert, 1982). Geodynamic constants such as the

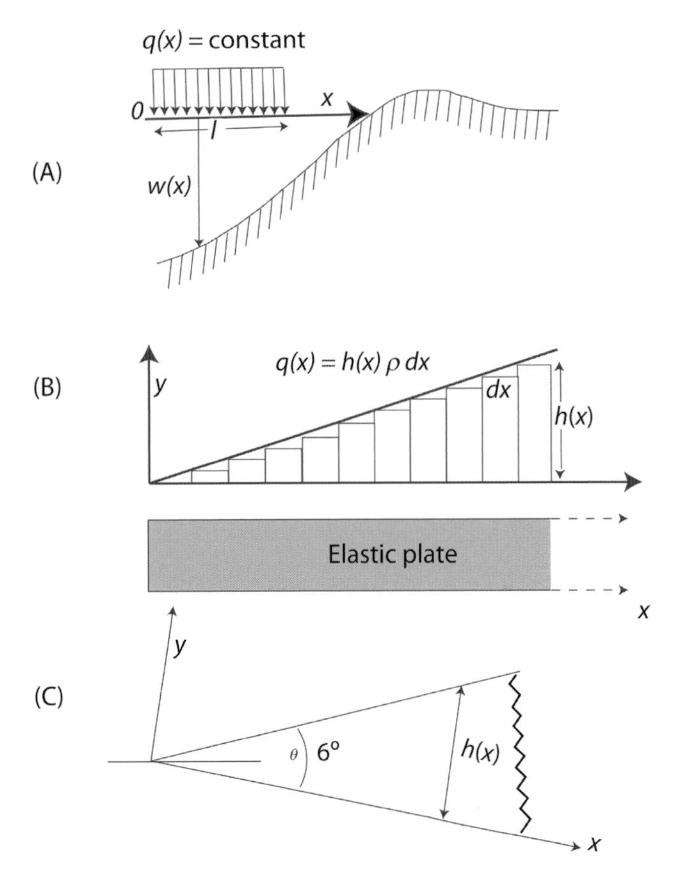

Figure 6. Flexural modeling of the tectonic load. (A) The tectonically induced deflection is forward modeled iteratively to best match the geometry of the tectonic load and the effective elastic thickness. (B) During modeling, the hinterland end of the tectonic wedge coincides with the edge of the semiinfinite elastic plate in our flexural models. We divide the wedge body into discrete, 1 km (1 mi) long rectangles whose heights are a function of the thrust belt taper angle (B, C), taken from regional cross sections.

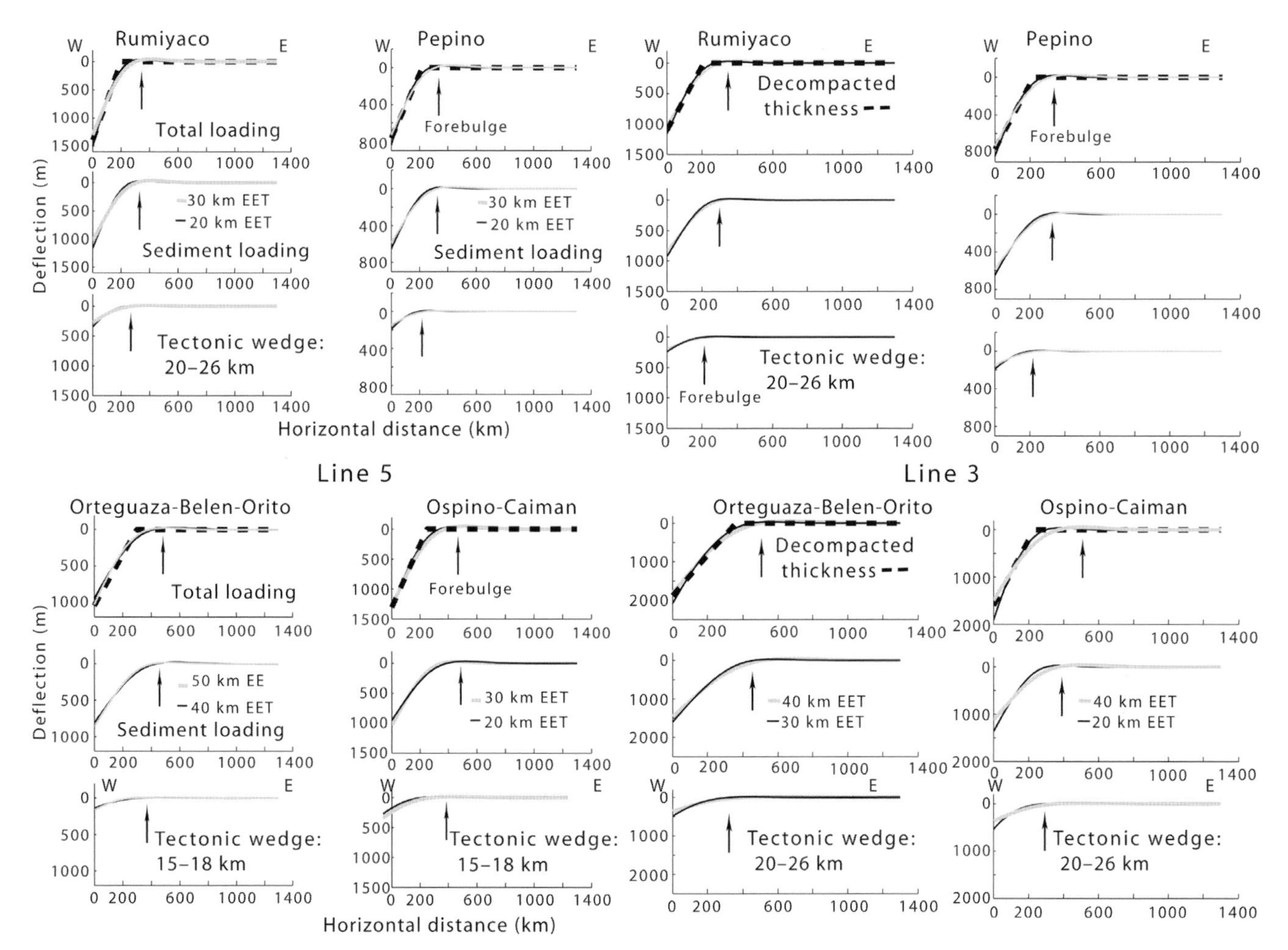

Figure 7. Flexural modeling results for composite seismic lines 5 (left) and 3 (right). For each tectonostratigraphic unit, the upper panel represents total flexure, the middle panel represents sedimentary flexure, and the lower panel represents tectonic flexure. Dashed lines represent the decompacted thickness of tectonostratigraphic units; gray lines represent total flexure; and solid black lines represent tectonic and sedimentary flexures. Note how the location of the forebulge changes concordantly with the evolving geometry of the sedimentary cover and thrust belt. EET = effective elastic thickness.

Poisson's ratio and Young's modulus (which constrains the flexural rigidity (D) and/or effective elastic thickness of the plate) are shown in Table 1. For an ideal elastic plate, the flexural subsidence is treated as occurring instantaneously with each loading phase.

FLEXURAL MODELING RESULTS

Flexural modeling using a semi-infinite homogeneous elastic plate is able to reproduce the subsidence history of the Putumayo foreland basin (Figure 7). The elastic thickness does not change during the entire evolution of the basin but remains constant at an average of 30 ± 10 km (19 ± 6 mi). During the late Oligocene to the middle Miocene, the models along line 5 match better a slightly stronger elastic lithosphere (40 km [25 mi]). However, this value returns to 30 ± 10 km (19 ± 6 mi) in the late Miocene–Holocene. This change is difficult to explain because no reported thermal events that might have affected the crust exist. The thermal age of the crust exceeds 300 m.y. (Cordoba et al., 1997). Therefore, we favor to use an average of 30 ± 10 km (19 ± 6 mi) during the entire evolution of the basin.

The first-order flexural deflection ranges between approximately 200 km (~125 mi) because of tectonic loading during the late Eocene and approximately 400 km (~250 mi) because of sediment loading since the Miocene. According to our results, approximately 450 km (~280 mi) of elastic lithosphere were flexed during the evolution of the PRB.

The geometry of the loads has a strong control on the final geometry of the basin as is manifested in our results. The location of the forebulge changes concordantly with the evolving geometry of the sedimentary

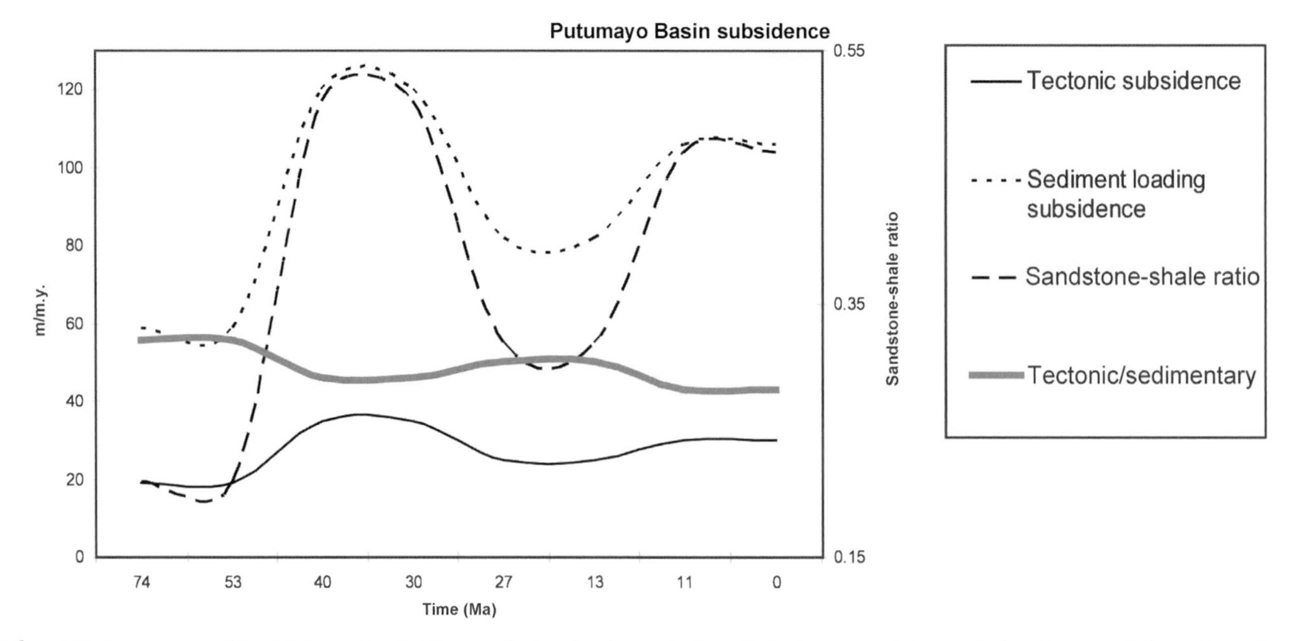

Figure 8. Summary of the history of subsidence during basin evolution. Note how sandstone-shale ratio parallels sedimentary loading.

cover and thrust belt. The geometry of the forebulge reached tens of kilometers in width but only tens of meters in height (Figure 7). Invariably, tectonic loads produce forebulges that lie closer to the hinterland and deflections that are narrower and deeper than those created by sedimentary loads. Tectonic-related forebulges during the history of the basin tend to move in the foreland direction (Figure 7).

Total foreland-sediment pileup above point-of-zero deflection reaches between 1100 ± 150 m (3608 ± 492 ft) and 1350 ± 100 m (4428 ± 328 ft; Figure 7). Tectonic deflection during the entire foreland period reaches approximately 1200 m (~3900 ft; line 3) and approximately 1000 m (~3000 ft; line 5). The difference between sediment pileup and tectonic deflection is between 150 and 250 m (492–820 ft). This difference is of the same order of magnitude as that of the present-day topography (between 500 and 250 m [1640–820 ft] in the basin), which quantitatively validates the results of the flexural model. The continental character of the basin during its foreland period indicates that it has been an area above sea level during its evolution because of sediment accumulation.

The width of the estimated effective tectonic wedge can range between 15 and 30 km (9–19 mi) for a critical taper angle of 7°. This geometry is within the parameter range of tectonic loads (thrust belts) reported in this tectonic setting around the world (Nemèok et al., 2005). The range in width values of the wedge could be a function of the uncertainty in the effective elastic thickness, which can be as high as approximately 30% in this chapter (25% for elastic thickness around the world; Burov and Diament, 1995). The height of the wedge ranges between 3 km (1.86 mi, in the Oligocene) and 1.5 km (0.93 mi, in the Eocene). Today, the Andes Mountains, contiguous to the Putumayo foreland, reach approximately 2 km (~1.24 mi) in elevation, approximately 40 km (~20 mi) from the foothills. It is noteworthy that the entire wedge is not necessarily above any datum such as sea level. Given the nature of thrust belt tectonics, it is reasonable to assume that only a part of this wedge would be subaerially exposed.

Figure 8 summarizes the flexural modeling result in terms of flexural subsidence rates at or near the deepest part of the basin. These values tend to zero at the edge of the basin (i.e., zero values in isochore maps). The Putumayo Basin reached a maximum rate of subsidence of 155 m/m.y. (508 ft/m.y.) during the late Eocene (Pepino Unit: 35 m/m.y. [114.8 ft/m.y.] resulting from tectonic loading and 120 m/m.y. [394 ft/m.y.] resulting from sediment loading) and a minimum rate of 78 m/m.y. (256 ft/m.y.) during the Paleocene: (19 m/m.y. [62 ft/m.y.] resulting from tectonic loading and 59 m/m.y. [194 ft/m.y.] resulting from sediment loading). On average, the tectonic loading subsidence is responsible for approximately 23% of the total subsidence, whereas sediment loading is responsible for the remaining 77%. The sandstone-shale ratio tends to be higher with increasing sediment-related subsidence rate: in the PRB, the highest ratio was reached during the late Eocene (Pepino Formation; Figure 8). However, it is also highest when the ratio of tectonic-related subsidence to sediment-related subsidence tends to be the lowest (Figure 8).

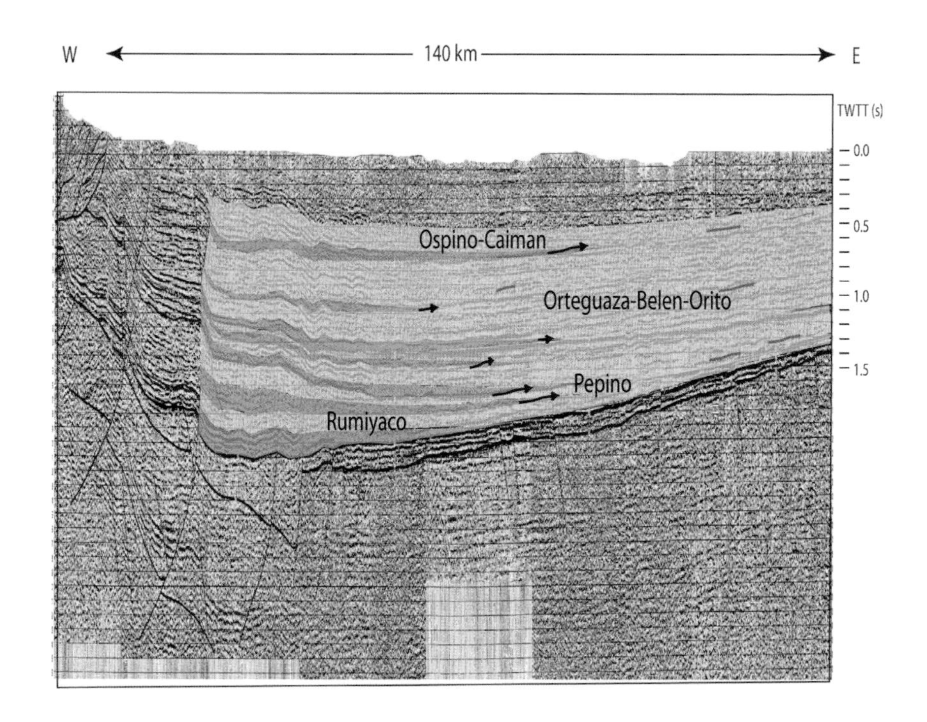

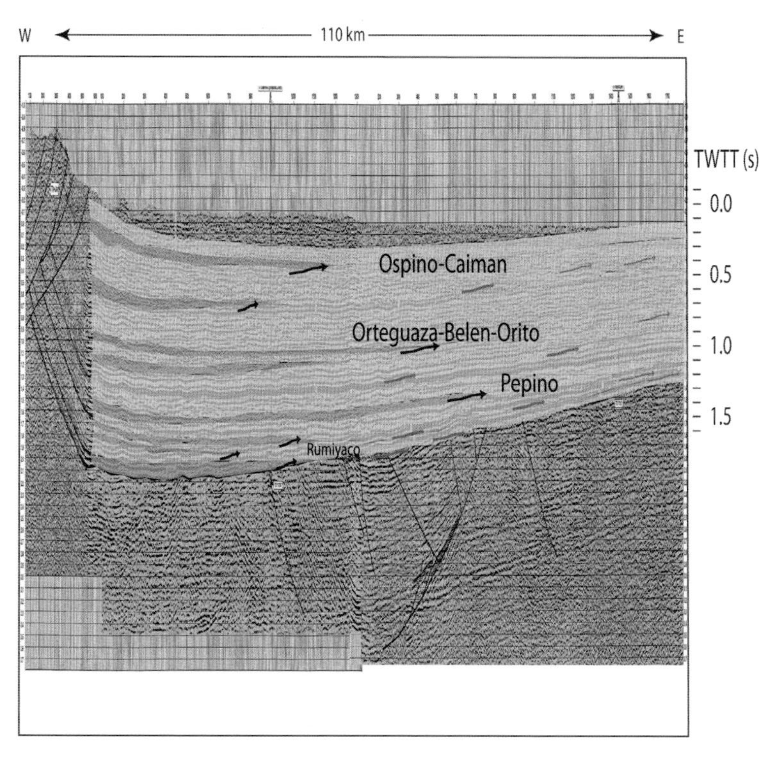

Figure 9. Seismic sections showing the results based on flexurally induced onlap and onlap-shift relationship. As much as eight regional onlap shifts (blue) are found in the seismostratigraphic data. They are interpreted renewed tectonic flexure caused by loading via thrust advancements. The onlap-shift events are overlaid by continuous foreland onlap events (yellow). TWTT = two-way traveltime. 110 km (68.3 mi); 140 km (86.9 mi).

CAUSAL LINK BETWEEN FLEXURE, ACCOMMODATION, AND BASE-LEVEL CHANGES

Despite that the seismic sequences in the Putumayo foreland exhibit a rather monotonous aggradational facies pattern, detailed seismic interpretation leads us to recognize two main reflector geometries along the entire foreland basin: a continuous onlap toward the foreland and some onlap shifts toward the hinterland (Figure 3). The maximum onlap shift, within the data boundaries, reaches approximately 75 km (~46.6 mi, during the Miocene; Figure 3). We find as much as eight of these shifts exceeding 50 km (31 mi) in the foreland sequences. Given their regional extent, these shifts indicate that accommodation is created toward the hinterland end of the depocenter (Figure 9). One

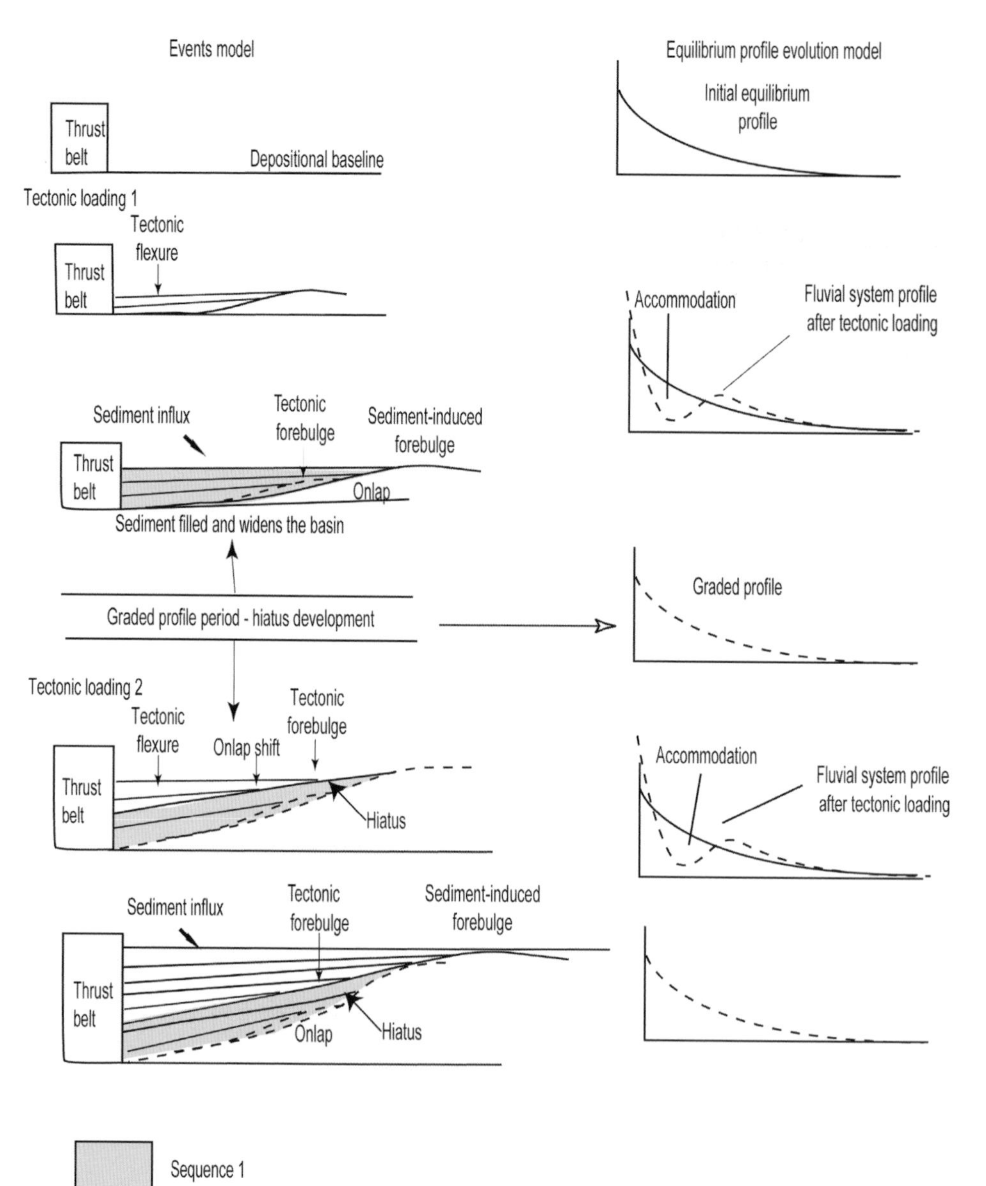

Figure 10. Foreland basin regional development model, integrating flexure, caused by tectonic and sedimentary loads, base-level and long-term hiatuses. The model initiates with a period of tectonic loading (regional hinterland onlap shifts), followed by a period of sedimentation that deepens and widens the basin resulting from sediment loading until either a new tectonic event loads the basin at the hinterland end or the fluvial system reaches the graded stage. If the latter is the case, a period of nondeposition (hiatus) follows and continues until a new loading event, via flexure, creates further accommodation.

likely cause for the deepening toward the hinterland is tectonic loading via thrust advancements. In this case, the new space for sedimentation is created through flexure of the lithosphere under these loads (Figures 3, 9). During periods of tectonic loading, the rate of creation of accommodation (flexure) is greater than the rate of sediment supply because deflection occurs instantaneously (in geologic terms). In addition, the rates of tectonic uplift may be as much as eight times those of denudation (Lawrence and Williams, 1987; Blair and McPherson, 1994). The onlap-shift events are overlaid by continuous foreland onlap events that represent continuous sedimentation without tectonic reactivation. Plate deflection tends to widen in the direction where the sediments are transported as they cover a part of the flexure created by their own weight (Figure 9).

In continental deposits, base level has been commonly considered as the limiting elevation that controls equilibrium between the processes of aggradation and degradation. This abstract definition lacks a physically equivalent boundary in the sedimentary record (Chul, 2006). We cannot always be sure whether an onlap termination coincides with the base-level elevation at that particular time of sedimentation. When a basin is flexed because of sedimentary and tectonic loads, as is the case in the Putumayo foreland basin, the depositional surface subsides continuously, creating more accommodation, although the standard base level may remain unchanged (Figure 10). Sediment aggradation proceeds up to base level and eventually tends to reach the equilibrium profile (Schumm, 1993). During the equilibrium stage, no net sediment deposition or erosion is observed and, therefore, no stratigraphic record remains nor is an erosional scar generated (Figure 10). One possible explanation for the hiatuses reported in the Putumayo Basin is that they represent periods of an equilibrium phase, where the prevailing fluvial-system profiles are near graded stage at base level. Conversely, within this context, a sequence within the continental foreland record may be defined as a sedimentary succession formed during the adjustment of the fluvial system toward an assumed base level during an equilibrium phase (Figure 10).

In addition, and following previous studies (Milana, 1998; Chul, 2006), in the Putumayo Basin, sedimentary sequences can be divided into the following according to the subsidence regime: (1) tectonically induced high-subsidence facies, with probably low sandstone-shale ratio sediments, that are predicted to be associated with onlap shifts toward the hinterland recognizable in seismic data, and (2) sediment-loading-induced low-subsidence facies, containing probably high sandstone-shale ratios, that are predicted to be associated with continual onlap events toward the foreland bounded by onlap shifts and/or unconformities recognizable in seismic data (Figure 3). In the Putumayo Basin, some sequences contain both facies or a succession of them, suggesting that it is not necessary to reach the equilibrium stage before a change in subsidence regime occurs.

As predicted in models (Plint et al., 2001; Miall, 2002, 2006) the sandstone-shale ratio increases with the increasing amount of sediment loading (Figures 2, 8) and its ratio to tectonic load.

DISCUSSION

The results obtained in this study disagree with the current evolutionary model proposed for the Putumayo foreland basin in two fundamental aspects. First, it is unclear whether a causal link exists between tectonic episodes and resultant unconformities (Cordoba et al., 1997). It is also not clear whether the three dated tectonostratigraphic unconformities described in the Putumayo Basin (Geotec, 1992; Cordoba et al., 1997) represent single or composite uplift events followed by the sedimentation of the corresponding tectonostratigraphic unit. The interpreted unconformities show no evidence of erosion and/or structural deformation across them, and so, little evidence to explain the development of paraconformities by tectonic episodes exists. On the contrary, tectonic episodes in foreland basins have been commonly related to the abrupt deposition of coarse-grained facies (Jordan, 1995). A tectonic episode creates, necessarily, a corresponding flexural event followed by the initiation of a high-subsidence phase (Figures 7, 9, 10). The initial tectonic event is followed by a low-subsidence phase controlled mostly by sediment-induced flexure. Our model, however, indicates that these paraconformities may represent hiatuses during periods of equilibrium or graded stage of the prevailing fluvial systems and total tectonic quiescence (Figure 9). Second, we recognize as much as eight regional onlap shifts toward the hinterland (exceeding 50 km [31 mi]) that represent tectonic loading events (Figure 9). Our results support a model with more than three tectonic events to match the seven or more onlap shifts. We interpret that seven periods of thrust belt advancements are recorded in the seismostratigraphic record, and although the resolution in the data does not allow us to assess the geodynamic model for flexure for each one of these tectonic-loading related units, our model consistently explains the development of the paraconformities caused during the tectonic quiescence stage and explains the seismostratigraphic architecture found in the foreland sequences.

Figure 9 shows how at least one onlap shift or tectonic episode is interpreted during the Paleocene early in the sequence, within the limits of our seismic data. Most of the Rumiyaco Formation sequence is dominated by sediment-loading facies, represented by continuous onlap toward the foreland, before reaching the equilibrium stage that produces the succeeding 13 m.y.-time gap unconformity. However, the tectonic-related flexure is the highest at this period compared to the total subsidence and, as predicted in the models, this sequence has the lowest sandstone-shale ratio.

Within the Eocene Pepino Formation, at least two pulses of tectonic episodes can be inferred in the equivalent-age seismic data. Following the second episode, the fluvial system appears to reach the equilibrium stage that generates a hiatus of approximately 6 m.y. Also during the second period, flexural analysis suggests that the maximum subsidence rate is attained for both tectonic and sedimentary loads.

During the late Oligocene to the middle Miocene, at least three episodes of tectonic reactivation are interpreted from onlap shifts in the seismic data. These episodes occurred in the lower part of the seismic sequence: two in the lower half of the section and one toward the middle. The rates of both tectonic-related and sedimentary-related subsidence appear to be slower than for the previous Oligocene section.

Finally, since the late Miocene, at least two tectonic reactivation pulses are interpreted in the lower half of the section. Noisy data quality precludes a reliable interpretation of the upper half of the remaining seismic section. Rates of subsidence appear to increase again, similar to those in the Oligocene, although the tectonic-related subsidence is relatively low compared to the total subsidence. As expected, a sharp increase of the sandstone-shale ratio is observed.

In our interpretation, we assume that the basin is always overfilled, for instance, the fluvial systems run transversally to the tectonic belt, which may not always be the case. We do not have evidence of an underfilled basin, in which case we expect the fluvial system to be parallel with the thrust belt, probably creating a different type of seismic facies, as described by Milana (1998).

As predicted in models (Plint et al., 2001; Miall, 2002, 2006), the sandstone-shale ratio increases with the increasing amount of sediment loading (Figures 2, 9) and its ratio to tectonic load. These results support the idea of a time gap between tectonic uplifting and denudation because they show an increasing amount of fine-grained sediments related to tectonic activity, as the amount of accommodation appears to be a lot higher than that of sediment supply. At the same time, and for the same reason, the results contradict the idea of direct temporal relationship between coarse-grained sediment and tectonic activity.

Although our models determine the maximum width and thickness of the tectonic load or the segment of the thrust belt reactivated during each studied period, we cannot establish with certainty a unique position for the load. Balance cross sections and palinspastic reconstructions extending through the mountain belt are necessary to assess the geologic position of the effective load. However, if we consider that the location of the predicted forebulges associated with each tectonic episode moves toward the foreland during the basin evolution, it is possible to infer a forward-breaking thrust sequence in the Putumayo Basin.

Flexural models predict the change of location and dimension of forebulges along the plate in response to the change in nature and magnitude of the loads throughout the basin history (DeCelles and Giles, 1996). In the Putumayo Basin, our flexural models predicted at least eight forebulges (four tectonic-related and four sedimentary-related) at different locations and with different dimensions (Figure 7). However, we interpret at least seven tectonic reactivation pulses from the seismic data. If the corresponding sediment-related forebulges are added, at least 14 forebulges should be predicted according to the foreland system models (DeCelles and Giles, 1996; DeCelles and Currie, 1996). Yet, when we consider each individual loading event, the scale of these forebulge features is subtle, only a few tens of kilometers wide and, more importantly, only tens of meters high. Distinguishing these forebulges from elevations created by autogenic processes using only stratigraphic records is an extremely difficult task (Figure 9). From the perspective of conventional and heavy oil exploration, one important implication of this assessment is that many stratigraphic traps may exist because of the larger number of tectonic pulses than identifiable in the seismic data. The location and scale of these expected traps can be predicted from flexural analysis. Exploring for hydrocarbons using a single forebulge model for the entire history of the basin could result in a costly mistake.

According to the results obtained in this study, the effective elastic thickness (30 ± 10 km [19 ± 6 mi]) does not change at a scale of 10^7 yr, similar to the results of Sinclair et al. (1991) in the Alps. No viscous-elastic relaxation seems to be necessary to explain the flexure in the Putumayo Basin. Neither the plate curvature nor the sedimentary cover, via blanketing (Lavier and Steckler, 1997), are sufficient to weaken the plate during the evolution of the basin. Cretaceous back-arc rifting processes appear to have not affected the lithosphere in Putumayo and, therefore, the thermal age of the plate is probably Precambrian to Cambrian (Sarmiento, 2002). The effective elastic thickness found in this work is similar to, or slightly lower than, the estimated thickness of the crust in the area (35–40 km [22–25 mi]; Sarmiento, 2002; Watts, 2001) during the latest episode of the Andean orogeny in Colombia during the late Miocene.

The geometry of the flexural deflection during the evolution of the PRB can be explained using tectonic and sedimentary loads only. The geometry of the modeled thrust belt (maximum width, ~30 km [~19 mi]) is within the range of similar systems around the world (Nemèok et al., 2005). In addition, because the wavelength of the first-order flexural deflection did not exceed 450 km (280 mi) at any point during the evolution of the basin, it becomes difficult to endorse dynamic topography as an acting downward force in the PRB (Mitrovica et al., 1989).

CONCLUSIONS

The regional geometry of stratigraphic sequences in retroarc foreland basins appears to be controlled primarily by the flexure of the lithosphere in response to tectonic (thrust sheets) and sediment loads. A tectonic event produces a narrow but deep depocenter, with a high subsidence rate (in relation to sediment supply) and a low sandstone-shale ratio. In the seismostratigraphic record, a tectonic loading event could be recognized by regional (tens of kilometers) onlap shifts from the foreland toward the hinterland. Conversely, a sediment-related subsidence period (controlled mostly by the weight and dispersion of sediments) would produce a wide but relatively shallow depocenter, with a high sediment supply rate compared to the subsidence rate, and a high sandstone-shale ratio. During these periods, the flexure widens and sediments propagate toward the foreland. Seismically, these facies could be recognized by continuous foreland onlap. The end of these periods is marked in the seismostratigraphic record by the regional onlap shifts toward the hinterland that mark the initiation of a new tectonic pulse.

Retroarc foreland sequences can be seen as base-level cycles that move through several stages that include a period of tectonic reactivation, a period of sedimentary-related subsidence, and a period of graded-stage development. Base-level cycles also explain the origin of regional long-live paraconformities within a geodynamic flexural framework. A sequence, in a continental foreland setting, may be defined as a sedimentary succession formed during the adjustment of a fluvial system to the equilibrium stage at base level.

Paraconformities may represent periods of tectonic quiescence and a graded stage of the prevailing fluvial systems in the Putumayo Basin. If a sedimentary unit deposited during a period dominated by sediment loading is topped by a paraconformity, we may interpret that no tectonic reactivation immediately following its deposition (no new accommodation was made available) was observed. Without new accommodation and no new sedimentation and/or erosion, the fluvial profile would be essentially at base level and at graded stage.

In the Putumayo Basin, the highest rate of subsidence (155 m/m.y. [508 ft/m.y.]: 35 m/m.y. [114.8 ft/m.y.] because of tectonic subsidence and 120 m/m.y. [394 ft/m.y.] because of sediment-related subsidence) was attained during the Eocene.

In the Putumayo Basin, at least seven pulses of tectonic reactivation are identified in the seismostratigraphic record of the Tertiary section. These pulses are recognized as regional (>50 km [>31 mi]) onlap shifts toward the hinterland. One tectonic pulse is identified during the Paleocene (Rumiyaco Formation), at least two during the Eocene (Pepino Formation), two during the early to middle Miocene (Orteguaza-Belen-Orito Formation) and, finally, at least two since the late Miocene (Ospino-Caiman Formation). Regional unconformities and hiatuses in the Putumayo represent tectonic quiescence period where fluvial systems reached, or were near, graded stage.

The mechanical models used in this work predicted the occurrence of at least 14 forebulges, coincident with seven periods of tectonic reactivation and seven periods of sediment-controlled subsidence. The scale of these forebulges reaches tens of kilometers in width and tens of meters of height.

The subsidence history of the basin can be reproduced with an elastic semiinfinite plate, whose effective elastic thickness is 30 ± 10 km (19 ± 6 mi). Elastic plate thickness does not need to change during the evolution of the basin to match the observed data. During the history of the basin, the total deflection exceeds 5500 m (18,045 ft) vertically and over approximately 450 km (~280 mi) in width.

ACKNOWLEDGMENTS

We thank Empresa Colombiana de Petroleos (Ecopetrol) and Geotec Colombia for allowing us to use proprietary data and the Landmark Graphics Corporation for allowing us to use their software through an educational grant to the Department of Geology and Geophysics, Louisiana State University, Baton Rouge. The originally submitted manuscript was greatly benefited by the detailed constructive comments from reviewers H. Luo and D. Harry.

REFERENCES CITED

Allen, J., 1978, Studies in fluvial sedimentation: An exploratory quantitative model for the architecture of avulsion-controlled alluvial suites: Sedimentary Geology, v. 21, p. 129–147, doi:10.1016/0037-0738(78)90002-7.

Allen, P., and J. Allen, 1990, Basin analysis: Principles and applications: Oxford, United Kingdom, Blackwell Scientific Publications, 451 p.

Aspden, J. A., W. McCourt, and M. Brook, 1987, Geometrical control of subduction-related magmatism: The Mesozoic and Cenozoic plutonic history of western Colombia: Journal of the Geological Society (London), v. 144, p. 893–905, doi:10.1144/gsjgs.144.6.0893.

Balkwill, H., G. Rodriguez, F. Paredes, and J. Almeida, 1995, Northern parts of Oriente Basin, Ecuador: Reflection seismic expression of structures *in* A. Tankard, R. Soruco, and H. Welsink, eds., Petroleum basins of South America: AAPG Memoir 62, p. 559–571.

Beamont, C., 1981, Foreland basins: Geophysics Journal Research of Astronomical Society, v. 65, p. 471–498.

Blair, T., and J. McPherson, 1994, Historical adjustment by Walker River to lake-level fill over a tectonically tilted half-graben floor, Walker Lake Basin, Nevada: Sedimentary Geology, v. 92, p. 7–16, doi:10.1016/0037-0738(94)00058-1.

Blum, M. D., and T. E. Törnqvist, 2000, Fluvial responses to climate and sea level change: A review and look forward: Sedimentology, v. 47, p. 2–48, doi:10.1046/j.1365-3091.2000.00008.x.

Burov, E. B., and M. Diament, 1995, The effective elastic thickness (T_e) of continental lithosphere: What does it really mean?: Journal of Geophysical Research, v. 100, no. B3, p. 3905–3927, doi:10.1029/94JB02770.

Cardozo, N., and T. Jordan, 2001, Causes of spatially variable tectonic subsidence in the Miocene Bermejo foreland basin, Argentina: Basin Research, v. 13, p. 335–358, doi:10.1046/j.0950-091x.2001.00154.x.

Catuneanu, O., 2004, Retroarc foreland systems evolution through time: Journal of African Earth Sciences, v. 38, p. 225–242, doi:10.1016/j.jafrearsci.2004.01.004.

Chul, W., 2006, Conceptual problems and recent progress in fluvial sequence stratigraphy: Geoscience Journal, v. 10, p. 433–443, doi:10.1007/BF02910437.

Clark, M., and L. Royden, 2000, Topographic ooze, building the eastern margin of Tibet by lower crustal flow: Geology, v. 28, p. 703–706, doi:10.1130/0091-7613(2000)28<703:TOBTEM>2.0.CO;2.

Cordoba, F., E. Kairuz, J. Moros, W. Calderón, F. Buchelli, C. Guerrero, and L. Magoon, 1997, Proyecto Evaluación Regional Cuenca del Putumayo: Definición de Sistemas Petrolíferos: Ecopetrol Internal Report, 119 p.

Daly, M. C., 1989, Correlations between Nazca/Farallon plate kinematics and forearc basin evolution in Ecuador: Tectonics, v. 8, p. 769–790, doi:10.1029/TC008i004p00769.

Davis, D., J. Supper, and F. Dahlia, 1983, Mechanics of fold and thrust belts and accretionary wedges: Journal of Geophysical Research, v. 88, p. 1153–1172, doi:10.1029/JB088iB02p01153.

DeCelles, P., and B. Currie, 1996, Long-term sediment accumulation in the Middle Jurassic–early Eocene Cordilleran retroarc foreland basin system: Geology, v. 24, no. 7, p. 591–594, doi:10.1130/0091-7613(1996)024<0591:LTSAIT>2.3.CO;2.

DeCelles, P., and K. Giles, 1996, Foreland basin systems: Basin Research, v. 8, p. 105–123, doi:10.1046/j.1365-2117.1996.01491.x.

Dengo, C. A., and M. C. Covey, 1993, Structure of the Eastern Cordillera of Colombia: Implications for trap styles and regional tectonics: AAPG Bulletin, v. 77, p. 1315–1337.

Emery, D., and K. Myers, 1997, Sequence stratigraphy: Oxford, England, Blackwell Science Publications, 297 p.

Geotec, 1992, Facies distribution and tectonic setting through the Phanerozoic of Colombia: A regional synthesis combining outcrop and subsurface data presented in 17 consecutive rock-time slices: Bogotá, Colombia, Geotec Limitada, 100 p.

Gomez, E., T. E. Jordan, R. W. Allmendinger, and N. Cardozo, 2005, Development of the Colombian foreland-basin system as a consequence of diachronous exhumation of the northern Andes: Geological Society of America Bulletin, v. 117, p. 1272–1292.

Henderson, W., 1979, Cretaceous to Eocene volcanic arc activity in the Andes of northern Ecuador: Journal of the Geological Society (London), v. 136, p. 367–378.

Hetenyi, M., 1946, Beams of elastic foundation: The University of Michigan Press, 257 p.

Higley, D., 2001, The Putumayo-Oriente-Maranon Province of Colombia, Ecuador, and Peru: Mesozoic–Cenozoic and Paleozoic petroleum systems: U.S. Geological Survey Digital Data Series 63, 20 p.

Horton, B. K., K. N. Constenius, and P. G. DeCelles, 2004, Tectonic control on coarse-grained foreland-basin sequences: An example from the Cordilleran foreland basin, Utah: Geology, v. 32, p. 637–640.

Jimenez, C., 1997, Structural styles of the Andean foothills, Putumayo Basin, Colombia: Master's thesis, University of Texas at Austin, Austin, Texas, 73 p.

Jordan, T., 1995, Retroarc foreland and related basins, *in* C. Busby and R. Ingersoll, eds., Tectonics and sedimentation: Blackwell Science Publications, p. 331–362.

Lavier, L., and M. Steckler, 1997, The effect of sedimentary cover on the flexural strength of continental lithosphere: Nature, v. 389, p. 476–479, doi:10.1038/39004.

Lawrence, D., and B. Williams, 1987, Evolution of drainage systems in response to Acadian deformation: The Devonian Battery Point Formation, eastern Canada, *in* F. Ethridge, R. Flores, and M. Harvey, eds., Recent developments in fluvial sedimentology: SEPM Special Publication 39, p. 287–300.

Liu, S., and D. Nummedal, 2004, Late Cretaceous subsidence in Wyoming: Quantifying the dynamic component: Geology, v. 32, no. 5, p. 397–400, doi:10.1130/G20318.1.

Londono, J., 2004, Foreland basins: Lithospheric flexure, plate strength and regional stratigraphy: Ph.D. dissertation, Louisiana State University, Baton Rouge, Louisiana, 175 p.

Mackin, J., 1948, Concept of graded river: Geological Society of America Bulletin, v. 59, p. 463–512.

Miall, A., 1997, The geology of stratigraphic sequences: Berlin, Germany, Springer-Verlag, 433 p.

Miall, A., 2002, Architecture and sequence stratigraphy of Pleistocene fluvial systems in the Malay Basin, based on seismic time-slice analysis: AAPG Bulletin, v. 86, p. 1201–1216.

Miall, A., 2006, Reconstructing the architecture and sequence stratigraphy of the preserved fluvial record as a tool for reservoir development: A reality check: AAPG Bulletin, v. 90, p. 989–1002, doi:10.1306/02220605065.

Mora, C., M. P. Torres, and J. Escobar, 1997, Potencial generador de hidrocarburos de la Formación Chipaque y su relación estratigráfica secuencial en la zona axial de la Cordillera Oriental (Colombia): VI Simposio Bolivariano Exploración Petrolera en las Cuencas Subandinas, Cartagena, Colombia, p. 217–237.

Michal, N., S. Steven, and R. Gayer, 2005, Thrust belts: Structural architecture, thermal regimes and petroleum systems: Cambridge, United Kingdom, Cambridge University Press, 541 p.

Milana, J., 1998, Sequence stratigraphy in alluvial settings: A flume-based model with applications to outcrop seismic data: AAPG Bulletin, v. 82, p. 1736–1753.

Mitrovica, J. X., C. Beaumont, and G. T. Jarvis, 1989, Tilting of continental interiors by the dynamical effects of subduction: Tectonics, v. 8, p. 1079–1094, doi:10.1029/TC008i005p01079.

Nemcok, M., S. Schamel, and R. Gayer, 2005, Thrustbelts: structural architecture, thermal regimes and petroleum systems: Cambridge University Press, 541 p.

Plint, G., P. McCarthy, and U. Faccini, 2001, Nonmarine sequence stratigraphy, updip expression of sequence boundaries and systems tracts in a high-resolution framework, Cenomanian Dunvegan Formation, Alberta foreland basin, Canada: AAPG Bulletin, v. 85, p. 1967–2001.

Portilla, O., E. Ch. Kairuz, C. A. Y Lombo, and H. Garzón, 1993, Informe Final Proyecto Putumayo Oeste Fase III: Bogotá, Colombia, Ecopetrol, 153 p.

Posamentier, H., and P. Vail, 1988, Eustatic controls on clastic deposition II: Sequence and systems tract models, *in* C. Wilgus, B. S. Hasting, C. G. Kendall, H. W. Posamentier, C. A. Ross, and J. C. Van Wagoner, eds., Sea level changes: An integrated approach: SEPM Special Publication 42, p. 125–154.

Sarmiento, L., 2002, Mesozoic rifting and Cenozoic basin inversion history of the Eastern Cordillera, Colombian Andes: Inferences from tectonic models: Ph.D. dissertation, Vrije Universiteit, Amsterdam, The Netherlands, 295 p.

Schumm, S., 1993, River response to base-level change: Implication for sequence stratigraphy: Journal of Geology, v. 101, p. 279–294, doi:10.1086/648221.

Schumm, S., J. Dumont, and J. Holbrook, 2000, Active tectonics and alluvial rivers: Cambridge, United Kingdom, Cambridge University Press, 276 p.

Steckler, M., and A. Watts, 1978, Subsidence of an Atlantic-type continental margin off New York: Earth Planetary Sciences, v. 7, p. 1–13, doi:10.1016/0012-821X(78)90036-5.

Sclater, J. G., and P. A. F. Christie, 1980, Continental stretching: An explanation of the post–mid-Cretaceous subsidence of the central North Sea Basin: Journal of Geophysical Research, v. 85, p. 3711–3739, doi:10.1029/JB085iB07p03711.

Shanley, M., and P. McCabe, 1994, Perspectives on sequence stratigraphy of continental strata: AAPG Bulletin, v. 78, p. 654–568.

Sinclair, H., B. Coakly, P. Allen, and A. Watt, 1991, Simulation of foreland basin stratigraphy using a diffusion model of mountain belt erosion: An example from the Alps of eastern Switzerland: Tectonics, v. 10, p. 599–620, doi:10.1029/90TC02507.

Sloss, L. L., 1962, Stratigraphic models in exploration: AAPG Bulletin, v. 74, p. 93–113.

Turcotte, D., and G. Schubert, 1982, Geodynamics: Application of continuum physics to geological problems: New York, Wiley, 448 p.

Vail, P., R. Mitchum, and S. Thompson, 1977, Seismic stratigraphy and global changes of sea level: Relative changes of sea level from coastal onlap, *in* C. E. Payton, ed., Seismic stratigraphy: Applications to hydrocarbon exploration: AAPG Memoir 26, p. 63–81.

Watts, A. B., 2001, Isostasy and flexure of the lithosphere: Cambridge, United Kingdom, Cambridge University Press, 472 p.

Wescott, W., 1993, Geomorphic thresholds and complex response of fluvial systems: Some implications for sequence stratigraphy: AAPG Bulletin, v. 77, p. 1208–1218.

17

Luo, Hongjun, and Dag Nummedal, 2012, Forebulge migration: A three-dimensional flexural numerical modeling and subsurface study of southwestern Wyoming, *in* D. Gao, ed., Tectonics and sedimentation: Implications for petroleum systems: AAPG Memoir 100, p. 377–395.

Forebulge Migration: A Three-dimensional Flexural Numerical Modeling and Subsurface Study of Southwestern Wyoming

Hongjun Luo

BP Exploration, Chertsey Rd., Sunbury-on-Thames, TW16 7LN, United Kingdom (e-mail: luo.hongjun@bp.com)

Dag Nummedal

Colorado Energy Research Institute, Colorado School of Mines, 1500 Illinois St., Golden, Colorado, 80401, U.S.A. (e-mail: nummedal@mines.edu)

ABSTRACT

The recognition of a forebulge in the subsurface is difficult because of its low amplitude and wide extent. It is further complicated by the subsequent tectonic modification (by the Laramide orogeny in this case) that may have overprinted the forebulge with complex younger structural patterns. Three-dimensional (3-D) flexural numerical modeling provides a strong supportive tool to help predict forebulge locations and focus subsurface search on their subtle isopach expression. Based on detailed well-log correlations and good outcrop control, three regional cross sections were established to identify Late Cretaceous forebulges in southwestern Wyoming. Along these sections in the Greater Green River Basin (two east–west and one northwest–southeast), the existence of forebulges was only recognized in the southern section. In response to the progressive eastward movement of the Crawford, early Absaroka, and late Absaroka thrusts, the forebulge migrated eastward to the Moxa Arch, the Rock Springs Uplift, and the Washakie Basin, respectively. The 3-D flexural modeling indicates that the forebulge was limited in its extent only to the southern part of the basin because of the distribution of thrust loads. The forebulge shifted southeastward over time because of the migration of these loads. The 3-D flexural modeling is critical to understanding Late Cretaceous forebulge migration across southwestern Wyoming.

INTRODUCTION

A forebulge is a small uplift on the distal margin of a foreland basin maintained by the strength of the lithosphere in response to loading on the proximal side. Unless crustal rigidity and thrust load changes abruptly in space and time, the separation distance between the leading edge of an active thrust belt and the forebulge should not change much during the migration of the orogen. This hypothesis, however, has rarely been tested because very few studies have been able to track forebulge migration through space and time. This article provides such a test and related 3-D numerical modeling through analysis of an extensive subsurface data set on

DOI:10.1306/13351563M1003538

the stratigraphy of the Late Cretaceous basin in southwestern Wyoming, a foreland basin immediately east of the Late Cretaceous Sevier fold and thrust belt.

This Late Cretaceous foreland basin represents a small part of the Cretaceous Western Interior basin that formed as a consequence of Mesozoic subduction along the western margin of North America. The Greater Green River Basin is an intermontane basin that was superimposed on the preceding Cretaceous foreland basin during the Laramide orogeny (post-80 Ma). The Wind River Mountains, Idaho-Wyoming fold and thrust belt, Uinta Mountains, Sierra Madre, and Rawlins uplift define the northwestern, western, southwestern, eastern, and northeastern margins of the basin, respectively. The Greater Green River Basin contains four sub-basins: the Green River Basin, Great Divide Basin, Washakie Basin, and Sand Wash Basin, separated by the Rock Springs Uplift, Wamsutter Arch, and Cherokee Ridge, respectively (Figure 1).

Cretaceous rocks are among the most prolific oil, gas, and coal producers in Wyoming as well as in surrounding Rocky Mountain states. Thanks to active oil and gas exploration, plenty of well-logs and seismic data are available throughout the basin. In the past several decades, the Greater Green River Basin has been extensively studied for its tectonics (Vietti, 1977; Armstrong and Oriel, 1986; Oriel and Armstrong, 1986; Kraig et al., 1987; Royse, 1993; Pang and Nummedal, 1995; Liu and Nummedal, 2004), lithostratigraphy (Roehler, 1965, 1990; Lawrence, 1992; Dyman et al., 1994; Martinsen et al., 1998), chronostratigraphy (Gill and Cobban, 1973; Miller, 1977; Nichols and Jacobson, 1982; Jacobson and Nichols, 1982; Kauffman et al., 1993), sequence stratigraphy (Devlin et al., 1993; Nummedal et al., 2002), and tectonostratigraphy (Elliott, 1977; Devlin et al., 1993; DeCelles, 1994, 2004; Liu et al., 2005). The abundance of geologic data, extensive research results, and easy access to outcrops make the Greater Green River Basin an ideal location to conduct research on foreland basins.

To document more precisely the Upper Cretaceous stratigraphy in southwestern Wyoming, Luo (2005) constructed three regional and one local cross section across the basin using detailed well-log correlations as well as published and new outcrop observations along the basin margins. Only after such detailed correlations were completed was it possible to define and model forebulge migration in response to progressive eastward thrusting in the Sevier thrust belt.

TIMING OF MAJOR THRUSTING EVENTS

To set the stage for forebulge identification and numerical modeling, major tectonic elements and their timing in Southwestern Wyoming are briefly reviewed (Figure 2).

The Uinta Mountains are a major east–west-trending range in northeastern Utah and northwestern Colorado, located just south of, and roughly parallel with, the Utah-Wyoming border (Figure 1). The major Laramide uplift of the Uinta Mountains occurred in two distinct pulses (Bradley, 1995). The first period of uplift occurred during latest Cretaceous to early Paleocene, and the second uplift occurred in late early to early middle Eocene.

The Wind River Range is the largest basement-cored uplift in the broken foreland of Wyoming, exposing Precambrian crystalline rocks in west-central Wyoming (Figure 1; Steidtmann and Middleton, 1991). Initial displacement on the Wind River fault, along the southwestern margin of the range, probably began at approximately 90 Ma (Shuster and Steidtmann, 1988) and ceased at approximately 49 to 50 Ma (Steidtmann and Middleton, 1991).

The Sierra Madre is the northern extension of the Park Range of Colorado and forms the eastern boundary of the Greater Green River Basin (Figure 1). The Sierra Madre also formed during the Laramide orogeny (Mears, 1998).

The Idaho-Wyoming fold and thrust belt is a remnant of the much more extensive Late Cretaceous–early Paleocene Sevier orogenic belt. According to Liu et al. (2005) and Royse (1993), the major thrust-fault systems and the time intervals of major displacement are (from west to east) Paris-Willard (Barremian?–Aptian), Meade-Laketown (Aptian–Turonian), Crawford (Coniacian–Santonian), Absaroka (Santonian–early Paleocene), and Hogsback (Paleocene–early Eocene) (Figure 1).

The Rock Springs Uplift is located in the middle part of the Greater Green River Basin (Figure 1). During the Laramide orogeny, the Rock Springs area was greatly uplifted and Upper Cretaceous strata were breached and partially eroded.

The Moxa Arch is a broad, gently folded basement uplift in the western part of the basin (Figure 1). Devlin et al. (1993), building on the tectonic analysis of the Moxa Arch by Kraig et al. (1987, 1988), concluded that the Moxa Arch uplift caused major unconformities. Based on the age of these unconformities, the timing of the Moxa Arch uplift is set to 79.5 to 72.4 Ma (middle and early Late Campanian).

FOREBULGE ANALYSIS

Identification of Forebulges

The Upper Cretaceous stratigraphic units in southwestern Wyoming, from bottom to top within the target interval of study in this chapter, are the Frontier Formation,

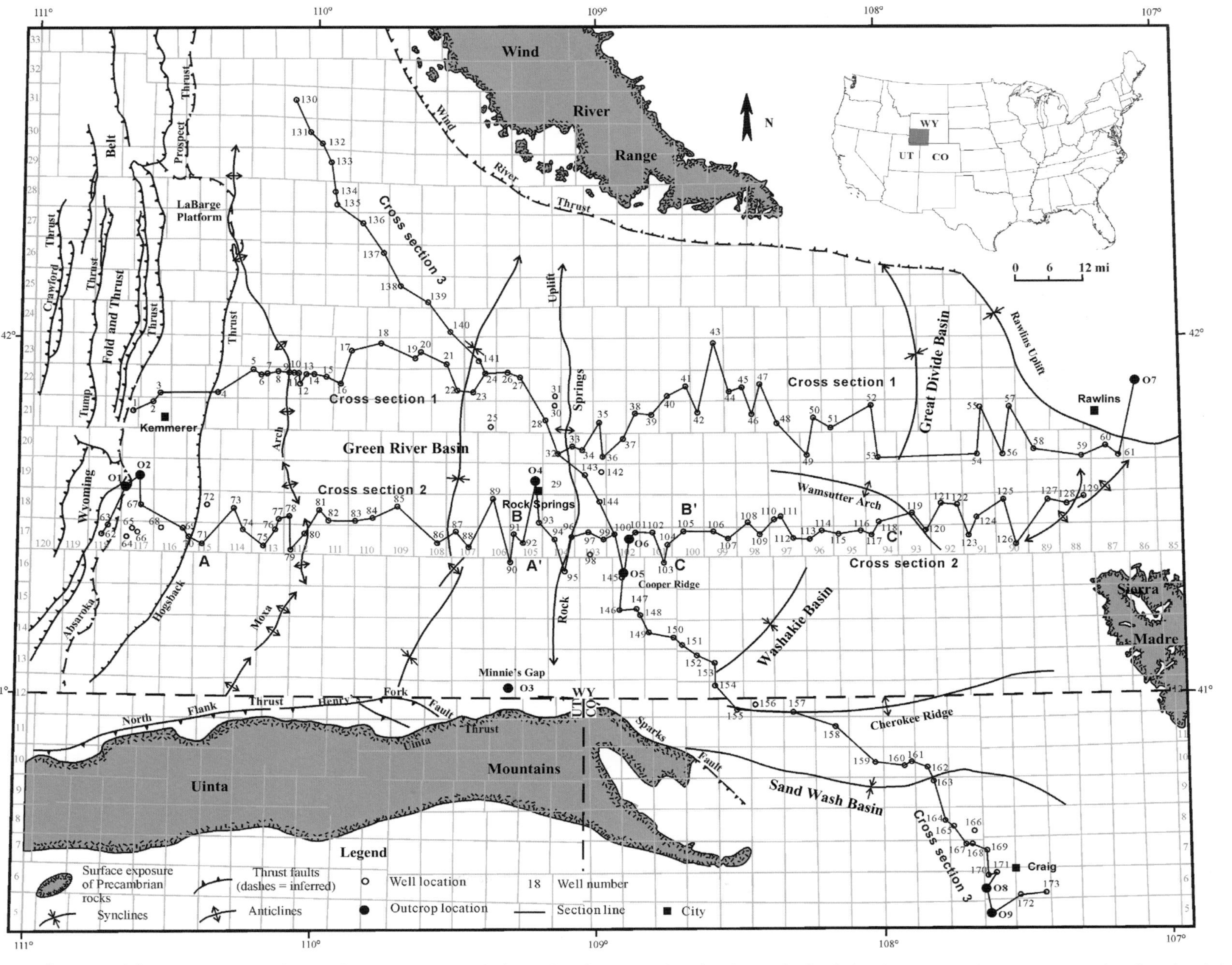

Figure 1. Index map of the Greater Green River Basin. Cross section 2 is the regional cross section that forms the basis for Figure 3. Section segments AA′, BB′, and CC′ along cross section 2 show the section locations for Figures 4, 5, and 6, respectively. WY = Wyoming; UT = Utah; CO = Colorado. 12 mi (19.3 km).

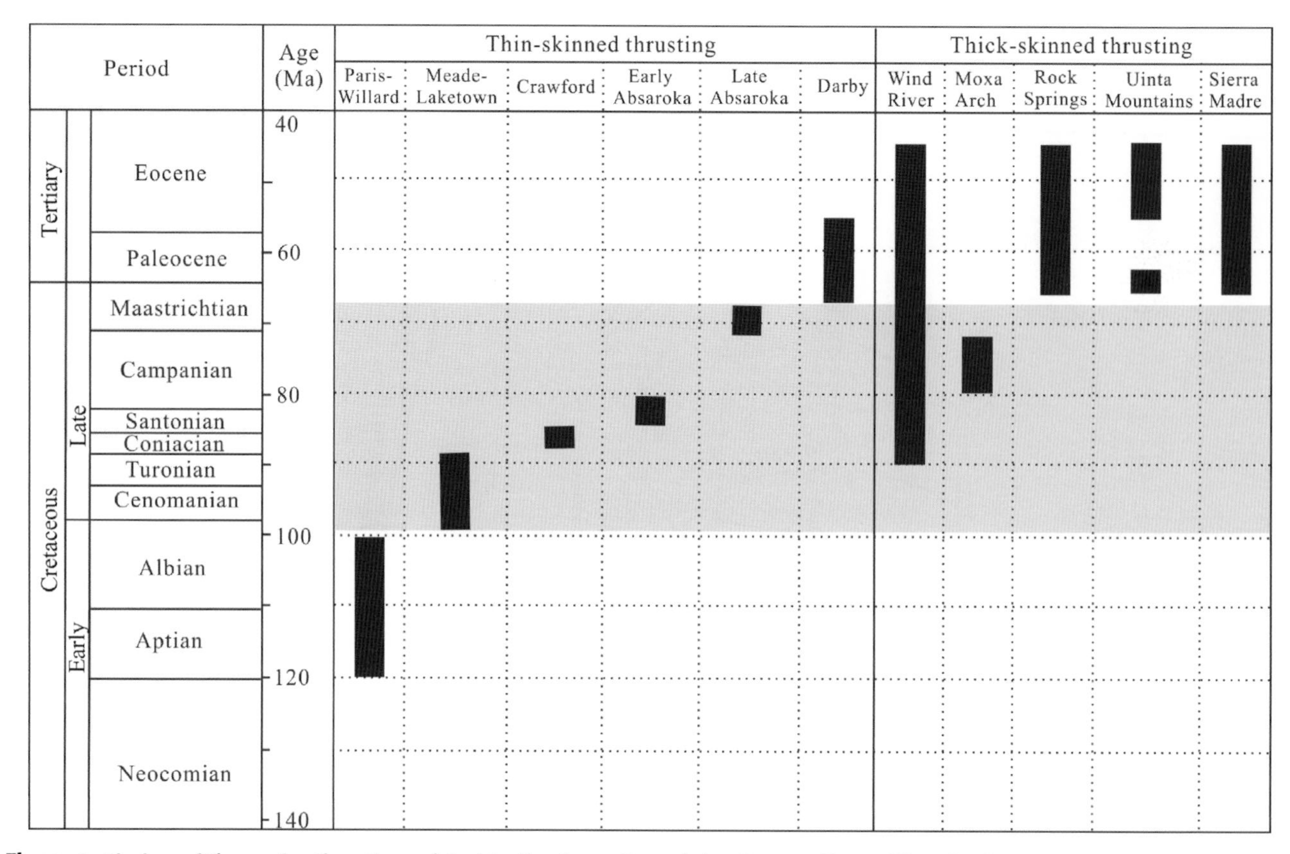

Figure 2. Timing of the major thrusting related to the formation of the Greater Green River Basin.

Baxter-Steele shales, Niobrara, Blair, Adaville, Rock Springs, Haystack Mountain, Ericson, Allen Ridge, Almond formations, and Lewis shale. Three regional cross sections across the Greater Green River Basin are included in Figures 3–5.

Among these three regional cross sections, only along cross section 2 (Figures 1, 4) did the subsurface correlations reveal three thinning and partly erosional stratigraphic intervals. Moreover, these thinning intervals shifted location with age. They are interpreted as stratigraphic expressions of migrating forebulge, based on the following geologic observations: (1) the amplitude of the thinning and erosion is low, commonly no more than 80 m (240 ft); (2) the area of thinning is wide, 30 to 50 km (20–30 mi); (3) the timing of thinning is coeval with a thrusting event, and thinning does not persist with time; (4) the distance between the individual thinning area and the coeval thrust front is 160 to 200 km (100–125 mi), which corresponds to the modeled width of the associated thrust-belt foredeep. The cross sections that provide evidence for these three forebulges are given in Figures 6, 7, and 8, respectively.

The forebulge located at the Moxa Arch (Figure 6) was formed by the Crawford thrusting, which was active from the Coniacian–Santonian (Figure 2). The cross section in Figure 6 emphasizes part of the Hilliard shale, just above the Frontier Formation, in the Moxa Arch area and uses a datum at the FS-1 regional bentonite bed. The amplitude of thinning is approximately 40 m (~120 ft), and the width is approximately 50 km (~30 mi).

The late Santonian–early Campanian, early Absaroka thrusting (Figure 2) caused forebulge development at the Rock Springs Uplift (Figure 7). The datum for this section is the regional bentonite bed FS-1. The major stratigraphic intervals are the Airport Sandstone and surrounding shales (Figure 4). The amplitude of thinning is approximately 70 m (~220 ft), and the width is approximately 30 km (~20 mi). The width of the early Absaroka forebulge at the Rock Springs Uplift (30 km [19 mi]) is less than that of the Crawford-Meade forebulge at the Moxa Arch (50 km [31 mi]).

The late Absaroka thrusting, active from the late Campanian to Maastrichtian (Figure 2), contributed to forebulge development at the western margin of the Washakie Basin (Figure 8). The cross section in Figure 8 includes the stratigraphic interval of the Lewis shale, Almond Formation, and Ericson Sandstone in Washakie Basin. The datum for this section is the Asquith marker of the Lewis shale. The thinning of the underlying Lewis shale and Almond Formation indicates the existence of a forebulge (Figure 8), with an amplitude of thinning of approximately 50 m (~150 ft) and a width of approximately 30 km

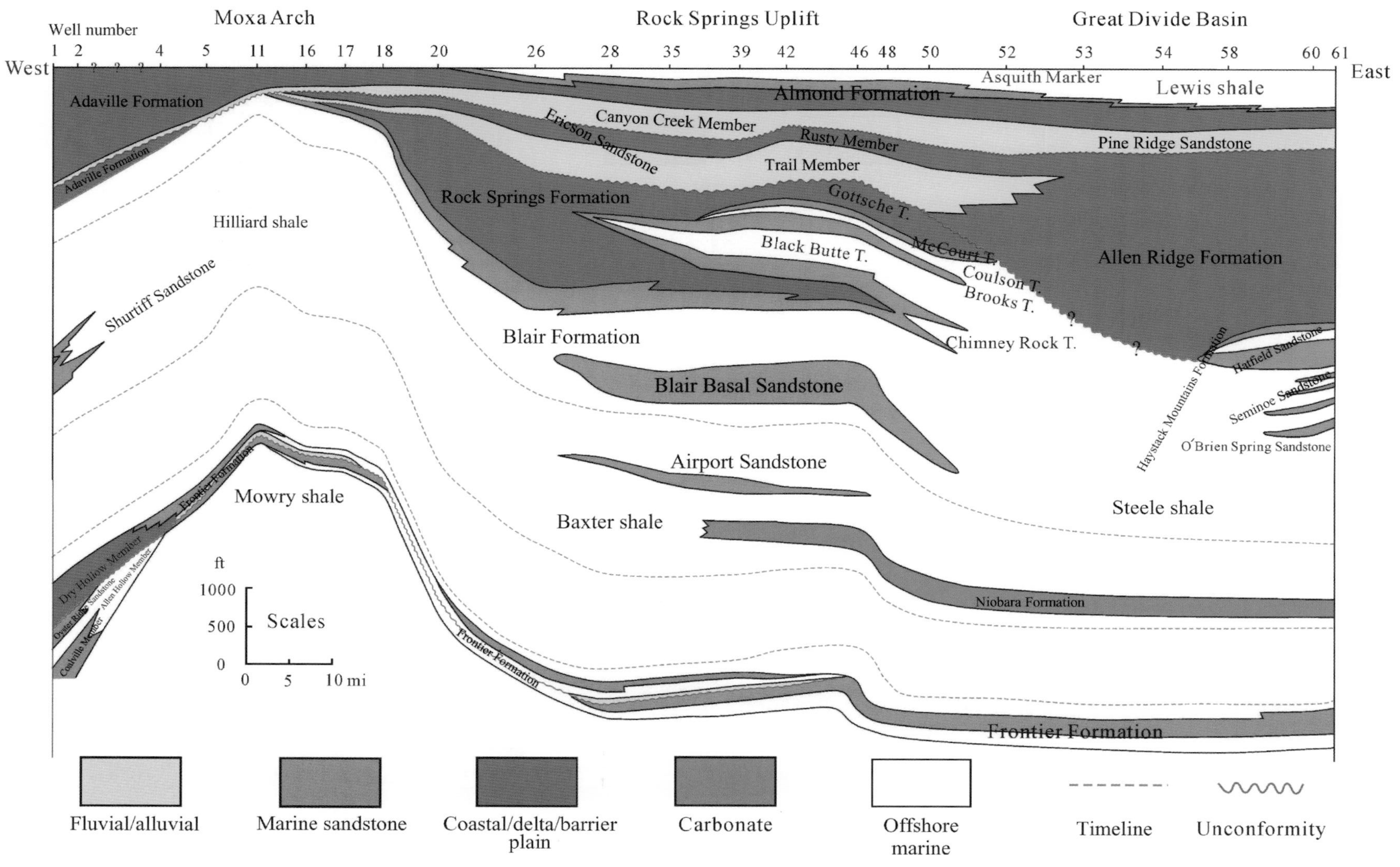

Figure 3. Cross section 1 showing the restored Upper Cretaceous from the west to the east across southwestern Wyoming. The location is cross section 1 in Figure 1. T. = Tongue.

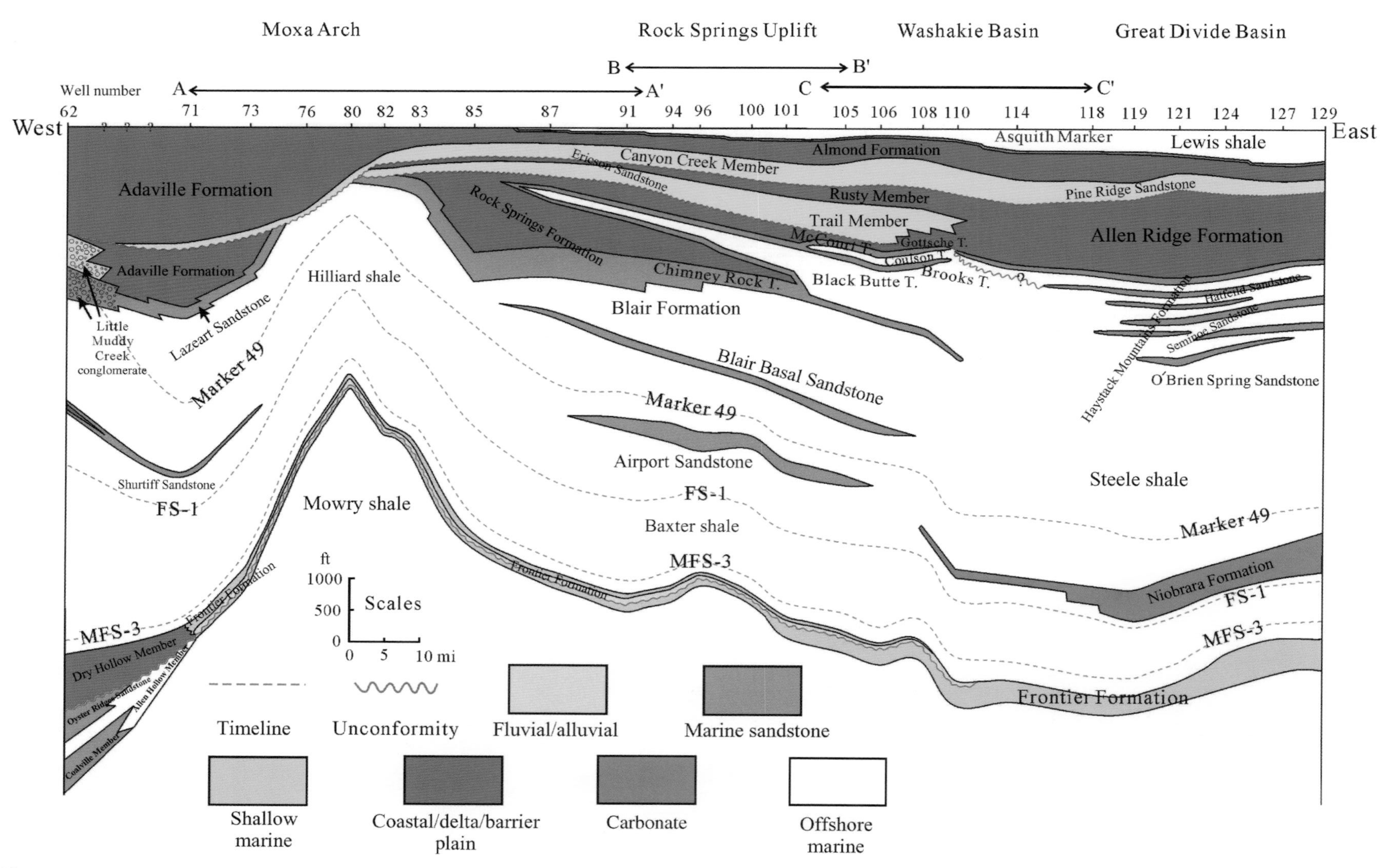

Figure 4. Cross section 2 showing the restored Upper Cretaceous stratigraphy from the west to the east across southwestern Wyoming. The location is cross section 2 in Figure 1. Section segments AA′, BB′, and CC′ show the section locations for Figures 4, 5, and 6, respectively. T. = Tongue.

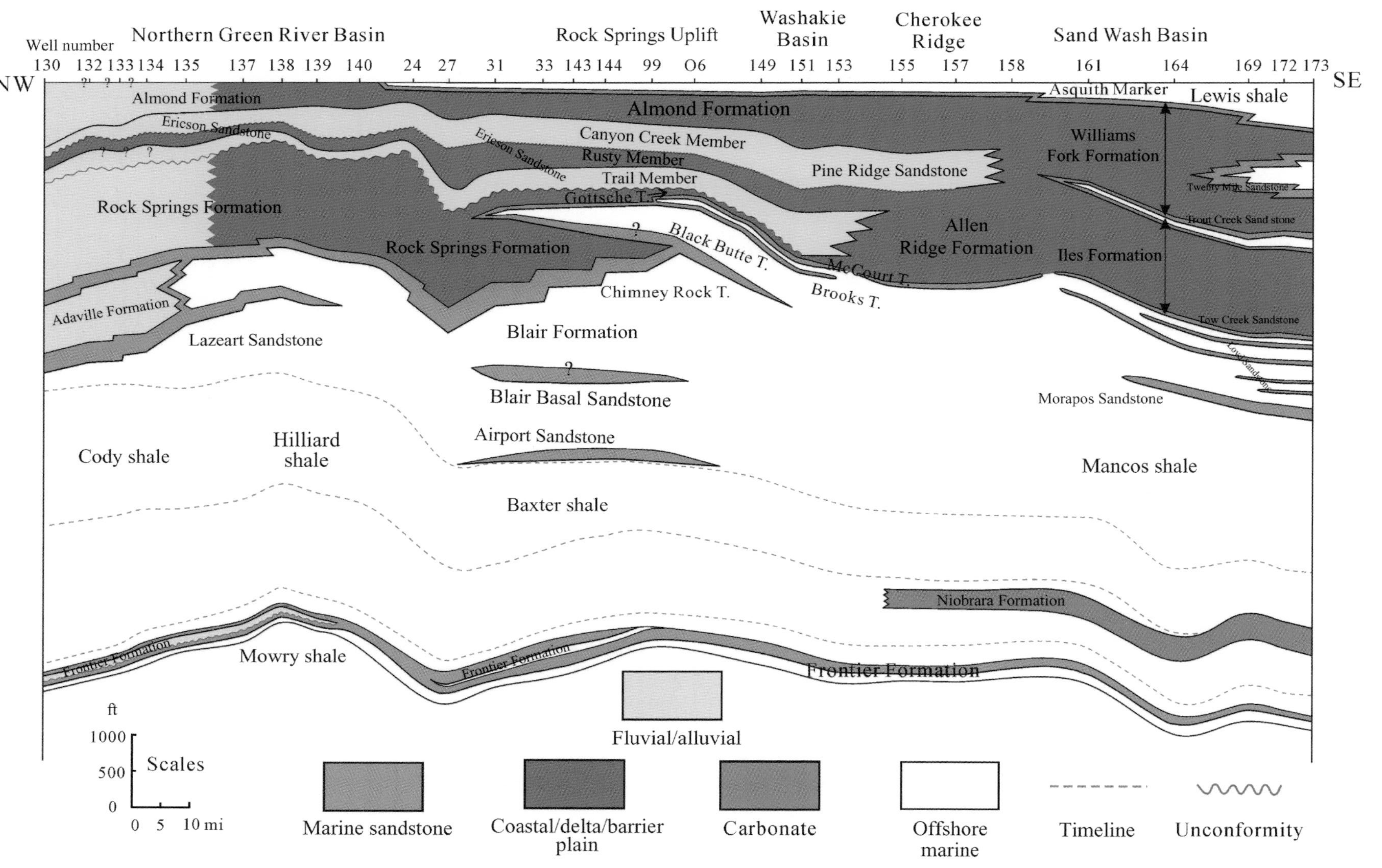

Figure 5. Cross section 3 showing the restored Upper Cretaceous from the northwest to the southeast across the Greater Green River Basin. The location is cross section 3 in Figure 1. T. = Tongue.

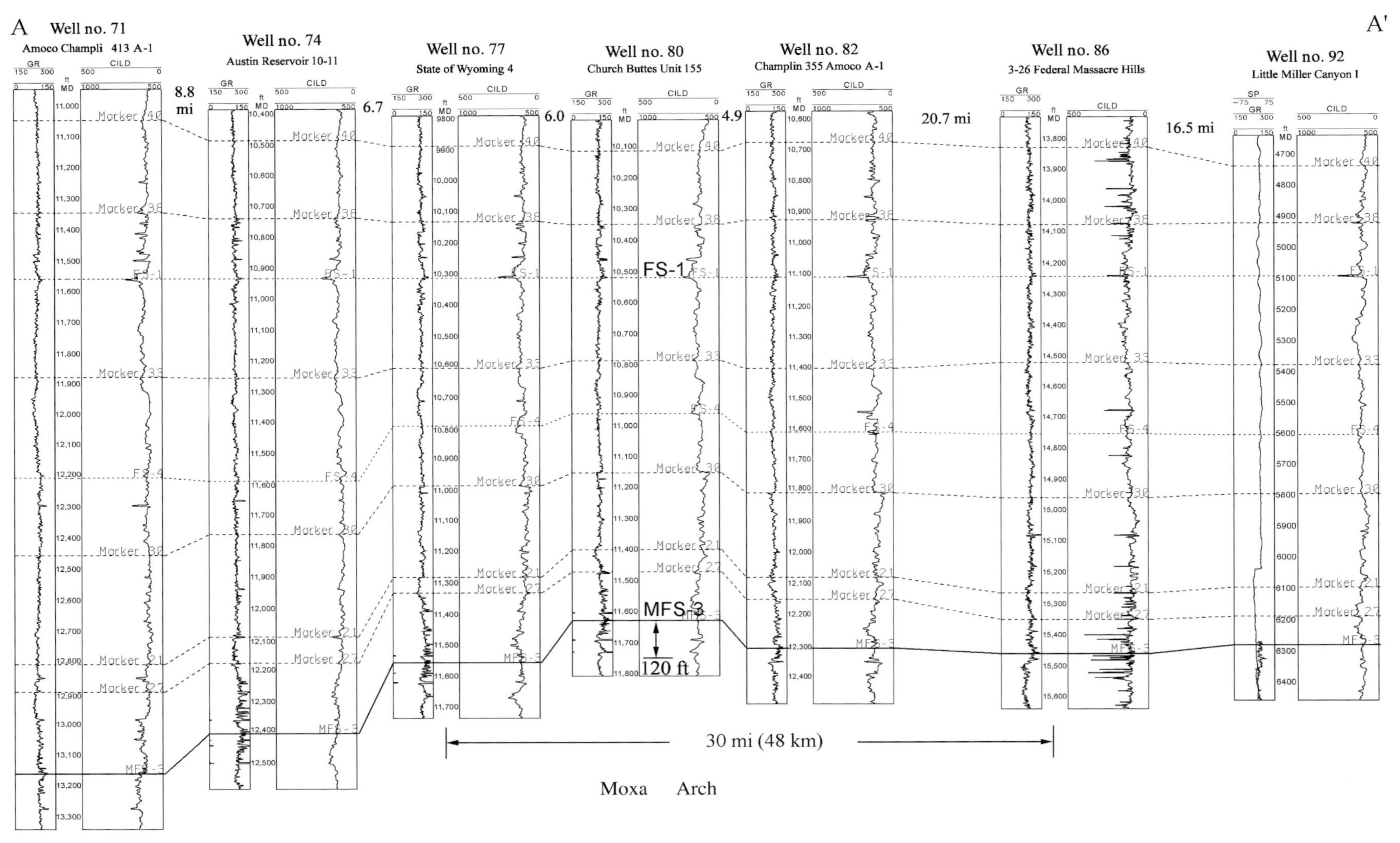

Figure 6. Subsurface cross section showing evidence for the forebulge formed by the Crawford thrusting. FS = flooding surface; MFS = maximum flooding surface; GR = gamma ray; sp = spontaneous potential; CILD = conductivity, induction log deep.

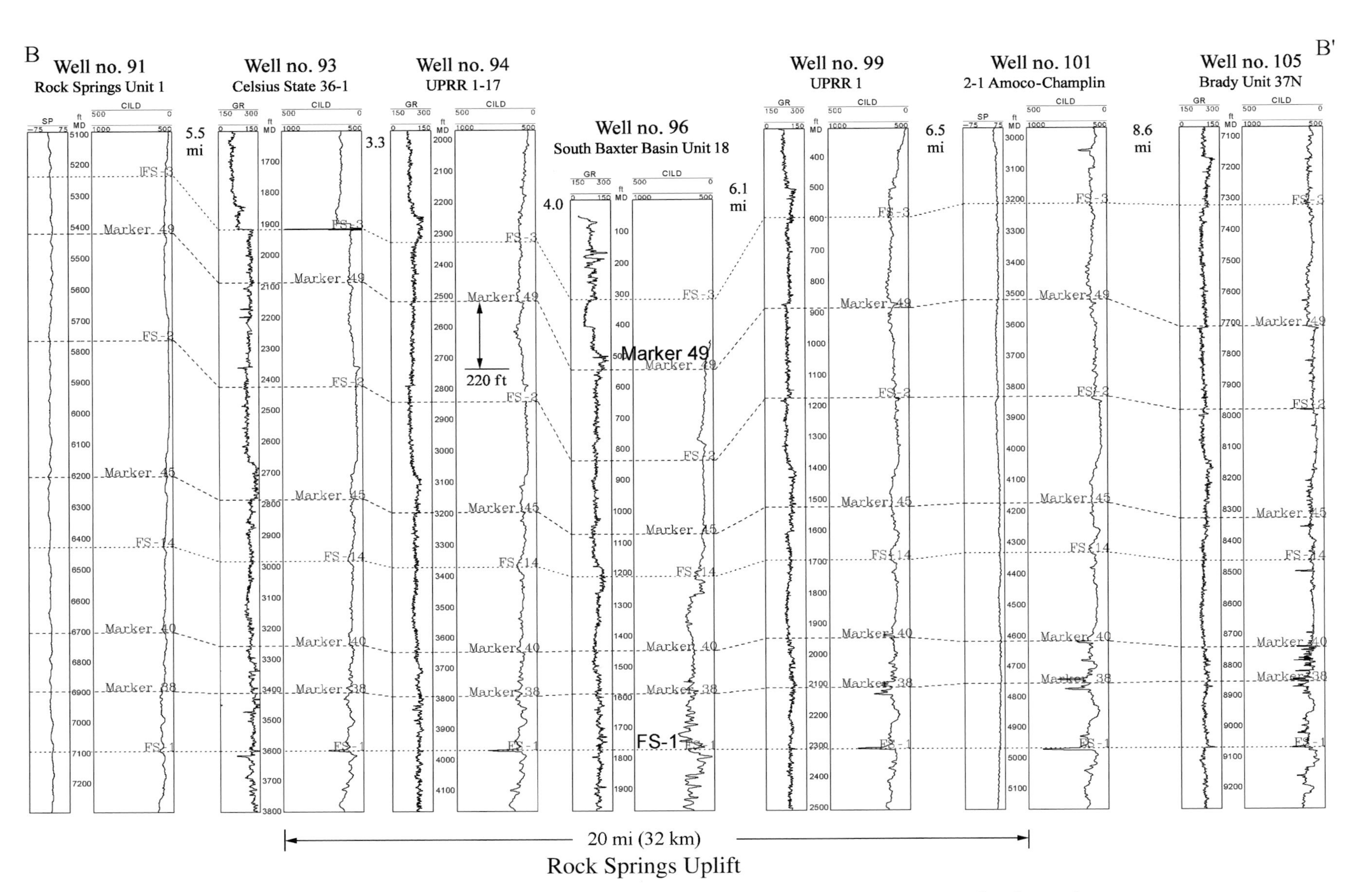

Figure 7. Subsurface cross section showing evidence for the forebulge formed by the Early Absaroka thrusting. FS = flooding surface; GR = gamma ray; sp = spontaneous potential; CILD = conductivity, induction log deep.

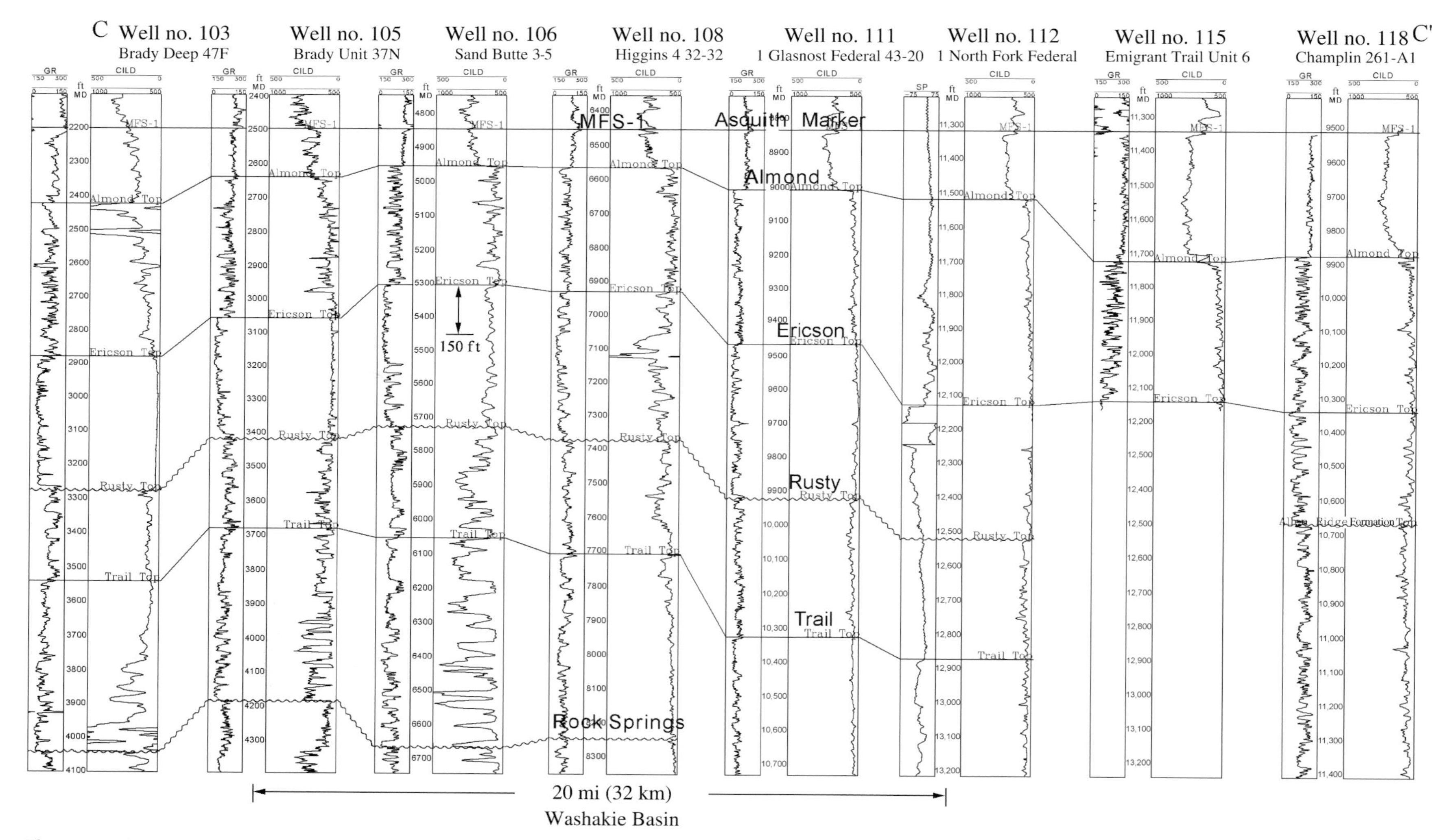

Figure 8. Subsurface cross section showing evidence for the forebulge formed by the Late Absaroka thrusting. GR = gamma ray; sp = spontaneous potential; CILD = conductivity, induction log deep.

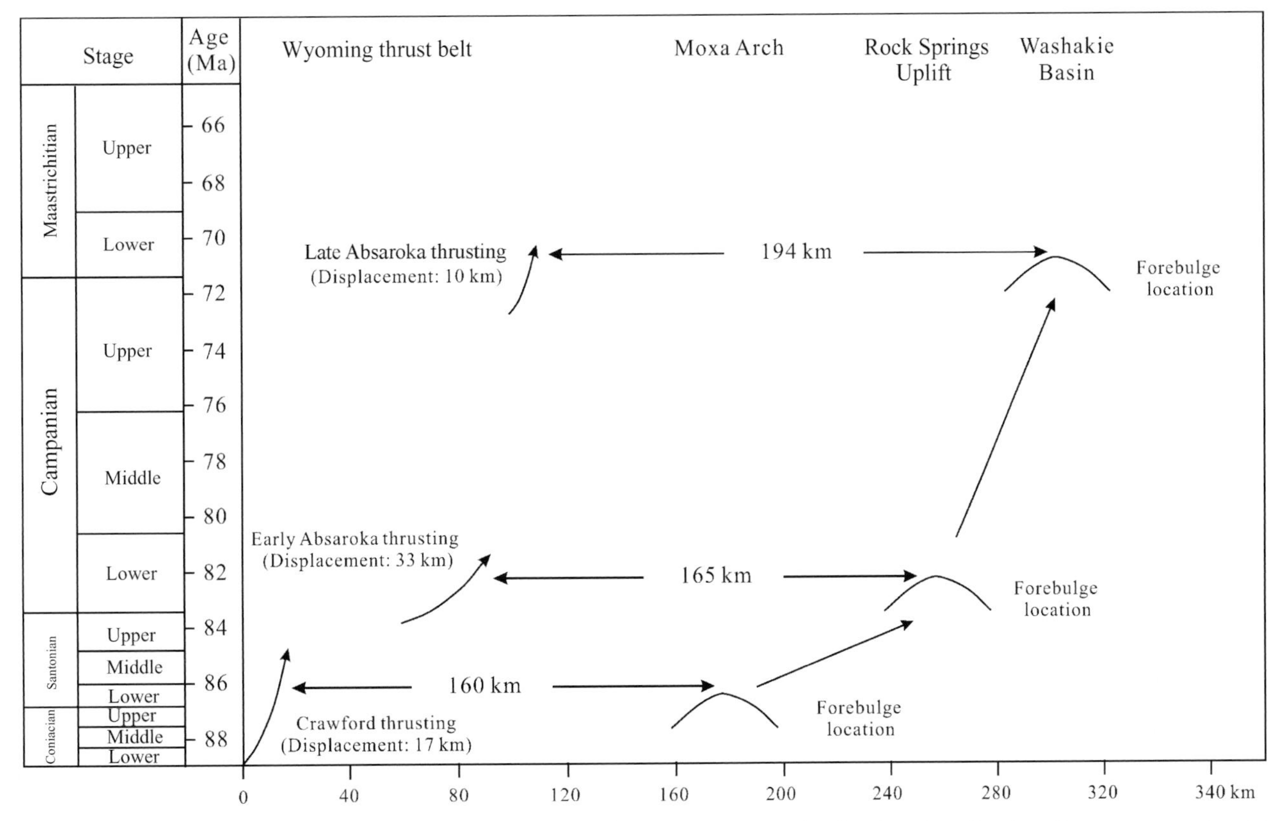

Figure 9. Schematic chart tracing forebulge migration across the Late Cretaceous foreland basins (the thrusting was approximately restored to the original locations). Note that the distance from the thrust front to the forebulge changed very little over time.

(~20 mi). The width of this forebulge is the same as that of the older forebulge at the Rock Springs Uplift.

Forebulge Migration

From the data just presented, it is clear that the forebulge migrated basinward (eastward) with progressively younger thrusting in the Wyoming thrust belt (Figure 9). The widths of the foredeep (the distance from the thrust front to the associated forebulge) formed by the Crawford and early Absaroka thrusting are very similar. They are 160 km (100 mi) and 165 km (103 mi), respectively (Figure 9). The width of the foredeep for the late Absaroka thrusting is 194 km (121 mi), approximately 30 km (19 mi) more than the previous two foredeeps. This width increase may be caused by a lesser 3-D mountain load by the late Absaroka thrusting.

THREE-DIMENSIONAL NUMERICAL MODELING

Three-dimensional flexural numerical modeling was conducted to support the above observation on forebulge migration.

The rheological model for the lithosphere is a key factor in the numerical modeling of any basin. Generally, three rheological models for the lithosphere exist: elastic, viscoelastic, and temperature-dependent viscoelastic models (Quinlan and Beaumont, 1984). Beaumont (1978, 1981) first advocated the viscoelastic lithosphere and used the North Sea Basin and the Alberta foreland basin of Western Canada as examples to test the model, which was supported later by an example from the mid-Cretaceous foreland basin of central Utah (Gardner, 1995a). The model by Beaumont et al. evolved to a more complex temperature-dependent viscoelastic model (Quinlan and Beaumont, 1984; Stockmal et al., 1986; Stockmal and Beaumont, 1987; Beaumont et al., 1988). Most other geologists preferred the elastic model for the lithosphere because of its simplicity (Jordan, 1981; Flemings and Jordan, 1989; Angevine et al., 1990; Flemings, 1990; Sinclair et al., 1991; Waschbusch and Royden, 1992a, b; White et al., 2002) and found it sufficient to describe the configuration of foreland basins, because the time scale for viscous relaxation is too short to have a significant impact on overall stratigraphic patterns. Jordan (1981), working on the Idaho-Wyoming foreland, found that an elastic lithosphere was sufficient to describe the subsidence history of that basin. For these reasons, this article will use an elastic lithosphere model in 3-D numerical modeling of the Late Cretaceous foreland basin across the southwestern Wyoming.

Three-dimensional Elastic Plate Flexure

The differential equation for the vertical deflection $w(x, y)$ of a thin elastic plate overlying a fluid substratum is (Ranalli, 1995):

$$D\frac{\partial^4 w(x,y)}{\partial x^4} + D\frac{\partial^4 w(x,y)}{\partial y^4} + 2D\frac{\partial^4 w(x,y)}{\partial x^2 \partial y^2} + (\rho_m - \rho_s)gw = p(x,y) \quad (1)$$

where $w(x, y)$ = vertical deflection; x, y = location coordinates; D = flexural rigidity = $Eh^3/12(1-v^2)$; E = Young's modulus (7×10^{10} N/m^2); v = Poisson's ratio (0.25); h = elastic-plate thickness; g = acceleration of gravity (9.8 ms^{-2}); $p(x, y)$ = applied surface load; ρ_m = density of mantle and asthenosphere (3300 kg/m^3); ρ_s = density of sediments or water that may fill the depression ($w(x, y) < 0$) or eroding materials ($w(x, y) > 0$) (density of sediments fill in: 2200 kg/m^3).

The same equation as equation 1 has been solved for the specified load $p(x, y)$ and appropriate boundary conditions by Lambeck and Nakiboglu (1980). For a load of finite dimensions, specified by a boundary $r = a(x, y)$, commonly used boundary conditions are as follows: w and d_w/d_r are finite at $r = 0$; w and d_w/d_r are continuous at $r = a$; moments and shears are continuous at $r = a$; and w and $r^{-1}/d_w/d_r$ vanish at infinity. This last condition follows from the requirement that the moments vanish at infinity.

For disc loads with various radii, analytical solutions of the deflection at a radius r from the center of the load are given by Lambeck and Nakiboglu (1980), using the above boundary conditions. The solutions are related to Bessel-Kelvin functions:

$$w(r) = -\frac{\rho h}{\Delta\rho}\left[\frac{a}{\alpha}\,\mathrm{ker}'\!\left(\frac{a}{\alpha}\right)\mathrm{ber}\!\left(\frac{r}{\alpha}\right) - \frac{a}{\alpha}\,\mathrm{kei}'\!\left(\frac{a}{\alpha}\right)\mathrm{bei}\!\left(\frac{r}{\alpha}\right) + 1\right] (r < a)$$

$$w(r) = -\frac{\rho h}{\Delta\rho}\left[\frac{a}{\alpha}\,\mathrm{ber}'\!\left(\frac{a}{\alpha}\right)\mathrm{ker}\!\left(\frac{r}{\alpha}\right) - \frac{a}{\alpha}\,\mathrm{bei}'\!\left(\frac{a}{\alpha}\right)\mathrm{kei}\!\left(\frac{r}{\alpha}\right)\right] (r \geq a) \quad (2)$$

where $\alpha = [D/\Delta\rho g]^{1/4}/1000$ (km); ρ = density of load (basement) (2800 kg/m^3); a = radius of disc load; ker, kei, ker$'$, kei$'$, ber, bei, ber$'$, bei$'$ are Bessel-Kelvin functions.

For a point load ($a = 0$), these reduce to $w(r) = -\frac{Q\alpha^2}{2\pi D}\mathrm{kei}(r)$
where Q = point load (N).
At origin $r = 0$, $w(0) = -\frac{Q\alpha^2}{2\pi D}\mathrm{kei}(0) = -\frac{Q\alpha^2}{8D}$.

Figure 10 shows the results of modeling a disc load with a height of 3000 m (9843 ft). The disc load with a 50 km (31 mi) radius is located with its center at (0, 0). The forebulge caused by the disc load surrounds the load symmetrically (Figure 10a, b). Along $x = 0$ or $y = 0$ (Figure 10c, d), the foredeep is 860 m (2822 ft) deep at its maximum and has a width of 140 km (87 mi), and the forebulge is 200 km (124 mi) wide, with an amplitude of 21 m (656 ft), for a crustal rigidity of 1×10^{23} N·m.

The Matlab codes written to simulate the flexure formed by disc loads are published by Luo (2005).

Three-dimensional Load Estimate

Three-dimensional flexural numerical modeling requires an estimation of the amount of crustal loading that occurred during each thrusting. This chapter follows the method of Jordan (1981) to estimate load. The basic data for load estimation at the Wyoming thrust belt come from eight cross sections of Royse (1993). After measuring the load increase along each section, this study interpolated the load increase in a 3-D extent according to the general trend of thrust distribution described by Royse (1993). The location of each thrust has been approximately restored to the original location of thrusting. The average crustal shortening of Meade-Laketown, Crawford, early Absaroka, late Absaroka, and Darby thrusts are 40 km (25 mi), 17 km (11 mi), 33 km (21 mi), 10 km (6 mi), and 10 km (6 mi), respectively.

The basic data for the Wind River thrust come from the article by Steidtmann et al. (1986). The total displacement time of the Wind River thrust was 40 m.y. (90–50 Ma). The crustal shortening was 20 km (12 mi). The dip of Wind River thrust is 40° (Shuster and Steidtmann, 1988). The shortening that occurred during 90 to 70 Ma (the target time interval for this article) was 10 km (6 mi), assuming that the shortening rate was constant. The total load width was approximately 40 km (25 mi) during 90 to 70 Ma.

Modeling Results

Computed 3-D flexural subsidence results, for the time intervals of 88.7 to 86.0 Ma, 86.0 to 79.5 Ma, and 72.4 to 67.9 Ma are presented in Figures 11, 12, and 13, respectively. These time intervals are associated with the Crawford, early Absaroka, and late Absaroka thrustings, respectively (Figures 2, 9). The Wind River thrust was assumed to have been uplifted at a constant rate during the modeling time interval beginning at 90 Ma. A range of flexural rigidities—10^{21}, 10^{22}, 5×10^{22}, 6×10^{22}, 7×10^{22}, 8×10^{22}, and 10^{23} N·m—was analyzed in the modeling, and 5×10^{22} N·m provided consistently the best match to the forebulge locations identified from the stratigraphic cross sections. The locations of sections 1, 2, and 3 are shown in the modeling results of Figures 11, 12, and 13.

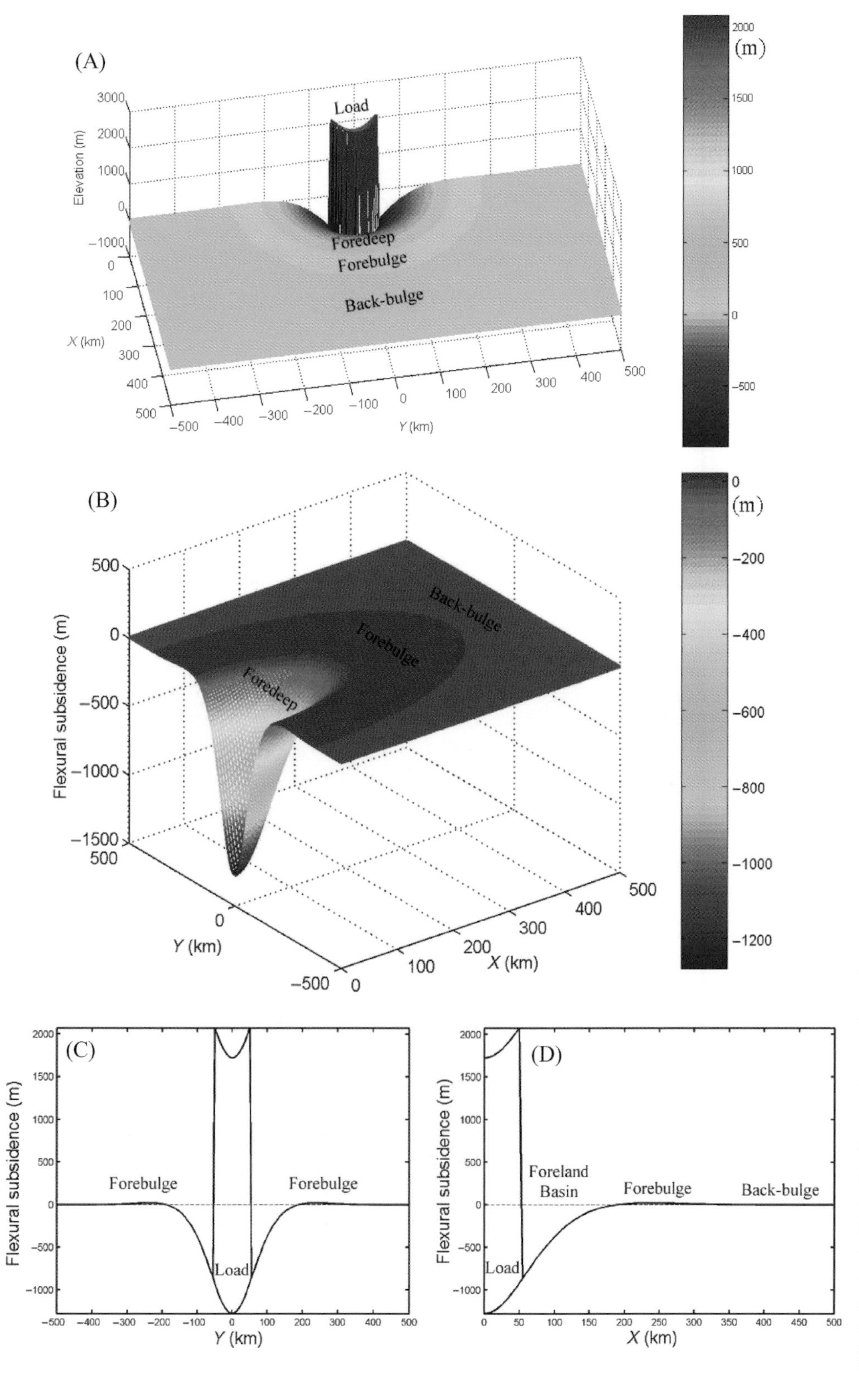

Figure 10. Flexural modeling for one disc load. Modeling parameters are as follows: flexural rigidity (D) = 1×10^{23} N·m; disc-load height = 3000 m (9843 ft); disc-load radius = 50 km (31 mi). (A) Overall three-dimensional topographic response (including the load). (B) Three-dimensional vertical flexure formed by the disc load (not including the load). (C) Two-dimensional section along $x = 0$ from A. (D) Two-dimensional section along $y = 0$ from (A).

From 88.7 to 86.0 Ma, the Crawford thrusting generated a major load increase. Meanwhile, the Wind River thrust continued its uplift (Figure 11). The major subsidence center was located in front of the Wyoming thrust belt, and the minor one, in front of the Wind River thrust (Figure 11). The modeled amplitude of the

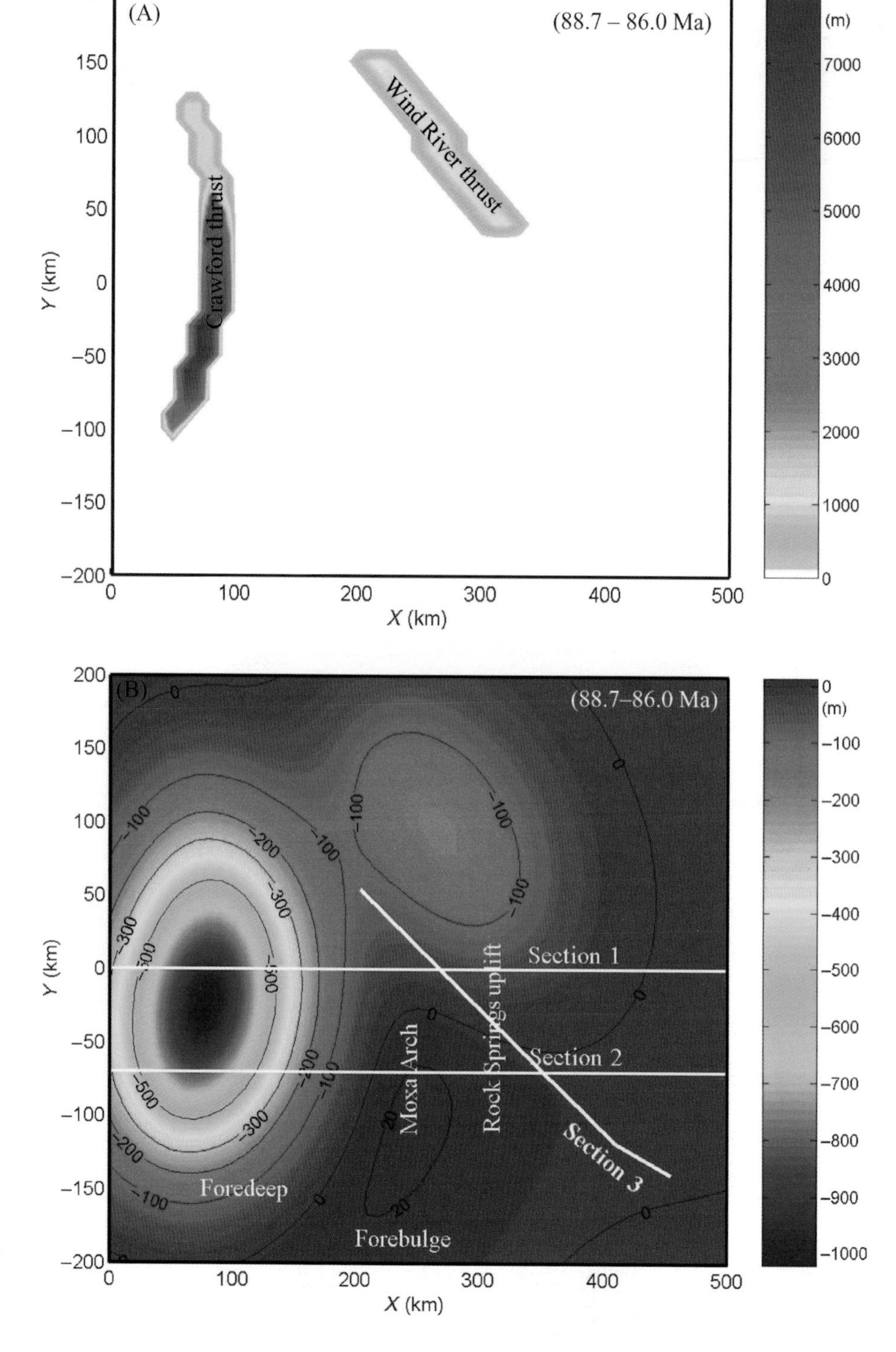

Figure 11. Three-dimensional flexural modeling results from 88.7 to 86.0 Ma (Crawford thrusting). (A) Three-dimensional load distribution. (B) Three-dimensional flexural subsidence.

forebulge is approximately 20 m (~66 ft). The location and geometry of the forebulge were very similar to the current basement-involved Moxa Arch.

From 86.0 to 79.5 Ma, the early Absaroka thrusting caused the major load increase and the Wind River thrust continued to add load at a constant rate. The major subsidence center remained in front of the Wyoming thrust belt, and the minor one in front of the Wind River thrust gradually became more important (Figure 12). The subsidence area in front of the Wind River thrust experienced a very gentle change of subsidence (Figure 12b), which is quite different from the quick subsidence change in the pure proximal foredeep. This gentle change of subsidence is the result of combined elastic flexure in response to both the Wyoming thrust belt and the Wind River Range. With the eastward growth of the Wyoming thrust belt and continuous load increase by the Wind River thrust, the forebulge migrated southeastward and became more dome shaped. The crest of the forebulge was located at the current Rock Springs Uplift, with an amplitude of more than 20 m (>66 ft) at this time (Figure 12).

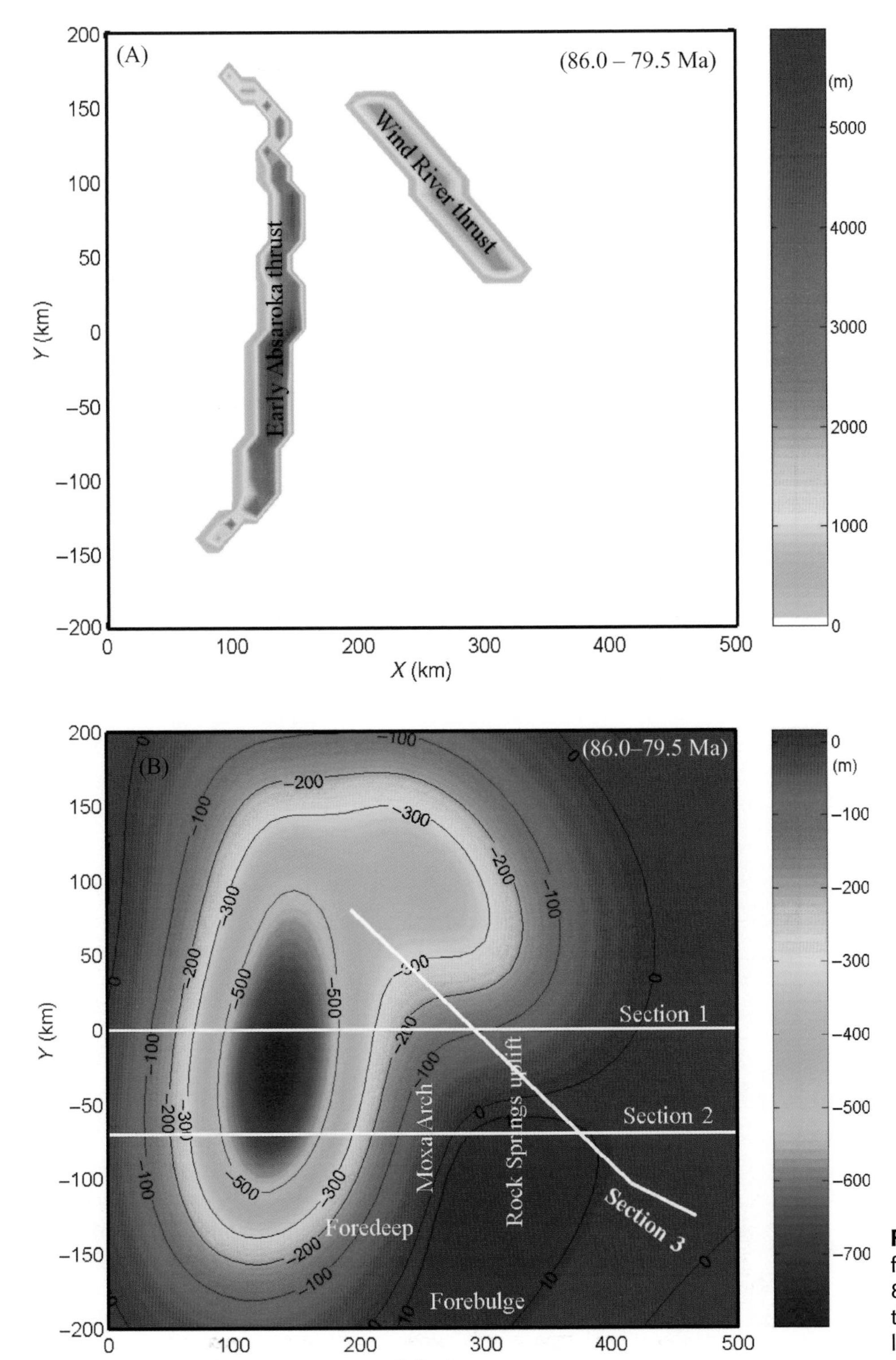

Figure 12. Three-dimensional flexural modeling results from 86.0 to 79.5 Ma (early Absaroka thrusting). (A) Three-dimensional load distribution. (B) Three-dimensional flexural subsidence.

From 72.4 to 67.9 Ma, the Absaroka and Wind River thrusts added new loads (Figure 13). The load increase was much less than that in the previous two stages. The new load added by the Wind River thrust was more important. With the continued eastward growth of the Wyoming thrust belt and continuous load increase by the Wind River thrust, the forebulge continued its southeastward migration. The migration

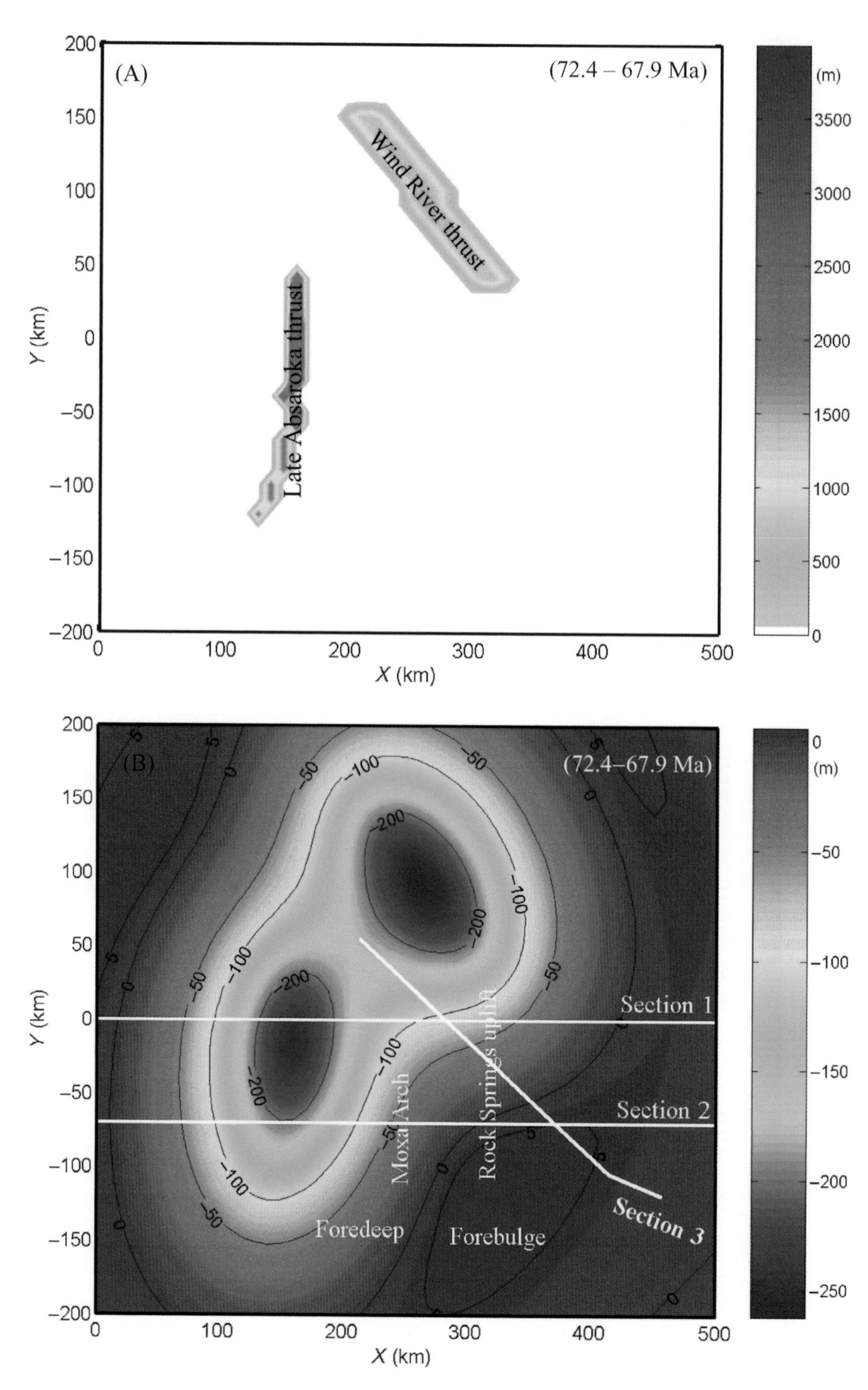

Figure 13. Three-dimensional flexural modeling results from 72.4 to 67.9 Ma (late Absaroka thrusting). (A) Three-dimensional load distribution. (B) Three-dimensional flexural subsidence.

distance was greater at the northern margin of the forebulge crest than that at the southern margin (Figure 13b). The amplitude of this late forebulge was only approximately 10 m (33 ft).

This 3-D flexural modeling exercise makes it easier to understand the 3-D migration of the forebulge. The forebulge migration along section 2 is only a profile across the 3-D forebulge evolution. As mentioned before, no apparent forebulge migration was found along section 1. The 3-D flexural modeling results provide a reason for this geologic observation. Section 1 does not cross the forebulge area for all modeled time intervals. The forebulge migration along section 2 is quite obvious in the 3-D modeling results from 88.7 to 67.9 Ma (Figures 11–13). The forebulge was located at the Moxa Arch from 88.7 to 86.0 Ma (Figure 11), migrated to the Rock Springs Uplift from 86.0 to 79.5 Ma (Figure 12) and to the western flanks of the Washakie Basin from 72.4 to 67.9 Ma (Figure 13). The modeling results are consistent with the geologic observations of forebulge migration and explain the complexity of the 3-D distribution of forebulge-related unconformities in the subsurface of the Greater Green River Basin.

CONCLUSIONS

1) Late Cretaceous forebulge migration is recognized in southwestern Wyoming. In response to the eastward progressive movement of the Crawford, early Absaroka, and late Absaroka thrusts, the forebulges migrated eastward to the Moxa Arch, the Rock Springs Uplift, and the Washakie Basin, respectively.
2) Three-dimensional flexural modeling indicates that the forebulges were most prominently expressed in the southern parts of the basin because of the 3-D geometry of the thrust load. The forebulges shifted southeastward over time because of the migration of this 3-D load.
3) Understanding forebulge formation and migration is still a subject of developing interest, and different interpretations exist. Consequently, what we have reported here may reflect the manifestation of individual bulges that move out with the advance of each successive thrust sheet (our preferred interpretation), or it may reflect the manifestation of a single bulge that migrated progressively more cratonward with the advance of each thrust, each time reactivating a more distal basement structure. The modeling results are consistent with either interpretation.
4) Three-dimensional flexural modeling results are consistent with the geologic observations and are critical to the understanding of the Late Cretaceous forebulge migration across southwestern Wyoming.

ACKNOWLEDGMENTS

We thank Ronald Steel, Paul Heller, Art Snoke, Shaofeng Liu, Peigui Yin, Randi Martinsen and James Steidtmann for discussions. We also thank James Drever for his help. This project was part of Hongjun Luo's dissertation funded by the Institute for Energy Research-Enhanced Oil Recovery Institute (IER–EORI) and the Department of Geology and Geophysics at the University of Wyoming during 2001 to 2005. Additional financial support came from the 2003 to 2005 Walter and Constance Spears Memorial Scholarship, the Phillips Geology Scholarship, the 2004 Weimer Family AAPG grant, and a 2004 SEPM travel grant. We thank peer reviewers Frank Ettensohn and John Londoño for the constructive suggestions to improve this paper.

REFERENCES CITED

Angevine, C. L., P. L. Heller, and C. Paola, 1990, Quantitative sedimentary basin modeling: AAPG Continuing Education Course Note Series 32, 247 p.

Armstrong, F. C., and S. S. Oriel, 1986, Tectonic development of the Idaho-Wyoming thrust belt, *in* J. A. Peterson, ed., Paleotectonics and sedimentation in the Rocky Mountain region, United States: AAPG Memoir 41, p. 243–280.

Beaumont, C., 1978, The evolution of sedimentary basins on a viscoelastic lithosphere: Theory and examples: Geophysical Journal of the Royal Astronomical Society, v. 55, p. 471–497, doi:10.1111/j.1365-246X.1978.tb04283.x.

Beaumont, C., 1981, Foreland basins: Geophysical Journal of the Royal Astronomical Society, v. 65, p. 291–329, doi:10.1111/j.1365-246X.1981.tb02715.x.

Beaumont, C., G. Quinlan, and J. Hamilton, 1988, Orogeny and stratigraphy: Numerical models of the Paleozoic in the Eastern Interior of North America: Tectonics, v. 7, p. 389–416, doi:10.1029/TC007i003p00389.

Bradley, M. D., 1995, Timing of the Laramide rise of the Uinta Mountains, Utah and Colorado, *in* R. W. Jones, ed., Resources of southwestern Wyoming: Wyoming Geological Association Field Conference Guidebook, p. 31–44.

DeCelles, P. G., 2004, Late Jurassic to Eocene evolution of the Cordilleran thrust belt and foreland basin system, western U.S.A.: American Journal of Science, v. 304, p. 105–168, doi:10.2475/ajs.304.2.105.

DeCelles, P. G., 1994, Late Cretaceous–Paleocene synorogenic sedimentation and kinematic history of the Sevier thrust belt, northeast Utah and southwest Wyoming: Geological Society of America Bulletin, v. 106, p. 32–56, doi:10.1130/0016-7606(1994)106<0032:LCPSSA>2.3.CO;2.

Devlin, W. J., K. W. Rudolph, C. A. Shaw, and K. D. Ehman, 1993, The effect of tectonic and eustatic cycles on accommodation and sequence-stratigraphic framework in the Upper Cretaceous foreland basin of southwestern Wyoming: International Association of Sedimentologists Special Publication, v. 18, p. 501–520, doi:10.1002/9781444304015.ch25.

Dyman, T. S., E. A. Merewether, C. M. Molenaar, W. A. Cobban, J. D. Obradovich, R. J. Weimer, and W. A. Bryant, 1994, Stratigraphic transects for Cretaceous rocks, Rocky Mountains and Great Plains regions, *in* M. V. Caputo, J. A. Peterson, and K. J. Franczyk, eds., Mesozoic systems of the Rocky Mountain region, U.S.A.: Rocky Mountain Section Society for Sedimentary Geology, Denver, Colorado, p. 365–392.

Elliott, D., 1977, Structural and stratigraphic consequences for the foreland of tectonic activity in the thrust belt: Wyoming Geological Association 29th Annual Field Conference, Teton Village, Wyoming, September 14–17, 1977.

Flemings, P. B., 1990, Synthetic stratigraphy of foreland basins and bed-load transport in a graded stream: Ph.D. dissertation, Cornell University, New York, 134 p.

Flemings, P. B., and T. E. Jordan, 1989, A synthetic stratigraphic model of foreland basin development: Journal of Geophysical Research, v. 94, p. 3851–3866.

Gardner, M. H., 1995a, The stratigraphic hierarchy and tectonic history of the mid-Cretaceous foreland basin of central Utah, *in* S. Dorobek and J. Ross, eds., Stratigraphic evolution of foreland basins: SEPM Special Publication, v. 52, p. 243–283.

Gill, J. R., and W. A. Cobban, 1973, Stratigraphy and geologic history of the Montana Group and equivalent rocks, Montana, Wyoming, and North and South Dakota: U.S. Geological Survey Professional Paper 776, 37 p.

Jacobson, S. R., and D. J. Nichols, 1982, Palynological dating of syntectonic units in the Utah-Wyoming thrust belt: The Evanston Formation Echo Canyon Conglomerate and Little Muddy Creek Conglomerate, *in* R. D. Powers, ed., Geologic studies of the Cordilleran thrust belt, 1982: Rocky Mountain Association of Geologists, p. 735–750.

Jordan, T. E., 1981, Thrust loads and foreland basin evolution, Cretaceous, western United States: AAPG Bulletin, v. 65, p. 2506–2520.

Kauffman, E. G., B. B. Sageman, J. I. Kirkland, W. P. Elder, P. J. Harries, and T. Villamil, 1993, Molluscan biostratigraphy of the Cretaceous Western Interior Basin, North America, *in* W. G. E. Caldwell and E. G. Kauffman, eds., Evolution of the Western Interior Basin: Geological Association of Canada Special Paper 39, p. 397–434.

Kraig, D. H., D. V. Wiltschko, and J. H. Spang, 1988, The interaction of the Moxa Arch (La Barge Platform) with the Cordilleran thrust belt, south of Snider Basin, southwestern Wyoming: Geological Society of America Memoir 171, p. 395–410.

Kraig, D. H., D. V. Wiltschko, and J. H. Spang, 1987, Interaction of basement uplift and thin-skinned thrusting, Moxa Arch and the western overthrust belt, Wyoming: A hypothesis: Geological Society of America Bulletin, v. 99, p. 654–662, doi:10.1130/0016-7606(1987)99<654:IOBUAT>2.0.CO;2.

Lambeck, K., and S. M. Nakiboglu, 1980, Seamount loading and stress in the ocean lithosphere: Journal of Geophysical Research, v. 85, p. 6403–6418, doi:10.1029/JB085iB11p06403.

Lawrence, D. T., 1992, Primary controls on total reserves, thickness, geometry, and distribution of coal seams: Upper Cretaceous Adaville Formation, southwestern Wyoming: Geological Society of America Special Paper 267, p. 69–100.

Liu, S., and D. Nummedal, 2004, Late Cretaceous subsidence in Wyoming: Quantifying the dynamic component: Geology, v. 31, p. 397–400, doi:10.1126/10.1130/G20318.1.

Liu, S., D. Nummedal, P. Yin, and H. Luo, 2005, Linkage of Sevier thrust episodes and Late Cretaceous megasequences across southern Wyoming (U.S.A.): Basin Research, v. 17, p. 487–506, doi:10.1111/j.1365-2117.2005.00277.x.

Luo, H., 2005, Tectonostratigraphy of foreland basins: The Upper Cretaceous in southwestern Wyoming: Ph.D. dissertation, University of Wyoming, Laramie, Wyoming, 284 p.

Martinsen, O. J., R. J. Steel, R. S. Martinsen, and L. L. Krystinik, 1998, The Mesaverde Group, Rock Springs Uplift, Wyoming: AAPG Annual Convention Field Trip Guidebook 25, Salt Lake City, Utah, May 17–20, 1998.

Mears Jr., B., 1998, Neogene normal faulting superposed on a Laramide uplift: Medicine Bow Mountains, Sierra Madre, and intervening Saratoga Valley, Wyoming and Colorado: Contributions to Geology, v. 32, p. 181–185.

Miller, F. X., 1977, Biostratigraphic correlation of the Mesaverde Group in southwestern Wyoming and northwestern Colorado: Rocky Mountain Association of Geologists, p. 117–137.

Nichols, D. J., and S. R. Jacobson, 1982, Cretaceous biostratigraphy in the Wyoming thrust belt: The Mountain Geologist, v. 19, p. 73–78.

Nummedal, D., P. Yin, and R. J. Steel, 2002, Sequence stratigraphy of coal-bearing strata in the Washakie and Sand Wash basins, Wyoming and Colorado: Derived from outcrops: Institute for Energy Research/Anadarko Coalbed Methane Project Final Report, Laramie, University of Wyoming, p. 1–23.

Oriel, S. S., and F. C. Armstrong, 1986, Tectonic development of the Idaho-Wyoming thrust belt: Author's commentary, *in* J. A. Peterson, ed., Paleotectonics and sedimentation in the Rocky Mountain region, United States: AAPG Memoir 41, p. 243–280.

Pang, M., and D. Nummedal, 1995, Flexural subsidence and basement tectonics of the Cretaceous Western Interior Basin, United States: Geology, v. 23, p. 173–176, doi:10.1130/0091-7613(1995)023<0173:FSABTO>2.3.CO;2.

Quinlan, G., and C. Beaumont, 1984, Appalachian thrusting, lithospheric flexure, and the Paleozoic stratigraphy of the Eastern Interior of North America: Canadian Journal of Earth Sciences, v. 21, p. 973–996, doi:10.1139/e84-103.

Ranalli, 1995, Rheology of the Earth, 2nd ed.: London, Chapman and Hall, 413 p.

Roehler, H. W., 1965, Summary of Pre-Laramide Late Cretaceous sedimentation in the Rock Springs uplift area, *in* R. H. Devoto and R. K. Bitter, eds., Sedimentation of Late Cretaceous and Tertiary outcrops, Rock Springs Uplift: Wyoming Geological Association 19th Field Conference Guidebook, Casper, Wyoming, April 2–3, 1965, p. 10–12.

Roehler, H. W., 1990, Stratigraphy of the Mesaverde Group in the center and eastern Greater Green River Basin, Wyoming, Colorado, and Utah: U.S. Geological Survey Professional Paper 1508, 52 p.

Royse Jr., F., 1993, An overview of the geologic structure of the thrust belt in Wyoming, northern Utah, and eastern Idaho, *in* A. W. Snoke, J. R. Steidtmann, and S. M. Roberts, eds., Geology of Wyoming: Geological Survey of Wyoming Memoir 5, p. 273–311.

Shuster, M. W., and J. R. Steidtmann, 1988, Tectonic and sedimentary evolution of the northern Green River Basin, western Wyoming: Geological Society of America Memoir 171, p. 515–529.

Sinclair, H. D., B. J. Coakley, P. A. Allen, and A. B. Watts, 1991, Simulation of foreland basin stratigraphy using a diffusion model of mountain belt uplift and erosion: An example from the central Alps, Switzerland: Tectonics, v. 10, p. 599–620, doi:10.1029/90TC02507.

Steidtmann, J. R., and L. T. Middleton, 1991, Fault chronology and uplift history of the southern Wind River Range, Wyoming: Implications for Laramide and post-Laramide deformation in the Rocky Mountain foreland: Geological Society of American Bulletin, v. 103, p. 472–485, doi:10.1130/0016-7606(1991)103<0472:FCAUHO>2.3.CO;2.

Steidtmann, J. R., L. T. Middleton, R. J. Bottjer, K. E. Jackson, L. C. McGee, E. H. Southwell, and S. Lieblang, 1986, Geometry, distribution, and provenance of tectogenic conglomerates along the southern margin of the Wind River Range, Wyoming, *in* J. A. Peterson, ed., Paleotectonics and sedimentation in the Rocky Mountain region, United States: AAPG Memoir 41, p. 321–332.

Stockmal, G. S., C. Beaumont, and R. Boutilier, 1986, Geodynamic models of convergent margin tectonics: Transition from rifted margin to overthrust belt and consequences for foreland-basin development: AAPG Bulletin, v. 70, p. 181–190.

Stockmal, G. S., and C. Beaumont, 1987, Geodynamic models of convergent margin tectonics: The southern Canadian Cordillera and the Swiss Alps, *in* C. Beaumont and A. J. Tankard, eds., Sedimentary basins and basin-forming mechanisms: Canadian Society of Petroleum Geologists Memoir 12, p. 393–411.

Vietti, J. S., 1977, Structural geology of the Ryckman Creek anticline area, Lincoln and Uinta counties, Wyoming: Wyoming Geological Association 29th Annual Field Conference Guidebook, p. 517–522.

Washbusch, P. J., and L. H. Royden, 1992a, Spatial and temporal evolution of foredeep basins: Lateral strength variations and inelastic yielding in continental lithosphere: Basin Research, v. 4, p. 179–196, doi:10.1111/j.1365-2117.1992.tb00044.x.

Waschbusch, P. J., and L. H. Royden, 1992b, Episodicity in foredeep basins: Geology, v. 20, p. 915–918, doi:10.1130/0091-7613(1992)020<0915:EIFB>2.3.CO;2.

White, T., K. Furlong, and M. Arthur, 2002, Forebulge migration in the Cretaceous Western Interior Basin of the central United States: Basin Research, v. 14, p. 43–54, doi:10.1046/j.1365-2117.2002.00165.x.

18

Yang, Fengli, Dengliang Gao, Zhuan Sun, Zuyi Zhou, Zhe Wu, and Qianyu Li, 2012, The evolution of the South China Sea Basin in the Mesozoic–Cenozoic and its significance for oil and gas exploration: A review and overview, *in* D. Gao, ed., Tectonics and sedimentation: Implications for petroleum systems: AAPG Memoir 100, p. 397–418.

The Evolution of the South China Sea Basin in the Mesozoic–Cenozoic and Its Significance for Oil and Gas Exploration: A Review and Overview

Fengli Yang, Zhuan Sun, Zuyi Zhou, and Zhe Wu

School of Ocean and Earth Science, Tongji University, 1239 Siping Rd. Shanghai, 200092, China (e-mails: yangfl@tongji.edu.cn; sun_zhuan2008@163.com; zhouzy@tongji.edu.cn)

Dengliang Gao

Department of Geology and Geography, West Virginia University, 98 Beechurst Ave. Morgantown, West Virginia, 26506, U.S.A. (e-mail: dengliang.gao@wvu.edu)

Qianyu Li

State Key Laboratory of Marine Geology, Tongji University, 1239 Siping Rd. Shanghai, 200092, China (e-mail: qli01@tongji.edu.cn)

ABSTRACT

The greater South China Sea (SCS) Basin is composed of basins of different generations and styles. These polyhistory basins formed in complicated geologic settings and evolved through different tectonic regimes. Based on a classical basin classification scheme and data from previous studies, we summarize the evolution of tectonic environments of the SCS in the Mesozoic–Cenozoic into a Late Triassic–middle Eocene divergent-convergent cycle and a late Eocene–present divergent-convergent cycle. The two cycles are in turn composed of four evolutionary phases, which are (1) Late Triassic–Middle Jurassic divergent continental margin setting, (2) Late Jurassic–middle Eocene convergent intracontinental setting, (3) late Eocene–Miocene divergent continental margin setting, and (4) Pliocene–present convergent continental margin setting. We identify temporal sequence and spatial distribution of major polyhistory basins in the SCS associated with the four basin evolutionary phases in the two tectonic cycles. Each basin corresponds to a specific pressure, space, and temperature, and overprinting of the basin caused changes in pressure, space, and temperature with time. Unraveling this complex and dynamic nature of the polyhistory basins can be instrumental in assessing the hydrocarbon potential and exploration risk in the SCS.

DOI:10.1306/13351562M1003528

INTRODUCTION

The area now called the "South China Sea" (SCS) (Figure 1) was located at a convergent plate boundary between the Tethys and the circum-Pacific zones in the Mesozoic. As a result of the interaction between the Eurasian, the Indian-Australian, and the Pacific plates, the tectonic setting in the greater SCS region underwent a complicated history during the Mesozoic. In the Cenozoic, the tectonic setting of the SCS changed to a divergent regime through three major stages: a prespreading stage before the Eocene (>37 Ma), a spreading stage during the late Eocene–middle Miocene (37–16 Ma), and a postspreading stage since the late Miocene (<16 Ma) (Hsu et al., 2004). Therefore, the present-day SCS basin (Figure 1) has experienced multiphase deformation and is composed of polyhistory basins of different generations as described by Kingston et al. (1983a, b). This review and overview chapter is primarily based on previously published studies of the SCS and adopts the global basin classification scheme of Kingston et al. (1983a, b), which emphasizes the polyhistory and cyclic nature of sedimentary basins.

With its uniquely complicated geohistory and great hydrocarbon potential, the SCS has become a hot spot attracting geoscientists from both academia and the petroleum industry. Over the years, many authors have published their studies on the classification schemes and hydrocarbon systems of the SCS Basin (e.g., Taylor and Hayes, 1980, 1983; Tapponnier et al., 1982, 1986; Zhang, 1985; Liu, 1988; Briais et al., 1993; Liu, 1993; Gong and Li, 1997, 2004; Xia and Huang, 2000; Gong et al., 2001; Lee et al., 2001; Packham, 2003; Hao et al., 2004; Hutchison, 2004; Xia et al., 2004; Yao et al., 2004; Wang et al., 2005; Wei et al., 2005; Zhou et al., 2005b, c; Metcalfe, 2006; Li and Li, 2006; Droust and Sumner, 2007; Wang and Li, 2009; Cullen et al., 2010). Many investigators of the SCS Basin have primarily focused on the Cenozoic sedimentary basins in the Cenozoic plate tectonic setting instead of on the polyhistory and dynamic nature of the basins generated and overprinted at different geologic times (Kingston et al., 1983a, b). Application of the basin classification scheme developed by Kingston et al. (1983a), which emphasizes the polyhistory and dynamic nature of sedimentary basins, could help better understand the basin structures and lead to new play concepts in assessing the hydrocarbon potential of the greater SCS Basin.

As a geologic entity, a sedimentary basin is a low-lying area with long-time settlement from subsidence to sediment infill (Kingston et al., 1983a, b; Zhu, 1985). It becomes petroliferous once a working petroleum system becomes mature for the accumulation of oil and gas (Kingston et al., 1983b). However, the subsidence and sediment infill in the basin are time and space dependent and are subjected to changes in tectonic regimes. Basins, especially large basins, are always complicated by different structural styles. A simple basin is characterized by finite settlement structures and sedimentary bodies in a given generation, whereas a basin complex features a combination of different structural styles (Kingston et al., 1983a, b; Zhu, 1985). Ever since the implications of basin analogy for oil and gas exploration were demonstrated in the 1950s (Weeks, 1952), a variety of basin classification schemes have been proposed based on plate tectonic setting, crustal composition, and tectonic-thermal regime (Kingston et al., 1983a; Miall, 1984; Zhu, 1985; Leckie, 1992; Zhang, 1997, 2010; Allen and Allen, 2005). The global basin classification system of Kingston et al. (1983a) is among the most popular and widely recognized and has been adopted by many authors (e.g., Zhang, 1997, 2010) (Figure 2). Laboratory modeling by Tirel et al. (2006) confirmed that these basins are indeed different in terms of tectonic-thermal mechanisms. Such differences could affect source and reservoir rocks in petroleum systems (Zhang, 1997). This chapter analyzes the polyhistory of basins of different generations and their distribution in the SCS Basin. Our geohistory analysis of the basins within the SCS should have important implications for reconstructing the evolution of the greater SCS Basin as an integrated basin complex and for revealing the dynamic interplay among tectonics, sedimentation, and hydrocarbon systems in the SCS and other petroliferous sedimentary basins worldwide.

The SCS lies at the southeastern edge of the Asian continent, which formed through the convergence of ancient China landmasses. Against the background of the Cathaysia-Indochina mainland, the SCS evolved from a divergent-convergent system in the Mesozoic into a marginal sea in the Cenozoic. Cyclic divergence-convergence phases of plate tectonics played an important function in the formation and evolution of the SCS and its peripheral sedimentary basins. According to previous studies, the evolution of the SCS can be divided into two divergent-convergent cycles (Taylor and Hayes, 1980, 1983; Tapponnier et al., 1982, 1986; Zhu, 1985; Briais et al., 1993; Liu, 1993; Northrup et al., 1995; Schlüter et al., 1996; Gong and Li, 1997, 2004; Hilde et al., 1997; Qiu, 1997; Gong et al., 2001; Morley et al., 2003; Hsu et al., 2004; Hutchison, 2004; Sibuet and Hsub, 2004; Zhou et al., 2005a, b; Li and Li, 2006; Metcalfe, 2006, 2007). Overall, the two tectonic cycles are composed of four basin evolutionary phases: (1) divergent continental margin environment in the Late Triassic–Middle Jurassic, (2) convergent intracontinental environment in the Late Jurassic–middle Eocene, (3) recurrence of the divergent continental margin environment in the late

Eocene–Miocene, and (4) convergent continental margin setting in the Pliocene to the present (Table 1). Although they lasted for a very short period, the latter two phases corresponding to the late divergent-convergent cycle are of special significance for the continental margin basin evolution and for hydrocarbon exploration.

THE EARLY TECTONIC CYCLE: LATE TRIASSIC–MIDDLE EOCENE

Late Triassic–Middle Jurassic Divergent Continental Margin

The SCS and its surrounding area are divided into several tectonic units, such as the South China (Huanan) block, the Indochina block, the Sibumasu block, the West Burma block, the Sikuleh block, and the southwest Borneo block (Qiu, 1997; Qiu and Zhang, 2000). After splitting from Gondwanaland, these tectonic blocks were amalgamated into one continent (Figure 3A). For example, the Huanan block was formed by the amalgamation and accretion of the Yangtze-Cathaysia blocks and allochthonous crustal fragments during the late early Paleozoic and joined by the Indochina block and Kontum massif in the late Paleozoic. Thus, the basement of the Xisha (XS) and Nansha (DS) islands and the East China Sea belongs to the Asia continent, and within it, rifts were developed (Qiu, 1997). The SCS area was in a divergent continental margin setting between the Panthalassa and the southeastern mainland, where a Late Triassic–Middle Jurassic transgressive system developed to drape over much of the Southeast Asian continent (Zhou et al., 2005b, c). The latter suture formed by the closure of the Tethys in the south of the Indochina–East Malaysia–southwest Borneo terrain (Figure 3A), as delineated by Metcalfe (2006), indicates a close relationship between the subducted Panthalassa and rotated Borneo terrane to the north. From the East Borneo terrain, it was connected with the Pacific crust through transform faults in the Triassic–Jurassic period (Hilde et al., 1977) (Figure 3B).

Late Jurassic–Middle Eocene Convergent Continental Margin

In the Late Jurassic–Early Cretaceous (Figure 3B), the SCS experienced a phase of intracontinental convergence, which was associated with the opening and closing of the Tethys Ocean in the heyday of Pacific subduction (Huang et al., 1984; Zhang, 1997; Hall and Wilson, 2000). It has been found (Kudrass et al., 1985; Wakita, 2000; Metcalfe, 2006; Pang et al., 2007) that the suture of the Tethys extended from the Woyla belt of Sumatra Island to the Luk Uio in Java and the Meratus in the southwest of Kalimantan Island and connected with eastern Palawan, Taidong, and the southwestern accretionary wedge of Kyushu by transform faults. Magnetic lineaments in the Pacific oceanic crust east of transform faults (Hilde et al., 1977) exhibit characteristics of accelerated northward subduction. Accordingly, with western convergence and eastward repulsion, the original tectonic framework was overprinted by the left-lateral twisting, leading to the formation of northeast–north-northeast-trending faults and en-echelon folds (Figure 3B). During this time, the eastern continental crust of China features intermediate–acid volcanic rocks extending from southeastern China to South Korea, suggesting an upwelling and eruption of lower crustal material in response to an intracontinental geothermal effect.

During the Late Cretaceous–middle Eocene (Figure 3C), the tectonic framework of the Asian continent underwent major changes caused by the convergence between the Indian plate and the Pacific plate toward the Asian continent. This period saw the Indian plate drifting northward to the Asian continent at approximately 15 to 17 cm/a (Packham, 2003). The Tethys Ocean closed off in eastern Tibet, the eastern margin of the Southeast Asian mainland was continuously subducted by the Pacific oceanic crust into Panthalassa, and the Australian plate drifted northward to collide with Southeast Asia (Hall et al., 1995). As part of the plate accretion and amalgamation processes, the Borneo terrane rotated counterclockwise by 45 to 50° at this time and obducted onto the present remaining ancient oceanic crust of the southern Nansha (NS) Island, thereby creating an uplifted orogenic fold belt along the northern edge of Borneo (Fuller et al., 1999; Soeria-Atmadja et al., 1999; Morley et al., 2003; Ingram et al., 2004; Hall et al., 2008). Compared to the East Asian continental convergence, the collision of India to southwest Asia at this time was stronger, which caused the reversal in the sense of shear from the Mesozoic left lateral to the Late Cretaceous–middle Eocene right lateral (Zhu, 1980, 1985; Gong et al., 2001). Accordingly, massive northeast-trending extensional faults and half grabens formed in eastern China including the SCS (Figure 3C).

THE LATE TECTONIC CYCLE: LATE EOCENE TO THE PRESENT

Late Eocene–Miocene Divergent Continental Margin

The late Eocene–Miocene was a period when the tectonic framework of East Asia changed to form the basis

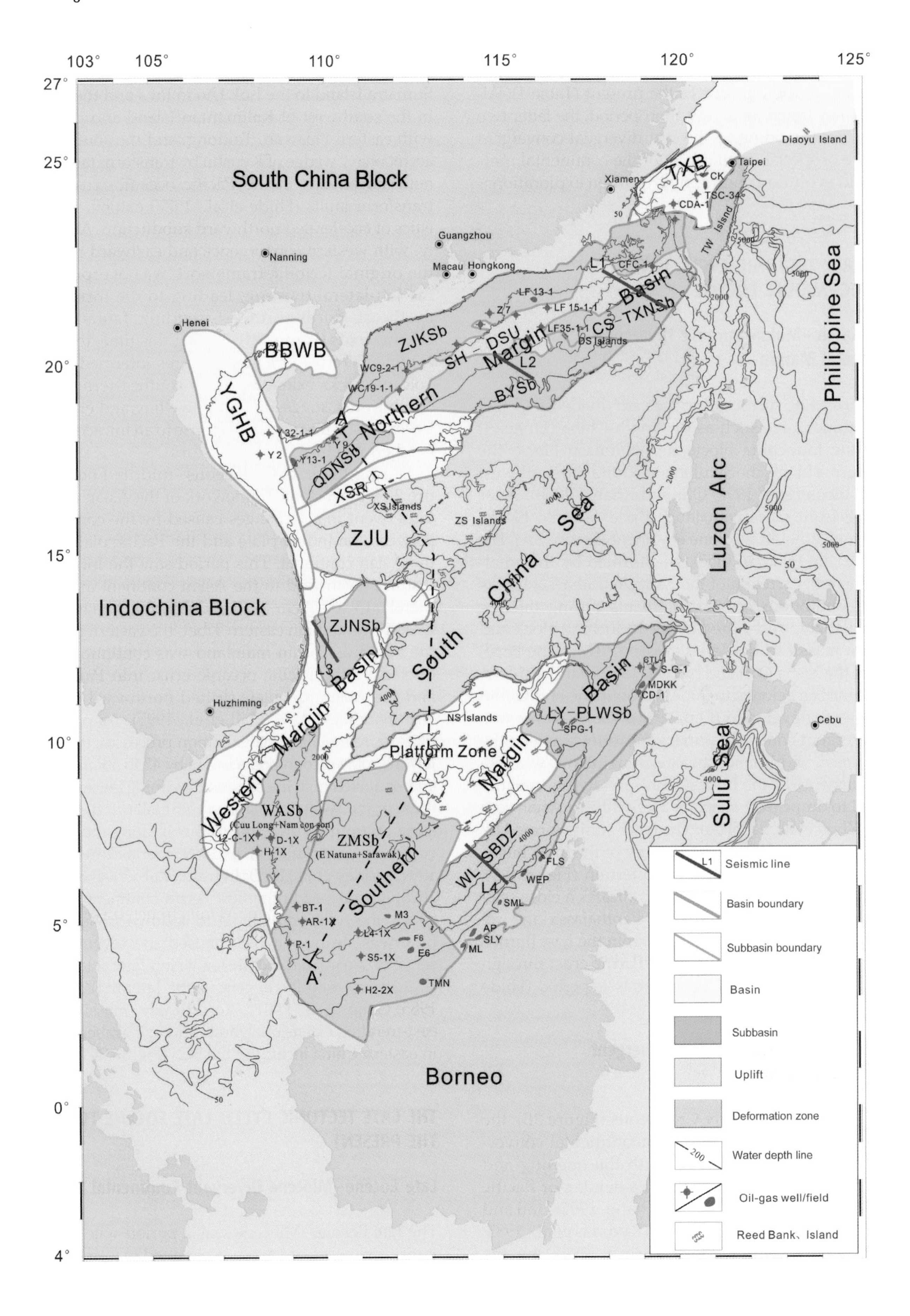
103°
105°
110°
115°
120°
125°
27°
25°
20°
15°
10°
5°
0°
4°
South China Block
Indochina Block
Borneo
Philippine Sea
Luzon Arc
Sulu Sea
South China Sea
Diaoyu Island
Taipei
Xiamen
TXB
CK
TSC-34
CDA-1
TW Islnd
Guangzhou
Nanning
Macau
Hongkong
Henei
Huzhiming
Cebu
L1
CFC-1
Basin
CS-TXNSb
DS Islands
LF 13-1
LF 15-1-1
LF35-1-1
Z 7
ZJKSb
SH-DSU
Margin
L2
BYSb
Northern
BBWB
YGHB
WC9-2-1
WC19-1-1
Y 32-1-1
Y 2
Y13-1
Y 9
QDNSb
XSR
XS Islands
ZS Islands
ZJU
ZJNSb
L3
Basin
Margin
Western
WASb
(Cuu Long+Nam con son)
12-C-1X
D-1X
H-1X
ZMSb
(E Natuna+Sarawak)
Southern
BT-1
AR-1X
P-1
A
A'
L4-1X
S5-1X
H2-2X
M3
F6
E6
TMN
NS Islands
Platform Zone
Margin
LY-PLWSb
SPG-1
Basin
GTL-1
S-G-1
MDKK
CD-1
WL-SBDZ
L4
FLS
WEP
SML
AP
SLY
ML
Seismic line
Basin boundary
Subbasin boundary
Basin
Subbasin
Uplift
Deformation zone
Water depth line
Oil-gas well/field
Reed Bank、Island

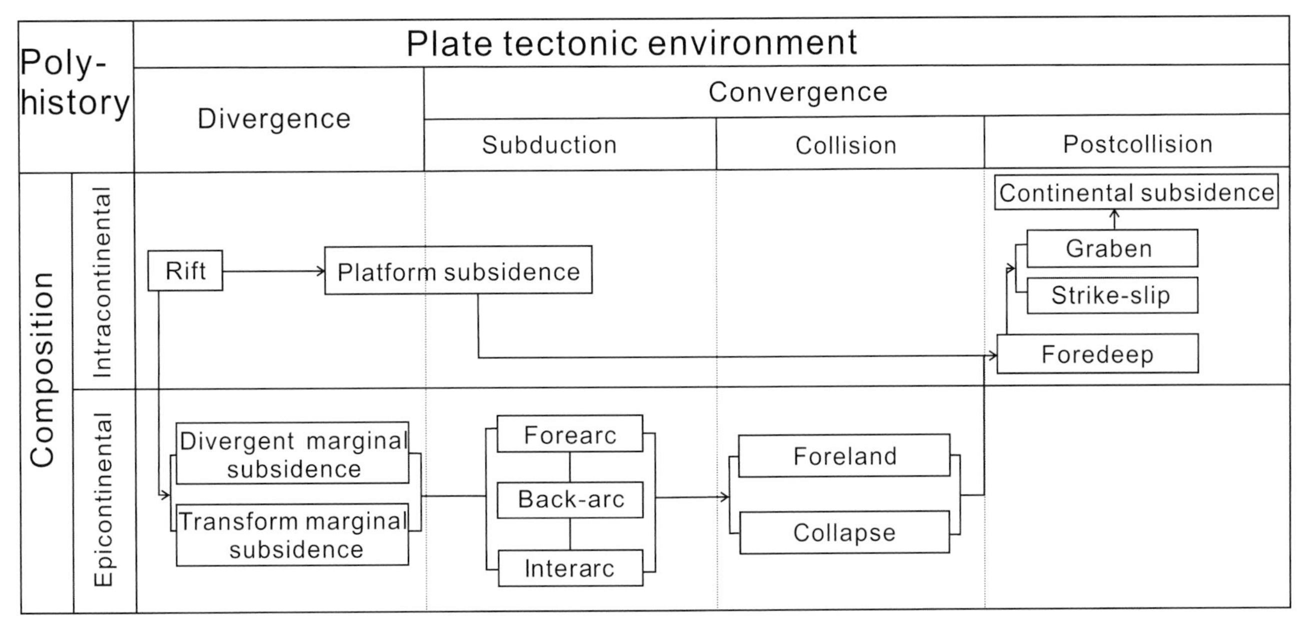

Figure 2. The polyhistory basin classification scheme of the South China Sea Basin (modified from Kingston et al., 1983a; Zhang, 1997, 2010).

of the modern tectonic style (Figure 3D, E). In the west and southwest of continental China, the collision between the Indian plate and Tibet caused large-scale intracontinental deformation and displacement in East Asia (Tapponnier et al., 1982, 1986). The oblique subduction of the Indian Ocean along the Java trench (Packham, 2003) caused intracontinental shearing in the southern Indochina continental block.

In eastern continental China, because of the abrupt eastward expansion and clockwise rotation of the northern side of the Philippine oceanic crust (Hall et al., 1995; Lee and Lawver, 1995), subduction of the Pacific plate turned toward the west northwest from north northwest, and its convergence toward the Eurasian plate, accordingly, became relatively weakened (Northrup et al., 1995). These changes not only affected the difference in stress regime between both sides of the Asian continent, but also caused sea-floor spreading and formation of the SCS (Taylor and Hayes, 1980, 1983; Briais et al., 1993; Barckhausen and Roeser, 2004; Hsu et al., 2004). The SCS continental crust breakup might be associated with the thermal creep of the deep mantle toward eastern continental China (Zhang, 1997). At this time, a series of rifts was created in the intracontinental margin during the late Eocene–Oligocene (Figure 3D).

Pliocene–Present Convergent Continental Margin

Since the late Miocene (Figure 3E), the SCS has evolved again to a convergent tectonic setting, which was associated with the continued collision of the Australian continent with East Timor (Butcher, 1990; Hilde et al., 1997; Pubellier et al., 2003; Sibuet and Hsub, 2004) and the clockwise rotation and northward obduction of the Philippine Sea plate (Li et al., 2004). The convergence not only formed the Manila trench (Li et al., 2004), closed the SCS, and created the orogenic belts in Taiwan area (Huang et al., 1997; Sibuet and Hsub, 2004), but also started a new tectonic-thermal event around the entire SCS and surrounding land and island arc (Gong et al., 2001; Zhou et al., 2005a). At this time, intense volcanic activity occurred along the residual oceanic crust and continental margin in the SCS region (Gong et al.,

Figure 1. Distribution of sedimentary basins in the South China Sea. B = basin; BBW = Beibuwan; BY = Baiyun; CS-TXN = Chaoshan-Taixinan; DS = Dongsha; DZ = deformation zone; LY-PLW = Liyue-Palawan; NP = North Palawan; NS = Nansha; QDN = Qiongdongnan; R = rift; RB = Reed Bank; Sb = subbasin; SH-DS = Shenhu-Dongsha; SL = Sulawesi Sea; SP = South Palawan; SU = Sulu Sea; TX = Taixi; U = uplift; WA = Wan'an (Cuu Long+Nam Con Son); WL-SB = Wenlai-Shaba; XS = Xisha; YGH = Yinggehai; ZJ = Zhongjian; ZJK = Zhujiangkou; ZJN = Zhongjiannan; ZM = Zengmu (East Natuna+Sarawak); ZS = Zhongsha. L_1, L_2, L_3, L_4, and AA′ are seismic sections or geologic traverses.

Table 1. Tectonic stages and evolutionary sequences in the polyhistory basins of the South China Sea in the Mesozoic–Cenozoic (see Figure 1 for the abbreviations).

Age of bottom (Ma)			Evolution stage		Tectonic environment	BBWB*	Northern margin basin				TXB*	YGHB*	Western margin basin		Southern margin basin	
							QDNSb*	ZJKSb*	BYSb*	CS-TXNSb*			ZJNSb*	WASb* (Cuu Long+Nam Con Son)	LY-PLWSb*	ZMSb* (E Natuna+Sarawak)
Quaternary	$N_{p\text{-}h}$	2.60	C	C_2	Convergence	Continental subsidence	Divergent marginal subsidence				Foreland	Rift	Divergent marginal subsidence		Divergent marginal subsidence	
Pliocene	N_2	5.33														
Miocene	N_1^3	11.63		C_1	Divergence						Transform marginal subsidence	Strike-slip				
	N_1^2	15.97														
	N_1^1	23.03														
Oligocene	E_3^1	28.40				Graben	Rift	Rift	Rift	Rift			Strike-slip	Strike-slip	Rift	Rift
	E_3^2	33.90														
Eocene	E_2^3	37.20														
	E_2^2	48.60	B	B_2	Convergence		Graben	Graben	Graben	Graben	Graben	Graben	Graben	Graben	Graben	Graben
	E_2^1	55.80														
Paleocene	E_1	65.50														
Cretaceous	K_2	99.60														
	K_1	145.5		B_1		Strike-slip										
Jurassic	J_3	161.2														
	J_2	175.6	A		Divergence	Divergent marginal subsidence										
	J_1	199.6														
Triassic	T_3	228.0														
	T_2	245.0				Accretion-collage of the Southeast Asia's block										
	T_1	251.0														
Permian	P	299.0														

*BBWB = Beibuwan Basin; QDNSb = Qiongdongnan subbasin; ZJKSb = Zhujiangkou subbasin; BYSb = Baiyun subbasin; CS-TXNSb = Chaoshan - Taixinan subbasin; TXB = Taixi Basin; YGHB = Yinggehai Basin; ZJNSb = Zhongjiannan subbasin; WASb = Wa'an (Cuu Long+Nam Con Son) subbasin; LY-PLWSb = Liyue-Palawan subbasin; ZMSb = Zengmu (East Nabuna + Sarawak) subbasin.

2001; Zhou et al., 2005a; Yan et al., 2006, 2008) (Figure 3E). For example, the regional thermal subsidence in the continental margin and the XS rift (XSR) became more active (He et al., 1980; Qiu and Ye, 2001; Liu and Wu, 2006) with renewed uplifts (Gao and Chen, 2006).

SEQUENCE OF MESOZOIC–CENOZOIC POLYHISTORY BASINS

The identification of unconformities helps us to distinguish the various polyhistory basins (Kingston et al., 1983a) that formed because of multiphase deformation. Take the top-Oligocene breakup unconformity of Falvey (1974) as an example; it is an important boundary between the earlier rift and the later divergent marginal subsidence (Table 2). The unconformity not only represents an important change in tectonic regime, but also separates two distinct, vertically stacked basins with different hydrocarbon potential.

Since the Triassic, at least four major unconformities and six secondary unconformities have been identified corresponding to regional tectonic events and sedimentary cycles (Lin et al., 2007; Xie et al., 2008). The four major unconformities developed at the top of the Mesozoic, top of the middle Eocene, top of the Oligocene (the breakup unconformity), and top of the middle Miocene, which have been termed as the "Liyue event," "Xiwei event," "Nanhai event," and "Nansha event" (or Wan'an event), respectively (Table 2). They are relatively easier to track on regional seismic profiles, especially the unconformities at the top of the Mesozoic and the top of the Oligocene (Figure 4). Regionally, small differences exist among them in the time and intensity of development, as some of them are not completely synchronous based on recent stratigraphic results (Zhan et al., 2006; Xie et al., 2008).

Generally, the Mesozoic and older strata are difficult to identify and track because of the poor quality of seismic data (Figures 4–8) and the limited number of deep wells. However, according to Xia and Huang (2000) and Hao et al. (2001), these strata are sequentially bounded by several unconformities. Stratigraphic and facies data from the coastal outcrops of Guangzhou to the Chaoshan–Taixinan (TXN) subbasin (Figure 1) indicate diachroneity for Mesozoic unconformities, which become younger in age from north to south (Li et al., 2008). Seismic interpretation reveals that the strata below the top Mesozoic unconformity are incomplete, whereas the Cenozoic strata above the unconformity are widespread and controlled by faults and half-graben structures (Xia and Huang, 2000; Hao et al., 2001).

DISTRIBUTION OF MESOZOIC–CENOZOIC POLYHISTORY BASINS

Basins of the Earlier Tectonic Cycle (Late Triassic–middle Eocene)

Basins in the Late Triassic–Middle Jurassic

Shallow-marine deposits of Triassic–Lower Jurassic occur in outcrops along the coast of Guangzhou. They extend into the sea as found by drilling in the East China Sea and at outcrops on the footwall of thrust structures in Taiwan (Zhou et al., 2005c; Shao et al., 2007; Li et al., 2008). Many residual shallow-marine strata of the Upper Triassic–Lower Jurassic also occur along the continental margins of Palawan and Mindoro islands (Zhou et al., 2005c). This distribution pattern indicates that the eastern Asian continent was occupied by widely distributed basins of the divergent marginal subsidence. During the late Mesozoic, the neritic facies in these basins were deformed in a convergent tectonic setting. At the southern margin, they were overprinted in the Late Jurassic–Early Cretaceous by strike-slip tectonics before merging into the orogenic belt of the North Borneo associated with the subduction of Panthalassa in the Late Cretaceous–middle Eocene (Zhou et al., 2005c) (Figure 9A) (Table 1).

Basins in the Late Jurassic–middle Eocene

During the Late Jurassic–Early Cretaceous, the SCS was located within the continent of Southeast Asia (Figure 3B). Influenced by subduction and accretion in a convergent tectonic setting, the continental margin had been uplifted, whereas the southern margin of the SCS was associated with the orogenic belt because of subduction of the oceanic crust and collision with North Borneo. Further influenced by the northward subduction of the Pacific, clastic and volcanic deposition in a continental setting was interrupted by occasional or cyclical marine deposition in a wrench tectonic setting (Wang et al., 2000; Zhou et al., 2005a; Yan et al., 2006). In the TXN region of the SCS, marine deposits can reach a thickness of more than 5000 m (16,404 ft) (Feng et al., 2001), and their northward extensions are exposed along the coast of Guangzhou. In the TXN region, seismic data indicate folds and inversion structures in the Mesozoic sequences. Some seismic profiles show that the uplifted Mesozoic sequences were overprinted by graben subsidence in the Paleogene (Shi et al., 2008). Based on these characteristics in a regional contractional or transpressional tectonic setting, we see it as a major strike-slip basin in the Late

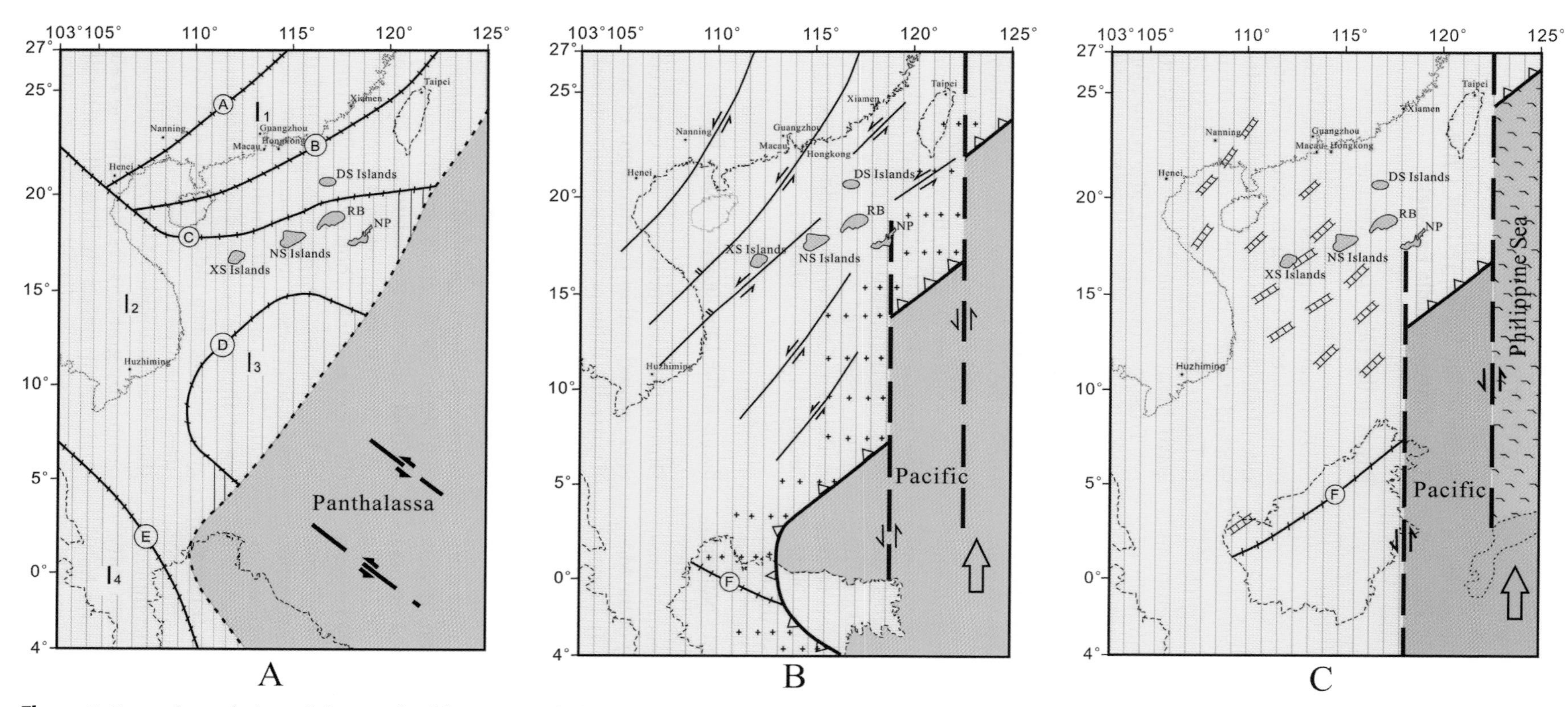

Figure 3. Tectonic evolution of the South China Sea Basin in Late Triassic–Middle Jurassic (A), Late Jurassic–Early Cretaceous (B), Late Cretaceous–middle Eocene (C), late Eocene–Oligocene (D), and since Miocene (E) (see Figure 1 for the abbreviations). 1 = continental block; 2 = Panthalassa; 3 = pristine oceanic basin; 4 = Tertiary oceanic basin; 5 = island; 6 = continent-ocean boundary; 7 = suture and collision orogens; 8 = subducted zone; 9 = orogenic revived zone; 10 = strike-slip fault; 11 = transtension fault; 12 = hypothetical transform fault; 13 = center of spreading; 14 = rift zone; 15 = moving direction of oceanic plate; 16 = calc-alkalic magmatic arc; 17 = magmatic rock; I_1 = South China continent; I_2 = Indochina continent; I_3 = Dongma–southwest Borneo plate; I_4 = Zhenghe-Zengmu (Luconia) plate; I_5 = Sulu-Borneo accretionary series; I_6 = southwestern subbasin; I_7 = east subbasin; I_8 = Luzon arc; I_9 = Philippine Sea plate; A = Qinzhou-Fangcheng suture; B = South China offshore suture; C = Heishui River–Yinggehai-Xisha rift suture; D = northwest Nansha suture; E = Dongma-Gujin suture; F = Kalimantan suture; G = East Natuna–Lupar subducted zone; H = Bukit-Mersing subducted zone; I = Palawan suture; J = Tingjia fault belt; K = Red River–Yuedong shear zone; L = Luzon Trough subducted zone; M = Manila trench–Negros-Gothabator subducted zone; N = Philippine trench subducted zone; O = Taiwan collision orogens.

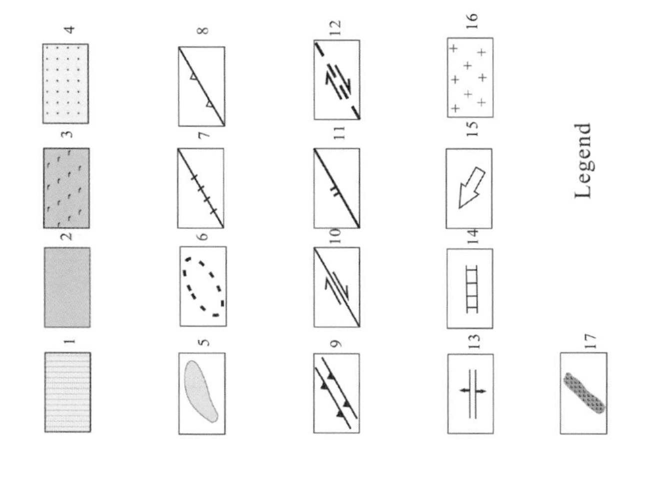

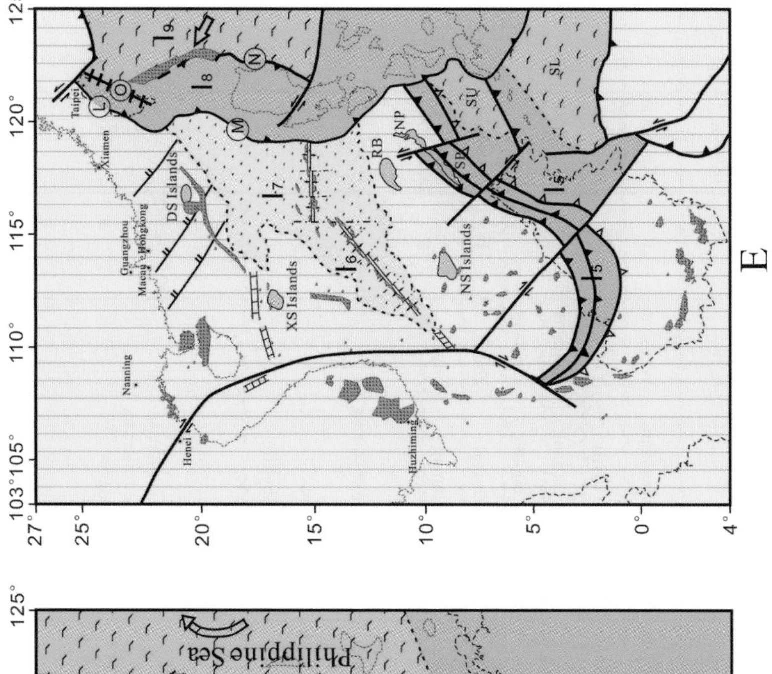

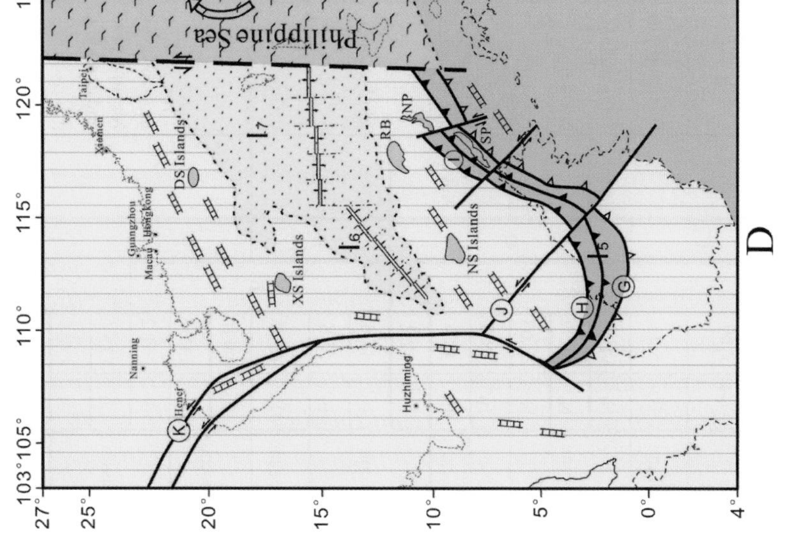

Figure 3. (cont.).

Table 2. Stratigraphy and correlation of the polyhistory basins on the continental shelf and slope of the greater South China Sea Basin (see Figure 1 for the abbreviations).

Age of Bottom (Ma)			Evolution Stage		Tectonic Event	BBWB	Northern Margin Basin				YGHB	Western Margin Basin		Southern Margin Basin		WL-SBDZ
							QDNSb	ZJKSb	BYSb	CS-TXNSb		ZJNSb	WASb (Cuu Long + Nam con son)	LY-PLWSb	ZMSb (E Natuna + Sarawak)	
Quaternary	$N_{p\text{-}h}$	2.60	C	C_2	Nansha		Ledong T_2			A T_0	Ledong T_2	A T_1		A		A T_1
Pliocene	N_2	5.33				Wanglougang	Yinggehai T_3	Wanshan	Wanshan	B T_2	Yinggehai T_3	B T_2	Guangya T_2	T_2	Beikang T_1	B T_2
Miocene	N_1^3	11.63		C_1		Dengloujiao T_0	Huangliu T_4	Yuehai T_2	Yuehai T_2	C T_4	Huangliu T_4	C T_3	Kunlun T_3	B T_3	Nankang T_2	C T_3
	N_1^2	15.97			Nanhai	Jiaowei T_1	Meishan T_5	Hanjiang T_4	Hanjiang T_4	D	Meishan T_5	D	Lizhun $T_3{}'$	C	Haining	D
	N_1^1	23.03				Xiayang T_2	Sanya T_6	Zhujiang T_5	Zhujiang T_5	T_6	Sanya T_6	T_4	Wanan T_4	T_4	Lidi	
Oligocene	E_3^2	28.4			Xiwei	Weizhou	Lingshui T_7	Zhuhai T_7	Zhuhai T_7	E	Lingshui T_7					T_4
	E_3^1	33.9				T_4	Yacheng T_8	Enping T_8	Enping T_8	F	Yacheng T_8	E	Xiwei	D	Zengmu	E
Eocene	E_2^3	37.2								T_7		T_5	T_5	T_5		T_5
	E_2^2	48.6	B	B_2	Liyue	Liushagang	?	Wenchuang	?	G						
	E_2^1	55.8				T_7						F	Renjun	E	Larang	F
Paleocene	E_1	65.5				Changliu		Shenhu			?					
Cretaceous	K_1	99.6		B_1			T_g	T_g	T_g	T_g H		T_g	T_g	T_g		T_g G
	K_2	145.5								T_m						H
Jurassic	J_3	161.2				limestone, and Metamorphite, and granide etc.	?	Granide, and sedimentary rocks ?	?	?			Granide, and metamorphite	Clastic rocks, and metamorphite	Metamorphite, limestone, and clastic rocks	Crystalline rocks
	J_2	175.6	A													
	J_1	199.6														
Triassic	T	251.0														
Permian	P	299.0														
Data sources	International stratigraphic chart (2004)					Gong and Li, 1997						Zhong and Gao, 2005	Xie et al., 2008	Gong et al., 2001		

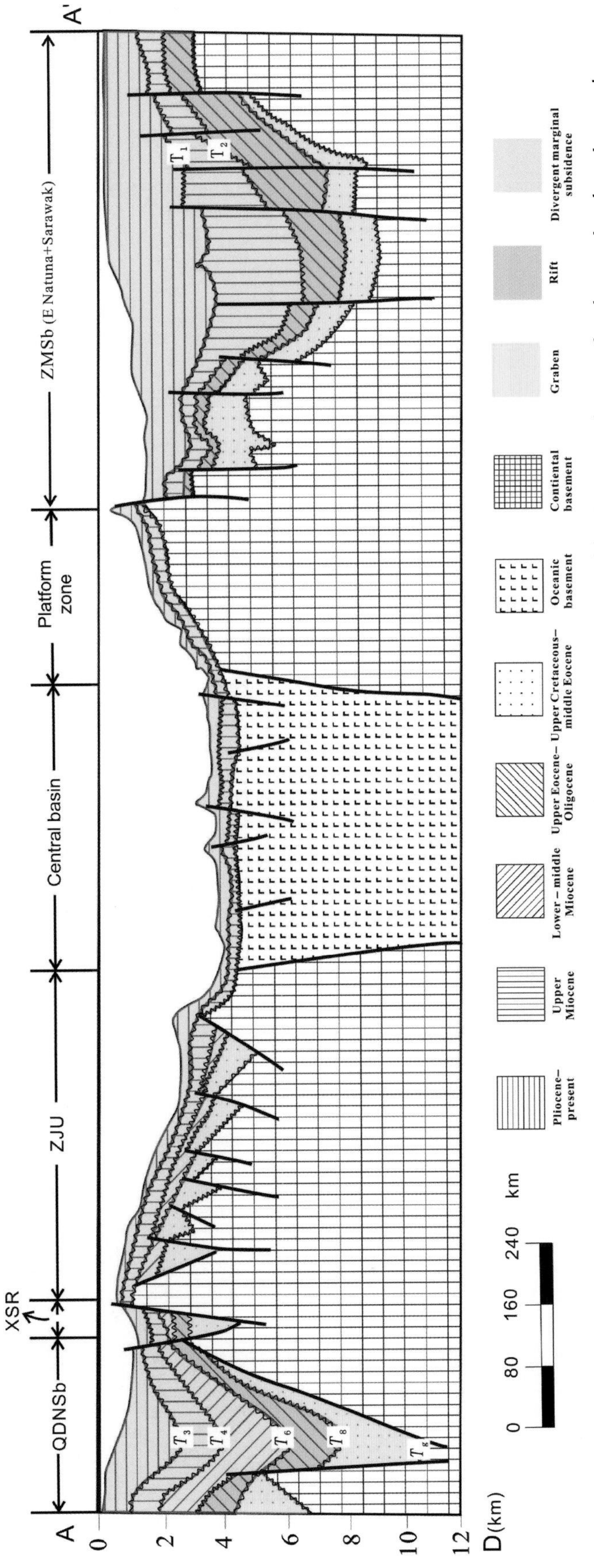

Figure 4. Geologic section AA′ across the western South China Sea showing various unconformities and basins (see Figure 1 for the section location and abbreviations and Table 2 for the symbols) (modified from Wu et al., 2003). 80 km (49.7 mi).

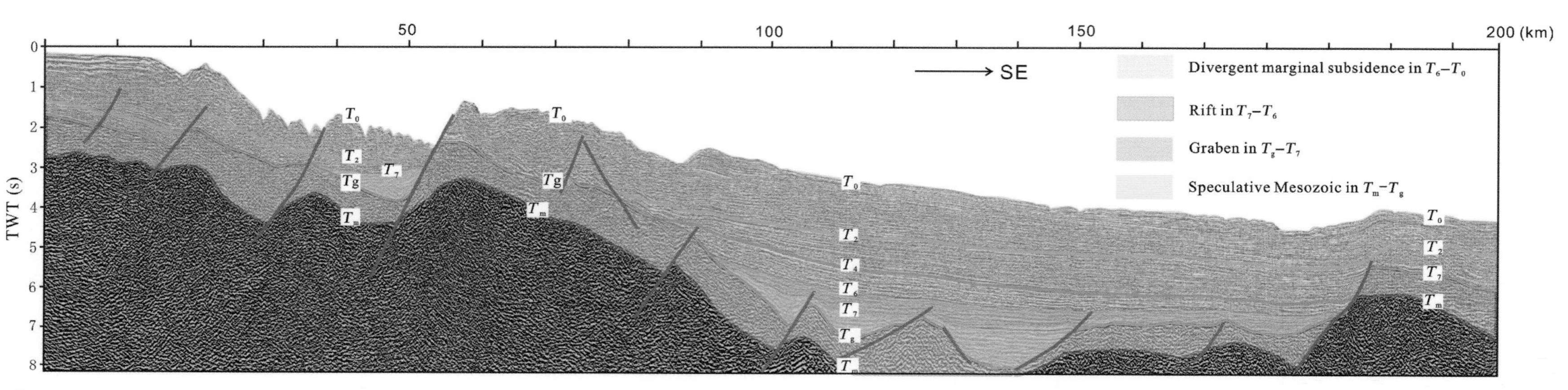

Figure 5. Seismic section L_1 from the Shenhu-Dongsha uplift to the Chaoshan-Taixinan subbasin in the northern margin (see Figure 1 for the location and Table 2 for the symbols) (modified from Chen and Wen, 2010). TWT = two-way traveltime. 50 km (31 mi).

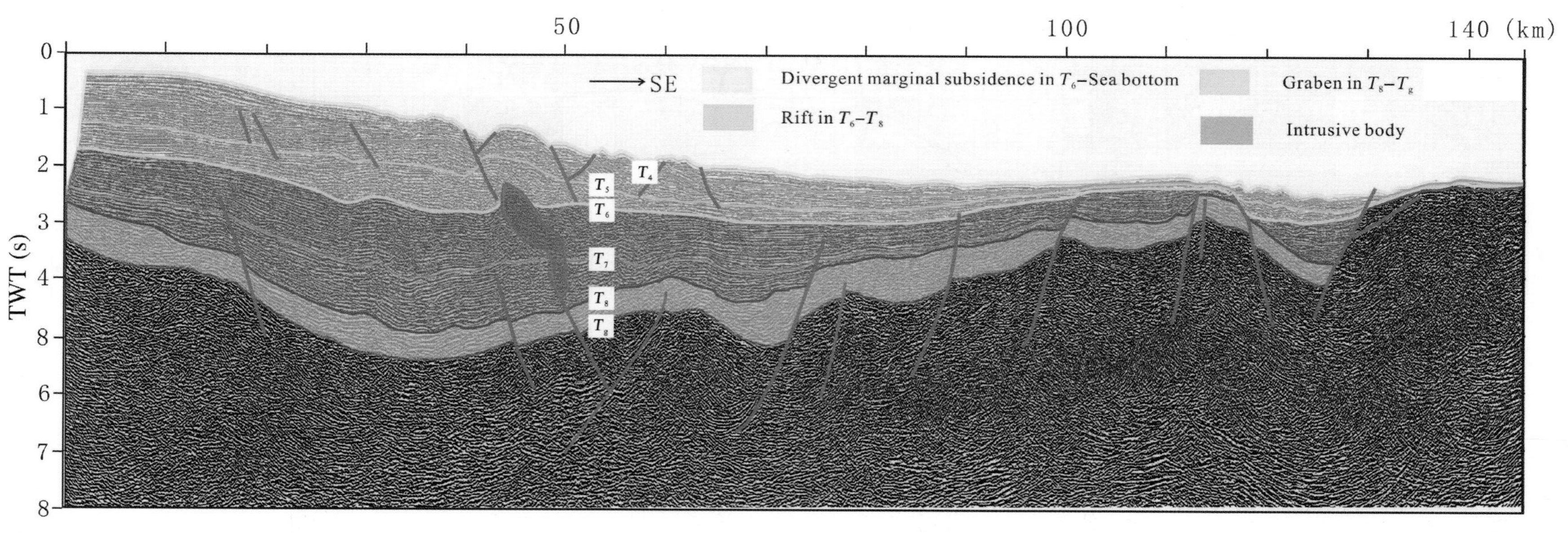

Figure 6. Seismic section L_2 across the Baiyun subbasin in the northern margin (see Figure 1 for the location and Table 2 for the symbols) (modified from Chen and Wen, 2010). TWT = two-way traveltime. 50 km (31 mi).

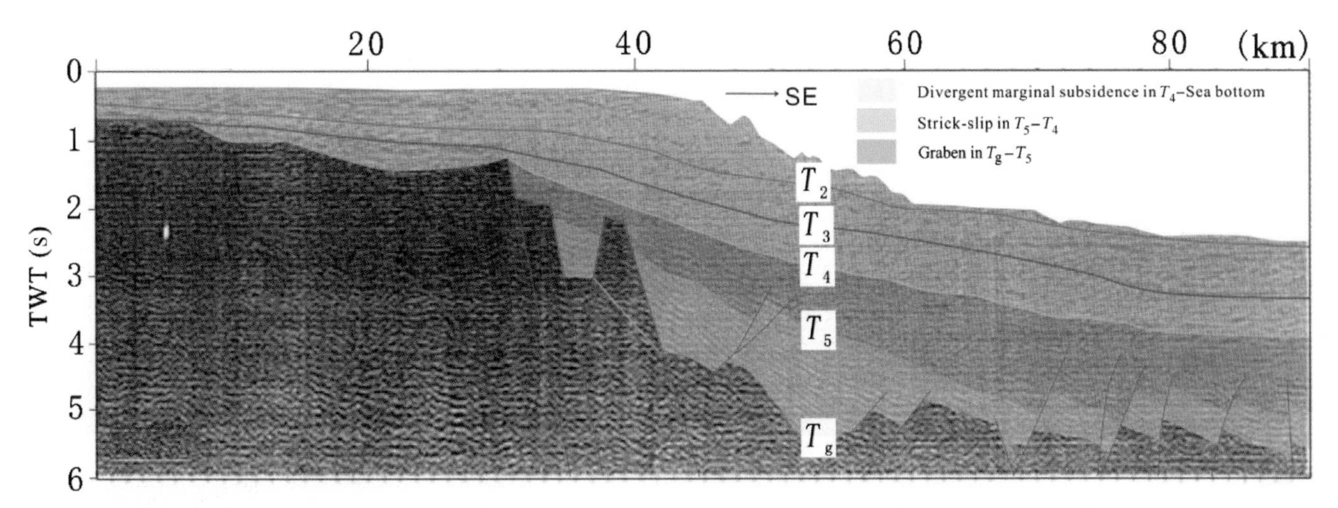

Figure 7. Seismic section L_3 across the Zhongjiannan subbasin in the western margin (see Figure 1 for the location and Table 2 for the symbols) (modified from Clift et al., 2008). TWT = two-way traveltime. 20 km (12.4 mi).

Jurassic–Early Cretaceous (Figure 9A) (Table 1). Recently, oil and gas fields have been discovered along a northeastern structural trend in the TXN strike-slip basins formed in the Late Jurassic–Early Cretaceous (Feng et al., 2001; Xia et al., 2004; Wei et al., 2005). However, details of contemporary basins in the northern SCS need further investigation.

In the southern SCS, Early Cretaceous shallow-marine clastic rocks interbedded with limestone lignite and coal seams were found by drilling in the offshore regions of Palawan, Reed Bank, and Nansha islands (Schliiter et al., 1996). Being superimposed on the residual coastal delta system, these deposits formed in an earlier divergent marginal subsidence, again indicating transformation of the basins in the late Mesozoic (Figure 9A) (Table 1).

As part of the China-Indochina continent during the Late Cretaceous–middle Eocene, a series of basins was developed and characterized by a half-graben structural style. Many geologists have studied them in the SCS and would call them "rift basins" (e.g., Gong et al., 1989; Liu, 1993; Gong and Li, 1997, 2004; Packham, 2003; Hutchison, 2004; Yao et al., 2004; Wang et al., 2005; Li and Li, 2006; Droust and Sumner, 2007). Based on their plate tectonic configuration, we define them as the kind of graben basins formed in a convergent tectonic setting (Figure 9B) (Table 1). This type of basin was characterized by a structural style of half graben (Figures 5–7) and a nonmarine sediment infill in a contractional environment (Gong and Li, 1997, 2004; Li and Li, 2006; Xie et al., 2008). It overlies the earlier basins of strike-slip or divergent marginal subsidence, and at the northern margin of the SCS, it gradually becomes smaller southward.

Basins of the Later Tectonic Cycle (Late Eocene–Present)

Basins in the Late Eocene–Miocene

After the birth of the oceanic crust in the SCS, a series of new basins was developed from the earlier basins. At the northern margin, the so-called rift basins actually experienced two stages (e.g., Gong et al., 1989; Liu, 1993; Gong and Li, 1997, 2004; Zhang and Bai, 1998; Lee et al., 2001; Packham, 2003; Hutchison, 2004; Yao et al., 2004; He et al., 2005; Qiu et al., 2005; Wang et al., 2005; Wei et al., 2005; Li and Li, 2006; Droust and Sumner, 2007), which are an earlier rift stage associated with the structural style of half graben in the late Eocene–Oligocene and a later regional thermal subsidence stage since the Miocene. Here, we classify them as rift basins in the late Eocene–Oligocene and as divergent marginal subsidence since the Miocene in a divergent tectonic setting. The rift basin developed in the late Eocene–Oligocene was characterized by a domino structural style (seaward) and superimposed with divergent marginal subsidences since the Miocene (Figure 9C, D). The earlier rift basin can be easily identified from the later basin of the divergent marginal subsidence by a breakup unconformity (Figures 4–8).

At the northern margin, the breakup unconformity T6 is a seismic marker (Table 2) between the Zhuhai and Zhujiang Formations in the Zhujiangkou subbasin or the Pearl River Mouth Basin (Gong and Li, 1997, 2004), and the unconformity represents a transformation from rift to divergent marginal subsidence. The structural style of the rift basin was characterized by a half graben. Major faults were sliding above the

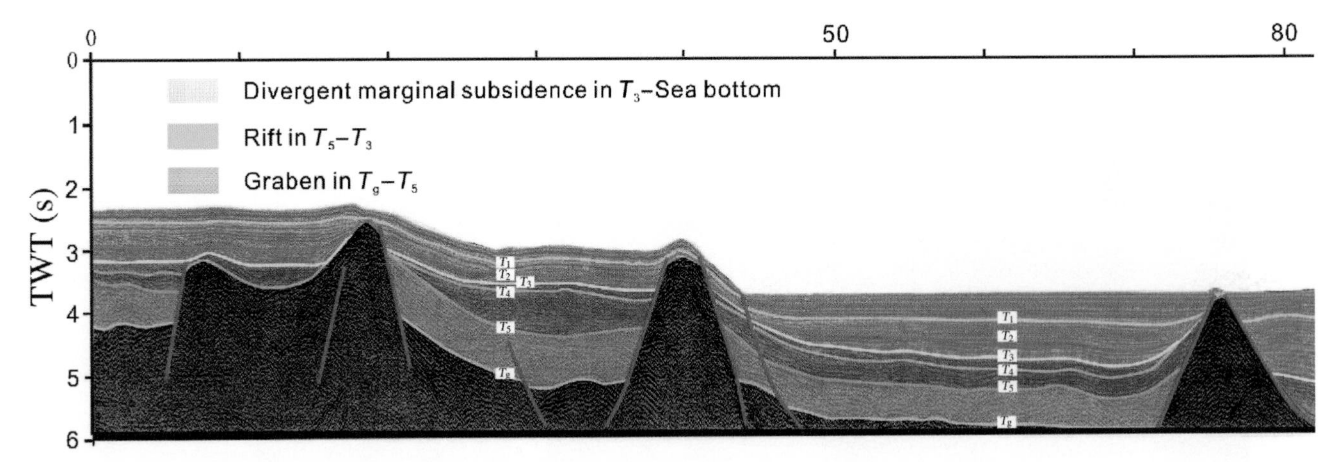

Figure 8. Seismic section L_4 across the Wenlai-Shaba deformation zone in the southern margin (see Figure 1 for the location and Table 2 for the symbols) (modified from Chen and Wen, 2010). TWT = two-way traveltime. 20 km (12.4 mi).

thinning crust and show a parallel domino style with the main faults dipping seaward (Figures 4, 6, 9C). Generally, the major part of the rift basin developed on the marginal slope. Although the sequences above and below the unconformity T8 (Table 2) are different (Table 1), they are easy to be confused with the same structural style in the half graben. The major fault is generally dipping to the south (seaward) in the rift basin, whereas the major fault in the graben basin commonly dips to the north (landward) (Figures 4–6; 9B, C).

At the western margin (Figure 9C), the strike-slip basins were created at the same time as the rift basins were formed at the northern margin in the late Eocene–Oligocene. The flower-shape inversion structure formed in a strike-slip setting can be identified on two-dimensional seismic sections. This structural style can be distinguished from the underlying Eocene graben basins (Rangin et al., 1995a, b; Clift et al., 2008). The strike-slip basins were later overlain by basins of the divergent marginal subsidence associated with thermal subsidence similar to the northern margin since the Miocene (Figures 7; 9C, D) (Table 1). The divergent marginal subsidence on the shelf (<200 m [656 ft]) is relatively narrow, possibly inheriting the strike-slip fault zone in Indochina (Rangin et al., 1995a, b; Pubellier et al., 2003).

At the southern margin, some authors suggested three stages of basin development (Zhong et al., 1995; Madon et al., 1999; Jin and Li, 2000; Chen, 2002). The first stage is marked by a peripheral foreland basin during the late Eocene–early Miocene; the second one, by a strike-slip or transtensional pull-apart basin during the middle and late Miocene; and the third one, by a passive continental margin basin since the middle Miocene. However, many regional seismic reflection profiles (Qiu et al., 2005; Clift et al., 2008) do not show the style of the flexural subsidence, which characterizes the foreland basin (Figure 2). Instead, they show the structural style of graben or half graben during the late Eocene–Oligocene and the seaward subsidence since the Miocene, indicating the rift and divergent marginal subsidence, respectively. We suggest that the thrusts or folds and flower structures during the late Eocene–Oligocene represent only a later deformational event or inversion of the earlier extensional faults caused by the collisional orogen of the Zengmu block with the West Borneo since the Miocene (Figure 8). Seismic surveys further reveal that basin change can be tracked from the NS Islands eastward to the Reed Bank and even Palawan (Williams, 1997), indicating that, similar to the northern margin, the rift basins have been transformed into the basins of divergent marginal subsidence since the Miocene (Figure 9C, D).

The Miocene deltaic deposits outside the Rajang Estuary in northern Sarawak were sourced from the Borneo orogen in the south of the SCS (Hutchison, 2004). The sedimentary sequence established by exploration at the northwestern coast of Palawan shows shallow-marine sandstone and shale with limestone intercalations in the Oligocene rift basin, overlying the riverine and lacustrine sandstone and mudstone in Eocene and Paleocene graben basins. In the divergent marginal subsidence basins, Miocene limestone developed on the carbonate platform to form the Nido Formation, which was overlain by clasolites in the Galoc Formation of deep-marine environments. Drilling also proved that mature source rocks exist in the rift basin underlain by the divergent marginal subsidence basin, and oil migrated up to the limestone across sequence unconformities (Williams, 1997).

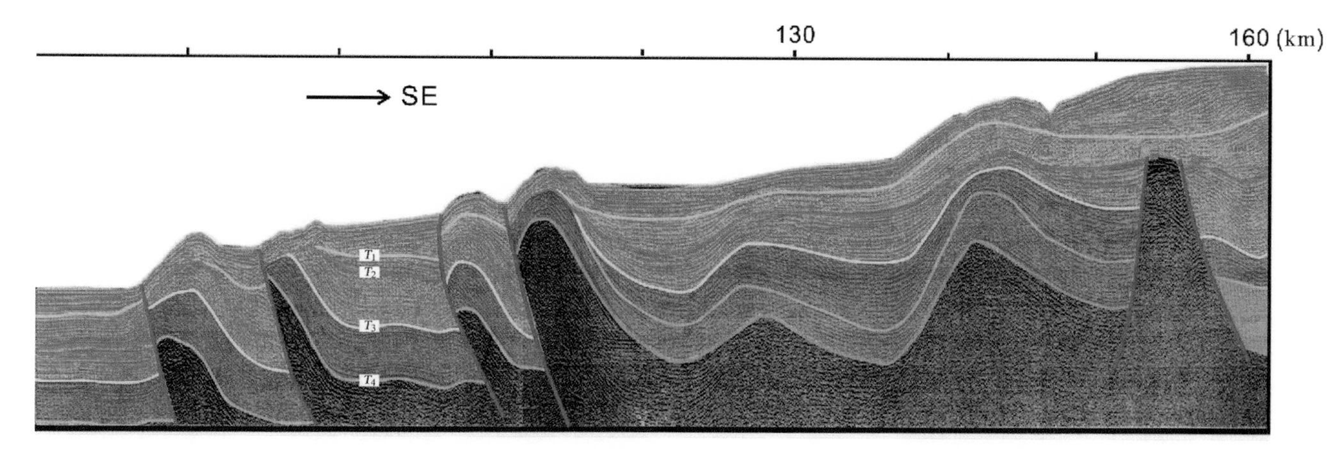

Figure 8. (cont.).

Basins in the Pliocene to the Present

Since the Pliocene, in an overall convergent tectonic setting associated with the Philippine island arc obduction onto the SCS oceanic crust and movement toward the Chinese mainland, a series of southeast–northwest-trending sinistral strike-slip faults developed (Gong and Li, 1997; Yan et al., 2008) (Figure 3E). The divergent marginal subsidence was overprinted by strike-slip faulting in the northern margin (Figure 9D). These strike-slip faults have been considered as the primary migration pathways for oil (Gong and Li, 1997).

At the same time, the east–west-trending XSR (Qiu and Ye, 2001), the northwest-trending Yinggehai (YGH) rift (Gong and Li, 1997), and the Zhongjian uplift were also formed (Qiu et al., 1997; Gao and Chen, 2006) (Figures 1, 9D). The XSR bends west to link with the north–south-trending faults. Refraction-seismic and gravity-magnetic data (Qiu and Ye, 2001) indicate that a new generation of rift basins was created by mantle upwelling and crustal stretching, probably representing one of the triple rifts that extend eastward to form the new oceanic crust. The northwest-trending YGH rift basin developed in the continental crust since 5.5 Ma by superimposing on a strike-slip basin formed by the earlier Red River fault (Table 1). In the rift basin, the thicknesses of the Pliocene Huangliu Formation and the Quaternary deposits together reach more than 8000 to 10,000 m (26,247–32,808 ft) (Table 2), in sharp contrast to the much thinner sequence in the divergent marginal subsidence of the Qiongdongnan region (Gong and Li, 1997). Drilling has confirmed a very thick sequence with deposition rates of 3000 m/Ma (9843 ft/Ma) or more in the YGH rift basin (Gong and Li, 1997; Sun et al., 2003). The thermal upwelling caused the present geothermal gradient of 42.5°C/km and the measured heat-flow value of 84.1 MW/m^2, indicating an anomalous thermal structure spatially and temporally associated with the multiphase extension (He and Xiong, 2000). The Pliocene quartz tholeiite at a depth of 100 m (328 ft) in well Ya 32-1-1 from the northern YGH Basin (Figure 1) was dated at 3.85 Ma (Gong and Li, 2004), proving a close relation between rifting and thermal upwelling from the mantle. This geothermal condition provided an optimal setting for the generation and accumulation of oil and gas in the SCS Basin. In addition, whenever and wherever the mantle heat flow was blocked, the sedimentary wedges in the western and eastern deep zones of the northern marginal subsidence were invaded by basic magma. Heat flow increases from 66 ± 9.8 MW/m^2 (thermal gradient at 30.4 ± 4.9°C/km) to 77.5 ± 14.8 MW/m^2 (thermal gradient at 35.5 ± 6.4°C/km) toward the deep-marine setting, which has been considered by Gong and Li (2004) as an important thermal condition for oil and gas generations in the divergent marginal subsidence basin. At the southern margin (Figure 9D), the divergent marginal subsidence basin was overprinted by tectonic inversion and thrusting. Deposition continued only in troughs in the front of thrust faults, forming reservoir lithofacies as indicated by oil and gas discoveries in the associated inversion structures of the upper Pliocene (Gong et al., 2001).

IMPLICATIONS OF POLYHISTORY BASINS FOR OIL AND GAS EXPLORATION

As one of the major petroliferous sedimentary basins globally, the SCS has been the focus of a lot of attention in the past and will continue to be an exploration hot spot in the future. Unlike other offshore sedimentary

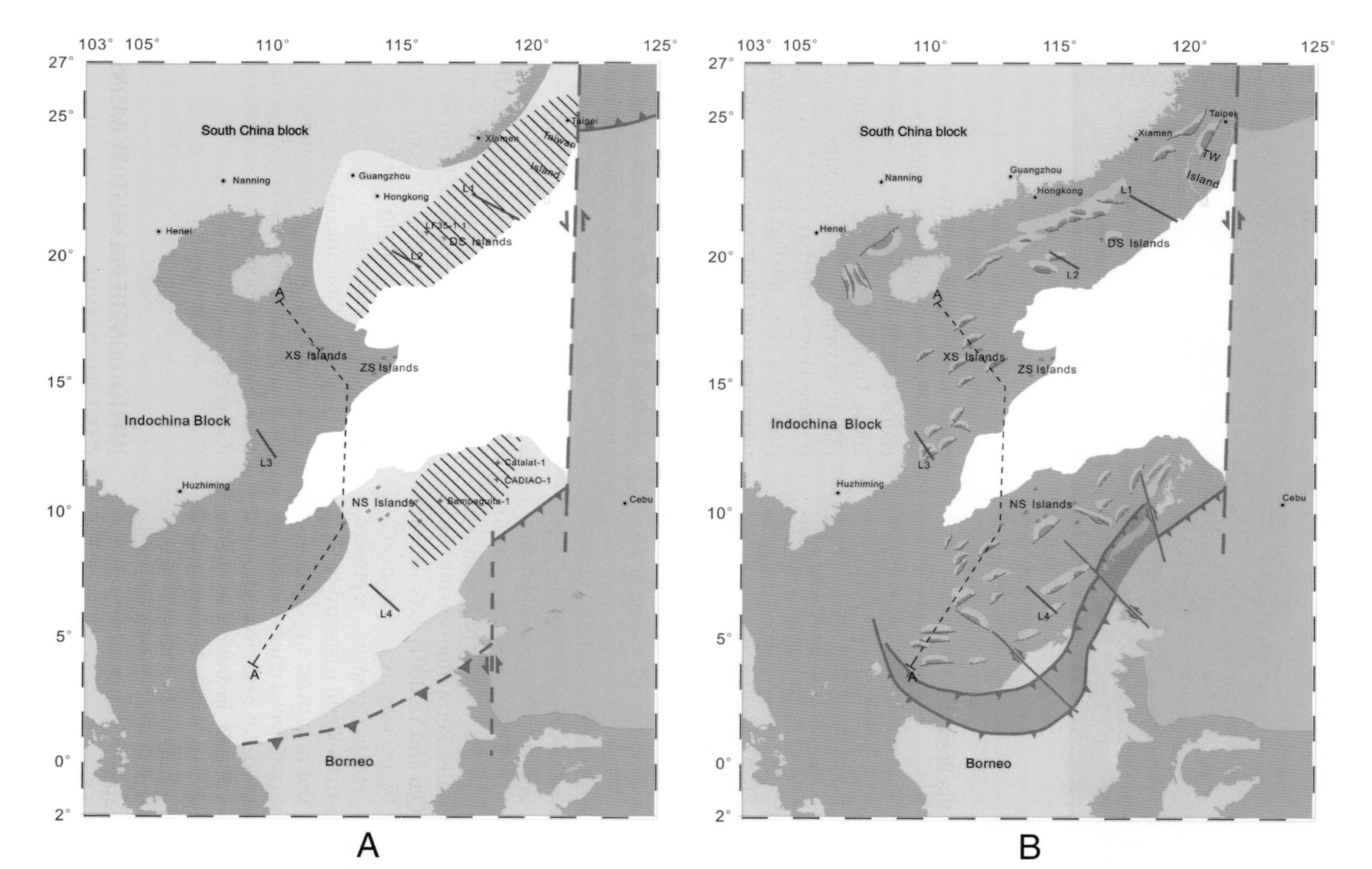

Figure 9. Evolution and distribution of the depositional sequences in polyhistory basins of the South China Sea in Late Jurassic–Early Cretaceous (A), Late Cretaceous–middle Eocene (B), late Eocene–Oligocene (C), and Miocene to the present (D) (see Figure 1 for the abbreviations).

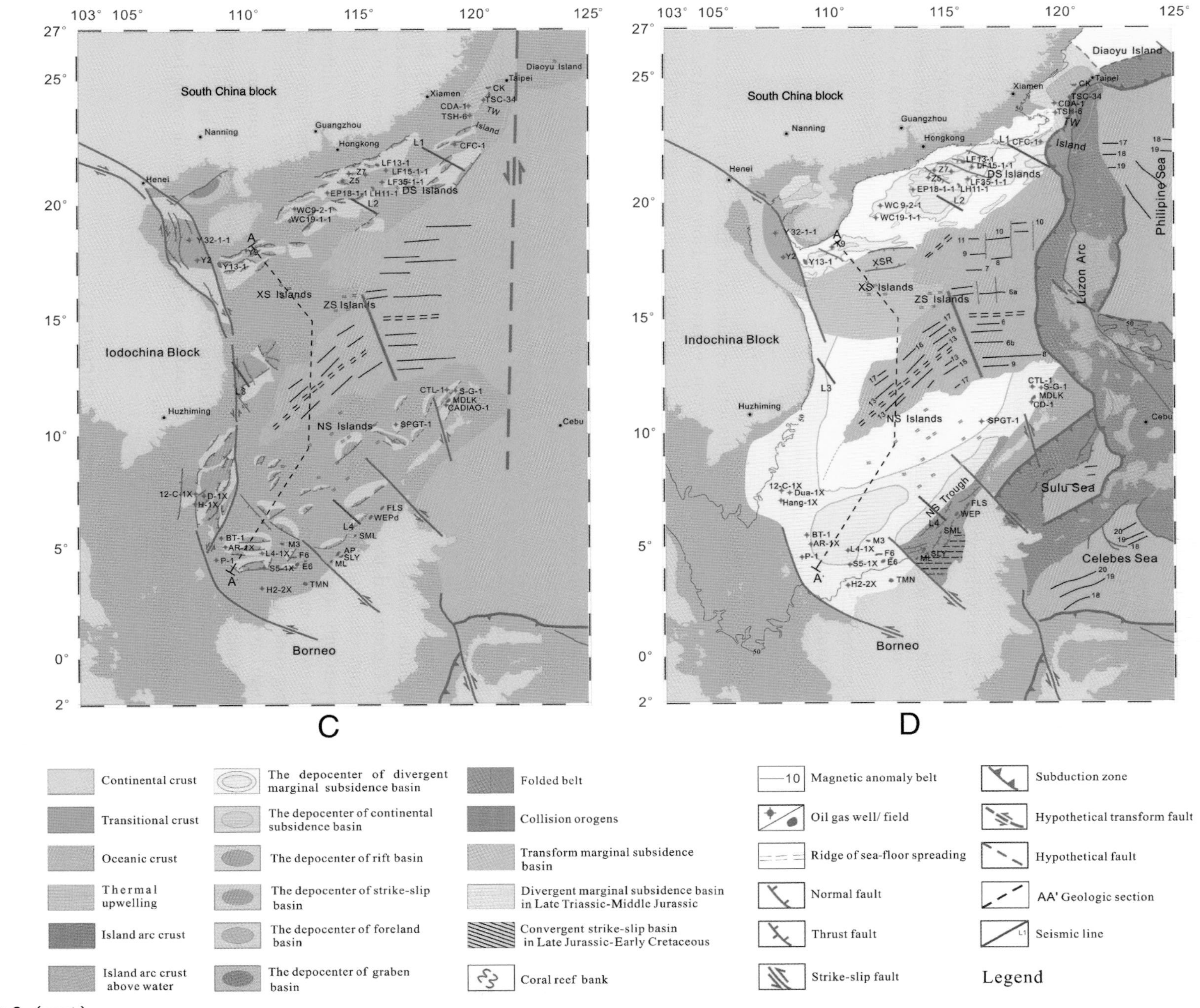

Figure 9. (cont.).

basins such as in west Africa and east Brazil, the SCS has the most complicated spatial variation and temporal evolution in tectonic setting, basin structure style, and petroleum systems. Part of the reason for such complexities is that it is located in a composite geologic setting featuring complicated plate interactions and tectonic-thermal events. Because of these complexities, it is particularly difficult to characterize and classify the polyhistory basins, making it highly risky and challenging in the exploration for hydrocarbons in the basins. It is important to use a basin classification scheme that is most representative of and relevant to the greater SCS Basin to capture the dynamic nature of the polyhistory basins and petroleum systems. Eventually, a robust geohistory analysis of the basins might lead to new play concepts and to future exploration successes in the SCS.

Unraveling the polyhistory of the greater SCS Basin is helpful to track the spatial and temporal variations in physical conditions (Zhang, 1997, 2010) and overprinting relationship of basins. Each polyhistory basin has a specific pressure (P), space (V), temperature (T), and history (t). Overprinting of different basins will cause changes in P, S, and T with t. If the basin type changed at time t, the P, V, and T would become different, leading to changes (either enhancement or reduction) in the hydrocarbon potential of petroleum systems in the basin. Therefore, polyhistory basin analysis is instrumental in evaluating hydrocarbon potential from a dynamic and historic perspective in the SCS.

In the Zhujiangkou subbasin in the northern margin (Figure 1), for example, exploration wells have found reservoirs in a working hydrocarbon system. In the system, V controlled the distribution of the deeply buried source rock and reservoirs of turbidites formed in the deep-marine basin slope, P and T controlled the maturity of the source rock, whereas t controlled the hydrocarbon expulsion that could happen only after the rift basin was overlapped by the divergent marginal subsidence basin. Exploration drilling confirmed that oil and gas generated in the graben basin have migrated to reservoirs in the superposed divergent marginal subsidence basin, primarily in carbonate rocks with good porosity and permeability deposited on the uplifted zone of the continental shelf. Subsurface data indicate that the high-potential reservoirs were spatially associated with the migration pathway created by the northwest-trending strike-slip fault (Gong and Li, 1997). The exploration successes in this region demonstrate that the spatial and temporal correlations (P-V-T-t) of polyhistory basins help identify exploration targets of highest potential in the northern margin of the SCS.

The SCS Basin is one of the major offshore petroliferous basins at the southeastern China continental margin. In these basins, major oil and gas fields have been discovered. Although these oil and gas fields occur in diverse geologic settings, many of them are associated with major strike-slip faults. For example, in the Bohai Bay Basin, oil and gas fields are associated with the Tan-Lu strike-slip fault (Xu, 1993; Hsiao et al., 2004). Given that many other major strike-slip faults exist in southeastern China (Xu, 1993) and that an enhanced hydrocarbon potential along strike-slip faults is observed (Gao and Milliken, 2012), it is particularly important to investigate the polyhistory basins and their dynamic relationship to strike-slip structure and tectonics in the SCS and the southeastern China continental margin, where wrench tectonics have played a major function in the Mesozoic–Cenozoic (Xu, 1993).

SUMMARY

The tectonic evolution of the SCS in the Mesozoic–Cenozoic can be generalized into an early (Late Triassic–middle Eocene) divergent-convergent cycle and a later (late Eocene–present) divergent-convergent cycle. The two tectonic cycles are composed of four phases of evolution, which are (1) Late Triassic–Middle Jurassic divergent continental margin setting, (2) Late Jurassic–middle Eocene convergent intracontinental setting, (3) late Eocene–Miocene divergent continental margin setting, and (4) Pliocene–present convergent continental margin setting. Through the two tectonic cycles and the four basin evolutionary phases, the greater SCS Basin has assembled different polyhistory basin types, which primarily include divergent marginal subsidence basin, strike-slip basin, graben basin, and rift basin. Each polyhistory basin features a specific P, V, T, and t. As the basin evolved, overprinting of different basins would cause changes in P-V-T conditions with t, making hydrocarbon systems dynamic and four dimensional. Unraveling the spatial and temporal relationships among different basins formed in divergent, convergent, and wrench tectonic settings is the key to finding oil and gas in geologically challenging basins such as the SCS.

ACKNOWLEDGMENTS

This study was supported by the 863 China National High-Tech Research and Development Key Projects (2006AA09A101) and by the National Natural Science Foundation of China (40631007 and 41076017). We thank

Yuchang Zhang in Wuxi Research Institute of Petroleum Geology of SINOPEC (China Petroleum and Chemical Corporation) China, for his comments and suggestions, and Guangzhou Marine Geological Survey in China, for providing the South China Sea data. We also thank Paul Reemst and Taizhong Duan for their constructive peer reviews. Colin P. North served as the guest editor for this chapter. We thank Harry Droust and Christopher Morley for providing additional peer reviews that helped further improve the quality of the chapter.

REFERENCES CITED

Allen, P. A., and J. R. Allen, 2005, Basin analysis: Principles and applications: Oxford, United Kingdom, Blackwell Publishing, 549 p.

Barckhausen, U., and H. A. Roeser, 2004, Sea-floor spreading anomalies in the South China Sea revisited, *in* P. Clift, P. Wang, W. Kuhnt, and D. Hayes, eds., Continent-ocean interactions within the East Asian marginal seas: American Geophysical Union Geophysical Monograph, v. 149, p. 121–125.

Briais, A., P. Patriat, and P. Tapponnier, 1993, Update interpretation of magnetic anomalies and sea-floor spreading stages in the South China Sea: Implications for the Tertiary tectonics of southeast Asia: Journal of Geophysical Research, v. 98, no. B4, p. 6299–6328, doi:10.1029/92JB02280.

Butcher, B. P., 1990, Northwest Shelf of Australia, *in* J. D. Edwards and P. A. Santogrossi, eds., Divergent/passive margin basins: AAPG Memoir 48, p. 81–115.

Chen, J., and N. Wen, 2010, Geophysical atlas in the South China Sea (in Chinese): Beijing, China, Science Press, 137 p.

Chen, L., 2002, Geologic structural feature in west of Zengmu Basin, Nansha Sea area: Oil Geophysical Prospecting (in Chinese with English abstract), v. 37, no. 4, p. 354–362.

Clift, P., G. H. Lee, N. A. Duc, U. Barckhausen, H. V. Long, and Z. Sun, 2008, Seismic reflection evidence for a Dangerous Grounds miniplate: No extrusion origin for the South China Sea: Tectonics, v. 27, p. 1–16.

Cullen, A., P. Reemst, G. Henstra, S. Gozzard, and A. Ray, 2010, Rifting of the South China Sea: New perspectives: Petroleum Geoscience, v. 16, p. 273–282, doi:10.1144/1354-079309-908.

Droust, H., and H. S. Sumner, 2007, Petroleum systems in rift basins: A collective approach in Southeast Asian basins: Petroleum Geoscience, v. 13, p. 127–144, doi:10.1144/1354-079307-746.

Falvey, D. A., 1974, The development of continental margins in plate tectonic theory: Journal of Australian Petroleum Exploration Association, v. 14, p. 95–106.

Feng, X. J., C. Y. Zhang, C. X. Wang, and L. Gao, 2001, Mesozoic in the East China Sea shelf and Taixinan Basin and its petroleum potential: China Offshore Oil and Gas (in Chinese with English abstract), v. 15, no. 5, p. 306–310.

Fuller, M., J. Li, and S. Moss, 1999, Paleomagnetism of Borneo: Journal of Asian Earth Science, v. 17, p. 3–24, doi:10.1016/S0743-9547(98)00057-9.

Gao, D., and J. Milliken, 2012, Cross-regional intraslope lineaments on the lower Congo Basin slope, offshore Angola (west Africa): Implications for tectonics and petroleum systems at passive continental margins, *in* D. Gao, ed., Tectonics and sedimentation: Implications for petroleum systems: AAPG Memoir 100, p. 229–248.

Gao, H., and L. Chen, 2006, An analysis of structural framework and formation mechanism of Zhongjiannan Basin in the west of South China Sea: Oil and Gas Geology (in Chinese with English abstract), v. 27, no. 4, p. 512–516.

Gong, M., T. G. Li, and Y. J. Wu, eds., 2001, Characteristics of structures and evolution of basins in Nansha Sea area (in Chinese): Wuhan, China, China University of Geosciences Press, 88 p.

Gong, Z. S., and S. T. Li, 1997, Continental margin basin analysis and hydrocarbon accumulation of the northern South China Sea (in Chinese): Beijing, China, Science Press, 510 p.

Gong, Z. S., and S. T. Li, 2004, Dynamic research of oil and gas accumulation in northern marginal basin of South China Sea (in Chinese): Beijing, China, Science Press, 339 p.

Gong, Z. S., Q. Jin, Z. Qiu, S. Wang, and J. Meng, 1989, Geology tectonics and evolution of the Pearl River Mouth Basin, *in* X. Zhu, ed., Chinese sedimentary basins: Amsterdam, Netherlands, Elsevier, p. 181–196.

Hall, R., and M. E. J. Wilson, 2000, Neogene sutures in eastern Indonesia: Asian Earth Sciences, v. 18, p. 781–808, doi:10.1016/S1367-9120(00)00040-7.

Hall, R., J. G. Ali, C. D. Anderson, and S. J. Baker, 1995, Origin and motion history of the Philippine Sea plate: Tectonophysics, v. 251, p. 229–250.

Hall, R., W. A. Marco, H. Van, and S. Wim, 2008, Impact of India-Asia collision on SE Asia: The record in Borneo: Tectonophysics, v. 451, p. 366–389, doi:10.1016/j.tecto.2007.11.058.

Hao, H. J., H. M. Lin, M. X. Yang, H. Y. Xue, and J. Chen, 2001, The oil-gas exploration of Chaoshan depression: A new domain of petroleum exploration: China Offshore Oil and Gas (Geology) (in Chinese with English abstract), v. 15, no. 3, p. 157–163.

Hao, H. J., R. L. Wang, and X. T. Zhang, 2004, Mesozoic marine sediment identification and distribution in the eastern Pearl River Mouth Basin: China Offshore Oil and Gas (in Chinese with English abstract), v. 16, no. 2, p. 84–88.

He, L. J., and L. P. Xiong, 2000, The simulation of tectonic and thermal evolution of the Yinggehai Basin: Science in China (D) (in Chinese with English abstract), v. 30, no. 4, p. 415–419.

He, L. S., G. Y. Wang, and X. C. Shi, 1980, Xisha trough: A Cenozoic rift: Geology Review (in Chinese), v. 26, p. 486–489.

He, Q., Z. G. Tong, and G. C. Hu, 2005, Sediment filling and its effect on hydrocarbon accumulation in Wan'an Basin: China Offshore Oil and Gas (Geology) (in Chinese with English abstract), v. 17, no. 2, p. 80–88.

Hilde, T. W., C. S. Uyeda, and L. Kroewke, 1977, Evolution

of the western Pacific and its margin: Tectonophysics, v. 38, no. 1–2, p. 145–165.

Hsiao, L. Y., S. A. Graham, and N. Tilander, 2004, Seismic reflection imaging of a major strike-slip fault zone in a rift system: Paleogene structure and evolution of the Tan-Lu fault system, Liaodong Bay, Bohai, offshore China: AAPG Bulletin, v. 88, p. 71–97, doi:10.1306/09090302019.

Hsu, S. K., Y. C. Yeh, W. B. Doo, and C. H. Tsai, 2004, New bathymetry and magnetic lineations identifications in the northeasternmost South China Sea and their tectonic implications: Marine Geophysical Researches, v. 25, p. 29–44, doi:10.1007/s11001-005-0731-7.

Huang, C. Y., W. Y. Wu, C. P. Chang, S. Tsao, P. B. Yuan, C. W. Lin, and K. Y. Xia, 1997, Tectonic evolution of accretionary prism in the arc-continent collision terrane of Taiwan: Tectonophysics, v. 281, p. 31–51, doi:10.1016/S0040-1951(97)00157-1.

Huang, J. Q., G. M. Chen, and B. W. Chen, 1984, Preliminary analysis of the Tethys-Himalayan tectonic domain: Acta Geologica Sinica (in Chinese with English abstract): v. 1, p. 1–17.

Hutchison, C. S., 2004, Marginal basin evolution: The southern South China Sea: Marine and Petroleum Geology, v. 21, p. 1129–1148, doi:10.1016/j.marpetgeo.2004.07.002.

Ingram, G. M., T. J. Chisholm, C. J. Grant, C. A. Hedlund, P. Stuart-Smith, and J. Teasdale, 2004, Deep-water northwest Borneo: Hydrocarbon accumulation in an active fold and thrust belt: Marine and Petroleum Geology, v. 21, no. 7, p. 879–887, doi:10.1016/j.marpetgeo.2003.12.007.

Jin, Q. H., and T. G. Li, 2000, Regional geologic tectonics of the Nansha Sea area: Marine Geology and Quaternary Geology (in Chinese with English abstract), v. 20, p. 1–8.

Kingston, D. R., C. P. Dishroon, and P. A. Williams, 1983a, Global basin classification system: AAPG Bulletin, v. 67, p. 2175–2193.

Kingston, D. R., C. P. Dishroon, and P. A. Williams, 1983b, Hydrocarbon plays and global basin classification: AAPG Bulletin, v. 67, p. 2194–2198.

Kudrass, H. R., M. Wiedicke, and P. Capek, 1985, Mesozoic and Cenozoic rocks dredged from the South China Sea (Reed Bank area) and Sulu Sea and their significance for plate tectonic reconstructions: Marine and Petroleum Geology, v. 3, p. 19–30.

Leckie, D. A., 1992, Regional setting evolution and depositional cycles of the western Canada foreland basin: AAPG Memoir 55, p. 9–45.

Lee, G. H., K. Lee, and J. S. Watkins, 2001, Geologic evolution of the Cuu Long and Nam Con Son basins, offshore southern Vietnam, South China Sea: AAPG Bulletin, v. 85, p. 1055–1082.

Lee, T. Y., and L. A. Lawver, 1995, Cenozoic plate reconstruction of Southeast Asia: Tectonophysics, v. 251, p. 85–138, doi:10.1016/0040-1951(95)00023-2.

Li, C. F., Z. Y. Zhou, H. J. Hao, H. J. Chen, and J. L. Wang, 2008, Last Mesozoic tectonic structure and evolution along the present-day northeastern South China Sea continental margin: Journal of Asian Earth Sciences, v. 31, p. 546–561, doi:10.1016/j.jseaes.2007.09.004.

Li, J. B., X. L. Jin, and A. G. Ren, 2004, The diapir tectonic in the middle segment of accretionary wedge of the Manila Trench: Chinese Science Bulletin (in Chinese with English abstract), v. 49, no. 10, p. 1000–1008.

Li, W. Y., and D. X. Li, 2006, Tectonic characteristics on the sedimentary basins with different plate margins in the South China Sea: Geoscience (in Chinese with English abstract), v. 20, no. 1, p. 19–29.

Lin, C. S., F. Y. Chu, and J. Y. Gao, 2007, On tectonic movement in the South China Sea during the Cenozoic: Acta Oceanologica Sinica (in Chinese with English abstract), v. 29, no. 14, p. 87–96.

Liu, F. L., and L. S. Wu, 2006, Topographic and morphologic characteristics and genesis analysis of Xisha Trough area in the South China Sea: Marine Geology and Quaternary Geology (in Chinese with English abstract), v. 26, p. 7–14.

Liu, G. D., ed., 1993, Map series of geology and geophysics of China seas and adjacent regions: Chinese Earth Institute of Physics Miscellaneous Investigations Map 7-116-01072-6, scale 1:5,000,000, 9 sheets.

Liu, Z. S., 1988, Geologic structure and the continental margin extension in the South China Sea: Marine Science Bulletin (in Chinese), v. 47, no. 4, p. 13–14.

Madon, M. B. H., R. B. Abd. Karim, and R. W. H. Fatt, 1999, Tertiary stratigraphy and correlation schemes, *in* K. M. Leong, ed., The petroleum geology and resources of Malaysia: Kuala Lumpur, Malaysia, Petronas, p. 113–137.

Metcalfe, I., 2006, Paleozoic and Mesozoic tectonic evolution and paleogeography of East Asian crustal fragments: The Korean Peninsula in context: Gondwana Research, v. 9, p. 24–46, doi:10.1016/j.gr.2005.04.002.

Metcalfe, I., 2007, Conodont index fossil *Hindeodus changxingensis* Wang fingers greatest mass extinction event: Palaeoworld, v. 16, p. 202–207, doi:10.1016/j.palwor.2007.01.001.

Miall, A. D., 1984, Principles of sedimentary basin analysis: New York, Springer-Verlag, 490 p.

Morley, C. K., S. Back, P. Crevello, P. van Rensbergen, and J. J. Lambiase, 2003, Characteristics of repeated, detached, Miocene–Pliocene tectonic inversion events: In a large delta province on an active margin, Brunei Darussalam, Borneo: Journal of Structural Geology, v. 25, p. 1147–1169, doi:10.1016/S0191-8141(02)00130-X.

Northrup, C. J., L. H. Royden, and B. C. Burchfiel, 1995, Motion of the Pacific plate relative to Eurasia and its potential relation to Cenozoic extension along the eastern margin of Eurasia: Geology, v. 23, p. 719–722, doi:10.1130/0091-7613(1995)023<0719:MOTPPR>2.3.CO;2.

Packham, H. G., 2003, Plate tectonics and the development of sedimentary basins of the dextral regime in western Southeast Asia: Southeast Asian Earth Sciences, v. 8, p. 497–511, doi:10.1016/0743-9547(93)90048-T.

Pang, X., C. M. Chen, and D. J. Peng, 2007, The Pearl River deep-water fan system and petroleum in South China Sea (in Chinese): Beijing, China, Science Press, 360 p.

Pubellier, M., F. Ego, N. Chamot-Rooke, and C. Rangin, 2003, The building of pericratonic mountain ranges : Structural and kinematic constraints applied to GIS-based reconstructions of SE Asia: Bulletin Societe Geologique France, v. 174, no. 6, p. 561–584, doi:10.2113/174.6.561.

Qiu, X. L., and S. Y. Ye, 2001, Crustal structure across the Xisha Trough, northwestern South China Sea: Tectonophysics, v. 341, p. 179–193, doi:10.1016/S0040-1951(01)00222-0.

Qiu, Y., B. C. Yao, and T. G. Li, 1997, Geologic and tectonic features and hydrocarbon potential of the Zhongjiannan Basin, South China Sea: Geological Research of China Sea (in Chinese with English abstract), v. 9, p. 37–53.

Qiu, Y., G. Chen, X. Xie, L. Wu, X. Liu, and T. Jiang, 2005, Sedimentary filling evolution of Cenozoic strata in Zengmu Basin, southwestern South China Sea: Journal of Tropical Oceanology (in Chinese with English abstract), v. 24, no. 5, p. 43–52.

Qiu, Y. X., 1997, Some problems about Cathaysia (Cathaysia block): Scientia Geologica Sinica, v. 16, no. 3, p. 325–335.

Qiu, Y. X., and B. Y. Zhang, 2000, On eastern extension of the paleo-Tethys in South China: Regional Geology of China (in Chinese with English abstract), v. 19, no. 2, p. 175–180.

Rangin, C., M. Klein, D. Roques, X. Le Pichon, and L. V. Trong, 1995a, The Red River fault system in the Tonkin Gulf, Vietnam: Tectonophysics, v. 243, p. 209–222, doi:10.1016/0040-1951(94)00207-P.

Rangin, C., P. Huchon, X. Le Pichon, H. Bellon, C. Lepvrier, D. Roques, N. D. Hoe, and P. V. Quynh, 1995b, Cenozoic deformation of central and south Vietnam: Tectonophysics, v. 251, p. 179–196, doi:10.1016/0040-1951(95)00006-2.

Schliiter, H. U., K. Hinz, and M. Block, 1996, Tectonostratigraphic terranes and detachment faulting of the South China Sea and Sulu Sea: Marine Geology, v. 130, p. 39–78, doi:10.1016/0025-3227(95)00137-9.

Shao, L., H. Q. You, H. J. Hao, G. X. Wu, P. J. Qiao, and Y. C. Lei, 2007, Petrology and depositional environments of Mesozoic strata in the northeastern South China Sea: Geological Review (in Chinese with English abstract), v. 53, no. 2, p. 164–169.

Shi, X. B., H. H. Xu, X. L. Qiu, K. Y. Xia, X. Q. Yang, and Y. M. Li, 2008, Numerical modeling on the relationship between thermal uplift and subsequent rapid subsidence: Discussions on the evolution of the Tainan Basin: Tectonics, v. 27, p. TC6003, doi:10.1029/2007TC002163.

Sibuet, J. C., and S. K. Hsub, 2004, How was Taiwan created?: Tectonophysics, v. 379, p. 159–181, doi:10.1016/j.tecto.2003.10.022.

Soeria-Atmadja, R., D. Noeradi, and B. Priadi, 1999, Cenozoic magmatism in Kalimantan and its related geodynamic evolution: Journal of Asian Earth Sciences, v. 17, p. 25–45, doi:10.1016/S0743-9547(98)00062-2.

Sun, Z., Z. H. Zhong, and D. Zhou, 2003, Deformation mechanism of Red River fault zone during Cenozoic and experimental evidences related to Yinggehai Basin formation: Journal of Tropical Oceanology (in Chinese with English abstract), v. 22, no. 2, p. 1–9.

Tapponnier, P., G. Peltzer, and A. Y. Le Dain, 1982, Propagating extrusion tectonics in Asia: New insights from simple experiments with plasticine: Geology, v. 10, p. 611–616, doi:10.1130/0091-7613(1982)10<611:PETIAN>2.0.CO;2.

Tapponnier, P., G. Peltzer, and P. Armijo, 1986, On the mechanics of the collision between India and Asia: Geological Society, v. 19, p. 115–157.

Taylor, B., and D. E. Hayes, 1980, The tectonic evolution of the South China Sea Basin, *in* D. E. Hayes, ed., The tectonic and geologic evolution of southeast Asian seas and islands: American Geophysical Union Geophysical Monograph, v. 23, p. 89–104.

Taylor, B., and D. E. Hayes, 1983, Origin and history of the South China Sea basin, *in* D. E. Hayes, ed., The tectonic and geologic evolution of Southeast Asian seas and islands II: American Geophysical Union Geophysical Monograph, v. 27, p. 23–56.

Tirel, C., J. P. Brun, and D. Sokoutis, 2006, Extension of thickened and hot lithospheres: Inferences from laboratory modeling: Tectonics, v. 25, p. TC1005, doi:10.1029/2005TC001804.

Wakita, K., 2000, Cretaceous accretionary-collision complexes in central Indonesia: Journal of Asian Earth Science, v. 18, p. 739–749, doi:10.1016/S1367-9120(00)00020-1.

Wang, J. Q., B. C. Yao, L. Wan, and Z. H. Liu, 2005, Characteristics of tectonic dynamics of the Cenozoic sedimentary basins and the petroleum resource in the South China Sea: Marine Geology and Quaternary Geology (in Chinese with English abstract), v. 25, no. 2, p. 92–99.

Wang, P., K. Y. Xia, and C. L. Huang, 2000, Distribution and geological and geophysical characteristics of Mesozoic marine strata in northeastern part of the South China Sea: Journal of Tropical Oceanology (in Chinese with English abstract), v. 19, no. 4, p. 28–35.

Wang, P. X., and Q. Y. Li, eds., 2009, The South China Sea: Paleoceanography and sedimentology: Amsterdam, Netherlands, Springer, 506 p.

Weeks, L. G., 1952, Factors of sedimentary basin development that control oil occurrence: AAPG Bulletin, v. 36, no. 11, p. 2071–2124.

Wei, X., J. F. Deng, and Y. H. Chen, 2005, Distribution characters and exploration potential of Mesozoic sea facies sedimentary strata in the South China Sea Basin: Earth Science (in Chinese with English abstract), v. 35, no. 4, p. 456–468.

Williams, H. H., 1997, Play concepts: Northwest Palawan, Philippines: Asian Earth Sciences, v. 15, no. 2–3, p. 251–273.

Wu, N. Y., W. J. Zeng, H. B. Song, Z. Y. Zhou, D. L. Du, and L. Wan, 2003, Tectonic subsidence of the South China Sea: Marine geology and quaternary geology (in Chinese with English abstract), v. 23, no. 1, p. 55–65.

Xia, K. Y., and C. L. Huang, 2000, The sedimentary basin and the petroleum prospect in Tethys period of Mesozoic: Earth Science Frontiers (in Chinese with English abstract), v. 7, no. 3, p. 227–238.

Xia, K. Y., C. L. Huang, and Z. M. Huang, 2004, Upper Triassic–Cretaceous sediment distribution and hydrocarbon potential in South China Sea and its adjacent areas: China Offshore Oil and Gas (in Chinese with English abstract), v. 16, no. 2, p. 73–83.

Xie, J. L., C. Huang, and F. Y. Xiang, 2008, Evolution of Cenozoic tectonic paleogeography and its petroleum significance in the western South China Sea: Chinese Journal of Geology (in Chinese with English abstract), v. 43, no. 1, p. 133–153.

Xu, J., 1993, Basic characteristics and tectonic evolution of the Tancheng-Lujiang fault zone, *in* J. Xu, ed., The Tancheng-Lujiang strike-slip fault system: Chichester, England, John Wiley and Sons Ltd., p. 17–50.

Yan, P., H. Deng, H. L. Liu, Z. R. Zhang, and Y. K. Jiang, 2006, The temporal and spatial distribution of volcanism in the South China Sea region: Journal of Asian Earth Science, v. 27, p. 647–659, doi:10.1016/j.jseaes.2005.06.005.

Yan, P., Y. Wang, and H. L. Liu, 2008, Postspreading transpressive faults in the South China Sea Basin: Tectonophysics, v. 450, p. 70–78, doi:10.1016/j.tecto.2008.01.015.

Yao, B. C., L. Wan, and Z. H. Liu, 2004, Tectonic dynamics of Cenozoic sedimentary basins and hydrocarbon resources in the South China Sea: Earth Science (in Chinese with English abstract), v. 29, no. 5, p. 543–549.

Zhan, W. H., Z. Y. Zhu, L. T. Sun, Z. X. Sun, and Y. T. Yao, 2006, The epoch and diversities of neotectonic movement in the South China Sea: Acta Geologica Sinica (in Chinese with English abstract), v. 80, no. 4, p. 491–496.

Zhang, G. X., and Z. L. Bai, 1998, The characteristics of structural styles and their influences on oil and gas accumulation of the Wa'an Basin in the southwestern South China Sea: Petroleum Geology and Experiment (in Chinese with English abstract), v. 20, no. 3, p. 210–216.

Zhang, Y. C., 1997, The prototype analysis in Chinese petroleum basins (in Chinese): Nanjing, China, Nanjing University Press, 450 p.

Zhang, Y. C., ed., 2010, Dynamical basin and petroleum (in Chinese): Beijing, China, Petroleum Industry Press, 480 p.

Zhang, Y. X., 1985, The mantle heat activities in the Cenozoic of East Asia and the formation of the marginal sea in West Pacific: The Second National Conference of Marine Geology Conference Papers (in Chinese), v. 19, p. 10–13.

Zhong, G. F., H. Xu, and L. Wang, 1995, Structure and evolution of Cenozoic basins in the southwest area of South China Sea: Marine Geology and Quaternary Geology (in Chinese), v. 15, p. 87–94.

Zhou, D., H. L. Liu, and H. Z. Chen, 2005a, Mesozoic–Cenozoic magmatism in southern South China Sea and its surrounding areas and its implication to tectonics: Geotectonica et Metallogenia (in Chinese with English abstract), v. 29, no. 3, p. 354–363.

Zhou, D., H. Z. Chen, Z. Sun, and H. H. Xu, 2005b, Three Mesozoic sea basins in eastern South China Sea and their relation to Tethys and paleo-Pacific domains: Journal of Tropical Oceanology (in Chinese with English abstract), v. 24, no. 2, p. 16–25.

Zhou, D., Z. Sun, H. Z. Chen, and Y. X. Qiu, 2005c, Mesozoic lithofacies, paleogeography, and tectonic evolution of the South China Sea and surrounding areas: Earth Science Frontiers (in Chinese with English abstract), v. 12, no. 3, p. 204–218.

Zhu, X., ed., 1985, The tectonic of petroliferous basins in China (in Chinese): Beijing, China, Petroleum Industry Press, 300 p.

Zhu, X., 1980, Some ideas on the basin research: Petroleum Geology and Experiment (in Chinese), v. 2, no. 3, p. 100–108.